FIRST EDITION

GUIDE TO

PACIFIC NORTHWEST MARINAS

Powered by Datastract ™

Cathy Haden, Chris Haden & Beth Adams-Smith

Richard Y. Smith & Beth Adams-Smith, Editors

Jerawyn Publishing Inc.

www.AtlanticCruisingClub.com

Atlantic Cruising Club's
Guide to Pacific Northwest Marinas

FIRST EDITION

Copyright © 2007 Jerawyn Publishing Inc.

ISBN 0-0664028-9-8
ISBN 978-0-9664028-9-6
Library of Congress Catalog Number (PCN) 2006932517

Front Cover Photo by Chris Haden
Back Cover Photos by Beth Adams-Smith, Chris Haden & Richard Y. Smith

Front Cover: Friday Harbor, WA
Back Cover: Coal Harbour, Lund Harbour, Victoria Inner Harbour, Sequim Bay State Park

Senior Editor and Chief Researcher — Irina Adams
Assistant Editor & Communication Manager — Catherine Rose
Assistant Editor — Alice Picon
Executed Cover Designs — Jessica Gibbon, Spark Design
Initial Cover Design — Rob Johnson
Book Design — Spark Design
Photo Retouching — Karina Tkarch
Cartography — Amy Rock
Research Assistants — Brittney Taylor, Jessica Swanston and Sarah Newgaard
Printed by Walsworth Commerical Press

Atlantic Cruising Club's Guide to Pacific Northwest Marinas
is written, compiled, edited and published by the Atlantic Cruising Club, an imprint of:

Jerawyn Publishing Inc.
PO Box 978; Rye, New York 10580

www.AtlanticCruisingClub.com

Table of Contents

Table of Contents

INTRODUCTION

FIRST EDITION

Preface to the First Edition of the
Atlantic Cruising Club's Guide to Pacific Northwest Marinas

With this volume, the Atlantic Cruising Club expands its marina coverage to the Pacific Coast for the first time. For 10 years, we have been publishing independent, "boater-biased" information, ratings, reviews and photos based on personal visits backed by painstaking research — until now focused on 2,000 marinas on the Atlantic Coast. We believe the Atlantic Cruising Club's Guides to Marinas are now the most detailed and accurate source of marina and "what's-nearby" information available on both coasts. We hope you will agree as you cruise the Pacific Northwest's breathtakingly beautiful waters — and refer to these pages again and again for help in deciding which marinas, harbors and towns provide the services and attractions you are seeking.

We've seen a lot of magnificent country in our boating and publishing pursuits, and we believe the Pacific Northwest regions covered in this Guide provide some of the most glorious cruising grounds in all of North America. Better yet, most are in relatively protected areas — although huge tidal ranges and their associated currents can provide some interesting moments. We learned the pleasures of precipitation — whether it's a light mist, dense fog, brief shower, or teeth-chattering deluge — that is the price of all that intense greenery. It also made us appreciate the sun. When we were blessed with clear skies, the views were unmatched — anywhere. From most places on the water, there are dramatic vistas of small rock-bound archipelagos, densely layered forests, and, when they are "out," soaring snow-capped mountains. This can be raw country with wilderness close at hand — but it can also be exciting cosmopolitan cities, funky fishing camps, charming, laid-back little villages, world-class resorts, bustling harbor towns. We were bewitched by all of it.

We also ventured outside the comfort zone of most cruising guides. The Washington and Oregon coasts, south of the Puget Sound region, are often bypassed by recreational boaters — with good reason. Notorious bars, Alaskan-spawned storms, and the United States as a lee shore are powerful incentives to stay well offshore and hurry southward or northward to gentler waters before making landfall. This dramatic coastline, however, has some real rewards for the careful skipper who elects to stay inshore and explore the rivers and harbors.

Of course, there are no secrets; even the most remote regions have been discovered. To wit, over a third of the northernmost marinas covered in this Guide had changed hands in the previous year — many are evolving from fishing camps to full-service yachting facilities. Thankfully, most new owners seemed deeply committed to preserving the environment that had attracted them in the first place. On that same note, it is heartening to see that more "no-discharge" zones are being designated, pump-out facilities installed and "Clean Marinas" certified — these programs are vital to the health of these fragile waters (please see page 298 for a complete list). As new "Clean Marinas" and "No Discharge Zones/Areas" are christened, and new pump-put services inaugurated, ACC will post the information on its website — www.AtlanticCruisingClub.com.

This volume is also the Atlantic Cruising Club's first foray into a region that includes cruising grounds outside the United States. More than forty percent of the *Atlantic Cruising Club's Guide to Pacific Northwest Marinas* covers facilities in the Canadian province of British Columbia. An international border may seem to be just a dashed line on the chart, but some very real differences exist when crossing it in either direction. It might prove useful to understand the distinctions before setting off. Coast Guard regulations differ from one side to the other, and clearing customs requires some preplanning. For some assistance, please see the box on page 21 for current U.S. and Canadian Ports of Entry and www.AtlanticCruisingClub.com for the U.S. and Canadian Custom Offices' Boaters Fact Sheets — which also describes the all-important import restrictions. Be aware, too, that U.S. and Canadian nautical charts use different units of measure and a different tide datum. Even some of the boating language is unique: Looking for the Harbormaster or Harbourmaster? In B.C., ask for the Wharfinger. If you are new to the whole region, "moorage" means "dockage" — not "moorings." And, save your "loonies" — they are often required for a hot shower.

In this Guide, all prices in U.S. waters are shown in U.S. dollars. All prices in Canadian waters are shown in Canadian dollars. As of this writing, Canadian dollars are worth a little bit less than American dollars, and U.S. boaters may think they are getting a good deal after conversion. However, they then find out about the Canadian taxes, which often offset any apparent price advantage!

We wish you wonderful cruising and exciting adventures — whether you are discovering these waters for the first time or are revisiting them for the tenth. We can't wait for our next trip.

Chris Haden
Cathy Haden
Beth Adams-Smith
Richard Y. Smith

December 15, 2006

Talk To Us

The Atlantic Cruising Club's writer/reviewers, editors and publisher all want to hear from you. Our goal is to "get it right." If we didn't get something right, if we missed an important stop, if you know of a recent change, or you want to share a regional boating or marina experience, please contact us. Either send us an email at guide@AtlanticCruisingClub.com, call us at (888) 967-0994 or log onto the website, find the marina you wish to write about, and then select "Cruiser Comments." Your feedback is very, very important.

ACC's Guide to Pacific Northwest Marinas' Twenty Sub-Regions

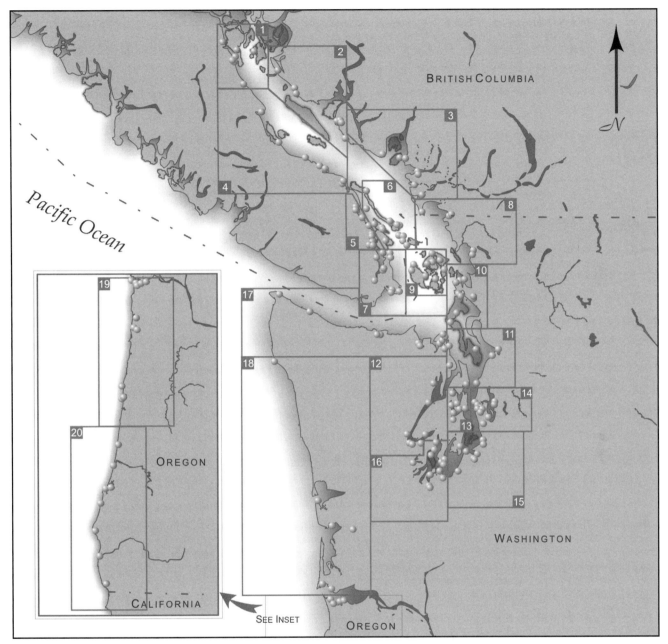

Atlantic Cruising Club's Ratings

The Bell Ratings generally reflect the services and amenities available for the captain and crew rather than the services available for the boat. By their nature, ratings are subjective and may also reflect certain biases of the writers, editors, and other reviewers. It is important to note that a five-bell marina will not always be a boater's best choice. There tends to be a correlation between higher bell ratings and higher overnight transient rates. Many of the resort-type amenities available at four- and five-bell marinas may be of little interest to boaters arriving late in the day and planning an early start the next morning. Similarly, a facility which has a one- or two-bell rating, good security, convenient transportation and a service-oriented staff, may be the best place to "leave the boat" for a period of time between legs of a longer cruise.

The Boatyard Ratings, on the other hand, are less subjective. They simply indicate the extent of the boatyard services and the size of yachts that the facility can manage. To receive a boatyard rating at all (one-travelift), a facility must have a haul-out mechanism — a travelift or marine railway (or, in some cases, a heavy-duty crane or forklift) plus, at a minimum, a standard array of basic boatyard services. To receive a two-travelift rating, a facility will generally have haul-out capacity in excess of 70 tons and a full complement of all possible boatyard services. Facilities that are primarily dry-stack storage operations are not given boatyard designations.

The Sunset Rating is the most subjective rating of all. This symbol indicates remarkable places with special appeal — like a pristine, untouched mooring field with no other facilities in sight, a marina that is, itself, so exquisitely turned out that it is truly beautiful, a view from the docks of a skyline or distant vista that is simply breathtaking, above and beyond services or amenities that are more than the basic rating would suggest, or a marina that offers the possibility of a unique experience. A Sunset means that, in our view, there is more here than the first two ratings can convey — and only you can determine if the additional notation is valid for you and your crew. We'd be very interested in hearing your collective views.

The Bell Ratings

Outlined below are some of the facilities and amenities one might generally find at a marina or boating facility within a given Bell-Rating category. Please note that some marinas within a particular category may not have all of the facilities listed here and some may have more. (The word "Marina" is used generically here, and throughout the *Guide*, to denote all types of marina facilities, including mooring fields and single, unattended piers.)

One Bell: The marina comfortably accommodates vessels over thirty feet in length, accepts overnight transients at docks or on moorings, and generally has heads. These are the "basic requirements" for ACC inclusion. Most facilities that are strictly mooring fields with a dinghy dock or basic docks with limited or no power pedestals or other services fall into this category.

Two Bells: In addition to meeting the basic ACC requirements, the marina generally has docks with power pedestals or a mooring field served by a launch (or a "dinghy loaner"). It has a dedicated marina staff, offers docking assistance, monitors the VHF radio, holds mail, and has an available fax machine. There are heads, showers, and, perhaps, a laundry. It likely has dock carts, a picnic area, and grills.

Three Bells: With attractive and convenient facilities, the marina significantly exceeds basic requirements in many physical and operational categories. In addition to the two-bell services described above, the docks will usually have finger piers, and there will usually be a restaurant on-site or adjacent. A pool, beach, or major recreational amenity, (i.e. a sport fishing center, a museum or sightseeing venue, a nature preserve, a significant "downtown") may also be nearby. The marina usually offers docking assistance and other customer-oriented services, a ships' store, cable TV, some kind of Internet access, and, hopefully, a pump-out facility.

Four Bells: Worth changing course to visit, the marina significantly exceeds requirements in most physical and operational categories and offers above average service in well-appointed, appealing, and thoughtfully turned-out facilities. In addition to the three-bell services described above, it will have a restaurant on-site, as well as a pool or beach and other desirable amenities like tennis courts, sport fishing charter, or historic or scenic sites. The marina will generally offer concierge services and have a particularly inviting environment.

Five Bells: A renowned, "destination" facility, the marina is worth a special trip. It has truly superior facilities, services, and atmosphere. A five-bell marina is located in a luxurious, impeccably maintained environment and provides absolutely everything a cruising boater might reasonably expect, including room service to the boat. It offers all that is promised in a four-bell marina, plus outstanding quality in every respect.

Bell ratings reflect both subjective judgment and objective criteria. The ratings are intended to reflect the overall boater experience and are significantly impacted by a marina's setting and general ambiance. An ACC bias is discovering interesting and distinctive waterfront destinations which may not have all the standard marina services, but which provide a unique experience. These may be given ratings higher than their facilities would suggest. Similarly, maritime museums, which most boaters find particularly compelling, are usually given a Sunset Rating to indicate that they offer maritime buffs more than just services. Ratings are also geographically specific and reflect the general level of available services in a given region. In other words, a five-bell marina in Florida (with a year-round season) will usually offer significantly more services and facilities than a five-bell marina in British Columbia with a much shorter season.

A Tour of a Marina Report

Photos: *One for each marina in the printed Marina Report, up to 17 in full color on the enclosed DVD and on the website. Most were taken by ACC personnel during periodic visits. They are intended to provide a noncommercial, visual sense of each facility.*

Ratings: *Bells (1 – 5) reflect the quality of on-site marina facilities plus location, recreation, dining, lodgings, etc. Travelifts (1 – 2) indicate the extent of the boatyard services. A Sunset notes a particularly beautiful, special, unique, or interesting place.*

Top Section: *Facts, facts and more facts, including VHF channels, phone numbers, e-mail/website addresses, number of slips, moorings, power options (rates for all), and much more. The format of this section is identical for every Marina Report for easy reference and comparison.*

Marina Name: *238 marinas and marine facilities are included in the Pacific Northwest volume.*

Sub-Regions: *The Pacific Northwest volume includes 20 sub-regions. For quick reference, these sub-region tabs are visible on the outside page edges. In each sub-region, Marina Reports are ordered North to South.*

Middle Section: *What's available, where and how to find it. Marine Services & Boat Supplies, Boatyard Services (including rates), Restaurants and Accommodations, Recreation and Entertainment, Provisioning and General Services, Transportation and Medical Services, all classified by distance — On-Site, Nearby, Within 1 mile, 1-3 miles or beyond. Names, phone numbers, price ranges, and more.*

Photos on DVD: *Indicates the number of full color photos of this facility that are on the DVD.*

Bottom Section: *The "Setting" commentary portrays a sense of the marina's surroundings and location. "Marina Notes" provides important and useful facts about the marina and its operations that may not have been covered in either of the earlier sections. "Notable" addresses anything the writers/reviewers feel is noteworthy about this facility, from special events or services to interesting side trips and/or local lore.*

Harbor: *Harbor or major body of water on which the marina resides.*

Semiahmoo Marina

Navigational Information
Lat: 48°59.196' Long: 122°46.366' Tide: 10 ft. Current: 0 kt. Chart: 18424
Rep. Depths (MLW): Entry 30 ft. Fuel Dock 10 ft. Max Slip/Moor 12 ft./-
Access: First marina to starboard on entering Drayton Harbor

Marina Facilities *(In Season/Off Season)*
Fuel: Gasoline, Diesel
Slips: 294 Total, 10 Transient Max LOA: 120 ft. Max Beam: n/a
Rate (per ft.): Day $0.80 Week n/a Month $4.85
Power: 30 amp $0.08/kwh, 50 amp $0.08/kwh, 100 amp n/a, 200 amp n/a
Cable TV: No Dockside Phone: Yes
Dock Type: Floating, Alongside, Concrete
Moorings: 0 Total, 0 Transient Launch: n/a
Rate: Day n/a Week n/a Month n/a
Heads: 10 Toilet(s), 8 Shower(s)
Laundry: 4 Washer(s), 4 Dryer(s), Iron, Iron Board Pay Phones: Yes
Pump-Out: OnSite, 1 Central, 1 Port Fee: n/a Closed Heads: Yes

Marina Operations
Owner/Manager: Lou Herrick Dockmaster: Same
In-Season: Jun-Sep, 8:30am-6pm Off-Season: Oct-May, 8:30am-5pm
After-Hours Arrival: Call ahead
Reservations: Yes, Preferred Credit Cards: Visa/MC, Amex
Discounts: None
Pets: Welcome, Dog Walk Area Handicap Access: Yes, Heads, Docks

Marina Services and Boat Supplies
Services - Docking Assistance, Room Service to the Boat, Dock Carts Communication - Mail & Package Hold, Data Ports (Wi-Fi - Broadband Xpress), FedEx, DHL, UPS, Express Mail (Sat Del) Supplies - OnSite: Ice (Block, Cube), Propane 3+ mi: West Marine (332-1918, 8 mi.)

Boatyard Services
OnSite: Travelift (35T), Engine mechanic (gas, diesel), Electrical Repairs, Hull Repairs, Rigger, Bottom Cleaning, Brightwork, Compound, Wash & Wax, Yacht Interiors

Restaurants and Accommodations
Near: Restaurant (Packers Lounge 318-2000), L & D Sun-Thu 11am-Mid, Fri-Sat 11am-1am, B Sat 9am-1pm), (Stars 318-2000, B Mon-Fri 6:30-11:30am, till Noon winds, D daily 5-10pm), (Pierside 318-2000), (Blue Heron 371-5745, at Semiahmoo Golf, casual and to-go), Hotel (Semiahmoo Resort 318-2010, $139-409) 3+ mi: Restaurant (Stephanie's 371-7033, B $5-9, L $7-28, D $10-24, 5 mi.), Motel (Anchor Inn 332-5539, $60-125, 8 mi.)

Recreation and Entertainment
OnSite: Picnic Area, Grills, Boat Rentals (kayaks, paddle boats), Roller Blade/Bike Paths (rentals too) Near: Heated Pool (at resort), Spa, Beach, Tennis Courts, Golf Course (shuttle to Semiahmoo Golf 371-7005 & Loomis Trail 332-1725), Fitness Center, Jogging Paths, Video Arcade (at resort) 1-3 mi: Park (Semiahmoo Park 733-2900), Museum (Drayton Harbor Museum at Semiahmoo Park - local history) 3+ mi: Fishing Charter (Eagle Point - Bellingham 966-3334, 25 mi.), Video Rental (Blockbuster 332-2441, 9 mi.), Cultural Attract (Peace Arch Playhouse 332-4678, 8 mi.), Special Events

(Blaine: Ski to Sea Fest - May, Hands Across the Border - Jun, Jazz Fest - Jul, Summer Aire Art - Jul, Galleries (Peace Arch Park International Sculpture Exhibition 322-7165, 8 mi.)

Provisioning and General Services
OnSite: Convenience Store (snacks, deli items, sundries, gifts) Near: Beauty Salon (at resort 318-2009), Barber Shop 1-3 mi: Retail Shops (gifts, antiques) 3+ mi: Market (Linda's Maxi Market 371-2804, 6 mi.), Supermarket (Cost Cutters 332-5909, 8 mi.), Liquor Store (8 mi.), Green Market (Grace Harbor Farms 371-9060, 4 mi.), Bank/ATM (4 mi.), Post Office (332-7184, 8 mi.), Catholic Church (8 mi.), Protestant Church (8 mi.), Library (Blaine 332-8146, 8 mi.), Dry Cleaners (Biz Center Plus 332-2030, 8 mi.), Bookstore (Book Warehouse 366-5354, 10 mi.), Hardware Store (True Value 332-4077, 8 mi.), Clothing Store (Bells Fair Mall, BC 360-734-5022; by bus, 22 mi.), Department Store (Goff's 332-6663, 8 mi.)

Transportation
OnSite: Bikes ($12 at store), Ferry Service (Plover Foot Ferry to Blaine, Fri-Sun May-Sep) OnCall: Rental Car (Enterprise 714-0243), Taxi (City Cab 332-8294) Near: Courtesy Car/Van (to golf & casino) 3+ mi: Local Bus (Wilcom Transportation Authority 676-7433, 8 mi.) Airport: Blaine Municipal/ Bellingham Int'l. (9 mi/18 mi.)

Medical Services
911 Service OnCall: Ambulance Near: Holistic Services (full-service European spa) 3+ mi: Doctor (Bay Medical Clinic 332-6327, 8 mi.), Dentist (Blaine Harbor Dental 332-2400, 8 mi.), Veterinarian (Cat & Dog Clinic 332-280, 9 mi.) Hospital: St. Joseph 734-5400 (25 mi.)

9450 Semiahmoo Parkway; Blaine, WA 98230

Tel: (360) 371-0440 VHF: Monitor Ch. 68 Talk n/a
Fax: (360) 371-0200 Alternate Tel: n/a
Email: semimarina@bbxmail.net Web: www.semiahmoomarina.com
Nearest Town: Blaine, WA (8 mi.) Tourist Info: (360) 671-3990

Setting – Located on the west side of the entrance to Drayton Harbor, quiet, clean, and well-groomed Semiahmoo Marina is situated at the end of a long spit of undeveloped land. Approaching the marina, to starboard, the facilities of the world-class Semiahmoo Resort sprawl along the shoreline. There are stunning harbor views with snowcapped Mount Baker lying to the south and the Peace Arch - marking the border between Canada and the U.S. - to the north.

Marina Notes – Stay in the channel, as a shoal parallels the fuel dock. Call ahead for slips or check in at the fuel dock. Overlooking the docks are the marina office, chandlery, picnic tables (some covered) and a large charcoal grill. A lift is on-site and Blaine Marine provides full boatyard services (371-5700). Kayak, bike, and roller blade rentals are available, as well as nature hikes and beachcombing. Modern tile bathhouse and laundry. Semiahmoo Yacht Club also makes its home here (reciprocal, 371-0440). The 32-foot wooden ferry "Plover" runs in season between the marina and Blaine Harbor.

Notable – The adjacent Inn at Semiahmoo, on the site of an old salmon cannery, offers three outstanding restaurants, a European-style spa & salon, and entry to the two top-rated golf courses in Washington. A fee of $15/day per adult ($5 per child) provides access to the Inn's ammenities: heated pool, fitness center, tennis courts, indoor track, racquetball court, and aerobics classes. Dining options include informal Packers Lounge, fine dining at Stars, and Sunday brunch buffet at Pierside - with views out to the bay (9am-1pm, under 5 free). Shuttle service to golf courses, the Skagit Casino, and room service to the boats are provided by the Inn. Several antique/gift shops and casual eateries are on this side of town. Blaine is a short dinghy/ferry ride, or about 8 miles by road.

Semiahmoo Marina

2 & WA - PT. ROBERTS to BELLINGHAM

PHOTOS ON CD-ROM: 15

DRAYTON HARBOR   134

The Marina Reports

In the book, individual Marina Reports are presented in a one-page, easy-reference format to make the *Guide* as user-friendly as possible. On the DVD, as well as on the website, second and third pages contain up to sixteen additional full color photos.

Each Report provides over 300 items of information, grouped into the following sections and categories:

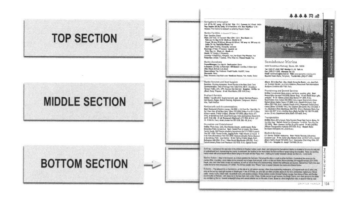

TOP SECTION

Name	Primary & Toll-Free Phone Numbers	VHF channels — Talk & Monitor	Nearest Town (distance)
Address	Alternate Phone — After-Hours	E-mail Address	Tourist Office Phone Number
	Fax Number	Web URL	

Navigational Information

Harbor (*bottom of page*)	Tidal Range & Maximum Current	Access — General Directions	Entry & Fuel Dock Depths
Latitude/Longitude	Chart Number	MLW Depths (*reported by the marinas*)	Deepest Slip/Mooring

Marina Facilities (In Season/Off Season) — *Note: the terms "dockage" and "moorage" are used interchangeably.*

Fuel: Availability and Brand	*Dock Type*	*Heads:*	*Pump-out Availability & Fees*
Diesel or Gasoline	Fixed or Floating	Number of Toilets	On-Site or OnCall
Slip-Side or OnCall Fueling	Alongside/Side-Tie	Number of Showers	Full Service or Self Service
High-Speed Pumps	Short or Long Fingers, Pilings	Dressing Rooms	Number of Central Stations
Maximum Boat LOA & Beam	*Dock Material*	Hair Dryers & Other Amenities	Number of Portable Stations
Slips: Number of Total/Transient	Wood, Concrete, Vinyl,	*Laundry:*	In-Slip Dedicated Units
Rates: Daily, Weekly, Monthly	Aluminum or Composite	Number of Washers & Dryers	Pump-Out Boats
Power (Availability & Rates):	*Moorings: Total/Transient*	Irons and Ironing Boards	
30amp, 50 amp, 100 amp, 200 amp	Rates: Daily, Weekly, Monthly	Book Exchange	*Closed Head Requirements*
Cable TV: Availability, Terms, Rates	Launch Service — Terms and Fees	Pay Phones	
Dockside Phone: Terms, Rates	Dinghy Dock — Terms and Fees	*Services to Anchored Boats*	

The number of transient slips or moorings does not necessarily mean dedicated transient slips or moorings. Many facilities rent open slips and moorings to transient cruisers when their seasonal tenants are not in port. The number of Transient Slips/Moorings indicated is the facility's guesstimate, based on past experience, of the number generally available at any given time. In-Season and Off-Season rates are listed as $2.50/1.75. The parameters of those seasons are outlined in "Marina Operations" — Dates and Hours. If rates are complicated, which is becoming the norm, then the daily rate for a 40-foot boat is selected, followed by an asterisk, and a complete explanation of the rate structure is given in the "Marina Notes" section. The availability of 300 amp, European Voltage and 3-Phase is listed under "Marina Services and Boat Supplies." If there is alongside dockage, rather than individual slips, then the total number of alongside feet is divided by an average boat length of 40 feet and that number is displayed next to the Total/Transient Slips heading — followed by another asterisk. Dock Type is then listed as "Alongside" and the specifics are explained in "Marina Notes." The lack of finger piers — long or short — may signal a "Stern-To, Med-Mooring" approach. Since this can be a critical factor in choosing a marina, this will be highlighted in "Marina Notes." Since Book Exchanges or Lending Libraries, along with Pay Phones, are traditionally in the laundry room, these are itemized in the Laundry section. If there is Launch Service to the mooring field, the hours are usually included in Marina Notes, too.

A discussion of No Discharge Zones, Certified Clean Marinas, Pump-Out Facilities and the current U.S. and Canada Federal and State/Provincial regulations pertaining to British Columbia, Washington and Oregon can be found in the Addendum, beginning on page 298.

Marina Operations

Marina Owner/Manager	After Hours Arrival Procedure	Discount Programs	Credit Cards Accepted:
Harbormaster/Wharfinger	Reservation Policies	Boat-US, Marina Life, Safe/Sea	Visa, MasterCard, Discover,
Dockmaster	Pets: Welcome?	Dockage Fuel & Repair Discounts	Diners Club, American Express
Dates & Hours of Operation	Dog Walk Area, Kennel	Handicap Accessibility	

For municipal facilities the Harbormaster or Wharfinger is listed under "Marina Owner/Manager" with the designation "(Harbormaster)" following his/her name. Dates and Hours for both in-season and off-season are provided and indicate the requisite time frames for the In-season and Off-season rates. In-season precedes off-season, separated by a (/) slash.

MIDDLE SECTION

Most of the information in this section is classified by "Proximity" — the distance from the marina to the service or facility, as follows:

On-Site — at the marina

OnCall — pick-up, delivery or slipside service

Nearby — up to approximately 4/10 of a mile — a very easy walking distance

Within 1 mile — a reasonable, though more strenuous, walking distance

1 – 3 miles — a comfortable biking distance, a major hike or a cab ride

3+ miles — a taxi, courtesy car or rental car distance — generally included is the approximate distance from the marina

(FYI: In this section, telephone area codes are included only if they are different from the area codes in the marina's contact information.)

Marina Services and Boat Supplies

General Services:	*Megayacht Facilities:*	*Communications:*	*Supplies: (Listed by Proximity)*
Docking Assistance	Additional Power Options:	Mail and Package Hold	Ice — Block, Cubes, Shaved
Dock Carts	300 Amps	Courier Services	Ships' Stores — Local Chandlery
Trash Pick-up	Three-Phase	FedEx. Airborne, UPS	West Marine, Boat-U.S.,
Security — Type & Hours	European Voltage	Express Mail, Saturday Delivery	Boaters World,
Concierge Services	Crew Lounge	Phone Messages	Other Marine Discount Stores
Room Service to the Boat		Fax In and Out — Fees	Bait & Tackle
		Internet Access/Data Ports	Live Bait
		Type, Location & Fees	Propane & CNG

Under Services are additional power options beyond the basic amperage covered in the "Marina Facilities" section. Communications covers Internet access, specifying the presence of Wi-Fi as well as the locations of broadband or dial-up dataports — plus the fees. Additional information is in "Marina Notes." (A discussion of Wi-Fi systems can be found on page 300 in the Addendum.) Communications also lists couriers that service the marina's area; this does not imply that the marina will manage the process (unless concierge services are offered). Assume that dealing with couriers will be up to the individual boater. The notation Megayacht Facilities means a marina accommodates at least 100 feet LOA with 100 or higher amperage service. Listed here, too, are additional power options, beyond the 30, 50, 100 & 200 amp services covered in the "Marina Facilities" section. Under Supplies are, among many other items, resources for galley fuel. As those who rely on CNG know, it is becoming harder and harder to find. (Please share any sources you discover with your fellow boaters.) Note, too, that West Marine now owns the Boat/US stores, but continues to operate them under the Boat/US name — so they are listed separately.

Boatyard Services

Nearest Boatyard (If not on-site):	Air Conditioning	Metal Fabrication	*Yard Rates:*
Travelift (including tonnage)	Refrigeration	Divers	General Hourly Rate
Railway	Rigger	Bottom Cleaning	Haul & Launch (Blocking included)
Forklift	Sail Loft	Compound, Wash & Wax	Power Wash
Crane	Canvas Work	Inflatable Repairs	Bottom Paint (Paint included?)
Hydraulic Trailer	Upholstery	Life Raft Service	*Boat Storage Rates:*
Launching Ramp	Yacht Interiors	Interior Cleaning	On Land (Inside/Outside)
Engine Mechanics — Gas & Diesel	Brightwork	Yacht Design	In the Water
Electrical Repairs	Painting	Yacht Building	*Memberships & Certifications:*
Electronic Sales	Awlgrip (or similar finish)	Total Refits	ABBRA — No. of Cert. Techs.
Electronic Repairs	Woodworking	Yacht Broker	ABYC — No. of Cert. Techs.
Propeller Repairs	Hull Repairs	*Dealer For:* (Boats, Engines, Parts)	Other Certifications

If the facility does not have a boatyard on-site, then the name and telephone number of the nearest boatyard is provided. In most cases, the services listed as "Nearby" or "Within 1 mile" will be found at that facility. "Dealer For" lists the manufacturers that the Boatyard services and its Authorized Dealerships. "Memberships and Certifications" refers to the two maritime trade organizations (ABBRA — American Boat Builders & Repairers Association & ABYC — American Boat and Yacht Council) which have programs that train and certify boatyard craftspeople and technicians. Several of the other professional maritime organizations and many manufacturers also offer rigorous training and certification on their particular product lines. These are included under "Other Certifications." A brief description of ABBRA and ABYC, and their certification programs, as well as the other major marine industry organizations is provided in the Addenda section on page 10.

Restaurants and Accommodations

Restaurants	Snack Bars	Fast Food	Motels
Seafood Shacks	Coffee Shops	Pizzeria	Inns/B&Bs
Raw Bars	Lite Fare	Hotels	Cottages/Condos

Since food is a major component of cruising, considerable attention has been given to both restaurants and provisioning resources. Eateries of all kinds are included (with phone numbers); full-service restaurants are listed simply as Restaurants. If delivery is available it is either noted or the establishment is listed as "OnCall." An attempt has been made to provide a variety of dining options and, whenever possible, to include the meals served Breakfast, Lunch, Dinner, Sunday Brunch (B, L, and/or D) plus the price range for entrées at each meal. If the menu is Prix Fixe (table d'hôte — one price for 3 - 4 courses), this is indicated in the commentary. On rare occasion, if a restaurant has received very high marks from a variety of reviewers, that will be noted, too. If we are aware of a children's menu, the listing will indicate "Kids' Menu." Often the hours of on-site restaurants are included in "Marina Notes" or "Notable."

Price ranges have been gathered from menus, websites, site visits, marina notes and phone calls. Although these change over time, the range should give you an idea of the general price point. In large cities, the list generally consists of a handful of the closest restaurants — we expect that you will supplement this with a local restaurant guide. In small towns, the list provided may be "exhaustive" — these may be all there are and they may not be close.

Frequently, the need for local off-boat overnight accommodations arises — either for guests, crew changes, or just because it's time for a real shower or a few more amenities or a bed that doesn't rock. We have attempted to list a variety of local lodgings and, whenever possible, have included the room rate, too. The rates listed generally cover a 12-month range. So, if you are cruising in high season, expect the high end of the range. If the lodgings are part of the marina, then there is often a "package deal" of some sort for marina guests wishing to come ashore. We have asked the question about "package deals" and included the answers in "Marina Notes."

Recreation and Entertainment

Pools (heated or not)	Tennis Courts	Bowling	Park
Beach	Golf Course	Sport Fishing Charter	Museum
Picnic Areas & Grills	Fitness Center	Group Fishing Boat	Galleries
Children's Playground	Jogging Paths	Movie Theater	Cultural Attractions
Dive Shop	Horseback Riding	Video Rentals	Sightseeing
Volleyball	Roller Blade & Bicycle Paths	Video Arcade	Special Events

What there is to do, once you're tied up and cleaned up, is often a driving force in choosing a harbor or a particular marina. If you are choosing a facility to spend a lay-day or escape foul weather, the potential land-based activities become even more important. We have created a list of the possible major types of recreation and entertainment activities and have organized them, again, by proximity; if they are more easily reached by dinghy we note that, too.

A public golf course is almost always listed unless it is farther than 25 miles. Boat Rentals include kayaks, canoes and small boats. Group Fishing Boats are sometimes known as "head boats." Museums cover the gamut from art and maritime to historic houses and districts to anthropological and environmental. Cultural Attractions can range from local craft ateliers to aquaria to live theaters to all manner of musical concerts. Sightseeing can range from whale watching to historical walking tours. Special Events usually covers the major once-a-year local tourist extravaganzas — and almost all require significant advance planning (often these also appear in the "Notable" section). Galleries, both fine art and crafts, are listed under Entertainment rather than General Services, since we view them more as opportunities for enlightenment than the shops that they really are. Admission prices are provided for both Adults and Children, when available, and are listed with the Adult price first, followed by the Children's price, i.e., $15/7. Occasionally, there is a family package price, which is also listed. Most entertainment and recreation facilities also offer Student and Senior Citizen pricing and other discounts; unfortunately, we don't have space to note them, but we do provide a phone number, so call and ask.

Provisioning and General Services

Complete Provisioning Service	Bakery	Houses Of Worship	Bookstore
Convenience Store	Farmers' Markets	Catholic Church	Pharmacy
Supermarket — usually major chain	Green Grocer	Protestant Church	Newsstand
Market — smaller, local store	Fishmonger	Synagogue	Hardware Store
Gourmet Shop	Lobster Pound	Mosque	Florist
Delicatessen	Meat Market	Beauty Salon	Retail Shops
Health Food Store	Bank/ATMs	Barber Shop	Department Store
Wine/Beer Purveyor	Post Office	Dry Cleaners	Copy Shops
Liquor/Package Store	Library	Laundry	Buying Club

As noted previously, we think that most boaters travel on their stomachs, so knowing how to find local provisioning resources is very important. In addition, there is a fairly constant need for all kinds of services and supplies. When delivery is available for any of the provisioning resources or services, we've either noted that in the commentary or listed it as "OnCall."

For major provisioning runs, we have tried to identify the closest outlet for a regional supermarket chain. If a smaller, but fairly well supplied, market is close by, we include it as well as the more distant chain supermarket. Most people, we've discovered, really prefer to find interesting, local purveyors, so the presence of a Farmers' Market is notable. Usually these are a one or two-day-a-week events, so the exact days, times and locations are included. To differentiate Farmers' Markets from produce markets and farm stands, the latter are listed as Green Grocers. We've also tried to locate full Provisioning Services that will "do it all" and deliver dockside. And Fishmonger is just another name for fish sellers — these could be regular fish markets or directly "off-the-boat."

In the "General Services" category, we've included the nearest libraries because they can be wonderful sources of all kinds of "local knowledge," and can provide a welcome port on a foul weather day. They also usually have children's programs during the "season" and, with growing frequency, offer data ports or public Internet access on their own PCs. The Laundry in this section should not to be confused with the washers and dryers at the marina. Laundries are usually combination "do it for you" drop-off and self-service operations — and are frequently near restaurants or recreation or entertainment venues.

Transportation

Courtesy Car or Van	Rental Car — Local and Nat'l	Intercity Bus	Airport Limo
Bikes	Taxi	Rail — Amtrak & Commuter	Airport — Regional & Nat'l
Water Taxi	Local Bus	Ferry Service	

Once most cruisers hit land, they are on foot (except for those fortunate souls with sufficient on-board storage to travel with folding bikes). So transportation, in all its guises, becomes a very important consideration. We've divided transportation into two categories — getting around while in port and the longer-range issue of getting to and from the boat. If the marina or boatyard provides some form of courtesy car or van service, it's noted first. These services can include unlimited use of a car (very rare), scheduled use of a car (often 2 hours), an on-demand "chauffeured" van service, a scheduled van service, or a marina manager or employee willing to drive you to "town" at a mutually convenient time. The guests' use of this service is sometimes completely unrestricted; other times, it is reserved exclusively for provisioning or restaurant trips. If the details of this arrangement are simple, they are explained in the commentary; if complicated, they are explained in the "Marina Notes." Courtesy cars and/or vans are one of the most volatile of the marina services so, if this is important to you, call ahead to confirm that it's still available and to ask about the terms.

The Airport Limo services are either individual car services or vans. Rail covers Amtrak as well as commuter services that connect to a city, an Amtrak stop, or an airport. Local Buses also include the seasonal trolleys that are becoming more common (and extraordinarily useful) in more tourist-oriented ports. Local, regional and inter-city ferry services are listed. Rates, when included, are usually for both Adults and Children, and indicate if one-way or round-trip; they are listed with the Adult price first, followed by the Children's price, i.e., $25/17RT. Note that there are usually Senior Citizen and Student prices, but space has precluded their inclusion. Don't forget to ask.

For those of us cruising the coast less than full time, the logistics of going back and forth to the boat is often the stuff of nightmares. Rental cars have a variety of uses — local touring or long distance (back to where you left your car, to the airport, or back home). We list local rental car agencies (for day rents), and regional ones (like Enterprise). We tend not to list the national ones (where pick-up and drop-off may be restricted to airports or downtown locations) because these are obvious and often very difficult to get to. Because Enterprise delivers and picks up — remembering that you have to return the driver to his/her office — we always include the nearest Enterprise office, if one exists, as "OnCall." (If another agency advertises pick-up/delivery, that information is included as "OnCall," too).

Note that some franchise auto rentals, because the outlets are locally or regionally owned, seem to have a wide range of "drop-off" policies. Sometimes, if the region is large enough, it is possible to pick the car up at the current marina and drive to the marina where you left your own car, and leave the rental right there. When available, this service is just great. Call and check. The Airport listing often includes both the nearest regional and the nearest international. Because, as noted, long-distance, one-way car rentals are often based at airport locations, a marina's distance from an airport takes on a larger meaning than just catching a plane.

Medical Services

911 Service	Dentist	Holistic Service	Ambulance	Hospital
Doctor	Chiropractor	Veterinarian	Optician	

The data in this section is provided for informational purposes only and does not imply any recommendation. ACC is simply listing the nearest practitioner in each category. The first listing is the availability of 911 service; this service is surprisingly not ubiquitous — so it is important to know if dialing 911 will "work" in a given area. In the listings for Doctors, preference is given to walk-in clinics, then group practices and then general practitioners or internists. Dentists, Chiropractors, Veterinarians, and Holistic Services are also chosen in a similar fashion. A single practitioner is listed only if a "group" is not nearby. Holistic services will generally list massage therapists, but may also include acupuncturists, energy healers, and yoga classes when we find them. Hospital is usually the nearest major facility; if this is very far away and we are aware of a satellite, that will be noted — especially if there are no physicians nearby.

BOTTOM SECTION

Setting

This section provides a description of the location, environment, and ambiance of each marina, boatyard or mooring field, including its views both landside and waterside, and any easily identifiable landmarks.

Marina Notes

Marina-specific information not included in the middle and top sections is detailed here. The source for this data includes interviews with marina staff, marina literature, marina comments provided to ACC, and surveyor/reviewer observations during site visits. If the rate structure is too complicated to detail in the "Facilities" section, a thorough explanation will be included here, preceded by an asterisk. Anything that is noteworthy or interesting about the facility is described in this paragraph. This includes history, background, recent changes, damage, renovations, new facilities, new management, comments on heads and showers (if they are, for instance, below par for the rating or particularly nice), and general marina atmosphere. If there is a restaurant or some form of lodging on-site, it is noted here, including any special deals for marina guests. If the marine services are part of a resort or a private yacht club with extensive recreation facilities, the level of access to those facilities is also detailed here or in "Notable." The facilities and services available to the visiting cruiser are also reflected in the rating. Certified Clean Marinas are indicated by CCM (see pages 298 – 299 for a complete description).

Notable

This section focuses on what is special or unique about the/each facility — additional items of interest related to the marina itself or the surrounding area. Details on special events, nearby attractions, the local community, ways of "getting around," the best beaches, special things to do, or other noteworthy facilities and/or services are listed here. Occasionally, there's also an elaboration on the on-site or nearby restaurants, amenities, or accommodations. Or, perhaps the marina is near an airport, in a secure basin, the owner lives on-site, or the rates are low — any or all of which might make this a good place to leave the boat for while.

For a complete explanation of the Data Gathering and Review Process, please see page 318.

The Atlantic Cruising Club WebSite and Boaters' Comments

Marina Search — Find a Marina: The ACC website houses the complete 2,500 marina database which underlies the *Atlantic Cruising Club's* regional *Guides to Marinas*. There are two ways for boaters to use the website to find the *right* marina:

▸ **The Map/Chart or Geographic Interface** — Beginning with a map of North America, a boater can "drill down" through increasingly more detailed maps — 1. Choose a specific region; 2. Choose a specific geographic area; then 3. Choose a marina. It's all point & click simple. Pass the cursor over the numbered button and the marina name appears; click and the complete full color Marina Report is displayed — with up to 17 photos.

▸ **The Marina Search Screen Interface** — The same tabbed Marina Search screen that's on the DVD is also on the website. Select up to 100 different search criteria – either singly or in any combination – and the program will display all the marinas that meet those criteria. Or simply choose a harbor to see all the facilities located there; or just type in the name of a marina. The complete full color Marina Report — with all of its photos — will display.

Boaters' Comments – A new section has been added to each online Marina Report allowing boaters to share their experiences. At the top of each online Marina Report are the words, "Add a comment/Read Other Comments." Please share your experiences.

Please Rate the Following on a Scale of 1 (basic) to 5 (world class)

▸ **Facilities & Services** (Fuel, Reservations, Concierge Services and General Helpfulness)

▸ **Amenities** (Pool, Beach, Internet Access, including Wi-Fi, Picnic/Grill Area, Boaters' Lounge)

▸ **Setting** (Views, Design, Landscaping, Maintenance, Ship-Shapeness, Overall Ambiance)

▸ **Convenience** (Access (including delivery services) to Supermarkets, other Provisioning Sources, Shops, Services, Attractions, Entertainment, Sightseeing, Recreation, including Golf and Sport Fishing & Medical Services)

▸ **Restaurants/Eateries/Lodgings** (Availability of Fine Dining, Snack Bars, Lite Fare, OnCall food service, and Lodgings ashore)

▸ **Transportation** (Courtesy Car/Vans, Buses, Water Taxis, Bikes, Taxis, Rental Cars, Airports, Amtrak, Commuter Trains)

▸ **Please Add Any Additional Comments** *(such as ... "Good place to leave the boat unattended for a week/month," "Accommodates catamarans with no up-charge." "Snug spot in a storm," "Excellent fuel prices," "Caters to Sportfish/sailboats/trawlers, etc.," "Boats over 50 ft. LOA/beam may find tight maneuvering," "Many liveaboards," "Megayacht haven," "Hurricane damage still not repaired," "Terrific spot for kids," "Excellent mechanics/electricians/carpenters/varnishers," or "Watch out for alligators.")*

Marina Updates: The Reports in the *Atlantic Cruising Club's Guide to Pacific Northwest Marinas – Book & CD-ROM* are current when published, but marinas are in a fairly constant state of evolution. ACC updates the Marina Reports as facts change, as new information is received from the facilities, or as new site visits by ACC Reviewers provide added insights. These updates are posted on the ACC Website. Reports for new marinas are posted after data has been gathered and site visits made.

Cruising Info and Articles: ACC writers and reviewers share their experiences in articles on destinations, harbors, cruise itineraries, vacations for boaters (on and off boats), provisioning, cruising lifestyles, etc.

Suggested Reading: For every region, there is an extensive reference list of books that cruisers might find interesting. This is an expanded version of the Suggested Reading section in this *Guide* – with photos of the covers, pricing, and direct links for purchases. Additional recommendations are always welcome.

Cruising Links: These are hot URLs to resources on the web that ACC reviewers, writers and editors have found particularly useful. Recommendations are welcome.

Fleet Captains/Cruise Planners: A discussion board just for those planning group events, along with articles and cruise suggestions.

A complimentary six-month Silver Membership is included with the purchase of this *Guide*. Select "Join Now," then "Silver," then follow the instructions.

The Digital Guide on DVD

The enclosed DVD contains the *Atlantic Cruising Club's Digital Guide to Pacific Northwest Marinas*; it includes all of the data and Marina Reports that are in this print version, but in full color, with more than 3,000 color photographs, all searchable on over 100 datafields.

▸ **Installation**

Simply insert the enclosed DVD into your computer's DVD drive. The installation program starts automatically. If it doesn't, click "Start/Run" and enter "d:\ACCsDigitalGuidetoPacificNorthwestMarinas.exe" (assuming your DVD drive is "d"). Or, select "Explore" or "My Computer," click on the DVD drive, then the red & blue ACC icon. During the installation process, you are asked to choose which of the three components you wish to copy to your hard drive — this affects the amount hard disk space the program uses: Program (8.6 MB), Database (27 MB), and Photos file (878 MB). If you have the hard disk capacity, it is best to install all three components. The program will run somewhat faster, and you will not need to have the DVD always at hand.

▸ **The Digital Guide is built around four screens:**

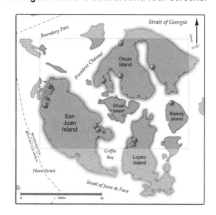

Region and Sub-Region Maps — Click on the "Maps" button to view a map of the Pacific Northwest Region with each of the 20 sub-regions outlined. Click on one of the outlines to access the map for that particular Sub-Region. The locations of each of the marinas in that geographic area are displayed on the map. The location "points" and the marina names are "hot." Click on them to display the Marina Report for that marina. If you would prefer to add additional criteria to a marina search (or to skip the graphics interface entirely), the "search" button on the Pacific Northwest Regional map screen and on each of the Sub-Region map screens takes you directly to the Marina Search screen.

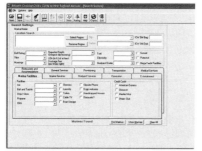

Marina Search Settings — This screen permits the user to enter up to 100 different search criteria — either singly or in combination. You may search for a particular marina by name (or part of a name), for all of the marinas in a particular geographic region, city, or body of water, for marinas able to accommodate your vessel's LOA and draft, etc., etc., etc. If you arrived at this screen from one of the Sub-Region maps, then that Sub-Region is already entered in the "Location Search" box. You may add more sub-regions using the "Select Sub-Region" button. Once you have set the criteria for your search, click the "Find Marinas" button at the bottom of the screen (or the "Find" button in the toolbar). The search result (the number of marinas meeting your criteria) is indicated in the "Marinas Found" field. At this point, either refine your search to generate more or fewer marina choices or click the "Show Marinas" button to proceed to the next screen.

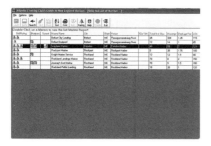

List of Selected Marinas — All of the marinas that meet your criteria are displayed on this screen. Next to each Marina Name are several items of information, including: Bell, Boatyard, and Sunset Ratings, city, state, and harbor. If, during the search, you set a criterion for "Slips," "Moorings," "LOA" or "Dockage Fee," a column for each selection also appears in the "List of Selected Marinas." The List may be sorted either geographically (default) or alphabetically. To view a full Marina Report for any of the marinas on the List, simply double-click on its name.

Marina Report — The Marina Reports in *ACC's Digital Guide* are identical to the Marina Reports in this printed version of the *Guide* but with some enhancements: The Reports are in color and each contains up to to 17 additional full color marina photos. The Marina Reports Screen also lists the names of the other marinas that met your most recent search criteria. Clicking on one of those marina names displays its Marina Report. Finally, there is a box in the lower left hand corner of the Marina Report for you to enter your comments and observations about a given marina. This data is stored within the program and becomes part of the record for that marina. You can also print Marina Reports (and your comments, if you wish) using the Print command.

Note: For proper operation of the Digital Guide, Internet Explorer 6.0 (or higher) must be installed on your computer. Each regional volume of the ACC Guides will automatically install into a separate, clearly labeled folder. They will operate independently.

Acknowledgments

The authors, editors and publisher would like to express their deep appreciation to a number of people who have contributed to the compilation and production of past and present editions of the Atlantic Cruising Club's Guides to Marinas:

Senior Editor: Irina Adams for gracefully managing a very complex process, for her tenacity and commitment to "getting it right," and for her impeccable research and editorial skills. And Matthew Codrin Adams, for his patience, thoughtfulness and perfect timing. **Assistant Editor & Communications Manager:** Catherine Rose for her dogged persistence in acquiring the data and reconfirming its accuracy. **Assistant Editor:** Alice Picon for her early initiatives in defining the scope of the Guide. **Book & Page Design, Cover Execution, Back Cover Design & Execution:** Jessica Gibbon & Rupert Edson at Spark Design for an elegant volume and DVD that has stood the test of many volumes; they were truly the midwives of the new editions. **DVD:** Surendranath Reddy at iRUS Infotech for developing the new DVD format. **Icon Design:** Jennifer Grassmeyer for the pitch perfect icons. **Maps:** Cartographer Amy Rock for her technical skills and elegant graphic interpretations of the GPS data. **Research:** Next Meridian for locating and mapping all the "what's nearby." **Photo Retouching:** Karina Tkach for 238 crisp B&W images and 1,500 sharp and colorful DVD and Web images. Gunshe Ramchandani for an initial 1,500 DVD and Web images. **Proofreading:** Claudia Volkman for being our last line of defense. **Research Assistants:** Jessica Swanston, Brittney Taylor and Sarah Newgaard for their enthusiasm, commitment, and excellent technical skills.

We also want to thank the **members of the Atlantic Cruising Club** for their notes and emails describing their marina experiences. And, most important, the **facilities owners, managers and dockmasters** who provided enormous quantities of detailed information on their marina and boatyard operations, (and reviewed ACC's interpretation at each step along the way), despite their discomfort with their inability to control the final Marina Report. We are grateful to the original **ACC Founder** John Curry, and editors Nancy Schilling and Jennifer Wise, for their impressive initial ground work and to Jason C. Smith for his invaluable vision and technical support in taking the JPI/ACC digital publishing house to the next level.

The authors and editors owe an extraordinary debt of gratitude to the multitude of **tourism and government agencies** (and a few innkeepers) that provided invaluable help along the way. A very special thanks to Doug Treleaven at BC Ferries, Jennifer Huitema & Alexa Tammerle at Vancouver, Coast & Mountains Tourism, Robin Jacobson at San Juan Islands Visitors Bureau, Heidi Wesling & Lana Kingston at Tourism Vancouver Island, Teresa Davis & Carmen Berry at Tourism North Central Island, Ian Hobbs & Donna Kaye at Sevilla Island Resort, Lund, and Robin Richardson at Small Craft Harbours, Fisheries and Oceans Canada for going above and beyond in helping to research and complete this volume.

And our deep appreciation to the many others at the Chambers of Commerce, Convention and Visitors Bureaus and U.S. and Canadian Government offices who provided vital assistance in the research and fact-checking of this volume. They are listed alphabetically by the name of the agency: Sue Keller, Anacortes Chamber of Commerce; Joanne Hansen, Anacortes Visitors Center; Paulette McCoy, Astoria-Warrenton Chamber of Commerce; John Brown, Bandon Chamber of Commerce; Karen Scott & Loreena Lund, City of Gig Harbor; Deanna Hendricks, Chamber of Commerce of Nehalem; Kathy Duncan, Cowichan Chamber of Commerce & Visitor Info Centre; Jennifer Ford, Destination Nanaimo; Al Cannon, Peter Binner & Rich Tamboline, Docks & Wharf Info Victoria; Cari McEdwards & D'Ann Schmit, Forks Chamber of Commerce; Connie Boice, Fort Flagger; Michaela Sugars, Gibsons Chamber of Commerce; Holly Lenk, Greater Victoria Visitors and Convention Bureau; Linda Franz & Leslie Taylor, Harbour Authority Association of British Columbia; Sally Christy, Kingston Chamber of Commerce; Claudia Young, La Conner Chamber of Commerce; Erika Von-Poser, Ladysmith Chamber of Commerce; Becky Smith, Lopez Island Chamber of Commerce; Andre Drummer, Olympia City Parks & Recreation Department; Tamara Garcia, Olympia-Lacey-Tumwater CVB; Martin Callery, Oregon International Port of Coos Bay; Janine Belleque, Oregon State Marine Board; Natasha Kammerle, Pender Harbor Chamber of Commerce; Kerry Thompson, Pender Island Chamber of Commerce; Maureen Edelblutt, Port Ludlow Chamber of Commerce; Anne Grimm, Port of Bellingham Chamber of Commerce; Brian I. Sauer, Ginger McKenzie Waye & Cec Kolb, Port of Bremerton; Vivian Blossom, Port of Everett Tourism; Susan Browne, Port of Gold Beach Chamber of Commerce; Cindy Weigardt; Port of ilwaco Chamber of Commerce; Megan Deinas & Laurie Lohrer, Port of Seattle; Susan Grantham, Port Townsend Chamber of Commerce; Jenny Docking, Powell River Visitor's Center; Barbara Carroll, Quilcene/Brinnon Chamber of Commerce; Perry Ruehlen, Salt Spring Island Visitor Information Centre; Louise Lightfoot, Seabeck Chamber of Commerce; David Blandford, Seattle's Convention & Visitors Bureau; Laurie Dunlap, Seattle Parks and Recreation; Kasey Cronfist, Shelton Chamber of Commerce; Reg Mooney, Sidney Chamber of Commerce; Julie Gangler, Tacoma Regional Convention & Visitor Bureau; Louise Dreyer, Thea Foss Development Authority; Janice Greenwood, Tourism British Columbia; Wendy Klyne, Tourism Cowichan; Hannah King, Tourism Nanaimo; Emily Armstrong, Tourism Vancouver; Heather McGillivray, Tourism Victoria; Louise Carlow, Tourism Victoria Visitor Info Centre; Wendy Underwood, Vancouver Chamber of Commerce; Virginia Painter, Washington State Dept. of Tourism; Linda Burnett, Washington State Parks; Carrie Wilkinson-Tuma, Washington Tourism; Julie Clement, West Vancouver Chamber of Commerce; Cheryl Collins, Whatcom County Tourism; Anne Steele, Willapa Chamber of Commerce. Also our appreciation to Jean & Dick Billingsly, Sechelt Inlet B&B, Sechelt, BC; Betsy & Gerry Parker, Seascape B&B, Gibsons; John Douglas at Poets Cove, and Janeen Jennings, Best Western Friday Harbor Suites.

Photo Credits: Our appreciation for the many wonderful images generously provided to enhance those taken at the time of our visits: Tom Woltjer's photos (tomsimages.com) enrich Reports in Poulsbo, Port Orchard, Bremerton, Silverdale, Sequim, Gig Harbor and Bainbridge Island. Aerial images of the Harbour Authorities at Campbell River, Comox Harbour, Cowichan, Crofton, Deep Bay, Discovery, Gibsons, Ladysmith, Lund, Maderia, Powell, Salt Spring (Fulford, Ganges & Vesuvius), and Whaler Bay were reproduced with permission of the Minister of Public Works and Government Services Canada, 2006. Nanaimo Harbour image courtesy of Chris Cheadle, Gig Harbor by Jim Nelson and Port Townsend by Paul Boyer & Valerie Henschel. Other images were courtesy of Tacoma Region Convention & Visitors Bureau, Deddeda Stemler & Tourism Victoria, Olympia-Lacey-Tumwater Visitor & Convention Bureau, Walla Walla College Marine Station, and BrewBook. Marinas that provided images include Pacific Playground, Stones Marina, Maple Bay Marina, Hope Bay, Butchart Gardens, Canoe Cove, Goldstream Boathouse, Oak Bay Marina, Van Isle Marina, Islands Marine, Eagle Harbor, Port of Brownsville, Nautical Landing, Olson's Resort, Charleston Harbor – Coos Bay, Poulsbo, Foss Waterway Development Authority, John Wayne Marina, April Point, Fisherman's Resort, Sunshine Coast Resort, Irvines Landing & Alderbrook.

ATLANTIC CRUISING CLUB'S

GUIDE TO
PACIFIC NORTHWEST MARINAS

THE
MARINA REPORTS

Geographical Listing of Marinas

Marina Name	Harbor	City, State	Page No.
5. BC – LADYSMITH to COWICHAN BAY			**73**
Ladysmith Maritime Society	Ladysmith Harbour	Ladysmith, BC	74
Page Point Inn & Marina	Ladysmith Harbour	Ladysmith, BC	75
Chemainus Public Wharf	Stewart Channel	Chemainus, BC	76
Crofton Public Wharf	Osborne Bay	Crofton, BC	77
Maple Bay Public Wharf	Maple Bay	Duncan, BC	78
Bird's Eye Cove Marina	Maple Bay/Bird's Eye Cove	Duncan, BC	79
Maple Bay Marina	Maple Bay/Bird's Eye Cove	Duncan, BC	80
Cherry Point Marina	Cowichan Bay/Cherry Point	Cowichan Bay, BC	81
Genoa Bay Marina	Cowichan Bay/Genoa Bay	Duncan, BC	82
Fishermen's Wharf Association	Cowichan Bay	Cowichan Bay, BC	83
Cowichan Bay Marina	Cowichan Bay	Cowichan Bay, BC	84
Pier 66 Marina	Cowichan Bay	Cowichan Bay, BC	85
Bluenose Marina	Cowichan Bay	Cowichan Bay, BC	86
6. BC – GULF ISLANDS			**87**
Silva Bay Resort & Marina	Silva Bay	Gabriola Island, BC	88
Page's Resort & Marina	Silva Bay	Gabriola Island, BC	89
Telegraph Harbour Marina	Telegraph Harbour	Thetis Island, BC	90
Thetis Island Marina	Telegraph Harbour	Thetis Island, BC	91
Montague Harbour Marine Park	Montague Harbour	Galiano, BC	92
Montague Harbour Public Wharf	Montague Harbour	Galiano, BC	93
Montague Harbour Marina	Montague Harbour	Galiano, BC	94
Whaler Bay Public Wharf	Whaler Bay	Galiano, BC	95
Sturdies Bay Public Float	Sturdies Bay	Galiano, BC	96
Vesuvius Bay Public Wharf	Vesuvius Bay	Saltspring Island, BC	97
Centennial Wharf	Ganges Harbour	Saltspring Island, BC	98
Kanaka Visitors Wharf	Ganges Outer Harbour	Saltspring Island, BC	99
Ganges Marina	Ganges Outer Harbour	Saltspring Island, BC	100
Salt Spring Marina	Ganges Outer Harbour	Saltspring Island, BC	101
Fulford Harbour Wharf	Fulford Harbour	Saltspring Island, BC	102
Fulford Harbour Marina	Fulford Harbour	Saltspring Island, BC	103
Port Washington Public Wharf	Grimmer Bay	North Pender Island, BC	104
Otter Bay Marina	Otter Bay	North Pender Island, BC	105
Hope Bay Public Wharf	Hope Bay	North Pender Island, BC	106
Browning Harbour Public Wharf	Browning Harbour	North Pender Island, BC	107
Port Browning Marina	Browning Harbour	North Pender Island, BC	108
Poets Cove Resort & Spa	Bedwell Harbour	South Pender Island, BC	109
7. BC – SAANICH PENINSULA & VICTORIA			**111**
Deep Cove Marina	Saanich Inlet/Deep Cove	North Saanich, BC	112
Mill Bay Marina	Saanich Inlet/Mill Bay	Mill Bay, BC	113
Compass Rose Cabins & Marina	Saanich Inlet/Brentwood Bay	Brentwood Bay, BC	114
Brentwood Bay Lodge & Spa Marina	Saanich Inlet/Brentwood Bay	Brentwood Bay, BC	115
Butchart Gardens	Saanich Inlet/Brentwood Bay	Brentwood Bay, BC	116
Goldstream Boathouse Marina	Saanich Inlet/Finlayson Arm	Victoria, BC	117
Swartz Bay Public Wharf	Swartz Bay	Sidney, BC	118
Canoe Cove Marina	Canoe Cove	Sidney, BC	119
Westport Marina	Tsehum Harbour	Sidney, BC	120
Van Isle Marina	Tsehum Harbour	Sidney, BC	121
Port Sidney Marina	Sidney Harbour	Sidney, BC	122
Oak Bay Marina	Oak Bay	Victoria, BC	123
Fisherman's Wharf	Victoria Middle Harbour	Victoria, BC	124

Continued on the next page

Marina Name	Harbor	City, State	Page No.
12. WA – HOOD CANAL			**175**
Quilcene Marina	Hood Canal/Quilcene Bay	Quilcene, WA	176
Pleasant Harbor State Park	Hood Canal/Pleasant Harbor	Brinnon, WA	177
Pleasant Harbor Marina	Hood Canal/Pleasant Harbor	Brinnon, WA	178
Port of Hoodsport	Hood Canal	Hoodsport, WA	179
Hood Canal Marina	Hood Canal	Union, WA	180
Alderbrook Resort and Spa	Hood Canal	Union, WA	181
Twanoh State Park	Hood Canal	Union, WA	182
Port of Allyn North Shore Dock	Hood Canal	Allyn, WA	183
13. WA – WEST CENTRAL PUGET SOUND			**185**
Bainbridge Island Marina	Eagle Harbor	Bainbridge Island, WA	186
Eagle Harbor Marina	Eagle Harbor	Bainbridge Island, WA	187
Eagle Harbor Waterfront Park	Eagle Harbor	Bainbridge Island, WA	188
Winslow Wharf Marina	Eagle Harbor	Bainbridge Island, WA	189
Harbour Marina	Eagle Harbor	Bainbridge Island, WA	190
Bremerton Marina	Sinclair Inlet	Port Orchard, WA	191
Port Orchard Marina	Sinclair Inlet	Port Orchard, WA	192
Port Washington Marina	Washington Narrows	Bremerton, WA	193
Port of Silverdale	Dyes Inlet	Silverdale, WA	194
Illahee State Park	Port Orchard Bay	Bremerton, WA	195
Port of Illahee	Port Orchard Bay	Bremerton, WA	196
Port of Brownsville	Burke Bay	Bremerton, WA	197
Port of Poulsbo	Liberty Bay	Poulsbo, WA	198
14. WA – SEATTLE AREA			**199**
Shilshole Bay Marina	Shilshole Bay	Seattle, WA	200
Ballard Mill Marina	Lake Washington Ship Canal	Seattle, WA	201
Salmon Bay Marina	Lake Wash. Ship Canal/Salmon Bay	Seattle, WA	202
Fishermen's Terminal	Lake Wash. Ship Canal/Salmon Bay	Seattle, WA	203
Ewing Street Moorings	Lake Washington Ship Canal	Seattle, WA	204
Lee's Landing	Lake Union	Seattle, WA	205
Commercial Marine	Lake Union	Seattle, WA	206
Nautical Landing	Lake Union	Seattle, WA	207
AGC Marina	Lake Union	Seattle, WA	208
Fairview Marinas	Lake Union	Seattle, WA	209
Kirkland Marina Park	Lake Washington	Kirkland, WA	210
Carillon Point Marina	Lake Washington	Kirkland, WA	211
Lakewood Moorage	Lake Washington/Andrews Bay	Seattle, WA	212
Elliott Bay Marina	Elliott Bay	Seattle, WA	213
Bell Harbor Marina	Elliott Bay	Seattle, WA	214
South Park Marina	Duwamish River	Seattle, WA	215
15. WA – SOUTHEAST PUGET SOUND			**217**
Des Moines Marina	East Passage	Des Moines, WA	218
Saltwater State Park	Poverty Bay	Des Moines, WA	219
Jerisich Dock	Gig Harbor	Gig Harbor, WA	220
Arabella's Landing	Gig Harbor	Gig Harbor, WA	221
Point Defiance Boathouse Marina	Dalco Passage/Commencement Bay	Tacoma, WA	222
Breakwater Marina	Dalco Passage/Commencement Bay	Tacoma, WA	223
Ole and Charlie's Marina	Hylebos Waterway	Tacoma, WA	224
Chinook Landing Marina	Hylebos Waterway	Tacoma, WA	225

Continued on the next page

Marina Name	Harbor	City, State	Page No.
15. WA – SOUTHEAST PUGET SOUND, continued			
Foss Waterway Marina	Thea Foss Waterway	Tacoma, WA	226
Johnny's Dock	Thea Foss Waterway	Tacoma, WA	227
Dock Street Marina	Thea Foss Waterway	Tacoma, WA	228
Foss Landing Marina & Storage	Thea Foss Waterway	Tacoma, WA	229
Steilacoom Marina	Puget Sound/Gordon Point	Steilacoom, WA	230
16. WA – SOUTHWEST PUGET SOUND			**231**
Penrose Point State Park	Carr Inlet/Mayo Cove	Lakebay, WA	232
Lakebay Marina	Carr Inlet/Mayo Cove	Lakebay, WA	233
Longbranch Marina	Filucy Bay	Lakebay, WA	234
Tolmie State Park	Nisqually Reach	Olympia, WA	235
Zittel's Marina	Johnson Point	Olympia, WA	236
Joemma Beach State Park	Case Inlet/Whiteman Cove	Lakebay, WA	237
Jarrell's Cove Marina	Pickering Passage/Jarrell Cove	Shelton, WA	238
Jarrell Cove Marine State Park	Pickering Passage/Jarrell Cove	Shelton, WA	239
Fair Harbor Marina	Case Inlet/Fair Harbor	Grapeview, WA	240
Port of Allyn Waterfront Park	Case Inlet/North Bay	Allyn, WA	241
Port of Shelton	Oakland Bay	Shelton, WA	242
Boston Harbor Marina	Boston Harbor	Olympia, WA	243
West Bay Marina	Budd Inlet	Olympia, WA	244
Port Plaza Guest Dock	Budd Inlet	Olympia, WA	245
Percival Landing Moorage	Budd Inlet	Olympia, WA	246
Swantown Marina & Boatworks	Budd Inlet	Olympia, WA	247
17. WA – STRAIT OF JUAN DE FUCA			**249**
Makah Marina	Neah Bay	Neah Bay, WA	250
Olson's Resort and Marina	Clallam Bay	Sekiu, WA	251
Port Angeles Boat Haven	Port Angeles Harbor	Port Angeles, WA	252
Port Angeles City Pier	Port Angeles Harbor	Port Angeles, WA	253
John Wayne Marina	Sequim Bay	Sequim, WA	254
Sequim Bay State Park	Sequim Bay	Sequim, WA	255
Fort Worden State Park	Admiralty Inlet/Point Wilson	Port Townsend, WA	256
Point Hudson Resort and Marina	Port Townsend Bay	Port Townsend, WA	257
Port Townsend Boat Haven	Port Townsend Bay	Port Townsend, WA	258
Old Fort Townsend State Park	Port Townsend Bay	Port Townsend, WA	259
Port Hadlock Marina	Port Townsend Bay/Port Hadlock Bay	Port Hadlock, WA	260
18. WA – SOUTH COAST			**261**
Quileute Marina	Quillayute River	La Push, WA	262
Westport Marina	Grays Harbor	Westport, WA	263
Tokeland Marina	Willapa Harbor	Tokeland, WA	264
Port of Willapa Harbor	Willapa Harbor	Raymond, WA	265
Port of Ilwaco	Columbia River/Ilwaco Harbor	Ilwaco, WA	266
Port of Chinook	Columbia River/Baker Bay	Chinook, WA	267
19. OR – COLUMBIA RIVER to YAQUINA BAY			**269**
Hammond Marina	Columbia River/Hammond Harbor	Hammond, OR	270
Skipanon Landing	Columbia River/Skipanon Waterway	Warrenton, OR	271
Warrenton City Mooring Basin	Columbia River/Skipanon Waterway	Warrenton, OR	272
Port of Astoria — West Basin	Columbia River/Port of Astoria	Astoria, OR	273

Continued on the next page

Marina Name	Harbor	City, State	Page No.
19. OR – COLUMBIA RIVER to YAQUINA BAY, continued			
Port of Astoria — East Basin	Columbia River/Port of Astoria	Astoria, OR	274
Wheeler Marina	Nehalem River	Wheeler, OR	275
Garibaldi Marina	Tillamook Bay	Garibaldi, OR	276
Port of Garibaldi	Tillamook Bay	Garibaldi, OR	277
Depoe Bay Marina	Depoe Bay	Depoe Bay, OR	278
Port of Newport Marina & RV Park	Yaquina River/Yaquina Bay	Newport, OR	279
Embarcadero Resort Hotel & Marina	Yaquina River/Yaquina Bay	Newport, OR	280
20. OR – SIUSLAW RIVER to BROOKINGS			**281**
Port of Siuslaw RV Park & Marina	Siuslaw Bay	Florence, OR	282
Salmon Harbor Marina	Umpqua River/Salmon Harbor	Winchester Bay, OR	283
Charleston Marina on Coos Bay	Coos Bay/South Slough	Charleston, OR	284
Coos Bay City Dock	Coos Bay Harbor	Coos Bay, OR	285
Port of Bandon	Coquille River/Bandon Harbor	Bandon, OR	286
Port of Port Orford	Port of Port Orford	Port Orford, OR	287
Port of Gold Beach	Rogue River/Gold Beach Harbor	Gold Beach, OR	288
Port of Brookings Harbor	Chetco River/Chetco Cove	Brookings, OR	289

CUSTOMS, IMMIGRATION & PLEASURE CRAFT PORTS OF ENTRY

Customs and immigration regulations on both sides of the border can change without much notice so it is important to keep abreast of them. Currently, U.S. citizens returning from Canada will have to present their passports and, after June 1, 2009, Canada will also require passports in order to cross its border. For Canadian citizens, passports are currently required to enter the U.S. and will be required when returning after June 2009. These regulations apply to all aboard your vessel, so remind your guests as well. In the border region Marina Reports, ACC has noted the nearest customs clearance office and its contact info. If everyone aboard has a one-year I-68 form or a very handy five-year NEXUS card, then it is possible to clear-into the U.S. by phone (800-562-5943) after entering U.S. waters. Otherwise it is necessary to appear in person. For boaters traveling from the U.S. to Canada, NEXUS and CANPASS offer expedited clearance — except the call must be made up to four hours before entering Canadian waters (888-226-7277).

United States Customs and Border Protection
Designated U.S. Small Boat Ports of Entry:

Mid-September to Mid-May 9am-5pm
Mid-May to Mid-September 8am-8pm

- ▸ **Friday Harbor:** (360) 378-2080
- ▸ **Roche Harbor:** (360) 378-2080
- ▸ **Port Angeles:** (360) 457-4311
- ▸ **Point Roberts:** (360) 945-2314
- ▸ **Anacortes:** (360) 293-2331

Canada Customs and Revenue Agency
Designated Canadian Small Boat Ports of Entry:

May to September 8am-8pm
October to April 8am-5pm

- ▸ **Campbell River:** Discovery Harbor; Coast Marina
- ▸ **Nanaimo:** Brechin Pt. Marina; Nanaimo Harbour Commission; Townsite Marina (CANPASS only)
- ▸ **Pender Islands:** Bedwell Harbour (seasonal); Port Browning Marina
- ▸ **Sidney:** Canoe Cove Marina; Port Sidney Marina; RVYC-Tseum Harbour; Van Isle Marina
- ▸ **Ucluelet:** 52 Steps Dock
- ▸ **Vancouver:** All Coal Harbour/Burrard marinas; False Creek Public Docks
- ▸ **Victoria:** Oak Bay Marina; Royal Victoria Yacht Club (CANPASS only); Victoria Customs Float-Inner Harbour
- ▸ **White Rock:** White Rock Public Dock; Crescent Beach Marina

Please visit www.AtlanticCruisingClub.com for the U.S. and Canadian Custom Offices' Boaters Fact Sheet. In addition, you will find the U.S. and Canadian Custom Offices' Boaters Fact Sheet — which also describes the all-important import restrictions. You will also find details on how to secure an I-68 form, NEXUS card or CANPASS card and more information on these Alternative Inspection Systems Programs — including a list of Centers a list of all issuing sites and reporting centers.

Boaters' Notes

_____

_____

_____

_____

_____

_____

_____

_____

_____

_____

_____

_____

_____

_____

_____

_____

_____

_____

_____

_____

_____

_____

_____

Add Your Ratings and Reviews at www.AtlanticCruisingClub.com

A complimentary six-month Silver Membership is included with the purchase of this _Guide_. Select "Join Now," then "Silver," then follow the instructions.

The AtlanticCruisingClub website provides updated Marina Reports, Destination and Harbor Articles and much more — including an option within each online Marina Report for boaters to add their ratings and comments regarding that facility. Please log on frequently to share your experiences — and to read other boaters' comments.

On the website, boaters may rate marinas in one or more of the following categories — on a scale of 1 (basic) to 5 (world class) — and also enter additional commentary.

▸ **Facilities & Services** (Fuel, Reservations, Concierge Services and General Helpfulness)

▸ **Amenities** (Pool, Beach, Internet Access, including Wi-Fi, Picnic/Grill Area, Boaters' Lounge)

▸ **Setting** (Views, Design, Landscaping, Maintenance, Ship-Shapeness, Overall Ambiance)

▸ **Convenience** (Access — including delivery services — to Supermarkets, other Provisioning Sources, Shops, Services, Attractions, Entertainment, Sightseeing, Recreation, including Golf and Sport Fishing & Medical Services)

▸ **Restaurants/Eateries/Lodgings** (Availability of Fine Dining, Snack Bars, Lite Fare, OnCall food service, and Lodgings ashore)

▸ **Transportation** (Courtesy Car/Vans, Buses, Water Taxis, Bikes, Taxis, Rental Cars, Airports, Amtrak, Commuter Trains)

▸ **Please Add Any Additional Comments**

1. BC – Desolation Sound

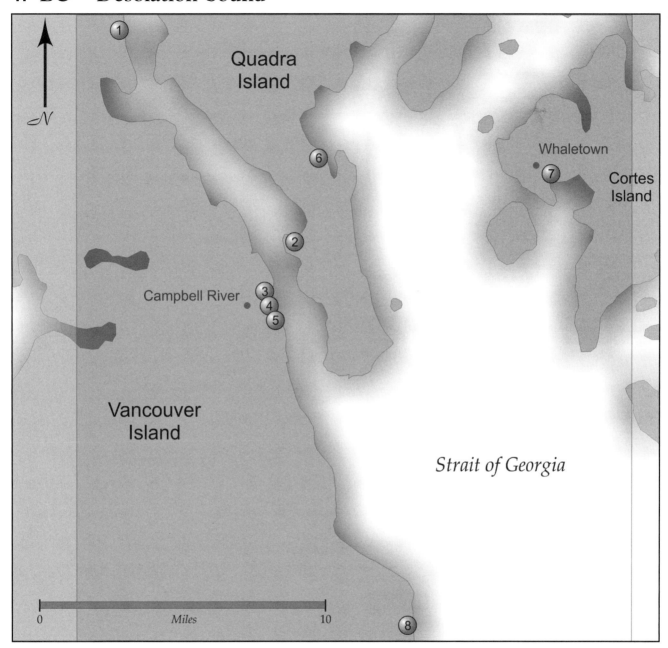

MAP	MARINA	HARBOR	PAGE	MAP	MARINA	HARBOR	PAGE
1	Brown's Bay Marina	Discovery Pas./Brown Bay	24	5	Campbell River Harbour Auth.	Discovery Pas./Campbell River	28
2	April Point Resort, Spa & Marina	Gowlland Harbor/April Point	25	6	Heriot Bay Inn & Marina	Drew Harbour/Heriot Bay	29
3	Discovery Harbour Marina	Discovery Pas./Campbell River	26	7	Gorge Harbour Marina & Resort	Gorge Harbour	30
4	The Coast Discovery Inn & Marina	Discovery Pas./Campbell River	27	8	Pacific Playgrounds Int'l. Marina	Strait of Georgia	31

▶ **Currency** — In Canadian Marina Reports, all prices are in Canadian dollars. In U.S. Marina Reports, all prices are in U.S. dollars.

▶ **"CCM"** — Denotes a Certified Clean Marina, a state/provincial award for environmental excellence. See page 298 for an explanation and page 299 for a list of Pump-Out facilities.

▶ **Ratings & Reviews** — An overview of the Atlantic Cruising Club's rating system is on page 6 and details on the content of each Marina Report are on pages 7 – 11.

▶ **Marina Report Updates** — Comments from boaters and new information from ACC reviewers and marinas are posted regularly on www.AtlanticCruisingClub.com.

Brown's Bay Marina

15021 Brown's Bay Road; Campbell River, BC V9W 7H6

Tel: (250) 286-3135 **VHF: Monitor** Ch. 66A **Talk** Ch. 66
Fax: (250) 286-0951 **Alternate Tel:** n/a
Email: n/a **Web:** www.brownsbayresort.com
Nearest Town: Campbell River *(12 mi.)* **Tourist Info:** (250) 287-4636

Navigational Information
Lat: 50°09.715' **Long:** 125°22.511' **Tide:** 15 ft. **Current:** 5 kt. **Chart:** 3312
Rep. Depths (*MLW*): Entry 100 ft. **Fuel Dock** 100 ft. **Max Slip/Moor** 100 ft./-
Access: 2 miles north of Seymour Narrows on Discovery Passage

Marina Facilities *(In Season/Off Season)*
Fuel: *Chevron* - Gasoline, Diesel
Slips: 90 Total, 20 Transient **Max LOA:** 100 ft. **Max Beam:** 50 ft.
 Rate *(per ft.):* **Day** $0.85/0.64 **Week** $4.83/3.50 **Month** $12/9
 Power: 30 amp $4*, **50 amp** $5, **100 amp** n/a, **200 amp** n/a
 Cable TV: No **Dockside Phone:** No
 Dock Type: Floating, Long Fingers, Alongside, Wood
Moorings: 0 Total, 0 Transient **Launch:** n/a
 Rate: Day n/a **Week** n/a **Month** n/a
Heads: 2 Toilet(s), 1 Shower(s)
Laundry: 2 Washer(s), 2 Dryer(s) **Pay Phones:** Yes, 3
Pump-Out: OnSite **Fee:** n/a **Closed Heads:** No

Marina Operations
Owner/Manager: John Dawson/Mike Sparks **Dockmaster:** Same
In-Season: May-Sep, 6am-6pm **Off-Season:** Oct-Apr, 7:30am-3pm
After-Hours Arrival: Call ahead
Reservations: Yes, Preferred **Credit Cards:** Visa/MC, Amex, Debit
Discounts: None
Pets: Welcome, Dog Walk Area **Handicap Access:** No

Marina Services and Boat Supplies
Services - Docking Assistance, Boaters' Lounge, Trash Pick-Up, Dock Carts **Communication -** Phone Messages, Fax in/out *(Free)* **Supplies -** OnSite: Ice *(Block, Cube)*, Ships' Store, Bait/Tackle, Live Bait **3+ mi:** West Marine *(12 mi.)*, Propane *(12 mi.)*

Boatyard Services
3+ mi: Railway *(12 mi.)*, Engine mechanic *(diesel)*. **Nearest Yard:** Ocean Pacific (250) 285-1011

Restaurants and Accommodations
OnSite: Restaurant *(The Narrows 287-3512, B $7-10, L $8-15, D $12-18, floating restaurant; 7 days)* **3+ mi:** Restaurant *(San Marcos Steak House 287-7066, L $7-12, D $16-26, 12 mi.)*, *(Chan's Kitchen 286-6776, 12 mi.)*, *(Royal Coachman Pub 286-0231, 12 mi.)*, Seafood Shack *(Joey's Only Seafood 287-4482, 12 mi.)*, *(Crabby Bob's 923-4674, 12 mi.)*, Fast Food *(McDonald's, Subway, Wendy's 12 mi.)*, Lite Fare *(Boston Pizza 286-6120, 12 mi.)*, Motel *(Ramada 286-1131, 12 mi.)*, *(Town Centre Inn 287-8866, 12 mi.)*, *(Painters Lodge 286-1108, 12 mi.)*, Hotel *(Coast Discovery 800-663-1144, 12 mi.)*, Inn/B&B *(Rivers Ridge B&B 286-9696, 12 mi.)*

Recreation and Entertainment
OnSite: Beach, Group Fishing Boat *(Hook & Reel, White Wolf, Discovery Ocean Tours)*, Sightseeing *(over 300 Alaska-bound vessels pass by each year)* **3+ mi:** Heated Pool *(12 mi.)*, Golf Course *(Sequoia Springs 287-4970, 12 mi.)*, Movie Theater *(Galaxy 286-1744, 12 mi.)*, Museum *(Maritime Heritage Centre 286-3161, Campbell River 287-3103, 12 mi.)*, Cultural Attract *(Tidemark Theatre 287-7465, 12 mi.)*, Galleries *(Pier Street 286-9717, 12 mi.)*

Provisioning and General Services
OnSite: Clothing Store *(Foul weather gear, fleeces, T-shirts, etc - many logo products)* **3+ mi:** Supermarket *(Safeway 287-4900,, 12 mi.)*, Farmers' Market *(Campbell River Pier St. from Apr-Sep Sun 9am-2pm, 12 mi.)*, Pharmacy *(Peoples Drug Mart 287-8311, 12 mi.)*, Hardware Store *(Home Hardware 287-7147, 12 mi.)*, Retail Shops *(Real Canadian Superstore 830-2736, 12 mi.)*, Department Store *(Zellers 287-7922, 12 mi.)*

Transportation
3+ mi: Rental Car *(Budget, 12 mi.)*, Taxi *(Bee Line 287-8383, 12 mi.)*, Ferry Service *(Campbell River/Quadra Island, 12 mi.)* **Airport:** Campbell River/YVR *(18 mi./150 mi.)*

Medical Services
911 Service **OnCall:** Ambulance **3+ mi:** Doctor *(River City Medical 287-2111, 12 mi.)*, Dentist *(Arnet 287-7343, 12 mi.)*, Holistic Services *(Sense of Wellness 830-8012, 12 mi.)*, Veterinarian *(Greenwood 286-1129, 12 mi.)* **Hospital:** Campbell River 287-7111 *(12 mi.)*

Setting -- Located just north of Seymour Narrows on the Vancouver Island side of Discovery Passage, Brown's Bay offers quiet moorage in a deep half-moon bay protected by a breakwater of railroad tankers. The docks form a floating "Y" - the two arms host slips with finger piers and the stem links them to the shore. At the nexus, perched on floating barges, are the main office, a small marine store, a fuel dock, adventure businesses and the Narrows Floating Restaurant.

Marina Notes -- *15 & 20 amp service available. Closed weekends & holidays off season. Brown's Bay sponsors a salmon fishing derby in mid-October where 50+ pounders are not unusual. The marina shares the cove with the attractive, forest green buildings of the Brown's Bay Packing Company - an aqua culture operation and fish packing plant directly north of the docks. The laundry room is in a separate building beyond the parking lot. It has a nicely varnished pine table and benches, two washers, two dryers and extensive Formica countertops for folding.

Notable -- The on-site Narrows Floating Restaurant serves breakfast, lunch and dinner seven days a week in season - mid-May through October. Three dining venues feature pine tables and benches - some upholstered in a fresh bright blue Naugahyde. Eat inside, or in a covered porch area directly adjacent to the marina. There is a separate floating dining area protected by large transparent panels that create a cozy spot during bad weather. Ripple Rock RV Park's arc of water-view concrete pads hovers on a ridge above the cove; the Octagon Clubhouse, and its adjacent indoor hot tub, overlooks the passage. Almost all other services are in Campbell River - it's 5 km. from the marina to Route 19 and then 12 miles due south to Campbell River - it's faster by boat.

PHOTOS ON DVD: 14

1. BC - DESOLATION SOUND

Navigational Information
Lat: 50°03.709' **Long:** 125°13.690' **Tide:** 13 ft. **Current:** 4 kt. **Chart:** 3312
Rep. Depths *(MLW)*: **Entry** 9 ft. **Fuel Dock** n/a **Max Slip/Moor** 35 ft./-
Access: Discovery Pas. to Gowlland Harb., hug red spar buoy on starboard

Marina Facilities *(In Season/Off Season)*
Fuel: No
Slips: 100 Total, 10* Transient **Max LOA:** 150 ft. **Max Beam:** n/a
  Rate *(per ft.)*: **Day** $1.00 **Week** n/a **Month** $25.50
  Power: 30 amp $5** **, 50 amp** $12, **100 amp** n/a, **200 amp** n/a
Cable TV: Yes, $5 **Dockside Phone:** No
Dock Type: Floating, Pilings, Alongside, Wood
Moorings: 0 Total, 0 Transient **Launch:** n/a, Dinghy Dock
  Rate: Day n/a **Week** n/a **Month** n/a
Heads: 2 Toilet(s), 2 Shower(s) *(with dressing rooms)*
Laundry: 3 Washer(s), 3 Dryer(s), Iron, Iron Board **Pay Phones:** Yes, 2
Pump-Out: No **Fee:** n/a **Closed Heads:** No

Marina Operations
Owner/Manager: Oak Bay Marine Group/Lindsay Humber **Dockmaster:** n/a
In-Season: May-Oct, 7am-9pm **Off-Season:** Closed
After-Hours Arrival: Call ahead
Reservations: Yes, Preferred **Credit Cards:** Visa/MC, Amex
Discounts: None
Pets: Welcome, Dog Walk Area **Handicap Access:** Yes, Docks

April Point Resort & Marina

PO Box 248; Campbell River, BC V9W 4Z9

Tel: (250) 285-2222; (800) 663-7090 **VHF: Monitor** Ch. 66A **Talk** Ch. 66A
Fax: (250) 598-1361 **Alternate Tel:** (250) 598-3366
Email: aprilpoint@obmg.com **Web:** www.obmg.com
Nearest Town: Quathiaski Cove *(1.5 mi.)* **Tourist Info:** (866) 830-1113

Marina Services and Boat Supplies
Services - Docking Assistance, Concierge, Room Service to the Boat, Security, Trash Pick-Up, Dock Carts **Communication -** Mail & Package Hold, Phone Messages, Fax in/out, Data Ports *(Rear office or Wi-Fi Broadband Xpress)*, Express Mail *(Sat Del)* **Supplies - OnSite:** Ice *(Block, Cube)* **1-3 mi:** Ships' Store, Bait/Tackle, Propane *(Petro Canada)*

Boatyard Services
OnSite: Launching Ramp **OnCall:** Engine mechanic *(gas, diesel)*, Rigger, Sail Loft, Canvas Work, Bottom Cleaning, Brightwork, Air Conditioning, Refrigeration, Divers, Compound, Wash & Wax, Interior Cleaning, Propeller Repairs, Inflatable Repairs **Nearest Yard:** Cape Mudge (250) 285-2155

Restaurants and Accommodations
OnSite: Restaurant *(April Point Resort B $8-12, L $9-14, D $18-30, kids' menu $8, jazz Tue nights)*, Hotel *(April Point Resort 285-2222, $145-319, 57 rooms)*, Condo/Cottage *($189-429, 1-4 bedroom suites/houses)*
1-3 mi: Restaurant *(Legends at Painter's Lodge via free water shuttle; Sun brunch $25)*, Coffee Shop *(Dockers Espresso 285-3338, coffee, hot dogs, snacks)*, Lite Fare *(Tybee Pub at Painter's Lodge)*, Pizzeria *(Joe's Pizza and Subs 285-2122)*, Hotel *(Painters Lodge $175-299)*

Recreation and Entertainment
OnSite: Picnic Area, Grills, Boat Rentals *(ocean kayaks)*, Fishing Charter *(Boston whaler 2/$285, Cruiser 3/$459, Traditional Tyee rowboats 2/$159)*, Sightseeing *(whale watching $169/105, wildlife adventure or ocean rapids $105/85,)* **Under 1 mi:** Fitness Center *(Quadra Fitness 285-2144)*
1-3 mi: Pool *(Painters Lodge - comp. shuttle)*, Beach, Playground, Dive

Shop, Tennis Courts, Golf Course *(Sequoia Springs near Painters 287-4970)*, Video Rental *(Quadra Station 285-3222)*, Park *(Community Centre Trails)*, Museum *(Kwagiulth Museum 285-3733)*, Galleries *(Chris Rose Soapstone 285-3245, James Pottery 285-3108, Batoche Gallery 285-2044)*

Provisioning and General Services
OnCall: Dry Cleaners **Near:** Newsstand **1-3 mi:** Market *(Quadra Foods 285-3391, organic produce, deli, liquor; call in orders for pick-up or delivery)*, Delicatessen, Health Food, Liquor Store, Farmers' Market, Green Market *(Mapleview Farm 285-2508)*, Fishmonger, Bank/ATM, Catholic Church, Beauty Salon, Barber Shop, Pharmacy *(Peoples 285-2275)*, Hardware Store *(Quadra Island Building Supply 285-3221)*, Florist, Retail Shops *(Island Treasures 285-2456; Explore 285-3293)*, Copies Etc. *(Explore 285-3293)*
3+ mi: Gourmet Shop *(Quadra Island Market 285-3223, groceries, produce, fresh meat, deli items, liquor, 5 mi.)*, Post Office *(4 mi.)*, Library *(Quadra Island 285-2216, 5 mi.)*

Transportation
OnSite: Bikes *(& mopeds)*, Water Taxi *(Free shuttle to Painter's Lodge)* **OnCall:** Taxi *(Quadra 205-0505)* **1-3 mi:** Ferry Service *(to Campbell River, 1.5 mi./Cortes Island, 5.5 mi.)* **Airport:** Campbell R./YVR - NW Seaplanes - Seattle to Painters Lodge *(10 mi./150 mi.)*

Medical Services
911 Service **OnSite:** Holistic Services *(Aveda Spa at April Point 285-2668, full-service - 1 hr. massage $110)* **OnCall:** Ambulance **1-3 mi:** Doctor *(Quadra Is. Medical Clinic 285-3540)*, Dentist **Hospital:** Campbell River 287-7111 via water taxi *(4 mi.)*

Setting -- Off Discovery Passage, the striking northwest architecture of April Point dominates the entrance to Quadra Island's Gowlland Harbour. Well-landscaped, sage green buildings trimmed in deep red are an upscale riff on a fishing lodge. The dramatic Asian-inspired spa building, reached by a single walkway, seems to float in its own lagoon. Deeper into the harbor, just past the red buoy, the marina is tucked into a protected, densely forested cove.

Marina Notes -- *4,500 linear ft. **20, 30 & 50 amp service available. 5 sets of alongside floating docks, paralleling the shore, easily accommodate large yachts. A single main wharf ties them together, ending at twin gray metal sheds, and a two-story cinderblock and forest green-sided office. The 50-year-old resort, owned by Oak Bay Marine since '98, is part of a group of 7 fishing-oriented BC resorts & 5 marinas. Every 30 minutes, a free water taxi runs between April Point & Painter's Lodge (on Vancouver Island, 5 km. north of Campbell River). All facilities are shared. Painter's Lodge's dock is available to April Point guests. Off season, Oak Bay's enormous floating fishing "lodges" line the docks. A well-used helipad is adjacent to the Spa building. The bathhouse has serviceable full baths (showers $2) and a bright laundry room. Note: Cross when Seymour Narrows is slack to avoid 14 kt. current.

Notable -- The Resort's well-appointed lodge is a short walk from the marina. A spectacular all-window dining room, sushi bar and an outside sundeck feature stunning views of Discovery Passage. 57 guest rooms and ten 1-4 bedroom houses also sport water views. On-site are adventure, fishing & ecotourism excursions plus trail bikes & mopeds for touring Quadra Island's artist colony. Luxurious neo-Victorian Painter's has a pool, hot tub, tennis courts & 2 eateries.

Navigational Information

Lat: 50°02.096' **Long:** 125°14.726' **Tide:** 14 ft. **Current:** 5 kt. **Chart:** 3312
Rep. Depths *(MLW)*: **Entry** 20 ft. **Fuel Dock** 20 ft. **Max Slip/Moor** 20 ft./-
Access: Northernmost breakwater facility on west shore at Campbell River

Marina Facilities *(In Season/Off Season)*

Fuel: *Esso* - Gasoline, Diesel
Slips: 150 Total, 50 Transient **Max LOA:** 200 ft. **Max Beam:** n/a
  Rate *(per ft.)*: **Day** $1.10/0.75* **Week** n/a **Month** $13.50/6.00
  Power: 30 amp Incl.** **50 amp** Incl., **100 amp** Incl., **200 amp** n/a
  Cable TV: No **Dockside Phone:** No
  Dock Type: Floating, Long Fingers, Alongside
Moorings: 0 Total, 0 Transient **Launch:** n/a
  Rate: Day n/a **Week** n/a **Month** n/a
Heads: 8 Toilet(s), 5 Shower(s) *(with dressing rooms)*
Laundry: 3 Washer(s), 3 Dryer(s) **Pay Phones:** Yes
Pump-Out: No **Fee:** n/a **Closed Heads:** No

Marina Operations

Owner/Manager: Vicky Lagos **Dockmaster:** Tara Henderson
In-Season: May-Sep, 8am-8pm **Off-Season:** Oct-Apr, 9am-4:30pm
After-Hours Arrival: Call in advance
Reservations: Yes, Preferred **Credit Cards:** Visa/MC, Amex, Debit
Discounts: None
Pets: Welcome **Handicap Access:** Yes, Heads

Discovery Harbour Marina

1434 Island Highway, Suite 392; Campbell River, BC V9W 8C9

Tel: (250) 287-2614 **VHF: Monitor** Ch. 66A **Talk** Ch. 73
Fax: (250) 287-8939 **Alternate Tel:** n/a
Email: tara@discoveryharbourmarina.com **Web:** See below
Nearest Town: Campbell River **Tourist Info:** (250) 287-4636

Marina Services and Boat Supplies

Services - Docking Assistance, Security *(24/7, video)*, Dock Carts, Megayacht Facilities, 3 Phase **Communication -** Phone Messages, Fax in/out, Data Ports *(Wi-Fi - Broadband Xpress)*, FedEx, DHL, UPS, Express Mail *(Sat Del)* **Supplies -** OnSite: Ice *(Block, Cube)*, Ships' Store *(Ocean Pacific 286-9600, near Tyee 287-2641)* **Under 1 mi:** Propane

Boatyard Services

OnSite: Launching Ramp **OnCall:** Air Conditioning, Refrigeration, Divers, Propeller Repairs **Near:** Electronics Repairs *(Sea-Com 286-3717)*.

Restaurants and Accommodations

OnSite: Restaurant *(Moxie's 830-1500, L&D $15-25)*, *(Riptide 830-0044, L $8-14, D $8-24, delivers)*, *(Joey's 287-4422)*, *(Harbour Grill 287-4143, D $20-36, water views)*, Seafood Shack *(Patti Finn's 287-3957, L & D $7-25 "G" Dock)*, Coffee Shop *(Starbucks)*, Fast Food *(A&W)*, Inn/B&B *(Marina Inn 923-7255)* **Near:** Restaurant *(Reef 492-0997)*, *(Quinsam 287-7127)*, *(Jager's 286-1661)*, *(San Marcos 287-7066)*, Coffee Shop *(Java Shack 287-9881)*, Fast Food *(Wendy's, Quizno's)*, Pizzeria *(Boston Pizza 286-6120)*

Recreation and Entertainment

OnSite: Fishing Charter, Group Fishing Boat *(JZ's 850-1050)*, Video Rental *(Blockbuster 830-0800)*, Cultural Attract *(Gildas Box 287-7310)*, Sightseeing *(Eagle Eye Zodiac Adventures 286-0809)*, Galleries *(Wei Wai Kum House of Treasures 286-1440)* **Near:** Boat Rentals *(CR Kayaks 287-2278)*, Park **Under 1 mi:** Beach, Picnic Area, Playground, Movie Theater *(Galaxy 286-1744)*, Museum *(Campbell River 287-3103, Maritime Heritage Centre 286-3161)*, Special Events *(Oceans Day - Jun, Children's Festival - Jul 1,*

Logger Sports - Aug) **1-3 mi:** Dive Shop *(Beaver 287-7652)*, Golf Course *(Sequoia Springs 287-4970)*, Fitness Center *(Nautilus 286-3623)*

Provisioning and General Services

OnSite: Supermarket *(Real Canadian, Thrifty Foods)*, Liquor Store *(Riptide)*, Bank/ATM *(Riptide)*, Post Office, Pharmacy *(Superstore)*, Clothing Store *(Mark's Work,)*, Retail Shops *(Innovations 286-9986)*, Department Store *(Canadian Tire, Zellers 287-7922)*, Copies Etc. *(Staples 286-3500)*
OnCall: Dry Cleaners *(Mr. One Hour 286-6631)* **Near:** Bakery *(Plaza 286-1916)*, Library *(Campbell River 287-3655)*, Beauty Salon *(Parlor 287-2177)*, Bookstore *(Page Eleven 286-6476)*, Newsstand **Under 1 mi:** Gourmet Shop *(On Line 286-6521)*, Delicatessen *(Pumpernickles 287-7414)*, Health Food *(Eat Well 830-1088)*, Farmers' Market *(Pier St. Apr-Sep, Sun 9am-2pm)*, Fishmonger *(Susi's 287-2457)*, Catholic Church, Barber Shop *(Ron's 286-9678)*, Hardware Store *(Home 287-7147)*

Transportation

OnSite: Local Bus *(CRT - Rt.8 to Tyee Plaza)* **OnCall:** Water Taxi *(Discovery Launch 828-7577)*, Taxi *(Bee Line 287-8383)* **Near:** Ferry Service *(to Quadra Island)* **Under 1 mi:** Bikes *(Spokes 286-0500)*, Rental Car *(Budget/Rent-a-Wreck 287-8353)* **Airport:** Campbell River/YVR *(7 mi./150 mi.)*

Medical Services

911 Service **OnSite:** Optician *(Maycock)* **OnCall:** Ambulance **Under 1 mi:** Dentist *(Arnet 287-7343)*, Holistic Services *(Sense of Wellness 830-8012)*, Veterinarian *(Greenwood 286-1129)* **1-3 mi:** Doctor *(River City 287-2111)* **Hospital:** Campbell River 287-7111 *(2 mi.)*

Setting -- Heading south on Discovery Passage, this is the first, and largest, of the three Campbell River recreational marinas; it's snugged within one long curved arm of the breakwater. A main boardwalk closely parallels the shore and ten docks radiate out - each with slips on either side - with the exception of "G" and "I" which are designed for alongside, megayacht tie-ups. At the head of "K" dock is the well-marked tan and chocolate brown two-story floating office.

Marina Notes -- *Over 100 ft., $1.50/ft. **20 amp also and 3-phase service. Monthly power is metered. Owned by the 600-member First Nation Campbell River Indian Band (CRIB). Built on reserve land in 1998, this $60-million project is a joint venture with Northwest Real Estate Developments and is staffed by Band members. CRIB operates House of Treasures, a large art gallery and gift shop. The northern sector of the marina is dedicated to commercial vessels. Customs clearance is available. Fish cleaning stations are on floats "D" and "F." A pleasant tile-floored bathhouse has Formica vanities with drop-in sinks and tile showers with dressing rooms. URL: www.discoveryharbourmarina.com. Note: First few docks roll in most weather.

Notable -- Along the inside of the main wharf, a funky assortment of floating buildings house all manner of charters and expeditions plus the red & white Patti Finn's Seafood Patio. The onshore promenade is lined with the waterfront side of the attractive 400,000 square foot Discovery Harbour Shopping Centre - 40 tenants include 6 eateries, major department and food stores, and a "welcome pole." Past the office, the contemporary, stone & driftwood Riptide Marine Pub Grill features a large mahogany, mostly glass, dining room with a central, copper-hooded fireplace plus two waterfront patios, dockage and Internet access.

PHOTOS ON DVD: 17

Navigational Information
Lat: 50°01.635' **Long:** 125°14.481' **Tide:** 14 ft. **Current:** 5 kt. **Chart:** n/a
Rep. Depths *(MLW)*: **Entry** n/a **Fuel Dock** n/a **Max Slip/Moor** 10 ft./-
Access: Campbell River west shore next to ferry dock

Marina Facilities *(In Season/Off Season)*
Fuel: No
Slips: 60 Total, 6-10 Transient **Max LOA:** 150 ft. **Max Beam:** n/a
  Rate *(per ft.)*: **Day** $1.50/1.25 **Week** n/a **Month** n/a
  Power: 30 amp Incl., **50 amp** Incl., **100 amp** Incl., **200 amp** n/a
  Cable TV: Yes, Incl. **Dockside Phone:** No
  Dock Type: Floating, Long Fingers, Concrete
Moorings: 0 Total, 0 Transient **Launch:** n/a
  Rate: Day n/a **Week** n/a **Month** n/a
Heads: 2 Toilet(s), 2 Shower(s) *(with dressing rooms)*, Hair Dryers
Laundry: None **Pay Phones:** Yes
Pump-Out: No **Fee:** n/a **Closed Heads:** No

Marina Operations
Owner/Manager: Sukhy Bains **Dockmaster:** Same
In-Season: Year-Round, 24 hrs. **Off-Season:** n/a
After-Hours Arrival: Go to Front Desk
Reservations: Preferred **Credit Cards:** Visa/MC, Dscvr, Din, Amex, JCB
Discounts: None
Pets: Welcome, Dog Walk Area **Handicap Access:** No

The Coast Discovery Marina

975 Shoppers Row; Campbell River, BC V9W 2C4

Tel: (250) 287-7455 **VHF: Monitor** Ch. 66A **Talk** Ch. 66A
Fax: (250) 287-2213 **Alternate Tel:** (250) 287-7155
Email: n/a **Web:** www.coasthotels.com/home/sites/campbellriver
Nearest Town: Campbell River **Tourist Info:** (250) 287-4636

Marina Services and Boat Supplies
Services - Room Service to the Boat, Trash Pick-Up, Dock Carts
Communication - Mail & Package Hold, Fax in/out, Data Ports *(Wi-Fi -
Broadband Xpress)*, FedEx, UPS **Supplies - OnSite:** Ice *(Block, Cube)*
Under 1 mi: Ships' Store *(Ocean Pacific 286-9600)*, Bait/Tackle *(Tyee
Marine 287-2641)*

Boatyard Services
Near: Travelift. **Nearest Yard:** Ocean Pacific (250) 286-1011

Restaurants and Accommodations
OnSite: Restaurant *(The Brasserie B $10, L $10-15, D $15-30, Thu
Schnitzel Mania $12, Fri Prime Rib Buffet $21, Sat Steak & Lobster $21,
Sun $12 Omelette Bar)*, Seafood Shack *(Dick's Fish & Chips 287-3336,
L $4-11, D $4-11)*, Hotel *(Discovery Coast 287-7155, $130, 6-story tower)*
Near: Restaurant *(Friends Bistro 250-0312)*, *(Banner's 286-6711, 22 ice
cream flavors)*, *(Baan Thai 286-4853)*, *(Duke's Dockside Grill 286-6650)*,
Coffee Shop *(Java Shack)*, Fast Food *(McDonald's, Subway)*, Pizzeria
(Boston Pizza 286-6120, delivers), Motel *(Town Centre Inn 287-8866)*
Under 1 mi: Motel *(Passage View 286-1156)*, Inn/B&B *(Pierhouse 287-2943)*

Recreation and Entertainment
OnSite: Spa *(whirlpool 7am-11pm)*, Dive Shop *(Aqua Shack 923-0848)*,
Fitness Center, Fishing Charter *(Sea Beyond 287-4497, Fast Water 923-
2302, Robert Nuttle 287-340, Coastal Wilderness 287-3427)*, Park *(Robert
V. Ostler Park)*, Sightseeing *(Nature's Best Pacific Wilderness Tours)*
Near: Beach *(Foreshore Park)*, Picnic Area, Playground, Jogging Paths,
Movie Theater *(Galaxy 286-1744)*, Video Rental *(Blockbuster 830-0800)*,

Special Events *(Oceans Day - Jun, Children's Fest - Jul 1, Logger Sports -
Aug, Haig-Brown Fes - Sep)*, Galleries *(Pier St. 286-9717)* **Under 1 mi:** Museum *(Campbell River 287-3103 Tue-Sun 12-5pm)*, Cultural Attract *(Tidemark Theatre 287-7465, Gildas Box of Treasures 287-7310 - Laichwiltach
performances)* **1-3 mi:** Golf Course *(Sequoia Springs 287-4970)*

Provisioning and General Services
OnSite: Liquor Store *(Wine & Spirit)* **Near:** Market *(SuperValu 287-4410)*,
Gourmet Shop *(On-Line Gourmet 286-6521, homemade B&L to go; Internet)*,
Health Food *(Eat Well 830-1088)*, Bakery *(Plaza Bakery 286-1916)*, Farmers'
Market *(Pier St. Apr-Sep, Sun 9am-2pm,)*, Fishmonger *(seasonal at fishing
pier)*, Bank/ATM, Post Office, Catholic Church, Library *(287-3655)*, Beauty
Salon *(Tangles 286-6291)*, Barber Shop *(Mike's 287-4226)*, Dry Cleaners
(Mr. One Hour 286-6631), Laundry *(C.R. Coin-Op, Internet access)*,
Pharmacy *(Shoppers Drug Mart 286-1166)*, Hardware Store *(Home
Hardware 287-7147)*, Clothing Store *(Intersport 286-1760, Incognito 286-
1903)*, Department Store *(Zellers 287-7922)*

Transportation
OnSite: Water Taxi *(Discovery Launch 828-7577, Charters, Tours &
Freight)*, Ferry Service *(Quadra Island Ferry)* **OnCall:** Taxi *(Bee Line 287-
8383)* **Near:** Rental Car *(Budget /Rent-a-Wreck 287-8353, 1 mi.)*
Airport: Campbell River/YVR *(7 mi./150 mi.)* Helijet-Seattle 800-665-4354

Medical Services
911 Service **OnCall:** Ambulance **Near:** Dentist *(Arnet 287-7343)* **Under 1
mi:** Doctor *(Alder Medical 287-7441)*, Chiropractor *(Shaw 287-7429)*, Holistic
Services *(Sense of Wellness 830-8012)* **Hospital:** C.R. 287-7111 *(1.5 mi.)*

Setting -- Inside the center set of jetties, the small boutique Coast Marina shares the basin with the Quadra Island Ferry. Upscale floating cement docks have individual slips with full length finger piers and high-end Seatech pedestals. A contemporary two-story, gray-sided office floats on a barge. Directly adjacent is Foreshore Park with a waterfront path lined with Rigosa roses; downtown Campbell River is a short walk.

Marina Notes -- New owners as of October 2005. Limited transient dockage, so reserve ahead. Part of the Coast Hotel complex, facilities are shared and discounts offered to boaters wishing to come ashore for a night or two. The heads and showers for marina guests are located in the hotel's fitness center (7am-11pm), which also has a whirlpool and a nice assortment of equipment. Fully tiled bathroom with fiberglass shower stalls, toiletries and towels - very comfortable and intimate. Note: There's a HeliJet to Seattle.

Notable -- Tyee Plaza, directly across Island Highway from the docks, is home to the Coast Discovery Hotel, a large number of shops and a tourism kiosk. In the hotel, the Brassiere Restaurant offers bistro or dinner entrees - inside or on the patio (and delivers boxed lunches to the marinas). The inn's six-story tower provides most rooms with views of the passage. Everything is nearby on Shoppers Row -- movies, billiards, a library and stores of every description. Just south is Fisherman's Wharf and the Campbell River museums. The ferry to Quadra Island is adjacent to the marina docks. The Cortes Connection, a van service that links Campbell River with the Quadra Island and Cortes Island ferries, makes a great daytrip. An airport shuttle links the ferry dock to C.R. Airport.

Campbell River Harbour Auth.

705 Island Highway; Campbell River, BC V9W 2L3

Tel: (250) 287-7931 **VHF:** Monitor Ch. 66A **Talk** Ch. 66A
Fax: (250) 287-8495 **Alternate Tel:** n/a
Email: fishermans@telus.net **Web:** n/a
Nearest Town: Campbell River **Tourist Info:** (866) 830-1113

Navigational Information
Lat: 50°01.342' **Long:** 125°14.347' **Tide:** 14 ft. **Current:** 5 kt. **Chart:** 3312
Rep. Depths (*MLW*): Entry 20 ft. **Fuel Dock** n/a **Max Slip/Moor** 20 ft./-
Access: Campbell River, southernmost breakwater on west shore

Marina Facilities (*In Season/Off Season*)
Fuel: No
Slips: 100 Total, 20 Transient **Max LOA:** 100 ft. **Max Beam:** n/a
  Rate (*per ft.*): **Day** $0.50 **Week** $1.75 **Month** $3.12
Power: 30 amp $2.50*, **50 amp** $5, **100 amp** n/a, **200 amp** n/a
Cable TV: No **Dockside Phone:** No
Dock Type: Floating, Short Fingers, Alongside, Concrete
Moorings: 0 Total, 0 Transient **Launch:** n/a
  Rate: Day n/a **Week** n/a **Month** n/a
Heads: 2 Toilet(s), 2 Shower(s)
Laundry: None **Pay Phones:** No
Pump-Out: Onsite (*Dock "C"*), 1 Central **Fee:** $10 **Closed Heads:** No

Marina Operations
Owner/Manager: CRHA **Dockmaster:** Linda Franz
In-Season: Summer, 8am-5pm** **Off-Season:** Winter, 8am-4pm
After-Hours Arrival: Call ahead
Reservations: Yes **Credit Cards:** Visa/MC, Debit
Discounts: None
Pets: Welcome, Dog Walk Area **Handicap Access:** Yes

Marina Services and Boat Supplies
Services - Docking Assistance, Security (*night foot patrol*), Trash Pick-Up, Dock Carts **Communication** - Mail & Package Hold, Data Ports (*Wi-Fi*), FedEx, DHL **Supplies - OnSite:** Ships' Store (*Ocean Pacific 286-9600*), CNG **Near:** Ice (*Cube*), Marine Discount Store (*River Sportsman 268-1017*), Bait/Tackle (*Tyee Marine 287-2641*) **1-3 mi:** Propane

Boatyard Services
OnSite: Travelift, Railway (*2 railways - Ocean Pacific Marine 286-1011*), Hull Repairs, Rigger, Bottom Cleaning, Compound, Wash & Wax, Woodworking, Metal Fabrication, Painting **OnCall:** Engine mechanic (*gas, diesel*), Electrical Repairs, Refrigeration, Divers, Propeller Repairs, Inflatable Repairs

Restaurants and Accommodations
OnSite: Seafood Shack (*Crabby Bob's 923-4674, 11am-5pm*), Snack Bar (*Espresso 101 cheescake bar*), Lite Fare (*Campbell River Discovery Pier B $2-5, L $2-10*) **OnCall:** Restaurant (*Best Wok 287-2831*), Pizzeria (*Boston Pizza 286-6120*) **Near:** Restaurant (*Pier Street Café 287-2772*), (*Koto Japanese 286-1422*), Seafood Shack (*Dick's Fish & Chips 287-3336*), Motel (*Discovery Coast Inn 287-7155, $90*), Inn/B&B (*Pier House B&B 287-2943*) **Under 1 mi:** Restaurant (*Lookout 286-6812*), (*San Marcos Steak House 287-7066, L $7-12, D $16-26*), Fast Food, Hotel (*Ancho 286-1131*)

Recreation and Entertainment
OnSite: Dive Shop (*Beaver Aquatics 287-7652*), Fishing Charter (*850-1050 charters & tours*) **Near:** Beach, Picnic Area, Grills, Playground, Tennis Courts, Jogging Paths (*Paved shoreline bike/jog trail*), Park, Museum

(*Maritime Heritage Centre 286-3161, Campbell River 287-3103*), Cultural Attract (*Tidemark Theatre 287-7465*), Special Events, Galleries (*Pier St. Gallery*) **Under 1 mi:** Fitness Center (*Fitness First 287-4011*), Movie Theater (*Galaxy 286-1744*), Video Rental (*Blockbuster 830-0800*)
1-3 mi: Golf Course (*Sequoia Springs 287-4970, plus Mini Pitch & Putt*)

Provisioning and General Services
OnSite: Farmers' Market (*Pier St. Parking Lot May-Oct, Sun 9am-1pm*), Fishmonger (*at docks 2 & 3, plus High Tide Seafood - will process your catch*) **Near:** Convenience Store (*Pier St. 286-3113*), Market (*Jolly Giant 287-3882*), Gourmet Shop (*On Line 286-6521, B&L to go; Internet*), Liquor Store (*Wine & Spirit*), Bank/ATM, Post Office, Beauty Salon (*Tangles 286-6291*), Dry Cleaners (*One Hour 286-6631*), Bookstore (*Book Bonanza 287-3212*), Pharmacy (*Seymour 287-8311*), Hardware Store (*Home Hardware 287-7147*) **Under 1 mi:** Health Food (*Eat Well 830-1088*), Bakery (*Plaza 286-1916*), Catholic Church, Protestant Church, Library (*287-3655*)
1-3 mi: Department Store (*Zellers 287-7922*)

Transportation
OnCall: Water Taxi (*$50/hr.*), Taxi (*Bee Line 287-8383*) **Near:** Rental Car (*Budget*), Local Bus **Under 1 mi:** Bikes (*Spokes 286-0500*), Ferry Service (*BC Ferries to Quadra Island*) **Airport:** Campbell River/YVR (*7 mi./ 150 mi.*)

Medical Services
911 Service **OnCall:** Ambulance **Near:** Dentist (*Arnet 287-7343*), Holistic Services (*Sense of Wellness 830-8012*) **Under 1 mi:** Doctor (*Alder 287-7441*) **Hospital:** Campbell River 287-7111 (*1 mi.*)

Setting -- Discovery Passage's third set of high, rock jetties protects Fisherman's Wharf's egalitarian mix of pleasure craft and working vessels - mingling at (mostly) alongside floating docks. Parallel to the jetty, directly outboard of the marina, is bustling Campbell River Discovery Pier and flanking the docks are two parks - Sequoia to the south and Foreshore to the north. Vibrant, colorful totem poles signal the important First Nation heritage and presence.

Marina Notes -- *20 amp, too. **Mon-Fri 8am-5pm, Sat & Sun 8am-4pm. Established in 1997as a commercial facility, it now welcomes pleasure craft. The main stationary wharf hosts the facilities (office & bathhouse) and divides the basin into North Harbour (docks "A", "B" & "C" - with pumpout) and South Harbour (Docks "1"-"5"). All are alongside tie-ups with some vessels rafted - except "1" with slips on one side. Ocean Pacific provides boatyard services.Three tidal grids. Boats on "2" & "3" sell fish direct as does Crabby Bob's on "6" & High Tide Seafood. Deep Bay Harvest (830-1415) will freeze your salmon catch.

Notable -- In keeping with Campbell River's international renown as a salmon & sport fish capital, a festive atmosphere awaits on Discovery Pier - tables & chairs surround the snack bar which sells sandwiches, fish & chips (Cod, Lingcod, Halibut, Snapper), plus ice cream and fishing licenses. Benches, high fishing chairs and fish cleaning stations line the wharf - all occupied by avid fishers and sightseers on the lookout for wildlife. Right across 19A is the Maritime Heritage Centre, a large gray vertical sided building, topped by a contemporary yellow lighthouse. A highlight of the Campbell River Museum is the collection of First Nation's masks and early artifacts. Thriving downtown Campbell River's galleries feature contemporary visual arts and First Nation's crafts.

Navigational Information
Lat: 50°06.154' **Long:** 125°12.645' **Tide:** 14 ft. **Current:** n/a **Chart:** 3539
Rep. Depths *(MLW)*: **Entry** 20 ft. **Fuel Dock** 20 ft. **Max Slip/Moor** 150 ft./-
Access: East side of Quadra Island in Heriot Bay near ferry dock

Marina Facilities *(In Season/Off Season)*
Fuel: Gasoline, Diesel
Slips: 20 Total, 20 Transient **Max LOA:** 150 ft. **Max Beam:** n/a
  Rate *(per ft.)*: **Day** $0.70/0.50* **Week** n/a **Month** n/a
  Power: 30 amp Incl., **50 amp** Incl., **100 amp** n/a, **200 amp** n/a
  Cable TV: Yes **Dockside Phone:** No
  Dock Type: Fixed, Floating, Alongside, Wood
Moorings: 0 Total, 0 Transient **Launch:** n/a, Dinghy Dock
  Rate: Day n/a **Week** n/a **Month** n/a
Heads: 2 Toilet(s), 4 Shower(s)
Laundry: 1 Washer(s), 1 Dryer(s), Book Exchange **Pay Phones:** Yes
Pump-Out: No **Fee:** n/a **Closed Heads:** No

Marina Operations
Owner/Manager: Lorraine Wright **Dockmaster:** Roisin Doran
In-Season: Jun16-Sep16, 7am-9pm **Off-Season:** Sep17-Jun15, 9am-5pm
After-Hours Arrival: Call ahead
Reservations: Yes, Required **Credit Cards:** Visa/MC, Amex
Discounts: Hotel guests **Dockage:** $0.10-20/ft. **Fuel:** n/a **Repair:** n/a
Pets: Welcome, Dog Walk Area **Handicap Access:** Yes

Heriot Bay Inn & Marina

PO Box 100; 673 Hotel Road; Heriot Bay, BC V0P 1H0

Tel: (250) 285-3322; (800) 605-4545 **VHF: Monitor** Ch. 66A **Talk** Ch. 66
Fax: (250) 285-2708 **Alternate Tel:** n/a
Email: info@heriotbayinn.com **Web:** www.heriotbayinn.com
Nearest Town: Heriot Bay **Tourist Info:** 866-830-1113

Marina Services and Boat Supplies
Services - Docking Assistance, Room Service to the Boat **Communication** - Mail & Package Hold, Phone Messages, Fax in/out, Data Ports *(Wi-Fi, Free)* **Supplies - OnSite:** Ice *(Block, Cube)*, Bait/Tackle, Propane

Boatyard Services
Nearest Yard: Cape Mudge Boat Works (250) 285-2155

Restaurants and Accommodations
OnSite: Restaurant *(Herons 285-3322, B $6-12, D $15-25, Sun brunch 10am-3pm, $13; afternoon tea $7)*, (HBI Pub 285-3539, L $8-13, D $8-13, live music 2 nts/wk), Inn/B&B *(Heriot Bay Inn 285-3322, $59-109, rooms, suites, cottages, RV & tent sites)*, Condo/Cottage *(Heriot Bay $119-239, 1-2 bedrooms)* **Near:** Snack Bar *(Yum Yum Tree 285-2941)*, Lite Fare *(Gateway Café 285-2600, Internet access)* **Under 1 mi:** Inn/B&B *(Quadra Island Harbour House B&B 285-2556)* **1-3 mi:** Restaurant *(Island Grill 285-2995)* **3+ mi:** Lite Fare *(Café Aroma 285-2404, 4 mi., Internet café)*, Pizzeria *(Lovin' Oven 285-2262, 4 mi.)*

Recreation and Entertainment
OnSite: Beach, Picnic Area, Grills, Playground, Boat Rentals *(kayaks - rentals $40/day, lessons, guided tours $39-84)*, Fishing Charter *(Salmon Fishing $500 - 5 hrs/4 fishers)*, Sightseeing *(Island Adventure Centre 877-285-2007, Eco & Whale watching $159/129, Grizzly bear expeditions $299, 4-hr. Quadra Island van tours $69, hiking tours $59, Marshwood Winery 285-2068 - tours/tastings 1.5 mi.)* **Near:** Jogging Paths, Roller Blade/Bike Paths, Video Rental, Galleries *(Earth Light Pottery 285-3931, Quadra Crafts 285-2184, Spirit Boxes 285-3108; James Pottery 285-3110, Chris Rose Soapstone 285-3245)* **Under 1 mi:** Park *(Rebecca Spit Marine 248-9460, dinghy)* **1-3 mi:** Fitness Center *(Quadra 285-2144)* **3+ mi:** Museum *(Kwagiulth 285-3733, First Nation artifacts, 4 mi.)*

Provisioning and General Services
OnSite: Fishmonger *(Sea Angel 285-3876, 830-1272 floating store)*, Bank/ATM, Retail Shops *(Sea Chest Gift Shop - books, some clothing & convenience items)* **Near:** Convenience Store *(Island Market's Heriot Bay Store 285-3223 - produce, meats, liquor - delivers)*, Delicatessen, Wine/Beer *(Heriot Bay)*, Liquor Store, Bakery, Post Office, Library *(Quadra Is. 285-2216)*, Barber Shop, Bookstore, Newsstand **Under 1 mi:** Market *(Quadra Island 285-3223, groceries, meat, liquor)* **1-3 mi:** Green Market, Catholic Church, Florist, Clothing Store **3+ mi:** Gourmet Shop *(Quadra Foods 285-3391, organic produce, deli, wines; deliveries to 8pm, 4 mi.)*, Beauty Salon *(Village Hair 285-3848, 4 mi.)*, Pharmacy *(Peoples 285-2275, 4 mi.)*, Hardware Store *(Quadra Is. Building Supply 285-3221, 5 mi.)*

Transportation
OnSite: Bikes *(Island Cycle 285-2637 $14/2 hrs, $20/day)*, Local Bus *(shuttle, fee)*, Ferry Service *(Cortes Island 45-min/ Campbell River 4 mi.)* **OnCall:** Taxi *(Quadra 24 hrs. 205-0505)* **1-3 mi:** Water Taxi *($140/hr.)* **Airport:** Campbell River/YVR *(12 mi./150 mi.)*

Medical Services
911 Service **1-3 mi:** Dentist **3+ mi:** Doctor *(Quadra Is. Medical 285-3540, 4 mi.)*, Holistic Services *(Balance 285-3080, Plum Blossom Acupuncture 285-2107, 4 mi.)* **Hospital:** Campbell River 287-7111, via ferry *(6 mi.)*

Setting -- Sitting on a short rise overlooking a network of five recently renovated parallel floating docks, the classic, 2-story, early 1900's white clapboard inn sports blue shutters, a gray metal roof and a newly enclosed dinning wing with an inviting deck. A natural, varnished railing fronts the inn & edges the deck; contemporary sling chairs ring tables & tall heaters ward off the evening chill. A dozen and a half Adirondack-style chairs line the well-trimmed waterfront.

Marina Notes -- *Dockage discount for Inn guests - $0.50/ft. Jun-Sep, $0.40/ft. Jan-May & Oct-Dec. Whale watch owner Lorraine Wright purchased the Inn in Nov '04 and upgraded the docks in Spring '05 & '06. Hosea Arminis Bull first established an inn on this site in 1895; the current edifice was rebuilt in 1912 after a fire. The original inn is now the lobby, lounge and office. The porch has been enclosed and added to the dining room; a new deck overlooks the water. In addition to the modest, but newly refurbished, guest rooms, one & two bedroom cabins with bay-view porches are strung along the ridge (5 new ones planned). Hidden away are 45 RV spaces. Bathhouse: rough-sawn knotty-pine dividers with a cement floor. Additional dockage may be available next door at HBHA.

Notable -- On-site dining is at casual, charming Heron's (inside & deck) and HBI Pub with inside or covered porch seating - plus a pool table, big screen TV for Canucks games and live music (Fri for easy rock; Sun jazz at 6:30pm). Beneath the dining deck is the Island Adventure Center: its new 700 hp, 12-seater "Coast Explorer" searches for orca whales & takes expeditions to see grizzly bears in the Orford River Valley. Other options include van tours of Quadra Island, guided kayak, eco and hiking tours, salmon fishing charters, fresh & saltwater fly fishing. Six ferries a day to Cortes Island leave right from the Inn's docks.

Navigational Information
Lat: 50°05.419' **Long:** 125°02.227' **Tide:** 14 ft. **Current:** 2 kt. **Chart:** 3538
Rep. Depths *(MLW):* **Entry** 20 ft. **Fuel Dock** n/a **Max Slip/Moor** 20 ft./-
Access: North to Blind Channel, Big Bang, and Stewart Island

Marina Facilities *(In Season/Off Season)*
Fuel: Gasoline, Diesel, High-Speed Pumps
Slips: 45 Total, 25* Transient **Max LOA:** 200 ft. **Max Beam:** n/a
  Rate *(per ft.):* **Day** $0.95/0.60** **Week** n/a **Month** $2.95
  Power: 30 amp $5*** , **50 amp** n/a, **100 amp** n/a, **200 amp** n/a
  Cable TV: No **Dockside Phone:** No
  Dock Type: Floating, Alongside, Wood
Moorings: 0 Total, 0 Transient **Launch:** n/a, Dinghy Dock
  Rate: Day n/a **Week** n/a **Month** n/a
Heads: 2 Toilet(s), 4 Shower(s) *(with dressing rooms)*
Laundry: 6 Washer(s), 6 Dryer(s), Book Exchange **Pay Phones:** Yes
Pump-Out: No **Fee:** n/a **Closed Heads:** Yes

Marina Operations
Owner/Manager: G. Clarke & B. Hansen **Dockmaster:** T. Stewart-Webb
In-Season: Jun-Sep, 8am-7pm **Off-Season:** Oct-May, 9am-5:30pm
After-Hours Arrival: Check in at store in morning
Reservations: Yes, Preferred **Credit Cards:** Visa/MC, Debit
Discounts: Lodge or RV guests **Dockage:** $0.50 **Fuel:** n/a **Repair:** n/a
Pets: Welcome **Handicap Access:** No

Gorge Harbour Marina & Resort

PO Box 89; 1374 Hunt Road; Whaletown, BC V0P 1Z0

Tel: (250) 935-6433 **VHF: Monitor** Ch. 66A **Talk** Ch. 66A
Fax: (250) 935-6402 **Alternate Tel:** (250) 935-6433
Email: gorgehar@oberon.ark.com **Web:** www.gorgeharbour.com
Nearest Town: Campbell River *(1.5 hrs.)* **Tourist Info:** (866) 830-1113

Marina Services and Boat Supplies
Services - Docking Assistance, Dock Carts **Communication -** Mail &
Package Hold, Phone Messages, Fax in/out **Supplies - OnSite:** Ice *(Block,
Cube)*, Ships' Store *(charts)*, Bait/Tackle, Propane

Restaurants and Accommodations
OnSite: Restaurant *(Floathouse 935-6631, D $35, B, L & D Jul-Aug, D year-
round on weekends 5:30-9pm, Sun brunch May-Sep 10am-2pm)*, Inn/B&B
(Gorge Harbour Resort $65-75, trailers $65/nt, $395 weekly) **Near:** Snack
Bar *(ice cream, coffee, snacks)*, Lite Fare *(Trude's Café)*, Condo/Cottage
*(Cortes Island Vacation Rentals 800-939-6644, private cottages and homes
around the island)* **3+ mi:** Restaurant *(Sunset Cortes 935-8555, 6 mi.)*,
(Cortes Café 6 mi., at Mason's Hall, open Mon, Wed, Fri)

Recreation and Entertainment
OnSite: Picnic Area, Grills, Boat Rentals *(14 ft. runabout $20/hr, $95/day;
kayak rentals)*, Fishing Charter, Video Rental **Near:** Beach *(freshwater lake
or sandy saltwater beach)*, Galleries *(Cortes Craftshop at Squirrel Cove plus
artists' studios throughout the island - potters, painters, carvers; Old Schoolhouse Gallery)* **1-3 mi:** Sightseeing *(bird and wildlife watching, Wolf Bluff
Castle)* **3+ mi:** Park *(Mansons Landing Provincial Park, 117 acres -
includes Hague Lake and sandy beaches, 6 mi.)*, Museum *(Cortes Island
Museum 935-6340, Wed-Sun 10am-4pm, 6 mi.)*, Cultural Attract *(community
theatre in summer, 6 mi.)*, Special Events *(Cortes Day, Music Festival,
Sandcastle Day, 6 mi.)*

Provisioning and General Services
OnSite: Market, Wine/Beer, Liquor Store, Post Office *(open Mon, Wed, Fri -
in Whaletown)* **Near:** Convenience Store *(Whaletown General Store)*,
Protestant Church, Library *(private, open Fri only)*, Retail Shops *(gift shop)*
1-3 mi: Green Market *(roadside stands, plus Cortes Natural Food Co-op
935-8577, 6 mi.)* **3+ mi:** Farmers' Market *(Fri in season at Mason's Hall, 6
mi.)*, Fishmonger *(Cortes Island Wild Harvest at Manson's Landing 935-
6939, 6 mi.)*, Bank/ATM *(Quadra Credit at Mason's Landing, 6 mi.)*,
Bookstore *(6 mi.)*

Transportation
OnSite: Bikes *(Scooter Rentals)*, Rental Car *($125/day)*, InterCity Bus
*("Cortes Connection" to Campbell River 9:25am Mon-Sat $14/9 1-way,
$24/14 RT plus ferry fares)* **Near:** Ferry Service *(45 min. to Quadra Island,
6 daily crossings, $5.50 RT/pp, $14/car, 888-BCFERRY)* **Airport:** Campbell
River/VYR *(15 mi./150 mi.)*

Medical Services
911 Service **OnCall:** Ambulance **Near:** Doctor, Dentist
Under 1 mi: Holistic Services *(Yoga at Gorge Hall in Whaletown Mon 2pm $6;
or 11 mi. Hollyhock Spa 800-933-6339, bodywork, yoga, meditation)*
Hospital: Campbell River *(15 mi.)*

Setting -- A small white clapboard-sided dock house with brown trim welcomes boaters to this rustic island oasis. Five linear floating docks parallel the shore, flanking both sides of a white-railed main wharf that leads to the wooded resort. A path passes the bark brown restaurant and ends at the cedar-sided main building that houses the lodge, store & office. Scattered about the deeply wooded grounds are 36 RV sites plus tent sites.

Marina Notes -- *1,800 ft. of linear moorage. **Dockage is $0.50/ft. for guests of RV park or lodge. ***15 amp service $3. Established in 1959, new owners Richard & Michelle Glickman took the helm in October '05. Look for many improvements over the next few seasons. Gazebo and BBQ for club/group cruises. The on-site market is well provisioned and ideal for boaters headed to Desolation Sound. 36 RV sites. Bathhouse & laundry were serviceable but basic; new ones were completed in fall '06. Additional moorage at government docks at Manson's Landing, Cortes Bay, and Squirrel Cove.

Notable -- The casual , licensed Floathouse Restaurant perches on a bank overlooking the docks. Despite its seasonality (May-September plus year-round weekends), the chef has developed quite a following. The resort's rustic lodge offers four guest rooms with private baths (one with shared), a lounge with satellite TV, and a group kitchen. Consider a dinghy adventure to Manson's Lagoon to gather oysters - reputed to be the best in North America. Nearby parks are designed for hikes and the island's protected inlets invite exploration. Two km. away, six ferries head to Quadra Island daily, landing at the Heriot Bay Inn (two eateries & a wide variety of adventure tours). Alternatively, hike or dinghy to sweet little Whaletown or take the "Cortes Connection" van to Campbell River.

Navigational Information

Lat: 45°52.285' **Long:** 125°06.959' **Tide:** 14 ft. **Current:** 0 kt. **Chart:** 3513
Rep. Depths (*MLW*): Entry 1 ft. **Fuel Dock** 10 ft. **Max Slip/Moor** 40 ft./-
Access: Strait of Georgia to private channel

Marina Facilities (*In Season/Off Season*)

Fuel: Gasoline
Slips: 240 Total, 10 Transient **Max LOA:** 40 ft. **Max Beam:** 15 ft.
 Rate (*per ft.*): **Day** $0.63* **Week** $168 **Month** n/a
 Power: 30 amp Incl.**, **50 amp** n/a, **100 amp** n/a, **200 amp** n/a
 Cable TV: No **Dockside Phone:** No
 Dock Type: Floating, Long Fingers, Wood
Moorings: 0 Total, 0 Transient **Launch:** n/a
 Rate: Day n/a **Week** n/a **Month** n/a
Heads: 10 Toilet(s), 10 Shower(s)
Laundry: 8 Washer(s), 8 Dryer(s), Book Exchange **Pay Phones:** Yes, 2
Pump-Out: No **Fee:** n/a **Closed Heads:** Yes

Marina Operations

Owner/Manager: Kelly Purden **Dockmaster:** Same
In-Season: Year-Round, 9am-9pm **Off-Season:** n/a
After-Hours Arrival: Check in at office or resort house
Reservations: Yes, Preferred **Credit Cards:** Visa/MC, Debit
Discounts: With Cottage Rental **Dockage:** n/a **Fuel:** n/a **Repair:** n/a
Pets: Welcome, Dog Walk Area **Handicap Access:** No

Pacific Playgrounds Int'l Marina

9082 Clarkson Drive; Black Creek, BC V9J 1B3

Tel: (250) 337-5600 **VHF: Monitor** n/a **Talk** n/a
Fax: (250) 337-5979 **Alternate Tel:** (250) 337-1744
Email: info@pacificplaygrounds.com **Web:** www.pacificplaygrounds.com
Nearest Town: Coutenay (13 mi.) **Tourist Info:** (250) 830-1115

Marina Services and Boat Supplies

Services - Boaters' Lounge, Security (*24 hrs.*), Trash Pick-Up, Dock Carts
Communication - Mail & Package Hold, Phone Messages, Fax in/out
($0.50), Data Ports (*In-Store or Wi-Fi***, $1*) **Supplies - OnSite:** Ice (*Block, Cube*), Ships' Store (*Some basic marine supplies at convenience store*), Bait/Tackle (*at convenience store - a large fishing tackle section*), Live Bait (*anchovies*), Propane

Boatyard Services

OnSite: Launching Ramp, Divers **OnCall:** Engine mechanic (*gas, diesel*)

Restaurants and Accommodations

OnSite: Snack Bar (*coffee, ice cream and snacks at the store*), Condo/Cottage (*Pacific Playgrounds 337-5600, $120-125, 20 2-bedroom cottages - ask for the riverside ones*) **Near:** Restaurant (*Black Creek Café 337-4022, B $4-10, L $7-12, D $7-12*), Snack Bar (*Ivy Cottage Sweets 'n Treats 337-5221*), Motel (*Sandpiper 923-4281*), Inn/B&B (*Ocean Air 337-5795, fronts Saratoga Beach*) **Under 1 mi:** Lite Fare (*Salmon Point Pub 923-7272, L, D, Sun brunch on the beach, 7 days - walk north through the provincial park*) **1-3 mi:** Restaurant (*Sound Point L $8-17, D $10-20*)

Recreation and Entertainment

OnSite: Heated Pool (*and kiddie pool*), Picnic Area (*covered - the "Cook-Out"*), Grills (*charcoal*), Playground, Volleyball (*plus basketball, horseshoes, ball field*), Tennis Courts, Roller Blade/Bike Paths, Fishing Charter (*Blue Orca, JZ's*) **Near:** Golf Course (*Saratoga Beach GC 337-8212 - 9-hole

executive, driving range, putting green, 18-hole miniature, pro shop plus Miracle Beach Adventure Mini Golf and 8 other courses within 15 mi.*), Jogging Paths, Video Rental, Park (*Saratoga Beach, Miracle Beach Provincial Park*) **1-3 mi:** Sightseeing (*Oyster River Hatchery -- open to public Tue 8am-2pm, call ahead 337-5967*), Special Events (*Outdoor concerts in summer at Saratoga Speedway 337-8106*), Galleries (*Little House Gallery*)

Provisioning and General Services

OnSite: Convenience Store (*good sized basic grocery section, ice cream, dairy and other basic picnic time essentials*), Post Office, Clothing Store
OnCall: Dry Cleaners **Near:** Market (*Black Creek Country Market 337-5514 - 7am-10pm daily; Miracle Beach Country Junction 337-5158 8am-9pm daily*), Delicatessen (*Miracle Beach*), Wine/Beer, Liquor Store, Bank/ATM, Newsstand, Retail Shops (*Heart & Home 337-8522; Marzanaz Consignments 337-5118*) **Under 1 mi:** Beauty Salon (*Black Creek Cuts 337-0015*) **1-3 mi:** Farmers' Market, Protestant Church (*Mennonite*)
3+ mi: Department Store (*Driftwood Mall in Courtnay 338-1071, 16 mi.*)

Transportation

OnCall: InterCity Bus **Airport:** Campbell River/YVR (*17 mi./130 mi.*)

Medical Services

911 Service **3+ mi:** Veterinarian (*Campbell River 287-7111, 19 mi.*)
Hospital: Washington Park Medical Clinic in Courtenay 334-9241 / Willow Point in Cambell River 923-6144 (*16 mi./19 mi.*)

Setting -- Just off the Strait of Georgia, a dredged channel leads into the very protected, man-made harbor. On the port side are big boat docks and, to starboard, trailer boat docks line the northern shore. These are backed by a small, sparsely treed spit of land, populated with RV sites, that separates the marina basin from the Oyster River. The majority of the 40-acre resort is a well-done, amenity-rich RV park nestled among grassy fields and old growth trees.

Marina Notes -- *Flat rate: $24-36/day. Moorage discounts with cottage stays. **15 amp power. Home of Campbell River Y.C. - reciprocity. ***Wi-Fi: $1/day, $5/week or on-line computer in store $1.50 /15 min. (printing $.05/10pp). Staffed 24/7. New management in '05 - expect more changes. A gated floating boardwalk parallels the south edge of the basin; radiating from it are four low freeboard main docks - each with floating slips. Expect max LOA to increase to 45-50 ft. in near future. Time arrival to the tides - at some points in channel there's 0 water on 1.5 tide. At the top of the boardwalk, a large wooden platform hosts a couple of picnic tables and adjacent is the "Look-Out," a small boaters' lounge with games & tables. The bathhouse is located in the RV Park.

Notable -- The resort boasts a 4-acre recreation area with volleyball, bocce, tennis, horseshoes, playground, badminton, baseball diamond, basketball, & heated pools. The Oyster River offers excellent fly-fishing (home of the Sea-Run Cutthroat Trout), swimming, and birding. A wood path meanders along the river to the strait and then parallels the shore. Reportedly great salmon & bottom fishing "right there." A two minute walk, Saratoga Beach promises sheltered, warm water, and the golf course is nearly adjacent. Most general services are 13 miles north (in Campbell River) or 16 miles south (in Courtenay).

Boaters' Notes

Add Your Ratings and Reviews at www.AtlanticCruisingClub.com

A complimentary six-month Silver Membership is included with the purchase of this *Guide*. Select "Join Now," then "Silver," then follow the instructions.

The AtlanticCruisingClub website provides updated Marina Reports, Destination and Harbor Articles and much more — including an option within each online Marina Report for boaters to add their ratings and comments regarding that facility. Please log on frequently to share your experiences — and to read other boaters' comments.

On the website, boaters may rate marinas in one or more of the following categories — on a scale of 1 (basic) to 5 (world class) — and also enter additional commentary.

▸ **Facilities & Services** (Fuel, Reservations, Concierge Services and General Helpfulness)

▸ **Amenities** (Pool, Beach, Internet Access, including Wi-Fi, Picnic/Grill Area, Boaters' Lounge)

▸ **Setting** (Views, Design, Landscaping, Maintenance, Ship-Shapeness, Overall Ambiance)

▸ **Convenience** (Access — including delivery services — to Supermarkets, other Provisioning Sources, Shops, Services, Attractions, Entertainment, Sightseeing, Recreation, including Golf and Sport Fishing & Medical Services)

▸ **Restaurants/Eateries/Lodgings** (Availability of Fine Dining, Snack Bars, Lite Fare, OnCall food service, and Lodgings ashore)

▸ **Transportation** (Courtesy Car/Vans, Buses, Water Taxis, Bikes, Taxis, Rental Cars, Airports, Amtrak, Commuter Trains)

▸ **Please Add Any Additional Comments**

2. BC – Sunshine Coast

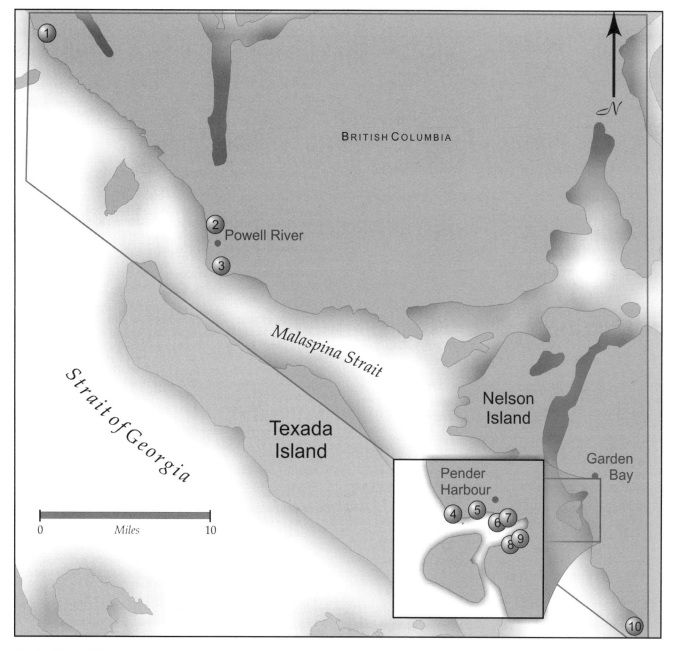

MAP	MARINA	HARBOR	PAGE	MAP	MARINA	HARBOR	PAGE
1	Lund Harbour Authority Public Docks	*Finn Bay*	34	6	Fisherman's Resort & Marina	*Pender Harbour/Hospital Bay*	39
2	Powell River Westview Harbour	*Strait of Georgia/Powell River*	35	7	Garden Bay Restaurant & Marina	*Pender Harbour/Garden Bay*	40
3	Beach Gardens Resort & Marina	*Malaspina Strait/Grief Point*	36	8	Madeira Park Public Wharf	*Pender Harbour/Welbourn Cove*	41
4	Irvines Landing Hotel Resort & Marina	*Pender Harbour/Joe Bay*	37	9	Sunshine Coast Marina & Resort	*Pender Harbour/Gunboat Bay*	42
5	Pender Harbour Resort	*Pender Harbour/Duncan Cove*	38	10	Secret Cove Marina	*Malaspina Strait/Secret Cove*	43

▸ **Currency** — In Canadian Marina Reports, all prices are in Canadian dollars. In U.S. Marina Reports, all prices are in U.S. dollars.

▸ **"CCM"** — Denotes a Certified Clean Marina, a state/provincial award for environmental excellence. See page 298 for an explanation and page 299 for a list of Pump-Out facilities.

▸ **Ratings & Reviews** — An overview of the Atlantic Cruising Club's rating system is on page 6 and details on the content of each Marina Report are on pages 7 – 11.

▸ **Marina Report Updates** — Comments from boaters and new information from ACC reviewers and marinas are posted regularly on *www.AtlanticCruisingClub.com*.

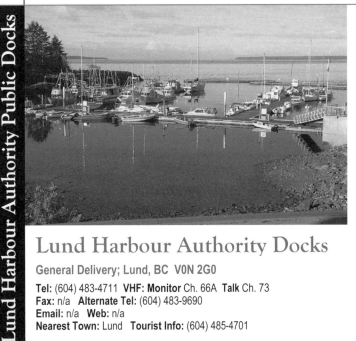

Lund Harbour Authority Docks

General Delivery; Lund, BC V0N 2G0

Tel: (604) 483-4711 **VHF: Monitor** Ch. 66A **Talk** Ch. 73
Fax: n/a **Alternate Tel:** (604) 483-9690
Email: n/a **Web:** n/a
Nearest Town: Lund **Tourist Info:** (604) 485-4701

Navigational Information
Lat: 49°58.848' **Long:** 124°45.687' **Tide:** 14 ft. **Current:** n/a **Chart:** 3311
Rep. Depths (*MLW*): **Entry** n/a **Fuel Dock** n/a **Max Slip/Moor** 14 ft./-
Access: Finn Bay past Lund Hotel docks; leave breakwaters to starboard

Marina Facilities *(In Season/Off Season)*
Fuel: *Lund Fuel Dock* - Gasoline, Diesel
Slips: 17 Total, 12 Transient **Max LOA:** 75 ft. **Max Beam:** n/a
  Rate *(per ft.):* **Day** $0.55* **Week** n/a **Month** n/a
  Power: 30 amp $6.25** **50 amp** n/a **100 amp** n/a **200 amp** n/a
Cable TV: No **Dockside Phone:** No
  Dock Type: Floating, Alongside, Concrete
Moorings: 0 Total, 0 Transient **Launch:** n/a
  Rate: Day n/a **Week** n/a **Month** n/a
Heads: 4 Toilet(s), 2 Shower(s)
Laundry: None **Pay Phones:** No
Pump-Out: OnSite **Fee:** n/a **Closed Heads:** No

Marina Operations
Owner/Manager: Neil Gustafson **Dockmaster:** Rosemarie O'Neill
In-Season: Jul-Aug, 7am-7pm **Off-Season:** Winter, 7-8am & 4-5pm
After-Hours Arrival: Call ahead
Reservations: No **Credit Cards:** Visa/MC
Discounts: None
Pets: Welcome **Handicap Access:** No

Marina Services and Boat Supplies
Services - Trash Pick-Up **Communication -** Data Ports *(Lund Hotel)*, DHL, UPS **Supplies - Near:** Propane *(Lund Fuel Dock)*

Boatyard Services
OnSite: Launching Ramp *(wide, concrete - 5 ft. above 0 tide)* **Near:** Travelift *(30T, to 60 ft. Jack's BY - right across the harbor)*, Engine mechanic *(gas, diesel)*. **Nearest Yard:** Jack's Boat Yard (604) 483-3566

Restaurants and Accommodations
Near: Restaurant *(Lund Hotel 414-0479, L $8-12, D $12-26, 11:30am-9:30pm, 7 days, D after 5pm; sandwiches, burgers, chicken, pasta, seafood)*, *(Nancy's Bakery 483-4180, B $4-8, L $4-9, D $7-14, soups, pizza, sandwiches, quiches; Tapas Night Fri & Sat; Sun-Thu 7am-6pm, Fri-Sat to 10pm)*, Snack Bar *(Lund Scoop, ice cream & snacks)*, Lite Fare *(Flo's 414-0304, Fri-Sun only, noon-8pm; burgers, salads, soups, dinner specials)*, Hotel *(The Historic Lund Hotel 414-0474, $80-135, 27 rooms)*
Under 1 mi: Inn/B&B *(Sevilla Island Resort 414-6880, $165-185, plus $50/day for the superb, locally-sourced meals; right across the harbor - the Hobbs pick up guests in their 20 ft. enclosed boat & specialize in eco-adventure packages)* **1-3 mi:** Restaurant *(The Laughing Oyster Restaurant 483-9775, D $18-26, noon-8:30pm)*

Recreation and Entertainment
OnSite: Boat Rentals *(Rock Fish Kayak 414-9355 $30-55/day)*, Fishing Charter *(Perfedia Charters)*, Video Rental *(Lund General Store)*, Sightseeing *(Terracentric Coastal Adventures 483-7900, high speed Zodiac tours:*

$79/3 hr., $125/6 hr. w/lunch; Island Explorer Boat Tours 414-6880), Galleries *(Debra Bevaart's Tug-Guhm Gallery & Studio features regional artists & soapstone carvng workshops; Skipped Gallery)* **Near:** Dive Shop *(Linda & Wayne Lewis 414-9595)*, Jogging Paths *(buy a hiking trail map at Nancy's Bakery $2)* **1-3 mi:** Beach *(white sand beaches on Savary Island via water taxi 483-9749, 5 times/day $8.50/pp)*, Golf Course *(Glen Rose Golf Course -- Craig Road in Lund, $10 or Les Furber's Myrtle Point Golf Club overlooking the Strait & the Comox Glaciers)*, Park *(Okeover Arm Provincial Park by kayak or dinghy)*

Provisioning and General Services
Near: Market *(Lund General Store 414-0471 Sat 10:30am-12:30pm, Sun 12:30-2:30pm, produce, dairy, meats, deli, prepared foods)*, Wine/Beer, Farmers' Market *(Sat 10:30am-12:30pm, Sun 12:30-2:30pm, Apr-Oct)*, Fishmonger, Bank/ATM, Post Office *(at hotel)*, Laundry *(Lund Hotel)*, Bookstore *(some books and magazines at the general store)*, Newsstand, Hardware Store *(Lund General Store)*, Clothing Store *(Limited)*

Transportation
OnCall: Water Taxi *(Glen Water Taxi 489-9749, $15 to Savory Island and other islands)* **Near:** Bikes *(Savary Island Bike Shop 414-4079, rentals $15/day, kids trailer: $10/day, Jul-Sep 15 9:30am-6pm, May-Jun weekends to 3:30pm, call in winter)* **Airport:** Powell River *(17 mi.)*

Medical Services
911 Service **OnCall:** Ambulance **3+ mi:** Veterinarian *(Westview 485-5111, 17 mi.)* **Hospital:** Powell River General 485-3211 *(17 mi.)*

Setting -- Three floating concrete breakwaters protect two long docks at the foot of historic, charming Lund. After passing the landmark Lund Hotel, fuel dock and boat launch, the first dock behind the breakwater is for pleasure craft, the second for commercial vessels. At the head of the main wharf is the tan shingled harbormaster's office, with brown metal roof & rust trim. Rising above is the strikingly handsome multistory contemporary signed "Nancy's Bakery."

Marina Notes -- *Flat rate: alongside moorage: 30' $15.02, 40' $20.04 45' $22.54, 50' $25.04, 55' $27.55, 58' $29.05. 680 feet of linear dockage. Moorage also on either side of the three breakwaters (dinghy ashore) and at a dock behind Sevilla Island - same rates. Boats are often rafted. **20 amp: $3.20 (purchase adaptors for $20). Fee for garbage drop. The Lund Hotel fuel dock is open 7 days, 8am-8pm, Ch. 73 or 414-0474. Two very nice all-tile full bathrooms plus two powder rooms; fiberglass shower stalls with colorful curtains - two loonies.

Notable -- Lund, on the shore of the Malaspina Peninsula, is the northern terminus of the 9,000 mile Pan-American Highway, the last town on the Sunshine Coast and the "Gateway to Desolation Sound." Just east of the docks, the venerable Lund Hotel anchors the hamlet; it was renovated and re-opened in 2000 by local businessman David Formosa and the Sliammon First Nation. The first floor hosts a restaurant and a café/bar, post office and the very well-supplied Lund General Store; below are galleries and, on the top floor, 27 rooms. Its basic moorage, unprotected by the breakwater, is dock 'n dine and overflow from the public docks. Don't miss Nancy's Bakery, which is a bistro & café, and across the harbor Donna Kaye & Ian Hobb's delightful, postmodern Sevilla Island Resort.

Navigational Information
Lat: 49°50.142' **Long:** 124°31.759' **Tide:** 16 ft. **Current:** 2 kt. **Chart:** 3311
Rep. Depths (*MLW*): **Entry** 15 ft. **Fuel Dock** 8 ft. **Max Slip/Moor** 8 ft./-
Access: Opposite the northern tip of Texada Island, south of ferry terminal

Marina Facilities (*In Season/Off Season*)
Fuel: *Chevron* - Gasoline, Diesel
Slips: 80 Total, 40 Transient **Max LOA:** 60 ft. **Max Beam:** n/a
Rate (*per ft.*): **Day** $0.60/same **Week** n/a **Month** $3.35/same
Power: 30 amp Incl., **50 amp** n/a, **100 amp** n/a, **200 amp** n/a
Cable TV: No **Dockside Phone:** No
Dock Type: Floating, Concrete, Wood
Moorings: 0 Total, 0 Transient **Launch:** n/a
Rate: Day n/a **Week** n/a **Month** n/a
Heads: 2 Toilet(s), 2 Shower(s)
Laundry: Yes **Pay Phones:** No
Pump-Out: No **Fee:** n/a **Closed Heads:** No

Marina Operations
Owner/Manager: City of Powell River **Dockmaster:** Jim Parsons
In-Season: Jun-Sep, 8am-9pm **Off-Season:** Oct-Apr, Mon-Fri, 8am-4:30pm
After-Hours Arrival: Call ahead
Reservations: No **Credit Cards:** Visa/MC, Amex, Debit
Discounts: None
Pets: Welcome **Handicap Access:** No

Powell River Westview Harbour

6910 Duncan Street; Powell River, BC V8A 1V4

Tel: (604) 485-5244 **VHF: Monitor** Ch. 68 **Talk** n/a
Fax: (604) 485-5286 **Alternate Tel:** n/a
Email: n/a **Web:** n/a
Nearest Town: Powell River **Tourist Info:** (604) 485-4701

Marina Services and Boat Supplies
Services - Docking Assistance, Security (*locked, coded gates*)
Communication - Data Ports (*Wi-Fi - Broadband Xpress*), DHL, UPS
Supplies - OnSite: Ice (*Block, Cube*), Ships' Store (*Marine Traders 485-4624; Kolszar Marine & Power*), Bait/Tackle **Near:** Marine Discount Store (*Powell Rvr. Outdoor 485-2555, delivers to docks*) **3+ mi:** Propane (*5 mi.*)

Boatyard Services
OnCall: Electronics Repairs, Refrigeration, Propeller Repairs
Near: Travelift, Engine mechanic (*gas, diesel*), Launching Ramp, Electrical Repairs, Hull Repairs, Bottom Cleaning, Divers, Total Refits.

Restaurants and Accommodations
Near: Restaurant (*Pier Pub & Dockside Café 485-6281*), (*Thaidal Zone 485-4807*), (*Chang Mai 485-0883, L $7-8, D $8-16*), (*Donna's Diner 485-0984, B & L*), (*Deli Truce Dutch Café 485-9224*), (*Marine Inn 485-4242*), (*Old Jailhouse Café 485-0926*), Lite Fare (*Westview Hotel Pub Café 485-0821, entertainment*), Pizzeria (*Rocky Mountain Pizza & Bakery 485-9111, Wi-Fi*), (*Westview 485-6162*), Motel (*Marine Inn 485-4242*), (*Westview Centre 485-4023*), (*Powell River Harbour Guesthouse hostel with showers, coin-op laundry*) **Under 1 mi:** Fast Food (*McDonald's, Tomatoe Rice, Dairy Queen*), Hotel (*Coast Town Centre 485-3000*)

Recreation and Entertainment
OnSite: Picnic Area, Fishing Charter, Park (*or Seawalk*) **Near:** Bowling (*Westview Bowling & Billiards 485-2033*), Cultural Attract (*Powell River Theatre, Evergreen Theatre 485-2891*), Sightseeing (*sea kayak tours 866-617-4444*), Special Events (*Seafair - Jul, Blackberry Fest - Aug, Sunshine Music - LabDay*) **Under 1 mi:** Beach, Playground, Fitness Center (*Avid Fitness 485-9580*), Jogging Paths, Video Rental (*Select Video 485-0333*), Museum (*Powell River Museum*), Galleries (*Wind Spirit 485-7572*)
1-3 mi: Heated Pool, Movie Theater (*Patricia Theatre 483-9345*)
3+ mi: Golf Course (*Myrtle Point 487-4653, 6 mi.*)

Provisioning and General Services
OnSite: Fishmonger (*South Basin, off the boats*), Clothing Store (*Powell River Outdoor 485-2555, delivers*) **Near:** Delicatessen (*Deli Truce 485-9224*), Wine/Beer, Liquor Store (*Westview 485-9599*), Bakery, Bank/ATM (*Scotia*), Library (*485-4796*), Beauty Salon (*Eclips 485-9797*), Retail Shops (*Blue Cat Mall 485-0904*) **Under 1 mi:** Supermarket (*Safeway 485-1233*), Catholic Church, Protestant Church, Dry Cleaners (*Westview Drycleaners 485-2616*), Bookstore (*Breakwater Books 489-0010*), Pharmacy (*Shoppers 485-2844*), Hardware Store (*Building Supply 485-2791*), Department Store (*Wal-Mart 485-9811*) **1-3 mi:** Market (*Overwaitea Foods 485-4751*), Farmers' Market (*McLeod Rd. Sat 10:30am-12:30pm, Sun 12:30-2:30pm*), Post Office, Copies Etc.

Transportation
OnSite: Courtesy Car/Van (*Jul-Aug*), Ferry Service (*Comox &Texada Is.*)
OnCall: Taxi (*Powell River 483-3666*) **Near:** Local Bus **Under 1 mi:** Rental Car (*Budget 485-4131*), Airport Limo **Airport:** Powell River (*1 mi.*)

Medical Services
911 Service **OnCall:** Ambulance **Under 1 mi:** Dentist (*Kean 485-9887*), Chiropractor **1-3 mi:** Doctor (*Medical Clinic 485-6261*), Veterinarian (*Powell River Vet 485-6333*) **Hospital:** Powell River General 485-2208 (*1.5 mi.*)

Setting -- Divided by the BC Ferry dock, two high rock jetties create two nearly landlocked, protected basins. The terraced Westview section of Powell River rises above. Recreational boats populate the northern harbor and commercial vessels the southern one, which is hugged by an impressive superstructure that supports the ferry landing and entrance ramps. Steps lead up the high bank above the north basin to a blue railed, nicely grassed picnic area and mini park.

Marina Notes -- *The fuel dock is at the entrance to the southern commercial basin. The floating cement docks have lights and full pedestals - rafting may be required in high season. Seasonal courtesy car makes the community very accessible. In season, the commercial fishermen sell their catch right from the boats. The on-site ferries create a lot of car and foot traffic. The bathhouse is at the southern end of the rec docks - the heads and showers are full bathrooms with fiberglass shower enclosures and terrazzo floors and tile walls - quite pleasant and a step up from municipal.

Notable -- Powell River is the region's supply depot and commercial and cultural hub - so just about anything needed will be found here. The nearby streets have an eclectic air and are lined with shops and eateries. About a mile away is Town Centre Mall with a Wal-Mart, clothing shops, supermarket, liquor/wine store, post office, The Source by Circuit City and Building Centre - a big hardware store. The Powell River Historical Museum (Jun-Aug 9am-5pm daily, Sep-May 9am-4:30pm Mon-Fri) features local logging and paper mill history and First Nation Sliammon's famous baskets and tribal history. Texada Island is 5 miles offshore; consider taking the ferry to Blubber Bay or the boat to Harbour at Van Anda - the only all-weather port.

Beach Gardens Resort & Marina

7074 Westminster Avenue; Powell River, BC V8A 1C5

Tel: (604) 485-6267; (800) 663-7070 **VHF: Monitor** Ch. 66A **Talk** n/a
Fax: (604) 485-2343 **Alternate Tel:** (604) 485-6267
Email: beachgardens@shaw.ca **Web:** www.beachgardens.com
Nearest Town: Powell River (2.5 mi.) **Tourist Info:** (604) 485-4701

Navigational Information
Lat: 49°48.107' **Long:** 124°31.113' **Tide:** 16 ft. **Current:** n/a **Chart:** 3535
Rep. Depths (MLW): Entry 6 ft. **Fuel Dock** 6 ft. **Max Slip/Moor** 6 ft./-
Access: East of Grief Point, east end of Westview

Marina Facilities (In Season/Off Season)
Fuel: Gasoline, Diesel
Slips: 90 Total, 30 Transient **Max LOA:** 110 ft. **Max Beam:** n/a
 Rate (per ft.): **Day** $0.85 **Week** n/a **Month** n/a
 Power: 30 amp $7, **50 amp** n/a, **100 amp** n/a, **200 amp** n/a
 Cable TV: No **Dockside Phone:** Yes at motel
 Dock Type: Floating, Wood
Moorings: 0 Total, 0 Transient **Launch:** n/a
 Rate: Day n/a **Week** n/a **Month** n/a
Heads: 2 Toilet(s), 2 Shower(s)
Laundry: 2 Washer(s), 2 Dryer(s) **Pay Phones:** No
Pump-Out: No **Fee:** n/a **Closed Heads:** Yes

Marina Operations
Owner/Manager: Joan Barszczewski **Dockmaster:** Same
In-Season: MidMay-MidSep, 7am-9pm **Off-Season:** Sep-May, 8am-4:30pm
After-Hours Arrival: Come to front desk
Reservations: Yes, Preferred **Credit Cards:** Visa/MC, Amex, Debit
Discounts: None
Pets: Welcome **Handicap Access:** Yes

Marina Services and Boat Supplies
Services - Docking Assistance, Boaters' Lounge, Security (24 hrs., onsite), Trash Pick-Up, Dock Carts **Communication -** Mail & Package Hold, Phone Messages, Fax in/out ($1.50), DHL, UPS, Express Mail **Supplies - OnSite:** Ice (Block, Cube) **OnCall:** West Marine **1-3 mi:** Ships' Store (Marine Traders 485-4624), Bait/Tackle, Propane

Boatyard Services
OnCall: Engine mechanic (gas, diesel), Electrical Repairs, Electronics Repairs, Hull Repairs, Canvas Work, Bottom Cleaning, Refrigeration, Divers, Compound, Wash & Wax, Interior Cleaning, Propeller Repairs, Woodworking, Inflatable Repairs, Life Raft Service, Metal Fabrication, Painting **Nearest Yard:** Valley Marine (604) 485-9257

Restaurants and Accommodations
OnSite: Motel (Beach Gardens 485-6267, $79-135) **Near:** Motel (Seaside Villa 485-2911), Inn/B&B (Ocean Point 485-5132) **1-3 mi:** Restaurant (Rene's Pasta 485-4555), (Chang Mai 485-0883), (Gourmet Canton 485-2885), (Casita 485-7220, Mexican), (Old Jailhouse Café 485-0926), (Pier Pub & Dockside Café 485-6281), Snack Bar (Robin's Donuts 485-5033), Fast Food (A&W), Pizzeria (Westview 485-6162), (Rocky Mountain Pizza & Bakery 485-9111), Motel (Marine Inn 485-4242)

Recreation and Entertainment
OnSite: Beach, Playground, Fitness Center, Jogging Paths, Boat Rentals (Mitchell's Canoes & Kayaks 487-1609) **Near:** Dive Shop (Alpha Dive 485-6939) **1-3 mi:** Pool, Golf Course (Myrtle Point 487-4653), Bowling, Movie Theater (Patricia Theatre 483-9345), Video Rental (Select Video 485-0333), Park, Museum (Powell River Museum), Cultural Attract (Evergreen Theatre 485-2891), Special Events, Galleries (Artique Co-op 485-4837)

Provisioning and General Services
OnSite: Liquor Store (11am-10pm) **Near:** Convenience Store (Thunder Bay Store 485-6633), Bank/ATM, Newsstand **1-3 mi:** Supermarket (Safeway 485-1233), Delicatessen (Killey), Farmers' Market (McLeod Rd. Sat 10:30am-12:30pm, Sun 12:30-2:30pm, Apr-Oct), Fishmonger (Powel River Seafood), Post Office, Catholic Church, Protestant Church, Library (485-4796), Beauty Salon (Eclips 485-9797), Laundry (Westview Cleaners & Laundry 485-2616), Bookstore (Breakwater Books 489-0010), Pharmacy (Shoppers 485-2844), Hardware Store, Clothing Store (Powell River Outdoor 485-2555), Retail Shops, Department Store (Towne Centre Mall, Wal-Mart 485-9811), Copies Etc. (Visitor Centre - Crossroads Village)

Transportation
OnSite: Courtesy Car/Van (van service to shopping center Jul-Aug), Local Bus **OnCall:** Taxi **1-3 mi:** Bikes, Rental Car (Rent a Wreck 485-7929, Budget 485-4131), Ferry Service (Comox &Texada) **Airport:** Powell River (3 mi.)

Medical Services
911 Service **OnCall:** Ambulance **1-3 mi:** Doctor (Medical Clinic Associates 485-6261), Dentist (Kean 485-9887), Chiropractor, Holistic Services, Veterinarian (Powell River Vet 485-6333) **Hospital:** Powell River General 485-2208 (4 mi.)

Setting -- Nestled among pine trees, the recently updated Beach Gardens Resort and Marina promises acres of quiet and gorgeous, wide-open views out to Malaspina Strait. A row of two and four-story waterfront, balconied motel units rim the edge of a hill just above the marina. Sheltered within the arms of two long rock jetties, the main pier parallels the northern jetty and hosts six long docks with side-tie moorage. A floating fuel dock greets boaters at the entrance.

Marina Notes -- Fuel in season only. Rafting is not permitted except in a storm situation. In season, free shuttle to town for provisioning, shopping or, in the evening, to restaurants. Resort offers concierge services and can arrange fishing charters, kayak or canoe tours, golf tee times or scuba diving. In early 2006, a new upscale four-story motel unit with 48 waterfront, well-equipped rooms with balconies joined the basic, dated two-story motel units and 12 housekeeping cottages - and stepped up the whole atmosphere. A restaurant is planned for 2007 and more improvements are on the way. This beautiful spot seems to be finally getting the TLC it deserves. We look forward to seeing the final result.

Notable -- Beach Gardens' proximity to town and its very useful seasonal van service makes it a good provisioning stop on the way to Desolation Sound. The Powell River area is a top scuba diving destination and has been called Canada's Dive Capital; Jacques Cousteau compared P.R.'s exceptionally clear water with that of the Red Sea. Nearly 100 dive sites have been identified by local guides with visibility to more than 30 meters, in winter. A highlight is the "HMCS Chaudiere," sunk to form an artificial reef. Nearby Alpha Diving (485-6939) is the central clearing house for guides and provides equipment.

Navigational Information
Lat: 49°37.937' **Long:** 124°03.455' **Tide:** 16 ft. **Current:** n/a **Chart:** 3535
Rep. Depths (MLW): Entry n/a **Fuel Dock** n/a **Max Slip/Moor** -/-
Access: Strait of Georgia to Pender Harbour to Joe Bay

Marina Facilities (In Season/Off Season)
Fuel: No
Slips: 50 Total, 12 Transient **Max LOA:** 150 ft. **Max Beam:** n/a
 Rate (per ft.): **Day** $1.00/0.85 **Week** $3.75 **Month** n/a
 Power: 30 amp n/a, **50 amp** Yes, **100 amp** n/a, **200 amp** n/a
 Cable TV: No **Dockside Phone:** No
 Dock Type: Floating, Wood
Moorings: 0 Total, 0 Transient **Launch:** n/a
 Rate: Day n/a **Week** n/a **Month** n/a
Heads: 2 Toilet(s), 2 Shower(s)
Laundry: 4 Washer(s), 4 Dryer(s) **Pay Phones:** No
Pump-Out: No **Fee:** n/a **Closed Heads:** Yes

Marina Operations
Owner/Manager: Peter Kwong **Dockmaster:** n/a
In-Season: May-Sep, 11am-11pm **Off-Season:** Oct-Apr, 11am-9pm
After-Hours Arrival: Call ahead
Reservations: No **Credit Cards:** Visa/MC, Amex
Discounts: None
Pets: Welcome, Dog Walk Area **Handicap Access:** Yes, Heads, Docks

Irvines Landing Resort & Marina

RR #1 Irvines Landing Road; Garden Bay, BC V0N 1S0

Tel: (778) 863-0311 **VHF: Monitor** n/a **Talk** n/a
Fax: n/a **Alternate Tel:** n/a
Email: info@irvineslandingresort.com **Web:** www.irvineslandingresort.com
Nearest Town: Garden Bay (2 mi.) **Tourist Info:** (604) 883-2561

Marina Services and Boat Supplies
Services - Trash Pick-Up **Communication -** Mail & Package Hold, Data
Ports (Inquire), FedEx, UPS **Supplies - OnSite:** Ice (Cube) **Under 1 mi:**
Ships' Store, Bait/Tackle **1-3 mi:** Propane (John Henry's 883-2253)

Boatyard Services
OnSite: Launching Ramp ($5) **OnCall:** Divers **Near:** Hull Repairs. **Under
1 mi:** Canvas Work, Bottom Cleaning. **1-3 mi:** Travelift, Electrical Repairs,
Electronic Sales, Woodworking, Inflatable Repairs, Upholstery, Metal
Fabrication. **3+ mi:** Engine mechanic (gas, diesel), Propeller Repairs.
Nearest Yard: Madeira Marina (604) 883-2261

Restaurants and Accommodations
OnSite: Restaurant (Irvines Landing Pub - a new 2000 square foot
restaurant will open by August '07), Motel (57 water-view, efficiencies with
terraces; in clusters of 8-12 units) **Under 1 mi:** Motel (Pender Harbor
Resort 883-2424, $80-220, motel rooms and cottages) **1-3 mi:** Restaurant
(Garden Bay Restaurant & Pub 883-9919, L & D: pub fare $6-19; formal
dining after 5pm $20-30), Snack Bar (John Henry's 883-2253, pizza, coffee,
ice cream, snacks), Lite Fare (LaVerne's Grill 883-2451, seasonal; takeout),
Condo/Cottage (Fisherman's Resort 883-2336, $100-120)

Recreation and Entertainment
OnSite: Pool (2008), Picnic Area, Playground, Volleyball (convertible to
basketball), Tennis Courts (2008), Fitness Center (2008) **Near:** Beach,
Jogging Paths, Boat Rentals **1-3 mi:** Dive Shop (Georgia Strait Diving &
Eco Tours 883-9120 - Pender Harbor is most popular dive spot on Sechelt
Peninsula), Horseback Riding, Roller Blade/Bike Paths, Fishing Charter

(Pender Harbour Charters 883-1181 - fishing, sightseeing, & eco tours;
Alpha Adventures 885-8838 - kayak rentals and tours, plus tennis lessons
771-1007), Video Rental (John Henry's 883-2253), Video Arcade (Garden
Bay), Park (Pender Hill Park; trail from Lee Bay Rd. to top of hill, panoramic
views), Cultural Attract (Pender Harbour Music School 883-2689),
Sightseeing, Special Events (May Day, Art Fest, Chamber Music - 3rd wknd
Aug, Jazz Fest - 3rd wknd Sep, Blues Festival), Galleries (Flying Anvil 883-
3660) **3+ mi:** Golf Course (Pender Harbour 883-9541, 5 mi.)

Provisioning and General Services
OnSite: Market (new onsite 1,500 sq. ft. market, or Oak Tree Market 883-
2411, by boat), Wine/Beer **1-3 mi:** Convenience Store (John Henry's 883-
2253), Supermarket (IGA Madeira Park 883-9100), Liquor Store (BC Liquor
883-2737, across bay), Bakery (IGA), Fishmonger, Bank/ATM, Post Office,
Catholic Church, Protestant Church, Beauty Salon, Bookstore (Harbour
Paper Mill 883-9911), Pharmacy (Marina Pahrmacy 883-2888, by boat),
Newsstand, Clothing Store, Retail Shops (gift shops, Madeira Beach)

Transportation
OnCall: Water Taxi (High Tide 883-9220) **1-3 mi:** Rental Car (PH
Transportation 883-2040, delivers), InterCity Bus (Malaspina Coach Lines
885-2217) **Airport:** Sechelt/YVR (22 mi./65 mi.)

Medical Services
911 Service **OnCall:** Ambulance **1-3 mi:** Doctor (Pender Harbour Health
Centre 883-2764, Mon-Fri 8am-4pm), Dentist, Veterinarian (Madeira
Veterinary Hospital 883-2488) **Hospital:** St. Mary's 885-2224 (30 mi.)

Setting -- The new Irvines Landing Hotel Resort & Marina complex sits at the mouth of Pender Harbour. A cluster of three-story, wood-framed, beachcomber/west coast-style hotel units will perch above the docks with gorgeous views of the rockbound shores of Joe Bay. The Government Dock fishing pier, lined with flower-filled planters, dominates the cove. Adjacent, but not accessible from the pier, will be Irvine Landing's newly refurbished 50-slip marina.

Marina Notes -- Formerly Irvines Landing Pub & Marina. The original structure is being demolished and rising in its stead, beginning early '07, will be a new complex developed by Peter Kwong of Vancouver. The restaurant and 30 motel units are scheduled for completion by August '07; the amenities and the remaining units by December '07. The facility will feature all new docks, a 2,000 square foot restaurant, a 1,500 square foot on-site grocery store, a heated pool, courts usable for tennis, basketball or volleyball, a playground, meeting/event room for 50. And, for a night ashore, 57 efficiency hotel rooms (500 sq. ft. each) with water views and decks will cluster in units of 8-12 rooms. The new bathhouse will have four boater-only full bathrooms and a coin-op laundry.

Notable -- In 1865, a trading post set up shop on this site followed, in early 1900s, by a hotel, saloon and store. By 1929, Union steamships were carrying loggers, settlers and merchants to & from Vancouver three times a week. In the mid-30s, the Sechelt Highway ended the era. Until fall '06, Irvines Landing Marine Pub was a beloved magnet for locals & boaters; it appears that this new, more upscale facility will be as well. Pender Hill rises 758 feet above Irvines Landing - the two-mile round-trip Pender Hill trail promises a good hike (Irvines Landing Rd. to Lee Rd.). More shopping is across the harbor in Madeira Park.

PHOTOS ON DVD: 8

Navigational Information
Lat: 49°38.062' **Long:** 124°02.639' **Tide:** 15 ft. **Current:** n/a **Chart:** 3535
Rep. Depths (MLW): Entry 12 ft. **Fuel Dock** n/a **Max Slip/Moor** 14 ft./-
Access: Sechelt Peninsula, Duncan Cove in Pender Harbour

Marina Facilities (In Season/Off Season)
Fuel: No
Slips: 75 Total, 10 Transient **Max LOA:** 48 ft. **Max Beam:** n/a
  Rate (per ft.): **Day** $0.90 **Week** $6.30 **Month** $9
  Power: 30 amp $6, **50 amp** n/a, **100 amp** n/a, **200 amp** n/a
  Cable TV: No **Dockside Phone:** No
  Dock Type: Floating, Wood
Moorings: 0 Total, 0 Transient **Launch:** n/a, Dinghy Dock
  Rate: Day n/a **Week** n/a **Month** n/a
Heads: 5 Toilet(s), 4 Shower(s)
Laundry: 2 Washer(s), 1 Dryer(s), Iron, Iron Board **Pay Phones:** Yes
Pump-Out: No **Fee:** n/a **Closed Heads:** Yes

Marina Operations
Owner/Manager: Adam Boothby **Dockmaster:** n/a
In-Season: May-Sep, 8am-8pm **Off-Season:** Oct-Apr, 9am-5pm
After-Hours Arrival: Call ahead
Reservations: Preferred **Credit Cards:** Visa/MC, Interac or Debit
Discounts: None
Pets: Welcome **Handicap Access:** Yes, Heads, Docks

Pender Harbour Resort

PO Box ; 4686 Sinclair Bay Road; Garden Bay, BC V0N 1S0

Tel: (604) 883-2424; (877) 883-2424 **VHF: Monitor** n/a **Talk** n/a
Fax: (604) 883-2414 **Alternate Tel:** n/a
Email: info@penderharbourresort.com **Web:** penderharbourresort.com
Nearest Town: Garden Bay (1.5 mi.) **Tourist Info:** (604) 883-2561

Marina Services and Boat Supplies
Services - Docking Assistance **Communication -** Phone Messages, Data Ports (Wi-Fi - Broadband Xpress), DHL, UPS **Supplies - OnSite:** Ice (Block, Cube), Bait/Tackle, Live Bait (Herring) **1-3 mi:** Ships' Store (limited supplies at Fisherman's Resort), Propane (John Henry's 883-2253)

Boatyard Services
OnSite: Launching Ramp **Under 1 mi:** Travelift, Engine mechanic (gas, diesel), Electronic Sales, Electronics Repairs, Hull Repairs, Bottom Cleaning, Brightwork, Divers, Compound, Wash & Wax, Interior Cleaning, Propeller Repairs, Woodworking, Metal Fabrication. **1-3 mi:** Electrical Repairs, Inflatable Repairs. **Nearest Yard:** Madeira Marina (604) 883-2261

Restaurants and Accommodations
OnSite: Motel (Pender Harbour $85, efficiency suites), Condo/Cottage (Pender Harbour $110-250, Garden Bay Chalet sleeps 8) **Under 1 mi:** Restaurant (Irvines Landing reopens August '07) **1-3 mi:** Restaurant (Garden Bay Restaurant & Pub 883-9919, L & D: pub fare $6-19; formal dining after 5pm, $20-30), Snack Bar (John Henry's 883-2253, pizza, coffee, ice cream, snacks), Lite Fare (LaVerne's Grill 883-2451, seasonal; burgers, fish & chips, ice cream), Pizzeria (Pizza Pantry 883-2543, across bay)

Recreation and Entertainment
OnSite: Heated Pool ($10), Picnic Area, Grills (included in pool fee), Playground, Jogging Paths, Boat Rentals, Roller Blade/Bike Paths
Under 1 mi: Park (Pender Hill Park; steep trail to top of hill) **1-3 mi:** Dive Shop (Georgia Strait Diving & Eco Tours 883-9120), Fishing Charter

(Pender Harbour Charters 883-1181, fishing, sightseeing, & ecotours; Alpha Adventures 885-8838, kayak rentals and tours, plus tennis lessons 771-1007), Video Rental (John Henry's 883-2253), Cultural Attract (Pender Harbour Music School 883-2689), Sightseeing, Special Events (May Day, Art Fest, sailing events - Jul & Aug, music events Aug & Sep), Galleries (Flying Anvil 883-3660) **3+ mi:** Golf Course (Pender Harbour 883-9541, 5 mi.), Horseback Riding (Malaspring Ranch)

Provisioning and General Services
Under 1 mi: Bank/ATM, Post Office **1-3 mi:** Convenience Store (John Henry's 883-2253), Market (IGA Madeira Park 883-9100/Oak Tree Market 883-2411, by boat), Wine/Beer (John Henry's 883-2253), Liquor Store (BC Liquor 883-2737), Catholic Church, Protestant Church, Beauty Salon, Bookstore (Harbour Paper Mill 883-9911), Pharmacy (Marina Pharmacy 883-2888, by boat), Newsstand, Clothing Store, Retail Shops (gifts & souvenirs: Knitting Zen 883-2922, Water & the Wind Gifts 883-9911, UE-Urban Eclectics)

Transportation
OnCall: Water Taxi (High Tide 883-9220) **1-3 mi:** Rental Car (PH Transportation 883-2040), InterCity Bus (Malaspina Coach Lines 885-2217)
Airport: Sechelt/YVR (22 mi./65 mi.)

Medical Services
911 Service **OnSite:** Holistic Services (Pender Harbour Resort Spa 883-0602) **OnCall:** Ambulance **1-3 mi:** Doctor (Pender Harbour Health Centre 883-2764, Mon-Fri, 8am-4pm), Dentist, Chiropractor, Veterinarian (Madeira Veterinary Hospital 883-2488) **Hospital:** St. Mary's 885-2224 (30 mi.)

Setting -- Tucked into Duncan Cove, a network of quality docks floats just below a cluster of cottages and buildings that hug the craggy shoreline; a screened gazebo perches above the kayak launch. Just up the hill a series of modest resort buildings -- a motel, cabins, yurts, RV sites and a large chalet -- are all festooned with well-tended flowers and greenery. Little picnic areas are tucked in and about - some awaiting the arrival of an RV.

Marina Notes -- A marina, formerly named Duncan Cove Resort, has been on this site since the '70s. For 13 years it was owned and operated by the current managers' parents. Recently, a non-family Vancouver-based group purchased the facility, but is still managed by the previous owners' knowledgeable next generation - Alexis and John McLeod - who live on-site. Outstation for Vancouver's Thunderbird Yacht Club and Vancouver Island's Schooner Cove Yacht Club. A new spa opened in 2005 and further development is planned for the area to the right of the marina, facing the water. Pool and grills are available for $10. Book exchange and Internet access. Ideal for group or club cruises is a private event space with a caterer-friendly kitchen. The pool table is available to all. Tiled-floor cinder-block bathhouses have coin-operated fiberglass showers with dressing rooms.

Notable -- The focus is more on the resort than on the marina and the amenities are plentiful. A lovely pool area, with a solar heat pool cover, is surrounded by attractive teak tables and chairs with green market umbrellas and brown strapped chaises. Visible from the pool area is a wide grassy area, which is actually a bocci court; a fire pit and a new playground are at the far end. Nestled in the trees, the diminutive new spa is housed in an inviting yurt with a sundeck.

Navigational Information
Lat: 49°37.857' **Long:** 124°01.943' **Tide:** 15 ft. **Current:** n/a **Chart:** 3535
Rep. Depths (*MLW*): **Entry** 80 ft. **Fuel Dock** 30 ft. **Max Slip/Moor** 60 ft./-
Access: Pender Harbour to starboard side of Hospital Bay

Marina Facilities *(In Season/Off Season)*
Fuel: *At John Henry's* - Gasoline, Diesel
Slips: 35 Total, 25 Transient **Max LOA:** 80 ft. **Max Beam:** n/a
  Rate *(per ft.)*: **Day** $1.00/0.75 **Week** $7/4.90 **Month** $5.25
  Power: 30 amp $6*, **50 amp** n/a, **100 amp** n/a, **200 amp** n/a
  Cable TV: No **Dockside Phone:** No
  Dock Type: Floating, Long Fingers, Pilings, Alongside, Wood
Moorings: 0 Total, 0 Transient **Launch:** n/a
  Rate: Day n/a **Week** n/a **Month** n/a
Heads: 2 Toilet(s), 2 Shower(s)
Laundry: 1 Washer(s), 1 Dryer(s), Book Exchange **Pay Phones:** Yes
Pump-Out: No **Fee:** n/a **Closed Heads:** Yes

Marina Operations
Owner/Manager: David Pritchard **Dockmaster:** Same
In-Season: Apr-Sep, 7am-9pm **Off-Season:** Oct-Mar, Closed
After-Hours Arrival: Call Ch. 66A or 604-883-2336
Reservations: Yes, Required **Credit Cards:** Visa/MC, Interac
Discounts: None
Pets: Welcome, Dog Walk Area **Handicap Access:** No

Fisherman's Resort & Marina

PO Box 68; 4890 Pool Road; Garden Bay, BC V0N 1S0

Tel: (604) 883-2336 **VHF: Monitor** Ch. 66A **Talk** n/a
Fax: (604) 883-2336 **Alternate Tel:** (604) 817-2336
Email: fishermans@dccnet.com **Web:** www.fishermansresortmarina.com
Nearest Town: Garden Bay *(0.3 mi.)* **Tourist Info:** (604) 883-2561

Marina Services and Boat Supplies
Services - Docking Assistance, Security *(owner onsite)*, Dock Carts
Communication - Phone Messages, Data Ports *(Wi-Fi - Broadband Xpress)*, FedEx, UPS **Supplies - OnSite:** Ice *(Cube)*, Ships' Store *(limited)*, Bait/Tackle **Near:** Live Bait *(herring)*, Propane *(John Henry's 883-2253)*

Boatyard Services
OnSite: Launching Ramp, Brightwork **OnCall:** Engine mechanic *(gas, diesel)*, Electrical Repairs, Air Conditioning, Refrigeration, Divers, Compound, Wash & Wax, Interior Cleaning, Propeller Repairs, Woodworking, Inflatable Repairs, Upholstery, Metal Fabrication
Under 1 mi: Travelift. **Nearest Yard:** Madeira Marina (604) 883-2261

Restaurants and Accommodations
OnSite: Condo/Cottage *($100-135, waterfront - add 10% for just one night)*
Near: Restaurant *(Garden Bay Restaurant & Pub 883-9919, L & D: pub fare $6-19, formal dining from 5pm $20-30)*, Snack Bar *(John Henry's 883-2253, pizza slices, coffee, ice cream, snacks)*, Lite Fare *(LaVerne's Grill 883-2451, seasonal fare & takeout)* **1-3 mi:** Motel *(Pender Harbour 883-2424, cottages too)*

Recreation and Entertainment
OnSite: Picnic Area, Grills, Boat Rentals *(Kayaks $22-30/2 hrs., $50-68/day; 14' open boats w/ 10 hp; 16' & 17' covered boats w/ 50-60 hp)* **Near:** Jogging Paths, Fishing Charter *(Pender Harbour Charters 883-1181 - fishing & ecotours; Alpha Adventures 885-8838 - kayak rentals, tours, & tennis lessons 771-1007)*, Video Rental *(John Henry's 883-2253)*

Under 1 mi: Dive Shop, Park *(Garden Bay Provincial Marine Park, hiking trails to Mount Daniel)*, Cultural Attract *(Pender Harbour Music School 883-2689)*, Special Events *(May Day, Chamber Music Festival - 3rd wknd Aug, Pender Harbour Jazz Fest - 3rd wknd Sep, Blues Festival)*, Galleries *(Flying Anvil 883-3660)* **1-3 mi:** Beach, Sightseeing *(Skookumehuc Rapids)*
3+ mi: Pool *(Aquatic Center 883-2612, call for hours; classes, instructors & kids' programs, 5 mi.)*, Golf Course *(Pender Harbour G.C. 883-9541, 5 mi.)*

Provisioning and General Services
Near: Convenience Store *(John Henry's 883-2253)*, Wine/Beer *(John Henry's 883-2253)*, Post Office **Under 1 mi:** Market *(IGA Madeira Park 883-9100, by boat /Country Roads)*, Liquor Store *(BC Liquor 883-2737 by boat)*, Bakery, Bank/ATM, Catholic Church, Protestant Church, Bookstore *(Harbour Paper Mill 883-9911)*, Pharmacy *(Marina Pharmacy 883-2888)* **1-3 mi:** Beauty Salon, Copies Etc. **3+ mi:** Hardware Store *(10 mi.)*

Transportation
OnSite: Courtesy Car/Van *(14' runabout for provisioning in Madeira Park)*
OnCall: Water Taxi *(High Tide 883-9220)* **Under 1 mi:** Rental Car *(PH Transportation 883-2040)* **1-3 mi:** InterCity Bus *(Malaspina Coach Lines 885-2217)* **3+ mi:** Taxi *(10 mi.)* **Airport:** Sechelt, 22 mi./YVR, 65 mi.; floatplane service to dock *(On-Site)*

Medical Services
911 Service **OnCall:** Ambulance **Under 1 mi:** Doctor *(Pender Harbour Health Centre 883-2764, 8am-4pm Mon-Fri)*, Dentist, Chiropractor
1-3 mi: Holistic Services, Veterinarian *(Madeira Veterinary Hospital 883-2488)* **Hospital:** St. Mary's 885-2224 *(30 mi.)*

Setting -- Entering Hospital Bay, the Fisherman's Resort docks are immediately to starboard. The three sets of floating docks share a single ramp that leads through an arbor to a "sunset deck," and then to the upland. There, amid well-maintained park-like grounds, a picnic grove, office, tackle shop and cottages share spectacular views. Just beyond, bright blue awnings on a gray half-timbered structure signal John Henry's - a small store, pub and fuel dock.

Marina Notes -- *20 amp $3.50. Established in 1961, it has maintained its "60s west-coast village" ambiance through three owners. As of late '05, owners/managers David Pritchard and Jennifer Love (from Bowen Island) are now at the helm; watch for many improvements that won't disturb the coveted casual atmosphere. Broadband Xpress Wi-Fi covers all 2,300 feet of dockage. Four quaint cabins plus limited camping and RV sites. The bathhouse consists of two recently updated complete bathrooms - with tile floors and fiberglass shower stalls. FYI: 75% are visitors from the U.S., many headed to Alaska.

Notable -- The two French blue buildings perched on the promontory above the docks were once loggers' float homes that have been permanently installed and renovated into charming rental units - an expansive deck hosts bright red chairs. Up the harbor is the permanent berth of a crayon yellow 1943 Curtiss N-7 floatplane (owned by Bill Thompson). Hopefully, you will see the plane take its 45-minute tour of the harbor and do a stalled drop right before the docks - it is quite a production. On the hill at the head of the bay, past John Henry's, is former Sundowners Inn - a cream clapboard building with bright green trim and a brick-red roof-signed "Restaurant." It's closed, but rumors abound. The area supports many music events including blues, chamber music and jazz festivals.

Garden Bay Restaurant & Marina

PO Box 90; 4958 Lyons Road; Garden Bay, BC V0N 1S0

Tel: (604) 883-2674 **VHF: Monitor** Ch. 66A **Talk** Ch. 66
Fax: (604) 883-2674 **Alternate Tel:** n/a
Email: gbhm@dccnet.com **Web:** www.gardenbaypub.com
Nearest Town: Garden Bay **Tourist Info:** (604) 883-2561

Navigational Information
Lat: 49°37.930' **Long:** 124°01.667' **Tide:** 15 ft. **Current:** 0 kt. **Chart:** 3535
Rep. Depths (*MLW***): Entry** 100 ft. **Fuel Dock** n/a **Max Slip/Moor** 30 ft./-
Access: Sunshine Coast, Pender Harbour to Garden Bay

Marina Facilities *(In Season/Off Season)*
Fuel: No
Slips: 30 Total, 12* Transient **Max LOA:** 140 ft. **Max Beam:** 20 ft.
 Rate *(per ft.):* **Day** $0.90 **Week** n/a **Month** n/a
 Power: 30 amp $6** , **50 amp** n/a, **100 amp** n/a, **200 amp** n/a
 Cable TV: No **Dockside Phone:** No
 Dock Type: Floating, Alongside, Wood
Moorings: 0 Total, 0 Transient **Launch:** Yes, Dinghy Dock
 Rate: Day n/a **Week** n/a **Month** n/a
Heads: 2 Toilet(s), 2 Shower(s)
Laundry: 1 Washer(s), 1 Dryer(s) **Pay Phones:** Yes, 2
Pump-Out: No **Fee:** n/a **Closed Heads:** Yes

Marina Operations
Owner/Manager: Ron Johnston **Dockmaster:** Same
In-Season: Year-Round, 11am-Mid **Off-Season:** n/a
After-Hours Arrival: Call ahead
Reservations: Preferred **Credit Cards:** Visa/MC, Amex
Discounts: None
Pets: Welcome, Dog Walk Area **Handicap Access:** No

Marina Services and Boat Supplies
Services - Docking Assistance, Trash Pick-Up, Dock Carts
Communication - Mail & Package Hold, Data Ports *(Wi-Fi - Broadband Xpress)* **Supplies - Near:** Ice *(Block, Cube)*, Bait/Tackle, Propane *(John Henry's)*

Boatyard Services
OnSite: Air Conditioning **OnCall:** Engine mechanic *(gas, diesel)*, Divers **Near:** Launching Ramp, Hull Repairs, Metal Fabrication. **Under 1 mi:** Travelift, Electrical Repairs, Electronic Sales, Bottom Cleaning, Upholstery, Yacht Interiors. **Nearest Yard:** Madeira Marina (604) 883-2261

Restaurants and Accommodations
OnSite: Restaurant *(Garden Bay 883-9919, D $20-30, from 5pm)*, Lite Fare *(Garden Bay Pub L $6-12, D $8-19, sandwiches, burgers, grilled dinners; jazz Fri & Sat nights, jam session Sun afternoon)* **Near:** Snack Bar *(John Henry's 883-2253, snacks, ice cream, pizza)*, Lite Fare *(LaVerne's Grill 883-2451, seasonal; burgers, fish & chips, ice cream)* **1-3 mi:** Restaurant *(Irvine's Landing 883-1145, reopens in August 2007)*, Motel *(Sunshine Coast Resort 883-9177, $100-260, across the bay)*, *(Pender Harbour Resort 883-2424, $80-220, motel rooms & cottages)*

Recreation and Entertainment
OnSite: Boat Rentals *(Alpha Adventures 885-8838 - sea kayak & sit-on-top rentals and tours, plus tennis lessons 771-1007)*, Fishing Charter *(Pender Harbour Charters 883-1181 - fishing, sightseeing, & ecotours)* **Near:** Video Rental *(John Henry's 883-2253)*, Video Arcade **Under 1 mi:** Dive Shop *(Georgia Strait Diving & Eco Tours 883-9120)*, Jogging Paths, Park *(Garden Bay Provincial Marine Park, hiking trails to Mount Daniel)*, Cultural Attract *(Pender Harbour Music School 883-2689)*, Special Events *(May Day, Art Fest, Sailing Races, Chamber Music - 3rd wknd Aug, Jazz Fest - 3rd wknd Sep, Blues Festival)*, Galleries *(Flying Anvil 883-3660)* **1-3 mi:** Beach, Picnic Area **3+ mi:** Pool *(Aquatic Center 883-2612, call for hours; classes, instructors and kids programs, 5 mi.)*, Golf Course *(Pender Harbour Golf Club 883-9541, 5 mi.)*

Provisioning and General Services
OnSite: Wine/Beer, Bank/ATM **Near:** Convenience Store *(John Henry's 883-2253)*, Post Office, Newsstand **Under 1 mi:** Market *(IGA Madeira Park 883-9100/Oak Tree Market 883-2411, by boat)*, Liquor Store *(BC Liquor 883-2737 by boat)*, Fishmonger *(Madeira Park)*, Catholic Church, Protestant Church, Beauty Salon, Bookstore *(Harbour Paper Mill 883-9911)*, Pharmacy *(Marina Pharmacy 883-2888)*, Clothing Store *(The Clothing Lounge)*, Retail Shops *(Water & the Wind 883-9911, Urban Eclectics, Knitting Zen 883-2922)*

Transportation
OnCall: Water Taxi *(High Tide 883-9220)* **Under 1 mi:** Rental Car *(PH Transportation 883-2040)*, InterCity Bus *(Malaspina Coach Lines 885-2217)* **Airport:** Sechelt/YVR *(22 mi./65 mi.)*

Medical Services
911 Service **Under 1 mi:** Doctor *(Pender Harbour Health Centre 883-2764, Mon-Fri 8am-4pm)*, Dentist, Chiropractor **1-3 mi:** Holistic Services, Veterinarian *(Madeira Veterinary Hospital 883-2488)* **Hospital:** St. Mary's 885-2224 *(30 mi.)*

Setting -- At the head of the long, narrow bay, Garden Bay Restaurant's L-shaped gray clapboard building with natural trim and brick red roof perches on a rocky rise above two floating wharves. It is surrounded by a dining deck protected by glass windscreens, and, just below, picnic tables sprawl on the grassy slope. At the water level, between the two piers, a low-slung dockhouse hosts several services. To port, the lovely gray shingle, multistory contemporary with white trim and glass-topped pavilion is a Royal Vancouver Yacht Club substation. To starboard, Sportsmen's Marina hosts a Seattle Yacht Club outstation.

Marina Notes -- *700 ft. "U" shaped alongside moorage. **Also 15 amp. Established in 1932. Dockhouse hosts Alpha Adventures, Pender Harbour Tours and a mini-florist. Adjacent Sportsmen's Marina, which may also have some transient dockage, has heads and showers that are available to boaters docking at Garden Bay Marina. Royal Vancouver Y.C. (604-224-1344; garden_bay@royalvan.com) and Seattle Y.C. (206-325-1000) may offer reciprocal privileges.

Notable -- Garden Bay Restaurant has 3,000 square feet of dining and entertainment space. A cozy bar with large screen TV and outdoor waterfront deck serves a pub menu and a respected, glass-walled fine dining restaurant features seafood specialties and an interesting wine list. Well-known throughout B.C. for its sponsorship of jazz groups and live bands, it offers jam sessions on Sundays. Entertainment Friday-Sunday plus Thursdays (and some Wednesdays) during the summer. The "Meat Draw" on Friday is a cultural experience. Pender Harbour has been called Canada's Venice; the labyrinth of bays, waterways and coves makes it a dive and kayak destination. Prized sites - Nelson Rock, Anderson & Charles Islands and Fearney Bluffs - can only be reached by boat.

2. BC - SUNSHINE COAST

PHOTOS ON DVD: 17

Navigational Information
Lat: 49°37.363' **Long:** 124°01.525' **Tide:** 15 ft. **Current:** n/a **Chart:** 3535
Rep. Depths (*MLW*): **Entry** 12 ft. **Fuel Dock** n/a **Max Slip/Moor** 12 ft./-
Access: Pender Harbour to Welbourn Cove

Marina Facilities *(In Season/Off Season)*
Fuel: No
Slips: 50 Total, 10 Transient **Max LOA:** 100 ft. **Max Beam:** n/a
  Rate *(per ft.)*: **Day** $0.65* **Week** n/a **Month** $3.40
  Power: 30 amp $6.45**, **50 amp** $10.75, **100 amp** n/a, **200 amp** n/a
  Cable TV: No **Dockside Phone:** No
  Dock Type: Floating, Long Fingers, Alongside, Wood
Moorings: 0 Total, 0 Transient **Launch:** n/a, Dinghy Dock
  Rate: Day n/a **Week** n/a **Month** n/a
Heads: 2 Toilet(s), 2 Shower(s)
Laundry: None **Pay Phones:** Yes
Pump-Out: OnSite, Full Service, 1 Central **Fee:** $12 **Closed Heads:** Yes

Marina Operations
Owner/Manager: Diana Pryde **Dockmaster:** Same
In-Season: Year-Round, 8am-7pm **Off-Season:** n/a
After-Hours Arrival: Self-service pay boxes
Reservations: No **Credit Cards:** Visa/MC, Amex
Discounts: None
Pets: Welcome **Handicap Access:** Yes, Heads

Madeira Park Public Wharf

PO Box 118; Madeira Park, BC V0N 2H0

Tel: (604) 883-2234 **VHF: Monitor** Ch. 66A **Talk** Ch. 66A
Fax: (604) 883-2234 **Alternate Tel:** (604) 883-9878
Email: n/a **Web:** n/a
Nearest Town: Madeira Park **Tourist Info:** (604) 883-2561

Marina Services and Boat Supplies
Services - Docking Assistance, Trash Pick-Up, Dock Carts
Communication - Phone Messages, Express Mail **Supplies - Near:** Ice
(Cube), Boat/US, Bait/Tackle **Under 1 mi:** Propane

Boatyard Services
OnSite: Launching Ramp **OnCall:** Divers, Propeller Repairs,
Woodworking **Near:** Crane *(1.5T)*, Engine mechanic *(gas, diesel)*, Electrical
Repairs, Hull Repairs, Bottom Cleaning, Brightwork, Air Conditioning,
Compound, Wash & Wax, Interior Cleaning, Upholstery, Metal Fabrication,
Painting, Yacht Broker *(Pacific Northwest Yacht 883-9430)*. **Under 1 mi:**
Inflatable Repairs. **Nearest Yard:** Madeira Marine (604) 883-2261

Restaurants and Accommodations
OnCall: Pizzeria *(Pizza Pantry 883-2543, del)* **Near:** Coffee Shop *(Java
Dock B $1-3, L $2-8, sandwiches, sausage rolls and hot entrees)*, Lite Fare
(Triple B's Buda-Bing-Burgers 883-9655, B $2-4, L $3-6), *(Jill's Place 883-
2355, B $2-7, L $3-9, all homemade, croissants, soups, sandwiches)*
Under 1 mi: Restaurant *(Garden Bay 883-9919, L $6-12, D $20-30, Pub
$8-19)*, *(Garden Hill)* **1-3 mi:** Lite Fare *(Grasshopper Pub 883-9013,
L $10-16, D $10-16)*, Hotel *(Pender Harbour Hotel 883-9013)*

Recreation and Entertainment
OnSite: Picnic Area **Near:** Dive Shop *(Georgia Strait Diving & Eco Tours
883-9120)*, Fishing Charter *(Pender Harbour Charters - fishing, sightseeing
and ecotours 883-1181)*, Video Rental *(Sean & Sarah's Internet Café - DVDs
and digital photo processing 883-9505)*, Park *(Seafarer's Park adjacent to
marina; gazebo, picnic tables, benches)*, Cultural Attract *(Pender Harbour

Music School 883-2689), Sightseeing *(Rain Coast Adventures 883-9239, 3
hr. harbor tours aboard 60' sailing hitch; Eco Tours 740-2486)*, Special
Events *(all summer)*, Galleries *(The Harbour Gallery 883-2807, Flying Anvil
883-3660)* **Under 1 mi:** Beach, Grills, Playground, Boat Rentals
1-3 mi: Pool *(Aquatic Center at Secondary School 883-2612, call for public
hours and programs)*, Golf Course *(Pender Harbour 883-9541)*

Provisioning and General Services
OnSite: Fishmonger *(seasonal 740-7942; fish, shrimp, crabs)*
Near: Convenience Store, Market *(IGA 883-9100, Oak Tree 883-2411-
produce, fresh meats, deli)*, Liquor Store *(BC Liquor 883-2737)*, Bakery,
Bank/ATM, Post Office, Catholic Church, Protestant Church *(St. Andrew's)*,
Beauty Salon *(Miss Sunny's 883-2715)*, Laundry, Bookstore *(Harbour Paper
Mill 883-9911)*, Pharmacy *(Guardian Marina 883-2888)*, Newsstand, Florist,
Clothing Store *(The Clothing Lounge)*, Retail Shops *(UE-Urban Eclectics,
Knitting Zen 883-2922, souvenirs and T-shirts, Water & the Wind Gifts 883-
9911)*, Copies Etc. *(Pender Harbour Paper Mill 883-9911)*

Transportation
OnSite: Ferry Service *(Slo-Cat, local shuttle)* **OnCall:** Water Taxi
(Malaspina 740-2486), Taxi *(Sunshine Coast 604-740-0703)* **Near:** Rental
Car *(PH Transportation 883-2040, delivers)*, InterCity Bus *(Malaspina Coach
885-2217)* **Airport:** Sechelt - floatplane pickup/YVR *(OC/65 mi.)*

Medical Services
911 Service **OnCall:** Ambulance **Under 1 mi:** Doctor *(P.H. Health Centre
883-2764, Mon-Fri, 8am-4pm)*, Dentist *(P.H. Health)*, Veterinarian *(Madeira
Vet. Hosp. 883-2488)* **Hospital:** St. Mary's 885-2224 *(20 mi.)*

Setting -- On the south side of beautiful Pender Harbour, in Wellbourne Cove, a pair of floating pontoon alongside side-tie docks leads up to a large
stationary wharf. On the shore is a classic, fieldstone and natural-sided dock house with a forest green steel roof. Flanking the dockhouse are two pretty,
carefully planted pocket Seafarer's Parks - one features a catelievered gazebo, brick paths, benches and stone tables - overlooking the "Govie." Madeira Park,
the region's modest supply depot, is a very short walk.

Marina Notes -- *Average rate - a chart spells out the rate for each LOA. 1 hr. grace time; 2-4 hrs., half-day rate; over 4 hrs., full-day rate. **15 amps
$4.30/day. On-site Mobile Marine Repairs provides boatyard services. Working fishing boats usually have halibut ($10/lb.), salmon, shrimp, cod ($6/lb.) prawns,
crab and halibut cheeks ($10/lb.) Inviting fully tiled bathrooms are keyed for boaters only.

Notable -- 150 meters away is the village of Maderia Park - Pender Harbour's center for services, supplies, activities, events and transportation. Boaters push
their Marketplace or Oak Tree provisions to the docks in grocery carts and then leave the buggies in an assigned spot. The Pender Harbour Cultural Centre
houses the library, PH School of Music and PH Gallery. Breakfast and lunch are available at Java Docks (7:30am-6pm, 5pm Sat, 4pm Sun), Jill's Place (5am-
4pm), and Triple B's Buda-Bing-Burgers. For dinner, buy fresh seafood off the boats. A handful of shops offer clothing and gifts. Malaspina's Powell River to
Vancouver bus stops twice daily. Events include May Day, an Arts Festival, sailing regattas, chamber music (3rd wknd Aug) and Jazz Fest (3rd wknd Sep).

Sunshine Coast Marina & Resort

12695 Sunshine Coast Hwy; Madeira Park, BC V0N 2H0

Tel: (604) 883-9177 **VHF: Monitor** Ch. 66A **Talk** n/a
Fax: (604) 883-9171 **Alternate Tel:** (604) 883-9177
Email: vacation@sunshinecoast-resort.com **Web:** sunshinecoastresort.com
Nearest Town: Madeira Park *(0.5 mi.)* **Tourist Info:** (604) 883-2561

Navigational Information
Lat: 49°37.458' **Long:** 124°01.123' **Tide:** 15 ft. **Current:** n/a **Chart:** 3535
Rep. Depths *(MLW)*: **Entry** 50 ft. **Fuel Dock** n/a **Max Slip/Moor** 50 ft./-
Access: Head of Pender Harbour

Marina Facilities *(In Season/Off Season)*
Fuel: No
Slips: 30 Total, 8 Transient **Max LOA:** 100 ft. **Max Beam:** 12 ft.
  Rate *(per ft.)*: **Day** $0.75 **Week** n/a **Month** $6.00
  Power: 30 amp $3, **50 amp** n/a, **100 amp** n/a, **200 amp** n/a
  Cable TV: Yes, $3, 30 channels **Dockside Phone:** No
  Dock Type: Fixed, Floating, Short Fingers, Pilings, Wood
  Moorings: 0 Total, 0 Transient **Launch:** n/a, Dinghy Dock
  Rate: Day n/a **Week** n/a **Month** n/a
Heads: 6 Toilet(s), 4 Shower(s) *(with dressing rooms)*, Hair Dryers
Laundry: 2 Washer(s), 1 Dryer(s), Iron, Iron Board **Pay Phones:** Yes
Pump-Out: No **Fee:** n/a **Closed Heads:** Yes

Marina Operations
Owner/Manager: Ralph Linnmann **Dockmaster:** Same
In-Season: Year-Round, 24/7 **Off-Season:** n/a
After-Hours Arrival: Call 604-883-9177
Reservations: Preferred **Credit Cards:** Visa/MC, Amex, Debit
Discounts: None
Pets: Welcome, Dog Walk Area **Handicap Access:** No

Marina Services and Boat Supplies
Services - Docking Assistance, Concierge, Security *(24 hrs., surveillance cameras)*, Trash Pick-Up, Dock Carts **Communication -** Mail & Package Hold, Phone Messages, Data Ports *(Wi-Fi)*, FedEx, UPS **Supplies -** **OnSite:** Ice *(Block, Cube, Shaved)*, Bait/Tackle, Live Bait *(herring)* **Near:** Ships' Store *(Madeira Marina)* **Under 1 mi:** Propane

Boatyard Services
OnCall: Divers **Near:** Railway *(50T)*, Engine mechanic *(gas, diesel)*, Electrical Repairs, Electronic Sales, Canvas Work, Bottom Cleaning, Brightwork, Air Conditioning, Refrigeration, Compound, Wash & Wax, Interior Cleaning, Propeller Repairs, Woodworking, Upholstery, Metal Fabrication, Painting, Awlgrip. **Nearest Yard:** Madeira Marina (604) 883-2261

Restaurants and Accommodations
OnSite: Hotel *(Sunshine Coast Resort 883-9177, $100-260)* **OnCall:** Pizzeria *(Pizza Pantry 883-2543, del)* **Near:** Restaurant *(Garden Bay Hotel 883-2674, L $10, D $17-32, by boat or shuttle)*, Coffee Shop *(Java Docks)*, Lite Fare *(Jill's Place 883-2355)*, *(Triple B's Buda-Bing-Burgers 883-9655)* **1-3 mi:** Restaurant *(Irvines Landing 883-1145, by boat - reopening August '07)*, Lite Fare *(Grasshopper Pub 883-9013, L $10-16, D $10-16, at Pender Harbour Hotel, shuttle)*, Motel *(Pender Harbour Resort 883-2424, $80-220, across the bay; motel rooms and cottages)*

Recreation and Entertainment
OnSite: Heated Pool *(new in 2006)*, Spa *(oceanside, open 24 hrs.)*, Picnic Area, Grills, Playground, Boat Rentals, Fishing Charter, Video Rental, Sightseeing *(883-2280 - Princess Luisa Inlet Tour - 6 hrs. $130/90,* Skookumchuk Narrows - 45 min. $40, Highlights - 2 hrs. $70) **Near:** Dive Shop, Jogging Paths, Roller Blade/Bike Paths, Video Arcade, Park *(Seafarer's Park - gazebo, picnic tables, benches)*, Cultural Attract *(Pender Harbour Music School 883-2689)*, Special Events *(May Day, Art Festival, Chamber Music - 3rd wknd Aug, Jazz Fest - 3rd wknd Sep, Blues Festival)*, Galleries *(Flying Anvil 883-3660)* **Under 1 mi:** Horseback Riding **1-3 mi:** Tennis Courts, Golf Course *(Pender Harbour 604-883-9541, shuttle)*

Provisioning and General Services
Near: Convenience Store, Market *(IGA Madeira Park 883-9100/Oak Tree 883-2411)*, Delicatessen *(IGA Market Place)*, Wine/Beer, Liquor Store *(BC 883-2737)*, Farmers' Market, Fishmonger, Bank/ATM, Post Office, Catholic Church, Protestant Church *(St. Andrew's)*, Beauty Salon *(Miss Sunny's 883-2715)*, Laundry, Bookstore *(Harbour Paper Mill 883-9911)*, Pharmacy *(Marina Pharmacy 883-2888)*, Newsstand, Florist, Clothing Store *(The Clothing Lounge)*, Retail Shops *(Knitting Zen 883-2922, souvenirs & T-shirts, Water & the Wind 883-9911, Urban Eclectics)*, Copies Etc.

Transportation
OnSite: Courtesy Car/Van *(to golf course & eateries)*, Bikes *($8/hr., $80/day)*, Water Taxi *($95/hr.; free shuttle to restaurants 741-3796)* **Near:** Rental Car *(PH Transportation 883-2040, delivers)*, InterCity Bus *(Malaspina Coach 885-2217)* **Airport:** Sechelt/YVR *(22 mi./65 mi.)*

Medical Services
911 Service **OnCall:** Ambulance **Near:** Veterinarian *(Madeira Vet Hospital 883-2488)* **Under 1 mi:** Doctor *(Pender Harbour Health Centre 883-2764, Mon-Fri 8am-4pm)*, Dentist **Hospital:** St. Mary's 885-2224 *(20 mi.)*

Setting -- Tumbling down the gorgeous hillside at the head of Pender Harbor are pale gray-shingled, multistory, contemporary buildings - an architectural spin on Bavarian - B.C. design - that feature natural varnished trim, deep overhangs and decks protected by glass windshields. Continuing toward the docks below are terraced expanses of lush lawn and decks that host a hot tub or a gaggle of picnic tables or a pocket garden - all snuggled behind more glass windbreaks and sporting spectacular views of the cove. Wide, comfortable wooden steps wind their way up from the water - level after level.

Marina Notes -- Owners designed this beautifully executed, first class facility in 1993 and live on-site. An upscale RV park is adjacent - but not visible. A new heated pool opened in mid-2006. Hot tub available 24 hrs. Wireless Internet on docks. No fee for moorings if a room is booked. The resort rents 12-18 ft. boats at $20-55/hr., $125-330/day, kayaks $15-18/hr., $90-110/day. Fishing charters $95/hr., $450/half day, $800/day. For kids there's ping-pong and various outdoor games. Complimentary water taxi "Slo-Cat" to Grasshopper Pub and Garden Bay, and a shuttle to Pender Harbour Golf Course. New coin-operated showers, laundromat and book exchange. Nanaimo Yacht Club maintains an outstation here and boatyard services are at Madeira Marina.

Notable -- In addition to this luxury hotel's well-appointed rooms and suites (some with kitchens), there is a penthouse and a two bedroom rental cottage on the grounds. During the summer this beautiful inlet near Gunboat Bay attracts 60 to 80 boats. Just around the bend, and dinghy-able, is Madeira Park for provisioning and services - a good place for a casual meal. And if you aren't taking your own boat to Princess Louisa Inlet, consider the Resort's excursion.

Navigational Information
Lat: 49°31.620' **Long:** 123°58.102' **Tide:** 11 ft. **Current:** 2 kt. **Chart:** 3512
Rep. Depths (*MLW*): **Entry** 160 ft. **Fuel Dock** 60 ft. **Max Slip/Moor** 70 ft./-
Access: Malaspina Strait to Secret Cove

Marina Facilities (*In Season/Off Season*)
Fuel: Gasoline, Diesel, High-Speed Pumps
Slips: 160 Total, 35 Transient **Max LOA:** 160 ft. **Max Beam:** 30 ft.
Rate (*per ft.*): **Day** $1.00 **Week** $10.50 **Month** $5.25
Power: 30 amp $5*, 50 amp n/a, 100 amp n/a, 200 amp n/a
Cable TV: No **Dockside Phone:** No
Dock Type: Floating, Wood
Moorings: 0 Total, 0 Transient **Launch:** n/a
Rate: Day n/a **Week** n/a **Month** n/a
Heads: 3 Toilet(s), 3 Shower(s)
Laundry: None **Pay Phones:** No
Pump-Out: No **Fee:** n/a **Closed Heads:** Yes

Marina Operations
Owner/Manager: Scott Rowland **Dockmaster:** Same
In-Season: Apr-Jun, 8am-6pm **Off-Season:** Jul-Sep, 8am-10pm
After-Hours Arrival: Ring the buzzer
Reservations: Yes **Credit Cards:** Visa/MC
Discounts: Fuel **Dockage:** n/a **Fuel:** 500 & 1000 lts **Repair:** n/a
Pets: Welcome **Handicap Access:** No

Secret Cove Marina

PO Box 1118; Sechelt, BC V0N 1Y2

Tel: (604) 885-3533; (866) 885-3533 **VHF: Monitor** Ch. 66A **Talk** n/a
Fax: (604) 885-6037 **Alternate Tel:** n/a
Email: scottrowland@dccnet.com **Web:** www.secretcovemarina.com
Nearest Town: Sechelt (*9 mi.*) **Tourist Info:** (604) 883-2561

Marina Services and Boat Supplies
Services - Docking Assistance, Security (*owner onsite*), Trash Pick-Up,
Dock Carts **Communication -** Mail & Package Hold, Phone Messages, Fax
in/out ($2), Data Ports (*Wi-Fi - Broadband Xpress*), FedEx, UPS **Supplies -**
OnSite: Ice (*Block, Cube*), Ships' Store (*limited*), Bait/Tackle (*also at
Buccaneer Marina*), Live Bait, Propane

Boatyard Services
OnSite: Electrical Repairs, Divers **OnCall:** Electronic Sales, Electronics
Repairs, Rigger, Sail Loft, Canvas Work, Bottom Cleaning, Brightwork, Air
Conditioning, Compound, Wash & Wax, Propeller Repairs, Inflatable
Repairs, Yacht Interiors, Metal Fabrication **Near:** Railway, Launching
Ramp. **Nearest Yard:** Buccaneer Marina (604) 885-7888

Restaurants and Accommodations
OnSite: Restaurant (*Upper Deck Café 885-3533, D $20-26*), Condo/Cottage
(*cottages*) **1-3 mi:** Restaurant (*Rockwater Secret Cove Resort L $10-16,
D $15-24, Tapas Menu, Sun brunch 11am-1pm; free shuttle*), Inn/B&B
(*Rock-Water 885-7038, $129-334*), (*Jolly Roger Inn 885-7860*)

Recreation and Entertainment
OnSite: Picnic Area, Boat Rentals, Group Fishing Boat **Near:** Beach,
Tennis Courts, Golf Course (*Sechelt Golf & Country Club, public 885-4653 +
golf academy*), Fitness Center (*Dolphin Fitness 885-2969*), Jogging Paths,
Horseback Riding, Roller Blade/Bike Paths **1-3 mi:** Park (*Bucanner Bay or
Smugglers Cove Marine Park - accessible by boat - north end of Welcome
Pass - watch for eyebolts along shore for stern lines*), Sightseeing (*Wood
Bay Salmon Farm; Homesite Creek Trail to waterfall and limestone caves*)

Provisioning and General Services
OnSite: Convenience Store (*sundries, snacks, drinks, produce, meat,
clothing*), Liquor Store, Bank/ATM, Post Office **Under 1 mi:** Market
(*Halfmon Bay General Store 885-8555; 3 mi - Fawn Rd. Market 885-3115*)
3+ mi: Delicatessen (*Big Mac's Superette & Delicatessen 885-9414, 13 mi.*),
Health Food (*Good Stuff Health Foods 885-9063, 13 mi.*), Bakery
(*Wheatberries, Pearl's 885-3395, 13 mi.*), Farmers' Market (*Farmers' and
Artisans Market at theatre parking lot, 8:30am-1pm Sat in season, 885-2276,
13 mi.*), Library (*Sechelt 885-3260, 12 mi.*), Beauty Salon (*Headquarters
885-3616, 13 mi.*), Dry Cleaners (*Care-A-Lot Cleaners 885-9166, 13 mi.*),
Bookstore (*Books & Stuff 885-2625, Talewind Books 885-2527, 13 mi.*),
Pharmacy (*Drugstore 740-5765, 13 mi.*), Hardware Store (*Home Hardware
885-9828, 13 mi.*), Clothing Store (*Maribel's Fine Fashions 885-2029, 13
mi.*), Retail Shops (*Trail Bay Source For Sports 885-2512, Radio Shack 885-
2568, 13 mi.*)

Transportation
OnSite: Courtesy Car/Van (*shuttle to 3 golf courses*), Bikes (*trail bikes &
mopeds, too*), Water Taxi (*740-0703; Thormanby Island 740-4317; MV
Grackle 885-1212 $10/pp*) **1-3 mi:** InterCity Bus (*Malaspina Coach Lines
885-2217*) **3+ mi:** Rental Car (*National 885-9120, 13 mi.*), Taxi (*Sunshine
Coast 885-3666, 13 mi.*) **Airport:** Sechelt/YVR (*13 mi./55 mi.*)

Medical Services
911 Service **1-3 mi:** Holistic Services (*Rockwater Spa at Lord Jim's Resort
885-7038, daily 10am-8pm: massage, facials, reflexology, Reiki, meditation*)
3+ mi: Dentist (*Midcoast Dental 885-2246, 13 mi.*), Veterinarian (*Atlas
Animal Hosp. 740-8208, 13 mi.*) **Hospital:** St. Mary's 885-2224 (*13 mi.*)

Setting -- Tucked into the northern arm of stunning Secret Cove on the Sechelt Peninsula, the marina's docks sprawl along a rockbound shoreline. The first
set of docks hosts a beige two-story floating dockhouse with a pale teal barrel roof; inside are a restaurant and general store and adjacent is the fuel dock.
Beyond are 150 upscale floating slips - some open and some beneath teal-colored boathouses.

Marina Notes -- *15 amp & 30 amp service. Owners live on-site. Both covered and open slips are available to transients. Fuel dock, general store, liquor
store and bathrooms are open from April 1 to Canadian Thanksgiving. Moped, canoe, paddleboat and mountain bike rentals. Impeccable rose and chocolate
marble tiled full bathrooms (hours same as store). Nearby Buccaneer Cove Marina, largely a seasonal facility, provides boatyard services.

Notable -- Secret Cove's Upper Deck restaurant, on the second floor of the floating office, features a canopied deck with spectacular long views of the cove
(open from May long weekend through mid-October). Its limited West Coast inspired menu is fresh, local and seasonal. Below, on the first floor, the well-
stocked general store sells provisions - including fresh produce, meat and seafood, groceries, liquor, wine & beer -- plus books, crafts, hardware, boating
supplies, fishing tackle, bait, marine wear and foul weather gear. Thormanby Jedediah Island Water Taxi makes its home here for an easy trip to Thormanby or
to the marine parks at Bucanneer Bay or Smuggler Cove. The area is popular for diving, year-round fishing, hiking and biking. It's 4 miles to the small town of
Halfmoon Bay, about 13 miles to Sechelt with Tem-Swiya Museum 885-8991 and Raven's Cry Theatre 885-4597 at House of Chiefs.

Boaters' Notes

Add Your Ratings and Reviews at www.AtlanticCruisingClub.com

A complimentary six-month Silver Membership is included with the purchase of this *Guide*. Select "Join Now," then "Silver," then follow the instructions.

The AtlanticCruisingClub website provides updated Marina Reports, Destination and Harbor Articles and much more — including an option within each online Marina Report for boaters to add their ratings and comments regarding that facility. Please log on frequently to share your experiences — and to read other boaters' comments.

On the website, boaters may rate marinas in one or more of the following categories — on a scale of 1 (basic) to 5 (world class) — and also enter additional commentary.

‣ **Facilities & Services** (Fuel, Reservations, Concierge Services and General Helpfulness)

‣ **Amenities** (Pool, Beach, Internet Access, including Wi-Fi, Picnic/Grill Area, Boaters' Lounge)

‣ **Setting** (Views, Design, Landscaping, Maintenance, Ship-Shapeness, Overall Ambiance)

‣ **Convenience** (Access — including delivery services — to Supermarkets, other Provisioning Sources, Shops, Services, Attractions, Entertainment, Sightseeing, Recreation, including Golf and Sport Fishing & Medical Services)

‣ **Restaurants/Eateries/Lodgings** (Availability of Fine Dining, Snack Bars, Lite Fare, OnCall food service, and Lodgings ashore)

‣ **Transportation** (Courtesy Car/Vans, Buses, Water Taxis, Bikes, Taxis, Rental Cars, Airports, Amtrak, Commuter Trains)

‣ **Please Add Any Additional Comments**

3. BC – Howe Sound & Vancouver

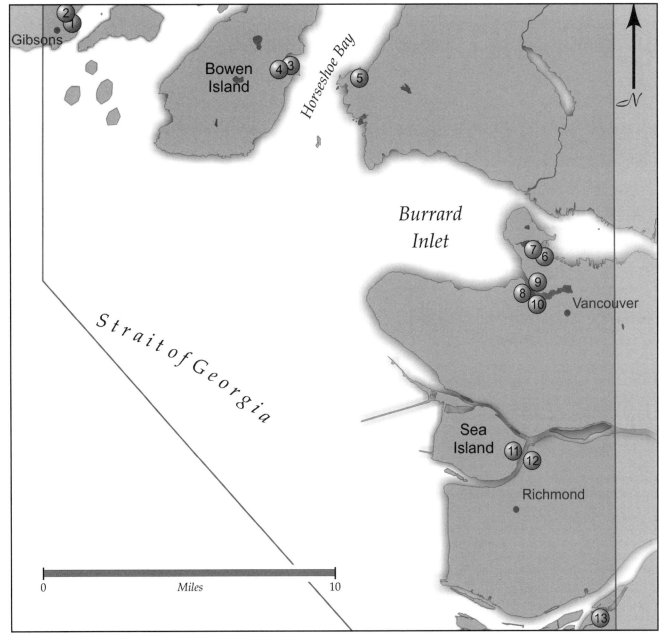

MAP	MARINA	HARBOR	PAGE	MAP	MARINA	HARBOR	PAGE
1	Gibsons Marina	Howe Sound/Gibsons	46	8	Cooper Boating	False Creek	53
2	Gibsons Landing Harbour Authority	Howe Sound/Gibsons	47	9	False Creek Yacht Club	False Creek	54
3	Bowen Island Marina	Manion Bay/Snug Harbour	48	10	Pelican Bay Marina	False Creek	55
4	Union Steamship Company Marina	Manion Bay/Snug Harbour	49	11	Delta Marina	Fraser River Middle Arm	56
5	Sewell's Marina	Horseshoe Bay	50	12	Vancouver Marina Limited	Fraser River Middle Arm	57
6	Coal Harbour Marina	Coal Harbour	51	13	Captain's Cove Marina	Fraser River/Deas Harbour	58
7	Bayshore West Marina	Coal Harbour	52				

▸ **Currency** — In Canadian Marina Reports, all prices are in Canadian dollars. In U.S. Marina Reports, all prices are in U.S. dollars.

▸ **"CCM"** — Denotes a Certified Clean Marina, a state/provincial award for environmental excellence. See page 298 for an explanation and page 299 for a list of Pump-Out facilities.

▸ **Ratings & Reviews** — An overview of the Atlantic Cruising Club's rating system is on page 6 and details on the content of each Marina Report are on pages 7 – 11.

▸ **Marina Report Updates** — Comments from boaters and new information from ACC reviewers and marinas are posted regularly on *www.AtlanticCruisingClub.com*.

Gibsons Marina

PO Box 1520; 675 Prowse Road; Gibsons, BC V0N 1V0

Tel: (604) 886-8686 **VHF: Monitor** Ch. 66A **Talk** Ch. 68
Fax: (604) 886-8686 **Alternate Tel:** n/a
Email: n/a **Web:** n/a
Nearest Town: Gibsons **Tourist Info:** (604) 886-2325

Navigational Information
Lat: 49°23.870' **Long:** 123°30.420' **Tide:** 16 ft. **Current:** n/a **Chart:** 3534
Rep. Depths (MLW): Entry 7 ft. **Fuel Dock** n/a **Max Slip/Moor** 7 ft./-
Access: Southeastern end of Howe Sound

Marina Facilities *(In Season/Off Season)*
Fuel: *Hyak Marine* - Gasoline, Diesel
Slips: 380 Total, 30 Transient **Max LOA:** 80 ft. **Max Beam:** n/a
Rate *(per ft.)*: **Day** $0.62* **Week** n/a **Month** n/a
Power: 30 amp $2** **, 50 amp** n/a, **100 amp** n/a, **200 amp** n/a
Cable TV: No **Dockside Phone:** No
Dock Type: Floating, Long Fingers, Wood
Moorings: 0 Total, 0 Transient **Launch:** n/a
Rate: Day n/a **Week** n/a **Month** n/a
Heads: 3 Toilet(s), 4 Shower(s)
Laundry: 2 Washer(s), 2 Dryer(s), Book Exchange **Pay Phones:** Yes, 2
Pump-Out: OnSite, Full Service, 1 Port **Fee:** $20 **Closed Heads:** No

Marina Operations
Owner/Manager: Arthur McGinnis **Dockmaster:** Same
In-Season: Easter-Oct 15, 7am-8pm **Off-Season:** Oct 16-Easter, 8:30-4:30
After-Hours Arrival: Emergency # is posted
Reservations: Yes, Preferred **Credit Cards:** Visa/MC, Debit
Discounts: None
Pets: Welcome **Handicap Access:** Yes, Heads, Docks

Marina Services and Boat Supplies
Services - Security *(24 hrs., locked security gate)*, Trash Pick-Up, Dock Carts **Communication -** FedEx, UPS **Supplies - OnSite:** Ice *(Cube)*, Ships' Store, Bait/Tackle, Propane, CNG

Boatyard Services
OnSite: Launching Ramp *($5)* **Near:** Travelift, Engine mechanic *(gas, diesel)*.

Restaurants and Accommodations
Near: Restaurant *(Waterfront 886-2831, L $7-12, D $8-26)*, *(Molly's Reach 886-9710, L $6-13, D $9-25)*, *(Leo's Tapas 886-9424)*, *(Bayview Szechuan 886-1728, L $6, D $8-15)*, Coffee Shop *(Just-A-Cuppa)*, Lite Fare *(Mike's Place)*, Inn/B&B *(Maritimer 886-0664)*, *(Edgewater 886-4764, $100-200)*, *(Ritz Inn 886-3343)* **Under 1 mi:** Fast Food *(Wendy's, Subway)*, Lite Fare *(Robbie's Pancake House 886-9090)*, Pizzeria *(Patra 886-7671, del.)*

Recreation and Entertainment
OnSite: Picnic Area **Near:** Beach *(Amours or Georgia)*, Playground, Tennis Courts, Jogging Paths, Boat Rentals *(kayaks)*, Roller Blade/Bike Paths, Fishing Charter, Park *(Winegarden Park)*, Museum *(Elphinstone Pioneer Museum 886-8232 - maritime room, shell collection; 1:30-4:30pm, closed Sun & Mon)*, Cultural Attract *(Heritage Playhouse 886-8998)*, Sightseeing *(Heritage Walking Tour)*, Special Events *(Jazz Festival - Jun; Fibre Arts Festival, International Outrigger Canoe Iron Race, Sea Cavalcade - Jul)*, Galleries *(Gibsons Landing Gallery 886-0099; Artesia gallery tours 886-7300)* **Under 1 mi:** Movie Theater *(Gibsons Cinema 886-6843)*, Video Rental *(Movies N Stuff 883-1331)*

1-3 mi: Golf Course *(Sunshine Coast 885-9212)*

Provisioning and General Services
Near: Market *(Fong's Market & Gift Shop 886-8515)*, Wine/Beer, Bakery *(Wild Blueberry 886-1917)*, Farmers' Market *(Holland Park, seasonal)*, Fishmonger *(Suncoast Seafood 886-7157)*, Bank/ATM, Post Office, Protestant Church, Library *(886-2130)*, Bookstore *(Coast Books 886-7744)*, Pharmacy *(Dockside Peoples Drugmart 886-8158)*, Clothing Store *(Matthews 886-8801)* **Under 1 mi:** Delicatessen *(Mountainview Market & Deli 886-2249)*, Health Food *(Seaweeds Health Food 886-1522)*, Catholic Church, Beauty Salon *(QT's 886-6860)*, Barber Shop, Dry Cleaners *(Bergner 886-8564)*, Laundry *(Gibsons 886-0838)*, Hardware Store *(Home Hardware 886-2442)*, Florist, Retail Shops *(Saan 886-9413, Sportstraders, A Buck or Two 886-8946)*, Copies Etc. **1-3 mi:** Supermarket *(IGA 886-3487, Super Valu 886-2424)*, Liquor Store *(BC Liquor 886-2013)*

Transportation
OnCall: Taxi *(Sunshine Coast 887-7337)* **Near:** Water Taxi *(Navigator, Dolphin, & Gibson)*, Rental Car *(Skookum Chrysler 886-3493)*, Local Bus **1-3 mi:** InterCity Bus *(Malaspina Coach Lines 885-2217)*, Ferry Service *(Langdale to Horseshoe Bay)* **Airport:** Sechelt/YVR *(13 mi./30 mi.)*

Medical Services
911 Service **OnCall:** Ambulance **Near:** Dentist *(Chiasson 886-7308)*, Holistic Services *(Inner Moves 886-9737, yoga, massage, meditation)* **Under 1 mi:** Doctor *(Gisbons Medical Clinic 886-2868)*, Chiropractor *(Coast Back Care 886-3622)* **Hospital:** St. Mary's 885-2224 *(13 mi.)*

Setting -- Tucked behind a long rock breakwater, just south of the floating Hyak Marine Fuel Dock, Gibsons Marina's network of 7 docks hosts almost 400 slips for recreational vessels. A wide ramp edged with signature bright red and chrome yellow railings leads to a low white building with a red and white roof that houses the office, store and amenities - surrounded by pretty gardens and nicely landscaped grounds. Picnic tables and a red and white pavilion perch on the banks of a rushing stream crossed by a red and yellow foot bridge. The views seaward from the docks are of Keats Island.

Marina Notes -- *Rates vary by size, approximately $0.62/ft. **Power is 15 amp, $2. Launch ramp $5. Established 1984. Transient moorage is mostly on "A" dock at the marina entrance - between the rock jetty and the gazebo at the end of the GLHA's Breakwater Boardwalk. Fuel is available from the immediately adjacent Hyak Marine's Esso station (886-9011). Pump-out is located on the dock. Integrated into the main office is a large, well-stocked chandlery with boating supplies, fishing tackle and nautical clothing and gear. A commodious laundry room with folding tables is next to a comfortable bathhouse with tiled showers.

Notable -- The charming little village of Lower Gibson is an easy quarter mile stroll around the harbor along the seawalk. A path, opposite Hyak Marine, leads up to the museum on Gower Pt. Road which is lined with shops and eateries. Or continue on the seawalk, past Winegarden Park and the Molly's Lane Shops to the public wharf and Molly's Reach - once the set for the CBC's two-decade long "The Beachcomber" TV series. Across the road is Pioneer Park and the Visitors' Center. To the right, Marina Drive and Gibsons Way host shops, galleries, provisioning and more eateries. The seawalk continues along the shore.

Navigational Information

Lat: 49°24.030' **Long:** 123°30.324' **Tide:** 16 ft. **Current:** n/a **Chart:** 3534
Rep. Depths *(MLW)*: **Entry** 15 ft. **Fuel Dock** n/a **Max Slip/Moor** 8 ft./-
Access: Southwestern end of Howe Sound

Marina Facilities *(In Season/Off Season)*

Fuel: *Hyak Marine* - Gasoline, Diesel
Slips: 150 Total, 10 Transient **Max LOA:** 50 ft. **Max Beam:** n/a
 Rate *(per ft.)*: **Day** $0.77 **Week** n/a **Month** $4
 Power: 30 amp $4, **50 amp** n/a, **100 amp** n/a, **200 amp** n/a
 Cable TV: No **Dockside Phone:** No
 Dock Type: Fixed, Floating, Long Fingers, Pilings, Wood
Moorings: 0 Total, 0 Transient **Launch:** n/a
 Rate: Day n/a **Week** n/a **Month** n/a
Heads: 2 Toilet(s), 2 Shower(s)
Laundry: 2 Washer(s), 2 Dryer(s), Book Exchange **Pay Phones:** Yes
Pump-Out: OnSite, Self Service, 1 Central **Fee:** $5 **Closed Heads:** No

Marina Operations

Owner/Manager: Gibsons Landing H.A. **Dockmaster:** Bill Oakford
In-Season: Jun-Sep, 8am-8pm **Off-Season:** Oct-May, 8am-4pm
After-Hours Arrival: Call ahead
Reservations: No **Credit Cards:** Visa/MC, Interac
Discounts: None
Pets: Welcome **Handicap Access:** Yes, Heads, Docks

Gibsons Landing Harbour Auth.

PO Box 527; Gibsons, BC V0N 1V0

Tel: (604) 886-8017 **VHF: Monitor** Ch. 66A **Talk** Ch. 68
Fax: (604) 886-1347 **Alternate Tel:** n/a
Email: gibsons@telus.net **Web:** n/a
Nearest Town: Gibsons **Tourist Info:** (604) 886-2325

Marina Services and Boat Supplies

Services - Docking Assistance, Security *(cameras)*, Trash Pick-Up, Dock Carts **Supplies - OnSite:** Ice *(Cube)* **Near:** Ships' Store

Boatyard Services

OnSite: Launching Ramp **OnCall:** Engine mechanic *(gas, diesel)*, Bottom Cleaning, Brightwork, Air Conditioning, Refrigeration, Divers

Restaurants and Accommodations

OnSite: Snack Bar *(B $2-5, L $3-5, hot dogs, sandwiches; 9:30am-2:30pm)*, Lite Fare *(Gramma's Marine Pub)* **Near:** Restaurant *(Waterfront 886-2831, L $7-12, D $8-26)*, *(Bayview Szechuan 886-1728, L $6, D $8-15)*, *(Seaview Gardens 886-9219)*, *(Molly's Reach 886-9710, L $6-13, D $9-25)*, *(Leo's Tapas 886-9414, L $6-9, D $10-19)*, *(Opa Japanese 886-4023, L $9, D $11-17)*, Lite Fare *(Old Country Fish & Chips 886-9797)*, Inn/B&B *(Maritimer 886-0664)*, *(Howe's House $150-175)* **Under 1 mi:** Fast Food *(Subway, McDonald's)*, Pizzeria *(Angelo's 886-4142)*, Motel *(Sunnycrest 886-2419, $50-90)*, Inn/B&B *(Cedars Inn 886-3008)*, *(Ocean View 886-8700)*

Recreation and Entertainment

OnSite: Boat Rentals *(C-D Sailing 886-4972; Sunshine Kayaking 886-9760)* **Near:** Beach *(Amours/Georgia)*, Jogging Paths, Roller Blade/Bike Paths, Fishing Charter, Park *(Winegarden, Pioneer)*, Museum *(Elphinstone Pioneer Museum 886-8232, Tue-Sat 1:30-4:30pm)*, Cultural Attract *(Heritage Playhouse 886-8998)*, Sightseeing *(Lookout Adventure Tours 886-1655; Artesia tours for artists 886-7300; Gibsons Ferry 886-4910 harbor tour $10/5)*, Galleries *(Gibsons Landing Gallery 886-0099)*

Under 1 mi: Playground, Tennis Courts, Bowling *(Gibsons Lanes 886-2086)*, Movie Theater *(Gibsons 886-6843)*, Video Rental *(Movies N Stuff 883-1331)* **1-3 mi:** Golf Course *(Sunshine Coast 885-9212)*, Fitness Center

Provisioning and General Services

OnSite: Wine/Beer *(Gamma's Pub)*, Fishmonger, Florist *(Flora's Float)*
OnCall: Hardware Store *(or Home 886-2442)* **Near:** Market *(Fong's 886-8515 - a trip)*, Bakery *(Wild Blueberry 886-1917)*, Farmers' Market *(Holland Park)*, Bank/ATM, Post Office, Protestant Church, Library *(Gibsons 886-2130)*, Bookstore *(Coast 886-7744)*, Pharmacy *(Dockside Peoples 886-8158)*, Clothing Store *(Matthews 886-8801)* **Under 1 mi:** Delicatessen *(Mountainview 886-2249)*, Health Food *(Nancy's; Seaweeds 886-1522)*, Catholic Church, Beauty Salon *(QT's 886-6860)*, Dry Cleaners *(Bergner 886-8564)*, Laundry *(Gibsons 886-0838)*, Retail Shops **1-3 mi:** Supermarket *(IGA 886-3487, Super Valu 886-2424)*, Liquor Store *(BC 886-2013)*

Transportation

OnSite: Water Taxi *(Navigator 886-7626; Dolphin 740-1937; Gibsons Ferry 886-4910)* **OnCall:** Taxi *(Sunshine Coast 887-7337)* **Near:** Rental Car *(Skookum Chrysler 886-3493)*, Local Bus *(BC Transit)* **1-3 mi:** InterCity Bus *(Malaspina Coach Lines 885-2217)*, Ferry Service *(Langdale to Horseshoe Bay)* **Airport:** Sechelt/YVR *(13 mi./30 mi.)*

Medical Services

911 Service **OnCall:** Ambulance **Under 1 mi:** Doctor *(Gisbons Medical Clinic 886-2868)*, Dentist *(Bland 886-7020)*, Chiropractor *(Coast Back Care 886-3622)*, Holistic Services *(Inner Moves 886-9737, yoga, massage, meditation)* **Hospital:** St. Mary's 885-2224 *(13 mi.)*

Setting -- On the southwest side of Howe Sound, a series of floating docks are connected by two steel gangplanks that lead up to a concrete wharf. Perched on the wharf is a dramatic natural-shingled dockhouse and outbuildings. The teal green roof is supported by whimsical, hand-carved pillars and topped by a large cupola. A stone jetty protects the slips and supports a long boardwalk, furnished with benches, that extends out to the "vista" gazebo.

Marina Notes -- Fuel is available from Hyak Marine Services' floating station (886-9011). This is the government wharf and a working harbor; the egalitarian mix of commercial and pleasure craft - along with marine contractors and water-based tours - make this a marina rich with atmosphere. Many fishing vessels sell direct from their boats - including Beldi's Fish "Market" (open dawn to dark). Internet access at Intime Café. Adjacent to the dockhouse are pleasant tiled heads with vanities and separate fiberglass showers with dressing areas.

Notable -- Just up Molly's Lane from the docks is Molly's Reach Restaurant - for two decades the set for CBC's "The Beachcombers" TV series. Just beyond, a bronze statue of George Gibson oversees flower-bedecked pocket Pioneer Park - the centerpiece of the artistic, bustling little village of Lower Gibsons. Next door is the Visitor's Center and stretched along Marine Drive are markets, restaurants, galleries, unique shops, a cyber café and a bookstore. Stroll the Seawalk and Heritage Walk and end up at the Elphinstone Pioneer Museum with its Maritime Room and Bedford Shell Collection (admission by donation). The bird sanctuary at Arrowhead Park is also a short walk from both marinas. Gambier and Keats Islands make interesting day trips via either of the water taxis.

PHOTOS ON DVD: 17

Bowen Island Marina

PO Box 1; 375 Cardena Road; Bowen Island, BC V0N 1G0

Tel: (604) 947-9710 **VHF: Monitor** Ch. 16 **Talk** Ch. 67
Fax: (604) 947-9710 **Alternate Tel:** n/a
Email: norma@bowen-island.com **Web:** www.bowen-island.com
Nearest Town: Vancouver *(9 mi.)* **Tourist Info:** (604) 947-9024

Navigational Information
Lat: 49°22.801' **Long:** 123°19.899' **Tide:** 12 ft. **Current:** 5 kt. **Chart:** 3534
Rep. Depths *(MLW)*: **Entry** 12 ft. **Fuel Dock** n/a **Max Slip/Moor** 12 ft./12 ft.
Access: Queen Charlotte Channel to Manion Bay to Snug Habour

Marina Facilities *(In Season/Off Season)*
Fuel: No
Slips: 45 Total, 3 Transient **Max LOA:** 45 ft. **Max Beam:** 12 ft.
 Rate *(per ft.)*: **Day** $0.85 **Week** $5.25 **Month** $8
 Power: 30 amp $5*, **50 amp** n/a, **100 amp** n/a, **200 amp** n/a
 Cable TV: No **Dockside Phone:** No
 Dock Type: Floating, Long Fingers, Wood
 Moorings: 0 Total, 0 Transient **Launch:** n/a, Dinghy Dock
 Rate: Day n/a **Week** n/a **Month** n/a
Heads: None
Laundry: None **Pay Phones:** Yes, 1
Pump-Out: No **Fee:** n/a **Closed Heads:** No

Marina Operations
Owner/Manager: Norma Dallas **Dockmaster:** Same
In-Season: May-Oct, 8am-8pm **Off-Season:** Nov-Apr, 8am-6pm
After-Hours Arrival: Call ahead
Reservations: Yes, Preferred **Credit Cards:** Visa/MC
Discounts: None
Pets: Welcome, Dog Walk Area **Handicap Access:** No

Marina Services and Boat Supplies
Services - Docking Assistance, Security *(Office hours, office close to marina)*, Trash Pick-Up, Dock Carts **Communication -** Mail & Package Hold, Phone Messages, Fax in/out *($1)*, Data Ports *(Office)*, FedEx, UPS, Express Mail **Supplies - OnSite:** Ice *(Cube)* **Near:** Ships' Store *(at Union Steamship)* **Under 1 mi:** Propane

Boatyard Services
OnSite: Engine mechanic *(gas, diesel)*, Interior Cleaning **OnCall:** Divers
Nearest Yard: Thunderbird Marina (604) 921-7434

Restaurants and Accommodations
OnSite: Snack Bar *(homemade tacos, ice cream, coffee, snacks)*
Near: Restaurant *(Bowen Sushi To Go 947-6806, 11:30-8:30)*, *(Doc Morgan's 947-0808, Restaurant & Pub)*, *(Blue Eyed Mary's Bistro 947-2583)*, Lite Fare *(The Happy Italian 947-2140, takeout foods, deli, coffee, gelato)*, *(Neighbourhood Pub 947-2782, L $6-10, D $6-10, pub fare, bugrers, salads, pizzas $11-21; takeout Noon-Mid, delivery 5-9pm)*, *(Snug Café 947-0402, B & L)*, *(Village Baker Café 947-2869, B&L)*, Pizzeria *(Tuscany Wood Oven 947-0550)*, Inn/B&B *(On the Sea B&B 947-2471)*, *(Lodge at the Old Dorm 947-0947, $75-140)*, Condo/Cottage *(Doc Morgan's Inn 947-0808)*

Recreation and Entertainment
OnSite: Boat Rentals (Bowen Island Sea Kayaking 947-9266), Special Events *(live music in summer, plus Dog Days of Summer - 2nd Sun Aug; Bowfest & Parade - last Sat Aug; Round Bowen Kayak Race - early Jun)*, **Near:** Beach*(several)*, Picnic Area, Grills, Playground, TennisCourts *(BICS 947-9337)*, Fitness Center *(BICS 947-9337)*, Jogging Paths, Horseback Riding, Movie Theater *(Bowen Film Society 947-0450)*, Video Rental, Park *(Crippen Regional)*, Museum *(Bowen Island Museum & Archive 947-2655, Jul-Aug daily 10am-4pm, off-season Mon-Thu 10am-4pm)*, Cultural Attract *(Theatre on the Island in summer, Tirnanog Theatre School performances 947-9557)*, Galleries *(Artisan Square 947-2454 Wed-Sun 11am-5pm)* **Under 1 mi:** Roller Blade/Bike Paths, Group Fishing Boat **3+ mi:** Golf Course *(Bowen Island, 947-4653, Bruce Russell, 4 mi.)*

Provisioning and General Services
OnSite: Retail Shops *(Gifts on the Pier 947-0107, local crafts, gifts, souvenirs)* **Near:** Market *(Snug Cove General Store 947-9619)*, Gourmet Shop *(Cocoa West Chocolatier 947-2996)*, Delicatessen, Health Food *(Ruddy Potato 947-0998)*, Wine/Beer *(Bowen Island Beer & Wine 947-2729)*, Liquor Store *(General Store)*, Bakery *(The Village Baker 947-2869, The Oven Door 947-2509)*, Bank/ATM, Post Office, Catholic Church, Protestant Church, Library *(947-9788, closed Sun & Mon)*, Beauty Salon, Barber Shop, Dry Cleaners, Laundry, Pharmacy, Newsstand, Hardware Store, Florist, Clothing Store

Transportation
OnCall: Taxi *(Bowen Island Taxi 947-0000)* **Near:** Bikes, Water Taxi *(Cormorant 947-2243)*, Local Bus *(Comm. Shuttle 947-0229)*, Ferry Service *(to Horseshoe Bay)* **Airport:** Vancouver *(20 mi.)*

Medical Services
911 Service **OnCall:** Ambulance **Near:** Doctor, Chiropractor, Veterinarian
1-3 mi: Dentist, Holistic Services *(Twiggleberries Spa 947-2876)*
Hospital: Lions Gate 988-3131 *(15 mi.)*

Setting -- The first marina in Sung Harbor off beautiful Howe Sound, Bowen Island Marina sits to starboard of the ferry landing on Bowen Island aka "The Rock." A long breakwater pier protects four inner docks - three offer 45 alongside tie-ups. Up the ramp "The Pier" hosts a funky collection of small natural-sided "west coast" cottages that house useful businesses and services. Music and art shows enliven the scene during the summer.

Marina Notes -- *15 amp $2, 30 amp $5. On-site are a kayak rental shop (that occupies the fourth inner dock and also gives lessons and leads tours), a gift shop (native and regional crafts), bait & tackle (frozen bait, ice & fishing licenses), and a taco and ice cream stand (with fresh coffee, too). SeaTow berths here. The nearest boatyard services are at Thunderbird Marina in West Vancouver three miles across the bay.

Notable -- The Pier is always a lively spot: During July and August, there's "Dessert on The Pier" from 6-8pm (made from Bowen Island fruit); from May to September buskers provide a variety of entertainment and Sundays at 3pm it's the Lawn Dogs with bluegrass - plus Tecumseh native arts, crafts, jewelry, sacred sculptures and artifacts by Mohawk artisans. The adjacent Bowen Island Library has internet access and nearby is the Chamber of Commerce's Visitors' Info cottage. Across the road there are several eateries and shops within the Union Steamship complex, and up the hill, Cates Hill Village Square hosts supplies, services and more eateries. Further upland is Artisan Square. The 20 square-mile Rock is home to hundreds of artists, 300 of whom are represented by the Bowen Island Arts Council at The Gallery at Artisan Square. And just beyond, walking and hiking trails wander through beautiful Crippen Regional Park.

Navigational Information
Lat: 49°22.777' Long: 123°20.026' Tide: 12 ft. Current: n/a Chart: 3534
Rep. Depths (*MLW*): Entry 15 ft. Fuel Dock n/a Max Slip/Moor 8 ft./-
Access: Queen Charlotte Channel to Manion Bay to Snug Harbour

Marina Facilities (*In Season/Off Season*)
Fuel: No
Slips: 150 Total, 24 Transient Max LOA: 210 ft. Max Beam: n/a
 Rate (*per ft.*): Day $1.50/0.85* Week $7.88 Month n/a
 Power: 30 amp $5, 50 amp $10, 100 amp $20, 200 amp n/a
 Cable TV: No Dockside Phone: No
 Dock Type: Long Fingers, Pilings
Moorings: 0 Total, 0 Transient Launch: yes, Dinghy Dock
 Rate: Day 0 Week n/a Month n/a
Heads: 1 Toilet(s), 1 Shower(s)
Laundry: 4 Washer(s), 4 Dryer(s) Pay Phones: Yes, 2
Pump-Out: OnSite, Self Service, 1 Port Fee: $10 Closed Heads: No

Marina Operations
Owner/Manager: Rondy Dike Dockmaster: Same
In-Season: May-Sep, 9am-9pm Off-Season: Oct-Apr, 9am-4pm
After-Hours Arrival: Check in in the am or call to leave CC #
Reservations: Yes, Preferred Credit Cards: Visa/MC, Amex
Discounts: 4 days or more Dockage: 25% Fuel: n/a Repair: n/a
Pets: Welcome, Dog Walk Area Handicap Access: Yes, Heads

Union Steamship Co. Marina

PO Box 250; 431 Bowen Is. Trunk Rd.; Bowen Is., BC V0N 1G0

Tel: (604) 947-0707 VHF: Monitor Ch. 66A Talk n/a
Fax: (604) 947-0708 Alternate Tel: n/a
Email: ussc@shaw.ca Web: www.steamship-marina.bc.ca
Nearest Town: Vancouver (*9 mi.*) Tourist Info: (604) 947-9024

Marina Services and Boat Supplies
Services - Docking Assistance, Trash Pick-Up, Dock Carts, Megayacht Facilities Communication - Data Ports (*Wi-Fi - Broadband Xpress*) Supplies - OnSite: Ice (*Block, Cube*), Ships' Store, Bait/Tackle (*frozen anchovies & herring*) 1-3 mi: Propane

Boatyard Services
OnSite: Launching Ramp (*$10*) Nearest Yard: Thunderbird Marina (604) 921-7434

Restaurants and Accommodations
OnSite: Restaurant (*Doc Morgan's Inn & Pub 947-0808, dining room & outdoor patio; Sun-Thu 11am-9pm, Fri & Sat 'til 10pm; from 9am Jun-Sep*), Lite Fare (*Doc Morgan's Marine Pub Sun-Thu 11am-Mid, Fri & Sat 11am-1am*), Condo/Cottage (*Doc Morgan's Inn 947-0707, $130-250, suites, cottages, house overlooking the bay*) Near: Restaurant (*Bowen Sushi To Go 947-6806*), (*Blue Eyed Mary's Bistro 947-2583*), Snack Bar (*Snug Café 947-0402*), Lite Fare (*Neighbourhood Pub 947-2782, L $6-10, D $6-10, pub fare, bugrers, salads, pizzas $11-21; takeout Noon-Mid, delivery 5-9pm*), (*The Happy Italian 947-2140, deli, takeout foods, gelato*), Pizzeria (*Tuscany Wood Oven 947-0550*), Motel (*Bowen Lodge by the Sea 947-2129*), Inn/B&B (*On the Sea B&B 947-2471*)

Recreation and Entertainment
OnSite: Beach (*all around the island*), Jogging Paths Near: Picnic Area, Grills, Tennis Courts (*BICS 947-9337*), Fitness Center (*BICS*), Boat Rentals (*Bowen Island Sea Kayaking 947-9266*), Video Rental (*Village Video*), Park (*Crippen Regional Park*), Museum (*Bowen Island Museum & Archive*

947-2655, Jul-Aug 10am-4pm daily, off-season 10am-4pm Mon-Thu*), Cultural Attract (*summer theatre, Tirnanog Theatre School performances 947-9557*), Special Events (*Dog Day of Summer - 2nd Sun Aug, Island Fest & Parade - last Sat Aug*), Galleries (*Artisan Square 947-2454*)
1-3 mi: Sightseeing (*challenging 7-mile trail to the top of Mt. Gardner*)
3+ mi: Golf Course (*Bowen Island, 947-4653, Bruce Russell, 4 mi.*)

Provisioning and General Services
Near: Market (*Snug Cove General Store 947-9619*), Gourmet Shop (*Cocoa West Chocolatier 947-2996*), Delicatessen, Health Food (*Ruddy Potato 947-0998*), Wine/Beer (*Neighbourhood Pub*), Liquor Store (*General Store*), Bakery (*The Village Baker 947-2869, The Oven Door 947-2509*), Fishmonger, Bank/ATM, Post Office, Catholic Church, Library (*947-9788, closed Sun & Mon*), Beauty Salon, Barber Shop, Dry Cleaners, Laundry, Pharmacy, Newsstand, Florist, Clothing Store, Retail Shops (*Black Bear 947-9040, gifts, crafts, smoked salmon*), Copies Etc. Under 1 mi: Protestant Church

Transportation
OnSite: Courtesy Car/Van (*license required*) OnCall: Taxi (*Bowen Island Taxi 947-0000*) Near: Bikes (*The Bike Rentals 947-2243*), Water Taxi (*Cormorant 947-2243 to Horseshoe Bay*), Local Bus (*Comm. Shuttle 947-0229*), Ferry Service (*to Horseshoe Bay*) Airport: YVR (*20 mi.*)

Medical Services
911 Service OnCall: Ambulance Near: Doctor, Dentist, Chiropractor, Veterinarian 1-3 mi: Holistic Services (*Twiggleberries Spa 947-2876*)
Hospital: Lions Gate 988-3131 (*15 mi.*)

Setting -- Nestled in Bowen Island's diminutive Snug Cove, adjacent to the ferry wharf, Union Steamship Company's five main docks berth 150 slips that can accommodate vessels to 210 feet. Onshore is a manicured greensward with vibrant gardens and plantings - edged with restored turn-of-the-last-century cottages and a 3-story lookout turret. In the surrounding 600-acre park, lakes, hiking and biking trails, beaches, picnic groves and a charming village beckon.

Marina Notes -- *25% off over 4 days. 3 hr. moorage $10 to 45', $20 over 45'. Transients mainly on Docks "A" (with slips, alongside dockage & a seaplane float at the end) and "E" (with 4 side-tie floats that radiate from the main dock). Since '85, owners have restored, enhanced and embellished the original facility to create an appealing destination with character. Courtesy car provides access to the whole island. The chandlery has nautical supplies, fishing gear, a line of logo clothing, snacks and ice cream. Manages club cruises of 50-350. Heads being expanded for '07. A spacious laundry has commercial washers & dryers.

Notable -- Once part of the Union Steamship Company's Bowen Island Resort - in the early 1900s, steamers would bring guests from the mainland for dinner, dancing or an overnight in one of 200 cottages (fare $1). Today, the essence remains - a boardwalk, pocket gardens and alleyways surprise with seating areas and pretty vistas. Doc Morgan's Inn Restaurant & Pub is in the former home of a Vancouver barber famous for his parties - a deck and patio overlook the docks and feature frequent entertainment. Renovated cottages are available for a night or a week ashore. Across the street is the library and the Visitor's Info Cottage is nearby. Walk up to compact Cates Hill center and beyond to Artisan Square where artists specialize in jewlery, ceramics, stained glass,or pottery.

Sewell's Marina

Sewell's Marina

6409 Bay Street; West Vancouver, BC V7W 3H6

Tel: (604) 921-3473 **VHF: Monitor** n/a **Talk** n/a
Fax: (604) 921-7027 **Alternate Tel:** n/a
Email: info@sewellsmarina.com **Web:** www.sewellsmarina.com
Nearest Town: Horseshoe Bay **Tourist Info:** (604) 926-6614

Navigational Information
Lat: 49°22.552' **Long:** 123°16.529' **Tide:** 16 ft. **Current:** 2 kt. **Chart:** 3534
Rep. Depths (MLW): Entry 100 ft. **Fuel Dock** 35 ft. **Max Slip/Moor** 70 ft./-
Access: Queen Charlotte Channel to Horseshoe Bay

Marina Facilities (In Season/Off Season)
Fuel: Esso - Gasoline, Diesel
Slips: 300 Total, 6 Transient **Max LOA:** 60 ft. **Max Beam:** 18 ft.
 Rate (per ft.): **Day** $1.00 **Week** $7 **Month** n/a
 Power: 30 amp $5*, **50 amp** n/a, **100 amp** n/a, **200 amp** n/a
 Cable TV: No **Dockside Phone:** No
 Dock Type: Floating, Wood
Moorings: 0 Total, 0 Transient **Launch:** n/a
 Rate: Day n/a **Week** n/a **Month** n/a
Heads: None
Laundry: None **Pay Phones:** Yes
Pump-Out: No **Fee:** n/a **Closed Heads:** No

Marina Operations
Owner/Manager: Megan & Eric Sewell **Dockmaster:** Same
In-Season: Year-Round **Off-Season:** n/a
After-Hours Arrival: Call ahead
Reservations: Yes **Credit Cards:** Visa/MC, Amex
Discounts: None
Pets: Welcome **Handicap Access:** No

Marina Services and Boat Supplies
Services - Dock Carts **Supplies - OnSite:** Ice (Block), Live Bait
Near: Ships' Store **Under 1 mi:** Propane **3+ mi:** Marine Discount Store

Boatyard Services
OnSite: Engine mechanic (gas), Launching Ramp **Near:** Electrical Repairs, Electronics Repairs, Hull Repairs, Rigger, Sail Loft, Canvas Work, Brightwork, Air Conditioning, Interior Cleaning. **Under 1 mi:** Bottom Cleaning, Compound, Wash & Wax, Painting, Total Refits (Thunderbird Marina). **1-3 mi:** Travelift (at Race Rocks).

Restaurants and Accommodations
OnSite: Restaurant (The Boathouse 921-8188, L $9-25, D $15-40, kids' menu), Snack Bar (Lookout) **Near:** Restaurant (Bay Mooring Steak & Seafood House 921-8184), (Arigato Sushi 921-6300), (Greenhouse 921-8865), (Trolls at Horseshoe Bay 921-7755), Raw Bar (YaYa's Oyster Bar 921-8848), Lite Fare (Trollers Pub 921-7616), Pizzeria (Pudgies Chicken and Pizza 921-6866), Motel (Horseshoe Bay Motel 921-7454, $60)

Recreation and Entertainment
OnSite: Boat Rentals, Fishing Charter (guided salmon charters, 21-28' boats, $400-500/5 hrs.), Sightseeing (Sewell's Sea Safari high-speed eco-tours $59/29, 10am, 1pm & 4pm; Baden-Powell hiking trail starts near ferry terminal), Special Events (fishing derbys) **Near:** Playground (fun & unusual equipment), Fitness Center, Jogging Paths, Park (Horseshoe Bay Park) **Under 1 mi:** Video Rental **1-3 mi:** Golf Course (Gleneagles 921-7353), Roller Blade/Bike Paths (mountain biking trails)

3+ mi: Museum (West Vancouver 925-7295, 8 mi.), Cultural Attract (Kay Meek Arts Center 981-1175, 5 mi.)

Provisioning and General Services
Near: Convenience Store (Dayal's 921-9612), Market (Bay Market 921-7155), Wine/Beer, Bank/ATM, Post Office, Catholic Church, Beauty Salon, Newsstand, Retail Shops (Word of Mouth Gifts 921-9121)
Under 1 mi: Green Market, Clothing Store **3+ mi:** Supermarket (Safeway 926-2575, 4 mi.), Liquor Store (BC Liquor 922-8201, 4 mi.), Bakery (The Bread Garden 925-0181, 8 mi.), Fishmonger (Dundarave Fish Market 922-1155, 6 mi.), Pharmacy (Pharmasave 926-5378, 4 mi.), Hardware Store (West Van Hardware 926-9265, 3.5 mi.), Department Store (Park Royal Mall 925-9576,8 mi.)

Transportation
OnSite: Courtesy Car/Van (New Downtown shuttle 921-3474), Water Taxi (Mercury Launch) **OnCall:** Taxi (North Shore Taxi 922-2222), Airport Limo (Excel 433-3550) **Near:** Local Bus (921-7414, at ferry terminal), Ferry Service (Bowen Island) **3+ mi:** Rental Car (National 609-7166/Enterprise 988-5878, 11 mi.) **Airport:** YVR (20 mi.)

Medical Services
911 Service **OnCall:** Ambulance **3+ mi:** Doctor (Park Royal Medical 922-7390, 8 mi.), Dentist (West Van Dental 922-3232, 8 mi.), Chiropractor (Kapilano Chiropractic & Massage Therapy 921-7246, 8 mi.), Veterinarian (Dundarave Veterinary Hospital 925-1244, 6 mi.) **Hospital:** Lions Gate 988-3131 (11 mi.)

Setting -- Just off Howe Sound and snugged behind a floating breakwater, Sewell's two sets of high-end floating docks fill most of Horseshoe Bay. Adjacent is the Bowen Island ferry dock which berths three vessels; add commuter boats, water taxis, tugs, sightseeing tours, pleasure craft and working boats and this becomes a vey busy place. On the shore, the glass-fronted A-frame Boathouse overlooks the scene; further along "The Lookout" - a soft yellow turretted Victorian, trimmed in reddish brown, hosts the Sewell's Ocean Adventure Center.

Marina Notes -- *15A with 30A circuits. Est. in 1931 - fourth generation at the helm. Limited transient dockage, reserve in advance. Quality docks have long, wide finger piers plus alongside tie-ups to manage larger vessels. Ocean Adventure Center features Fishing Derbies, Sea Safari Eco Tours, Company Scavenger Hunts, Sea to Tee Safari (ecotour and golf $149), and a self-drive boat rental fleet. On the first floor is a lounge, registration desk and office, and on the second floor in the windowed turret is tourist information, a gift shop, and a coffee bar with outside dining deck. More moorage at the governnment dock.

Notable -- The Boathouse, part of a BC chain, offers white tablecloth dining and a seafood bar - with a cathedral ceiling and 2-story glass wall filled with views of the docks and bay (note the two traditional West Coast style front doors carved by Nisga'a artist Norman Tait). The 2-hour Boathouse Dinner & Safari, combines a meal and ecotour of Howe Sound. Sewell's is set in the midst of Horseshoe Bay, an energetic, sophisticated seaside village with lots of eateries, unique shops, B&Bs, and activities. Next to The Lookout is Horseshoe Bay Park with a playgoround and a fountain with a propeller centerpiece.

Navigational Information
Lat: 49°17.468' **Long:** 123°07.923' **Tide:** 14 ft. **Current:** 2 kt. **Chart:** 3601
Rep. Depths (*MLW*): **Entry** 11.5 ft. **Fuel Dock** n/a **Max Slip/Moor** 13 ft./-
Access: Burrard Inlet to First Narrows to Coal Harbour

Marina Facilities *(In Season/Off Season)*
Fuel: *Chevron - nearby Fuel Barge* - Gasoline, Diesel
Slips: 243 Total, 15 Transient **Max LOA:** 330 ft. **Max Beam:** 50 ft.
  Rate *(per ft.)*: **Day** $1.90/1.70* **Week** Inq. **Month** Inq.
  Power: 30 amp kwh., 50 amp kwh., 100 amp kwh., 200 amp kwh.**
  Cable TV: Yes - Shaw Cable 280-8818 **Dockside Phone:** Yes, 5/day
  Dock Type: Floating, Long Fingers, Pilings, Concrete
Moorings: 0 Total, 0 Transient **Launch:** n/a
  Rate: Day n/a **Week** n/a **Month** n/a
Heads: 4 Toilet(s), 4 Shower(s) *(with dressing rooms)*
Laundry: 2 Washer(s), 4 Dryer(s) **Pay Phones:** Yes, 2
Pump-Out: Onsite *(In Slip)* **Fee:** $15 **Closed Heads:** Yes

Marina Operations
Owner/Manager: Steven Varley **Dockmaster:** Same
In-Season: Year-Round, 9am-5pm*** **Off-Season:** n/a
After-Hours Arrival: Radio ahead Ch. 66A
Reservations: Yes, Required **Credit Cards:** Visa/MC, Amex
Discounts: None
Pets: Welcome, Dog Walk Area **Handicap Access:** Yes, Heads, Docks

Coal Harbour Marina

1525 Coal Harbour Quay; Vancouver, BC V6G 3E7

Tel: (604) 681-2628 **VHF: Monitor** Ch. 66A **Talk** Ch. 66A
Fax: (604) 681-4666 **Alternate Tel:** n/a
Email: info@coalharbourmarina.com **Web:** www.coalharbourmarina.com
Nearest Town: Vancouver **Tourist Info:** (604) 682-2222

Marina Services and Boat Supplies
Services - Docking Assistance, Concierge, Boaters' Lounge, Security *(24 hrs., patrol twice daily)*, Trash Pick-Up, Dock Carts, Megayacht Facilities, 3 Phase **Communication -** Phone Messages, Fax in/out *($0.50)*, Data Ports *(Wi-Fi - Broadband Xpress)*, DHL, UPS **Supplies - OnSite:** Ice *(Block, Cube)*, Ships' Store *(Wright Mariner Supply)*, Marine Discount Store **Near:** Propane *(gas station)* **1-3 mi:** Bait/Tackle *(Anglers West 875-5474)*

Boatyard Services
OnSite: Compound, Wash & Wax, Interior Cleaning, Woodworking, Yacht Broker *(Westerly)* **OnCall:** Engine mechanic *(diesel)*, Electronics Repairs, Hull Repairs, Rigger, Bottom Cleaning, Air Conditioning, Refrigeration, Divers, Propeller Repairs, Inflatable Repairs, Life Raft Service, Upholstery, Metal Fabrication **Nearest Yard:** Arrow Marine Service/Lynnwood

Restaurants and Accommodations
OnSite: Restaurant *(Cardero's 669-7666, L $9-19, D $17-29, live music in Live Bait Marine Pub Sun-Thu 8:30pm)* **Near:** Restaurant *(Cannery 254-9606, L $16-42, D $23-40, dock'n'dine)*, *(Lift L $3-16, D $23-37, raw bar $3-25, Sun brunch $9-18)*, Lite Fare *(Blue Edge Café 669-3376, 8am-8pm; Wi-Fi)*, *(Bojangles B $3-6, L $5-8, D $5-8)*, Pizzeria *(Flying Wedge 669-5035, delivers)*, Motel *(O Canada 688-0555, $235)*, Hotel *(Westin Bayshore 682-3377, $400, adjacent)*, Condo/Cottage *(Marina Guest House 680-2270)*

Recreation and Entertainment
OnSite: Boat Rentals **Near:** Heated Pool, Beach, Picnic Area, Playground, Park *(Stanley - 5 mi. circumnavigation)*, Sightseeing *(Seabus to Lauderdale Key in North Vancouver or 1 mi. Grouse Mountain 984-0661)*

Under 1 mi: Dive Shop *(Diver's World 732-1344)*, Tennis Courts *(Stanley Park)*, Golf Course *(Stanley Park Pitch & Putt 257-8400 or Meadow Gardens 888-895-2870, $45-65, 5 mi.)*, Fitness Center *(Fitness World 24-hrs. 662-7774)*, Horseback Riding *(Hidden Trails 323-1141)*, Bowling *(Commodore 681-1531)*, Movie Theater *(Granville 7 684-4051)*, Video Rental *(Rogers 431-5332)*, Video Arcade *(Lions Lair)*, Museum *(Maritime 257-8300 $8/5; McMillan Space Center 738-7827)*, Galleries *(Vancouver Art 662-4719)*

Provisioning and General Services
Near: Convenience Store *(7-Eleven, 24 hrs. 668-0531)*, Supermarket *(Safeway 669-5177)*, Gourmet Shop *(Capers 739-6676)*, Liquor Store *(BC 660-4572)*, Fishmonger, Bookstore *(Chapters 682-4066)*, Pharmacy *(Safeway)*, Copies Etc. *(Kinkos 685-3338)* **Under 1 mi:** Farmers' Market *(Bidwell & Comox, Sat 9:30am-2pm)*, Bank/ATM, Post Office *(800-267-1177)*, Catholic Church, Protestant Church, Synagogue, Library *(331-3600)*, Beauty Salon *(Suki's 687-8805)*, Dry Cleaners *(Denman 683-8914)*, Laundry *(Darie 682-2717)*, Hardware Store *(West End 633-1941)*, Retail Shops *(Robson St.)*, Buying Club *(Costco 596-1183)*, Mosque *(Masid Al-Hidava)*

Transportation
OnCall: Water Taxi, Rental Car *(Enterprise 872-1600)*, Taxi *(Yellow 681-1111)* **Near:** Bikes *(688-5141)*, Local Bus *(Translink 953-3333)* **Under 1 mi:** InterCity Bus *(Greyhound)*, Rail *(Pacific Term.)* **Airport:** YVR *(10 mi.)*

Medical Services
911 Service **Near:** Doctor *(Care Point 681-5338)*, Dentist *(Denman 688-3335)*, Chiropractor *(Denman 646-4645)*, Holistic Services *(689-0308)*, Veterinarian *(Urban Animal 684-3632)* **Hospital:** St. Paul's 682-2344 *(1 mi.)*

Setting -- Set in the center of Vancouver's downtown waterfront district, Coal Harbour's network of seven high-end docks - hosting over 240 slips - sits snugly within the arms of the Seawall Trail that edges the harbor. Two floating breakwaters protect the slips and provide additional side-tie dockage for megayachts. Rising above are striking aqua-toned glass residential towers and the Westin Bayshore Hotel and, across the harbor, the snow-capped Coastal Mountains.

Marina Notes -- *In-season prices 30-59' $1.90/ft., 60-89' $2.10, 90-119' $2.40, 120-149' $2.70, 150-179' $2.95, 180-209' $3.20, 210-330' $3.50. **Metered power 30-150 amp. ***Open year-round, staffed 24 hrs.; 3 shifts. Dockmaster lives on-site. 20 liveaboards. Megayachts on "C" & "E" docks. Built in 1995 and still looks new. Vendor guide to marine services. 6 floating homes. Bright laundry (washers & dryers $1.75) and all tile heads & showers.

Notable -- Right on-site - surrounded by docks - is Cadero's, a bustling bistro with an open kitchen, dining rooms and an alfresco deck. All along Coal Harbour Quay are shops and eateries, sculptures, water features and pocket parks. Follow the Seawalk east to the Tourist Info Centre on Burrard Street ("Tickets Tonight" booth, too) and take the 5-hour self-guided downtown walking tour - first stop the soaring Marine Building that documents Vancouver's maritime history. (The Skytrain covers most of the area.) Or three blocks inland find famous Robson Street - a magnet for shoppers with world-class stores, trendy, local boutiques and many hotels - mostly concentrated between Burrar and Jervis Streets (plus dancing Friday nights in Robson Sq.). Or walk a quarter mile west to the entrance to fabulous 1,000 acre Stanley Park. If you are planning to sightsee, check out the SmartVisit card for significant savings.

Bayshore West Marina

450 Denman Street; Vancouver, BC V6G 3J1

Tel: (604) 689-5331 **VHF: Monitor** n/a **Talk** n/a
Fax: (604) 689-5332 **Alternate Tel:** n/a
Email: franka@thunderbirdmarine.com **Web:** www.thunderbirdmarine.com
Nearest Town: Vancouver **Tourist Info:** (604) 682-2222

Navigational Information
Lat: 49°17.613' **Long:** 123°07.923' **Tide:** 14 ft. **Current:** 2 kt. **Chart:** 3601
Rep. Depths (*MLW*): **Entry** 11 ft. **Fuel Dock** n/a **Max Slip/Moor** 12 ft./-
Access: Burrard Inlet to First Narrows to Coal Harbour

Marina Facilities (*In Season/Off Season*)
Fuel: *Chevron - nearby Fuel Barge* - Gasoline, Diesel
Slips: 50 Total, 5 Transient **Max LOA:** 120 ft. **Max Beam:** 25 ft.
 Rate (*per ft.*): **Day** $1.70* **Week** n/a **Month** n/a
Power: 30 amp $.10/kw, **50 amp** $.10/kw, **100 amp** $.10/kw, **200 amp** n/a
Cable TV: No **Dockside Phone:** No
Dock Type: Fixed, Long Fingers, Concrete
Moorings: 0 Total, 0 Transient **Launch:** n/a
 Rate: Day n/a **Week** n/a **Month** n/a
Heads: 2 Toilet(s)
Laundry: None **Pay Phones:** No
Pump-Out: Onsite (*In-Slip*), Full Service **Fee:** $15 **Closed Heads:** Yes

Marina Operations
Owner/Manager: Frank Armitage **Dockmaster:** Same
In-Season: Year-Round, 9am-5pm **Off-Season:** n/a
After-Hours Arrival: Call ahead
Reservations: Yes, Required **Credit Cards:** Visa/MC
Discounts: None
Pets: Welcome, Dog Walk Area **Handicap Access:** No

Marina Services and Boat Supplies
Services - Docking Assistance, Security, Trash Pick-Up, Dock Carts, Megayacht Facilities, 3 Phase **Communication -** Data Ports (*Wi-Fi - Broadband Xpress*), FedEx, DHL, UPS **Supplies - Near:** Ice (*Block, Cube*), Ships' Store (*Wright Mariner*), Propane **1-3 mi:** West Marine (*730-4093*)

Boatyard Services
OnCall: Engine mechanic (*diesel*), Electrical Repairs, Electronic Sales, Electronics Repairs, Hull Repairs, Rigger, Compound, Wash & Wax, Interior Cleaning **Nearest Yard:** Arrow Marine Service/Lynnwood

Restaurants and Accommodations
OnSite: Restaurant (*Lift L $3-16, D $23-37*) **Near:** Restaurant (*Cardero's 669-7666, L $9-19, D $17-29*), (*Yoshi 738-8226*), (*Le Restaurant Gavroche 685-3924*), (*Tapas Tree 606-4680*), (*Francesco's 687-7210*), (*Cloud 9 662-8328, revolving, at Greenbriar Hotel*), (*Salonica 681-8141*), (*Ciao Bella 688-5771*), (*Currents at Bayshore B $13-16, daily market buffet $23 6:30-11:30am; Sun brunch $45*), (*Seawell Bar & Grill L $9-15, D $17-39, 11:30am-11pm*), Coffee Shop (*Coffee Grind 689-9444*), (*Stanley Perks B $2-4, L $5-6.50*), Pizzeria (*Luigi's 687-7771*), Hotel (*Westin 682-3377, $200-600*), (*Lord Stanley Suites 688-9299*), (*Greenbriar 683-4558*) **Under 1 mi:** Restaurant (*Top of Vancouver 669-2220, L $12-40, D $32-64, revolving*)

Recreation and Entertainment
Near: Beach, Picnic Area, Grills, Playground, Tennis Courts (*Stanley*), Golf Course (*Stanley 257-8400*), Fitness Center (*World Gym 915-3032*), Jogging Paths, Boat Rentals, Park (*Stanley*), Sightseeing (*Stanley Park horse-drawn tour 681-5115 $20/13*), Special Events **Under 1 mi:** Heated Pool

(*Stanley Park*), Dive Shop (*Diver's World 732-1344*), Bowling (*Commodore 681-1531*), Movie Theater (*Granville 7 684-4051*), Video Rental (*Rogers 431-5332*), Museum (*Canadian Craft 687-8266, Golden Museum 688-7735, Roedde House 684-7040*), Cultural Attract (*Aquarium 659-3474 $9-16; Back Alley Theatre 688-7013*), Galleries (*Potters 685-3919, Howe St. 681-5777*)

Provisioning and General Services
Near: Convenience Store (*City Park 682-6063*), Supermarket (*Safeway 669-5177*), Delicatessen (*Wrap Zone 331-1343*), Health Food (*Natures Energy 688-7575*), Wine/Beer (*Divino 683-8466*), Liquor Store (*BC Liquor 660-4572*), Bakery (*Pastries of Lisbon 688-3340*), Fishmonger, Bank/ATM, Post Office (*682-6063*), Library (*331-3600*), Beauty Salon (*BC Glamour 688-2888*), Dry Cleaners (*Pacific Rim 688-5718*), Laundry, Bookstore (*Chapters 682-4066*), Pharmacy (*Safeway*), Hardware Store (*West End 633-1941*) **Under 1 mi:** Farmers' Market (*Nelson Park Sat 9:30am-2pm*), Catholic Church, Protestant Church, Retail Shops (*Pacific Centre 683-6861*), Buying Club (*Costco 596-1183*), Copies Etc. (*Kinkos 685-3338*), Mosque

Transportation
OnCall: Bikes (*Bikes'n Blades 602-9899*), Water Taxi (*Harbor Ferries*), Taxi (*Yellow 681-1111*) **Near:** Rental Car (*Hertz 606-3784, Nat'l 738-6006*), Local Bus (*Translink 953-3333*) **Under 1 mi:** Rail (*Pacific Term.*), Ferry Service (*Seabus to N Vancouver 953-3333*) **Airport:** YVR (*10 mi.*)

Medical Services
911 Service **Under 1 mi:** Doctor (*Care Point 681-5338*), Dentist (*Denman Dental 688-3335*), Chiropractor (*Denman Chiro 646-4645*), Veterinarian (*Urban Aminal 684-3632*) **Hospital:** St. Paul's 682-2344 (*1 mi.*)

Setting -- Tucked up into a protected cove near the head of Coal Harbour, Bayshore's docks lie in the shadow of the Bayshore Westin Hotel and Tower. Views across the harbor are of Stanley Park, the Vancouver Rowing Club and the aqua peaks of the Royal Vancouver Yacht Club's covered docks - and the Coastal Mountains. A six-foot wide ramp leads from the network of high-end concrete docks to the bulkhead - edged by the ten-mile Seawall Trail.

Marina Notes -- *$65 min. Redeveloped in 2000 when condos were built. Managed by Thunderbird. Widest slips in harbor. Sunsail Charters (320-7245) is on-site. A concrete float system features 8-foot-wide main floats and 4-foot-wide fingers with a pump-out connection at each slip. Extra wide channels & berths make for easy maneuvering. Fuel Barge anchored in the harbor. All major rental cars within walking distance. Underground car park $12/day. Nicola Internet Café 408-1559 is nearby. No showers or laundry. Note: New marina being developed near the cruise ship dock.

Notable -- Bayshore West is in the heart of Vancouver's recently developed waterfront neighborhood. It sits at the edges of Stanley Park and the West End residential neighborhood - home to 40,000 people living in contemporary high-rises. Denman Street is the West End's main drag with lots of shopping and fine dining and ethnic eateries. The ten-mile Seawall Trail circumnavigates the city and runs through 1,000 acre Stanley Park - which has three beaches, a pitch & putt golf course, a pool, gardens, Children's Farmyard and Miniature Railway, Theatre under the Stars, soaring totems, the Vancouver Aquarium Marine Science Centre, miles of trails and much, much more. Take a tour in a horse-drawn carriage and don't miss the overlook at Prospect Point.

Navigational Information
Lat: 49°16.294' **Long:** 123°08.247' **Tide:** 13 ft. **Current:** n/a **Chart:** 3481
Rep. Depths *(MLW)*: **Entry** 15 ft. **Fuel Dock** n/a **Max Slip/Moor** 8 ft./-
Access: Burrard Inlet to English Bay to False Creek, NW side of Granville Is.

Marina Facilities *(In Season/Off Season)*
Fuel: No
Slips: 90 Total, 30 Transient **Max LOA:** 50 ft. **Max Beam:** n/a
   Rate *(per ft.)*: **Day** $1.75/1.50 **Week** n/a **Month** n/a
   Power: 30 amp $10, **50 amp** n/a, **100 amp** n/a, **200 amp** n/a
   Cable TV: No **Dockside Phone:** No
   Dock Type: Floating, Long Fingers, Wood
Moorings: 0 Total, 0 Transient **Launch:** n/a
   Rate: Day n/a **Week** n/a **Month** n/a
Heads: 3 Toilet(s), 3 Shower(s)
Laundry: None **Pay Phones:** Yes
Pump-Out: No **Fee:** n/a **Closed Heads:** Yes

Marina Operations
Owner/Manager: David West **Dockmaster:** Rafi Sheskin
In-Season: Year-Round, 9am-5pm **Off-Season:** n/a
After-Hours Arrival: See office next day
Reservations: Yes, Required **Credit Cards:** Visa/MC, Amex
Discounts: None
Pets: Welcome, Dog Walk Area **Handicap Access:** Yes, Heads, Docks

Cooper Boating

1620 Duranleau Street; Vancouver, BC V6H 3S4

Tel: (604) 687-4110; (888) 999-6419 **VHF: Monitor** n/a **Talk** n/a
Fax: (604) 687-3267 **Alternate Tel:** n/a
Email: crew@cooperboating.com **Web:** www.cooperboating.com
Nearest Town: Vancouver **Tourist Info:** (604) 682-2222

Marina Services and Boat Supplies
Services - Docking Assistance, Security *(CCTV)*, Trash Pick-Up, Dock Carts **Communication -** Mail & Package Hold, Phone Messages, Fax in/out, Data Ports *(Wi-Fi, $10)*, FedEx, DHL, UPS, Express Mail **Supplies -** **OnSite:** Ships' Store **Near:** Ice *(Cube)*, West Marine *(730-4093)*, Marine Discount Store *(Steveston Marine 733-7031)*, Bait/Tackle, Propane

Boatyard Services
OnSite: Travelift, Engine mechanic *(gas, diesel)*, Launching Ramp, Electrical Repairs, Electronic Sales, Electronics Repairs, Hull Repairs, Rigger, Canvas Work, Bottom Cleaning, Brightwork, Air Conditioning, Refrigeration, Divers, Propeller Repairs, Woodworking, Inflatable Repairs, Painting **OnCall:** Sail Loft, Metal Fabrication **Yard Rates:** Power Wash $2/ft.

Restaurants and Accommodations
Near: Restaurant *(Sand Bar 669-9030)*, *(Cat's Meow 647-2287)*, *(Bridges Seafood 687-7351)*, *(Rubina Grill 662-7778)*, *(Omi of Japan 685-8011)*, *(Sammy J Peppers 696-0739)*, *(Tony's Fish & Oyster Café 683-7127)*, Lite Fare *(Fraser Valley Juice & Salad 669-0727)*, *(Blue Parrot Coffee 688-5127)*, *(Celine Fish & Chips 669-8650)*, *(Siegel's Bagels 685-5670)*, Pizzeria *(Pizza Pzazz 682-9002)*, Hotel *(Granville Island 683-7373)*

Recreation and Entertainment
OnSite: Playground, Dive Shop *(Alabaster Star)*, Boat Rentals *(Cooper power & sail; EcoMarine Kayaks 689-7575)*, Fishing Charter *(Bonnie Lee 290-7447 5 hrs. $375-995)* **Near:** Tennis Courts, Fitness Center *(False Creek Comm. Ctr.)*, Video Rental *(A&F 688-2727)*, Video Arcade *(Circuit Circus 608-6699)*, Park *(Sutcliffe)*, Museum *(Granville Is. Sport Fishing,*

Model Ships & Trains 683-1939 $16/family), Cultural Attract *(Performance Works 666-8139; Granville Stage)*, Galleries **Under 1 mi:** Pool *(Aquatic Ctr. - F.C.Ferry)* **3+ mi:** Golf Course *(Langara 713-1816, 5 mi.)*

Provisioning and General Services
OnSite: Lobster Pound, Bank/ATM, Newsstand **OnCall:** Provisioning Service **Near:** Market *(Granville Is. Public Market 666-6477)*, Gourmet Shop *(Salmon Shop 682-7178)*, Delicatessen *(Kaisereck 685-8810)*, Wine/Beer *(Granville Is. Brewrey 687-2739 - tours)*, Bakery *(Stuart's 685-8816)*, Farmers' Market *(Thu 9am-2pm)*, Fishmonger *(Longliner 681-9016)*, Post Office, Beauty Salon *(Hairloft 684-6177)*, Hardware Store *(Steveston 733-7031)* **Under 1 mi:** Convenience Store *(Mac's 736-5115)*, Liquor Store *(Spirit of Howe St. 682-2586)*, Catholic Church, Protestant Church, Library *(Firehall 665-3970)*, Dry Cleaners *(Lexx 682-5399)*, Laundry *(Fletchers 731-9313)*, Bookstore *(Fireside 734-7323)*, Pharmacy *(Real Canadian 322-3704)* **1-3 mi:** Supermarket *(Super Value 688-0911)*

Transportation
OnSite: Local Bus *(Transliner)*, InterCity Bus *(Greyhound)* **OnCall:** Taxi *(MacLures 731-9211)* **Near:** Water Taxi *(Aquabus 689-5858 $2.50/1.25-$5/3 - day pass $11/8)*, Ferry Service *(False Creek 684-7781 $2.50/1.25-$5/3 or day pass $12/8)* **Under 1 mi:** Bikes *(Ride Off 738-7747)*, Rental Car *(Lo-Cost 689-9664)* **1-3 mi:** Rail *(VIA Rail 640-3700)* **Airport:** YVR *(8 mi.)*

Medical Services
911 Service **Near:** Doctor *(Bayswater 731-0091)*, Dentist *(Seaside 733-2777)*, Holistic Services *(Peaceful Moments 943-7070)*, Veterinarian *(Animal Emergency 734-5104)* **Hospital:** Vancouver General 875-4111 *(1.5 mi.)*

Setting -- Cooper Boating is located in the heart of lively Granville Island, a redeveloped industrial site that is now a thriving entertainment destination with a world-class market and arts scene. Its docks are filled with the bustle of a large charter operation and are surrounded by the Maritime Market - a compilation of dozens of merchants, tours, and rental agencies that sprawls across the island's southwest shore. This is the heart of all things nautical on Granville Island.

Marina Notes -- Fuel 0.5 mi.. Catamaran dockage extra fee. Cooper charters the newest and largest power and sail bare-boat fleet in Canada with additional locations in Sidney and Desolation Sound. They also offer flotilla charters, an innovative yacht sharing club, day sail rentals and Friday night recreational races. CYA courses - 4-levels. On-site is full-service Granville Island Boatyard 685-6924. Across the waterway are the commercial fishing vessels.

Notable -- Granville Island is a major tourism destination; dodge the sightseers by touring early mornings and weekdays. Cheek-by-jowl across the whole island are an impressive mix of fine art galleries, craft shops, an arts institute, theaters and all manner of retail stores in repurposed industrial buildings - and right in the middle a still working cement plant! On the northern tip, the fabulous Public Market is the core of Granville Island - the array of foodstuffs and crafts is staggering and a fun place to provision. Eateries of all persuasions abound. The two-story Kids' Market is magical: level one is wall-to-wall toy stores; upstairs are an arcade, adventure zone, wet store and magic shop. For off-island pursuits, take the bus to the Waterfront Skytrain station, the rainbow-colored Aquabus (between Market & Arts Club) to Science World, Yaletown or 5 other spots or the little blue False Creek Ferries (next to Bridge's) to six stops on the mainland.

False Creek Yacht Club

1661 Granville Street; Vancouver, BC V6Z 1N3

Tel: (604) 682-3292; (866) 677-2628 **VHF: Monitor** Ch. 66A **Talk** n/a
Fax: (604) 682-3614 **Alternate Tel:** (604) 868-4275
Email: fcyc@fcyc.com **Web:** www.fcyc.com
Nearest Town: Vancouver **Tourist Info:** (604) 682-2222

Navigational Information
Lat: 49°16.405' **Long:** 123°07.952' **Tide:** 16 ft. **Current:** n/a **Chart:** 3481
Rep. Depths (MLW): Entry 15 ft. **Fuel Dock** n/a **Max Slip/Moor** 15 ft./-
Access: English Bay to False Creek, north side under Granville Bridge

Marina Facilities (In Season/Off Season)
Fuel: No
Slips: 100 Total, 10 Transient **Max LOA:** 100 ft. **Max Beam:** n/a
  Rate (per ft.): **Day** $1.50 **Week** n/a **Month** $10
  Power: 30 amp $5, **50 amp** $5, **100 amp** n/a, **200 amp** n/a
  Cable TV: No **Dockside Phone:** No
  Dock Type: Floating, Concrete, Wood
Moorings: 0 Total, 0 Transient **Launch:** n/a
  Rate: Day n/a **Week** n/a **Month** n/a
Heads: 4 Toilet(s), 4 Shower(s)
Laundry: 1 Washer(s), 1 Dryer(s), Book Exchange **Pay Phones:** No
Pump-Out: OnSite, Full Service **Fee:** $15 **Closed Heads:** Yes

Marina Operations
Owner/Manager: False Creek Yacht Club **Dockmaster:** n/a
In-Season: May-Sep 15, 10am-8pm **Off-Season:** Sep 16-Apr, 9am-4:30pm*
After-Hours Arrival: Call 604-202-0020
Reservations: Yes, Preferred **Credit Cards:** Visa/MC, Amex
Discounts: None
Pets: Welcome **Handicap Access:** No

Marina Services and Boat Supplies
Services - Docking Assistance, Boaters' Lounge, Security (8 hrs., overnight), Trash Pick-Up, Dock Carts **Communication -** Mail & Package Hold, Fax in/out, Data Ports (Wi-Fi - Broadband Xpress, $10/d), FedEx, UPS
Supplies - OnSite: Ice (Block, Cube) **Near:** Bait/Tackle **Under 1 mi:** Ships' Store (Steveston Marine 732-6220), West Marine (730-4093), Live Bait, Propane **1-3 mi:** Marine Discount Store (Western Marine 253-3322)

Boatyard Services
Nearest Yard: Granville Boatyard

Restaurants and Accommodations
OnSite: Restaurant (Stonegrill 637-0388, L & D $10-30), (The FCYC Lounge Tue-Fri L 11am-5pm, Wed-Fri dinner 7pm+), (Nu 646-4668, L & D $15-22, Sat & Sun brunch - enRoute Mag's Best New Rest. in Canada for '06)
Near: Restaurant (Fernando's 684-8815), (Two Parrots 685-9657), (Lickerish 696-0725), (Wildfire 682-3473), (Nippon Ichi 643-2898), (Kettle of Fish 682-6853), (English Bay Kettle 408-2739), (Gin Kaku Sushi 685-8381), ("C" 681-1164), (Tony Roma's 669-7336), Pizzeria (Fresh Slice 444-7444), Hotel (Executive Inn 688-7678), (Ramada 685-1111), (Best Western 669-9888)

Recreation and Entertainment
Near: Picnic Area, Playground, Dive Shop, Fitness Center (Fitness World 681-3232), Jogging Paths (Seawall Promenade), Video Rental (A&F Video 688-2727), Park (George Wainborn), Special Events **Under 1 mi:** Heated Pool (Aquatic Centre 665-3424 $4.75/2.40), Tennis Courts, Boat Rentals, Fishing Charter, Movie Theater (Kitten Theatre 689-0786), Cultural Attract (Orpheum Theatre 665-3050, Arts Club Theatre 687-1644), Sightseeing,

Galleries (City Art 684-6034, Art Beatus 688-2633) **1-3 mi:** Beach, Grills, Golf Course (Stanley Park 257-8400), Museum (Canadian Craft 687-8266, Golden 688-7735, Vancouver Maritime 257-8300)

Provisioning and General Services
Near: Convenience Store (7-Eleven 601-2489), Market (Choices 633-2392, Helen's 689-5937), Gourmet Shop (McKiney's 685-5403), Health Food (Yo Organic Plus 681-0101), Liquor Store (Sprint Canada 273-3610), Bank/ATM, Post Office (688-2068), Beauty Salon (Hairport 688-9099), Laundry (Swan's 684-0323), Bookstore (Bonanza 684-3775), Newsstand, Florist (Flowers Unlimited 685-6078), Copies Etc. **Under 1 mi:** Supermarket (Super Valu 688-0911), Delicatessen (Green City 669-2206), Bakery (Sugar Refinery 331-1184), Farmers' Market (Nelson Park Sat 9:30am-2pm), Green Market (West Valley Produce 669-5612), Catholic Church, Protestant Church, Library (660-2910), Dry Cleaners (Frank's 684-6917), Pharmacy (Pharmasave 669-7700), Hardware Store (A&V 647-0990)

Transportation
OnCall: Taxi (Vancouver 255-5111) **Near:** Water Taxi (Aquabus 689-5858 $2.50/1.25-$5/3 - day pass $11/8 - from Hornby St.), Rental Car (Lo-Cost 689-9664), Local Bus (Translink 953-3333) **Under 1 mi:** InterCity Bus (Grayhound) **1-3 mi:** Rail (VIA Rail 640-3700) **Airport:** YVR (8 mi.)

Medical Services
911 Service **Near:** Doctor (Yaletown Medical 633-2474), Dentist (Aarm 683-5530) **Under 1 mi:** Chiropractor (Vitality 687-7678), Holistic Services (Absolute Spa 684-2772), Veterinarian (Animal Emergency 734-5104)
Hospital: St. Paul's 682-2344 (1 mi.)

Setting -- Located on the north, downtown side of False Creek, the FCYC docks lie in the shadow of the Granville Island Bridge directly across from Granville Island's Public Market. The clubhouse occupies the top floor of a dramatic four-story barrel fronted glass contemporary. The club's lounge, decorated in deep blues and natural woods, spills out onto a wide dining terrace. The views from both are simply spectacular.

Marina Notes -- *Closed Sat-Sun & holidays in the off-season. 200-member club in operation since 1980. Welcomes transients without reciprocity - rates lower for other YC members. 5 dedicated transient slips. Fuel (gas & diesel) nearby. Site of the in-water part of the Vancouver International Boat Show (2nd week in Feb). Currently, the dock office is in a trailer. The FCYC hosts the Vancouver Boating Welcome Centre - on a blue barge on the main dock. Hail on Ch. 66A (648-2628) for available transient slips in Vancouver or to secure a False Creek Anchoring Permit. The surrounding neighborhood tends toward residential.

Notable -- Two well-reviewed eateries occupy the lower floors of the glass-fronted clubhouse - Wu (which means naked) is on the first level with a waterside patio and Stonegrill, whose menu revolves around table-top grills. is in the dramatic double-height middle space. A walk north along Granville Street yields the center of Vancouver's entertainment district. A few blocks east is Yaletown - once a warehouse district, it is now a trendy area fronting False Creek that is chockablock with hip eateries, sidewalk cafés, nightlife and small hotels. A short walk south across the Granville Bridge is Granville Island. Alternatively walk two blocks east to Hornby Street and take the Aquabus (it stops at 6 other spots along the creek, including Yaletown, Science Center and Plaza of Nations).

Navigational Information
Lat: 49°16.155' **Long:** 123°07.838' **Tide:** 12 ft. **Current:** n/a **Chart:** 3481
Rep. Depths (*MLW*): **Entry** 15 ft. **Fuel Dock** n/a **Max Slip/Moor** 8 ft./-
Access: False Creek through Granville Bridge to south end of Granville Isl.

Marina Facilities (*In Season/Off Season*)
Fuel: No
Slips: 32 Total, 3 Transient **Max LOA:** 100 ft. **Max Beam:** 22 ft.
  Rate (*per ft.*): **Day** $2.00 **Week** n/a **Month** n/a
  Power: 30 amp $5, **50 amp** $10, **100 amp** n/a, **200 amp** n/a
Cable TV: Yes **Dockside Phone:** No
Dock Type: Fixed, Pilings, Alongside
Moorings: 0 Total, 0 Transient **Launch:** n/a
  Rate: Day n/a **Week** n/a **Month** n/a
Heads: 1 Toilet(s)
Laundry: None **Pay Phones:** No
Pump-Out: No **Fee:** n/a **Closed Heads:** No

Marina Operations
Owner/Manager: Elaine Jensen **Dockmaster:** Same
In-Season: Year-Round, 8am-8pm **Off-Season:** n/a
After-Hours Arrival: Call (604) 729-1442
Reservations: Yes, Preferred **Credit Cards:** Visa/MC, Amex
Discounts: None
Pets: Welcome, Dog Walk Area **Handicap Access:** Yes

Pelican Bay Marina

1780 W. 6th Avenue; Vancouver, BC V6J 5E8

Tel: (604) 682-7420 **VHF: Monitor** n/a **Talk** n/a
Fax: (604) 682-7433 **Alternate Tel:** (604) 729-1442
Email: mejens@shaw.ca **Web:** globalairphotos.com/pelican_bay_marina
Nearest Town: Vancouver (*0 mi.*) **Tourist Info:** (604) 682-2222

Marina Services and Boat Supplies
Services - Docking Assistance, Security (*locked gate*), Dock Carts
Communication - Fax in/out (*$5*), Data Ports (*Office*), FedEx, DHL, UPS
Supplies - OnSite: Ice (*Cube*) **Near:** Ice (*Block*), Ships' Store, West Marine
(*730-4093*) **Under 1 mi:** Propane, CNG

Boatyard Services
OnSite: Bottom Cleaning, Divers **OnCall:** Brightwork, Air Conditioning,
Propeller Repairs, Woodworking **Near:** Travelift, Engine mechanic (*gas,
diesel*), Electronics Repairs, Hull Repairs, Canvas Work, Metal Fabrication.

Restaurants and Accommodations
OnSite: Restaurant (*Dockside 685-7070, B $8-16, L $12-16, D $24-31,
brunch $9-15, buffet $20/11, kids' menu, open 7 days; sustainable fish only*),
Lite Fare (*Dockside Brewing Co. D $8-24, 8 varieties*), Hotel (*Granville Island
683-7373, $150-240*) **Near:** Restaurant (*Rubina Grill 662-7778*), (*Dinos
684-1714*), (*Sand Bar 669-9030*), (*Bridges' 687-4400*), (*Omi of Japan 685-
8011*), (*La Tortilleria 684-2820*), Pizzeria (*Pizza Pzazz 682-9002*)
Under 1 mi: Hotel (*Ramada 685-1111*)

Recreation and Entertainment
OnSite: Boat Rentals, Roller Blade/Bike Paths, Fishing Charter **Near:** Pool
(*YMCA*), Playground, Tennis Courts (*Sutcliffe*), Fitness Center (*F.C.
Community Ctr, Sutcliffe Park 257-8195*), Jogging Paths, Park (*Basford*),
Museum (*Charles H. Scott at Emily Carr Institute of Art 844-3811; Granville
Is. Brewing 687-2739; Granville Is. Sport Fishing, Model Ships & Model
Trains 683-1939 $16/family*), Cultural Attract (*Waterfront Theatre 685-6217,
Performance Works 666-8139, Vancouver Theatre 688-7013*), Galleries

(*Net Loft*) **Under 1 mi:** Dive Shop (*BC Dive & Kayak 732-1344*), Movie
Theater, Video Rental (*A&F 688-2727*) **1-3 mi:** Bowling (*5 Pin Bowling 738-
5412*) **3+ mi:** Golf Course (*Langara 713-1816, 5 mi.*)

Provisioning and General Services
OnSite: Copies Etc. (*in hotel*) **OnCall:** Dry Cleaners (*picks up at marina*),
Laundry (*picks up & delivers*) **Near:** Market (*Granville Island Public Market
666-6477*), Gourmet Shop (*La Baguette Et L'Echalote 684-1351*),
Delicatessen (*Kaisereck Deli 685-8810*), Wine/Beer (*Granville Island
Brewery*), Bakery (*Stuart's 685-8816*), Farmers' Market (*Thu 9am-2pm*),
Fishmonger (*Longliner Seafoods 681-9016*), Bank/ATM, Post Office, Beauty
Salon (*Hairloft 684-6177*), Newsstand, Hardware Store (*Steveston 733-
7031*), Florist (*G.I. 669-1228*) **Under 1 mi:** Liquor Store, Catholic Church,
Protestant Church, Library (*Firehall 665-3970*), Pharmacy (*Real Canadian
322-3704*) **1-3 mi:** Supermarket (*Super Valu 688-0911*), Synagogue

Transportation
OnSite: Bikes **OnCall:** Taxi (*Yellow Cab 681-1111*) **Near:** Water Taxi
(*Aquabus 689-5858 $2.50/1.25-$5/3 - day pass $11/8*), Local Bus (*Translink
953-3333*), InterCity Bus, Ferry Service (*False Creek 684-7781 $2.50/1.25-
$5/3 or day pass $12/8*) **Under 1 mi:** Rental Car (*Lo-Cost Rent-a-Car 689-
9664*) **1-3 mi:** Rail (*VIA Rail 640-3700*) **Airport:** YVR (*8 mi.*)

Medical Services
911 Service **OnCall:** Ambulance, Veterinarian **Near:** Doctor (*Bayswater
731-0091*), Dentist (*Seaside Dental 733-2777*)
Hospital: Vancouver General 875-4111 (*1.5 mi.*)

Setting -- Stretched along the quieter, artsy northeast side of Granville Island, past colorful barge houses and adjacent to Ron Basford Park, Pelican Bay
offers a less touristy alternative. Pots of flowers dot well-maintained docks that float below the Granville Island Hotel & Dockside Restaurant - its
architecture a bold mix of industrially-inspired styles. Across the parking lot are the stark edifices of the prestigious Emily Carr Institute of Art & Design.

Marina Notes -- The marina offers power, water, heads, cable, and phone hook-ups. A small turquoise steel building, right on the dock, houses the office
and amenities. The attractive 82-room, pet-friendly Granville Island Hotel offers a wide range of uniquely appointed rooms - many overlook the docks, creek and
Vancouver skyline. The Dockside restaurant features window walls, a 50-foot aquarium and a delightful waterside patio with gas heaters; the more casual
Dockside Brewing Co. pub has a lighter menu, frequent live entertainment and its own brewery (formerly Creek Brewery). Note: there are no showers at the
marina but these are available at the nearby False Creek Community Center (257-8195).

Notable -- The painted shells of old factories linked by inviting walkways rule the island's aesthetic. The arts district includes dozens of galleries, studios and
ateliers spawned by the Institute. Sutcliffe Park has tennis courts, a playground, water park and fitness center. A short walk is the famous Public Market, two
theaters, dozens of eateries, Kids' Market and the Maritime Market which houses nautically-oriented shops and services. Take a ferry or Aquabus across False
Creek to downtown Vancouver or to Vanier Park for the Shakespeare Fest (Jun-Sep 739-0559), Pacific Space Center and Maritime & Vancouver Museums.

Delta Marina at the Delta Vancouver Airport Hotel

Delta Marina

3500 Cessna Drive; Richmond, BC V7B 1C7

Tel: (604) 273-4211; (800) 661-7762 **VHF: Monitor** n/a **Talk** n/a
Fax: (604) 273-7531 **Alternate Tel:** n/a
Email: deltacharters@telus.net **Web:** www.deltacharters.com
Nearest Town: Richmond *(2 mi.)* **Tourist Info:** (604) 271-8280

Navigational Information
Lat: 49°11.400' **Long:** 123°08.439' **Tide:** 12 ft. **Current:** n/a **Chart:** 3491
Rep. Depths *(MLW)*: **Entry** 15 ft. **Fuel Dock** n/a **Max Slip/Moor** 15 ft./-
Access: North shore of Fraser River's Middle Arm

Marina Facilities *(In Season/Off Season)*
Fuel: No
Slips: 138 Total, 3 Transient **Max LOA:** 120 ft. **Max Beam:** 25 ft.
 Rate *(per ft.)*: **Day** $1.00 **Week** n/a **Month** $8.50/7.50
 Power: 30 amp Incl.*, **50 amp** n/a, **100 amp** n/a, **200 amp** n/a
 Cable TV: No **Dockside Phone:** No
 Dock Type: Floating, Wood
 Moorings: 0 Total, 0 Transient **Launch:** n/a
 Rate: Day n/a **Week** n/a **Month** n/a
Heads: 6 Toilet(s)
Laundry: None **Pay Phones:** Yes
Pump-Out: No **Fee:** n/a **Closed Heads:** Yes

Marina Operations
Owner/Manager: Rick Cockburn **Dockmaster:** Same
In-Season: Year-Round, 9am-5pm **Off-Season:** n/a
After-Hours Arrival: Call ahead
Reservations: Yes, Required **Credit Cards:** Visa/MC, Amex, Interac
Discounts: None
Pets: Welcome, Dog Walk Area **Handicap Access:** No

Marina Services and Boat Supplies
Services - Docking Assistance, Concierge, Room Service to the Boat, Boaters' Lounge, Security *(locked gates & night watchman)*, Trash Pick-Up, Dock Carts **Communication -** Mail & Package Hold, Phone Messages, Fax in/out, Data Ports *(Wi-Fi in Hotel Lounge & Biz Ctr.)*, FedEx, DHL, UPS *(Sat Del)* **Supplies - OnSite:** Ice *(Block, Cube)* **Under 1 mi:** West Marine *(233-2327)* **1-3 mi:** Ships' Store, Propane *(Autogas 276-9924)*

Boatyard Services
OnSite: Travelift *(50T)*, Engine mechanic *(gas, diesel)*, Electrical Repairs, Hull Repairs, Bottom Cleaning, Brightwork, Compound, Wash & Wax, Interior Cleaning, Yacht Broker *(Delta Yachts)*, Total Refits **OnCall:** Electronic Sales, Electronics Repairs, Canvas Work, Air Conditioning, Refrigeration, Propeller Repairs, Inflatable Repairs, Life Raft Service **Under 1 mi:** Painting, Awlgrip. **1-3 mi:** Launching Ramp. Other Certifications: BCYBA
Yard Rates: $65/hr., Haul & Launch $7/ft., Power Wash $65/hr.

Restaurants and Accommodations
OnSite: Restaurant *(Elephant Castle 276-1962, L $10-23, D $10-23, 11am-Mid - seafood & steak to light fare to real pub classics)*, *(Atmosphere North B $11-14, D $13-29, B buffet $14 & $17; D small plates $13-16, main plates $18-29)*, Lite Fare *(Lounge - tapas $8-13)*, Hotel *(Delta Vancouver Airport Hotel 278-1241, $140-195, pet friendly)* **OnCall:** Pizzeria *(Panago 310-0001)* **Near:** Restaurant *(KG Boathouse 273-7014, L $9-18, D $15-25)* **Under 1 mi:** Restaurant *(Dem Bones 273-3822, D $17-20)*, *(Richmond Sushi 207-7799)*, *(Gala Seafood 821-0943)*, *(Curry House 231-9887)*, *(Sea Harbour Seafood 232-0816)*, *(New India Buffet 244-8858)*, *(Brooklyn Steakhouse 270-6030)*, Hotel *(Four Points 214-0888)*

Recreation and Entertainment
OnSite: Pool, Fishing Charter **Near:** Park *(Miller)* **1-3 mi:** Playground, Tennis Courts *(Richmond 273-3631)*, Golf Course *(Marine Drive 261-8111)*, Movie Theater *(Famous Players 273-7173)*, Video Rental *(Entertainment One 279-9098)*, Cultural Attract *(Gateway Theatre 270-1812)*

Provisioning and General Services
OnSite: Bank/ATM, Newsstand, Copies Etc. *(Hotel Business Center)* **Under 1 mi:** Delicatessen *(Sam's 821-1421)*, Wine/Beer *(Richmond 273-5969)*, Beauty Salon *(Rainbow 248-1625)*, Pharmacy *(Park Pacific 273-9812)*, Buying Club *(Costco 668-8450)* **1-3 mi:** Supermarket *(Superstore 233-2435/ Safeway 263-7267)*, Health Food *(Gibo 278-8716)*, Liquor Store *(Liquor Depot 233-0354)*, Bakery *(De Fresh 273-1323)*, Farmers' Market *(Cambie Community Centre Sun 10am-2pm)*, Green Market, Fishmonger *(Nexus 278-2998)*, Post Office, Catholic Church, Protestant Church, Library *(231-6465)*, Dry Cleaners *(Airway 273-5564)*, Laundry *(Easy Kleen 273-6925)*, Bookstore *(Manna 303-1102)*, Hardware Store *(Ace 278-0100)*

Transportation
OnSite: Courtesy Car/Van *(shuttle to airport & Richmond)*, Bikes *(Bike Star - kiosk)* **Near:** Taxi *(Richmond 278-8444)*, Local Bus *(Translink 279-0600)* **Under 1 mi:** Rental Car *(Budget 273-5508)* **1-3 mi:** InterCity Bus *(Int'l. Stage Lines 270-6135)* **Airport:** YVR *(0.5 mi.)*

Medical Services
911 Service **OnCall:** Ambulance **1-3 mi:** Doctor *(Continental Medical Clinic 231-8970)*, Dentist *(Continental Dental 207-2127)*, Holistic Services *(Chinese Medicine 273-8284)* **Hospital:** Richmond 278-9711 *(3 mi.)*

Setting -- On the Sea Island side of the Fraser River, the Delta Marina's network of six main piers lies direclty in front of the Delta Vancouver Airport Hotel. The bright red octagonal roof of the Elephant & Castle pub contrasts against the tall, gray, V-shaped hotel; a kidney-shaped pool is tucked in between. Across the river are the bright blue roofs of the Vancouver Marina's boathouses.

Marina Notes -- *20 & 30 amp. The hotel & marina are separate operations but marina guests have access to the hotel pool, hotel heads, room service to the boat and airport check-in kiosks in the lobby. Owner lives on-site. 4-5 liveaboards. Moorage full most of the time, so reserve ahead. Full customs clearance. Hosts Vancouver's largest power charter fleet. A full on-site boatyard operation has a 50-Ton platform synchro lift (70 ft. & 22 ft. capacities) and most other services. The proximity of the airport (0.5 mi.) and the city of Richmond (2 mi.), coupled with onsite BY services, makes this an interesting place to leave the boat.

Notable -- The towering, contemporary 415-room hotel provides boaters with a variety of amenities including two restaurants, a lounge with Wi-Fi, a well-equipped gift shop, ATM, pool, bike rentals, and shuttle service. The mostly glass eight-sided Elephant & Castle Pub and Restaurant sits on a pier smack in the center of the docks, affording expansive views of the river, boats, and bridge. An outside porch, with vinyl shades for inclement weather, and an open deck are furnished with green wrought iron tables and chairs - some topped with umbrellas. The pool, surrounded by white chaises, looks over hedges to the marina (dawn-dusk, May-Oct). Bike or hike the more than 20 kilometers of trails that run along the riverside.

Navigational Information
Lat: 49°11.295' **Long:** 123°07.970' **Tide:** 15 ft. **Current:** 4 kt. **Chart:** 3491
Rep. Depths *(MLW)*: **Entry** 15 ft. **Fuel Dock** 15 ft. **Max Slip/Moor** 25 ft./-
Access: Up Fraser River's North Arm then turn South down Middle Arm

Marina Facilities *(In Season/Off Season)*
Fuel: Gasoline, Diesel, High-Speed Pumps, On Call Delivery
Slips: 160 Total, 5 Transient **Max LOA:** 85 ft. **Max Beam:** 20 ft.
  Rate *(per ft.)*: **Day** $0.60 **Week** $4 **Month** $7.75
  Power: 30 amp 0.10/kw, 50 amp 0.10/kw*, 100 amp n/a, 200 amp n/a
  Cable TV: No **Dockside Phone:** No
  Dock Type: Floating, Long Fingers, Concrete
Moorings: 0 Total, 0 Transient **Launch:** n/a
  Rate: Day n/a **Week** n/a **Month** n/a
Heads: 1 Toilet(s)
Laundry: None **Pay Phones:** Yes
Pump-Out: No **Fee:** n/a **Closed Heads:** No

Marina Operations
Owner/Manager: Michael Short **Dockmaster:** Same
In-Season: Jun-Sep, 7am-7pm **Off-Season:** Oct-May, 7am-5pm
After-Hours Arrival: Call ahead
Reservations: Yes **Credit Cards:** Visa/MC, Dscvr, Din, Amex, Interac
Discounts: Bulk fuel **Dockage:** n/a **Fuel:** n/a **Repair:** n/a
Pets: Welcome **Handicap Access:** Yes, Docks

Vancouver Marina Limited

8331 River Road; Richmond, BC V6X 1Y1

Tel: (604) 278-9787 **VHF: Monitor** Ch. 66A **Talk** n/a
Fax: (604) 278-9785 **Alternate Tel:** n/a
Email: mooring@vancouvermarina.com **Web:** www.vancouvermarina.com
Nearest Town: Richmond *(1 mi.)* **Tourist Info:** (604) 271-8280

Marina Services and Boat Supplies
Services - Docking Assistance, Security, Trash Pick-Up, Dock Carts
Communication - Phone Messages, Data Ports *(Fuel dock - FatPort)*,
FedEx, DHL, UPS **Supplies - OnSite:** Ice *(Block, Cube, Shaved)*, Ships'
Store *(Handy Boating Supply)*, West Marine *(233-2327)*, Bait/Tackle
(frozen) **Near:** Propane *(Autogas 276-9924)*

Boatyard Services
OnSite: Hydraulic Trailer *(40T)*, Engine mechanic *(gas, diesel)*, Launching
Ramp, Electrical Repairs *(Powerwave)*, Electronics Repairs, Hull Repairs,
Rigger, Canvas Work, Bottom Cleaning, Brightwork, Divers *(Alpha Maritime)*,
Compound, Wash & Wax, Interior Cleaning **Near:** Travelift, Painting,
Awlgrip. **Under 1 mi:** Propeller Repairs, Woodworking, Metal Fabrication.
Dealer for: Champion, Harbercraft, Yamaha, Mercruiser.

Restaurants and Accommodations
OnSite: Restaurant *(Boathouse 273-7014, L $9-18, D $9-40, kids' menu)*
Near: Restaurant *(Richmond Sushi 207-7799)*, *(Sea Harbour Seafood 232-0816)*, *(Shanghai Shin Ya 273-2881)*, *(Dem Bones 273-3822)*, *(Sam's Restaurant 821-1421)*, *(Taipei Gourmet 303-6787)*, *(Brooklyn Steakhouse 270-6030)*, *(Mambo Café 273-3669)*, *(Ikku Tel Japanese 276-0205)*, *(Brooklyn Steakhouse 270-6030)*, *(Winsor Palace 303-1938)*, Hotel *(Hampton Inn 888-488-0101, $150-200, 1.5 blocks)*, *(Comfort Inn 800-663-0974, $150-200, adjacent)*, *(Fairmont Vancouver 800-257-7544)*

Recreation and Entertainment
OnSite: Jogging Paths, Roller Blade/Bike Paths, Fishing Charter
  Near: Special Events **1-3 mi:** Pool *(Minoru Aquatic Centre 718-8020)*,
Beach, Picnic Area, Grills, Playground, Tennis Courts *(Richmond Tennis Club 273-3631)*, Golf Course *(Marine Drive Golf Course 261-8111)*, Fitness Center *(Fitness World 278-3831)*, Boat Rentals, Movie Theater *(Famous Players 273-7173)*, Video Rental *(Rogers Video 8446)*, Park *(Minoru Park)*, Cultural Attract *(Gateway Theatre 270-1812)*

Provisioning and General Services
OnSite: Convenience Store **Near:** Market, Wine/Beer *(Richmond Wine 273-5969)*, Bank/ATM, Beauty Salon *(Arcadia Day Spa 278-9566)*, Barber Shop *(Time For Haircuts 207-4266)*, Pharmacy *(Park Pacific 273-9812)*, Hardware Store *(Ace 278-0100)* **Under 1 mi:** Supermarket *(Superstore 233-2435)*, Delicatessen *(Edelweiss 278-4430)*, Health Food *(Gibo 278-8716)*, Liquor Store *(Liquor Depot 233-0354)*, Farmers' Market *(Cambie Community Centre Sun 10am-2pm)*, Green Market *(Orchards 233-1855)*, Fishmonger *(Daily Seafood 207-9268)*, Post Office, Protestant Church, Dry Cleaners *(Busy Bee 276-2665)*, Bookstore *(What A Book Store 303-0383)*, Buying Club *(Costco 668-8450)* **1-3 mi:** Catholic Church, Library *(Richmond 231-6465)*, Laundry *(Easy Kleen 273-6925)*

Transportation
OnCall: Water Taxi *($150/hr.)*, Taxi *(Richmond Cab 278-8444)*
Near: Rental Car *(Budget 273-5508)*, Local Bus *(Translink 279-0600)*
1-3 mi: InterCity Bus *(Pacific Lines 278-3772)* **Airport:** YVR *(1.5 mi.)*

Medical Services
911 Service **OnCall:** Ambulance **Under 1 mi:** Doctor *(Continental Medical Clinic 231-8970)*, Dentist *(Continental Dental 207-2127)*, Holistic Services *(Pacific Breeze Spa 247-8503)* **Hospital:** Richmond 278-9711 *(2 mi.)*

Setting -- Vancouver Marina's rows of shipshape bright blue boathouses and open slips sprawl along the southern shore of the Fraser River's Middle Arm. The marina's three sections are divided by the Boathouse restaurant and the Sea Island Route 99 Bridge. At the head of the main fairway is the flower-bedecked, 60-foot fuel dock backed by the office and convenience store. Across the river the wedge-shaped Delta Airport Hotel rises above the Delta Marina docks that flank the Elephant & Castle Pub, with its bright red, octagonal roof.

Marina Notes -- *Metered power - 15-80 amps. Haul-out via a 40-Ton hydraulic lift - only for power boats. Stable concrete docks. On-site are Handy Boating Supply Store and West Marine. State-of-the-art surveillance systems - can even watch from home. Airport Y.C. makes its home here - in front of the Boathouse restaurant. One washroom in the office which is only available during office hours; there are no showers or laundry.

Notable -- The Boathouse restaurant is a big waterfront seafood eatery, part of a small BC chain, with bright blue awnings, and inside or outside dining. (Monday-Friday 11:30am-10pm, Fridays-Saturdays 4-10:30pm, Sundays 4-9:30pm). The surrounding neighborhood is heavily commercial and largely Asian. A number of strip malls are an easy walk and a Canadian Tire (with big parts & services sections) three blocks. A three-minute walk is a bus system that makes all Vancouver accessible. And another mile upriver is the nine-acre River Rock Casino Resort - the largest in Western Canada with 9 restaurants and lounges, a full service spa, over 200 hotel suites, indoor pool, 950-seat theater, 7,000 square foot casino, and, soon, a 144-berth marina (247-8900).

PHOTOS ON DVD: 16

Captain's Cove Marina

6100 Ferry Road; Delta, BC V4K 3M9

Tel: (604) 946-1294 **VHF: Monitor** Ch. 16 **Talk** Ch. 72
Fax: (604) 946-1273 **Alternate Tel:** n/a
Email: info@captainscovemarina.ca **Web:** www.captainscovemarina.ca
Nearest Town: Ladner *(1 mi.)* **Tourist Info:** (604) 964-4232

Navigational Information
Lat: 49°06.807' **Long:** 123°04.414' **Tide:** 12 ft. **Current:** 0 kt. **Chart:** 3491
Rep. Depths *(MLW)*: **Entry** 8 ft. **Fuel Dock** 8 ft. **Max Slip/Moor** 15 ft./-
Access: Georgia Harbour to Fraser River to Deas Slough

Marina Facilities *(In Season/Off Season)*
Fuel: *Esso* - Gasoline, Diesel
Slips: 250 Total, 20 Transient **Max LOA:** 60 ft. **Max Beam:** 22 ft.
Rate *(per ft.)*: **Day** $0.75* **Week** n/a **Month** $7.75
Power: 30 amp $0.10/mwh, **50 amp** $0.10/mwh, **100 amp** n/a, **200 amp** n/a
Cable TV: No **Dockside Phone:** No
Dock Type: Floating, Long Fingers, Wood
Moorings: 0 Total, 0 Transient **Launch:** n/a
Rate: Day n/a **Week** n/a **Month** n/a
Heads: 2 Toilet(s), 2 Shower(s)
Laundry: 1 Washer(s), 1 Dryer(s) **Pay Phones:** No
Pump-Out: OnSite, 20 InSlip **Fee:** Inq. **Closed Heads:** Yes

Marina Operations
Owner/Manager: Elizabeth Modal **Dockmaster:** Ray Mann
In-Season: Year-Round, 8am-5pm **Off-Season:** n/a
After-Hours Arrival: Check in at the Rusty Anchor Pub
Reservations: No **Credit Cards:** Visa/MC, Dscvr, Amex
Discounts: Fuel **Dockage:** n/a **Fuel:** 5% over 500 **Repair:** n/a
Pets: Welcome, Dog Walk Area **Handicap Access:** No

Marina Services and Boat Supplies
Services - Docking Assistance, Security *(24 hrs., manager onsite)*, Trash Pick-Up, Dock Carts **Communication -** FedEx, DHL, UPS **Supplies - Under 1 mi:** Bait/Tackle, Propane **1-3 mi:** Ships' Store *(Massey's Marine Supply 946-4488)* **3+ mi:** West Marine *(233-2327, 10 mi.)*

Boatyard Services
OnSite: Travelift *(60T)* **OnCall:** Engine mechanic *(gas, diesel)*, Launching Ramp, Electrical Repairs, Hull Repairs, Rigger, Canvas Work, Bottom Cleaning, Brightwork, Compound, Wash & Wax, Interior Cleaning, Propeller Repairs, Woodworking, Inflatable Repairs **Under 1 mi:** Electronics Repairs, Sail Loft, Air Conditioning, Refrigeration, Divers, Life Raft Service, Metal Fabrication, Painting, Awlgrip. **Yard Rates:** Haul & Launch $5.75/ft. *(blocking incl.)*, Power Wash $2/ft. **Storage:** In-Water $5.25/ft./mo.

Restaurants and Accommodations
OnSite: Restaurant *(Rusty Anchor L $8-12, D $8-12, 11am-11pm, 7 days in season, burgers, sandwiches & entrees)* **1-3 mi:** Restaurant *(Niagara Pizza 946-1522)*, *(Sharkey's Seafood Bar & Grille 946-7793)*, *(Uncle Herbert's Fish & Chips 946-8222)*, *(Go-Go Sushi & Café 940-3323)*, *(La Strada Ristorante 946-2535)*, *(La Belle Auberge 946-7717, D $33, Tue-Sun from 6pm)*, *(Ladner Sushi 946-7781)*, Fast Food *(Subway, KFC)*, Pizzeria *(Panago 940-5850)*, Motel *(Town & Country 946-4404)*, Inn/B&B *(The Duck Inn 946-7521, $100-210, waterfront)*, *(Blue Heron Inn 946-2754)*, *(Our House B&B 946-2628)*

Recreation and Entertainment
OnSite: Golf Course *(Coves Links 946-1839, 9 holes $19-22 Clubs $13, Cart $4)*, Jogging Paths *(Millennium Trail - along the Fraser past Deas Island)* **OnCall:** Fishing Charter, Group Fishing Boat, Sightseeing *(river tours)* **Near:** Picnic Area, Grills, Playground, Park *(Ladner Harbour - commercial moorage plus a nature preserve)* **Under 1 mi:** Pool, Dive Shop **1-3 mi:** Tennis Courts *(Ladner T.C. 946-8114)*, Video Rental *(Videoland 946-5558)*

Provisioning and General Services
OnSite: Wine/Beer, Bank/ATM **Near:** Convenience Store, Fishmonger *(off the boats in Ladner Harbour - or in Ladner at Superior 946-2097)*, Catholic Church, Protestant Church, Newsstand, Copies Etc. **1-3 mi:** Market *(Ladner Grocery 946-8938)*, Supermarket *(Safeway 940-0053, Save-On-Foods 946-5251)*, Gourmet Shop, Delicatessen *(Lux's 946-2989)*, Health Food *(Parsley, Sage & Thyme 946-1022)*, Liquor Store *(Speed's 940-6463)*, Bakery *(Trenant Park Bakery & Cafe 940-1659)*, Farmers' Market *(Ladner 48th Ave. 2nd & 4th Sun, 10am-2pm 946-8590)*, Post Office *(946-4014)*, Library *(Ladner 946-6215)*, Beauty Salon *(Changes 940-6337)*, Barber Shop *(Delta 946-8726)*, Dry Cleaners *(Harbourside 946-1828)*, Laundry, Bookstore *(Bryan's 946-2678)*, Pharmacy, Hardware Store *(Ladner Village 946-4833)*, Florist, Clothing Store, Retail Shops

Transportation
OnSite: Local Bus **Under 1 mi:** InterCity Bus *(Coastal)* **3+ mi:** Rental Car *(National 231-1670, 10 mi.)*, Ferry Service *(Victoria, 3 mi.)*
Airport: YVR *(5 mi.)*

Medical Services
911 Service **OnCall:** Ambulance **1-3 mi:** Doctor *(Delta Medical 946-7661)*, Dentist *(A Smile Clinic 946-8555)*, Veterinarian *(Ladner Animal 946-9567)*
Hospital: Delta 964-1121 *(3 mi.)*

Setting -- Eight miles up the Fraser River, tucked into Deas Slough, ten main piers edge 217 acres of riverfront. Captain's Cove is part of a large-scale environmentally-sensitive development that includes a nine-hole golf course, hundreds of dwellings in discrete neighborhoods and the Rusty Anchor Pub. Across the water, Deas Island Regional Park is flanked by the Fraser and Deas Slough. Picnic areas abound and hiking trails crisscross the peninsula.

Marina Notes -- Established 1978. *Multi-hull rate 50% up charge. Fuel dock 9am-5pm, 7 days. Boatyard provides haul and launch only - all other services are on-call or do-it-yourself. The dockmaster has a list of suggested service providers. On-land or in-water storage rates are the same as moorage rates. The attractive wood-sided laundry room leads to two bathhouses - men & women. Lovely full bathrooms feature stainless steel sinks, composite walls, nautically inspired light fixtures and fiberglass shower stalls with glass doors and large dressing areas.

Notable -- Two-story Rusty Anchor Pub is adjacent to the haul-out basin. It's very cozy, with ship's lanterns hanging from the ceiling and walls of windows looking out across the Slough - indoor dining room, sports bar with dart boards and an outside covered deck. Since 1985, Captain's Cove has been developing the 217 acres that surround the docks and Deas Slough with deliberateness and environmental awareness. The master plan anticipates 1,000 residential units of which 500 are in place. The phases currently underway add another 100 houses. Adjacent is Ladner Harbour Marina & Park for biking, hiking, picnicking, bald eagle watching and commercial moorage. A bus runs into Ladner, a little over a mile away, and most everything necessary is there.

4. BC – Comox to Nanaimo

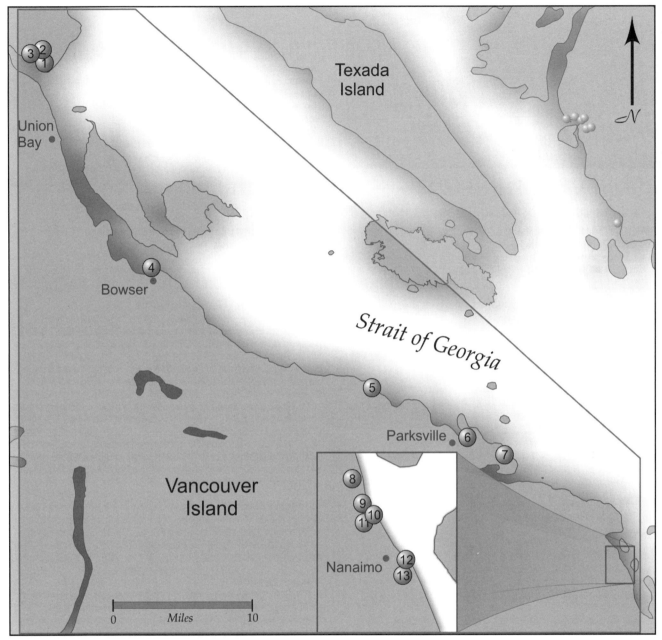

▸ **Currency —** In Canadian Marina Reports, all prices are in Canadian dollars. In U.S. Marina Reports, all prices are in U.S. dollars.

▸ **"CCM" —** Denotes a Certified Clean Marina, a state/provincial award for environmental excellence. See page 298 for an explanation and page 299 for a list of Pump-Out facilities.

▸ **Ratings & Reviews —** An overview of the Atlantic Cruising Club's rating system is on page 6 and details on the content of each Marina Report are on pages 7 – 11.

▸ **Marina Report Updates —** Comments from boaters and new information from ACC reviewers and marinas are posted regularly on www.AtlanticCruisingClub.com.

Navigational Information
Lat: 49°40.175' **Long:** 124°55.510' **Tide:** 17 ft. **Current:** n/a **Chart:** 3527
Rep. Depths *(MLW)*: Entry 15 ft. **Fuel Dock** n/a **Max Slip/Moor** 10 ft./-
Access: Off the Strait of Georgia north of Denman Island to Comox Harbour

Marina Facilities *(In Season/Off Season)*
Fuel: No
Slips: 200 Total, 35 Transient **Max LOA:** 150 ft. **Max Beam:** n/a
  Rate *(per ft.)*: **Day** $0.85 **Week** $4.37 **Month** n/a
  Power: 30 amp $7*, **50 amp** n/a, **100 amp** n/a, **200 amp** n/a
  Cable TV: No **Dockside Phone:** No
  Dock Type: Floating, Alongside, Wood
Moorings: 0 Total, 0 Transient **Launch:** n/a
  Rate: Day n/a **Week** n/a **Month** n/a
Heads: 7 Toilet(s), 5 Shower(s) *(with dressing rooms)*
Laundry: 2 Washer(s), 2 Dryer(s) **Pay Phones:** Yes, 2
Pump-Out: OnSite, Full Service, 1 Central **Fee:** $7 **Closed Heads:** No

Marina Operations
Owner/Manager: Elizabeth McLeod **Dockmaster:** Mo Nordstrom
In-Season: Jun-Aug, 9am-8pm **Off-Season:** Sep-May, 9am-5pm
After-Hours Arrival: Courtesy forms on the dock
Reservations: No **Credit Cards:** Visa/MC, Debit
Discounts: None
Pets: Welcome, Dog Walk Area **Handicap Access:** Yes, Heads, Docks

Comox Valley Harbour Authority

121 Port Augusta Street; Comox, BC V9M 3N8

Tel: (250) 339-6041 **VHF:** Monitor Ch. 66A **Talk** n/a
Fax: (250) 338-5325 **Alternate Tel:** n/a
Email: info@comoxfishermanswharf.com **Web:** comoxfishermanswharf.com
Nearest Town: Comox **Tourist Info:** (250) 334-3234

Marina Services and Boat Supplies
Services - Docking Assistance, Boaters' Lounge, Trash Pick-Up, Dock Carts **Communication -** Mail & Package Hold, Phone Messages, Fax in/out, Data Ports *(Office & Wi-Fi - Broadband Xpress)*, FedEx **Supplies - OnSite:** Ice *(Block, Cube)* **Near:** Ships' Store *(Crowsnest Chandlery 339-3676, Ted's 339-4942)*, CNG

Boatyard Services
OnSite: Forklift, Crane **Nearest Yard:** Nautech Industries (250) 338-0551

Restaurants and Accommodations
OnCall: Pizzeria *(Boston Pizza 334-2222, L $7-10, D $12-19)* **Near:** Restaurant *(Smitty's 339-3911, B $6-10, L $8-10, D $8-17)*, *(Edgewater Pub & Bistro 339-6151, B $8-11, L $5-18, D $9-21, dining deck, live entertainment Wed-Sat)*, *(Black Fin Pub 339-5030, L $8-14, D $10-23)*, Fast Food *(Subway)* **Under 1 mi:** Restaurant *(The Bamboo Inn 339-3500, L $7-12, D $11-23, delivers after 5pm)* **1-3 mi:** Inn/B&B *(Cope's Oceanfront B&B 339-1038, $65-110)*, *(Foskett House 339-4272, $60-110)*, *(Port Augusta Inn & Suites 339-2277, $55-140)*

Recreation and Entertainment
OnSite: Picnic Area, Grills, Playground **Near:** Beach *(Goose Spit)*, Video Rental *(Comox Videos 'n More 339-0112)*, Park *(Filberg Park)*, Museum *(Courtenay & District Museum 334-0888)*, Sightseeing *(Desolation Sound Discovery Tours 339-4914)*, Special Events *(Nautical Days - Aug, Father's Day Kite Fly, Canada Day)* **Under 1 mi:** Golf Course *(Comox Golf Course 339-4444)*, Cultural Attract *(Filberg Festival in Aug)*, Galleries *(Glass Expressions 339-7739)* **1-3 mi:** Fishing Charter *(G & M 800-577-6966)*

Provisioning and General Services
OnSite: Fishmonger *(fish sales area - buy direct from the fishermen)* **Near:** Market *(Extra Foods)*, Liquor Store *(Liquor Store 334-1335)*, Bakery *(Mcgavin's Bread Basket 338-6112)*, Meat Market *(Middleton Meats 339-5573)*, Bank/ATM *(CIBC 890-6820)*, Post Office, Library *(Comox 339-2971)*, Beauty Salon *(Anna's Coiffures 339-3944)*, Barber Shop *(Marty's Barber Shop and Art Studio 339-3395)*, Bookstore *(Blue Heron Books 339-6111)*, Pharmacy *(Rexall Drug 339-2235)*, Hardware Store *(Comox Hardware 339-2911)*, Florist *(Comox Valley Flowers 339-4141)*, Clothing Store *(Roxanne's Fashions Limited 339-6135)*, Retail Shops *(Comox Centre Mall 339-7344)* **Under 1 mi:** Convenience Store *(7-Eleven 334-3351)*, Supermarket *(Quality Foods 890-1005 - call before 10am for same day delivery to the boat)*, Health Food *(New Leaf Whole Foods 339-5911)* **3+ mi:** Farmers' Market *(Wed & Sat 9am-Noon in Courtenay, 334-1932, 4 mi.)*, Dry Cleaners *(Pressed for Time 334-3606, 4 mi.)*, Copies Etc. *(Staples 334-8357, 4 mi.)*

Transportation
OnCall: Taxi *(United Cab 339-7955)* **Near:** Local Bus *(BC Transit 339-5453)* **Under 1 mi:** Airport Limo *(Lady Driver 339-0606)* **1-3 mi:** Rental Car *(National 334-0202)*, Ferry Service *(888-724-5223)* **Airport:** Comox Valley *(6 mi.)*

Medical Services
911 Service **OnCall:** Ambulance *(897-1098)* **Near:** Doctor *(Comox Medical Clinic 339-2266)*, Dentist *(Wall 339-4044)*, Chiropractor *(Backworks 339-1148)* **Under 1 mi:** Veterinarian *(Shamrock Veterinary Clinic 339-2026)* **Hospital:** St. Joseph 339-2242 *(0.25 mi.)*

Setting -- The CVHA, also known as Comox Fisherman's Wharf is the first of four marinas inside the main harbor behind a floating metal breakwater. Homeport to the Comox fishing fleet, pleasure craft happily berth next to colorful gill netters and crabbers. The new two-story contemporary office, with its bright blue roof and signature burgundy trim, oversees the three long piers, with side-tie moorage, that are open to the public ("D," "F" & "H"). To the east, Goose Spit provides additional shelter and just to the north, via a wide flower-bedecked promenade, is a large public park with a picnic area and playground.

Marina Notes -- *15 and 20 amp $3.50. 30 amp is $7 and is limited to "H" dock. Hail the dockmaster for local knowledge and tie-up assistance. As you enter through the east side, "D" dock is the closest to shore, and "H" is the closest to the breakwater. Moorage may be available at the end of "F" dock as well. Very accommodating staff will make your stay pleasant. No anchoring is allowed outside the breakwater. A pump-out station is located at the head of the dock. The new marina office boasts internet service, a boaters' lounge, visitors information, plus new heads and showers.

Notable -- Several good restaurants, a few coffee shops, two chandleries, liquor stores, grocery stores, boutiques, galleries, and a golf course are within easy walking distance. Visitors and locals throng the docks to purchase fresh seafood right off the boats. Check the sign to see what is available - it sells out fast! Annual festivals include a Father's Day Kite Fly in June, Canada Day Celebrations on the First of July, and, in August, Comox Nautical Days (with "Build, Bail & Sail Your Own Creation," a parade, and Ceremony of the Flags) and the 4-day Filberg Festival, with arts, crafts, entertainment, concessions, and guest artists.

Navigational Information
Lat: 49°40.209' **Long:** 124°55.662' **Tide:** 17 ft. **Current:** n/a **Chart:** 3527
Rep. Depths (*MLW*): **Entry** 15 ft. **Fuel Dock** 9 ft. **Max Slip/Moor** 9 ft./-
Access: Off the Strait of Georgia north of Denman Island to Comox Harbour

Marina Facilities (*In Season/Off Season*)
Fuel: Gasoline, Diesel
Slips: 18 Total, 2 Transient **Max LOA:** 45 ft. **Max Beam:** n/a
  Rate (*per ft.*): **Day** $0.80 **Week** n/a **Month** n/a
  Power: 30 amp Incl., **50 amp** n/a, **100 amp** n/a, **200 amp** n/a
  Cable TV: No **Dockside Phone:** Yes
  Dock Type: Floating, Long Fingers, Alongside, Concrete, Wood
Moorings: 0 Total, 0 Transient **Launch:** n/a
  Rate: Day n/a **Week** n/a **Month** n/a
Heads: 4 Toilet(s), 2 Shower(s)
Laundry: 2 Washer(s), 2 Dryer(s) **Pay Phones:** Yes
Pump-Out: No **Fee:** n/a **Closed Heads:** No

Marina Operations
Owner/Manager: Joan Benda **Dockmaster:** Same
In-Season: Jun-Oct, 7am-9pm **Off-Season:** Nov-May, 8:30am-4:30pm
After-Hours Arrival: Call ahead
Reservations: Yes, Preferred **Credit Cards:** Visa/MC, Debit
Discounts: None
Pets: Welcome **Handicap Access:** No

Gas N Go Marina

PO Box 1296; 132 Port Augusta Street; Comox, BC V9M 7Z8

Tel: (250) 339-4664 **VHF: Monitor** Ch. 66A **Talk** n/a
Fax: (250) 339-4664 **Alternate Tel:** n/a
Email: n/a **Web:** n/a
Nearest Town: Comox **Tourist Info:** (250) 334-3234

Marina Services and Boat Supplies
Services - Docking Assistance, Security (*night patrol*), Trash Pick-Up
Communication - FedEx, UPS, Express Mail **Supplies - OnSite:** Ice
(*Cube*), Ships' Store

Boatyard Services
OnSite: Launching Ramp **Nearest Yard:** Nautech (250) 338-0551

Restaurants and Accommodations
OnSite: Restaurant (*Black Fin Pub 339-5030, L $8-14, D $10-23, burgers to fine dining*) **OnCall:** Pizzeria (*Boston Pizza 334-2222, L $7-10, D $12-19*)
Near: Restaurant (*Smitty's 339-3911, B $6-10, L $8-10, D $8-17*), (*Edgewater Pub & Bistro 339-6151, B $8-11, L $5-18, D $9-21, dining deck, live entertainment Wed-Sat*), Coffee Shop (*Comox Grind 339-2225*)
Under 1 mi: Restaurant (*The Bamboo Inn 339-3500, L $7-12, D $11-23, delivery from 5pm*) **1-3 mi:** Inn/B&B (*Alpine House 339-6181, $70-100*), (*Alan & Shirley Robb's Levenvale 339-3307, $55-70*), (*Port Augusta Inn 339-2277, $55-140*), Condo/Cottage (*Kairos Guest Suite 339-6573, $75-90*)

Recreation and Entertainment
Near: Playground, Golf Course (*Comox Golf Course 339-4444*), Jogging Paths, Video Rental (*Select Video 339-0112*), Park (*Filberg Heritage Lodge*), Sightseeing (*Desolation Sound Discovery Tours 339-4914*), Special Events (*Comox Nautical Days - Aug, Filberg Festival - Aug*), Galleries (*Artisans Courtyard 338-6564, Potters Place 334-4613, Muir Gallery 334-2983, Comox Valley Art Gallery 338-6211*) **Under 1 mi:** Tennis Courts, Cultural Attract
1-3 mi: Museum (*Comox Airforce Museum 339-8162, donations*)
3+ mi: Boat Rentals (*King Coho Resort and Boat Rentals 339-2039, 4 mi.*)

Provisioning and General Services
OnSite: Convenience Store (*Opened summer 2006*) **Near:** Health Food (*New Leaf 339-5911*), Bakery (*Strand 339-4322*), Fishmonger, Bank/ATM (*CIBC 890-6820*), Post Office, Library (*Comox 339-2971*), Beauty Salon (*Bellini Hair Studio 339-5150*), Barber Shop (*Marty's Barber Shop & Art Studio 339-3395*), Bookstore (*Blue Heron Books 339-6111*), Pharmacy (*Medicine Shoppe 339-5050*), Hardware Store (*Comox Hardware 339-2911*), Florist (*Comox Valley Flowers 339-4141*), Clothing Store (*Roxanne's Fashions 339-6133*) **Under 1 mi:** Supermarket (*Quality Foods 890-1005, call before 10am for same day delivery to the boat*), Liquor Store (*Liquor Store 334-1335*), Protestant Church, Retail Shops (*Comox Centre Mall 339-7344*) **3+ mi:** Farmers' Market (*Wed & Sat 9am-Noon in Courtenay, 334-1932, 4 mi.*), Dry Cleaners (*Pressed for Time 334-3606, 4 mi.*), Copies Etc. (*Staples 334-8357, 4 mi.*)

Transportation
OnCall: Taxi (*Designated Drivers 339-0997*) **Near:** Local Bus (*BC Transit 339-5453, to Courtenay*) **1-3 mi:** Rail (*E & N Dayliner 888-842-7245*)
3+ mi: Rental Car (*National 334-0202/Budget 338-7717, 3 mi./4 mi.*)
Airport: Comox Valley (*6 mi.*)

Medical Services
911 Service **OnCall:** Ambulance (*897-1098*) **Near:** Doctor (*Comox Medical Clinic 339-2266*), Dentist (*Fraser 339-2531*), Chiropractor (*Backworks 339-1148*) **1-3 mi:** Veterinarian (*Shamrock Veterinary Clinic 339-2026*)
Hospital: St. Joseph 339-2242 (*2 mi.*)

Setting -- Pass the breakwater and the Harbour Authority transient docks, then follow the channel. The Gas N Go fuel dock is straight ahead with the marina office at the end. The pier behind it berths several long slips on one side, with side-tie space on the other. The Comox Municipal Marina, which does not accept transients, is directly adjacent making the marina appear to be much larger than it is. Edgewater Pub & Bistro dominates the view toward shore.

Marina Notes -- Formerly the Black Fin Marina. This is the only fuel dock in the harbor with both gas & diesel. Call ahead - mostly permanent moorage, but transient spots available in summer. The marina office doubles as a small store with some marine supplies, cold drinks, snacks, and ice cream. Above the docks, the nautically inspired, upscale, Black Fin Pub features long harbor views from its dining deck - a Whitehall boat is suspended from its ceiling. Public restrooms and courtesy showers are at the top of the dock and to the right. A laundry is on the same level as Edgewater Pub & Bistro to the left.

Notable -- Comox Harbour, where pleasure boats share space with fishing vessels of all sizes, contributes to the town's authentic atmosphere. Restaurants & provisioning stops are within easy reach. The marine park next door has picnic tables, a gazebo, and a large playground. The Comox Air Force Museum (year-round, 10am-4pm daily) offers displays interpreting the history of Canadian military aviation intertwined with the history of Comox, plus a research library, gift shop, and an exhibit of numerous aircraft and military vehicles that have seen service. Comox Valley is alive with artists. In addition to the Filbert Gallery in Comox, nearby Courtenay is home to many galleries that are open year-round.

Comox Bay Marina

1805 Beaufort Avenue; Comox, BC V9N 1R9

Tel: (250) 339-2930 **VHF: Monitor** Ch. 66A **Talk** n/a
Fax: (250) 339-2930 **Alternate Tel:** n/a
Email: n/a **Web:** www.comoxbaymarina.com
Nearest Town: Comox **Tourist Info:** (250) 334-3234

Navigational Information
Lat: 49°40.208' **Long:** 124°55.740' **Tide:** 17 ft. **Current:** n/a **Chart:** 3527
Rep. Depths *(MLW):* **Entry** 15 ft. **Fuel Dock** n/a **Max Slip/Moor** 9 ft./-
Access: Off the Strait of Georgia north of Denman Island to Comox Harbour

Marina Facilities *(In Season/Off Season)*
Fuel: No
Slips: 247 Total, 7 Transient **Max LOA:** 135 ft. **Max Beam:** n/a
  Rate *(per ft.):* **Day** $0.75/0.50 **Week** n/a **Month** n/a
  Power: 30 amp $3*, **50 amp** $5, **100 amp** n/a, **200 amp** n/a
  Cable TV: No **Dockside Phone:** No
  Dock Type: Floating, Long Fingers, Alongside, Concrete, Wood
Moorings: 0 Total, 0 Transient **Launch:** n/a
  Rate: Day n/a **Week** n/a **Month** n/a
Heads: 2 Toilet(s), 2 Shower(s)
Laundry: 2 Washer(s), 2 Dryer(s) **Pay Phones:** Yes, 1
Pump-Out: No **Fee:** n/a **Closed Heads:** No

Marina Operations
Owner/Manager: Comox Bay Investments, Brad Jenkins **Dockmaster:** n/a
In-Season: Year-Round, 8am-4pm **Off-Season:** n/a
After-Hours Arrival: Call ahead
Reservations: Yes, Preferred **Credit Cards:** Visa/MC
Discounts: None
Pets: Welcome **Handicap Access:** Yes, Heads, Docks

Marina Services and Boat Supplies
Services - Docking Assistance, Security *(locked gates)*, Trash Pick-Up, Dock Carts **Communication -** Mail & Package Hold, Phone Messages, FedEx, UPS, Express Mail **Supplies - Near:** Ice *(Cube)*, Ships' Store *(Crowsnest Chandlery 339-3676, Ted's 339-4942)*

Boatyard Services
Near: Launching Ramp. **Nearest Yard:** Nautech (250) 338-0551

Restaurants and Accommodations
OnSite: Restaurant *(Edgewater Pub & Bistro 339-6151, B $8-11, L $5-18, D $9-21, patio dining, live entertainment Wed-Sat)* **OnCall:** Pizzeria *(Boston Pizza 334-2222, L $7-10, D $12-19)* **Near:** Restaurant *(Smitty's 339-3911, B $6-11, L $8-10, D $8-17)*, *(Black Fin Pub 339-5030, L $8-14, D $10-23)*, Coffee Shop *(Comox Grind 339-2225)*, Fast Food *(Subway)* **Under 1 mi:** Restaurant *(The Bamboo Inn 339-3500, L $7-12, D $11-23, delivery after 5pm)* **1-3 mi:** Inn/B&B *(Port Angusta Inn & Suites 339-2277, $55-140)*, *(Foskett House 339-4272, $60-110)*, *(Cope's Oceanfront B&B 339-1038, $65-110)* **3+ mi:** Restaurant *(Sandbar Grill 339-5570, 1 mi.)*

Recreation and Entertainment
OnSite: Picnic Area, Grills, Playground **Near:** Video Rental *(Select Video 339-0112)*, Park *(Filberg Park)*, Sightseeing *(Desolation Sound Discovery Tours 339-4914)*, Special Events *(Comox Nautical Days - Aug, Filberg Festival - Aug)* **Under 1 mi:** Golf Course *(Comox Golf Course 339-4444)*, Fitness Center *(Perfect Shapes 890-7519)*, Museum *(Comox Archives - maritime heritage; Tue-Sat 1-4pm, admission by donation 339-2885)*, Galleries *(Glass Expressions 339-7739)*

1-3 mi: Group Fishing Boat *(G & M Charters 800-577-6966)*

Provisioning and General Services
Near: Health Food *(New Leaf 339-5911)*, Bakery *(Strand 339-4322)*, Fishmonger *(off the boats 339-6041)*, Protestant Church, Pharmacy *(Rexall Drug 339-2235)*, Hardware Store *(Comox Hardware 339-2911)*
Under 1 mi: Supermarket *(Quality Foods 890-1005; call before 10am for same day delivery to the boat)*, Wine/Beer, Liquor Store *(334-1335)*, Bank/ATM *(CIBC 890-6820)*, Post Office, Library *(Comox 339-2971)*, Beauty Salon *(Anna's Coiffures 339-3944)*, Barber Shop *(Marty's Barber Shop and Art Studio 339-3395)*, Bookstore *(Blue Heron Books 339-6111)*, Florist *(Comox Valley Flowers 339-4141)*, Clothing Store, Retail Shops *(Comox Centre Mall 339-7344)* **3+ mi:** Convenience Store *(7-Eleven 334-3351, 4 mi.)*, Farmers' Market *(Courtenay 9am-Noon Wed & Sat, 334-1932, 4 mi.)*, Dry Cleaners *(Pressed for Time 334-3606, 4 mi.)*

Transportation
OnCall: Taxi *(Designated Drivers 339-0997)* **Near:** Local Bus *(to Courtenay 339-5453)* **Under 1 mi:** Bikes *(Simon's Cycles 339-6683)*, Airport Limo *(Lady Driver 339-0606)* **1-3 mi:** Rental Car *(National 334-0202)*, Ferry Service *(888-724-5223)* **Airport:** Comox Valley *(6 mi.)*

Medical Services
911 Service **OnCall:** Ambulance *(897-1098)* **Near:** Doctor *(Comox Medical Clinic 339-2266)*, Dentist *(Wall 339-4044)* **Under 1 mi:** Chiropractor *(Backworks 339-1148)*, Veterinarian *(Shamrock Veterinary Clinic 339-2026)*
Hospital: St. Joseph 339-2242 *(2 mi.)*

Setting -- Good breakwaters make small Port Augusta - popularly known as Comox Harbour - a well-protected, comfortable refuge. Pleasant grounds and parks back up four marinas - three with transient dockage. It's a beautiful location with the usually snow-capped Beaufort Mountains, home of the Comox Glacier, to the west, the Strait of Georgia to the east, and the village of Comox climbing the hill to the north. At the top of the Comox Bay Marina ramp, Edgewater Grill & Bistro promises gorgeous views from a panoramic window wall and dining deck -- plus a calendar of live entertainment.

Marina Notes -- *15 amp. power is also available. The dock and marina were renovated during 2004. Most moorage is permanent, so call ahead for availability. Despite the tremendous tides, the marina can accommodate up to 9 ft. drafts even at 0 tide. Tidal grids can be used at your own risk. The marina office is on the dock to the left as you clear the breakwater. Float plane operations are at the west end of the marina. Heads, showers, and laundry are at the top of the ramp. Be sure to carry some loonies ($1 Canadian coins) to operate the machines and showers.

Notable -- The Filberg Heritage Lodge and Park is only a short walk from the marina and well worth the stroll. Located on the beachfront east of the harbor, this 1930s era property is maintained in its original state and serves as home to one of the premier art events of the region. Each August, many of Canada's best artists display their work at the Filberg Festival. Tours are available in the summer - $3 for the lodge, $10 lodge & gardens, $15 with a stop at the teahouse. This is an area that loves festivals and has events nearly every weekend. The 9-hole Comox Golf Course is fun and within walking distance.

Deep Bay Harbour Authority

Navigational Information
Lat: 49°27.854' **Long:** 124°43.613' **Tide:** 15 ft. **Current:** n/a **Chart:** 3527
Rep. Depths (*MLW*): **Entry** 15 ft. **Fuel Dock** 10 ft. **Max Slip/Moor** 10 ft./-
Access: Strait of Georgia south of Denman Isl,. SE end of Baynes Sound

Marina Facilities (*In Season/Off Season*)
Fuel: Gasoline
Slips: 30 Total, 30* Transient **Max LOA:** 40 ft. **Max Beam:** n/a
 Rate (*per ft.*): **Day** $0.45 **Week** n/a **Month** $2.84
 Power: 30 amp $3.50**, **50 amp** n/a, **100 amp** n/a, **200 amp** n/a
 Cable TV: No **Dockside Phone:** Yes
 Dock Type: Floating, Long Fingers, Alongside, Wood
Moorings: 0 Total, 0 Transient **Launch:** n/a
 Rate: Day n/a **Week** n/a **Month** n/a
Heads: 4 Toilet(s), 1 Shower(s)
Laundry: 1 Washer(s), 1 Dryer(s) **Pay Phones:** Yes, 2
Pump-Out: Onsite ($7), OnCall, 1 Port **Fee:** n/a **Closed Heads:** No

Marina Operations
Owner/Manager: Deep Bay HA **Dockmaster:** Darrel Alexander
In-Season: Jun-Sep, 9am-9pm **Off-Season:** Oct-May, 10am-4pm
After-Hours Arrival: Call ahead
Reservations: No **Credit Cards:** Visa/MC, Amex, Debit
Discounts: None
Pets: Welcome **Handicap Access:** Yes

Deep Bay Harbour Authority

RR #1, Site 160C-10, 180 Burne Rd.; Bowser, BC V0R 1G0

Tel: (250) 757-9331 **VHF: Monitor** Ch. 66A **Talk** n/a
Fax: (250) 757-9319 **Alternate Tel:** n/a
Email: deepbay-mgr@shawcable.com **Web:** n/a
Nearest Town: Bowser (*1.5 mi.*) **Tourist Info:** (250) 752-9532

Marina Services and Boat Supplies
Services - Trash Pick-Up **Communication -** FedEx, UPS, Express Mail
Supplies - OnSite: Ice (*Cube*), Ships' Store (*Ship & Shore 757-8399*),
Bait/Tackle, Live Bait

Boatyard Services
OnSite: Launching Ramp (*at Ship & Shore*) **Nearest Yard:** Nautech
Industries (250) 338-0551

Restaurants and Accommodations
OnSite: Lite Fare (*Ship & Shore Café 757-8399, L $6-13, D $6-13, fish &
chips, burgers*) **1-3 mi:** Restaurant (*Henry's Kitchen 757-8288, L $6-19, D
$6-19*), Inn/B&B (*Lamplighter Cottage B&B 757-9394, $1200-1600/wk.*),
(*Shady Shores Beach Resort 757-8595, $80-130*), (*Mapleguard Resort 757-
9211, $70-90*)

Recreation and Entertainment
OnSite: Boat Rentals (*kayaks*), Fishing Charter (*Seaside Charters starting at
$30 757-2000*) **Near:** Group Fishing Boat (*G & M Fishing Charters

954-6500) **1-3 mi:** Video Rental (*Video Showcase 757-8353*), Park
(*Rosewall Creek, Wildwood Park*) **3+ mi:** Golf Course (*Arrowsmith Golf and
Country Club 752-9727, 10 mi.*), Horseback Riding (*10 mi.*), Sightseeing
(*Horne Lake Caves Park, info. 757-8687, reservations 248-7829, 14 mi.*)

Provisioning and General Services
OnSite: Convenience Store (*Ship & Shore 757-8399*) **1-3 mi:** Market
(*Tom's Food Village 757-8944*), Delicatessen (*Georgia Park Store 757-
8386*), Wine/Beer, Post Office, Protestant Church (*Wildwood Community
Church 757-8136*)

Transportation
3+ mi: Rental Car (*Budget 248-5341, 22 mi.*) **Airport:** Qualicum
Municipal (*15 mi.*)

Medical Services
911 Service **OnCall:** Ambulance (*897-1098*) **Hospital:** Trillium
248-8353 (*22 mi.*)

Setting -- Well protected Deep Bay lies at the south end of Baynes Sound, between Vancouver and Denman Islands. When at sea in this region, there is a certain sense of isolation - and little on the shoreline breaks up this impression. This is reinforced by that fact that this is the only marina between Comox and French Creek. The community of Deep Bay consists of a scattered collection of homes along the waterfront, adding to the rural air.

Marina Notes -- *About 1100 ft. of alongside dockage, no slips. **15 and 30 amp power. Deep Bay consists of a government facility run by the Harbour Authority, an adjacent private marina - also home to "reciprocal" Deep Bay Yacht Club, plus Ship & Shore - a land-side business. The public dock is shared by fishing and pleasure vessels; rafting is encouraged. Ship & Shore houses a fully-licensed café, laundry, plus a store with groceries, bait & tackle, and ice. It also provides fuel - pumped into a dolly-style tank at the store and rolled down to your boat. A small picnic area with tables and benches, and a fish-cleaning table are near the store. Cinderblock restrooms with fiberglass shower stalls are shared by all facilities and require a key from the Harbourmaster's office.

Notable -- North of the marina, Rosewall Creek Park provides great kayaking and bird watching opportunities. To the south, the small town of Bowser, (1.5 mi.) offers very limited services. It's about 15 mi. to Qualicum Beach, and another 6 to Parksville, along Hwy. 19A. This is a popular stretch among lighthouse lovers and clam diggers. The Horne Lake Caves Provincial Park offers tours of outstanding underwater caverns - ranging from an easy walk to an adrenalin pumping version that includes a seven-story rappel down a waterfall. Outfitters at the marina can supply everything required for a true aquatic adventure.

French Creek Boat Harbour

1055 Lee Road; Parksville, BC V9P 2E1

Tel: (250) 248-5051 **VHF: Monitor** n/a **Talk** n/a
Fax: (250) 248-5123 **Alternate Tel:** n/a
Email: hafc@frenchcreekharbour.com **Web:** n/a
Nearest Town: Parksville *(3 mi.)* **Tourist Info:** (250) 248-3613

Navigational Information
Lat: 49°20.922' **Long:** 124°21.454' **Tide:** 15 ft. **Current:** n/a **Chart:** 3512
Rep. Depths (*MLW*): Entry 12 ft. **Fuel Dock** n/a **Max Slip/Moor** 10 ft./-
Access: On the Strait of Georgia 3 miles north of Parksville

Marina Facilities *(In Season/Off Season)*
Fuel: Gasoline, Diesel - French Creek Seafood
Slips: 300 Total, 50 Transient **Max LOA:** 100 ft. **Max Beam:** n/a
  Rate *(per ft.)*: **Day** $0.71 **Week** n/a **Month** n/a
  Power: 30 amp $2.50*, **50 amp** n/a, **100 amp** n/a, **200 amp** n/a
  Cable TV: No **Dockside Phone:** No
  Dock Type: Floating, Alongside, Wood
Moorings: 0 Total, 0 Transient **Launch:** n/a
  Rate: Day n/a **Week** n/a **Month** n/a
Heads: 2 Toilet(s), 1 Shower(s)
Laundry: 1 washer(s), 2 dryer(s) **Pay Phones:** Yes, 2
Pump-Out: OnSite, Full Service, 1 Central **Fee:** Inq. **Closed Heads:** No

Marina Operations
Owner/Manager: Julie Blood **Dockmaster:** Same
In-Season: Year-Round, 8am-5:30pm **Off-Season:** n/a
After-Hours Arrival: Tie up and pay in the morning
Reservations: No **Credit Cards:** Visa/MC, Debit
Discounts: None
Pets: Welcome **Handicap Access:** No

Marina Services and Boat Supplies
Services - Security *(24 hr. video)*, Trash Pick-Up, Dock Carts
Communication - DHL, UPS **Supplies - OnSite:** Ships' Store **Near:** Bait/Tackle *(French Creek Store 248-8912, large inventory of fishing supplies and also a depot for St. Jean's Cannery which will process your catch)*

Boatyard Services
OnSite: Launching Ramp **Nearest Yard:** Nanaimo Shipyard (250) 753-1151

Restaurants and Accommodations
OnSite: Restaurant *(Wheelhouse Café 951-3301, B $4-8, L $7-10, D $7-12, 7am-9pm in summer)*, *(Creek House Restaurant 248-3214, L $9-12, D $15-20)* **1-3 mi:** Restaurant *(China Garden 954-2202)*, *(Captain Jim's Seafood Galley 248-4545)*, *(Heron B $3-10, L $10-16, D $18-32, ocean view, at Bayside Inn)*, Motel *(Travelodge 248-2232)*, *(Sandcastle Inn 248-2334)*, Hotel *(Best Western Bayside Inn 248-8333, $119-179)*

Recreation and Entertainment
OnSite: Group Fishing Boat *(Good Times 752-4221, Bald Eagle 248-4846)*, Special Events *(Fishing Fest - Aug 752-4221)* **Near:** Fishing Charter *(Oceans West 954-5266)* **1-3 mi:** Golf Course *(Qualicum Beach Memorial 752-6312)*, Fitness Center *(Body Sculptors 752-5553)*, Galleries *(Sea-Side Optical 248-1010, Potters Guild 954-1872)* **3+ mi:** Horseback Riding *(Tiger Lily Farm 248-2408, 3 mi.)*, Museum *(Craig Heritage Park 248-6966, Power House 752-6441, 5 mi.)*, Cultural Attract *(Village Theatre 752-3522, Bard to Broadway 752-6813, "Theatre in Tent" during summer, 5 mi.)*, Sightseeing *(Qualicum Fish Hatchery 757-8412, underwater viewing area, nature trails)*

Provisioning and General Services
OnSite: Convenience Store *(Little Mountain Grocery Store 954-2242)*
Near: Fishmonger *(French Creek Seafood 248-2888 - sells gas & diesel too)*, Bank/ATM *(Royal Bank 951-4000)*, Laundry *(Suds & Duds 248-5152)*
Under 1 mi: Wine/Beer, Beauty Salon *(Setting Trends 248-0526)*, Barber Shop *(Barber-Ret 248-3333)* **1-3 mi:** Market *(Shopppers Grocery Mart 248-3442)*, Supermarket *(Quality Foods 954-2262)*, Gourmet Shop *(Fore & Aft Foods 757-8682)*, Delicatessen, Health Food *(Vitalia by Rupert 248-9600, Heaven on Earth 752-3132)*, Bakery *(French Creek Bakery and Café 248-2080)*, Catholic Church *(Catholic Church of the Ascension 248-3747)*, Protestant Church *(Baptist 248-6322)*, Library *(Qualicum Beach 752-6121)*, Dry Cleaners *(Busy Bee 248-2551)*, Bookstore *(Fireside Books 248-5356)*, Pharmacy *(Arrowsmith 248-3162)*, Florist *(Home Hardware 248-9221)*, Clothing Store *(From A to Zebra 248-4164)* **3+ mi:** Farmers' Market *(Qualicum, Fir & Memorial St., Sat 9am-Noon in summer, 5 mi.)*

Transportation
OnCall: Taxi *(Alliance Taxi 954-5568)*, Local Bus *(Regional Transit 954-1001, Islandlink 954-3201)* **Near:** InterCity Bus *(Greyhound 800-661-8747)*, Ferry Service *(foot ferry to Lasqueti Isl.)* **1-3 mi:** Rental Car *(Rent-A-Wreck 248-4101, Budget 248-5341)* **Airport:** Qualicum Municipal/Nanaimo *(3 mi./35 mi.)*

Medical Services
911 Service **OnCall:** Ambulance *(741-0102)* **1-3 mi:** Doctor *(Nanoose Bay Medical 468-5939)*, Dentist *(Beaudoin 248-7088)*, Chiropractor *(Smith 248-6333)*, Holistic Services *(Ayurvedic Healing 954-1786)*, Veterinarian *(Parksville 248-8318)* **Hospital:** Trillium Lodge 248-8353 *(1 mi.)*

Setting -- Between Deep Bay and Northwest Bay this is the only place - on the Vancouver Island side of the Strait of Georgia - to duck behind a breakwater for shelter. 300 slips and side-ties berth a mix of recreational and working craft. At the top of the red-railed gangway, two eateries and fishing charter kiosks overlook the docks. Catch a taxi or rent a car and visit Parksville and Qualicum - their beach town atmospheres counterpoint the ruralness of French Creek.

Marina Notes -- *30 amp at some docks; 20 amp also available, $2. A small commercial area at the edge of the harbor provides limited services, and a few fishing charter operators maintain booths there. Fuel and fresh fish are next door at French Creek Seafood. A boat repair facility is located next to the launch ramp. The office plus additional showers, toilets and laundry are in the public building across the parking lot. Showers are $1/5 min.

Notable -- The stretch of coastline above and below French Creek, known as Oceanside, is a popular vacation spot and has many attractions such as sandy beaches, mini-golf, art galleries, museums, and a fish hatchery open to the public. To provision, head to Parksville 3 miles south, or Qualicum 5 miles north. The foot ferry to Lasqueti Island brings visitors to an unusual world where residents have chosen to forego the normal power grid connection and generate their electricity with solar, water and wind power. Tour the island by kayak or arrange for a taxi tour. Butterfly World and Gardens in Coombs (5 mi.) has achieved international acclaim for its collection of free-flying butterflies and tropical birds (248-7026, $7/3.50). There are no fewer than 5 golf courses in the area. A major sand castle competition is held in Qualicum each summer. Tiger Lily Farm offers horseback riding and an opportunity for children to interact with the animals.

Navigational Information
Lat: 49°18.185' **Long:** 124°12.132' **Tide:** 15 ft. **Current:** n/a **Chart:** 3459
Rep. Depths *(MLW):* **Entry** 14 ft. **Fuel Dock** 6 ft. **Max Slip/Moor** 27 ft./-
Access: Strait of Georgia to Northwest Bay , 8 miles south of Parksville

Marina Facilities *(In Season/Off Season)*
Fuel: Gasoline, Diesel
Slips: 90 Total, 2 Transient **Max LOA:** 45 ft. **Max Beam:** n/a
  Rate *(per ft.):* **Day** $1.00* **Week** n/a **Month** n/a
  Power: 30 amp $5*, **50 amp** n/a, **100 amp** n/a, **200 amp** n/a
  Cable TV: No **Dockside Phone:** No
  Dock Type: Floating, Alongside, Wood
Moorings: 0 Total, 0 Transient **Launch:** n/a
  Rate: Day n/a **Week** n/a **Month** n/a
Heads: 1 Toilet(s), 1 Shower(s)
Laundry: None **Pay Phones:** No
Pump-Out: No **Fee:** n/a **Closed Heads:** No

Marina Operations
Owner/Manager: Lesley Barnes **Dockmaster:** Same
In-Season: May-Sep, 8am-6pm **Off-Season:** Oct-Apr, 8:30am-12:30pm
After-Hours Arrival: Call ahead
Reservations: Yes, Preferred **Credit Cards:** Visa/MC, Debit
Discounts: None
Pets: Welcome, Dog Walk Area **Handicap Access:** Yes, Heads, Docks

Beachcomber Marina

7-1600 Brynmarl Road; Nanoose Bay, BC V9P 9E1

Tel: (250) 468-7222 **VHF: Monitor** Ch. 66A **Talk** n/a
Fax: (250) 468-5715 **Alternate Tel:** n/a
Email: lbarnes@shaw.ca **Web:** n/a
Nearest Town: Parksville *(8 mi.)* **Tourist Info:** (250) 248-3613

Marina Services and Boat Supplies
Services - Docking Assistance, Security *(24/7, locked gates & video)*, Trash Pick-Up, Dock Carts **Communication -** Mail & Package Hold, Phone Messages, FedEx, UPS, Express Mail **Supplies - OnSite:** Ice *(Cube)*, Bait/Tackle **1-3 mi:** Ships' Store, Propane

Boatyard Services
OnSite: Launching Ramp, Bottom Cleaning **OnCall:** Hydraulic Trailer, Engine mechanic *(gas, diesel)*, Electrical Repairs, Electronic Sales, Electronics Repairs, Hull Repairs, Rigger, Divers, Propeller Repairs
Nearest Yard: Marine Express (250) 468-7832

Restaurants and Accommodations
1-3 mi: Lite Fare *(Rocking Horse Pub 468-1735)* **3+ mi:** Restaurant *(Ichiban Sushi 248-3323, 8 mi.)*, *(Spinnaker Seafood 248-5532, 7 mi.)*, *(Bangkok Thai 248-4860, 7 mi.)*, *(British Bobby 954-3232, 6 mi.)*, *(The Landing 468-2400, D $13-30, 4 mi., oceanfront, at Pacific Shores Resort; gourmet pizzas, pasta, seafood, steaks)*, Fast Food *(Subway, 8 mi.)*, Pizzeria *(Pizza Connection 954-0088, 8 mi.)*, Motel *(Arbutus Grove 248-6422, $60-95, 6 mi.)*, Hotel *(Best Western Bayside Inn 248-8333, $119-179, 8 mi.)*, Condo/Cottage *(Beach Acres Resort 248-3424, $210-275, 8 mi.)*

Recreation and Entertainment
OnSite: Picnic Area, Grills, Playground, Sightseeing **Near:** Dive Shop *(Nanoose Bay - spectacular diving adventures with viewing distances up to 100 ft.)*, Park *(Jedediah Provincial Marine Park - by boat)* **1-3 mi:** Video Rental, Video Arcade **3+ mi:** Golf Course *(Fairwinds 468-7666, Qualicum Beach Memorial 752-6312, Arrowsmith Golf & Country Club 752-9729,*

Glengarry Golf Links 886-752-8787, 9 mi.)*, Fitness Center *(Fitness Connection 248-3144, 8 mi.)*, Museum *(Craig Heritage Park 248-6966 - historic buildings: church, log house, fire station, school house, two 19th-century post offices; daily 9am-5pm Jun-Sep 7, 5 mi.)*, Special Events *(Parksville Beach Festival: sand sculpting contests, kidfest - Aug, 8 mi.)*, Galleries *(Arrowsmith Potters Guild 954-1872, 8 mi.)*

Provisioning and General Services
Near: Convenience Store, Wine/Beer, Bank/ATM, Post Office *(Nanoose 468-7722)*, Protestant Church *(St. Mary's 468-5684)*, Bookstore, Newsstand **Under 1 mi:** Supermarket *(Quality Foods 468-7131)*, Beauty Salon *(Young's Hair Salon 468-5663)* **1-3 mi:** Library *(Nanoose 468-9977)* **3+ mi:** Market *(Shoppers Grocery 248-3442, 8 mi.)*, Delicatessen *(8 mi.)*, Liquor Store *(Arlington Liquor 468-7799, 6 mi.)*, Bakery *(Mom's Bakery 954-1887, 5 mi.)*, Fishmonger *(9 mi.)*, Catholic Church *(10 mi.)*, Dry Cleaners *(5th Generation 954-1551, 9 mi.)*, Laundry *(Craig St. Laundromat 954-2282, 10 mi.)*, Pharmacy *(Arrowsmith 248-3162, 8 mi.)*, Hardware Store *(Home Hardware 248-9221, 8 mi.)*, Department Store *(Walmart 390-2334, 12 mi.)*

Transportation
OnSite: Courtesy Car/Van *(when possible; small fee)* **3+ mi:** Rental Car *(Rent-A-Wreck 248-4101, Budget 248-5341, 8 mi.)*, Taxi *(Alliance Taxi 954-5568, 8 mi.)* **Airport:** Qualicum Municipal/Nanaimo *(9 mi./28 mi.)*

Medical Services
911 Service **OnCall:** Ambulance *(741-0102)* **3+ mi:** Doctor *(Waite 248-5677, 4 mi.)*, Dentist *(Shaw-Wood 248-1041, 8 mi.)*, Veterinarian *(Bellevue 248-2031, 7 mi.)* **Hospital:** Trillium Lodge 248-8353 *(9 mi.)*

Setting -- Beachcomber Marina lies in Northwest Bay off the Strait of Georgia. Watch for the striking white and gray-blue peaks of the contemporary condos on the bluff above the docks and the fluttering flags lining the white-railed main pier. This well-tended facility sports hanging baskets of flowers, ship-shape, mostly alongside, dockage and a pavilion with picnic tables overlooking the bay.

Marina Notes -- *15 amp also available. Originally a campground, the marina was established in the '60s and renovated in the '90s. The tidy, flower-trimmed office sells a few supplies, including ice, sodas, and fishing items. Manager Lesley Barnes offer critical transport for a small "gasoline fee." The pavilion and picnic tables are ideal for large group gatherings - make arrangements in advance. A fish cleaning station makes it easy to prepare the day's catch for the crowd. The bathhouse is near the office building. Note: This can be a useful shelter in bad weather, and a place to wait out the infamous "Qualicums."

Notable -- Beachcomber is 8 miles south of Parksville, in a wooded residential area - a peaceful place to relax and enjoy the tranquil harbor. Nearby Jedediah Island Provincial Marine Park, accessible only by boat, offers 640 acres of trails, sandy beaches, and campsites as well as some great kayaking and diving. Although it is no longer inhabited, there are remnants of an old homestead with its orchards and outbuildings. For golfers, 5 courses are available within 15 miles: Qualicum Beach Memorial (752-6312), Arrowsmith Golf & Country Club (752-9727), Glengarry Golf Links (866-752-8787), and Morningstar International (567-1320). Drop by the Station Gallery at the restored Parksville train station, where the Arrowsmith Potters Guild runs a hands-on studio.

Fairwinds Schooner Cove Marina

3521 Dolphin Drive; Nanoose Bay, BC V9P 9J7

Tel: (250) 468-5364; (800) 663-7060 **VHF:** Monitor Ch. 66A **Talk** n/a
Fax: (250) 468-5744 **Alternate Tel:** (250) 468-7691
Email: info@fairwindsbc.ca **Web:** www.fairwinds.ca
Nearest Town: Parksville *(12 mi.)* **Tourist Info:** (250) 248-3613

Navigational Information

Lat: 49°17.280' **Long:** 124°07.900' **Tide:** 15 ft. **Current:** n/a **Chart:** 3459
Rep. Depths *(MLW):* **Entry** 35 ft. **Fuel Dock** 8 ft. **Max Slip/Moor** 25 ft./-
Access: Off the Strait of Geogia south of Whiskey Golf

Marina Facilities *(In Season/Off Season)*

Fuel: Slip-Side Fueling, Gasoline, Diesel
Slips: 365 Total, 30 Transient **Max LOA:** 150 ft. **Max Beam:** n/a
  Rate *(per ft.):* **Day** $1.10/0.85 **Week** n/a **Month** $6.25
  Power: 30 amp $5, **50 amp** $5, **100 amp** n/a, **200 amp** n/a
  Cable TV: No **Dockside Phone:** No
  Dock Type: Floating, Long Fingers, Alongside, Wood
Moorings: 0 Total, 0 Transient **Launch:** n/a
  Rate: Day n/a **Week** n/a **Month** n/a
Heads: 6 Toilet(s), 6 Shower(s)
Laundry: 2 Washer(s), 2 Dryer(s) **Pay Phones:** Yes, 3
Pump-Out: OnSite, OnCall, Full Service **Fee:** $8 **Closed Heads:** Yes

Marina Operations

Owner/Manager: Ben Tall **Dockmaster:** Wayne Newport
In-Season: Jun-Sep, 6:30am-9pm **Off-Season:** Oct-May, 8am-4:30pm
After-Hours Arrival: Tie up and check in in the morning
Reservations: Yes, Preferred **Credit Cards:** Visa/MC, Amex, Debit
Discounts: Free moorage w/room **Dockage:** n/a **Fuel:** n/a **Repair:** n/a
Pets: Welcome, Dog Walk Area **Handicap Access:** Yes, Heads, Docks

Marina Services and Boat Supplies

Services - Docking Assistance, Concierge, Room Service to the Boat,
Security *(locked gate and 24 hr. patrol)*, Trash Pick-Up, Dock Carts
Communication - Mail & Package Hold, Phone Messages, Fax in/out, Data
Ports *(Wi-Fi - Broadband Xpress)*, FedEx, DHL, UPS, Express Mail *(Sat
Del)* **Supplies -** OnSite: Ice *(Block, Cube)*, Ships' Store, Bait/Tackle
(Frozen bait)

Boatyard Services

OnSite: Launching Ramp **OnCall:** Engine mechanic *(gas, diesel)*
Under 1 mi: Canvas Work *(Doyle Sails 468-9177)*.
Nearest Yard: Nanaimo Shipyard (250) 753-1151

Restaurants and Accommodations

OnSite: Restaurant *(The Laughing Gull Pub 468-7691, B $7-15, L $8-17,
D $10-30, seafood, steaks, pub fare)*, Coffee Shop *(Dockside Café 468-
7691, L $5-8, D $5-8, freshly baked goods daily)*, Hotel *(Fairwinds Schooner
Cove 468-7691, $69-139, oceanfront rooms with balconies; ask about 2 & 3
night Get-Away Packages $149-279, & Golf Get-Away Packages $215-505)*
OnCall: Pizzeria *(Oceanside Pizza & Pasta 248-9136)*
3+ mi: Seafood Shack *(The Landing 468-2400, D $13-30, 6 mi., at Pacific
Shores Resort, north side of peninsula)*, Lite Fare *(Rocking Horse Pub
468-1735, 5 mi.)*, Motel *(Arbutus Grove Hotel 248-6422, $40-80, 10 mi.)*,
Condo/Cottage *(Pacific Shores Resort 866-986-2222, 5 mi.)*

Recreation and Entertainment

OnSite: Heated Pool, Spa, Picnic Area, Volleyball *(Beach volleyball)*, Tennis
Courts, Fitness Center *(The Fairwinds Centre $10/day: circuit training,
aquacize*

$6, aerobics $6, bridge $3), Jogging Paths, Boat Rentals *(kayaks)*, Fishing
Charter *(4 hrs., max. 4 persons $300)* **Near:** Horseback Riding
3+ mi: Golf Course *(Fairwinds Golf and Country Club 468-7666, free shuttle;
18 holes$42-65, twilight $39, 9 holes $29-40, power cars $32, full set of clubs
$25, pull carts $5. Free clubs with full greens fee. Reserve Twilight Bite Tue
and Wed 4-6pm $40, incl. 9 holes of golf, power cart, and dinner, 5 mi.)*,
Museum *(Craig Heritage Park 248-6966, 12 mi.)*

Provisioning and General Services

OnSite: Convenience Store, Delicatessen, Wine/Beer, Liquor Store,
Crabs/Waterman *(Every Fri 3-6pm)*, Bank/ATM **Near:** Market *(Quality
Foods 468-7131, daily shuttle 11am)*, Post Office *(Nanoose 468-7722)*,
Protestant Church *(St. Mary's Anglican Church 468-5684)*
Under 1 mi: Beauty Salon *(Young's Hair Salon 468-5663)* **1-3 mi:** Library
(Nanoose 468-9977) **3+ mi:** Pharmacy *(Parksville Pharmasave 248-3521,
10 mi.)*, Hardware Store *(Albertsons 248-6888, 8 mi.)*

Transportation

OnSite: Courtesy Car/Van *(Shuttle to golf course & grocery store)*, Bikes
($6/hr. $22/half day), Rental Car *(marina staff will make arrangements with
Budget)* **OnCall:** Taxi **Airport:** Nanaimo *(25 mi.)*

Medical Services

911 Service **OnCall:** Ambulance *(741-0102)* **1-3 mi:** Holistic Services
(AquaTerre Spa 888-441-4442 at Pacific Shores, 5 mi.) **3+ mi:** Doctor
(Jensen Medical Clinic 248-6969, 11 mi.) **Hospital:** Nanaimo Regional
754-2141 *(16 mi.)*

Setting -- Sheltered by a high breakwater, the Schooner Cove Resort & Marina is part of the upscale residential development of Fairwinds which occupies
1,300 acres on the Nanoose Peninsula. Fine homes peek through the foliage on shore and the sprawling 3-story contemporary hotel and restaurant complex
overlooks the docks. Most resort amenities are available to visiting boaters and beautifully landscaped grounds provide ample space for picnics and relaxing.

Marina Notes -- Free moorage with hotel stay. 365 slips are distributed on 2 sides of several long piers. Transients are usually on the docks closer to the
breakwater, and the fuel dock is the farthest in, near the boat ramp. The marina boasts the cheapest gasoline on the island. Boat charters, kayak rentals and
sailing lessons can be arranged near the fuel dock. The on-site store carries convenience items, wine & beer, ice, bait & tackle. This is a first-class facility that
provides great customer service: staff are ready to take care of your every need. The central reservation office will also arrange rental cars; shuttle service is
provided to the grocery store, golf course, and Fairwinds Centre. The comfortable bathhouse sparkles. Note: watch the restrictions on the charts as you will be
passing the military torpedo range known as Whiskey Golf.

Notable -- An 18-hole championship golf course is part of the community, and the Fairwinds Fitness Centre, with indoor swimming pool, hot tub, exercise
room, billiards, etc., is available for a $10 drop-in fee. Learn the art of kayaking or sailing with a beginner's course or guided tour to observe the wildlife. Rent a
bike or hike Fairwinds' 1300 acres of trails. Hire a car and visit the nearby vacation destination Oceanside, or organize a scenic horseback ride.

Navigational Information
Lat: 49°11.235' **Long:** 123°56.938' **Tide:** 15 ft. **Current:** n/a **Chart:** 3447
Rep. Depths (*MLW*): **Entry** 12 ft. **Fuel Dock** n/a **Max Slip/Moor** 15 ft./-
Access: Newcastle Island Passage, west of Newcastle Island

Marina Facilities (*In Season/Off Season*)
Fuel: No
Slips: 300 Total, 13* Transient **Max LOA:** 100 ft. **Max Beam:** n/a
  Rate (*per ft.*): **Day** $1.00/0.60 **Week** n/a **Month** $6.60
  Power: 30 amp $2, **50 amp** n/a, **100 amp** n/a, **200 amp** n/a
  Cable TV: No **Dockside Phone:** No
  Dock Type: Floating, Long Fingers, Alongside, Wood
Moorings: 0 Total, 0 Transient **Launch:** n/a
  Rate: Day n/a **Week** n/a **Month** n/a
Heads: 9 Toilet(s), 8 Shower(s)
Laundry: 10 Washer(s), 12 Dryer(s) **Pay Phones:** Yes, 2
Pump-Out: No **Fee:** n/a **Closed Heads:** Yes

Marina Operations
Owner/Manager: Marc Stones **Dockmaster:** Carol Stones
In-Season: Jun-Sep, 9am-5pm **Off-Season:** Oct-May, 9am-5pm
After-Hours Arrival: Call ahead or go to the visitors' dock in front of Pub
Reservations: Yes, Required **Credit Cards:** Visa/MC, Debit
Discounts: None
Pets: Welcome **Handicap Access:** Yes, Heads, Docks

Stones Marina and Boatyard

1690 Stewart Avenue; Nanaimo, BC V9S 4E6

Tel: (250) 753-4232 **VHF: Monitor** n/a **Talk** n/a
Fax: (250) 753-4204 **Alternate Tel:** n/a
Email: email@stonesmarina.com **Web:** www.stonesmarina.com
Nearest Town: Nanaimo (*1 mi.*) **Tourist Info:** (250) 753-1191

Marina Services and Boat Supplies
Services - Docking Assistance, Security (*locked gate*), Trash Pick-Up, Dock Carts **Communication -** Mail & Package Hold, Phone Messages, Fax in/out, Data Ports, FedEx, DHL, UPS, Express Mail **Supplies - OnSite:** Ice (*Cube*), Bait/Tackle **Under 1 mi:** West Marine (*758-8048*)

Boatyard Services
OnSite: Travelift (*83T to 70 ft. LOA, 21 ft. beam*), Yacht Broker (*Passage Yacht Sales*)

Restaurants and Accommodations
OnSite: Restaurant (*Muddy Waters Pub 754-4220, L $7-12, D $7-12*), (*Beefeaters Chop House and Grill 753-2333, L $6-12, D $6-25*)
OnCall: Pizzeria (*Boston Pizza 751-0090*) **Near:** Restaurant (*The Granary 754-4899, B $5-10, L $6-13, D $11-19, Mon-Fri 8am-7:30pm, Sat to 3pm, closed Sun*), (*La Veranda 755-1124*), Inn/B&B (*Buccaneer Inn 753-1246, $60-120*) **Under 1 mi:** Motel (*Travelodge 754-6355, $87-110*) **1-3 mi:** Hotel (*Howard Johnson Harbourside 753-2241, $80-150*)

Recreation and Entertainment
OnSite: Dive Shop (*Ocean Explorers 753-2055*), Jogging Paths, Fishing Charter (*Nanaimo Yacht Charters & Sailing School 754-8601*) **Near:** Picnic Area, Boat Rentals (*Kayaks 754-6626*), Park (*Maffeo Sutton Park*) **1-3 mi:** Golf Course (*Pryde Vista 753-6188*), Fitness Center (*Pinnacle Health 753-0138*), Movie Theater (*Bay Theatre 754-3585*), Video Rental (*Tri Star 755-1817*), Museum (*Nanaimo District Museum 753-1821*), Cultural Attract (*Centre for the Arts 754-2264, Port Theatre 754-4555*), Sightseeing (*self-guided walking tours - harbor, coal, or railroad themed - map at CofC*),

Special Events (*Silly Boat Regatta, Bathtub Race, Dragon Boat Fest - Jul, Concerts in the Park - Sun in Aug, Bite of Nanaimo - Sep*), Galleries (*Artisans' Studio 753-6151*)

Provisioning and General Services
OnSite: Liquor Store (*at Muddy Waters*) **Near:** Convenience Store (*Wayne's Grocery 753-5133*) **Under 1 mi:** Supermarket (*Save-On Foods 753-8151*), Fishmonger (*Sea Drift Fish Market 754-4913*), Bank/ATM (*Bank of Montreal 754-0564*), Protestant Church, Beauty Salon (*Tresses 754-1194*), Pharmacy (*Central Drugs 753-5342*) **1-3 mi:** Gourmet Shop (*Mclean's 754-0100*), Delicatessen (*Delcado's Deli 753-6524*), Health Food (*Charlie Brown's 753-5211*), Bakery (*Mcgavin's Bread Basket 753-4487*), Farmers' Market (*Pioneer Plaza, Fri 10am-2pm, midApr-Oct*), Catholic Church, Library (*758-4697*), Dry Cleaners, Bookstore (*Bestsellers 755-1222*), Copies Etc. (*Staples 751-7770*)

Transportation
OnCall: Rental Car (*Enterprise 751-1200/National 758-3509, 1 mi.*), Taxi (*AC Taxi 753-1231*) **1-3 mi:** Ferry Service (*Protection Island 753-8244/ BC Ferries to Horseshoe Bay, Tsawwassen, & Islands 888-724-5223/ HarbourLynks to Vancouver 753-4443*) **Airport:** Nanaimo (*11 mi.*)

Medical Services
911 Service **OnCall:** Ambulance (*741-0102*) **Under 1 mi:** Dentist (*Malaspina 753-1181*), Chiropractor (*Mountainview 753-5351*) **1-3 mi:** Doctor (*Wellington Medical Clinic 753-9111*), Holistic Services (*Oceanview Massage Clinic 618-3191*), Veterinarian (*Island Veterinary 753-1288*) **Hospital:** Nanaimo Regional 248-2332 (*3 mi.*)

Setting -- The northernmost marina along Newcastle Island Passage, Stones' 13-acre facility features a central pier with nine docks. The mix of open and covered slips and alongside dockage berths 300 boats -- with wide comfortable fairways and views of Newcastle Island. Sprawled along the shore, the rustic land base is in the throes of change. Among the improvements a new boatyard operation including the installation of a travelift between the two eateries.

Marina Notes -- *Inner dock has 500 ft. alongside transient moorage. Gated 11 ft. wide docks with 5 ft. finger piers. Established in 1979. Boatyard construction, with installation of large 83T lift, during the summer of '06. Mostly do-it-yourself, but independent contractors are being invited to move on-site. Literally dockside is Muddy Waters marine pub; across the parking lot, Beefeaters Chop House & Grill, a full-service restaurant, offers a more subdued ambiance. On-site Ocean Explorers Dive Shop services the famous local artificial reefs, including HMCS Saskatchewan & Cape Breton. Charters to nearby islands and harbor tours are also based at the marina. The locked red-roofed bathhouse offers heads, shower and laundry.

Notable -- Nanaimo is built around the harbor. Downtown is a little over a mile, and most services are within easy walking distance. The Heritage Walking Tour passes several historic buildings, the waterfront Port Theatre, which presents Broadway productions and more experimental theater, and the Centre for the Arts with a dance academy, cinema, and galleries. Festivals run throughout the year; Adventure Games in September includes street luge, mountain biking, and wife carrying. Within walking distance is Brechin Point seaplane terminal. Further north, Vancouver ferries leave from the Departure Bay terminal.

PHOTOS ON DVD: 14

Newcastle Marina

Newcastle Marina

1300 Stewart Avenue; Nanaimo, BC V9S 4E1

Tel: (250) 754-1431 **VHF: Monitor** n/a **Talk** n/a
Fax: (250) 753-2974 **Alternate Tel:** n/a
Email: newcastle@shaw.ca **Web:** n/a
Nearest Town: Nanaimo *(1 mi.)* **Tourist Info:** (250) 753-1191

Navigational Information
Lat: 49°11.036' **Long:** 123°56.775' **Tide:** 15 ft. **Current:** n/a **Chart:** 3447
Rep. Depths *(MLW):* **Entry** 18 ft. **Fuel Dock** n/a **Max Slip/Moor** 12 ft./-
Access: Newcastle Island Passage, west of Newcastle Island

Marina Facilities *(In Season/Off Season)*
Fuel: No
Slips: 140 Total, 2 Transient **Max LOA:** 50 ft. **Max Beam:** n/a
Rate *(per ft.):* **Day** $1.00 **Week** n/a **Month** n/a
Power: 30 amp Incl.*, **50 amp** n/a, **100 amp** n/a, **200 amp** n/a
Cable TV: No **Dockside Phone:** No
Dock Type: Floating, Long Fingers, Wood
Moorings: 0 Total, 0 Transient **Launch:** n/a
Rate: Day n/a **Week** n/a **Month** n/a
Heads: 2 Toilet(s), 2 Shower(s)
Laundry: 2 Washer(s), 2 Dryer(s) **Pay Phones:** Yes, 1
Pump-Out: No **Fee:** n/a **Closed Heads:** Yes

Marina Operations
Owner/Manager: Gerald Chow **Dockmaster:** Tawnya Payne
In-Season: Year-Round, 9am-5pm **Off-Season:** n/a
After-Hours Arrival: Call ahead
Reservations: Yes, Required **Credit Cards:** Visa/MC, Debit
Discounts: None
Pets: Welcome **Handicap Access:** Yes, Docks

Marina Services and Boat Supplies
Services - Security *(night patrol, video, locked gate)*, Trash Pick-Up, Dock Carts **Communication -** Mail & Package Hold, Phone Messages, Fax in/out *($1/pg)*, FedEx, DHL, UPS, Express Mail **Supplies - Near:** Ice *(Cube)*, Ships' Store *(Shipyard)* **1-3 mi:** West Marine *(758-8048)*

Boatyard Services
OnSite: Travelift *(50T)*, Engine mechanic *(gas)*, Hull Repairs, Bottom Cleaning, Divers, Compound, Wash & Wax, Propeller Repairs, Painting, Yacht Broker *(Newcastle Boat Brokers 753-2626)* **OnCall:** Engine mechanic *(diesel)*, Electrical Repairs, Electronics Repairs, Canvas Work, Refrigeration, Interior Cleaning, Woodworking, Awlgrip **Near:** Launching Ramp, Inflatable Repairs, Life Raft Service, Upholstery.

Restaurants and Accommodations
OnSite: Restaurant *(Nauticals Seafood Bar & Grill 754-8881)* **Near:** Rest-aurant *(La Veranda 755-1124)*, *(The Granary 754-4899, B $5-10, L $6-13, D $11-19)*, Fast Food *(Subway, McDonald's)*, Hotel *(Moby Dick 753-7000, $69-130)* **Under 1 mi:** Restaurant *(Muddy Waters Pub 754-4220, L $7-12, D $7-12)*, *(Beefeaters 753-2333, B $6-12, L $6-25)*, *(White Spot 754-2241, B $6-9, L $7-14, D $7-19)*, Seafood Shack *(Cottage Fish & Chips 753-3944)* **1-3 mi:** Motel *(Best Western Dorchester 754-6835, $80-150, harbor views)*, Hotel *(Howard Johnson Harbourside 753-2241, $80-150)*

Recreation and Entertainment
OnSite: Picnic Area, Grills **Under 1 mi:** Fishing Charter, Park **1-3 mi:** Golf Course *(Pryde Vista 753-6188)*, Fitness Center *(Pinnacle 753-0138)*, Movie Theater *(Bay Theatre 751-3585)*, Video Rental *(Tri Star Video 755-1817)*,

Museum *(Nanaimo District Museum 753-1821)*, Cultural Attract *(Port Theatre 754-4555)*, Sightseeing *(downtown Nanaimo)*, Special Events *(festivals throughout the year)*, Galleries *(Art of the Siem 764-0074)* **3+ mi:** Tennis Courts *(Westwood Racquet Club 753-2866, 4 mi.)*

Provisioning and General Services
Near: Liquor Store *(Muddy Waters 754-4220)*, Fishmonger *(Sea Drift Fish Market 754-4913)*, Bank/ATM, Copies Etc. *(UPS Store 753-6245)*
Under 1 mi: Supermarket *(Save-On Foods 753-8151)*, Health Food *(L&M Natural 753-8413)*, Protestant Church, Beauty Salon *(Details Studio 753-4710)*, Pharmacy *(Central Drugs 753-5342)* **1-3 mi:** Gourmet Shop *(Mclean's 754-0100)*, Delicatessen *(Delcado's 753-6524)*, Bakery *(Mcgavin's 753-4487)*, Farmers' Market *(Pioneer Plaza, Fri 10am-2pm, midApr-Oct)*, Catholic Church, Library *(758-4697)*, Barber Shop *(His Hair 754-9933)*, Laundry (Boat *Basin 754-8654)*, Bookstore *(Bestsellers 755-1222)*, Clothing **3+ mi:** Department Store *(Walmart 390-2334, 6 mi.)*

Transportation
OnCall: Rental Car *(Enterprise 751-1200)*, Taxi *(AC Taxi 753-1231)*
1-3 mi: Ferry Service *(BC Ferries to Horseshoe Bay, Tsawwassen, & Islands 888-724-5223; HarbourLynks to Vancouver 753-4443)*
Airport: Nanaimo *(11 mi.)*

Medical Services
911 Service **OnCall:** Ambulance *(741-0102)* **Under 1 mi:** Dentist *(Malaspina 753-1181)*, Chiropractor *(Mountainview Chiropractic 753-5351)* **1-3 mi:** Doctor *(Wellington Medical Clinic 753-9111)*, Veterinarian *(Island Veterinary 753-1288)* **Hospital:** Nanaimo Regional 248-2332 *(2 mi.)*

Setting -- Newcastle Marina's boatyard atmosphere is softened by a two-story contemporary main office building - tan trimmed with chocolate brown. Colorful hanging plants mark the entrance to the office. At the foot of the docks, casual picnic tables and grills sit on a small rustic wooden deck overlooking the basin. Popular Nauticals Seafood Restaurant is on the right side of the parking lot.

Marina Notes -- *15 amp power is also available. Transient space is difficult to come by, so call ahead. Well-maintained wood piers have long finger slips on both sides with some covered slips. A yacht broker and working boatyard share the same parking lot as the marina, and a 50T travelift is on-site - expect an active, bustling yard. Note: If taking the approach from the north via Departure Bay, there are, reportedly, hazards in mid-channel south of the marina. Seaplanes operating from Brechin Point regularly land and take off in the passage - keep an eye on the sky. FYI: Anchorage Marina, just north, no longer accepts transients.

Notable -- The marina is just a dinghy ride or a short walk away from the heart of downtown Nanaimo. Boaters moored in the harbor are welcomed with Tourism Nanaimo's Come Ashore kits, filled with information on the port, local amenities, and things to do. Paddle into a secluded bay, or join the sailors for their annual S.I.N. (Snake Island-Nanaimo) Regatta. Downtown Nanaimo is divided into 3 distinct areas to suit every boater's tastes and needs: the Arts District, the Old City Quarter, and the Waterfront, all within walking distance of the marina, and offering marvelous dining, shopping, and entertainment opportunities.

Navigational Information

Lat: 49°10.898' **Long:** 123°56.579' **Tide:** 15 ft. **Current:** n/a **Chart:** 3447
Rep. Depths (*MLW*): **Entry** 18 ft. **Fuel Dock** n/a **Max Slip/Moor** 20 ft./-
Access: Newcastle Island Passage, west of Newcastle Island

Marina Facilities *(In Season/Off Season)*

Fuel: No
Slips: 20 Total, 2 Transient **Max LOA:** 80 ft. **Max Beam:** n/a
  Rate (*per ft.*): **Day** $4.50/2.00 **Week** n/a **Month** n/a
  Power: 30 amp $5, **50 amp** n/a, **100 amp** n/a, **200 amp** n/a
  Cable TV: No **Dockside Phone:** No
  Dock Type: Floating, Alongside, Wood
Moorings: 0 Total, 0 Transient **Launch:** n/a
  Rate: Day n/a **Week** n/a **Month** n/a
Heads: 5 Toilet(s)
Laundry: None **Pay Phones:** No
Pump-Out: No **Fee:** n/a **Closed Heads:** Yes

Marina Operations

Owner/Manager: Kevin Ratcliff **Dockmaster:** Same
In-Season: Year-Round, 8am-5pm **Off-Season:** n/a
After-Hours Arrival: Call ahead
Reservations: Yes, Preferred **Credit Cards:** Visa/MC
Discounts: None
Pets: Welcome **Handicap Access:** No

Nanaimo Shipyard

1040 Stewart Avenue; Nanaimo, BC V9S 4C9

Tel: (250) 753-1151 **VHF: Monitor** n/a **Talk** n/a
Fax: (250) 753-2235 **Alternate Tel:** n/a
Email: kevin@nanaimoshipyard.com **Web:** www.nanaimoshipyard.com
Nearest Town: Nanaimo (*1 mi.*) **Tourist Info:** (250) 753-1191

Marina Services and Boat Supplies

Services - Docking Assistance, Trash Pick-Up, Dock Carts
Communication - Mail & Package Hold, Phone Messages, Fax in/out,
FedEx, DHL, UPS, Express Mail (*Sat Del*) **Supplies - OnSite:** Ships'
Store **1-3 mi:** West Marine (*758-8048*)

Boatyard Services

OnSite: Railway (*three to 1,000T & 500 ft. LOA*), Engine mechanic (*gas,
diesel*), Launching Ramp, Electrical Repairs, Electronic Sales, Hull Repairs,
Rigger, Bottom Cleaning, Air Conditioning, Refrigeration, Compound, Wash
& Wax, Interior Cleaning, Propeller Repairs, Woodworking, Life Raft Service,
Upholstery, Yacht Interiors, Metal Fabrication, Painting, Awlgrip, Yacht
Design, Yacht Building, Total Refits **Yard Rates:** $62/hr.

Restaurants and Accommodations

Near: Motel (*Moby Dick Oceanfront Lodge 753-7111, $69-130*)
Under 1mi: Restaurant (*Muddy Waters Pub 854-4220, L $7-12, D $7-12*),
(*Beefeaters Chop House and Grill 753-2333, L $6-12, D $6-25*)
1-3 mi: Restaurant (*White Spot 754-2241, B $6-9, L $7-14, D $7-19*), (*La
Rosa Restaurant 754-5444, D $10-18, Greek*), Pizzeria (*Boston Pizza 751-
0090, will deliver*), Motel (*Blue Bird 753-4114, $49-64*), Hotel (*Best Western
Dorchester 754-6835, $90-160*), (*Coast Hotels 753-4155, $99-165*)

Recreation and Entertainment

1-3 mi: Golf Course (*Pryde Vista 753-6188*), Fitness Center (*Pinnacle Health
and Fitness 753-0138*), Movie Theater (*Bay Theatre 753-3585*), Video Rental
(*Tri Star 755-1817*), Park, Museum (*Nanaimo District Museum 753-1821,
Nanaimo Bastion 753-1821*), Cultural Attract (*Centre for the Arts 754-2264,*

Port Theatre 754-4555, Music in the Park - Sun eve in summer 745-8141),
Sightseeing (*Harbourside Walkway*), Special Events (*Cemetry Strolls every
2nd & 4th Thu Jul-Aug 7:30pm 753-1821, Bathtub Racing and Maple Sugar
Tasting Festival - Jul, Jazz Fest*), Galleries (*Art 10 753-4009, The Public
Hanging 756-1148, Barton & Leier 722-7140*)

Provisioning and General Services

Under 1 mi: Convenience Store (*Ocean Store 753-8151*), Market (*Townsite
Grocery 753-2233*), Health Food (*L&M Natural Food 753-8413*), Wine/Beer,
Fishmonger (*Sea Drift Fish Mkt. 754-4913*), Bank/ATM (*Bank of Montreal
754-0564*), Beauty Salon (*Tresses 754-1194*), Pharmacy (*Central Drugs
753-5342*) **1-3 mi:** Gourmet Shop (*Mclean's 754-0100*), Delicatessen
(*Delcado's 753-6524*), Liquor Store (*BC Liquor 741-5555*), Bakery (*Mcgavin's
Bread Basket 753-4487*), Farmers' Market (*Pioneer Plaza, Fri 10am-2pm,
midApr-Oct*), Library (*758-4697*), Dry Cleaners (*Pressed for Time 751-0005*),
Laundry (*Boat Basin 754-8654*), Bookstore (*Bestsellers 755-1222*)

Transportation

OnCall: Rental Car (*Enterprise 751-1200, Budget 754-7368, 1 mi.*), Taxi (*AC
Taxi 753-1231*) **1-3 mi:** Ferry Service (*Protection Island 753-8244, BC
Ferries - Horseshoe Bay, Tsawwassen, & Islands 888-724-5223,
HarbourLynks - Vancouver 753-4443*) **Airport:** Nanaimo (*11 mi.*)

Medical Services

911 Service **OnCall:** Ambulance (*741-0102*) **Under 1 mi:** Dentist
(*Malaspina 753-1181*), Chiropractor (*Mountainview 753-5351*) **1-3 mi:**
Doctor (*Wellington Medical Clinic 753-9111*) **Hospital:** Nanaimo Regional
248-2332 (*2 mi.*)

PHOTOS ON DVD: 8

Setting -- Nanaimo Shipyard's two main piers and three marine railways lie along Newcastle Island Passage. The upland hosts a sprawling blue work shed
with metal roofs of various heights. This is a major boatyard and it doesn't pretend to be anything else - all the docks are workdocks. The boatyard building and
store are sandwiched between Moby Dick's Oceanfront Lodge & Marina and the Nanaimo Harbour City Marina -- all located side-by-side on Stewart Avenue.

Marina Notes -- Nanaimo Shipyard is a full-service boatyard and marine center boasting a wide array of parts and services and a helpful staff prepared to
meet most any need. Three marine railways on-site manage up to 1,000 tons and 500 feet LOA. They specialize in repairs and refits of large ships, from fishing
boats to Coast Guard and Ferry vessels; work on the larger ships is conducted at the government-owned Esquimalt Graving Dock. Customs clearance available. Note: When approaching the shipyard, Passage Rock sits right in the entry - making its acquaintance will surely increase your boatyard bill.

Notable -- It's about a mile to downtown Nanaimo, where dining and entertainment opportunities abound. The Bathtub Racing and Maple Sugar Tasting
Festival has been held the fourth Sunday in July since 1967, and the event grows annually. Motorized bathtubs race from Nanaimo's harbor to the protected
waters of Departure Bay. Like to golf? There are at least 19 golf courses within an hour's drive. The arts scene is also busy, with a cultural events calendar
booked solid every month of the year and numerous art galleries throughout the town. Catch the foot ferry to Newcastle Island (800-663-7337) for a day of
hiking, picnicking, kite flying or a swim in one of its many secluded coves.

Moby Dick Lodge & Marina

1000 Stewart Avenue; Nanaimo, BC V9S 4C9

Tel: (250) 753-7111; (800) 663-2116 **VHF: Monitor** Ch. 67 **Talk** n/a
Fax: (250) 753-4333 **Alternate Tel:** n/a
Email: mobydicklodge@shaw.ca **Web:** www.mobydicklodge.com
Nearest Town: Nanaimo *(1 mi.)* **Tourist Info:** (250) 753-1191

Navigational Information
Lat: 49°10.831' **Long:** 123°56.686' **Tide:** 15 ft. **Current:** n/a **Chart:** 3447
Rep. Depths *(MLW):* **Entry** 6 ft. **Fuel Dock** n/a **Max Slip/Moor** 10 ft./-
Access: South end of Newcastle Island Passage, west of Newcastle Island

Marina Facilities *(In Season/Off Season)*
Fuel: No
Slips: 95 Total, 7 Transient **Max LOA:** 94 ft. **Max Beam:** n/a
 Rate *(per ft.):* **Day** $1.50/$15-20* **Week** n/a **Month** n/a
 Power: 30 amp Incl.**, 50 amp n/a, 100 amp n/a, 200 amp n/a
 Cable TV: No **Dockside Phone:** No
 Dock Type: Floating, Long Fingers, Alongside, Wood
Moorings: 0 Total, 0 Transient **Launch:** n/a
 Rate: Day n/a **Week** n/a **Month** n/a
Heads: 1 Toilet(s), 1 Shower(s)
Laundry: 1 Washer(s), 1 Dryer(s) **Pay Phones:** Yes, 1
Pump-Out: No **Fee:** n/a **Closed Heads:** Yes

Marina Operations
Owner/Manager: Alan Christensen **Dockmaster:** Same
In-Season: Jun-Sep, 7am-11:30pm **Off-Season:** Oct-May, 8am-11pm
After-Hours Arrival: Call ahead
Reservations: Yes, Preferred **Credit Cards:** Visa/MC, Dscvr, Amex, Debit
Discounts: None
Pets: Welcome **Handicap Access:** No

Marina Services and Boat Supplies
Services - Docking Assistance, Security *(24, video)*, Trash Pick-Up
Communication - Mail & Package Hold, Phone Messages, Fax in/out
($1/pg), Data Ports *(lobby)*, FedEx, DHL, UPS, Express Mail *(Sat Del)*
Supplies - 1-3 mi: West Marine (758-8048)

Boatyard Services
Nearest Yard: Nanaimo Shipyard (250) 753-1244

Restaurants and Accommodations
OnSite: Hotel *(Moby Dick Oceanfront Lodge & Marina 753-7111, $69-130)*,
Inn/B&B *(Elfin Yacht 753-5956)* **OnCall:** Pizzeria *(Boston Pizza 751-0090)*
Near: Restaurant *(Nauticals Seafood 754-8881)*, Fast Food *(Wendy's,
Burger King, Subway)*, Motel *(Port-O-Call 753-3421)* **Under 1 mi:** Rest-
aurant *(Muddy Waters Pub 754-4220, L $7-12, D $7-12)*, *(Beefeaters
753-2333, L $6-12, D $6-25)*, *(White Spot 754-2241, B $6-9, L $7-14,
D $7-19)*, *(La Rosa 754-5444, D $10-18, Greek)*, *(The Granary 754-4899,
B $5-10, L $6-13, D $11-19, Mon-Fri 8am-7:30pm, Sat 'til 3pm, closed
Sun)*, Seafood Shack *(Cottage Fish & Chips 753-3944)*
1-3 mi: Hotel (Howard *Johnson Harbourside 753-2241, $80-150)*

Recreation and Entertainment
OnSite: Picnic Area, Grills **Under 1 mi:** Park **1-3 mi:** Golf Course *(Pryde
Vista 753-6188)*, Fitness Center *(Pinnacle 753-0138)*, Bowling *(Fiesta Lanes
753-2451)*, Movie Theater *(Bay Theatre 754-3585)*, Video Rental *(Tri Star
755-1817)*, Museum *(Nanaimo District 753-1821)*, Cultural Attract *(Port
Theatre 754-4555, Centre for the Arts 754-2264)*, Sightseeing *(walking tours,
Cemetery Strolls Jul-Aug every 2nd & 4th Thu)*, Special Events

(Empire Days - May, Marine Fest - Jul), Galleries *(Artisans' Studio 753-
6151)* **3+ mi:** Tennis Courts *(Westwood Racquet Club 753-2866, 4 mi.)*

Provisioning and General Services
OnSite: Fishmonger *(on F dock)* **Under 1 mi:** Convenience Store *(Ocean
753-8151)*, Market *(Townsite 753-2233)*, Supermarket *(Save-On Foods 753-
8151)*, Health Food *(L&M Natural 753-8413)*, Wine/Beer, Bank/ATM *(Bank of
Montreal 754-0564)*, Protestant Church, Beauty Salon *(Tresses 754-1194)*,
Pharmacy *(Central 753-5342)*, Retail Shops **1-3 mi:** Gourmet Shop
(Mclean's 754-0100), Delicatessen *(Delcado's 753-6524)*, Liquor Store *(BC
Liquor 741-5555)*, Bakery *(McGavin's 753-4487)*, Farmers' Market *(Pioneer
Plaza, Fri 10am-2pm, MidApr-Oct)*, Catholic Church, Library *(758-4697)*,
Barber Shop *(Larry's 753-4469)*, Dry Cleaners *(Pressed for Time 751-0005)*,
Laundry *(Boat Basin 754-8654)*, Bookstore *(Bestsellers 755-1222)*, Clothing
Store *(Cynthia's 753-6336)*, Copies Etc. *(Staples 751-7770)*

Transportation
OnCall: Rental Car *(Enterprise 751-1200/Budget 754-7368, 1 mi.)*, Taxi *(AC
Taxi 753-1231)* **1-3 mi:** Ferry Service *(Protection Island 753-8244/ BC
Ferries - Horseshoe Bay, Tsawwassen, & Islands 888-724-5223/
HarbourLynks - Vancouver 753-4443)* **Airport:** Nanaimo *(11 mi.)*

Medical Services
911 Service **OnCall:** Ambulance *(741-0102)* **Under 1 mi:** Dentist
(Malaspina 753-1181), Chiropractor *(Mountainview 753-5351)*
1-3 mi: Doctor *(Wellington Medical Clinic 753-9111)*, Holistic Services
(Oceanview Massage 618-3191), Veterinarian *(Island Vet 753-1288)*,
Optician *(Q-Vision 753-2021)* **Hospital:** Nanaimo Regional 248-2332 *(2 mi.)*

Setting -- Moby Dick Lodge & Marina is located on Nanaimo's 4 km. Harbourside Walkway, near the southern end of Newcastle Island Passage. The four-story motel building, white with bright blue trim, sits high on the bluff separated from the docks by a parking area. A paved picnic area with tables and grills is to the left of the hotel lobby, and Nanaimo Shipyard is directly adjacent.

Marina Notes -- *Under 20' $1.00/ft.; off-season flat rate, $15-20/day. **15 and 20 amp power available. The hotel and marina are under 24-hour video surveillance, and a gate is at the bottom of the main ramp leading to the docks. The marina staff can provide traps for crabbing and arrange fishing charters, rental cars and other activities for guests. Next door, Nanaimo Shipyard provides a full range of services, should they be needed.

Notable -- The Harbourside Walkway leads 2.5 kilometers downtown, where most of Nanaimo's attractions are concentrated. The Nanaimo District Museum & Nanaimo Bastion (753-1821) hold several annual events, including: Summer Bastion Season with Noon Cannon Firing every day; Sundaes on Sundays Jul-Aug 1-4pm, $2.50 to build your own sundae; Miner's Cottage & Locomotive Summer Program, Sat in Jul-Aug 1-4pm; Cemetery Strolls every 2nd and 4th Thu in Jul-Aug at 7:30pm; Marine Festival activities in July, including the Bathtubs and Ballyho Street Fair (call for dates). The famous International Bathtub Race is on the fourth Sunday in July. Newcastle Island Marine Park, accessible only by boat or foot ferry, was named by officers of the Hudson's Bay Co. after the ancient coal city of Newcastle in Northumberland. Enjoy the hiking trails there and check out the tidal stream for a chance to see raccoons and river otters.

Navigational Information
Lat: 49°10.273' **Long:** 123°55.986' **Tide:** 15 ft. **Current:** n/a **Chart:** 3447
Rep. Depths *(MLW):* **Entry** 30 ft. **Fuel Dock** 26 ft. **Max Slip/Moor** 26 ft./-
Access: On the Strait of Georgia west of Gabriola Island

Marina Facilities *(In Season/Off Season)*
Fuel: Gasoline, Diesel
Slips: 225 Total, 10 Transient **Max LOA:** 590 ft. **Max Beam:** 60 ft.
　Rate *(per ft.):* **Day** $1.05/0.75* **Week** n/a **Month** $6.65
　Power: 30 amp $3, **50 amp** $6, **100 amp** $12, **200 amp** n/a
Cable TV: No **Dockside Phone:** No
Dock Type: Floating, Alongside, Wood
Moorings: 0 Total, 0 Transient **Launch:** n/a
　Rate: Day n/a **Week** n/a **Month** n/a
Heads: 3 Toilet(s), 3 Shower(s)
Laundry: 6 Washer(s), 6 Dryer(s), Book Exchange **Pay Phones:** Yes, 2
Pump-Out: OnSite, 1 Central **Fee:** $2 **Closed Heads:** Yes

Marina Operations
Owner/Manager: Andrew Pitcher, HM **Dockmaster:** David Mailloux
In-Season: Year-Round, 7am-11pm **Off-Season:** n/a
After-Hours Arrival: Tie up and pay in the morning
Reservations: Yes, Preferred **Credit Cards:** Visa/MC
Discounts: None
Pets: Welcome **Handicap Access:** Yes, Heads, Docks

Cameron Island

PO Box 131; 104 Front Street; Nanaimo, BC V9R 5K4

Tel: (250) 755-1216 **VHF: Monitor** Ch. 67 **Talk** n/a
Fax: (250) 753-4899 **Alternate Tel:** n/a
Email: marina@npa.ca **Web:** www.npa.ca
Nearest Town: Nanaimo **Tourist Info:** (250) 753-1191

Marina Services and Boat Supplies
Services - Docking Assistance, Security *(night watchman)*, Trash Pick-Up, Dock Carts **Communication -** Mail & Package Hold, Phone Messages, Fax in/out *($1.50)*, Data Ports *(Wi-Fi - Broadband Xpress)*, FedEx, UPS, Express Mail **Supplies -** OnSite: Ice *(Cube)* **Under 1 mi:** Ships' Store *(Harbour Chandler 753-2425)* **1-3 mi:** West Marine *(758-8048)*

Boatyard Services
OnSite: Crane *(1000 lb. hydraulic)* **Nearest Yard:** Nanaimo Shipyard (250) 753-1151

Restaurants and Accommodations
Near: Restaurant *(Penny's Palapa 753-6300, L $6-9, D $6-9, Apr-Oct)*, *(Grapevine on the Bay 746-0797, B $3-6, L $4-19, D $10-19)*, *(Lighthouse Bistro 754-3212, L $7-11, D $11-21, Sun brunch 10am-2pm $13/7)*, *(Cutter's Bistro 753-6601, D $17-19)*, *(Green Garden 753-2828)*, Seafood Shack *(Troller's Fish and Chips 741-7994, L, D)*, *(Charlies Seafood Bar 753-7044)*, Lite Fare *(Perkins Coffee Co. 753-2582)*, Motel *(Best Western Dorchester 754-6835, $80-150)*, Hotel *(The Coast Bastion Inn 753-6601, $99-200)*, *(Howard Johnson Harbourside 753-3241, $80-150)* **Under 1 mi:** Motel *(Travelodge 754-6355, $87-110)*

Recreation and Entertainment
Near: Bowling *(Fiesta Lanes 753-2451)*, Movie Theater *(Bay Theatre 754-3585)*, Museum *(Nanaimo District Museum 753-1821, Nanaimo Bastion 753-1821)*, Cultural Attract *(Centre for the Arts 754-2264, Theatre One 754-7587)* **Under 1 mi:** Video Rental *(Nichol 753-9441)*, Park, Special Events *(Jazz Fest, Synphony at the Harbour at Maffeo Sutton Park - Aug, Bite of*

Nanaimo - Sep)*, Galleries *(Crescent Moon 740-3955, Artisans' Studio 753-6151, Hill's Native Art 755-7873)* **1-3 mi:** Golf Course *(Pryde Vista 753-6188)*, Fitness Center *(Pinnacle 753-0138)*

Provisioning and General Services
Near: Supermarket *(Thrifty Foods 754-6273)*, Farmers' Market *(Pioneer Plaza, Fri 10am-2pm, midApr-Oct)*, Fishmonger *(on F dock)*, Bank/ATM *(Scotiabank 716-2300)*, Protestant Church *(St. Paul's Anglican 753-2523)*, Dry Cleaners, Laundry *(Boat Basin 854-8654)*, Pharmacy *(London Drugs 753-5566)*, Retail Shops *(Pioneer Waterfront Plaza 753-9900, Harbour Park Mall)* **Under 1 mi:** Convenience Store *(Superette 754-5741)*, Market *(J & J Market 754-3601)*, Delicatessen *(Delicado's 753-6524)*, Health Food *(Charlie Brown's 753-5211)*, Liquor Store *(BC Liquor 741-5555)*, Library *(758-4697)*, Beauty Salon *(Tan's 754-5132)*, Barber Shop *(His Hair 754-9933)*, Bookstore *(Downtown Books 753-0102)*, Clothing Store *(Man-Deez 716-0558)*, Copies Etc. **1-3 mi:** Bakery *(Bun's 753-1273)*, Catholic Church

Transportation
OnCall: Rental Car *(Avis 716-8898 at Best Western)*, Taxi *(AC Taxi 753-1231)* **Under 1 mi:** Ferry Service *(Protection Island 753-8244/ Gabriola Island/ BC Ferries - Horseshoe Bay, Tsawwassen, & Islands 888-724-5223/ HarbourLynks - Vancouver 753-4443)* **Airport:** Nanaimo *(11 mi.)*

Medical Services
911 Service **OnCall:** Ambulance *(741-0102)* **Under 1 mi:** Doctor *(Medical Arts Centre 741-0447)*, Holistic Services *(Oceanview Massage 618-3191)* **1-3 mi:** Dentist *(Harewood 754-1949)*, Chiropractor *(Mountainview 753-5351)* **Hospital:** Nanaimo Regional 248-2332 *(3 mi.)*

Setting -- Located off the north tip of Cameron Island, the marina consists of two docking areas: the 600 foot Visiting Vessel Pier, designed for megayachts and cruise ships - which also serves as a floating breakwater, and the 250 and 160 ft. piers for smaller pleasure craft. The facility is an easy dinghy ride or a pleasant stroll to the services of downtown Nanaimo. The BC Ferries dock is just around the corner and, overhead, seaplanes bustle to and fro.

Marina Notes -- *3 hrs. of moorage no charge twice per week. $90 deposit for a power cord converter. The wide, concrete docks at Cameron Island are very well maintained and the Visiting Vessel Pier is particularly well-suited to mega yachts. The ability to make reservations, as well as the location in the heart of downtown Nanaimo, makes this a very popular facility. It is adjacent to and operated by the Nanaimo Port Authority Boat Basin, whose facilities it shares - including a fuel dock, the Eco-Barge pump-out and a bathhouse with a laundry (showers $1/3min; laundry $2/load).

Notable -- Pioneer Waterfront Plaza's shops and restaurants edge the boardwalk. For a casual stroll with stunning Gulf Island views, take the Harbourside Walkway from Harbour Park Mall to Departure Bay Ferry Terminal. It passes many popular attractions - Georgia Park, the Bastion, Nanaimo District Museum, and Maffeo Sutton Park. A foot bridge over Swy-A-Lana Lagoon leads to the park and playground. Newcastle Island Provincial Marine Park, via the foot ferry, makes a good day trip - beaches, trails, wildlife viewing areas, forests, and an interpretive center in the historic pavilion. On Protection Island, the Dinghy Dock Floating Marine Pub (753-2373), is a popular spot; dock 'n' dine with great views of Nanaimo (or take the 10 minute ferry ride - to 11pm).

Nanaimo Port Authority

PO Box 131; 104 Front Street; Nanaimo, BC V9R 5KA

Tel: (250) 755-1216 **VHF: Monitor** Ch. 67 **Talk** n/a
Fax: (250) 753-4899 **Alternate Tel:** n/a
Email: marina@npa.ca **Web:** www.npa.ca
Nearest Town: Nanaimo **Tourist Info:** (250) 753-1191

Navigational Information
Lat: 49°10.256' **Long:** 123°56.032' **Tide:** 15 ft. **Current:** n/a **Chart:** 3447
Rep. Depths (MLW): Entry 15 ft. **Fuel Dock** 26 ft. **Max Slip/Moor** 26 ft./-
Access: On the Strait of Georgia west of Gabriola Island

Marina Facilities *(In Season/Off Season)*
Fuel: Gasoline, Diesel
Slips: 225 Total, 25 Transient **Max LOA:** 590 ft. **Max Beam:** 60 ft.
 Rate (per ft.): Day $0.90* **Week** n/a **Month** n/a
 Power: 30 amp $3**, **50 amp** $6, **100 amp** $12, **200 amp** n/a
 Cable TV: No **Dockside Phone:** No
 Dock Type: Floating, Long Fingers, Alongside, Wood
Moorings: 0 Total, 0 Transient **Launch:** n/a
 Rate: Day n/a **Week** n/a **Month** n/a
Heads: 3 Toilet(s), 3 Shower(s)
Laundry: 6 Washer(s), 6 Dryer(s), Book Exchange **Pay Phones:** Yes, 2
Pump-Out: OnSite, 1 Central **Fee:** $2 **Closed Heads:** Yes

Marina Operations
Owner/Manager: Andrew Pitcher, HM **Dockmaster:** David Mailloux
In-Season: Year-Round, 7am-11pm **Off-Season:** n/a
After-Hours Arrival: Tie up and pay in the morning
Reservations: No **Credit Cards:** Visa/MC, Debit
Discounts: None
Pets: Welcome **Handicap Access:** Yes, Heads, Docks

Marina Services and Boat Supplies
Services - Docking Assistance, Security *(night watchman and video)*, Trash Pick-Up, Dock Carts **Communication -** Mail & Package Hold, Phone Messages, Fax in/out *($1.50)*, Data Ports *(Wi-Fi - Broadband Xpress)*, FedEx, UPS **Supplies - OnSite:** Ice *(Cube)* **Near:** Ships' Store *(Harbour Chandler 753-2425, Dock Shoppe)* **1-3 mi:** West Marine *(758-8048)*

Boatyard Services
OnSite: Crane *(2T)* **Nearest Yard:** Nanaimo Shipyard (250) 753-1151

Restaurants and Accommodations
OnSite: Restaurant *(Penny's Palapa 753-6300, L & D $6-9, Apr-Oct)*, *(Troller's Fish & Chips 741-7994, L & D $8-13)* **Near:** Restaurant *(Lighthouse Bistro 754-3212, L $7-11, D $11-21, Sun brunch 10am-2pm $13/7)*, *(Cutter's Bistro 753-6601, D $17-19)*, *(Gina's 753-5411, L $6-12, D $10-16)*, Snack Bar *(Grapevine on the Bay 746-0797, B $3-6, L $4-19, D $10-19)*, *(Just Desserts 753-2264)*, Coffee Shop *(Javawocky)*, Motel *(Best Western 754-6835, $80-150)*, Hotel *(Howard Johnson Harbourside 753-2241, $80-150)*, *(The Coast Bastion Inn 753-6601, $99-200)*

Recreation and Entertainment
Near: Bowling *(Fiesta Lanes 753-2451)*, Movie Theater *(Bay Theatre 754-3585)*, Museum *(Nanaimo District Museum 753-1821)*, Cultural Attract *(Port Theatre 754-8550)*, Sightseeing *(The Nanaimo Bastion 10:30am-4pm Jun-LabDay 753-1821)*, Special Events *(Seafood Chowder Cook-Off 753-1821 - Aug)*, Galleries *(Artisans Studio 753-6151)* **Under 1 mi:** Video Rental *(Nichol Video 753-9441)*, Park *(Swy-A-Lana Park 765-5200)* **1-3 mi:** Golf Course *(Pryde Vista 753-6188)*, Fitness Center *(Pinnacle 753-0138)*

Provisioning and General Services
OnSite: Fishmonger *(On F dock)* **Near:** Supermarket *(Thrifty Foods 754-6273)*, Gourmet Shop, Health Food *(Charlie Brown's 753-5211)*, Wine/Beer, Liquor Store *(BC Liquor 741-5555)*, Farmers' Market *(Pioneer Plaza, Fri 10am-2pm, midApr-Oct)*, Bank/ATM *(Scotiabank 716-2300)*, Post Office, Protestant Church *(St. Paul's Anglican 753-2523)*, Library *(753-1154)*, Beauty Salon *(Hers Hair 754-5252)*, Barber Shop *(His Hair 754-9933)*, Dry Cleaners, Laundry *(Boat Basin 754-8654)*, Pharmacy *(London Drugs 753-5566)*, Florist *(Exclusively Yours 754-6363)*, Clothing Store *(Man-Deez Fashions 716-0558)*, Retail Shops *(Harbour Park Mall)*, Copies Etc. *(In Print 754-2542)* **Under 1 mi:** Convenience Store *(Superette Foods 754-5741)*, Market *(J & J Market 754-3601)*, Delicatessen *(Delicado's 753-6524)*, Bookstore *(Downtown Books 753-0102)* **1-3 mi:** Bakery *(Bun's Master Bakery 753-1273)*, Catholic Church

Transportation
OnSite: Ferry Service *(FastFerry to Vancouver 866-206-5969; HarbourLynks - 753-4443 $30/17, 3 trips a day. Also 1 mi: Protection Island 753-8244/Gabriola Island/ BC Ferries - Horseshoe Bay, Tsawwassen, & Islands)* **OnCall:** Taxi *(AC Taxi 753-1231)* **Near:** Rental Car *(Avis 716-8898 at Best Western)* **Airport:** Nanaimo *(11 mi.)*

Medical Services
911 Service **OnCall:** Ambulance *(741-0102)* **Near:** Optician *(Q-Vision 753-2021)* **Under 1 mi:** Doctor *(Medical Arts Centre 741-0447)*, Holistic Services *(Oceanview Massage Clinic 618-3191)* **1-3 mi:** Dentist *(Harewood Dental Clinic 754-1949)*, Chiropractor *(Mountainview Chiropractic 753-5351)* **Hospital:** Nanaimo Regional 248-2332 *(3 mi.)*

Setting -- Safely tucked behind Newcastle & Protection Islands and the 600-foot Visiting Vessels Pier, the NPA Boat Basin is surely one of the most secure marinas on the west coast. The well-sheltered docks lie right at the foot of downtown Nanaimo, with shopping, dining, theatres, and fine hotels only steps away. An attractive multistory contemporary pink building trimmed with turquoise (and rimmed with white tubular railings) houses the office and amenities.

Marina Notes -- *3 hrs. of moorage complimentary twice per week. 9,000 feet of dock. **15 and 20 amp power $3/day. $90 deposit for a power cord adapter. The friendly staff, clean accommodations, and surrounding activities make this marina a must to visit. No reservations are taken, so call ahead to inquire about availability. This is also a customs port-of-entry site. The unique floating Eco Barge with its solar powered reception facility is located on the central breakwater at the entrance to the Commercial Boat Basin. Showers $1/3 min. Wash/dryers $2/load.

Notable -- So much to do! First-class restaurants and pubs, gift shops, and galleries are grouped in the area surrounding the basin. Nearby 800-seat Port Theatre (754-8550) is a fabulous facility that celebrates the arts throughout the year with music, dance, and live theatre; it also offers a waterfront room for private receptions and conferences. Take the foot ferry to either Newcastle Island or Protection Island for some hiking and exploration, or kayak around Newcastle Island by full moon - a two-hour odyssey filled with brilliant phosphorescence and seals. Or take to the skies and see Vancouver Island by seaplane (from $49, 800-661-5599). Visit the saltwater lagoon in Swy-A-Lana Park where the natural ebb and flow of the tides creates a marine life habitat.

5. BC – Ladysmith to Cowichan Bay

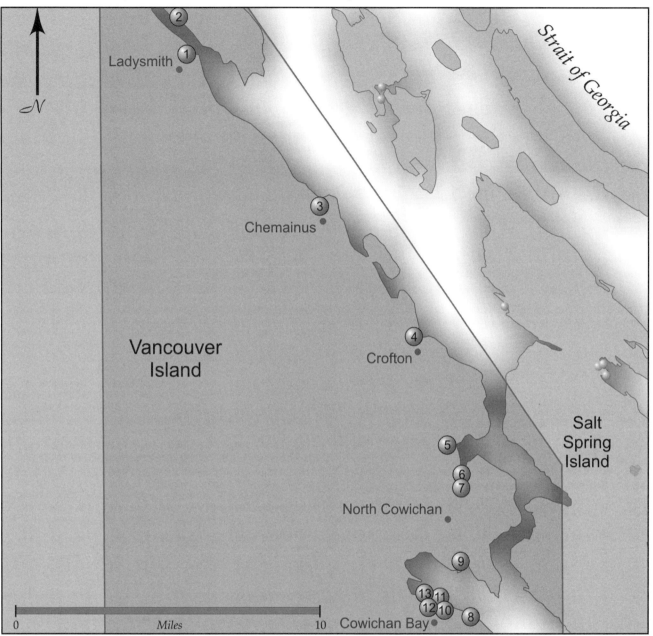

MAP	MARINA	HARBOR	PAGE	MAP	MARINA	HARBOR	PAGE
1	Ladysmith Maritime Society	Ladysmith Harbour	74	8	Cherry Point Marina	Cowichan Bay/Cherry Point	81
2	Page Point Inn & Marina	Ladysmith Harbour	75	9	Genoa Bay Marina	Cowichan Bay/Genoa Bay	82
3	Chemainus Public Wharf	Stewart Channel	76	10	Fishermen's Wharf Association	Cowichan Bay	83
4	Crofton Public Wharf	Osborne Bay	77	11	Cowichan Bay Marina	Cowichan Bay	84
5	Maple Bay Public Wharf	Maple Bay	78	12	Pier 66 Marina	Cowichan Bay	85
6	Bird's Eye Cove Marina	Maple Bay/Bird's Eye Cove	79	13	Bluenose Marina	Cowichan Bay	86
7	Maple Bay Marina	Maple Bay/Bird's Eye Cove	80				

▸ **Currency** — In Canadian Marina Reports, all prices are in Canadian dollars. In U.S. Marina Reports, all prices are in U.S. dollars.

▸ **"CCM"** — Denotes a Certified Clean Marina, a state/provincial award for environmental excellence. See page 298 for an explanation and page 299 for a list of Pump-Out facilities.

▸ **Ratings & Reviews** — An overview of the Atlantic Cruising Club's rating system is on page 6 and details on the content of each Marina Report are on pages 7 – 11.

▸ **Marina Report Updates** — Comments from boaters and new information from ACC reviewers and marinas are posted regularly on www.AtlanticCruisingClub.com.

Ladysmith Maritime Society

Ladysmith Maritime Society

PO Box 1030; 12335 Rocky Creek Rd.; Ladysmith, BC V9G 1A7

Tel: (250) 245-2248 **VHF: Monitor** n/a **Talk** n/a
Fax: (250) 245-9133 **Alternate Tel:** n/a
Email: aendacott@shaw.com **Web:** n/a
Nearest Town: Ladysmith *(0.25 mi.)* **Tourist Info:** (250) 245-2112

Navigational Information
Lat: 48°59.768' **Long:** 123°48.845' **Tide:** 12 ft. **Current:** n/a **Chart:** 3475
Rep. Depths *(MLW)*: **Entry** 7 ft. **Fuel Dock** n/a **Max Slip/Moor** 12 ft./-
Access: First set of docks on the south side of Ladysmith Harbour

Marina Facilities *(In Season/Off Season)*
Fuel: No
Slips: 130 Total, 8 Transient **Max LOA:** 100 ft. **Max Beam:** n/a
Rate *(per ft.)*: **Day** $0.75 **Week** n/a **Month** n/a
Power: 30 amp $5* **50 amp** n/a **100 amp** n/a **200 amp** n/a
Cable TV: No **Dockside Phone:** No
Dock Type: Floating, Long Fingers, Alongside, Wood
Moorings: 0 Total, 0 Transient **Launch:** n/a
Rate: Day n/a **Week** n/a **Month** n/a
Heads: None
Laundry: None **Pay Phones:** No
Pump-Out: No **Fee:** n/a **Closed Heads:** Yes

Marina Operations
Owner/Manager: Dianna & Cliff Fisher **Dockmaster:** Same
In-Season: Year-Round, 24 hrs.** **Off-Season:** n/a
After-Hours Arrival: Pay at the toll booth
Reservations: Yes, Preferred **Credit Cards:** No
Discounts: None
Pets: Welcome **Handicap Access:** No

Marina Services and Boat Supplies
Services - Dock Carts **Supplies - Near:** Ice *(Cube)* **Under 1 mi:** Bait/Tackle

Boatyard Services
Nearest Yard: Dalby's Service Ltd. (250) 245-2244

Restaurants and Accommodations
Near: Restaurant *(Dragon City Restaurant 245-3322)*, *(George's Tavern 245-2292, L $7-11, D $9-30, will deliver)*, Inn/B&B *(Sandpiper 245-5794, $70-95)* **Under 1 mi:** Restaurant *(The Sea Galley 245-4614)*, Lite Fare *(Renee's Soup & Sandwich 245-4198)*, Pizzeria *(Riailka's Greek and Mediterranean 245-1331)*, Motel *(Arcady Auto Court 245-3530, $55-95)*, *(Seaview Motel 245-3768, $50-90)*, Hotel *(Ladysmith Inn 245-8033, $52-60)* **1-3 mi:** Restaurant *(Northbrook Restaurant III 245-5000, L $7-15, D $8-22, Mon-Sat 11:30am-11pm, Sun 4-9pm)*, Fast Food *(KFC, Dairy Queen, McDonald's)*, Inn/B&B *(Hawley Place B&B 245-4431)*

Recreation and Entertainment
OnSite: Sightseeing *(Harbour Tours)* **Near:** Beach *(Transfer - notably warm water)*, Volleyball, Boat Rentals *(Sealegs Kayaking Adventure 245-4096 at Transfer Beach)*, Park *(Transfer Beach - picnic & BBQ, adventure playground, water spray park, kayaks)*, Museum *(Black Nugget Museum 245-2112, by app't.)*, Cultural Attract *(Summer performances at the outdoor amphitheatre at Transfer Beach)*, Special Events *(Oyster Fest - May/Jun, Ladysmith Days - Jul, Art on the Avenue - Aug, Fall Fair - Sep)*, Galleries *(Udderly Art 722-7140)*

Provisioning and General Services
Under 1 mi: Market *(49th Parallel Grocery 245-3221)*, Health Food *(Ladysmith Health Foods Store 245-2123)*, Wine/Beer, Liquor Store *(Ladysmith Inn 245-8868)*, Bakery *(Old Town Bakery 245-2531)*, Bank/ATM *(Royal Bank of Canada 245-7111)*, Protestant Church, Library *(245-2322)*, Beauty Salon *(Chez Christine 245-3832)*, Dry Cleaners *(Capri Dry Cleaners 245-2334)*, Pharmacy *(Pharmasave 245-3113)*, Hardware Store *(Ladysmith Home 245-3441)*, Florist *(Blossoms at the 49th 245-3221)*, Retail Shops **1-3 mi:** Convenience Store *(Timberland Convenience Store & Pizza 245-1001)*, Supermarket *(Carlo's Place)*, Catholic Church, Barber Shop *(Island Haircutting 245-8695)*, Laundry *(Coronation Laundromat & Dry Cleaning 245-1288)*, Bookstore *(Fraser and Naylor Booksellers 245-4726)*, Copies Etc. *(Ladysmith Printers 245-8121)*

Under 1 mi:
Heated Pool *(Frank Jameson Community Centre Pool)*, Tennis Courts, Fitness Center *(Frank Jameson)*, Video Rental *(Crazy Mike's 245-3151)* **1-3 mi:** Golf Course *(Ladysmith 245-7313)*

Transportation
Near: Local Bus *(Island Coach Lines 245-7332)* **1-3 mi:** Rail *(VIA Rail 888-842-7245)* **3+ mi:** Rental Car *(Budget 245-8733, 9 mi.)* **Airport:** Cassidy *(7 mi.)*

Medical Services
911 Service **OnCall:** Ambulance **Under 1 mi:** Doctor *(Hillside Medical Centre 245-2235)*, Chiropractor *(Samek Family Chiropractic 245-8778)* **1-3 mi:** Dentist *(Phelps 245-7151)* **Hospital:** Ladysmith 245-2221 *(0.5 mi.)*

Setting -- Floating boathouses edge the Ladysmith Maritime Society's network of rugged wood docks - Purple Martin birdhouses perch on the pilings. A rustic office bobs on one of the piers. The interesting collection of classic vessels berthed here warrants a stroll around the docks. Interrupting the peaceful tree-lined vista ashore is a large blue and brick-red kayak shop. The laid-back setting is a notable counterpoint to the bustling, pretty town nearby.

Marina Notes -- *15 amp power $3/night; call ahead for 30 amp power (limited number available). **Dockmaster is on-site daily from 9 -11am. The marina is intended primarily for permanent moorage, but there is usually room for guests so there are few amenities for boaters. But the town provides ample provisioning opportunities. FYI: Purple Martins are a species in distress in B.C.; a widespread nesting box program is attempting to bring them back.

Notable -- Free Harbor tours are conducted during the summer months (donations appreciated). The lovely town of Ladysmith is just across the highway (after a 100+ step climb) and offers many services and activities. Visit the Chamber of Commerce Info Centre for dates and times. There are art fairs, museums and the Ladysmith Heritage Walking Tour - featuring the town's turn-of-the-last-century architecture. If your tastes run to the unusual, they will steer you to the local skinny-dipping spot, recommend a bungee jumping outing, or, if you are looking for a truly unique experience, send you on a tour of the local sewage treatment plant. Nearby Transfer Beach Park hosts the largest amphitheater in B.C.- as well as many activiities In mid-June, the harbor hosts the Offshore Sailing & Heritage Boat and Marine Show (Contact Pat Samson at 250-709-0367).

Navigational Information
Lat: 49°00.701' Long: 123°49.349' Tide: 12 ft. Current: n/a Chart: 3475
Rep. Depths (MLW): Entry 10 ft. Fuel Dock 10 ft. Max Slip/Moor 15 ft./-
Access: On the eastern side of Ladysmith Harbour off Stuart Channel

Marina Facilities (In Season/Off Season)
Fuel: Gasoline, Diesel
Slips: 90 Total, 20 Transient Max LOA: 100 ft. Max Beam: n/a
 Rate (per ft.): Day $1.00 Week n/a Month $6.50
 Power: 30 amp $3, 50 amp n/a, 100 amp n/a, 200 amp n/a
Cable TV: Yes Dockside Phone: No
Dock Type: Floating, Alongside, Concrete
Moorings: 0 Total, 0 Transient Launch: n/a
 Rate: Day n/a Week n/a Month n/a
Heads: 2 Toilet(s), 2 Shower(s)
Laundry: 1 Washer(s), 1 Dryer(s), Iron Board Pay Phones: Yes, 1
Pump-Out: No Fee: n/a Closed Heads: Yes

Marina Operations
Owner/Manager: Lawrence & Lexie Lambert Dockmaster: Same
In-Season: Year-Round, 8am-5pm Off-Season: n/a
After-Hours Arrival: Call ahead
Reservations: Yes, Preferred Credit Cards: Visa/MC, Amex, Debit
Discounts: None
Pets: No Handicap Access: No

Page Point Inn & Marina
4760 Brenton-Page Road; Ladysmith, BC V9G 1L7

Tel: (250) 245-2312; (877) 860-6866 VHF: Monitor Ch. 66A Talk n/a
Fax: (250) 245-7546 Alternate Tel: n/a
Email: info@pagepointinn.com Web: www.pagepointinn.com
Nearest Town: Ladysmith (9 mi.) Tourist Info: (250) 245-2112

Marina Services and Boat Supplies
Services - Docking Assistance, Trash Pick-Up Communication - Mail & Package Hold, Phone Messages, Fax in/out, Data Ports (owner's office), FedEx, DHL, UPS, Express Mail (Sat Del)

Boatyard Services
Nearest Yard: Dalby's Service Ltd. (250) 245-2244

Restaurants and Accommodations
OnSite: Restaurant (Page Point Inn Restaurant 245-2312, B $7-9, L $9-14, D $25-30), Inn/B&B (Page Point Inn 245-2312, $85-125, includes continental breakfast - 11 nicely appointed rooms in the Inn & Old Quarter - some with water views & fireplaces) 1-3 mi: Restaurant (The Sea Galley 245-4614, dinghy), (Northbrook 245-5000, L $7-15, D $8-22, dinghy), (Juniper Café 245-3226, dinghy), Lite Fare (Renee's Soup & Sandwich 245-4198, dinghy), Pizzeria (Roberts Street Pizza 245-0168), Motel (Holiday House Motel 245-2231, $50-135)

Recreation and Entertainment
OnSite: Boat Rentals (Sailboats, kayaks; charter or gaze at 12-meter Dame Pattie, the 1967 Australian America's Cup challenger) Near: Jogging Paths Under 1 mi: Golf Course (Ladysmith Golf Club 245-7313), Video Rental (Crazy Mike's Video 245-3151) 1-3 mi: Park, Museum (Black Nugget Museum 245-2112, Jul-Aug 9am-3:30pm), Cultural Attract (Transfer Beach outdoor amphitheatre), Sightseeing (A Heritage Walk in Ladysmith, brochure at the Ladysmith & District Historical Society 245-7611),

Special Events (Oyster Fest - May or Jun, Ladysmith Celebration Days, downtown and at Transfer Beach - end of Jul, Art on the Avenue - Aug), Galleries (Mary Fox Pottery 245-3778)

Provisioning and General Services
1-3 mi: Market (49th Parallel Grocery 245-3221, 7:30am-9pm; phone ahead or preorder a prepackaged picnic basket), Supermarket (Carlo's Place, dinghy), Health Food (Ladysmith Health Foods Store 245-2123, dinghy), Liquor Store, Bakery (Old Town Bakery 245-2531), Post Office, Catholic Church (St. Mary's 245-3414), Protestant Church (Ladysmith Fellowship Baptist Church 245-5113), Library (245-2322), Beauty Salon (Chop Stix 245-5788), Barber Shop (Adele's Hair Salon 245-0555), Bookstore (Fraser & Naylor Booksellers 245-4726), Pharmacy (Rexall Drug Stores 245-7184), Retail Shops (Coronation Mall)

Transportation
3+ mi: Rental Car (Budget 245-8733, 9 mi.), Rail (VIA Rail 888-842-7245, 9 mi.) Airport: Cassidy (7 mi.)

Medical Services
911 Service OnCall: Ambulance (245-2157) 3+ mi: Doctor (Hillside Medical Centre 245-2235, 8 mi.), Dentist (8 mi.), Holistic Services (Cedar Massage Therapy 722-4652, 8 mi.), Veterinarian (Ladysmith Animal Hospital 245-2274, 8 mi.), Optician (Vision Arts 245-8203, 8 mi.) Hospital: Ladysmith & District Hospital 245-2221 (8 mi.)

Setting -- Beyond the Woods Islands on the north side of the 49th parallel, Page Point Inn & Marina nestles in the trees on a hillside overlooking Ladysmith Harbor. Picnic tables and canvas gazebos, sheltered by old pines, are scattered about the park-like grounds overlooking the clean and uncluttered. docks. The view from the lovely inn's dining deck takes in the forested shore and the town of Ladysmith on the far side of the harbor.

Marina Notes -- 1,000 ft. of new concrete docks add more space for residents and transient boaters. Established in 1948 - as Manana Lodge - sailors were attracted by the warm water and microclimate. Sailboat rentals on-site - plus kayaking and canoeing. Also on-site is the 12-meter Dame Pattie, available for charter. A shuttle bus may be available for groups. Distances to shopping are by boat; distances to medical services are by land. Note: No pets allowed.

Notable -- Page Point Inn, with its excellent wine list and devoted following, serves three meals daily in the main Dining Room and the Fireside Pub. The hardwood floors, antiques, stone fireplace, and outdoor seating overlooking the harbor complement the fine food; it's worth changing course for a meal. Ladysmith, named one of the 10 prettiest towns in Canada, is a 9 mile drive or a short boat ride across the harbor. It offers two distinct shopping experiences: the quaint First Avenue in the downtown area and the modern Coronation Mall on the Island Highway. Annual celebrations include Paddlefest in the spring for paddle sport enthusiasts - a two-day festival of boats, gear, food, and music. Canada Day is celebrated with birthday cake and music on the historic waterfront. Arts on the Avenue in late August and the Ladysmith Fall Fair in September both attract many visitors.

PHOTOS ON DVD: 14

Navigational Information
Lat: 48°55.509' **Long:** 123°42.873' **Tide:** 12 ft. **Current:** n/a **Chart:** 3475
Rep. Depths (*MLW*): Entry 20 ft. **Fuel Dock** n/a **Max Slip/Moor** 20 ft./-
Access: West of Salt Spring Island

Marina Facilities *(In Season/Off Season)*
Fuel: No
Slips: 17 Total, 10 Transient **Max LOA:** 130 ft. **Max Beam:** n/a
 Rate *(per ft.)*: **Day** $0.85/0.65 **Week** n/a **Month** n/a
 Power: 30 amp $5, **50 amp** n/a, **100 amp** n/a, **200 amp** n/a
 Cable TV: No **Dockside Phone:** No
 Dock Type: Floating, Long Fingers, Alongside, Wood
Moorings: 0 Total, 0 Transient **Launch:** n/a
 Rate: Day n/a **Week** n/a **Month** n/a
Heads: 2 Toilet(s), 2 Shower(s)
Laundry: None **Pay Phones:** Yes, 1
Pump-Out: No **Fee:** n/a **Closed Heads:** No

Marina Operations
Owner/Manager: Harmen Bootsma, HM **Dockmaster:** Same
In-Season: Year-Round, 9am-5pm **Off-Season:** n/a
After-Hours Arrival: Call ahead
Reservations: Yes, self-pay envelope at office **Credit Cards:** No
Discounts: None
Pets: Welcome **Handicap Access:** Yes, Heads, Docks

Chemainus Public Wharf

PO Box 193; Chemainus, BC V0R 1K0

Tel: (250) 246-4655 **VHF: Monitor** Ch. 66A **Talk** n/a
Fax: (250) 246-1595 **Alternate Tel:** (250) 715-8186
Email: n/a **Web:** n/a
Nearest Town: Chemainus **Tourist Info:** (250) 246-3944

Marina Services and Boat Supplies
Services - Docking Assistance, Security *(locked gates)* **Communication -** Mail & Package Hold, Phone Messages, Fax in/out **Supplies - Near:** Ice *(Cube)*, Bait/Tackle **1-3 mi:** Propane *(Propane Bottle 246-1838)*

Boatyard Services
Nearest Yard: Body Shop for Boats (888) 310-0222

Restaurants and Accommodations
Near: Restaurant *(Barnacle Barney's Fish & Chips 246-2710, B $5-8, L $5-8, D $5-10)*, *(Ding Ho 416-0338, L $6-9, D $6-24, will deliver to the boat until 9pm)*, *(The Harbourside Café 416-0107, B $4-7, L $5-10, D $8-17)*, Snack Bar *(Small Tall Treats 246-3001, sandwiches and desserts)*, Coffee Shop *(Chemainus Coffee House 246-3443)*, Pizzeria *(Mandolinos 246-2848, L $20-27, D $20-27)* **Under 1 mi:** Fast Food *(Subway 246-1101)*, Inn/B&B *(At The Sea-Breeze 246-4593, $55-75)* **1-3 mi:** Snack Bar *(Ma's Colectibles & Gelato 246-9583)*, Motel *(Chemainus Motel 246-3282, $55-90)*, Inn/B&B *(A Small World B&B 246-9962, $60-110)*

Recreation and Entertainment
Near: Beach, Picnic Area, Playground, Jogging Paths, Video Rental *(Pioneer's Video & Music 246-9944)*, Park *(Kin Park and Beach, Waterwheel Park)*, Museum *(Chemainus Valley Museum 246-2445, 10am-3pm daily)*, Cultural Attract *(Chemainus Theatre 246-9800)*, Special Events *(Chemainus Daze - Jun)* **Under 1 mi:** Sightseeing *(murals and sculptures; walk or call Chemainus Tours 246-5055 for local or Cowichan Valley tours)*, Galleries *(at Theatre, plus Oak Leaf Gallery 246-4990)* **1-3 mi:** Golf Course *(Mount Brenton 246-2588)*, Bowling *(Chemainus Lanes 246-3541)*

Provisioning and General Services
Near: Market *(49th Parallel Grocery 246-2850, Viking Foods 246-3012)*, Liquor Store *(BC Liquor 246-3531)*, Bakery *(Chemainus Bakery 246-4321)*, Post Office *(246-9531)*, Protestant Church *(Congregational 246-9119)*, Beauty Salon *(Shear Impressions 246-2141)*, Barber Shop *(Trout's Barber Shop 246-7443)*, Dry Cleaners *(Mrs. One Hour Dry Cleaners 246-1444)*, Laundry *(Mrs. One Hour)*, Pharmacy *(Chemainus Pharmacy 246-2151)*, Retail Shops *(Bonnie & Frank's Gifts 246-2254, Toad Hall Emporium 246-4400)* **Under 1 mi:** Convenience Store *(Old Road Market 246-2345)* **1-3 mi:** Delicatessen *(Galleto Market & Deli 246-4770)*, Wine/Beer, Bank/ATM *(CIBC 246-3257)*, Catholic Church *(St. Joseph's 246-3260)*, Library *(246-9471)*, Bookstore *(The Little Shop of Novels 246-5330)*, Hardware Store *(True Value 246-3631)*, Florist *(Sunshine 246-4354)*, Clothing Store *(Pennies From Heaven 246-1069)*

Transportation
OnCall: Rental Car *(Budget 245-8733)*, Taxi *(Saltair Taxi 246-4414)*, Ferry Service *(Thetis & Kuper Islands)* **Near:** Bikes *(Chemainus Tours 246-5055)*, Local Bus *(Island Coach Lines)* **Under 1 mi:** Water Taxi *(Gulf Island Explorers 246-7866)* **Airport:** Cassidy *(14 mi.)*

Medical Services
911 Service **OnCall:** Ambulance *(746-4058)* **Near:** Doctor *(Chemainus Medical 246-3261)*, Veterinarian *(Chemainus 246-1222)*, Optician *(Chemainus Family Eye Care 246-3405)* **Under 1 mi:** Dentist *(Rosengart 246-9921)* **1-3 mi:** Chiropractor *(Dares 246-4011)* **Hospital:** Cowichan District 746-4141 *(20 mi.)*

Setting -- The pocket-sized public docks at Chemainus are neatly wedged between the Thetis Island ferry dock and the Weyerhaeuser shipping pier. A small, yellow building sits right at the foot at the ramp, overlooking the basin - it houses the office, as well as the heads & showers. The ferries keep the water stirred up and the generators on the ocean-going freighters run all night. This apparent chaos is presided over by the able Harbormaster and he makes it work. Old Town Chemainus is only a few steps from the dock.

Marina Notes -- The harbormaster accepts U.S. dollars but advises visitors to convert to Canadian currency before arriving to get a better exchange rate. Use a self-pay envelope after hours. Well kept wooden docks; 30 amp power is available. This is one of the few public docks in the area to offer a bathhouse and docking assistance.

Notable -- Restaurants, ice cream parlors, and gift shops line the street down to the marina, making this a convenient stop. Chemainus is justly proud of its murals and sculptures, the first created in 1982. Professionally done, they tell the story of the town's pioneers and of the First Nations' people who were the area's original residents. Walking tours are available as well as tours by horse-drawn wagon or replica steam train - Chemainus Tours start at Waterwheel Park ($10/5 kids for 30 minutes). The Chemainus Theatre Festival is a year-round professional company guaranteed to entertain; a dining room with buffet menu and a gallery are in the same building. There are free, live Sunday afternoon concerts in the park during the summer season.

Navigational Information

Lat: 48°51.956' **Long:** 123°38.184' **Tide:** 11.5 ft. **Current:** n/a **Chart:** 3441
Rep. Depths (*MLW*): **Entry** 6 ft. **Fuel Dock** n/a **Max Slip/Moor** 12 ft./-
Access: Off Stuart Channel in Osborne Bay west of Salt Spring Island

Marina Facilities *(In Season/Off Season)*

Fuel: No
Slips: 13 Total, 2 Transient **Max LOA:** 45 ft. **Max Beam:** n/a
 Rate *(per ft.)*: **Day** $0.85/0.65 **Week** n/a **Month** n/a
 Power: 30 amp $5*, **50 amp** n/a, **100 amp** n/a, **200 amp** n/a
 Cable TV: No **Dockside Phone:** No
 Dock Type: Floating, Alongside, Wood
Moorings: 0 Total, 0 Transient **Launch:** n/a
 Rate: Day n/a **Week** n/a **Month** n/a
Heads: 2 Toilet(s)
Laundry: None **Pay Phones:** Yes, 1
Pump-Out: No **Fee:** n/a **Closed Heads:** No

Marina Operations

Owner/Manager: Harmen Bootsma, HM **Dockmaster:** Same
In-Season: Year-Round **Off-Season:** n/a
After-Hours Arrival: Call ahead
Reservations: No **Credit Cards:** No
Discounts: None
Pets: Welcome **Handicap Access:** No

Crofton Public Wharf

Chaplin Street; Crofton, BC V0R 1R0

Tel: (250) 246-4655 **VHF: Monitor** n/a **Talk** n/a
Fax: (250) 246-1595 **Alternate Tel:** (250) 715-8186
Email: n/a **Web:** n/a
Nearest Town: Crofton **Tourist Info:** (250) 748-1111

Marina Services and Boat Supplies

Supplies - 1-3 mi: Propane *(Crofton Auto Service 246-3115)*

Boatyard Services

OnSite: Launching Ramp **Nearest Yard:** Covey Marine (250) 748-3130

Restaurants and Accommodations

Near: Restaurant *(Crofton Grill 246-4827, B $5-11, L $7-10, D $5-12, specialize in Curry Chicken and bake their own cinnamon buns)*, Lite Fare *(Brass Bell Pub 246-1230, L $6-10, D $7-15, no kids; Sat wings $2/doz, Sat burger & fries $6, Sun roast beef $8; homemade desserts on wknds)*, Pizzeria *(Canadian 2 for 1 Pizza 246-1212)*, Motel *(Twin Gables Motel 246-3112, $40-80, 1 bd $40, 1 bd + kitchenette $60, 2 bd $80)*, *(Croft Inn 246-9222, $69-115, all one bedroom; ea. w/ queen-size and double bed; walking distance to the Wharf; fully supplied kitchen units in each room)* **Under 1 mi:** Restaurant *(Carol's Place 246-2414, L $5-8, D $7-12)*, Hotel *(Crofton Hotel 246-3122, $50-70)*, Condo/Cottage *(Osborne Bay Resort 246-4787, log cottages)* **1-3 mi:** Restaurant *(The Seaside Kitchen 537-2249, L $6-12, D $6-22)*, *(The Vesuvius Bar and Grill 537-2312, L $8-17, D $8-17)* **3+ mi:** Pizzeria *(Pizza Hut 310-1010, 8 mi.)*

Recreation and Entertainment

OnSite: Picnic Area, Park *(or Osborne Bay Regional Park 746-2500)*, Museum *(1905 Olde School House Museum 246-2456, local history;* home to the visitors' info center and a gift gallery) **Under 1 mi:** Playground, Fitness Center *(Crofton Community Fitness 246-5399)*, Jogging Paths, Video Rental *(Crofton Movies and More 246-4066)* **1-3 mi:** Beach *(Osborne Bay)* **3+ mi:** Golf Course *(Mount Brenton Golf Course 246-2588, 7 mi.)*, Galleries *(Genoa Bay Gallery 746-5506, 6 mi.)*

Provisioning and General Services

Near: Post Office *(246-9911)*, Beauty Salon *(Maryann's Hair Studio 246-4446)* **Under 1 mi:** Protestant Church **1-3 mi:** Convenience Store *(Shell Convenience Store)*, Market *(Crofton Foods 246-3471)*, Hardware Store *(Crofton Hardware 246-9239)*, Retail Shops **3+ mi:** Health Food *(Community Farm Store 748-6227, 8 mi.)*, Liquor Store *(BC Liquor 746-1323, 8 mi.)*

Transportation

OnSite: Ferry Service *(BC Ferry to Spring Island 888-223-3779)* **OnCall:** Taxi *(Duncan Taxi 746-4444)* **Airport:** Duncan *(12 mi.)*

Medical Services

911 Service **OnCall:** Ambulance *(746-4058)* **3+ mi:** Doctor *(Arbutus Medical Clinic 746-5115, 8 mi.)*, Dentist *(Island Dental 748-0301, 8 mi.)*, Chiropractor *(Hoshizaki 746-6171, 8 mi.)*, Veterinarian *(Duncan Animal Hospital 746-1966, 8 mi.)*, Optician *(Opti View 1 Hour 748-9229, 8 mi.)*
Hospital: Cowichan District Hospital 746-4141 *(8 mi.)*

PHOTOS ON DVD: 6

Setting -- Crofton sits on the west side of Stuart Channel. This well-maintianed public dock is adjacent to the Salt Spring Island ferry dock. At the top of the red-railed ramp, is a grassy park with picnic tables, backed by a combination visitor center, small museum and gift shop in a lovely restored schoolhouse - all nestled in a residential area. Fishing boats provide most of the clientele, particularly in the off-season months.

Marina Notes -- *20 amp $3/day. If you need some quiet downtime and are self-contained, this is the place. The only disturbance will be the wash of the ferries as they come and go. Harbormaster Harmen Bootsma also oversees the Chemainus and Maple Bay Public Docks. The heads are located behind the museum. Note: Paper and pulp are the mainstays of the local economy, which creates a setting that is decidedly industrial.

Notable -- Crofton is equidistant from Victoria and Nanaimo and 8 miles south of Chemanius and 7 miles north of Maple Bay. It is surrounded by the rolling hills of the Cowichan Valley, and is home to several wineries and a cidery that offer tours and tastings. Interesting free tours of the Crofton Pulp & Paper Mill - and its Japanese Garden - are scheduled on weekends June through August - call ahead 246-6006. The new seawalk starts at the wharf and leads to Osborne Bay Regional Park which is popular for fishing, clamming and wildlife watching. Board the ferry next door for a hop to Salt Spring Island. It lands at Vesuvius Bay, home to two restaurants: the Seaside Kitchen Restaurant (free dock for diners), and the Vesuvius Bar and Grill. Several art galleries are within walking distance of the ferry landing and make for a pleasant afternoon outing. Alternatively, hike Maple Mountain in nearby Maple Bay Centennial Park.

Maple Bay Public Wharf

Maple Bay Public Wharf

PO Box 278; Duncan, BC V9L 3X4

Tel: (250) 246-4655 **VHF: Monitor** n/a **Talk** n/a
Fax: (250) 246-1595 **Alternate Tel:** n/a
Email: n/a **Web:** n/a
Nearest Town: Duncan *(6 mi.)* **Tourist Info:** (250) 748-1111

Navigational Information
Lat: 48°48.878' **Long:** 123°36.598' **Tide:** 12 ft. **Current:** n/a **Chart:** 3478
Rep. Depths *(MLW)*: **Entry** 6 ft. **Fuel Dock** n/a **Max Slip/Moor** 6 ft./-
Access: At the end of Maple Bay off Stuart Channel

Marina Facilities *(In Season/Off Season)*
Fuel: No
Slips: 4 Total, 4 Transient **Max LOA:** 50 ft. **Max Beam:** n/a
 Rate *(per ft.)*: **Day** $0.55 **Week** n/a **Month** n/a
 Power: 30 amp n/a, **50 amp** n/a, **100 amp** n/a, **200 amp** n/a
 Cable TV: No **Dockside Phone:** No
 Dock Type: Floating, Alongside, Wood
Moorings: 0 Total, 0 Transient **Launch:** n/a
 Rate: Day n/a **Week** n/a **Month** n/a
Heads: None
Laundry: None **Pay Phones:** No
Pump-Out: No **Fee:** n/a **Closed Heads:** No

Marina Operations
Owner/Manager: Harmen Bootsma, HM **Dockmaster:** Same
In-Season: Year-Round, 24/7 **Off-Season:** n/a
After-Hours Arrival: Self register at the kiosk
Reservations: No **Credit Cards:** No
Discounts: None
Pets: Welcome **Handicap Access:** No

Marina Services and Boat Supplies
Supplies - Near: Ice *(Cube)* **1-3 mi:** Ships' Store *(Lindstrom's at Maple Bay Marina 746-8482)*, Propane *(Maple Bay Marina)*, CNG

Boatyard Services
Nearest Yard: Covey Marine (250) 748-3130

Restaurants and Accommodations
Near: Restaurant *(Brigantine Inn 746-5422, L $6-10, D $13-17, Mon 2-for-1 Steak)*, Lite Fare *(Grapevine on the Bay 746-0797, takeout only Wed-Sun)*, Condo/Cottage **1-3 mi:** Restaurant *(Quamichan Inn 746-7028, D $22-35, free ride to the Inn)*, Inn/B&B *(The Quamichan Inn 746-7028, $115-135, restored B&B)* **3+ mi:** Restaurant *(Low's Garden 748-1288, 6 mi., Chinese)*, *(Barnacles748-0758, L $6-11, D $7-20, 6 mi.)*, Fast Food *(McDonald's, Wendy's, etc. 6 mi.)*, Pizzeria *(Slice of Life 715-1020, B $3, L $4-8, D $5-10, 6 mi., quick service)*, Motel *(Best Western 748-2722, $105-121, 6 mi., seasonal rates)*, *(South Side Inn 748-0661, $60-70, 6 mi., quiet atmosphere)*,*(Travelodge 748-4311, $79-189, seasonal rates;kitchenette, Jacuzzi, fiireplace, conf. center, pub)*

Recreation and Entertainment
Near: Beach *(black sand)*, Picnic Area, Boat Rentals *(Wilderness Kayaking 746-0151)*, Park **Under 1 mi:** Jogging Paths **3+ mi:** Pool *(Aquannis Pool Cowichan Community 746-0450, 6 mi.)*, Golf Course *(Fir Meadows 709-2221, 5 mi./Duncan Meadows 746-8993, 6 mi.)*, Fitness Center *(Fit Co Cardio and Fitness Centre 748-2202, 6 mi.)*, Movie Theater *(Caprice Showcase 748-6231, 6 mi.)*, Video Arcade *(Rogers Video 715-1280, 6 mi.)*,

Museum *(Cowhichan Valley Museum 746-6612, $2, 6 mi.)*, Cultural Attract *(CowichanTheatre 748-7529, 6 mi.)*, Sightseeing *(BC Forest Discovery Centre 715-1170, 6 mi.)*

Provisioning and General Services
Near: Convenience Store *(Brigantine Inn 746-5422)*, Gourmet Shop *(Ann's Gourmet 748-7373)*, Liquor Store *(Brigantine Pub)* **1-3 mi:** Bank/ATM, Post Office **3+ mi:** Market *(Coronation 748-6656, 6 mi.)*, Supermarket *(Safeway 746-6122, Overweitea 748-6011, 6 mi.)*, Delicatessen *(Delicado's 701-0855, 6 mi.)*, Green Market *(Community Farm Store 748-6227, 6 mi.)*, Fishmonger *(Cowichan Bay 715-1167, 6 mi.)*, Protestant Church *(5 mi.)*, Library *(746-7661, 6 mi.)*, Beauty Salon *(Cutler & Co. 746-8667, 6 mi.)*, Barber Shop *(Bob's 746-8514, 6 mi.)*, Dry Cleaners *(Betty's One Hour 748-3341, 6 mi.)*, Laundry *(Seabreeze Laundry 748-3021, 6 mi.)*, Bookstore *(Volume One 748-1533, 6 mi.)*, Pharmacy *(Drugstore 746-0535, 6 mi.)*, Florist *(Floral Art 746-5185, 6 mi.)*, Clothing Store *(Suzanne's 748-2821, 6 mi.)*, Department Store *(Wal-Mart 448-2566, 6 mi.)*, Copies Etc. *(Staples 715-1922, 6 mi.)*

Transportation
OnCall: Rental Car *(National 748-4400, Budget 748-3221)*, Taxi *(Cobble Hill 743-5555)* **3+ mi:** Rail *(Rail 888-842-7245, 6 mi.)* **Airport:** Duncan *(20 mi.)*

Medical Services
911 Service **OnCall:** Ambulance *(746-4058)* **3+ mi:** Doctor *(Festubert Family Practice 746-7720, 6 mi.)*, Dentist *(Alderlee Dental Health 748-1842, 6 mi.)*, Chiropractor *(First Chiropractic 746-6229, 6 mi.)*, Veterinarian *(Longair 746-7178, 6 mi.)* **Hospital:** Cowichan District 746-4141 *(8 mi.)*

Setting -- Maple Bay, a popular boating area with several marinas, lies in the narrow waterway between Vancouver Island and Salt Spring Island. From the north, approach through Stuart Channel and from the south, through Sansum Narrows. The Public Wharf is in the northerly portion of the bay, at the foot of Maple Mountain. Signature bright blue and brick-red railings trim the gangplank that leads to the single floating wharf. The surrounding small neighborhood can be best described as rural residential.

Marina Notes -- This public wharf offers only 150 ft. of dock space, so mostly smaller boats tie up here. There are no boater amenities, no power on-site, and the wood docks are uneven yet uncluttered. Payment is on the honor system (use the red kiosk at the top of the ramp). Harmen Bootsma is harbormaster for this wharf and two others in Chemainus and Crofton; his cell phone number is posted in case you need to reach him. Heads are at the nearby park. Note: Reciprocal moorgae is available at Maple Bay Y.C. (746-4521).

Notable -- An undersea dive area lies 200 feet off the end of the dock. Near the top of the ramp, Grapevine on the Bay, advertises "comfort food to go." Just up the street is a public park with sports, picnic tables, a swim platform & beach area, and public restrooms. This is a great place to relax on the grass under the shade trees and let the kids swim. A little farther up the street is the Brigantine Inn with its restaurant, pub, and liquor store. Ice and other mainstays can be found there. Supermarkets, motels, and additional services are available 6 miles east in Duncan.

PHOTOS ON DVD: 10

Navigational Information
Lat: 48°47.995' **Long:** 123°36.068' **Tide:** 12 ft. **Current:** n/a **Chart:** 3478
Rep. Depths (*MLW*): **Entry** 36 ft. **Fuel Dock** 28 ft. **Max Slip/Moor** 28 ft./-
Access: Off the southwest corner of Maple Bay in Bird's Eye Cove

Marina Facilities *(In Season/Off Season)*
Fuel: Gasoline, Diesel
Slips: 57 Total, 8 Transient **Max LOA:** 60 ft. **Max Beam:** n/a
  Rate *(per ft.)*: **Day** $0.85 **Week** n/a **Month** n/a
  Power: 30 amp Incl.*, **50 amp** n/a, **100 amp** n/a, **200 amp** n/a
  Cable TV: No **Dockside Phone:** No
  Dock Type: Floating, Long Fingers, Alongside, Concrete, Wood
Moorings: 0 Total, 0 Transient **Launch:** n/a
  Rate: Day n/a **Week** n/a **Month** n/a
Heads: 1 Toilet(s)
Laundry: None **Pay Phones:** Yes, 1
Pump-Out: No **Fee:** n/a **Closed Heads:** No

Marina Operations
Owner/Manager: James Marshall **Dockmaster:** Same
In-Season: May-Sep, 8am-8pm **Off-Season:** Oct-Apr, 9am-5pm
After-Hours Arrival: Tie up by the store for the night
Reservations: Yes, Preferred **Credit Cards:** Visa/MC, Debit
Discounts: None
Pets: Welcome **Handicap Access:** No

Bird's Eye Cove Marina

6271 Genoa Bay Road; Duncan, BC V9L 5T8

Tel: (250) 746-5686 **VHF: Monitor** Ch. 66A **Talk** n/a
Fax: (250) 746-5685 **Alternate Tel:** n/a
Email: n/a **Web:** n/a
Nearest Town: Duncan *(6 mi.)* **Tourist Info:** (250) 748-1111

Marina Services and Boat Supplies
Services - Docking Assistance, Security *(owner lives onsite)*, Trash Pick-Up, Dock Carts **Communication -** Mail & Package Hold, Phone Messages, Fax in/out, FedEx, DHL, UPS, Express Mail *(Sat Del)* **Supplies - OnSite:** Ice *(Cube)*, Bait/Tackle **Near:** Ships' Store, Propane, CNG

Boatyard Services
OnSite: Yacht Broker *(746-5686)* **Near:** Travelift *(15T)*, Railway *(100T)*, Crane *(5T)*, Sail Loft *(Rose Sailmakers 746-7245)*. **Nearest Yard:** Cove Yachts (250) 748-8136

Restaurants and Accommodations
OnCall: Pizzeria *(Domino's 715-0715)* **Under 1 mi:** Restaurant *(The Shipyard Restaurant 746-1026, B $7-10, L $7-12, D $7-23)* **1-3 mi:** Restaurant *(The Quamichan Inn 746-7028, complimentary limo to marina)*, *(Grapevine Café 746-0797)* **3+ mi:** Motel *(Falcon Nest 748-8188, $60-90, 6 mi.)*, *(Days Inn 748-0661, $70-100, 6 mi.)*, *(Thunderbird $60-75, 6 mi., 877-748-8192)*, Hotel *(Best Western 748-2722, $80-120, 6 mi.)*, Condo/ Cottage *(Alderlea Escape 715-1251, 4 mi., 1-2 nights $109/night; 3-5 nights $99; 5 and over $89)*

Recreation and Entertainment
1-3 mi: Beach, Picnic Area, Park **3+ mi:** Pool *(Aquannis Pool at Community Center 746-0450, 6 mi.)*, Golf Course *(Fir Meadows 709-2221, Duncan Meadows 746-8993, 6 mi.)*, Fitness Center *(Fit Co-Cardio and Fitness Centre 748-2202, 6 mi.)*, Movie Theater *(Caprice 748-6231, 6 mi.)*, Video Rental *(Rogers Video 814-1280, 6 mi.)*, Museum *(Cowichan Valley Museum 746-6612 $2, BC Forest Discovery Centre 715-1170, Quw'utsun'* Cultural and Conference Centre 746-8119, 6 mi.)*, Cultural Attract *(Cowichan Theatre 748-7529, 6 mi.)*

Provisioning and General Services
Near: Convenience Store **1-3 mi:** Liquor Store *(Brigantine Inn 746-5422)*, Bank/ATM, Post Office **3+ mi:** Market *(Coronation Market 748-6655, 6 mi.)*, Supermarket *(Safeway 746-6122, Overweitea 748-6011, 6 mi.)*, Delicatessen *(Delicado's Deli 701-0855, 6 mi.)*, Bakery *(6 mi.)*, Farmers' Market *(Sat 10am-2pm at train station parking, 6 mi.)*, Fishmonger *(Cowichan Bay Fish Market 715-1167, 6 mi.)*, Catholic Church *(6 mi.)*, Protestant Church *(Baptist 746-7432, 6 mi.)*, Library *(746-7355, 6 mi.)*, Beauty Salon *(Vanity Fair 748-7355, 6 mi.)*, Barber Shop *(Bob's 746-8514, 6 mi.)*, Dry Cleaners *(Eco Cleaning Express 748-7703, 6 mi.)*, Laundry *(Seabreeze Laundry 748-3021, 6 mi.)*, Bookstore *(Volume One 748-1533, 6 mi.)*, Pharmacy *(Pharmasave 748-5252, 6 mi.)*, Clothing Store *(Suzanne's 715-1922, 6 mi.)*, Department Store *(Wal-Mart 448-2566, 6 mi.)*, Copies Etc. *(Staples 715-1922, 6 mi.)*

Transportation
OnCall: Rental Car *(National 748-4400, Budget 748-3221)*, Taxi *(Cobble Hill Taxi 753-5555)* **3+ mi:** Rail *(VIA Rail 888-842-7245, 6 mi.)*
Airport: Duncan *(20 mi.)*

Medical Services
911 Service **OnCall:** Ambulance *(746-4058)* **3+ mi:** Doctor *(Festubert Family Practice 746-7720, 6 mi.)*, Dentist *(Alderlee Dental 748-1842, 6 mi.)*, Chiropractor *(First Chiropractic 746-6229, 6 mi.)*, Veterinarian *(Longair 746-7178, 6 mi.)* **Hospital:** Cowichan District 746-4141 *(8 mi.)*

Setting -- Bird's Eye Cove is actually the southern tip of Maple Bay, where a small headland provides good protection. Pretty Bird's Eye Cove Marina sits comfortably among the forest-clad hills at the Cove's entrance. Hanging pots, overflowing with flowers, adorn the gray clapboard office - with red, white and blue roof - next to the fuel dock. And at the top of the ramp an archway opens to a wide, flower-trimmed deck that hosts chairs and dock carts. Clearly, a real effort has been made to create a ship-shape ambiance. The surrounding area is residential; this is a quiet spot where relaxation is king.

Marina Notes -- *15 amp power also available. Guest moorage is tight, so call ahead. The office/store is conveniently located on the deck next to the fuel dock. It offers some groceries, coffee, and fishing tackle. Ice is also available on the fuel dock. Cove Yachts Shipyard next door provides boatyard services.

Notable -- Duncan is known as The City of Totems with over 70 poles in and about town. Follow the painted footsteps in Duncan and you will be treated to a tour that includes the culture and First Nations' heritage depicted in the totems. Duncan is also home to "The World's Largest Hockey Stick" and the BC Forest Discovery Centre (715-1170), north of town - it offers exhibits, steam train rides, easy trails, picnic areas, and playgrounds. The Quw'utsun' Cultural and Conference Centre (746-8119) in Duncan also depicts the major role played by the First Nations in establishing the area. Take the time to really enjoy the numerous exhibits, artifacts, and Native history.

Maple Bay Marina

6145 Genoa Bay Road; Duncan, BC V9L 5T7

Tel: (250) 746-8482; (866) 746-8482 **VHF: Monitor** Ch. 66A **Talk** n/a
Fax: (250) 746-8490 **Alternate Tel:** (866) 746-8482
Email: info@maplebaymarina.com **Web:** www.maplebaymarina.com
Nearest Town: Duncan *(6 mi.)* **Tourist Info:** (250) 748-1111

Navigational Information
Lat: 48°47.721' **Long:** 123°36.121' **Tide:** 12.5 ft. **Current:** n/a **Chart:** 3478
Rep. Depths *(MLW)*: **Entry** 25 ft. **Fuel Dock** 25 ft. **Max Slip/Moor** 30 ft./-
Access: Off the SW corner of Maple Bay at the head of Bird's Eye Cove

Marina Facilities *(In Season/Off Season)*
Fuel: Gasoline, Diesel
Slips: 375 Total, 60 Transient **Max LOA:** 150 ft. **Max Beam:** n/a
  Rate *(per ft.)*: **Day** $0.95/0.75 **Week** n/a **Month** n/a
  Power: 30 amp $5.50*, **50 amp** $7.50, **100 amp** n/a, **200 amp** n/a
  Cable TV: No **Dockside Phone:** No
  Dock Type: Floating, Long Fingers, Alongside, Wood
Moorings: 0 Total, 0 Transient **Launch:** n/a
  Rate: Day n/a **Week** n/a **Month** n/a
Heads: 8 Toilet(s), 9 Shower(s) *(with dressing rooms)*
Laundry: 3 Washer(s), 4 Dryer(s), Book Exchange **Pay Phones:** Yes, 2
Pump-Out: No **Fee:** n/a **Closed Heads:** No

Marina Operations
Owner/Manager: David Messier **Dockmaster:** Same
In-Season: Jul-Aug, 8am-8pm **Off-Season:** Sep-Jun, 9am-5pm
After-Hours Arrival: Call ahead
Reservations: Yes, Preferred **Credit Cards:** Visa/MC, Debit
Discounts: Pre-registered groups **Dockage:** n/a **Fuel:** n/a **Repair:** n/a
Pets: Welcome **Handicap Access:** Yes, Heads, Docks

Marina Services and Boat Supplies
Services - Docking Assistance, Security *(video and patrol)*, Trash Pick-Up, Dock Carts, Megayacht Facilities **Communication -** Mail & Package Hold, Phone Messages, Fax in/out *($.85*)*, Data Ports *(Wi-Fi - Broadband Xpress)*, FedEx, DHL, UPS, Express Mail *(Sat Del)* **Supplies - OnSite:** Ships' Store *(Lindstrom's Marine Supply)*, Propane, CNG

Boatyard Services
OnSite: Travelift *(50T; services by Lindstrom's 748-9199)*, Engine mechanic *(gas, diesel)*, Yacht Broker *(Passage Yacht Sales 748-5004)* **Near:** Railway *(100T)*. **Dealer for:** Caterpiller, Perkins - Gabbren Jac Marine.

Restaurants and Accommodations
OnSite: Restaurant *(Shipyard Pub and Restaurant 746-1026, B $7-10, L $7-12, D $7-23, outdoor dining available)* **OnCall:** Pizzeria *(Romeo's Pizza 746-9944)* **Near:** Inn/B&B *(Gray's Gate 748-4729)* **1-3 mi:** Restaurant *(The Quamichan Inn 746-7028, D $22-35, complimentary limo to marina)*, *(Grapevine Café 746-0797)*, Inn/B&B *(The Quamichan Inn 746-7028, $125)* **3+ mi:** Motel *(Falcon Nest 748-8188, $60-90, 6 mi.)*, *(Days Inn 748-0661, $70-100, 6 mi.)*, Hotel *(Best Western 748-2722, $80-120, 6 mi.)*

Recreation and Entertainment
Near: Jogging Paths **Under 1 mi:** Pool **1-3 mi:** Beach, Picnic Area **3+ mi:** Golf Course *(Fir Meadows 709-2221, 5 mi./Duncan Meadows 746-8993, 6 mi./Cowichan Golf 746-5333, 7 mi.)*, Fitness Center *(Fit-Co Cardio & Fitness Centre 748-2202, 6 mi.)*, Movie Theater *(Caprice 748-6231, 6 mi.)*, Video Rental *(Rogers 715-1280, 6 mi.)*, Museum *(Cowhichan Valley 746-6612, 6 mi.)*, Cultural Attract *(Cowichan Theatre 748-7529, 6 mi.)*

Provisioning and General Services
OnSite: Convenience Store **1-3 mi:** Liquor Store *(Brigantine Inn 746-5422)*, Bank/ATM, Post Office **3+ mi:** Market *(Wing on Food 746-6632, Coronation Market 748-6655, 6 mi.)*, Supermarket *(Safeway 746-6122, Overweitea 748-6011, 6 mi.)*, Delicatessen *(Delicado's Deli 701-0855, 6 mi.)*, Wine/Beer *(Village Green Inn 746-4328, 6 mi.)*, Farmers' Market *(Sat 10am-2pm, 743-7055 at train station on Canada Ave., 6 mi.)*, Fishmonger *(Cowichan Bay Fish Market 715-1167, 6 mi.)*, Catholic Church *(6 mi.)*, Protestant Church *(Bethel Baptist 746-7432, 6 mi.)*, Library *(746-7661, 6 mi.)*, Beauty Salon *(Vanity Fair 748-7355, 6 mi.)*, Barber Shop *(Bob's Barber Shop 746-8514, 6 mi.)*, Dry Cleaners *(Eco Cleaning Express 748-7703, 6 mi.)*, Laundry *(Seabreeze Laundry 748-3021, 6 mi.)*, Bookstore *(Wishes 748-9411, 6 mi.)*, Pharmacy *(Pharmasave 748-5252, 6 mi.)*, Florist *(Floral Art Shop 746-5185, 6 mi.)*, Clothing Store *(Suzanne's 748-2821, 6 mi.)*, Department Store *(Wal-Mart 448-2566, 6 mi.)*, Copies Etc. *(Staples 715-1922, 6 mi.)*

Transportation
OnSite: Rental Car *(Budget 748-3221)* **OnCall:** Taxi *(Duncan Taxi 746-4444)* **1-3 mi:** Local Bus **3+ mi:** Rail *(VIA Rail 888-842-7245, 6 mi.)* **Airport:** Duncan, 20 mi./YVR; floatplane service to airports: Saltspring Air 877-537-9880, Harbour Air 800-665-0212

Medical Services
911 Service **OnCall:** Ambulance *(746-4058)* **3+ mi:** Doctor *(Festubert Family Practice 746-7720, 6 mi.)*, Dentist *(Alderlea Dental 748-1842, 6 mi.)*, Chiropractor *(First Chiropractic 746-6229, 6 mi.)*, Veterinarian *(Longair 746-7178, 6 mi.)*, Optician *(Pearle Vision 746-4322, 6 mi.)* **Hospital:** Cowichan District Hospital 746-4141 *(8 mi.)*

Setting -- At the head of Bird's Eye Cove off Maple Bay, this marina delivers beautiful, long views of Salt Spring Island, Sansum Narrows, and the bay. The cove is protected, but outside the bay winds can be strong due to restricted flow through the Narrows and the influence of the canyons intersecting the waterway. The marina, with its burgee-lined roof, has lovely grounds, and offers badminton and a large covered picnic area for groups - with tables, chairs, a sink, and electricity. Follow the signs to Nature's Corner for a secluded spot to enjoy the birds and flowers.

Marina Notes -- *15 amp $3/night; call ahead for 50 amp. **Fax charge $0.85/pp incoming, $0.95/pp outgoing. A 50T travelift is available, and do-it-yourself and outside trades are welcome (surcharge). Oil and lubes also on-site. A little city by itself - several businesses are part of the complex: Lindstrom's Marine Supply, Budget Rent A Car, Passage Yacht Sales, Dockside Gifts, a small market and Internet Café ($3/15 min., $5/30 min., $2 and $4 to use your laptop). Broadband Xpress Wireless, 800-729-4603. Buildings are aging but well-maintained. Remodeling planned for near future. Harbour Air Seaplanes provides daily flights to Vancouver ($80 one way). Maple Bay Marina boasts the "cleanest washrooms on the coast."

Notable -- The marina pub, store and surrounding businesses make this a self-sufficient and convenient stop. Rent a car and visit the Cowichan Valley Museum (Jun-Sep, Mon-Sat 10am-4pm; Oct-May, Wed-Fri 11am-4pm, Sat 1-4pm; $2/free up to age 14). Or plan a self-guided walking tour of Duncan's famous totems (maps at the Visitors Centre). Golf courses dot the surrounding area and are open most of the year - each offers unique challenges and scenery.

Navigational Information
Lat: 48°44.120' **Long:** 123°35.407' **Tide:** 11 ft. **Current:** n/a **Chart:** 3441
Rep. Depths (*MLW*): **Entry** 4 ft. **Fuel Dock** n/a **Max Slip/Moor** 5 ft./-
Access: Off Satellite Channel, southern entrance to Cowichan Bay

Marina Facilities *(In Season/Off Season)*
Fuel: No
Slips: 20 Total, 2** Transient **Max LOA:** 72 ft. **Max Beam:** n/a
 Rate *(per ft.):* **Day** $0.50* **Week** n/a **Month** n/a
 Power: 30 amp Incl., **50 amp** n/a, **100 amp** n/a, **200 amp** n/a
 Cable TV: No **Dockside Phone:** No
 Dock Type: Floating, Alongside, Wood
Moorings: 0 Total, 0 Transient **Launch:** n/a
 Rate: Day n/a **Week** n/a **Month** n/a
Heads: 1 Toilet(s)
Laundry: None **Pay Phones:** Yes, 1
Pump-Out: No **Fee:** n/a **Closed Heads:** No

Marina Operations
Owner/Manager: Kathy Blades **Dockmaster:** Same
In-Season: Year-Round, 8am-9pm **Off-Season:** n/a
After-Hours Arrival: Call ahead
Reservations: Yes **Credit Cards:** None
Discounts: None
Pets: Welcome **Handicap Access:** No

Cherry Point Marina

1241 Sutherland Drive; Cowichan Bay, BC V0R 1N2

Tel: (250) 748-0453; (250) 701-2674 **VHF: Monitor** n/a **Talk** n/a
Fax: n/a **Alternate Tel:** n/a
Email: n/a **Web:** n/a
Nearest Town: Cowichan Bay *(2.5 mi.)* **Tourist Info:** (250) 748-1111

Marina Services and Boat Supplies
Services - Security *(owner onsite)*, Dock Carts **Supplies - 1-3 mi:** Ships' Store *(Island Marine Centre 746-6000, Pier 66 748-8444)*, Bait/Tackle

Boatyard Services
OnSite: Launching Ramp *($5)* **Nearest Yard:** Cowichan Bay Shipyard *(250) 748-7285*

Restaurants and Accommodations
1-3 mi: Restaurant *(The Masthead 748-3714, D $17-35)*, *(Rock Cod Café 746-1550, L $5-8, D $11-14)*, *(Amici at the Bluenose 748-2841, L $10-16, D $17-28, pizzas $12-16)*, *(The Tides 746-0166, B $6-10, L $7-10, D $23-60, Sun brunch $20)*, Snack Bar *(The Udder Guys 746-4300, ice , cream, coffeesnacks)*, Hotel *(Oceanfront Grand Resort 800-663-7898, $100-205)*, Inn/B&B *(Wessex Inn 748-4214)* **3+ mi:** Fast Food *(McD's, Quiznos, Subway, 7 mi.)*, Motel *(Days Inn 748-0661, $69-99, 7 mi.)*

Recreation and Entertainment
OnSite: Picnic Area **Near:** Beach **1-3 mi:** Boat Rentals *(We Go Kayaking 800-434-9346)*, Fishing Charter *(Bluenose Marina 748-2222)*, Video Rental *(Pioneer Video 709-9944, 11am-10pm)*, Museum *(Cowichan Bay Maritime Centre 746-4955/ Cowichan Valley Museum 746-6612, BC Forest Discovery Centre 715-1113, 7 mi.)*, Galleries *(Timeless Co. 709-9985, Hillbank Pottery 743-6686, Salish Pride)* **3+ mi:** Pool *(Aquannis 746-0450, 7 mi.)*, Tennis Courts *(South Cowichan Lawn Tennis 746-7282, 5 mi.)*, Golf Course *(Fir Meadows Golf Club 709-2221, 6 mi.)*, Fitness Center *(Movement Unlimited 748-2223, 7 mi.)*, Movie Theater *(Caprice Showcase Theatre 748-6231, 7 mi.)*, Sightseeing *(Pacific Northwest Raptors 746-0372, 7 mi.)*

Provisioning and General Services
1-3 mi: Convenience Store *(Four Ways Convenience Store 746-8888)*, Liquor Store *(Pier 66 Liquor and Convenience Store 748-8444)*, Bank/ATM **3+ mi:** Market *(Country Grocery at Beverly Corners 715-2630, 7 mi.)*, Supermarket *(Overwaitea 748-6151, Safeway 746-3655, 7 mi.)*, Delicatessen *(7 mi.)*, Wine/Beer *(7 mi.)*, Bakery *(7 mi.)*, Farmers' Market *(Sat 10am-2pm at Duncan train station 743-7055, 7 mi.)*, Fishmonger *(Cowichan Bay Fish Market 715-1167, 7 mi.)*, Catholic Church *(7 mi.)*, Protestant Church *(United Church of Canada 746-6043, 7 mi.)*, Library *(746-7661, 7 mi.)*, Beauty Salon *(Studio One 709-2195, 7 mi.)*, Barber Shop *(First Choice 748-3833, 7 mi.)*, Dry Cleaners *(Betty's One Hour 748-3341, 7 mi.)*, Laundry *(Spiffys 746-9771, 7 mi.)*, Bookstore *(Volume One 748-1533, 7 mi.)*, Pharmacy *(Drugstore 746-0535, 7 mi.)*, Hardware Store *(Duncan Home 746-4456, 7 mi.)*, Clothing Store *(Wall Street 746-8832, 7 mi.)*, Copies Etc. *(Staples 715-1922, 7 mi.)*

Transportation
OnCall: Taxi *(Cobble Hill Taxi 743-5555)* **1-3 mi:** Local Bus *(Cowichan Valley Transit System 748-1230)* **3+ mi:** Rental Car *(National 748-4400, 7 mi.)* **Airport:** Duncan *(6 mi.)*

Medical Services
911 Service **OnCall:** Ambulance *(743-3232)* **1-3 mi:** Holistic Services *(Breeze Day Spa 748-2811)* **3+ mi:** Doctor *(Ingram Family Physicians 746-4401, 7 mi.)*, Dentist *(Alderlea Dental Health 748-1842, 7 mi.)*, Chiropractor *(First Chiropractic Clinic 746-6229, 7 mi.)*, Veterinarian *(Duncan Animal Hospital 746-1966, 7 mi.)*, Optician *(Pearle Vision 746-4322, 7 mi.)* **Hospital:** Cowichan District 746-4141 *(7 mi.)*

Setting -- Cherry Point Marina sits all by itself at the southern entrance to Cowichan Bay. Its position is somewhat exposed, with a floating breakwater consisting of miscellaneous odds and ends providing shelter. Approaching from the south, Satellite Channel leads directly to Cherry Point; otherwise run down Sansum Narrows to get there. The marina is a quiet little facility that includes an RV park surrounded by a large expanse of lawn and brightly colored flowers, a boat ramp for small boats, a small deck with tables and chairs overlooking the water, and a fish cleaning area for preparing the day's catch.

Marina Notes -- *Flat rate per night. **50 ft. of moorage for visitors. The breakwater is an interesting accumulation of eclectic items. One head, no showers.

Notable -- Nearby is a pleasant sand and pebble beach ideal for a leisurely stroll. The seaside village of Cowichan Bay, in the throes of rejuvenation, is less than 3 miles away. Several restaurants, an ice cream parlor, galleries, a maritime museum with boat building classes, kayak rentals and fishing charters make it worth the visit. A short drive north, Duncan has carved out a name for itself as the "City of Totems." Six local carvers were commissioned to create totems in 1985. Today the city has 70 majestic poles on display, and walking tours are available to guide visitors around. Also in Duncan, Pacific Northwest Raptors trains prey birds and their handlers and educates the public through exhibits and daily flying demonstrations.

Navigational Information

Lat: 48°45.518' **Long:** 123°35.915' **Tide:** 11.5 ft. **Current:** n/a **Chart:** 3478
Rep. Depths *(MLW)*: **Entry** 27 ft. **Fuel Dock** n/a **Max Slip/Moor** 24 ft./-
Access: At the south end of Sansum Narrows

Marina Facilities *(In Season/Off Season)*

Fuel: No
Slips: 70 Total, 60 Transient **Max LOA:** 120 ft. **Max Beam:** n/a
  Rate *(per ft.)*: **Day** $1.00/$20* **Week** n/a **Month** n/a
  Power: 30 amp $5.50** **, 50 amp** n/a, **100 amp** n/a, **200 amp** n/a
  Cable TV: No **Dockside Phone:** No
  Dock Type: Floating, Alongside
Moorings: 0 Total, 0 Transient **Launch:** n/a
  Rate: Day n/a **Week** n/a **Month** n/a
Heads: 2 Toilet(s), 2 Shower(s)
Laundry: 2 Washer(s), 2 Dryer(s) **Pay Phones:** Yes, 2
Pump-Out: No **Fee:** n/a **Closed Heads:** No

Marina Operations

Owner/Manager: Will & Ben Kiedaisch **Dockmaster:** Same
In-Season: Year-Round, 9am-5pm **Off-Season:** n/a
After-Hours Arrival: Call ahead
Reservations: Yes, Preferred **Credit Cards:** Visa/MC, Debit
Discounts: None
Pets: Welcome **Handicap Access:** Yes, Heads, Docks

Genoa Bay Marina

5100 Genoa Bay Road; Duncan, BC V9L 5Y8

Tel: (250) 746-7621; (800) 572-6481 **VHF: Monitor** Ch. 66A **Talk** n/a
Fax: (250) 746-7621 **Alternate Tel:** n/a
Email: reservations@genoabaymarina.com **Web:** genoabaymarina.com
Nearest Town: Duncan *(11 mi.)* **Tourist Info:** (250) 748-1111

Marina Services and Boat Supplies

Services - Docking Assistance, Security *(owner onsite)*, Trash Pick-Up, Dock Carts **Communication -** Mail & Package Hold, Phone Messages, Fax in/out, Data Ports *(in the store)*, FedEx, DHL, UPS, Express Mail *(Sat Del)* **Supplies - OnSite:** Ice *(Block, Cube)*, Bait/Tackle **OnCall:** Marine Discount Store *(Next day delivery)* **1-3 mi:** Propane, CNG

Boatyard Services

OnSite: Launching Ramp **OnCall:** Divers **Nearest Yard:** Coastal Shipyard and Marine (250) 746-4705

Restaurants and Accommodations

OnSite: Restaurant *(Genoa Bay Café 746-7621, L $8-13, D $17-25)*
3+ mi: Restaurant *(The Aerie 743-7115, L $14-18, D $24-30, 11 mi.)*, *(Friday's 743-5533, L $10-16, D $10-16, 11 mi., wknd brunch $5-8)*, *(Shipyard Pub and Restaurant 746-1026, B $7-10, L $7-12, D $7-23, 3 mi., at Maple Bay Marina)*, Fast Food *(Quiznos, Dairy Queen, 11 mi.)*, Lite Fare *(Brigantine Inn 746-5422, L $5-9, D $9-14, 5 mi.)*, Motel *(Days Inn 748-0661, $70-100, 11 mi.)*, *(The Quamichan Inn 746-7028, 6 mi.)*, Inn/B&B *(Grays' Gate 748-4729, $85-105, 11 mi.)*, Condo/Cottage *(Alderlea Escape 715-1251, $89-109, 11 mi.)*

Recreation and Entertainment

OnSite: Picnic Area, Grills, Video Rental, Galleries *(Genoa Bay Gallery 746-5506)* **Near:** Beach, Boat Rentals *(We Go Kayaking 748-5400, $25-85)*, Sightseeing *(Tom's Cruise 2 hr. sail $39.95/p; to Butchart Gardens $79/p 748-7526)* **1-3 mi:** Cultural Attract *(Cowichin Bay Maritime Centre 746-4955, 9am-dusk, donations; by boat)* **3+ mi:** Pool *(Aquannis 746-0450,*

11 mi.), Golf Course *(Fir Meadows 709-2221, 11 mi.)*, Fitness Center *(Fit-Co Cardio & Fitness 748-2202, 11 mi.)*, Movie Theater *(Caprice 748-6231, 11 mi.)*, Museum *(BC Forest Discovery Centre 715-1170, Quw'utsun Cultural Centre 746-8119, Cowhichan Valley 746-6612, 11 mi.)*

Provisioning and General Services

OnSite: Convenience Store **1-3 mi:** Bank/ATM **3+ mi:** Supermarket *(Safeway 746-6122, Overweitea 748-6011, 11 mi.)*, Delicatessen *(Delicado's 701-0855, 11 mi.)*, Liquor Store *(Brigantine Inn 746-5422, 5 mi.)*, Farmers' Market *(Sat 10am-2pm, 743-7055 at train station parking lot, 11 mi.)*, Green Market *(Community Farm Store 448-6227, 11 mi.)*, Fishmonger *(Cowichan Bay Fish Market 715-1167, 11 mi.)*, Library *(746-7661, 11 mi.)*, Beauty Salon *(Matisse Day Spa 715-1126, 11 mi.)*, Barber Shop *(Jubilee 748-3262, 11 mi.)*, Dry Cleaners *(Betty's One Hour 748-3341, 11 mi.)*, Laundry *(Seabreeze 748-3021, 11 mi.)*, Bookstore *(Wishes 748-9411, 11 mi.)*, Pharmacy *(Drugstore 746-0535, 11 mi.)*, Hardware Store *(Duncan Home 746-4456, 11 mi.)*, Florist *(Little Shop 746-4147, 11 mi.)*, Clothing Store *(Wall St. 746-8832, 11 mi.)*, Copies Etc. *(Staples 715-1922, 11 mi.)*

Transportation

OnSite: Rental Car *(call ahead)* **OnCall:** Taxi *(Cobble Hill 743-5555)* **3+ mi:** Rail *(VIA Rail 888-842-7245, 11 mi.)* **Airport:** Duncan *(13 mi.)*

Medical Services

911 Service **OnCall:** Ambulance *(746-4058)* **3+ mi:** Doctor *(Festubert Family Practice 746-7720, 11 mi.)*, Dentist *(Alderlee Dental 748-1842, 11mi.)*, Chiropractor *(First Chiro.746-6229, 11 mi.)* **Hospital:** Cowichan District 746-4141 *(8 mi.)*

Setting -- Genoa Bay lies on the north side of Cowichan Bay between Separation Point and Skinner Point. Skinner Bluff dominates the north side of the bay. The marina is surrounded by mountains. On the docks, the marina office and store features a variety of sundries and, outside, a picnic patio with tables and chairs hosts colorful potted flowers. A delightful surprise is the high-quality art gallery that also shares the docks.

Marina Notes -- *Off-season flat rate of $20/night. **15 amp power $3.50/night. Owned and operated by the Kiedaisch brothers, this family-oriented marina offers personal service. Although some of the buildings are showing their age, they are well kept; guests remarked that they return here for the friendly service and clean facilities. Heads and laundry in a building across the parking lot. Note: Drying shoals narrow the entrance channel to a hundred meters or so at low tide, but they are marked and should pose no problem. The channel itself is very deep.

Notable -- The delightful Geonoa Bay Café, reported to be one of Vancouver Island's best, serves elegant meals in a casual setting overlooking the marina. Quiet, untouched Genoa Bay is far from the madding crowds and few services are available in the area. But the quaint town of Duncan, about 11 miles away, offers everything from antique shops to supermarkets, with entertainment options including live theater, a billiards club, and interesting museums. Self-guided walking tours (or guided in summer) are available. Savor the native culture reflected in the totems throughout town, or at the Quw'utsun' Cultural Centre. Learn about the majestic birds of prey at Pacific Northwest Raptors (746-0372, daily from Apr-Oct), or trek through the woods at BC Forest Discovery Centre.

Navigational Information
Lat: 48°44.489' **Long:** 123°37.090' **Tide:** 11 ft. **Current:** n/a **Chart:** 3478
Rep. Depths (*MLW*)**: Entry** 15 ft. **Fuel Dock** n/a **Max Slip/Moor** 15 ft./-
Access: In Cowichan Bay off Satellite Channel

Marina Facilities *(In Season/Off Season)*
Fuel: No
Slips: 80 Total, 10* Transient **Max LOA:** 150 ft. **Max Beam:** n/a
  Rate *(per ft.)*: **Day** $1.00 **Week** n/a **Month** $5.00
  Power: 30 amp $3** **, 50 amp** n/a, **100 amp** n/a, **200 amp** n/a
  Cable TV: No **Dockside Phone:** Yes
  Dock Type: Floating, Alongside, Concrete, Wood
Moorings: 0 Total, 0 Transient **Launch:** n/a
  Rate: Day n/a **Week** n/a **Month** n/a
Heads: 2 Toilet(s), 2 Shower(s)
Laundry: 1 Washer(s), 1 Dryer(s) **Pay Phones:** Yes, 1
Pump-Out: OnSite **Fee:** $15 **Closed Heads:** No

Marina Operations
Owner/Manager: Chuck Von Haas **Dockmaster:** Asst. Jim & Vicky
In-Season: Summer, 8am-8pm **Off-Season:** Winter, 9am-4pm
After-Hours Arrival: Call (250) 755-6763
Reservations: Yes **Credit Cards:** Visa/MC, Amex, Debit
Discounts: Clubs **Dockage:** Inq. **Fuel:** n/a **Repair:** n/a
Pets: Welcome **Handicap Access:** No

Fishermen's Wharf Association

PO Box 52; 1699 Cowichan Bay; Cowichan Bay, BC V0R 1N0

Tel: (250) 746-5911 **VHF: Monitor** Ch. 66A **Talk** Ch. 66
Fax: (250) 701-0729 **Alternate Tel:** (250) 755-6763
Email: cbfwa@shaw.ca **Web:** www.haa.bc.ca/
Nearest Town: Cowichan Bay **Tourist Info:** (250) 748-1111

Marina Services and Boat Supplies
Services - Security *(video)*, Trash Pick-Up **Communication -** FedEx, DHL, UPS, Express Mail **Supplies - Near:** Ice *(Cube)*, Ships' Store *(Yacht Sales 746-6000)*, Bait/Tackle

Boatyard Services
Near: Crane *($25 plus labor)*, Yacht Broker *(Island Marine and Brokerage Centre 746-6000)*. **Nearest Yard:** Cowichan Bay (250) 748-7285

Restaurants and Accommodations
OnSite: Snack Bar *(The Bay Pub 748-2330, next door, offers L & D)*
Near: Restaurant *(The Tides 748-0166, B $6-10, L $7-10, D $23-60, Sun brunch $20)*, *(Amici at the Bluenose 748-2841, L $10-16, D $17-28, pizza $12-16)*, *(Rock Cod Café 746-1550, L $5-8, D $11-14)*, *(The Masthead 748-3714, D $17-35)*, Snack Bar *(The Udder Guys Ice Cream Parlour 746-4300, ice cream, organic coffee, and snacks)*, Motel *(Oceanfront Grand Resort & Marina 800-663-7898, $100-205)* **1-3 mi:** Inn/B&B *(Dream Weaver B&B 748-7688, $130-150)*, Condo/Cottage *(Anchor's Guest House 748-7206, $96-195)*

Recreation and Entertainment
Near: Boat Rentals *(We Go Kayaking 800-434-9346)*, Fishing Charter *(Bluenose Marina 748-2222)*, Video Rental *(Pioneer 709-9944, 11am-10pm)*, Sightseeing *(Cowichan Bay Whale Watching 746-0166, Cowichan Tours 743-9444)*, Special Events *(Cowichan Bay Regatta - Aug)*, Galleries *(Timeless Co. 709-9985, Salish Pride)* **Under 1 mi:** Park *(Coverdale Watson Park)*, Museum *(Cowichan Bay Maritime Centre 746-4955)*
1-3 mi: Tennis Courts *(South Cowichan Lawn Tennis 746-7282)*

3+ mi: Golf Course *(Cowichan GC 746-5333, 6 mi.)*

Provisioning and General Services
Near: Convenience Store *(Pier 66 Liquor and Convenience Store 748-8444)*, Liquor Store *(Pier 66)*, Bakery *(True Grain Bread 746-7664)*, Fishmonger *(Cowichan Bay Seafoods 748-0020)*, Bank/ATM *(ATMs at Amici & The Bay Pub)*, Beauty Salon *(Rumors 709-2313)*, Clothing Store *(Mixed Blessings 801-4225)* **Under 1 mi:** Gourmet Shop *(Hilary's Fine Cheeses 715-0563)*, Bookstore *(Starfish Books and Charts 746-7597)* **1-3 mi:** Market *(Country Grocer 743-5639)* **3+ mi:** Supermarket *(6 mi.)*, Catholic Church *(St. Edward's 746-6831, 6 mi.)*, Protestant Church *(6 mi.)*, Library *(746-7661, 6 mi.)*, Dry Cleaners *(Betty's One Hour 748-3341, 6 mi.)*, Laundry *(Seabreeze 748-3021, 6 mi.)*, Pharmacy *(Pharmasave 748-5252, 6 mi.)*, Hardware Store *(Duncan Home 746-4456, 6 mi.)*, Florist *(Flower Affairs 746-3977, 6 mi.)*

Transportation
OnCall: Taxi *(Duncan Taxi 746-4444)* **Under 1 mi:** Local Bus *(Cowichan Valley Transit System 746-9899)* **3+ mi:** Rental Car *(Budget 748-3221, 6 mi.)* **Airport:** Duncan *(6 mi.)*

Medical Services
911 Service **OnCall:** Ambulance *(743-3232)* **Near:** Holistic Services *(Breeze Day Spa 748-2811)* **1-3 mi:** Doctor *(Brookside Medical Clinic 749-0120)*, Chiropractor *(Valleyview Family Chiropractic 743-3775)* **3+ mi:** Dentist *(Alderlea Dental Health Centre 748-1842, 6 mi.)*, Veterinarian *(Duncan Animal Hospital 746-1966, 6 mi.)*, Optician *(Opti View 1 Hour Optical 748-9229, 6 mi.)* **Hospital:** Cowichan Valley District 746-4141 *(6 mi.)*

Setting -- Cowichan Bay is virtually a continuation of Satellite Channel running south of Salt Spring Island. At the head of the bay, near the mouths of the Cowichan and Koksilah Rivers lies the village of Cowichan Bay. The Fishermen's Association Wharf is located on the east end of the busy waterfront, closest to the tall piling breakwater, and berths many of the town's commercial vessels. The Cowichan Bay Shipyard is right next door and the Village is 300 feet away.

Marina Notes -- *About 400 ft. of transient moorage during July & August. Only 200 ft. are available to transients in winter - when the commercial fishing fleet is dockside. A new 300' by 250' floating breakwater provides ample space - up to a 150' boat can side-tie inside; the area behind the breakwater is used primarily by fishermen when in port. **20 amp power only, $3/day. When not in the office, the dockmaster can be reached by cell phone. A two-story office housing the restrooms and showers is conveniently located on the docks.

Notable -- The town, centered around the waterfront, is undergoing a major face-lift. It offers almost 90 shops and an abundance of activities, including the Cowichan Bay Maritime Center and a Wooden Boat Society open to the public. Tugboat Annie's Flea Market, located below The Starfish Emporium, has an eclectic assortment of "treasures" (701-8470, open Thu-Sun only). The Udder Guys Ice Cream Parlour is a great place to pause to enjoy a homemade treat. If you'd like to get out on the bay with an experienced guide and learn more about the area, group tours of the bay can be arranged with Cowichan Tours (877-743-9444). A visit to Quw'utsun' Cultural & Conference Centre in Duncan is a must to learn about the First Nation's history and traditions (open 7 days).

Cowichan Bay Marina

Cowichan Bay Marina

PO Box 2517; 1721 Cowichan Bay; Cowichan Bay, BC V0R 1N0

Tel: (250) 701-9033 **VHF: Monitor** n/a **Talk** n/a
Fax: n/a **Alternate Tel:** n/a
Email: cowichanbaymarina@canada.com **Web:** n/a
Nearest Town: Cowichan Bay **Tourist Info:** (250) 748-1111

Navigational Information
Lat: 48°44.443' **Long:** 123°37.188' **Tide:** 11 ft. **Current:** n/a **Chart:** 3478
Rep. Depths (*MLW*): **Entry** 6 ft. **Fuel Dock** n/a **Max Slip/Moor** 6 ft./-
Access: In Cowichan Bay off Satellite Channel

Marina Facilities (*In Season/Off Season*)
Fuel: No
Slips: 90 Total, 2 Transient **Max LOA:** 40 ft. **Max Beam:** n/a
  Rate (*per ft.*): **Day** $0.90 **Week** n/a **Month** n/a
  Power: 30 amp Incl., **50 amp** n/a, **100 amp** n/a, **200 amp** n/a
  Cable TV: No **Dockside Phone:** No
  Dock Type: Floating, Long Fingers, Alongside, Wood
Moorings: 0 Total, 0 Transient **Launch:** n/a
  Rate: Day n/a **Week** n/a **Month** n/a
Heads: None
Laundry: None **Pay Phones:** No
Pump-Out: No **Fee:** n/a **Closed Heads:** No

Marina Operations
Owner/Manager: Richard Parker **Dockmaster:** T. Gary Marshall
In-Season: Year-Round, 9am-5pm **Off-Season:** n/a
After-Hours Arrival: Call ahead
Reservations: Yes, Preferred **Credit Cards:** No
Discounts: None
Pets: Welcome **Handicap Access:** No

Marina Services and Boat Supplies
Supplies - Near: Ice (*Cube*), Ships' Store (*Island Marine Centre & Island Yacht Sales 746-6000*), Bait/Tackle

Boatyard Services
Near: Yacht Broker (*Island Marine and Brokerage Centre 746-6000*).
Nearest Yard: Cowichan Bay Shipyard (250) 748-7285

Restaurants and Accommodations
Near: Restaurant (*The Tides 701-0166, B $6-10, L $7-10, D $23-60, Sun brunch $20*), (*Amici at the Bluenose 748-2841, L $10-16, D $17-28, pizza $12-16*), (*Rock Cod Café 746-1550, L $5-8, D $11-14*), (*The Masthead 748-3714, D $17-35*), Snack Bar (*The Udder Guys Ice Cream Parlour 746-4300*), (*The Bay Pub 748-2330, L & D*), Motel (*Oceanfront Grand Resort & Marina 800-663-7898, $100-205*), (*Wessex Inn 748-4214*) **1-3 mi:** Inn/B&B (*Bay Watch B&B 748-0712, $95*)

Recreation and Entertainment
Near: Boat Rentals (*We Go Kayaking 800-434-9346*), Fishing Charter (*Bluenose Marina 748-2222*), Video Rental (*Pioneer Video 709-9944, 11am-10pm*), Museum (*Cowichan Bay Maritime Centre 746-4955/ Quw'utsun' Cultural & Conference Centre 746-8119, Cowichan Valley Museum 746-6612, 6 mi.*), Sightseeing (*Cowichan Bay Whale Watching 746-0166*), Special Events (*Cowichan Bay Regatta - Aug*), Galleries (*Coast Salish Journey 748-1399 at the Maritime Centre*) **Under 1 mi:** Group Fishing Boat **1-3 mi:** (*Excalibur Charters 246-0142*), Park (*Coverdale Watson Park*) Tennis Courts (*South Cowichan Lawn Tennis 746-7282*)
3+ mi: Golf Course (*Cowichan Golf Club 746-5333, 6 mi.*)

Provisioning and General Services
Near: Convenience Store (*Pier 66 Liquor & Convenience Store 748-8444*), Liquor Store (*Pier 66*), Bakery (*True Grain Bread 746-7664*), Fishmonger (*Cowichan Bay Seafoods 748-0020*), Bank/ATM (*at Amici and The Bay Pub*), Beauty Salon (*Rumors 709-2313*), Bookstore (*Starfish Books & Charts 746-7597*), Clothing Store (*Mixed Blessing 801-4225*) **Under 1 mi:** Gourmet Shop (*Hilary's Fine Cheeses 715-0563*) **1-3 mi:** Market (*Country Grocer 743-5639*) **3+ mi:** Supermarket (*Safeway 746-3655, 6 mi.*), Health Food (*Community Farm Store 748-6227, 6 mi.*), Catholic Church (*St. Edward's 746-6831, 6 mi.*), Protestant Church (*6 mi.*), Library (*746-7661, 6 mi.*), Dry Cleaners (*Betty's One Hour 748-3341, 6 mi.*), Laundry (*Seabreeze Laundry 748-3021, 6 mi.*), Pharmacy (*Pharmasave 748-5252, 6 mi.*), Hardware Store (*Duncan Home 746-4456, 6 mi.*), Florist (*Flower Affairs 746-3977, 6 mi.*), Department Store (*Wal-Mart 748-1226, 6 mi.*)

Transportation
OnCall: Taxi (*Cobble Hill Taxi 743-5555*) **Under 1 mi:** Local Bus (*Cowichan Valley Transit System 746-9899*) **3+ mi:** Rental Car (*Budget 748-3221, 6 mi.*) **Airport:** Duncan (*6 mi.*)

Medical Services
911 Service **OnCall:** Ambulance (*743-3232*) **Near:** Holistic Services (*Breeze Day Spa 748-2811*) **1-3 mi:** Doctor (*Brookside Medical Clinic 749-0120*), Chiropractor (*Valleyview Family Chiropractic 743-3775*) **3+ mi:** Dentist (*Alderlea Dental Centre 748-1842, 6 mi.*), Veterinarian (*Duncan Animal Hospital 746-1966, 6 mi.*), Optician (*Opti View 1 Hour Optical 748-9229, 6 mi.*) **Hospital:** Cowichan Valley District 746-4141 (*6 mi.*)

Setting -- In this small, crowded harbor, the marinas are all grouped together, making it difficult to tell them apart from the water - Cowichan Bay Marina sits just east of the center of the waterfront, squeezed between Pier 66 and Masthead. The rustic floating docks offer a mix of side-tie, slip and stern-to dockage and hosts a wide variety of vessels, including a couple of large, permanent houseboats. Shoreward, the view is of the back side of several stores lining the street.

Marina Notes -- Although the marina offers few boater amenities at the present time, there are plans for expansion and growth. There are no heads or showers so, for the time being, boaters can use heads and showers at the Fisherman's Wharf Association or Bluenose Marina. Most of the Cowichan Bay facilities are aging, and this one is no exception - but repairs are being made.

Notable -- A valuable asset for this marina, as well as the others in the harbor, is its proximity to the seaside village of Cowichan Bay: restaurants, an ice cream parlor, a coffee shop, and art galleries are a stone's throw away. The Village is pulling together to make this a fun place to visit. The Cowichan Bay Whale Watching Tours scour the area looking for pods of killer whales and are almost always successful in finding them. The Cowichan Bay Maritime Centre & Wooden Boat Society, a short walk away, is a must-see; its three unique buildings, called "pods" are situated along a 350' pier. If an extended stay is planned, you might be able to sign up for one of their boatbuilding programs. Nearby Rock Cod Café reportedly serves excellent fish and chips in a comfort food atmosphere, and a few other eateries are nearby.

PHOTOS ON DVD: 8

Navigational Information
Lat: 48°44.457' **Long:** 123°37.253' **Tide:** 11 ft. **Current:** n/a **Chart:** 3478
Rep. Depths (*MLW*): **Entry** 26 ft. **Fuel Dock** 26 ft. **Max Slip/Moor** 18 ft./-
Access: Cowichan Bay off Satellite Channel Island

Marina Facilities *(In Season/Off Season)*
Fuel: Gasoline, Diesel
Slips: 25 Total, 5 Transient **Max LOA:** 50 ft. **Max Beam:** n/a
 Rate *(per ft.)*: **Day** $1.00/0.75 **Week** $5.50 **Month** n/a
 Power: 30 amp $1.00*, **50 amp** n/a, **100 amp** n/a, **200 amp** n/a
 Cable TV: No **Dockside Phone:** No
 Dock Type: Floating, Long Fingers, Pilings, Wood
Moorings: 0 Total, 0 Transient **Launch:** n/a
 Rate: Day n/a **Week** n/a **Month** n/a
Heads: None
Laundry: None **Pay Phones:** Yes, 1
Pump-Out: No **Fee:** n/a **Closed Heads:** Yes

Marina Operations
Owner/Manager: Tom and Sharon Ingram **Dockmaster:** Same
In-Season: Year-Round, 8am-9pm **Off-Season:** n/a
After-Hours Arrival: Report to the store
Reservations: Yes, Preferred **Credit Cards:** Visa/MC
Discounts: Fuel **Dockage:** n/a **Fuel:** $.05/ltr** **Repair:** n/a
Pets: Welcome **Handicap Access:** No

Pier 66 Marina

1745 Cowichan Bay Road; Cowichan Bay, BC V0R 1N0

Tel: (250) 748-8444 **VHF: Monitor** n/a **Talk** n/a
Fax: (250) 748-8444 **Alternate Tel:** n/a
Email: pier66@shaw.ca **Web:** www.pier66marina.com
Nearest Town: Cowichan Bay **Tourist Info:** (250) 748-1111

Marina Services and Boat Supplies
Services - Docking Assistance, Security *(locked gate)*, Trash Pick-Up, Dock Carts **Supplies - OnSite:** Ice *(Block, Cube)*, Ships' Store, Bait/Tackle

Boatyard Services
OnSite: Railway *(10T)*, Engine mechanic *(gas, diesel)*, Hull Repairs
Near: Launching Ramp, Yacht Broker, Total Refits **1-3 mi:** Propeller Repairs
Nearest Yard: Covey Marine (250) 748-3130

Restaurants and Accommodations
OnSite: Lite Fare *(Rodney's 748-8444, takeout: soups, burgers, fish & chips)* **Near:** Restaurant *(The Masthead 748-3714, D $17-35)*, *(The Tides 701-0166, B $6-10, L $7-10, D $23-60)*, *(Amici at the Bluenose 748-2841, L $10-16, D $17-26, pizzas $12-16)*, *(Rock Cod Café 746-1550, L $5-8, D $11-14)*, Snack Bar *(The Udder Guys Ice Cream Parlour 746-4300)*, Hotel *(Oceanfront Grand Resort & Marina 800-663-7898, $100-205, docks for hotel guests only)* **1-3 mi:** Inn/B&B *(Dream Weaver 748-7688, $130-150)*

Recreation and Entertainment
Near: Heated Pool, Playground, Dive Shop, Boat Rentals *(We Go Kayaking 800-434-9346)*, Fishing Charter *(Bluenose Marina 748-2222)*, Video Rental *(Pioneer Video 709-9944, 11am-10pm)*, Park *(Coverdale Watson Park)*, Museum *(Cowichan Bay Maritime Centre 746-4955, Cowichan Valley Museum 746-6612, 6 mi.)*, Special Events *(Cowichan Bay Regatta - Aug, Father's Day Tractor Show)* **Under 1 mi:** Group Fishing Boat *(Excalibur 246-0142, Morning Mist 746-4300)* **1-3 mi:** Tennis Courts *(South Cowichan Lawn Tennis 746-7282)*, Galleries *(Catherine Fraser 748-2067)* **3+ mi:** Golf Course *(Cowichan Golf Club 746-5333, 6 mi.)*

Provisioning and General Services
OnSite: Convenience Store *(Pier 66 Liquor & Convenience Store: groceries & liquor)*, Wine/Beer, Liquor Store **Near:** Bakery *(True Grain Bread 746-7664)*, Fishmonger *(Cowichan Bay Seafoods 748-0020)*, Bank/ATM *(ATMs at Amici and The Bay Pub)*, Beauty Salon *(Rumors 709-2313)*, Clothing Store *(Mixed Blessings 801-4225)* **Under 1 mi:** Gourmet Shop *(Hilary's Fine Cheeses 715-0563)*, Bookstore *(Starfish Books and Charts 746-7597)* **1-3 mi:** Market *(Country Grocer 743-5639)* **3+ mi:** Supermarket *(Safeway 746-6122, 6 mi.)*, Health Food *(Community Farm Store 748-6227, 6 mi.)*, Catholic Church *(St. Edward's 746-6831, 6 mi.)*, Protestant Church *(6 mi.)*, Library *(746-7661, 6 mi.)*, Dry Cleaners *(Betty's One Hour 748-3341, 6 mi.)*, Laundry *(Seabreeze Laundry 748-3021, 6 mi.)*, Pharmacy *(Pharmasave 748-5252, 6 mi.)*, Hardware Store *(Duncan Home 746-4456, 6 mi.)*, Florist *(Flower Affairs 746-3977, 6 mi.)*, Department Store *(Wal-Mart 748-1226, 6 mi.)*

Transportation
OnCall: Taxi *(Cobble Hill Taxi 743-5555)* **Under 1 mi:** Local Bus *(Cowichan Valley Transit System 746-9899)* **3+ mi:** Rental Car *(Budget 748-3221, 6 mi.)* **Airport:** Duncan *(6 mi.)*

Medical Services
911 Service **OnCall:** Ambulance **Near:** Holistic Services *(Breeze Day Spa 748-2811)* **1-3 mi:** Doctor *(Brookside Medical Clinic 749-0120)*, Chiropractor *(Valleyview Family Chiropractic 743-3775)* **3+ mi:** Dentist *(Alderlea Dental Health Centre 748-1842, 6 mi.)*, Veterinarian *(Duncan Animal Hospital 746-1966, 6 mi.)*, Optician *(Opti View 1 Hour Optical 748-9229, 6 mi.)* **Hospital:** Cowichan District Hospital 746-4141 *(6 mi.)*

Setting -- Pier 66 Marina sits among several facilities on the Cowichan Bay waterfront; it is easily identified by its fuel dock fronting the bay. Behind the marina, facing the highway, the Pier 66 Market offers groceries and supplies. Just to the west, a small shopping mall juts out over the water.

Marina Notes -- *Only 15 amp power is available ($1). **$0.05/litre fuel discount on cash purchases of 300 liters or more. Gas, diesel, and 50:1 mixes are available, as well as ice and fishing tackle. The on-site market is also a fully-licensed liquor store offering a wide selection. Covey Marine is adjacent, and provides haul out for boats up to 10T and some boatyard services. An indoor mini-storage is also part of the marina. Check the sign out front for rates. Although currently lacking many boater amenities, changes and improvements are on the way as the townspeople are trying to encourage tourism.

Notable -- On-site Rodney's Restaurant offers light fare including burgers, fried fish, calamari, and daily specials to go - perfect for a picnic on the new waterfront promenade. Nearby Cowichan Bay Maritime Museum features exhibits on boatbuilding techniques and outboard motor history. The seaside village of Cowichan provides much to do. Try Ocean Eco Adventures for a day of water exploration (748-3800). Excalibur Charters offers fishing, sightseeing, and nature cruises. Morning Mist Charters provides Cowichan Bay and area ecotours and half or full day fishing charters. Inquire at the Udder Guys Ice Cream Parlour (ice cream is considered a staple here). Annual attractions include a Father's Day Tractor Show, traditional First Nations' war canoe races, and a Boat Festival in June. Visitors to the Village (as it is called by the locals) whale watch, scuba dive, fish, sail, paddle, eat or poke into the shops.

Bluenose Marina

Navigational Information
Lat: 48°44.456' **Long:** 123°37.361' **Tide:** 11 ft. **Current:** n/a **Chart:** 3478
Rep. Depths *(MLW)*: **Entry** 6 ft. **Fuel Dock** n/a **Max Slip/Moor** 6 ft./-
Access: Located in Cowichan Bay off Satellite Channel

Marina Facilities *(In Season/Off Season)*
Fuel: No
Slips: 35 Total, 3 Transient **Max LOA:** 88 ft. **Max Beam:** n/a
  Rate *(per ft.)*: **Day** $1.00/0.75 **Week** n/a **Month** $6.50
  Power: 30 amp $1*, **50 amp** n/a, **100 amp** n/a, **200 amp** n/a
  Cable TV: No **Dockside Phone:** No
  Dock Type: Floating, Alongside, Wood
Moorings: 0 Total, 0 Transient **Launch:** n/a
  Rate: Day n/a **Week** n/a **Month** n/a
Heads: 3 Toilet(s), 1 Shower(s)
Laundry: 2 Washer(s), 2 Dryer(s) **Pay Phones:** Yes, 1
Pump-Out: No **Fee:** n/a **Closed Heads:** No

Marina Operations
Owner/Manager: Deben Jones **Dockmaster:** Same
In-Season: Year-Round, 10am-6pm **Off-Season:** n/a
After-Hours Arrival: Tie up and pay in the morning
Reservations: Yes, Preferred **Credit Cards:** No
Discounts: None
Pets: Welcome **Handicap Access:** No

Bluenose Marina

PO Box 40; 1765 Cowichan Bay; Cowichan Bay, BC V0R 1N0

Tel: (250) 748-2222 **VHF: Monitor** n/a **Talk** n/a
Fax: (250) 748-8040 **Alternate Tel:** n/a
Email: alinef@shaw.ca **Web:** n/a
Nearest Town: Cowichan Bay **Tourist Info:** (250) 748-1111

Marina Services and Boat Supplies
Services - Docking Assistance, Trash Pick-Up, Dock Carts
Communication - FedEx, DHL, UPS, Express Mail **Supplies - Near:** Ice *(Cube)*, Ships' Store *(at Pier 66)*, Bait/Tackle

Boatyard Services
Near: Launching Ramp. **Nearest Yard:** Covey Marine (250) 748-3130

Restaurants and Accommodations
OnSite: Restaurant *(Bluenose II 748-2841, L $10-16, D $17-28, pizza $12-16; oceanview deck)* **Near:** Restaurant *(The Tides 701-0166, B $6-10, L $7-10, D $23-60, Sun brunch $20)*, *(Rock Cod Café 746-1550, L $5-8, D $11-14)*, Snack Bar *(The Udder Guys Ice Cream Parlour 746-4300)*, Motel *(Oceanfront Grand Resort & Marina 800-663-7898, $100-205, docks for hotel guests only)*, *(Wessex Inn 748-4214)* **Under 1 mi:** Restaurant *(The Masthead Restaurant 748-3714, D $17-35)*, Snack Bar *(The Bay Pub 748-2330, L & D)* **1-3 mi:** Inn/B&B *(Dream Weaver B&B 748-7688, $130-150)*, Condo/Cottage *(Anchor's Guest House 748-7206, $96-195)*

Recreation and Entertainment
OnSite: Fishing Charter *(Bluenose Marina and Charter 748-2222)*
Near: Beach, Picnic Area, Playground, Boat Rentals *(We Go Kayaking 748-5400)*, Video Rental *(Pioneer Video 709-9944, 11am-10pm)*, Museum *(Cowichan Bay Maritime Centre 746-4955)*, Special Events *(Cowichan Bay Regatta in Aug)*, Galleries *(Timeless Co. 709-9985, coffee shop and art gallery; Hillbank Pottery)* **Under 1 mi:** Park *(Coverdale Watson Park)* **1-3 mi:** Tennis Courts *(South Cowichan Lawn Tennis 746-7282)* **3+ mi:** Golf Course *(Cowichan 748-5333, Arbutus Ridge 743-5100, 6 mi.)*

Provisioning and General Services
OnSite: Bank/ATM *(ATMs at The Bay Pub)* **Near:** Convenience Store *(Pier 66 Liquor and Convenience Store 748-8444)*, Bakery *(True Grain Bread 746-7664)*, Fishmonger *(Cowichan Bay Seafoods 748-0020)*, Beauty Salon *(Rumors 709-2313)*, Clothing Store *(Mixed Blessings 801-4225)* **Under 1 mi:** Gourmet Shop *(Hilary's Fine Cheeses 715-0563)*, Bookstore *(Starfish Books and Charts 746-7597 - gifts and more)* **1-3 mi:** Market *(Country Grocer 743-5639)* **3+ mi:** Supermarket *(Safeway 746-6122, 6 mi.)*, Health Food *(Community Farm Store 748-6227, 6 mi.)*, Catholic Church *(St. Edward's 746-6831, 6 mi.)*, Protestant Church *(6 mi.)*, Library *(746-7661, 6 mi.)*, Dry Cleaners *(Betty's One Hour 748-3341, 6 mi.)*, Laundry *(Seabreeze Laundry 748-3021, 6 mi.)*, Pharmacy *(Pharmasave 748-5252, 6 mi.)*, Hardware Store *(Duncan Home 746-4456, 6 mi.)*, Florist *(Flower Affairs 746-3977, 6 mi.)*, Department Store *(Wal-Mart 748-1226, 6 mi.)*

Transportation
OnCall: Taxi *(Country Cabs 746-0009)* **Under 1 mi:** Local Bus *(Cowichan Bay Transit 746-9899)* **3+ mi:** Rental Car *(Budget 748-3221, 6 mi.)*
Airport: Duncan *(6 mi.)*

Medical Services
911 Service **OnCall:** Ambulance *(743-3232)* **Near:** Holistic Services *(Breeze Day Spa 748-2811)* **1-3 mi:** Doctor *(Brookside Medical Clinic 749-0120)*, Chiropractor *(Valleyview Family Chiropractic 743-3775)* **3+ mi:** Veterinarian *(Duncan Animal Hospital 746-1966, 6 mi.)*, Optician *(Opti View 1 Hour Optical 748-9229, 6 mi.)* **Hospital:** Cowichan District 746-4141 *(6 mi.)*

Setting -- Bluenose Marina is situated just north of the Cowichan Bay Maritime Centre dock on the west end of the Village's waterfront. Its upscale restaurant with open deck, pizza parlor, and amenities set it apart from the other Cowichan facilities. A large, wooden picnic deck, adjacent to the restaurant, hosts new bench swings and picnic tables. Across the bay, the commercial docks stay busy with forest product exports. An unusual community of stilt houses lies west of the marina, along with the boat launch ramp and picnic areas of Hecate Park.

Marina Notes -- *15 amp power available. New owners as of August '06. Look forward to many changes in the upcoming years. Manager lives on-site. Mostly side-tie moorage along wooden docks, and some slips. The marina office is at the top of the ramp, but the entrance is hidden around the back of the building. Expansion and renewal is taking place throughout Cowichan Bay, and this marina is no exception. Locked restrooms and shower facilities.

Notable -- Cowichan translates from the native tongue as "The Warm Land." Outdoor enthusiasts visit for the hiking and biking opportunities it offers. An assortment of activities is available in downtown Cowichan Bay. We Go Kayaking offers just what the name implies - so rent a boat and explore the bay with a picnic lunch. Or pamper yourself at the Breeze Day Spa where you can get the works with a massage and even a tattoo. The Cowichan Bay Maritime Centre is a must-see for those who love wooden boats. Housed in unique "pods" along a 350' of pier, the Centre provides interesting insights into the history of the bay. Stop by The Udder Guys for a handmade ice cream treat after a long day of sightseeing.

6. BC – Gulf Islands

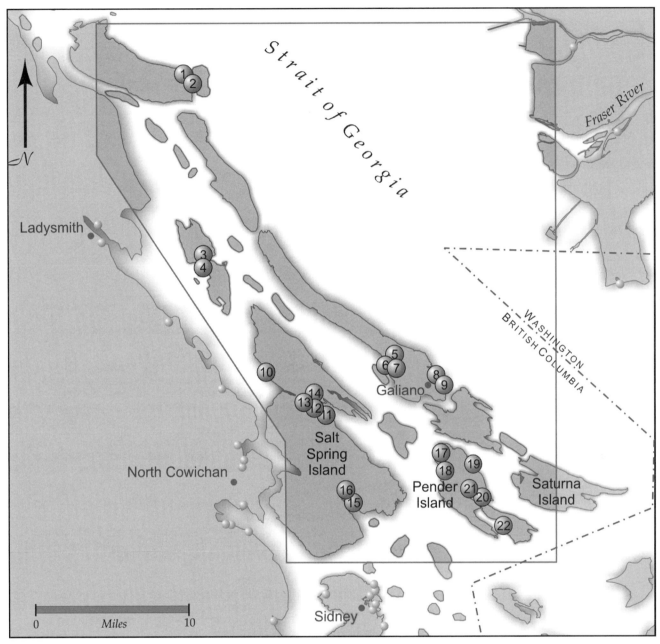

MAP	MARINA	HARBOR	PAGE	MAP	MARINA	HARBOR	PAGE
1	Silva Bay Resort & Marina	Silva Bay	88	12	Kanaka Visitors Wharf	Ganges Outer Harbour	99
2	Page's Resort & Marina	Silva Bay	89	13	Ganges Marina	Ganges Outer Harbour	100
3	Telegraph Harbour Marina	Telegraph Harbour	90	14	Salt Spring Marina	Ganges Outer Harbour	101
4	Thetis Island Marina	Telegraph Harbour	91	15	Fulford Harbour Wharf	Fulford Harbour	102
5	Montague Harbour Marine Park	Montague Harbour	92	16	Fulford Harbour Marina	Fulford Harbour	103
6	Montague Harbour Public Wharf	Montague Harbour	93	17	Port Washington Public Wharf	Grimmer Bay	104
7	Montague Harbour Marina	Montague Harbour	94	18	Otter Bay Marina	Otter Bay	105
8	Whaler Bay Public Wharf	Whaler Bay	95	19	Hope Bay Public Wharf	Hope Bay	106
9	Sturdies Bay Public Float	Sturdies Bay	96	20	Browning Harbour Public Wharf	Browning Harbour	107
10	Vesuvius Bay Public Wharf	Vesuvius Bay	97	21	Port Browning Marina	Browning Harbour	108
11	Centennial Wharf	Ganges Harbour	98	22	Poets Cove Resort & Spa	Bedwell Harbour	109

Silva Bay Resort & Marina

3383 South Road; Gabriola Island, BC V0R 1X7

Tel: (250) 247-8662 **VHF: Monitor** Ch. 66A **Talk** n/a
Fax: (250) 247-8663 **Alternate Tel:** n/a
Email: silvabay@canada.com **Web:** www.silvabay.com
Nearest Town: Nanaimo *(9 mi.)* **Tourist Info:** (250) 247-9332

Navigational Information
Lat: 49°08.983' **Long:** 123°41.936' **Tide:** 15 ft. **Current:** n/a **Chart:** 3475
Rep. Depths *(MLW)*: **Entry** 9 ft. **Fuel Dock** 20 ft. **Max Slip/Moor** 30 ft./-
Access: On Gabriola Island off Commodore Passage

Marina Facilities *(In Season/Off Season)*
Fuel: Gasoline, Diesel
Slips: 120 Total, 36 Transient **Max LOA:** 90 ft. **Max Beam:** n/a
  Rate *(per ft.)*: **Day** $1.10 **Week** n/a **Month** n/a
  Power: 30 amp $5, **50 amp** $10, **100 amp** n/a, **200 amp** n/a
  Cable TV: No **Dockside Phone:** No
  Dock Type: Long Fingers, Alongside, Wood
Moorings: 0 Total, 0 Transient **Launch:** n/a
  Rate: Day n/a **Week** n/a **Month** n/a
Heads: 6 Toilet(s), 6 Shower(s)
Laundry: 4 Washer(s), 2 Dryer(s) **Pay Phones:** Yes, 2
Pump-Out: No **Fee:** n/a **Closed Heads:** Yes

Marina Operations
Owner/Manager: Janice Fuller **Dockmaster:** Same
In-Season: Apr-Sep, 7am-9pm **Off-Season:** Oct-Mar, 8am-6pm
After-Hours Arrival: Call ahead
Reservations: Yes, Preferred **Credit Cards:** Visa/MC, Debit
Discounts: Fuel **Dockage:** n/a **Fuel:** Bulk disc. **Repair:** n/a
Pets: Welcome **Handicap Access:** Yes, Heads, Docks

Marina Services and Boat Supplies
Services - Docking Assistance, Trash Pick-Up **Communication -** Data Ports *(Wi-Fi - Broadband Xpress)* **Supplies - OnSite:** Ice *(Block, Cube)*, Bait/Tackle, Ships' Store *(Silva Bay Shipyard 247-8809)*

Boatyard Services
Nearest Yard: Silva Bay Boatyard (250) 247-8385

Restaurants and Accommodations
OnSite: Restaurant *(Silva Bay Bar & Grill 247-8662, L $8-12, D $16-24, B on wknds 9:30-11am)*, Snack Bar *(The Silva Bay Pub 247-8662, 7am-4pm)*
OnCall: Pizzeria *(Underground Pizza & Pasta 247-9622, free delivery "even when the power's out!")* **Under 1 mi:** Inn/B&B *(Silva Bay Inn 247-9351)*, Condo/Cottage *(Page's Resort & Marina 247-8931, $85-110)*
3+ mi: Restaurant *(Suzy's Restaurant & Deli 247-2010, 7 mi., family style & takeout)*, Inn/B&B *(Hummingbird Lodge 247-9300, $89-120, 6 mi.)*, *(Cherry Cottage 247-7912, $85, 5 mi.)*

Recreation and Entertainment
OnSite: Playground, Tennis Courts, Boat Rentals *(Jim's Kayaking 247-8335)*, Fishing Charter *(Silva Blue Charters 247-8807 - salmon fishing on 25.5 ft. Searay $280/half-day)*, Sightseeing *(Silva Bay Shipyard - the only full-time wooden boat building school in Canada 247-8809)*, Special Events *(Maritime Fest - Apr, Salmon Derby - Jun, Beat the Heat & Gabriola Non-Marine Boat Race - Aug)* **Under 1 mi:** Beach, Dive Shop **1-3 mi:** Park *(Drumbeg Provincial Park)* **3+ mi:** Golf Course *(Gabriola Golf and Country Club 247-8822, 4 mi.)*, Museum *(Gabriola Museum 247-9987, Jun-Aug Wed-Sun 10:30am-4pm, Sep-May wknds 1-4pm, $2/under 16 free, 7 mi.)*, Cultural Attract *(Gabriola Chamber Players 247-7858, 7 mi.)*, Galleries *(Pumphouse Gallery 247-9445, Ebbas Pottery 247-8152, Gabriola Artworks 247-7412, Taracotta Pottery 247-7583, Willcox Wildlife Photography 247-9043)*

Provisioning and General Services
OnSite: Wine/Beer *(Silva Bay Liquor, 7 days 10am-8pm)*, Liquor Store Farmers' Market *(Sun Market 10am-2pm Jul 4-LabDay - crafts, jewelry, art or Sat in town at Agi Hall)* **Near:** Market *(G&S Foods 247-8828)*, **3+ mi:** Convenience Store *(Village Quickstop 247-8755, 7 mi.)*, Supermarket *(Village Food Market 247-8755, 7 mi.)*, Gourmet Shop *(Gabriola Gourmet 7 mi.)*, Garlic 247-0132, Green Market *(Gabriola Greenhouse 247-8168, 7 mi.)*, Bank/ATM *(Direct Cash ATM, 7mi.)*, Post Office *(247-8862, closed Sun, 7 mi.)*, Library *(247-7878, 7 mi.)*, Pharmacy *(Medicine Centre 247-8310, 7 mi.)*, Hardware Store *(Village Paint and Hardware 247-9266, 7 mi.)*, Clothing Store *(Just Bliss Clothing 247-8233, 7 mi.)*, Retail Shops *(Folklife Village Shoping Centre 247-9676, 7 mi.)*

Transportation
OnCall: Bikes *(i-MoPed 247-0049)*, Rental Car, Taxi *(Gabriola Island 247-0049)* **3+ mi:** Ferry Service *(to Nanaimo, 7 mi.)* **Airport:** Tosino Air 800-665-2359/Nanaimo Airport via ferry *(On-Site/7 mi.)*

Medical Services
911 Service **OnCall:** Ambulance *(247-0001)* **3+ mi:** Doctor *(Medical Clinic 247-9922, 7 mi.)*, Dentist *(Aerie Dental Centre 247-8212, 7 mi.)*, Veterinarian *(Twin Cedars Veterinary Services 247-9185, 7 mi.)* **Hospital:** Nanaimo Regional 248-2332 *(13 mi.)*

Setting -- Silva Bay, located on the east end of Gabriola Island, is a small, somewhat cluttered bay almost completely surrounded by the Flat Top Islands - providing near total protection from the sometimes rambunctious Strait of Georgia. After rounding the shallows off Tugboat Island, newly renovated Silva Bay Resort is the northernmost of the marinas that line the western shore. A patio hosts tables, topped with gaily-colored umbrellas, that overlook the fan of immaculate slips and side-tie dockage. In addition to the eatery, there's a pub, coffee bar, liquor/wine store, pool table, dart boards, tennis courts, plus a variety of rentals.

Marina Notes -- Although there is a lot to see and do on the island, you need not leave the Silva Bay Marina to have a good time. The service-oriented staff ensures that boaters' needs are met. In the summer, there's live entertainment Saturday night in the restaurant and a seasonal Sunday Market from 10am-2pm. TheSilva Bay Fishing Derby takes place in June, and other events are held on-site. The marina can accommodate group cruises of 40-60 boats, supported by convenient direct flights from Vancouver Airport are available to get the gang together. On-site Silva Bay Shipyard offers marine repairs, a chandlery, and runs a wooden boat building school. Kayak, moped and car rentals are easily at hand. Beautifully maintained tiled heads & showers.

Notable -- Gabriola is just 8.6 miles long and 2 miles wide, so it's an easy moped ride to everything. There are three provincial waterfront parks and two regional waterfront parks. Besides the 4,500 human residents, you may run into seals, deer, herons, sea lions, raccoons, bald eagles, otters, and numerous birds. Stop by the Visitors' Information Centre for a detailed map, then hop on a moped and tour the island's many galleries.

Navigational Information
Lat: 49°08.899' **Long:** 123°41.807' **Tide:** 15 ft. **Current:** n/a **Chart:** 3475
Rep. Depths (MLW): Entry 11 ft. **Fuel Dock** 9 ft. **Max Slip/Moor** 11 ft./-
Access: On Silva Bay on Gabriola Island's west shore

Marina Facilities *(In Season/Off Season)*
Fuel: Gasoline, Diesel
Slips: 60 Total, 10 Transient **Max LOA:** 45 ft. **Max Beam:** n/a
  Rate *(per ft.):* **Day** $0.80 **Week** n/a **Month** $4.75*
  Power: 30 amp $3**, **50 amp** n/a, **100 amp** n/a, **200 amp** n/a
  Cable TV: No **Dockside Phone:** No
  Dock Type: Long Fingers, Alongside, Wood
Moorings: 0 Total, 0 Transient **Launch:** n/a
  Rate: Day n/a **Week** n/a **Month** n/a
Heads: 2 Toilet(s), 2 Shower(s)
Laundry: 2 Washer(s), 1 Dryer(s) **Pay Phones:** Yes, 1
Pump-Out: No **Fee:** n/a **Closed Heads:** Yes

Marina Operations
Owner/Manager: Ted & Phyllis Reeve **Dockmaster:** Same
In-Season: Year-Round, 8:30am-6pm **Off-Season:** n/a
After-Hours Arrival: Call ahead
Reservations: Yes, Preferred **Credit Cards:** Visa/MC
Discounts: None
Pets: Welcome, Dog Walk Area **Handicap Access:** Yes, Heads, Docks

Page's Resort & Marina

3350 Coast Road; Gabriola Island, BC V0R 1X7

Tel: (250) 247-8931 **VHF: Monitor** n/a **Talk** n/a
Fax: (250) 247-8997 **Alternate Tel:** n/a
Email: mail@pagesresort.com **Web:** www.pagesresort.com
Nearest Town: Nanaimo *(9 mi.)* **Tourist Info:** (250) 247-9332

Marina Services and Boat Supplies
Services - Security *(owners live onsite)*, Trash Pick-Up **Communication -**
Mail & Package Hold, Phone Messages **Supplies - OnSite:** Ice *(Cube)*
Under 1 mi: Ships' Store *(Silva Bay Resort)*, Bait/Tackle *(Sliva Bay)*

Boatyard Services
Nearest Yard: Silva Bay Boatyard (250) 247-8385

Restaurants and Accommodations
OnSite: Condo/Cottage *(Pages's Resort and Marina 247-8931, $85-105)*
Under 1 mi: Restaurant *(Silva Bay Bar & Grill 247-8662, L $8-12, D $16-24)*,
Snack Bar *(Silva Bay Pub 247-8662, 7am-4pm)*, Inn/B&B *(Silva Bay Inn 247-9351)* **3+ mi:** Restaurant *(Allycat Bistro 247-7698, 7 mi.)*, *Suzy's Restaurant & Deli 247-2010, 7 mi., steaks, seafood, pizza, sandwiches; takeout available)*, *(Café Provencal 247-9328, 7 mi.)*, *(Rapsberrys Jazz Café 247-9959, 7 mi.)*, Inn/B&B *(Hummingbird Lodge B&B 247-9300, $89-120, 6 mi.)*, Condo/Cottage *(Casa Blanca B&B 247-9824, $65-180, 5 mi.)*

Recreation and Entertainment
OnSite: Picnic Area, Cultural Attract *(art and book events in the owners' home, adjacent to the office)*, Galleries *(The Sandstone Studio 247-8931; studios throughout the island)* **Near:** Jogging Paths **Under 1 mi:** Boat Rentals *(Jim's Kayaking 247-8335)*, Fishing Charter *(Silva Blue Charters 247-8807 - salmon fishing on 25.5' Sea Ray $280/half-day)*, Sightseeing *(Silva Bay Shipyard - wooden boat building 247-8809)*, Special Events *(Salmon Derby - Jun, Potato Cannon - Jul, Beat the Heat - Aug, Sand Sculpture - Aug)* **1-3 mi:** Beach *(Drumbeg Park)*, Park *(Drumbeg Park)*
3+ mi: Golf Course *(Gabriola 247-8822, 4 mi.)*, Video Rental *(Village*

Quickstop 247-8755, 7 mi.), Museum *(Gabriola Museum 247-9987, Jun-Aug Wed-Sun 10:30am-4pm, Sep-May wknds 1-4pm, $2/under 16 free, 7 mi.)*

Provisioning and General Services
OnSite: Bakery *(plus, under 1 mi., Old Town Bakery 245-2531)*, Bookstore *(charts too)* **Under 1 mi:** Market *(Silva Bay Boatel 247-9351)*, Wine/Beer, Liquor Store *(Silva Bay Liquor Store at Silva Bay Marina 247-8662)*, Farmers' Market *(Weekly Market at Silva Bay on Sun 10am-2pm, or in town at Agi Hall on Sat 10am-Noon)* **3+ mi:** Convenience Store *(Village Quickstop and Video 247-8755, 7 mi.)*, Supermarket *(Village Food Market 247-8755, 7 mi.)*, Gourmet Shop *(Gabriola Gourmet Garlic 247-0132, 7 mi.)*, Bank/ATM *(Direct Cash ATM, 7 mi.)*, Post Office *(247-8862, closed Sun, 7 mi.)*, Protestant Church *(7 mi.)*, Library *(247-7878, 7 mi.)*, Pharmacy *(Medicine Centre 247-8310, 7 mi.)*, Hardware Store *(Village Paint and Hardware 247-9266, 7 mi.)*, Retail Shops *(Folklife Village Shoping Centre 247-9676, Twin Beaches Mall, 7 mi.)*

Transportation
OnCall: Taxi *(Gabriola Island Taxi 247-0049)* **Under 1 mi:** Bikes *(i-MoPed 247-2029)*, Rental Car *(at Silva Bay Resort)* **3+ mi:** Ferry Service *(BC ferry to Nanaimo 386-3431, 7 mi.)* **Airport:** Nanaimo - via ferry or Tofino Air 800-665-2359/ Amigo Air 753-1115 *(20 mi.)*

Medical Services
911 Service **OnCall:** Doctor *(Medical Clinic 247-9922)*, Ambulance *(247-0001)* **3+ mi:** Dentist *(Dental Centre 247-8212, 7 mi.)*, Veterinarian *(Twin Cedars Veterinary Services 247-9185, 7 mi.)* **Hospital:** Nanaimo Regional 754-2141 *(13 mi.)*

Setting -- Like Silva Bay Resort to the north, Page's Resort lies along the western shore of Silva Bay on Gabriola Island. It is the southern-most of the three facilities here. Much of Page's Resort is hidden among the trees that cover the shoreline, but a large welcoming sign and "Monique," a glamorous large wooden "Fogo" folk-art lady are clearly visible as one approaches. White gangways topped with blue rails lead up from the set of docks to the rustic resort.

Marina Notes -- *Monthly rate requires 3 month stay. **15 amp power only ($3/night). The Page family owned the marina from 1943-1987 and transformed it from a fishing harbor to a destination resort. The current owners, the Reeve family, carry on the tradition. The office is a delight: quality artwork hangs on the walls above a wide variety of books and charts, and home-baked goods are sold from the counter. Occasional summer concerts and author events are held in the owners' home - surrounded by their art collection and a view of the bay. Heads are servicable. Note: Silva Bay Boatel serves mainly smaller boats.

Notable -- Gabriola Island, otherwise known as The Isle of the Arts, is well represented by artists and artisans of all kinds, including a garlic chocolate maker. Rent a moped or a bicycle and go for a spin around the island: you'll see signs advertising painters, quilters, potters, jewelers, and garden ornament makers. In October, studios welcome visitors during the annual Thanksgiving Art Tour. In the summer, events include the Potato Cannon Contest on Canada Day, when contestants aim for Mudge Island across the Narrows. For natural beauty, visit one of the island's parks or ride to the end of Malaspina Drive to the famous Malaspina Galleries, a unique cave-like sandstone formation - also accessible by dinghy at Gabriola Sands. Good diving locations are also nearby.

Telegraph Harbour Marina

PO Box 710; Thetis Island, BC V0R 2Y0

Tel: (250) 246-9511; (800) 246-6011 **VHF: Monitor** Ch. 66A **Talk** n/a
Fax: (250) 246-2668 **Alternate Tel:** n/a
Email: n/a **Web:** www.telegraphharbour.com
Nearest Town: Chemainus *(4 mi.)* **Tourist Info:** (250) 246-3944

Navigational Information
Lat: 48°58.959' **Long:** 123°40.207' **Tide:** 10 ft. **Current:** n/a **Chart:** 3477
Rep. Depths *(MLW):* **Entry** 6 ft. **Fuel Dock** 8 ft. **Max Slip/Moor** 10 ft./-
Access: At the head of Telegraph Harbour on Thetis Island

Marina Facilities *(In Season/Off Season)*
Fuel: Gasoline, Diesel
Slips: 75 Total, 75** Transient **Max LOA:** 70 ft. **Max Beam:** n/a
  Rate *(per ft.):* **Day** $1.00/.80* **Week** n/a **Month** $7/3
  Power: 30 amp $4.50***, **50 amp** n/a, **100 amp** n/a, **200 amp** n/a
  Cable TV: No **Dockside Phone:** No
  Dock Type: Floating, Long Fingers, Alongside, Concrete, Wood
Moorings: 0 Total, 0 Transient **Launch:** n/a
  Rate: Day n/a **Week** n/a **Month** n/a
Heads: 4 Toilet(s), 4 Shower(s) *(with dressing rooms)*
Laundry: 2 Washer(s), 3 Dryer(s) **Pay Phones:** Yes, 2
Pump-Out: No **Fee:** n/a **Closed Heads:** Yes

Marina Operations
Owner/Manager: Ron & Barbara Williamson **Dockmaster:** Same
In-Season: Year-Round, 8am-8pm **Off-Season:** n/a
After-Hours Arrival: Call ahead
Reservations: Yes, Preferred **Credit Cards:** Visa/MC, Debit
Discounts: None
Pets: Welcome **Handicap Access:** Yes, Heads, Docks

Marina Services and Boat Supplies
Services - Docking Assistance, Trash Pick-Up **Communication -** Fax in/out, Data Ports *(Wi-Fi - Broadband Xpress)* **Supplies - OnSite:** Ice *(Cube)*, Bait/Tackle **1-3 mi:** Propane

Boatyard Services
Nearest Yard: Body Shop for Boats (888) 310-0222

Restaurants and Accommodations
OnSite: Lite Fare *(Burgees by the Bay 246-9511, B $2-6, L $4-16, D $4-16)*
Under 1 mi: Snack Bar *(Pot of Gold Coffee 246-4944)*, *(Blue Heron Donuts 246-9602, daily 'til sold out)* **1-3 mi:** Restaurant *(Thetis Island Marina Pub 246-3464, L $5-14, D $5-14, Continental B 8-11am)*, Inn/B&B *(Clam Bay Cottage 246-1016, $70-195)*, *(Overbury Farm Resort 246-9769, $95-150, wkly rates available in summer)*

Recreation and Entertainment
OnSite: Picnic Area, Grills, Playground, Volleyball, Special Events *(ice cream eating contests)* **Near:** Jogging Paths *(along the roads)*
1-3 mi: Boat Rentals *(Overbury Kayak Rentals 246-9769)*, Sightseeing *(Thetis Island Vineyards for vineyard tours & wine tasting 246-2258)*, Galleries *(Oracle Readings & Art by appt. 416-0638)*

Provisioning and General Services
OnSite: Convenience Store, Farmers' Market *(produce and art, Sun 10am-Noon)*, Bookstore **1-3 mi:** Wine/Beer, Liquor Store, Bakery *(Bread in the Bone)*, Bank/ATM, Post Office, Protestant Church, Library *(at the Community Hall - Forbes Centre 246-9626)*, Clothing Store *(Ellen's Fashions in Fleece 246-9528)*

Transportation
OnCall: Water Taxi *(Harbour Air Seaplanes 640-274-1277)*
Under 1 mi: Ferry Service *(BC ferry to Chemainus)* **Airport:** Cassidy *(19 mi.)*

Medical Services
911 Service OnCall: Ambulance **3+ mi:** Doctor *(Chemainus Medical 246-3261, ferry, 4 mi.)*, Dentist *(Rosengart 246-9921, ferry, 4 mi.)*
Hospital: Cowichan Bay Regional 746-4141 *(15 mi.)*

Setting -- At the head of Telegraph Harbour, Telegraph Harbour Marina's wide-open, parklike grounds anchor 3,000 feet of alongside dockage. Dotted with flowers, this marina is a delight to the eye; pots overflowing with colorful blooms line the walkway from the dock to the store. Popular with boaters and non-boaters alike, many families ride the ferry from Chemainus on Vancouver Island to enjoy the facilities & picnic for the day.

Marina Notes -- *$1/ft. in summer (May-midSep), $0.80/ft. fall & spring (midSep-earlyOct, midMar-May), $0.50/ft. in winter (earlyOct-midMar); monthly: summer $7, fall, winter & spring $3. **Approximately 3,000 ft. of side-tie dockage makes this a busy, bustling marina in the summer. ***15 and 30 amp power. Activities include shuffleboard, volleyball, tetherball, and horseshoes; a playground and a large group pavilion with a barbeque are also on-site. A fish cleaning station is available to prepare the day's catch. The store/restaurant boasts over 300 burgees from around the world and is looking for more. The effervescent staff serves sandwiches, pizza, salads, espresso, and - from a classic 1930s soda fountain - ice cream concoctions that have become famous among boaters: milk shakes, banana splits, hand-packed ice cream.

Notable -- The island is named after a 36-gun frigate that served the British Navy in this area from 1851 to 1853. It's only a short walk to Preedy Harbour to catch the ferry which runs regularly to Chemainus and its many attractions. Many boaters prefer the ferry, as dock space in Chemainus is limited. Thetis Island is fortunate, like many areas of the Gulf Islands, to fall under Vancouver Island's rain shadow. The annual rainfall here is under 35 inches.

Navigational Information
Lat: 48°58.660' **Long:** 123°40.220' **Tide:** 12 ft. **Current:** n/a **Chart:** 3477
Rep. Depths (MLW): Entry 6 ft. **Fuel Dock** 10 ft. **Max Slip/Moor** 10 ft./-
Access: In Telegraph Harbour on Thetis Island

Marina Facilities *(In Season/Off Season)*
Fuel: Slip-Side Fueling, Gasoline, Diesel
Slips: 75 Total, 75 Transient **Max LOA:** 150 ft. **Max Beam:** n/a
  Rate *(per ft.)*: **Day** $1.00/0.85 **Week** n/a **Month** $3.25
  Power: 30 amp $5.50** , **50 amp** n/a, **100 amp** n/a, **200 amp** n/a
  Cable TV: No **Dockside Phone:** No
  Dock Type: Floating, Long Fingers, Alongside, Wood
Moorings: 0 Total, 0 Transient **Launch:** n/a
  Rate: Day n/a **Week** n/a **Month** n/a
Heads: 5 Toilet(s), 3 Shower(s)
Laundry: 1 Washer(s), 1 Dryer(s) **Pay Phones:** Yes, 2
Pump-Out: No **Fee:** n/a **Closed Heads:** Yes

Marina Operations
Owner/Manager: Paul Deacon **Dockmaster:** Peter Jennings
In-Season: May-Sep, 8am-Mid **Off-Season:** Oct-Apr, 10am-7pm
After-Hours Arrival: Call ahead
Reservations: Yes, Preferred **Credit Cards:** Visa/MC, Amex, Debit
Discounts: Large groups & events **Dockage:** Inq. **Fuel:** n/a **Repair:** n/a
Pets: Welcome **Handicap Access:** Yes, Heads, Docks

Thetis Island Marina

General Delivery; Thetis Island, BC V0R 2Y0

Tel: (250) 246-3464 **VHF: Monitor** Ch. 66A **Talk** n/a
Fax: (250) 246-1433 **Alternate Tel:** (250) 246-3464
Email: marina@thetisisland.com **Web:** www.thetisisland.com
Nearest Town: Chemainus *(4 mi.)* **Tourist Info:** (250) 246-3944

Marina Services and Boat Supplies
Services - Docking Assistance, Trash Pick-Up **Communication -** Mail &
Package Hold, Phone Messages, Fax in/out, Express Mail **Supplies -**
OnSite: Ice *(Cube)*, Bait/Tackle, Propane **3+ mi:** Ships' Store *(Duncan, 20 mi.)*, West Marine *(Nanaimo, 25 mi.)*

Boatyard Services
OnCall: Engine mechanic *(gas, diesel)*, Divers **Near:** Launching Ramp.
3+ mi: Railway *(4 mi.)* **Nearest Yard:** Body Shop for Boats (888) 310-0222

Restaurants and Accommodations
OnSite: Restaurant *(The Pub 246-3464, L $5-14, D $5-14, open 8am 'til Mid, Continental breakfast 8-11am)* **1-3 mi:** Restaurant *(Burgees by the Bay 246-9511, B $2-6, L $4-16, D $4-16)*, Snack Bar *(Blue Heron Donuts 246-9602, daily, 9am 'til sold out in-season)*, *(Pot of Gold Coffee 246-4944)*, Lite Fare *(Afternoon Tea at Clam Bay Cottage 246-1016, Fri-Sun, call 2 hrs. ahead)*, Inn/B&B *(Telegraph Harbour Inn 246-2810, $125, per couple, wknd packages)*, *(Clam Bay Cottage 246-1016, $70-195)*, *(Overbury Farm Resort 246-9769, $135, wkly rates in summer)*, *(Arbutus View B&B 416-0347, $65)*, Condo/Cottage *(Thetis Cottage 246-4209, from $125)*

Recreation and Entertainment
OnSite: Picnic Area, Grills, Playground *(swings)*, Special Events *(Thetis Island Regatta - May: race around Thetis and Kuper Islands followed by a BBQ and party)* **Near:** Beach, Jogging Paths *(along the roads)*, Horseback Riding **1-3 mi:** Tennis Courts, Boat Rentals *(Overbury Kayak Rentals 246-9769)*, Galleries *(Solar Photography by appt. 416-0638)*

Provisioning and General Services
OnSite: Convenience Store *(well supplied with groceries, drinks, books, clothing)*, Liquor Store, Farmers' Market *(Sat mornings during the summer)*, Bank/ATM *(ATM)*, Post Office, Bookstore, Newsstand **Near:** Protestant Church **Under 1 mi:** Clothing Store *(Ellen's Fashions in Fleece 246-9528)* **1-3 mi:** Bakery *(Bread in the Bone)*, Library *(At the Community Hall - Forbes Centre 246-9626)* **3+ mi:** Pharmacy *(Chemainus 246-2151, 4 mi.)*

Transportation
OnCall: Water Taxi *(Seair 800-447-3247)* **1-3 mi:** Bikes *(Thetis Island Bike Shop and Touring 246-5281)*, Ferry Service *(BC ferry to Chemainus)*
Airport: Cassidy *(19 mi.)*

Medical Services
911 Service OnCall: Ambulance **3+ mi:** Doctor *(Chemainus Medical 246-3261, ferry, 4 mi.)*, Dentist *(Rosengart 246-9921, ferry, 4 mi.)*
Hospital: Cowichan Valley Regional 746-4141 *(15 mi.)*

Setting -- Thetis Island Marina is in sheltered Telegraph Harbour on the south end of Thetis Island on the opposite shore from Thetis Island Marina. A network of floating wood docks, managed by a small gray dockhouse, sits at the foot of a densely treed hillside. The shoreline is dominated by the long, low green-roofed "Pub" - brightly-colored umbrellas top tables that march along the waterfront deck. Note: The channel past the marina is somewhat narrow.

Marina Notes -- **15 amp power $3.50. *3,000 ft. of dockage. Pay at the fuel dock. Founded in the 1940s as a store and fishing re-fueling stop. Today, the marina, known for good food and a welcoming atmosphere, is particularly popular with groups, and offers facilities for them to gather. Summer weekend dinner specials at the on-site pub include prime rib and turkey, and the fully stocked convenience store offers a wide selection of goods, including basic foodstuffs, ice cream, books, soft drinks, tide tables and charts, souvenir T-shirts, sweats and hats, and liquor. Postal & faxing services are also available. Join other boaters in the covered picnic area for one of the marina's famous pig roasts (schedule ahead with staff) or for a game of horseshoes in the regulation size pit next to the picnic area. If you visit in May and wish to take part in the Thetis Island Regatta, details and an application are available on the marina website.

Notable -- Rent a bike and explore lovely, quiet Thetis Island. Several popular diving spots are nearby, and the protected waters are ideal for kayaking. Stroll to nearby Pot of Gold Coffee Roasters for a cup of java or to Telegraph Harbour Marina for ice cream. Visitors are encouraged to conserve water (no boat washing) and to take their garbage along when leaving the island. Chemainus, known as The City of Murals, is only a short ferry ride away.

Montague Harbour Marine Park

Galiano, BC V0N 1P0

Tel: (250) 539-2115; (877) 539-2115 **VHF: Monitor** n/a **Talk** n/a
Fax: n/a **Alternate Tel:** n/a
Email: n/a **Web:** www.gocampingbc.com
Nearest Town: Sturdies Bay *(5 mi.)* **Tourist Info:** (250) 539-2233

Navigational Information
Lat: 48°53.817' **Long:** 123°24.154' **Tide:** 11 ft. **Current:** n/a **Chart:** 3473
Rep. Depths *(MLW)*: **Entry** 8 ft. **Fuel Dock** n/a **Max Slip/Moor** 12 ft./-
Access: Southwest side of Montague Harbour on Galiano Island

Marina Facilities *(In Season/Off Season)*
Fuel: No
Slips: 6 Total, 6** Transient **Max LOA:** 36 ft. **Max Beam:** n/a
 Rate *(per ft.)*: **Day** $0.65* **Week** n/a **Month** n/a
 Power: 30 amp n/a, **50 amp** n/a, **100 amp** n/a, **200 amp** n/a
 Cable TV: No **Dockside Phone:** No
 Dock Type: Floating, Alongside, Concrete
Moorings: 39 Total, 39 Transient **Launch:** n/a, Dinghy Dock
 Rate: Day $10 **Week** $70 **Month** n/a
Heads: 20 Toilet(s)
Laundry: None **Pay Phones:** Yes
Pump-Out: No **Fee:** n/a **Closed Heads:** Yes

Marina Operations
Owner/Manager: K2 Park Services **Dockmaster:** Park Ranger
In-Season: Apr-Oct **Off-Season:** Nov-Mar
After-Hours Arrival: Drop box at the top of the stairs
Reservations: No **Credit Cards:** Visa/MC
Discounts: None
Pets: Welcome **Handicap Access:** No

Marina Services and Boat Supplies
Services - Security *(park ranger)*, Trash Pick-Up **Supplies - 1-3 mi:** Ice *(Cube)*, Bait/Tackle **3+ mi:** Propane *(Wells Propane 539-2516, Tue, Fri, Sat 10am-2pm, 5 mi.)*

Boatyard Services
OnSite: Launching Ramp **Nearest Yard:** Canoe Cove Marina (250) 656-5566

Restaurants and Accommodations
OnCall: Pizzeria *(Galiano Pizza 539-5544, delivery only)* **1-3 mi:** Restaurant *(The Harbour Grill 539-5733, B $5-10, L $6-12, D $10-20)*, *(La Berengerie 539-5392)*, Inn/B&B *(Holland House 539-5754)* **3+ mi:** Restaurant *(The Hummingbird Inn Pub 539-5472, L $8-11, D $8-11, 4 mi., live music Fri-Sat, Hummingbird Pub bus will pick up)*, *(Grand Central Emporium 539-9885, B $6-11, L $4-11, D $11-19, 5 mi., kids' menu $5-7)*, Inn/B&B *(Dragonfly 539-2084, 3 mi.)*, *(Galiano Inn 539-3388, from $249, 5 mi.)*, Condo/Cottage *(Sticks and Stones Country Cottage 539-3443, $120-240, 4 mi.)*

Recreation and Entertainment
OnSite: Beach, Picnic Area, Grills **1-3 mi:** Dive Shop *(Lead Foot Diving and Water Sports 539-5341)*, Boat Rentals *(Gulf Island Kayaking 539-2442)*, Group Fishing Boat *(Sporades Tours 539-2278)*, Sightseeing *(Sporades Tours 539-2278)*, Special Events *(Festival of Boats - Sep)* **3+ mi:** Golf Course *(Galiano Golf Club 539-5533 9 hole, restaurant, 5 mi.)*, Video Rental *(The Corner Store 539-2986, 4 mi.)*

Provisioning and General Services
1-3 mi: Convenience Store *(Montague Harbour Marina)*, Wine/Beer, Beauty Salon *(Keo's Kitchen and Hairstyling by appt. 539-3224)*, Barber Shop, Retail Shops *(Ixchel Crafts 539-9819, 10am-5pm)* **3+ mi:** Supermarket *(Galiano Garage and Groceries 539-5500, 5 mi.)*, Health Food *(Daystar Market and Café 539-2505, 4 mi.)*, Liquor Store *(The Corner Store/Liquor Agency 539-2986, 4 mi.)*, Farmers' Market *(539-3672, Sat 10am-3pm, 5 mi.)*, Bank/ATM *(at Hummingbird Pub & The Corner Store, 4 mi.)*, Post Office *(At the Daystar Market and Café, 4 mi.)*, Protestant Church *(St. Margaret's Church, 5 mi.)*, Bookstore *(Galiano Island Books 539-3340, 10am-6pm daily, 5 mi.)*, Pharmacy *(Galiano Garage and Groceries, 5 mi.)*, Hardware Store *(Galiano Trading Company 539-5529, 5 mi.)*

Transportation
OnSite: Local Bus *(Go Galiano Island Shuttle to golf course, the marina, ferry dock at Sturdies Bay 539-0202)* **OnCall:** Taxi *(539-0202)* **1-3 mi:** Bikes *(Galiano Mopeds 539-3443: 1-3 hrs. $18/hr.; 4-5 hrs. $15/hr.; full day $79; wknds 1-3 hrs. $20/hr.; 4-5 hrs. $17/hr.; day rate $99)* **3+ mi:** Water Taxi *(Gulf Islands Water Taxi 537-2510, $15 to Salt Spring Island, 5 mi.)*, Ferry Service *(BC Ferries to Swartz Bay or Vancouver 888-223-3779, 5 mi.)* **Airport:** Harbour Air Seaplane 800-665-0212 *(On-Call)*

Medical Services
911 Service **OnCall:** Ambulance **3+ mi:** Doctor *(Galiano Health Care Centre 539-3230, 5 mi.)*, Holistic Services *(Serenity by the Sea 800-944-2655 - massage, Reiki, yoga, 3 mi.; Galiano Inn & Spa 877-530-3939, 5 mi.)* **Hospital:** Lady Minto 538-4800 - Saltspring Isl. *(18 mi.)*

Setting -- Montague Harbour Marine Park covers both sides of Grey Peninsula on Galiano Island. Montague Harbour's position on the south side of the island, away from the Strait of Georgia, and behind Parker Island offers great protection. A large number of mooring buoys and a good concrete float - which also hosts the Floating Nature House - provide easy access to the park's attractions. Hiking in the park or along the beach is excellent.

Marina Notes -- *Dock fee is $2/meter. **No slips, side-tie on the main dock, for boats up to 36' (11 meters). A minimum fine of $50 is assessed for nonpayment. Rafting is not permitted at the dock or on the moorings. Garbage charges are $2/small bag and $3/large bag. Park regulations prohibit discharge of gray water. Pit toilets are scattered throughout the park. The Park's Floating Nature House, which is a particular hit with kids, features regularly scheduled programs (Noon to 5pm, Thu-Sun only). Other popular organized actives include nature walks and night shows - schedules are posted at information shelters. This is a very popular stop in the summertime. Sunsets are spectacular!

Notable -- Communing with nature is the primary activity at Montague Harbour Marine Park. Tide pooling is excellent here and the salt marsh environment in the lagoon attracts a wide variety of wildlife. Be sure to walk the white sand of Shell Beach. An ancient shell midden created by First Nation people who harvested shellfish here for 3,000 years has been broken up, ground to fine particles and deposited on the beach. Hence, the white "sand." The "Go Galiano" island shuttle stops here and will get you to the eastern side where several shops offer provisions.

Navigational Information
Lat: 48°53.548' **Long:** 123°23.524' **Tide:** 11 ft. **Current:** n/a **Chart:** 3473
Rep. Depths *(MLW)*: **Entry** 8 ft. **Fuel Dock** n/a **Max Slip/Moor** 25 ft./25 ft.
Access: Located on the southwestern side of Galiano Island

Marina Facilities *(In Season/Off Season)*
Fuel: No
Slips: 10 Total, 10 Transient **Max LOA:** 30 ft. **Max Beam:** 8 ft.
  Rate *(per ft.)*: **Day** $0.50* **Week** n/a **Month** $3.10
  Power: 30 amp n/a, **50 amp** n/a, **100 amp** n/a, **200 amp** n/a
  Cable TV: No **Dockside Phone:** No
  Dock Type: Floating, Alongside, Wood
Moorings: 20 Total, 20** Transient **Launch:** n/a, Dinghy Dock
  Rate: Day $15 **Week** n/a **Month** n/a
Heads: 2 Toilet(s)
Laundry: None **Pay Phones:** Yes, 1
Pump-Out: No **Fee:** n/a **Closed Heads:** Yes

Marina Operations
Owner/Manager: Wesley Gross, Wharfinger **Dockmaster:** Same
In-Season: Year-Round, 9am-5pm **Off-Season:** n/a
After-Hours Arrival: Use the drop box
Reservations: No **Credit Cards:** No
Discounts: None
Pets: Welcome **Handicap Access:** Yes, Docks

Montague Harbour Public Wharf

3451 Montague Road; Galiano, BC V0N 1P0

Tel: (250) 539-2488 **VHF: Monitor** Ch. 66 **Talk** n/a
Fax: n/a **Alternate Tel:** n/a
Email: n/a **Web:** n/a
Nearest Town: Sturdies Bay *(5 mi.)* **Tourist Info:** (250) 539-2233

Marina Services and Boat Supplies
Supplies - Near: Ice (Cube), Bait/Tackle *(Montague Harbour Marina 539-5733)* **3+ mi:** Propane *(Wells Propane 539-2516, Tue, Fri, Sat 10am-2pm, 4 mi.)*

Boatyard Services
Nearest Yard: Canoe Cove Marina (250) 656-5566

Restaurants and Accommodations
OnCall: Pizzeria *(Galiano Pizza 539-5544, delivery only)* **Near:** Restaurant *(The Harbour Grill 539-5733, B $5-10, L $6-12, D $10-20)*, *(La Berengerie 539-5392)* **1-3 mi:** Restaurant *(The Hummingbird Inn Pub 539-5472, L $8-11, D $8-11, live music Fri-Sat)*, Inn/B&B *(Holland House 539-5754)*, *(Woodstone Country Inn 539-2022, 199-249)* **3+ mi:** Restaurant *(Artevida 539-3388, D $12-30, 4 mi.)*, *(Grand Central Emporium 539-9885, B $6-11, L $4-11, D $11-19, 4 mi., kids' menu $5-7; pick up at the marina)*, Fast Food *(Madrona 539-3448, 4 mi., fresh takeout meals)*, Inn/B&B *(Galiano Inn 539-3388, from $249, 4 mi., free pick up at the marina)*, Condo/Cottage *(Sticks & Stones Country Cottage 539-3443, $120-240, 4 mi.)*

Recreation and Entertainment
Near: Beach, Picnic Area, Dive Shop *(Lead Foot Diving & Water Sports 539-5341)*, Boat Rentals *(Gulf Island Kayaking 539-2442)*, Group Fishing Boat *(Sporades Tours 539-2278)*, Sightseeing *(Sporades Tours 539-2278)*, Special Events *(Festival of Boats - Sep)* **1-3 mi:** Grills, Golf Course *(Galiano Golf Club 539-5533 9 hole & a restaurant)*, Video Rental *(The Corner Store 539-2986)*, Park *(Montague Harbour Provincial Marine Park)*

Provisioning and General Services
Near: Convenience Store *(Montague Harbour Marina 539-5733)*, Wine/Beer, Retail Shops *(Ixchel Crafts 539-9819, 10am-5pm)* **1-3 mi:** Market *(Daystar Market & Café)*, Liquor Store *(The Corner Store/Liquor Agency 539-2986)*, Bank/ATM *(The Hummingbird Pub and The Corner Store 539-2986)*, Post Office *(At the Daystar Market and Café)*, Beauty Salon *(Keo's Kitchen and Hairstyling by appt. 539-3224)*, Barber Shop **3+ mi:** Supermarket *(Galiano Garage and Groceries 539-5500, 4 mi.)*, Gourmet Shop *(St. Margaret's, 4 mi.)*, Farmers' Market *(Sat 10am-3pm 539-3672, 4 mi.)*, Bookstore *(Galiano Island Books 10am-6pm daily 539-3340, 4 mi.)*, Pharmacy *(Galiano Garage and Groceries 539-5500, 4 mi.)*, Hardware Store *(Galiano Trading Company 539-5529, 4 mi.)*

Transportation
OnSite: Local Bus *(Go Galiano Island Shuttle 539-0202)* **OnCall:** Rental Car *(Galiano Car Rentals 539-2488)*, Taxi *(539-0202)* **Near:** Bikes *(Galiano Mopeds 539-3443, 1-3 hrs. $18/hr.; 4-5 hrs. $15/hr.; 7 hrs. $79; wknds 1-3 hrs. $20/hr.; 4-5 hrs. $17/hr.; daily $99)* **3+ mi:** Water Taxi *(Gulf Islands Water Taxi 537-2510, $15 to Salt Spring Island, 4 mi.)*, Ferry Service *(BC Ferries to Swartz Bay or Vancouver 888-223-3779, 4 mi.)* **Airport:** Harbour Air 800-665-0212 - seaplanes to YVR *(On-Call)*

Medical Services
911 Service **OnCall:** Ambulance **3+ mi:** Doctor *(Galiano Health Care Centre 539-3230, 4 mi.)*, Holistic Services *(Galiano Inn & Spa 877-530-3939, Island Accupressure & Massage 539-9918, 4 mi.)*, Veterinarian *(Lady Minto 538-4800 - Saltspring Isl., 18 mi.)*

Setting -- At first glance, the boater might think this small public dock is part of the larger Montague Harbour Marina next door. The Public Wharf's dock and the marina's docks are only a good jump apart. Shore access, however, is separate. Boaters anchored or moored in the harbor use the float primarily as a dinghy dock.

Marina Notes -- *Alongside dockage. **20 new moorings planned for 2007. While Montague Harbour Public Wharf offers no boater amenities (no garbage disposal or water available), it does lie next door to Montague Harbour Marina where you will find a restaurant, a convenience store with ice cream, a craft shop, and moped and boat rentals. Dinghy over to enjoy the facilities and activities there. Note: Seasonal tenants bring cars to the island for transportation during their stay and the resulting "parking jam" creates a bit of a problem for users of the marina.

Notable -- A shuttle bus named "Go Galiano" stops at the top of the ramp and takes visitors to the ferry dock at Sturdies Bay, the golf course, as well as lodgings and restaurants. On the east side of the island, a couple of grocery stores provide supplies and takeout lunches. A bookstore and several galleries are in the same area. The Hummingbird Pub, with live music on weekends, is a popular spot. Galiano Inn & Spa is a popular a getaway destination, and offers a free shuttle; guests can enjoy treatments at the Madrona del Mar Spa, and dine at the oceanfront Artevida Restaurant (reserve ahead). Hiking trails, a beach, and nature center for kids are at Montague Harbour Provincial Marine Park about 1.5 miles from the public wharf.

Navigational Information

Lat: 48°53.544' **Long:** 123°23.497' **Tide:** 11 ft. **Current:** n/a **Chart:** 3473
Rep. Depths *(MLW):* **Entry** 8 ft. **Fuel Dock** 9 ft. **Max Slip/Moor** 10 ft./-
Access: Located on the southwestern side of Galiano Island

Marina Facilities *(In Season/Off Season)*

Fuel: Gasoline, Diesel
Slips: 90 Total, 30 Transient **Max LOA:** 80 ft. **Max Beam:** n/a
  Rate *(per ft.):* **Day** $0.99/0.55 **Week** n/a **Month** n/a
  Power: 30 amp $5* , **50 amp** n/a, **100 amp** n/a, **200 amp** n/a
  Cable TV: No **Dockside Phone:** No
  Dock Type: Floating, Long Fingers, Alongside, Wood
Moorings: 0 Total, 0 Transient **Launch:** n/a
  Rate: Day n/a **Week** n/a **Month** n/a
Heads: 2 Toilet(s)
Laundry: None **Pay Phones:** Yes, 1
Pump-Out: No **Fee:** n/a **Closed Heads:** Yes

Marina Operations

Owner/Manager: Marilyn & Graham Breeze **Dockmaster:** Same
In-Season: May-Sep **Off-Season:** n/a
After-Hours Arrival: Call ahead
Reservations: Yes, Preferred **Credit Cards:** Visa/MC, Debit
Discounts: None
Pets: Welcome **Handicap Access:** Yes, Docks

Montague Harbour Marina

3451 Montague; Galiano, BC V0N 1P0

Tel: (250) 539-5733 **VHF: Monitor** Ch. 68 **Talk** n/a
Fax: (250) 539-3593 **Alternate Tel:** n/a
Email: montaguemarina@gulfislands.com **Web:** montagueharbour.com
Nearest Town: Sturdies Bay *(5 mi.)* **Tourist Info:** (250) 539-2233

Marina Services and Boat Supplies

Services - Docking Assistance **Supplies - OnSite:** Ice *(Cube)*,
Bait/Tackle **3+ mi:** Propane *(Wells Propane 539-2516, Tue, Fri, Sat 10am-2pm, 4 mi.)*

Boatyard Services

Nearest Yard: Canoe Cove Marina (250) 656-5566

Restaurants and Accommodations

OnSite: Restaurant *(The Harbour Grill 539-5733, B $5-10, L $6-12, D $10-20, outdoor dining available)* **OnCall:** Pizzeria *(Galiano Pizza 539-5544, delivery only)* **Under 1 mi:** Restaurant *(La Berengerie 539-5392)* **1-3 mi:** Restaurant *(The Hummingbird Inn Pub 539-5472, L $8-11, D $8-11, pick-up availble; live music Fri-Sat)*, Inn/B&B *(Holland House 539-5754)*, *(Woodstone Country Inn 539-2022, $199-499, wknd rates)*, *(Galiano Inn 539-3388, from $249, free pick up at Montague Harbour Marina)* **3+ mi:** Restaurant *(Grand Central Emporium 539-9885, B $6-11, L $4-11, D $11-19, 4 mi., kids' menu $5-7; call for pick-up at the marina)*, Condo/Cottage *(Sticks and Stones 539-3443, $120-240, 4 mi.)*

Recreation and Entertainment

OnSite: Picnic Area, Boat Rentals *(Gulf Island Kayaking 539-2442, rentals & tours; Galiano Boat & Moped Rentals 539-3443)*, Group Fishing Boat *(Sporades Tours 539-2278)*, Sightseeing *(Sporades Eco Tours)* **Near:** Beach, Dive Shop *(Lead Foot 539-5341)*, Special Events *(Galiano Island Fun Triathalon 539-2600)* **1-3 mi:** Golf Course *(Galiano Golf Club 539-5533)*, Video Rental *(The Corner Store 539-2986)*, Park *(Montague Harbour Provincial Park)*

Provisioning and General Services

OnSite: Convenience Store *(groceries, books, gifts, camping & fishing supplies, ice cream)*, Wine/Beer, Retail Shops *(Ixchel Crafts 539-9819, 10am-5pm)* **1-3 mi:** Market *(Daystar Market and Café 539-2505)*, Liquor Store *(The Corner Store 539-2986)*, Bank/ATM *(The Corner Store 539-2986 and The Hummingbird Pub)*, Post Office *(at the Daystar Market & Café)*, Beauty Salon *(Keo's Kitchen & Hairstyling 539-3224 by appt.)*, Barber Shop **3+ mi:** Supermarket *(Galiano Garage and Groceries 539-5500, 4 mi.)*, Farmers' Market *(Sat 10am-3pm 539-3672, 5 mi.)*, Protestant Church *(St. Margaret's Church, 4 mi.)*, Bookstore *(Galiano Island Books 539-3340 10am-6pm daily, 4 mi.)*, Pharmacy *(Galiano Garage and Groceries 539-5500, 4 mi.)*, Hardware Store *(Galiano Trading Company 539-5529, 4 mi.)*

Transportation

OnSite: Bikes *(Galiano Mopeds "toadies, roadies, moped and scooter rentals" 1-3 hrs. $18/hr., 4-5 hrs. $15/hr., whole day-7 hrs. $79; wknds 1-3 hrs. $20/hr., 4-5 hrs. $17/hr., day rate $99)*, Rental Car **OnCall:** Taxi *(539-0202)* **Near:** Local Bus *(Go Galiano Island Shuttle 539-0202)* **3+ mi:** Water Taxi *(Gulf Islands Water Taxi 537-2510, $15 to Salt Spring Island, 4 mi.)*, Ferry Service *(BC Ferries to Swartz Bay or Vancouver 888-223-3779, 4 mi.)* **Airport:** Harbour Air Seaplanes 800-665-0212 *(On-Call)*

Medical Services

911 Service **OnCall:** Ambulance **3+ mi:** Doctor *(Galiano Health Care Centre 539-3230, 4 mi.)*, Holistic Services *(Galiano Inn & Spa 877-530-3939, Melody Linda 539-2661, massage therapist, 4 mi.)* **Hospital:** Lady Minto 538-4800 - Saltspring Isl. *(18 mi.)*

Setting -- Bustling Montague Harbour Marina is easy to spot as you enter the harbor; this is Galiano's largest boating facility, with a fuel dock, store and restaurant on-site. An extensive network of quality floats is backed by two crisp white clapboard buildings with ice-blue roofs that host The Harbour Grill and the other shoreside facilities. Overflowing flower boxes add a festive touch to the deck railings. The marina is the heart of one of only two commercial centers on a decidedly noncommercial island. The rural atmosphere throughout Galiano is one of the island's many charms.

Marina Notes -- *15 amp power also available. There is no trash pick-up on the island, so boaters must take their garbage with them (only recyclables can be left). The marina offers boat and kayak rentals, plus mopeds, a well equipped convenience store, crafts store, restaurant, and best of all, ice cream! The energetic staff are undaunted by the summer crowds. 16 foot boats rent for $34/hr. (1-7 hrs.) or $185/day; 17' boats cost $44/hr. (1-7 hrs.) or $225/day.

Notable -- The on-site Harbour Grill restaurant serves three meals daily in a casual atmosphere, with some seating on the long deck overlooking the water. Galiano, the driest of the Gulf Islands, is 14 nautical miles long and named after Dionisio Galiano who discovered the islands in 1792. To enjoy its scenic beauty, rent a moped and follow the country roads, or head out in a kayak. Orca sightings are frequent in the area - J Pod, a group of about 19 whales, is often seen at Active Pass. In May, the Galiano Fun Triathlon offers sweat and good times for participants and spectators: kayak 3 km., bicycle 13 km., and then run 5 km. Fun and food comes after. Start and finish in Montague Provincial Park.

Navigational Information
Lat: 48°52.993' **Long:** 123°19.514' **Tide:** 16 ft **Current:** n/a **Chart:** 3473
Rep. Depths *(MLW)*: **Entry** 2 ft. **Fuel Dock** n/a **Max Slip/Moor** 16 ft./-
Access: Located on the east side of Galiano Island north of Active Pass

Marina Facilities *(In Season/Off Season)*
Fuel: No
Slips: 9 Total, 9 Transient **Max LOA:** 45 ft. **Max Beam:** n/a
  Rate *(per ft.)*: **Day** $0.57* **Week** n/a **Month** $3.33
  Power: 30 amp $5** , **50 amp** n/a, **100 amp** n/a, **200 amp** n/a
  Cable TV: No **Dockside Phone:** No
  Dock Type: Floating, Alongside, Wood
Moorings: 0 Total, 0 Transient **Launch:** n/a
  Rate: Day n/a **Week** n/a **Month** n/a
Heads: None
Laundry: None **Pay Phones:** Yes
Pump-Out: No **Fee:** n/a **Closed Heads:** Yes

Marina Operations
Owner/Manager: Whaler Bay Harbour Authority **Dockmaster:** Joy Wilson
In-Season: Year-Round, 24/7 **Off-Season:** n/a
After-Hours Arrival: n/a
Reservations: No **Credit Cards:** No
Discounts: None
Pets: Welcome **Handicap Access:** No

Whaler Bay Public Wharf
Whaler Bay Road; Galiano, BC

Tel: (250) 539-5420 **VHF: Monitor** n/a **Talk** n/a
Fax: n/a **Alternate Tel:** n/a
Email: n/a **Web:** n/a
Nearest Town: Sturdies Bay *(3 mi.)* **Tourist Info:** (250) 539-2233

Marina Services and Boat Supplies
Supplies - 1-3 mi: Propane *(Wells Propane 539-2516 Tue, Fri, Sat 10am-2pm)*

Boatyard Services
Nearest Yard: Canoe Cove Marina (250) 656-5566

Restaurants and Accommodations
OnCall: Pizzeria *(Galiano Pizza 539-5544, delivery only)* **Under 1 mi:** Hotel *(Whaler Bay Lodge 539-2249, $225)*, Inn/B&B *(Woodstone Country Inn 539-2022, $199-499, wknd package rates available)*, Condo/Cottage *(Cain Beach Cottage 656-2035, $110-121)* **1-3 mi:** Restaurant *(The Hummingbird Inn Pub 539-5472, L $8-11, D $8-11, live music Fri-Sat, via its own bus $2)*, *(Atrevida 539-3388, D $12-30, res. req. Will pick up)*, Lite Fare *(Madrona, takeout)*, *(Max & Moritz 539-5888, B $4-7, L $5-7, D $5-7, Indonesian & hamburgers, etc.,at ferry parking lot; call ahead for takeout)*, *(Daystar Market and Café 539-2505, organic juice, coffee, gourmet L & D to go)*

Recreation and Entertainment
1-3 mi: Beach, Playground, Golf Course *(Galiano Golf Club 539-5533)*, Boat Rentals *(Chinook Key Boat and Eco Tours 539-3388)*, Video Rental *(Galiano Garage and Groceries 539-5500)*, Park *(Bluffs Park)*, Sightseeing *(Galiano Arts Studio Tours - map at the Visitor Centre)*, Special Events, Galleries *(Art and Soul Craft Gallery 539-2944, Galiano Art Gallery 539-3539, Cabra Gallery Museum 539-2507)*

Provisioning and General Services
Near: Laundry *(Sparkles)* **1-3 mi:** Market *(The Corner Store 539-2986)*, Supermarket *(Galiano Garage and Groceries 539-5500)*, Delicatessen *(Trincomali Bakery and Deli 539-2004)*, Health Food *(Daystar Market and Café 539-2505)*, Liquor Store *(Galiano Garage and Groceries/Liquor Agency)*, Bakery *(Trincomali)*, Farmers' Market *(539-3672, Sat 10am-3pm)*, Bank/ATM, Post Office *(Gulf Islands Insurance and Post Office 539-5615)*, Protestant Church *(St. Margaret's)*, Bookstore *(Galiano Island Books 539-3340, daily 10am-6pm)*, Pharmacy *(Galiano Garage and Groceries 539-5500)*, Clothing Store *(Thistledown 539-2592 - crafts, clothes, souvenirs)*

Transportation
OnCall: Taxi *(539-0202)* **Near:** Bikes *(Galiano Bicycle 539-9906, daily and multi-day rates)* **Under 1 mi:** Local Bus *(Go Galiano Island Shuttle 539-0202)* **1-3 mi:** Rental Car *(Galiano Mopeds and Boats 539-3443)*, Ferry Service *(BC Ferries to Swartz Bay or Vancouver 888-223-3779)*
Airport: Harbour Air Seaplanes 800-665-0212 *(On-Call)*

Medical Services
911 Service **OnCall:** Ambulance **1-3 mi:** Doctor *(Galiano Health Care Centre 539-3230)*, Holistic Services *(Island Acupressure and Massage 539-9918, Galiano Inn & Spa 877-530-3939)* **Hospital:** Lady Minto 538-4800 - Saltspring Isl. *(12 mi.)*

Setting -- Tiny Whaler Bay seems an unlikely spot for a public dock, but it provides excellent protection. Signature, bright red railings edge the main ramp that leads from the Public Wharf's small set of floats to a road that runs up the hill - there is nothing else here or nearby. Nevertheless, this is a popular launch spot for both recreational and commercial fishers.

Marina Notes -- *Nightly fees are flat rates based on $1.89/meter ($0.57/ft.): 30' $16.93, 40' $22.69, 50' $28.27; monthly $11/meter ($3.33/ft.) - 30' $100.10, 40' $134.20, 50' $137.20. 3 months or more $9.90/meter. **Power is 20 or 30 amp. Rafting mandatory. Priority is given to commercial vessels, so if one wants your spot, you will be obliged to find another or raft up to the fishing boat. Note: The entry is quite shallow at low tide, so make your approach at high tide.

Notable -- There are no facilities or attractions within easy walking distance of the float, for, like the rest of Galiano Island, the area around Whaler Bay is rural. The only real "downtown" is at Sturdies Bay - take the shuttle and then rent a bike or moped. Montague Harbour Provincial Marine Park, about 3 miles to the west, is a must-see for Galiano visitors. Its white shell beach makes for a nice stroll, or follow the path around Gray's Peninsula where there are several forested hiking trails for the more adventuresome. Also on the west side, Montague Harbour Marina offers boat, kayak, and moped rentals (valid driver's license required) - the Go Galiano bus will get you there for only $3. It will also take you to stores, the ferry terminal, and the park. The Hummingbird Pub Bus runs around the island from 5:50-10:30pm ($2), and Atrevida Restaurant (539-3388, reservations required) also offers a shuttle to its elegant dining room.

Navigational Information

Lat: 48°52.599' **Long:** 123°18.921' **Tide:** 16 ft. **Current:** n/a **Chart:** 3473
Rep. Depths (*MLW*): Entry 10 ft. **Fuel Dock** n/a **Max Slip/Moor** 12 ft./-
Access: On Galiano Island toward the eastern end of Active Pass

Marina Facilities *(In Season/Off Season)*

Fuel: No
Slips: 2 Total, 2 Transient **Max LOA:** 30 ft. **Max Beam:** n/a
 Rate *(per ft.)*: **Day** $0.60* **Week** n/a **Month** $3.40/ft.
 Power: 30 amp n/a, **50 amp** n/a, **100 amp** n/a, **200 amp** n/a
 Cable TV: No **Dockside Phone:** No
 Dock Type: Floating, Alongside, Wood
Moorings: 0 Total, 0 Transient **Launch:** n/a
 Rate: Day n/a **Week** n/a **Month** n/a
Heads: 2 Toilet(s)
Laundry: None **Pay Phones:** Yes, 1
Pump-Out: No **Fee:** n/a **Closed Heads:** Yes

Marina Operations

Owner/Manager: Jean Jones, Wharfinger **Dockmaster:** n/a
In-Season: Year-Round, 24/7 **Off-Season:** n/a
After-Hours Arrival: Use pay envelope at kiosk
Reservations: No **Credit Cards:** No
Discounts: None
Pets: Welcome **Handicap Access:** No

Sturdies Bay Public Float

PO Box ; Site 4,Comp 9, RR 1; Galiano, BC VON 1P0

Tel: (250) 539-5053 **VHF: Monitor** n/a **Talk** n/a
Fax: (250) 539-5053 **Alternate Tel:** n/a
Email: n/a **Web:** n/a
Nearest Town: Sturdies Bay **Tourist Info:** (250) 539-2233

Marina Services and Boat Supplies

Supplies - Near: Ice *(Cube)*, Propane *(Wells Propane 539-2516, Tue, Fri, Sat 10am-2pm)*

Boatyard Services

Nearest Yard: Canoe Cove Marina (250) 656-5566

Restaurants and Accommodations

OnCall: Pizzeria *(Galiano Pizza 539-5544, delivery only)* **Near:** Restaurant *(Galiano Grand Central Emporium 539-9885, B $6-11, L $4-11, D $11-19, kids' menu $5-7)*, *(Atrevida Restaurant 539-3388, D $12-30, reservations required, will pick up)*, Fast Food *(Madrona 539-3448, fresh takeout meals)*, *(Max and Moritz 539-5888, B $4-7, L $5-7, D $5-7, bring your own cup or plate and get $0.25 off)*, Inn/B&B *(Galiano Inn 530-3939)* **1-3 mi:** Restaurant *(The Hummingbird Inn Pub 539-5472, L $8-11, D $8-11, live music Fri-Sat, via bus)*, Hotel *(Whaler Bay Lodge 539-2249, $225)*, Inn/B&B *(Woodstone Country Inn 539-2022, $199-499, wknd package rates)*, Condo/Cottage *(Cain Beach Cottage 656-2035, $110-121)*

Recreation and Entertainment

Near: Beach, Jogging Paths *(Sturdies Trail)*, Boat Rentals *(Chinook Key Boat and Eco Tours 539-3388, whale watching, islands and nature tours in a 32' adventure boat)*, Video Rental *(Galiano Garage and Groceries 539-5500)*, Special Events, Galleries *(Art and Soul Gallery 539-2944, Galiano Art Gallery 539-3539, Cabra Gallery Museum 539-2507)* **Under 1 mi:** Picnic Area *(Bellhouse Provincial Park)* **1-3 mi:** Golf Course *(Galiano Golf Club*

539-5533, 9-hole course & restaurant), Park *(Bluffs Park)* **3+ mi:** Group Fishing Boat *(Sporades Tours 539-2278, fishing charters, ecotours, and sightseeing, 4 mi.)*, Cultural Attract *(Gulf Islands Film and Television School 539-5729, shuttle, 9 mi.)*

Provisioning and General Services

Near: Market *(Galiano Garage and Groceries 539-5500)*, Delicatessen *(Trincomali Bakery and Deli 539-2004)*, Wine/Beer, Liquor Store *(Galiano Garage and Groceries/Liquor Agency)*, Bakery *(Trincomali Bakery and Deli)*, Farmers' Market *(Sat 10am-3pm 539-3672)*, Bank/ATM, Post Office *(Gulf Islands Insurance and Post Office 539-5615)*, Bookstore *(Galiano Island Books 539-3340, daily 10am-6pm)*, Pharmacy *(Galiano Garage and Groceries)*, Clothing Store *(Thistledown 539-2592)* **1-3 mi:** Health Food *(Daystar Market and Café 539-2505)*, Protestant Church *(St. Margaret's)*

Transportation

OnSite: Ferry Service *(BC Ferries to Swartz Bay or Vancouver 888-223-3779)* **Near:** Water Taxi *(Gulf Island Water Taxi to Mayne, Pender, and Salt Spring Islands 539-2510)*, Local Bus *(Go Galiano 539-0202)* **1-3 mi:** Bikes *(Galiano Bicycle 539-9906)* **Airport:** Harbour Air Seaplanes 800-665-0212 *(On-Call)*

Medical Services

911 Service **OnCall:** Ambulance **Near:** Holistic Services *(Galiano Inn & Spa 877-530-3939)* **Under 1 mi:** Doctor *(Galiano Health Care Centre 539-3230)* **Hospital:** Lady Minto 538-4800 - Saltspring Isl. *(12 mi.)*

Setting -- This tiny public float is actually part of - and protected by - the ferry dock that serves Mayne Island and Swartz Bay on Vancouver Island. The terminal's three-peaked gray and blue office overlooks the single transient dock; a bright-blue railed ramp connects it to the large, main ferry dock. The view is out to Active Pass, aptly named because it is the busy main route between Vancouver and Victoria, and of Mt. Baker. Huge ferries pass by the entrance regularly, providing a bit of a show.

Marina Notes -- *First 4 hours are free; thereafter it's a flat rate: 4-12 hrs: 30' $7.50, 40' $10, 50' $12.50; 13-24 hrs: 30' $15, 40' $20, 50' $25. Leave payment in the drop box and place the back portion of the envelope on the boat so it is visible. No showers, power, or water. Public heads are available at the BC Ferries Wharf terminal at the top of the ramp. Note: The pass is subject to significant tidal currents and eddies, but the bay is generally quiet.

Notable -- The Sturdies Bay Dock is the most convenient moorage to the shops on Galiano Island. Just around the corner, you'll find a modest commercial area that passes for "downtown" on this rural island. Bike rentals are nearby and the island shuttle (Go Galiano) runs regularly in season. Near the docks, at the end of Jack Road, is Bellhouse Provincial Park. This day use area is one of the premier places in the islands to whale watch from shore. J Pod, a group of about 19 orcas, frequents Active Pass in search of the salmon that migrate through the pass. Take a picnic lunch and cross your fingers for whale sightings. The Gulf Islands Film & Television School, 9 miles north, offers one-week "boot camps" for children and adults ($1,000; free shuttle from ferry dock).

Navigational Information
Lat: 48°52.868' **Long:** 123°34.403' **Tide:** 11 ft. **Current:** n/a **Chart:** 3442
Rep. Depths *(MLW):* **Entry** 18 ft. **Fuel Dock** n/a **Max Slip/Moor** 10 ft./-
Access: Located on the northwestern side of Salt Spring Island

Marina Facilities *(In Season/Off Season)*
Fuel: No
Slips: 3 Total, 3 Transient **Max LOA:** 36 ft. **Max Beam:** n/a
  Rate *(per ft.):* **Day** $0.43* **Week** n/a **Month** $0.60
  Power: 30 amp n/a, 50 amp n/a, 100 amp n/a, 200 amp n/a
  Cable TV: No **Dockside Phone:** No
  Dock Type: Floating, Alongside, Wood
Moorings: 0 Total, 0 Transient **Launch:** n/a
  Rate: Day n/a **Week** n/a **Month** n/a
Heads: 1 Toilet(s)
Laundry: None **Pay Phones:** No
Pump-Out: No **Fee:** n/a **Closed Heads:** Yes

Marina Operations
Owner/Manager: Salt Spring Harbour Authority **Dockmaster:** n/a
In-Season: Year-Round, 24/7 **Off-Season:** n/a
After-Hours Arrival: Pay station near the dock
Reservations: No **Credit Cards:** No
Discounts: None
Pets: Welcome **Handicap Access:** No

Vesuvius Bay Public Wharf

Vesuvius; Saltspring Island, BC V8K1Z2

Tel: (250) 537-5711 **VHF: Monitor** n/a **Talk** n/a
Fax: (250) 537-5711 **Alternate Tel:** n/a
Email: n/a **Web:** n/a
Nearest Town: Ganges *(3.5 mi.)* **Tourist Info:** (250) 537-5252

Marina Services and Boat Supplies
Supplies - Near: Ice *(Cube)* **3+ mi:** Ships' Store *(Ganges or Salt Spring Marina, 4 mi.)*, Bait/Tackle *(4 mi.)*, Propane *(4 mi.)*

Boatyard Services
Nearest Yard: Canoe Cove Marina (250) 656-5533

Restaurants and Accommodations
OnSite: Restaurant *(Seaside Kitchen Restaurant 537-2249, L $5-14, D $11-20, dock'n'dine, no overnights; kids' menu, vegetarian)* **Near:** Coffee Shop *(Hon Hon's Espresso 645-2115)*, Inn/B&B *(Vesuvius Beach B&B 537-4123, $85-100)* **1-3 mi:** Inn/B&B *(Anchor Point B&B 538-0110, $160, every 6th night stay is free)* **3+ mi:** Restaurant *(House Picolo 537-1844, D $25-33, 4 mi.)*, *(Oystercatcher 537-5041, L $9-12, D $9-23, 4 mi.)*, *(Moby's Marine Pub 537-5559, L $10-14, D $10-17, 4 mi.)*, Pizzeria *(Luigi's 537-5660, 4 mi.)*, Motel *(Sea Breeze Motel 537-4145, $59-119, 4 mi.)*, *(Best Bays Lodging 645-3325, 4 mi.)*, Hotel *(Harbour House Hotel 537-5571, $59-295, 4 mi.)*

Recreation and Entertainment
OnSite: Fishing Charter *(Big Salmon Fishing 645-2374)* **Near:** Beach *(Vesuvius Beach)*, Cultural Attract *(Studio 12)*, Galleries *(Tina Louise Spalding Gallery 537-1741)* **1-3 mi:** Playground, Tennis Courts, Golf Course *(Salt Springs Golf and Country Club 537-2121)*, Movie Theater *(Central Hall Movies)*, Sightseeing *(Studio Tour, map at Visitor Info Centre 538-0140)* **3+ mi:** Fitness Center *(North End Fitness 537-5217, 4 mi.)*, Boat Rentals *(Sea Otter Kayaking 537-5678, 4 mi.)*, Special Events *(Fall Fair 3rd wknd Sep, Art Craft May-Sep, 4 mi.)*

Provisioning and General Services
Near: Convenience Store *(Vesuvius Village Store 537-1515)*, Beauty Salon *(Images 537-4712)* **Under 1 mi:** Gourmet Shop *(Salt Spring Cheese Co. / Old Salty Shop 537-5593, 4 mi.)* **1-3 mi:** Bank/ATM **3+ mi:** Market *(Ganges Village Market 537-4144, 4 mi.)*, Supermarket *(Thrifty Foods 537-1522, 4 mi.)*, Delicatessen *(Ganges Village Market, 4 mi.)*, Health Food *(Nature Works 537-2325, 4 mi.)*, Farmers' Market *(Saturdays in Centennial Park at Grace Point, 4 mi.)*, Fishmonger *(at Ganges Marina 537-4242, 4 mi.)*, Catholic Church *(4 mi.)*, Protestant Church *(4 mi.)*, Library *(537-4666, Mon-Sat 10am-5pm, Internet access, 4 mi.)*, Barber Shop *(Lock, Stock, and Barber Shop, 4 mi.)*, Pharmacy *(Pharmasave 538-0323, 4 mi.)*, Clothing Store *(Ganges Garment Company 537-8999, 4 mi.)*, Retail Shops *(4 mi.)*

Transportation
OnSite: Ferry Service *(BC Ferries to Crofton 888-223-3779)* **OnCall:** Water Taxi *(Gulf Islands Water Taxi 537-2510)* **Near:** Taxi *(Silver Shadow Taxi 537-3030)*, Local Bus *(The Ganges Faerie Mini Shuttle 357-6758, Pager 538-9007)* **Airport:** Saltspring Air 537-9880 Seaplanes only *(On-Call)*

Medical Services
911 Service **OnCall:** Ambulance **3+ mi:** Doctor *(4 mi.)*, Dentist *(Hayden 537-1400, 4 mi.)*, Holistic Services *(Acupressure Jin Shin Do 537-8525, 4 mi.)*, Veterinarian *(Gulf Islands Vet 537-5334, 4 mi.)* **Hospital:** Lady Minto Hospital 538-4800 *(4 mi.)*

Setting -- Salt Spring is the largest and most developed of the Gulf Islands. Vesuvius Bay, situated on the northwest side, is primarily a ferry stop, one of three on the island. The ferry to and from Crofton, just across the straight on Vancouver Island, docks here. The Vesuvius Bay Public Wharf is a small float inboard of the ferry terminal - a red-railed ramp leads to a stationary pier and the pier to a small "settlement." The adjacent Seaside Kitchen Restaurant overlooks the wharf and dominates the landside views from the dock.

Marina Notes -- *Flat rates range from $17 to $40/day depending on length. Use the pay station near the dock. Boats are restless when the ferry is at the adjacent terminal. There is no power at the dock and no boater amenities. A portable toilet is at the top of the ramp. Seaside Kitchen offers dock 'n' dine, but no overngihts. Note: the docks are relatively exposed to northwest weather coming out of Stuart Channel.

Notable -- A visit to Salt Spring Island is not complete without a trip to the primary town, Ganges. The Ganges Faerie Mini Shuttle transports visitors to this vibrant village where arts and culture are a way of life. The ArtCraft exhibit at Mahon Hall runs throughout the summer and is perfect for seeing local artwork and buying souvenirs. Hiking, biking, and kayaking are big year-round on all of the islands, and Salt Island is no exception. The Visitor Info Centre, located in downtown Ganges (open all year) will happily offer ideas on tours. The long list of activities available on the island includes swimming, fishing, golf, disk golf, tennis, horseback riding, hang gliding, scuba diving, kayaking, beachcombing, and bird-watching. Public Internet access is at Ganges' Mary Hawkins Library.

Centennial Wharf

127 Fulford Ganges Road; Saltspring Island, BC V8K 2C2

Tel: (250) 537-5711 **VHF: Monitor** Ch. 9 **Talk** n/a
Fax: (250) 537-5711 **Alternate Tel:** n/a
Email: ssha@Saltspring.com **Web:** n/a
Nearest Town: Ganges **Tourist Info:** (250) 537-4223

Navigational Information

Lat: 48°51.130' **Long:** 123°29.865' **Tide:** 11 ft. **Current:** n/a **Chart:** 3478
Rep. Depths *(MLW)*: **Entry** 12 ft. **Fuel Dock** n/a **Max Slip/Moor** 8 ft./-
Access: South side of Salt Spring Island behind the breakwater

Marina Facilities *(In Season/Off Season)*

Fuel: No
Slips: 100 Total, 4 Transient **Max LOA:** 45 ft. **Max Beam:** n/a
Rate *(per ft.)*: **Day** $0.75 **Week** n/a **Month** n/a
Power: 30 amp n/a, 50 amp n/a, 100 amp n/a, 200 amp n/a
Cable TV: No **Dockside Phone:** No
Dock Type: Floating, Concrete, Wood
Moorings: 0 Total, 0 Transient **Launch:** n/a
Rate: Day n/a **Week** n/a **Month** n/a
Heads: 2 Toilet(s), 1 Shower(s)
Laundry: None **Pay Phones:** No
Pump-Out: OnSite, OnCall **Fee:** $5 **Closed Heads:** Yes

Marina Operations

Owner/Manager: Salt Spring Harbour Authority **Dockmaster:** Bart Terwiel
In-Season: May-Sept, 9am-5pm **Off-Season:** Oct-May, 9am-1pm
After-Hours Arrival: n/a
Reservations: No **Credit Cards:** Visa/MC
Discounts: None
Pets: Welcome **Handicap Access:** No

Marina Services and Boat Supplies

Communication - FedEx, DHL, UPS **Supplies - Near:** Ice *(Cube)*, Ships' Store *(Thrifty)*, Bait/Tackle, Propane *(Save On Gas 537-5671)*

Boatyard Services

OnSite: Launching Ramp **Nearest Yard:** Harbour's End Marine & Equip. (250) 537-4202

Restaurants and Accommodations

OnSite: Restaurant *(Artists Bistro 537-1701)* **Near:** Restaurant *(Piccolo House 537-1344, D $25-50)*, *(Auntie Pesto's Café & Deli 537-4181, D $10-20)*, *(Café el Zocalo 537-9411, D $7-12)*, *(Calvin's Bistro 538-5551)*, *(Oystercatcher Seafood Bar & Grill 537-5041, L $9-12, D $9-23)*, *(La Cucina Italian Grill 537-5747)*, Pizzeria *(Luigi's 537-5660)* **Under 1 mi:** Hotel *(Anchorage Cove 537-5337, $125-155)* **1-3 mi:** Hotel *(Salt Spring Lodge & Spa 537-9522, $109-250, waterfront resort)*, Inn/B&B *(Hastings House Inn 537-2362)*

Recreation and Entertainment

OnSite: Playground, Park *(Centennial Park)*, Galleries *(J. Mitchell 537-8822, Thunderbird 537-8448/ Waterfront Galley 537-4525 - over 75 artists, nearby)* **Near:** Fitness Center *(Fitness Centre 537-5217)*, Video Rental *(Island Star Video 537-4477)* **Under 1 mi:** Beach, Picnic Area *(Mouat Provincial Park)*, Boat Rentals *(Sea Otter Kayaks 537-5678)*, Fishing Charter *(Lady Patricia 537-2154, Rope N Reel 537-9509)*, Cultural Attract *(ArtSpring 537-2102 - performing arts and exhibits)*, Special Events *(Sizzling Summer Nights Music Fest - Jul-Aug, Arts Festival - Jul)* **1-3 mi:** Tennis Courts

(Tennis Bubble $14/hr.; book at Apple Photo 537-4243), Golf Course *(Salt Spring Island 537-2121)*, Movie Theater *(Cinema Central 537-4656)*

Provisioning and General Services

OnSite: Liquor Store *(537-1919)*, Farmers' Market *(Market in the Park - Sat 8:30am-3pm midApr-midOct - produce & crafts)*, Bookstore *(Sabine's Used Books 538-0025 - good maritime collection/ Salt Spring Books 537-2812, nearby)*, Clothing Store *(Ganges Garment Co. 537-8999)* **Near:** Convenience Store, Supermarket *(Thrifty Foods 537-1522, free delivery with min. $25 order)*, Delicatessen *(Aunty Pesto's 537-4181)*, Health Food *(Nature Works 537-2325)*, Wine/Beer, Bakery *(Embe 537-5611, Barb's Buns 537-4491)*, Post Office *(537-232)*, Library *(537-4666, Mon-Sat 10am-5pm, Internet access)*, Beauty Salon *(Studio 103 537-2700)*, Barber Shop *(Lock, Stock, and Barber Shop)*, Laundry *(Mrs. Clean Laundromat 537-4133)*, Pharmacy *(Pharmasave 537-5534)*, Newsstand, Hardware Store *(Mouat's Trading Co. 537-5551)*, Florist *(Flowers & Wine 537-2231)*, Retail Shops *(Sports Traders 537-5148)*

Transportation

OnCall: Water Taxi *(Gulf Islands 537-2510)*, Taxi *(Silver Shadow 537-3030, Amber Cab 537-3277)* **Near:** Local Bus *(Ganges Faerie 537-6758)* **Under 1 mi:** Rental Car *(at SS Marina)* **1-3 mi:** Ferry Service *(BC Ferries 888-223-3779)* **Airport:** Saltspring Air 537-9880 - seaplanes *(0.1 mi.)*

Medical Services

911 Service **OnSite:** Holistic Services *(Skin Sensations Day Spa 537-8807 - Reiki, reflexology, spa treatments)* **Near:** Dentist *(Hayden 537-1400)*, Veterinarian **Under 1 mi:** Doctor **Hospital:** Lady Minto 538-4800 *(1 mi.)*

Setting -- Tucked in behind a stone breakwater at Grace Point, the four strings of floats at Centennial Wharf Boat Harbour - also known as Ganges Inner Harbour - host mostly seasonal tenants and the local fishing fleet. A boardwalk leads to a wide expanse of greensward with a creative playground, small decks high above the water, and a small open pavilion. Further inland is a shopping center and the rest of bustling, delightful Ganges.

Marina Notes -- Dockage available only when the fishing fleet is "out" or when a seasonal tenant is off cruising. Availability varies dramatically from year to year. A large rock mound breakwater provides some shelter from SE winds. There are heads and showers on-site.

Notable -- The ramp off the docks lead right into Centennial Park - site of the Market in the Park that takes place on Saturdays, 8:30am-3pm from midApr to midOct, rain or shine; it combines a farmers' market and crafts fair, and has become a major event, popular with the locals - and an attraction for visitors to the island. Salt Spring is known as Canada's organic gardening capital - often called the Banana Belt because of its mild microclimate. The park is also the site of concerts and many annual events. Just past the park is the Centennial Park Shopping Center, which hosts many useful services. Walk out of the shopping center and down the hill into the heart of the village of Ganges. The Ganges Faerie Mini Shuttle offers van service between the island towns and to the ferry terminals; reservations are available, and they will pick up at any location along the way (year-round, 7am-6pm; Jun-Sep 7 days, off-season no service on Wed; $7.50-12.50 depending on destination).

Navigational Information
Lat: 48°51.268' **Long:** 123°29.912' **Tide:** 11 ft. **Current:** n/a **Chart:** 3478
Rep. Depths (*MLW*): **Entry** 12 ft. **Fuel Dock** n/a **Max Slip/Moor** 10 ft./-
Access: Northeast of Grace Point including Breakwater docks

Marina Facilities *(In Season/Off Season)*
Fuel: No
Slips: 30 Total, 10 Transient **Max LOA:** 60 ft. **Max Beam:** n/a
  Rate *(per ft.)*: **Day** $0.43* **Week** n/a **Month** n/a
  Power: 30 amp n/a, **50 amp** n/a, **100 amp** n/a, **200 amp** n/a
  Cable TV: No **Dockside Phone:** No
  Dock Type: Floating, Long Fingers, Alongside, Wood, Composition
Moorings: 0 Total, 0 Transient **Launch:** n/a
  Rate: Day n/a **Week** n/a **Month** n/a
Heads: None
Laundry: None **Pay Phones:** Yes, 1
Pump-Out: No **Fee:** n/a **Closed Heads:** Yes

Marina Operations
Owner/Manager: Salt Spring Harbour Authority **Dockmaster:** Bart Terwiel
In-Season: Year-Round, 9-5 Sun-Tue, 8-8 Wed-Sat **Off-Season:** n/a
After-Hours Arrival: Call ahead
Reservations: No **Credit Cards:** No
Discounts: None
Pets: Welcome **Handicap Access:** No

Kanaka Visitors Wharf
Saltspring Island, BC V8K 2T2

Tel: (250) 537-5711 **VHF: Monitor** n/a **Talk** n/a
Fax: (250) 537-5711 **Alternate Tel:** n/a
Email: n/a **Web:** n/a
Nearest Town: Ganges **Tourist Info:** (250) 537-5252

Marina Services and Boat Supplies
Services - Dock Carts **Supplies - Near:** Ice *(Cube)*, Ships' Store
Under 1 mi: Bait/Tackle, Propane *(Save On Gas 537-5671)*

Boatyard Services
Nearest Yard: Harbour's End Marine & Equip. (250) 537-4202

Restaurants and Accommodations
OnSite: Restaurant *(Oystercatcher Seafood Bar & Grill 537-5041, L $9-12, D $9-23)* **Near:** Restaurant *(Treehouse Café 537-9644, B $3-8, L $4-8, D $8-10, umbrellaed tables under an enormous tree)*, *(La Cucina Italian Grill 537-5747, B $6-10, L $10-18, D $12-25)*, *(Calvin's Bistro 538-5551)* **Under 1 mi:** Restaurant *(Goden Island 537-2535)*, *(Piccolo 537-1844, D $25-33)*, *(Moby's Marine Pub 537-5559, L $10-14, D $10-17)*, Inn/B&B *(Anchorage Cove 537-5337, $125-155)* **1-3 mi:** Motel *(Sea Breeze 537-4145, $59-119)*, Inn/B&B *(Wisteria Guest House 537-5899, $55-99)*

Recreation and Entertainment
OnSite: Playground, Sightseeing *(Island Escapades 537-2553)*, Special Events *(Celebrate Art - Sep 1)*, Galleries *(Jill Louise Campbell 537-1589, Pegasus 537-2421)* **Near:** Fitness Center *(Fitness Centre 537-5217)*, Boat Rentals *(Sea Otter Kayaks 537-5678)*, Park *(Centennial Park)* **Under 1 mi:** Beach, Picnic Area *(Mouat Provincial Park)*, Jogging Paths, Bowling *(King's Lane)*, Fishing Charter *(Salmon Charters 653-4902, Lady Patricia 537-2154, Rope N Reel 537-9509)*, Video Rental *(Island Star Video 537-4477)*, Cultural Attract *(ArtSpring 537-2102)* **1-3 mi:** Golf Course *(Salt Spring Island Golf & Country Club 537-2121)*, Movie Theater *(Cinema Central 537-4656)*

Provisioning and General Services
Near: Supermarket *(Thrifty Foods 537-1522; free delivery with min. $25 order)*, Gourmet Shop *(Old Salty Shop 537-5551)*, Delicatessen *(Aunty Pesto's 537-4181)*, Wine/Beer, Liquor Store *(Licensed Liquor Store 537-1919)*, Bakery *(Embe Bakery 537-5611)*, Farmers' Market *(Saturdays in Centennial Park 8:30am-3pm)*, Bank/ATM, Post Office *(537-232)*, Library *(537-4666 Mon-Sat 10am-5pm, Internet access)*, Beauty Salon *(Studio 103 537-2700)*, Barber Shop *(Lock, Stock, and Barber Shop)*, Laundry *(Mrs. Clean Laundromat 537-4133)*, Bookstore *(Salt Spring Books 537-2812)*, Pharmacy *(Pharmasave 537-5534)*, Hardware Store *(Mouat's Trading Co 537-5551)*, Clothing Store *(Mouat's Clothing 537-5551)* **Under 1 mi:** Health Food *(Growing Circle 537-4247)*, Catholic Church, Protestant Church, Florist *(Flowers by Arrangement 537-9280)*, Copies Etc. *(Apple Photo)* **1-3 mi:** Market *(Ganges Village Market 537-4144)*

Transportation
OnCall: Water Taxi *(Gulf Islands Water Taxi 537-2510)*, Taxi *(Silver Shadow 537-3030, Amber Cab 537-3277)* **Near:** Rental Car *(at SS Marina)*, Local Bus *(The Ganges Faerie Mini Shuttle 537-6758)* **1-3 mi:** Ferry Service *(BC Ferries to Islands and Tsawwassen 888-223-3779)* **Airport:** Saltspring Air 537-9880; seaplanes only *(On-Call)*

Medical Services
911 Service **OnCall:** Ambulance **Near:** Holistic Services *(Acupressure Jin Shin Do 537-8525, Hastins House Spa 537-2362)*, Veterinarian *(Will pick up at Ganges Marina 537-5334)* **Under 1 mi:** Doctor, Dentist *(Hayden 537-1400)*, Chiropractor *(Richardson 537-9399)*, Optician *(Gulf Islands Optical 537-2648)* **Hospital:** Lady Minto 538-4800 *(0.5 mi.)*

Setting -- The town of Ganges is a bustling community that bellies up to this beautiful harbor. The Kanaka Public Visitors Wharf, consisting of three floats and a seaplane dock, is directly in front of the Oyster Catcher Restaurant & Waterside Pub. Guests at the dock have easy walking access to everything in town. A boardwalk edges the harbor and tree-shaded gravel paths snake from the pristine floats and red-railed ramps through carefully landscaped groups of eateries and shops. In the summer, both the harbor and the streets are crowded with visiting boaters. This is definitely where it's happening in the Gulf Islands.

Marina Notes -- *First 2 hrs. free (until 4pm). Overnight flat rates vary by size: 30' $13, 40' $17, 50' $21. Register at the Harbourmaster's Office two blocks away at Centennial Park. Grocery carts apparently double as dock carts here. There are no heads and showers but public washrooms are behind Pegasus Gallery and there is a bathhouse at Centennial Wharf. Mrs. Clean Laundromat down the street also has showers: wash your clothes while you wash yourself.

Notable -- This exciting, vibrant community is well worth a visit. Step off the dock and you are downtown, where groceries, shopping, and restaurants are all a stone's throw away. The atmosphere is bubbly and energetic - and very well-tended. On Friday nights, ten downtown galleries are open from 5pm-9pm. Every Saturday, a fabulous green market and arts and crafts fair is held at Centennial Wharf. There is much to do and the Visitors' Center will help sort out the choices. Ganges was actually named after a boat, the British naval ship "HMS Ganges." Launched in 1821, she was the last British sailing battleship commissioned for service in foreign waters.

PHOTOS ON DVD: 16

Ganges Marina

Ganges Marina

161 Lower Ganges Road; Saltspring Island, BC V8K 2T2

Tel: (250) 537-5242 **VHF: Monitor** Ch. 66A **Talk** n/a
Fax: (250) 538-1719 **Alternate Tel:** n/a
Email: gangesmarina@shaw.ca **Web:** www.gangesmarina.com
Nearest Town: Ganges **Tourist Info:** (250) 537-5252

Navigational Information
Lat: 48°51.321' **Long:** 123°29.998' **Tide:** 11 ft. **Current:** n/a **Chart:** 3478
Rep. Depths (MLW): Entry 12 ft. **Fuel Dock** 20 ft. **Max Slip/Moor** 25 ft./-
Access: West of Galiano Island and east of Crofton around Grace Point

Marina Facilities *(In Season/Off Season)*
Fuel: *Evergreen* - Gasoline, Diesel
Slips: 100 Total, 75 Transient **Max LOA:** 120 ft. **Max Beam:** n/a
 Rate *(per ft.):* **Day** $1.45/0.75 **Week** $8.70 **Month** n/a
 Power: 30 amp $3-5*, **50 amp** $5, **100 amp** n/a, **200 amp** n/a
 Cable TV: No **Dockside Phone:** No
 Dock Type: Floating, Long Fingers, Alongside, Wood
Moorings: 0 Total, 0 Transient **Launch:** n/a
 Rate: Day n/a **Week** n/a **Month** n/a
Heads: 5 Toilet(s), 6 Shower(s) *(with dressing rooms)*
Laundry: 2 Washer(s), 2 Dryer(s), Book Exchange **Pay Phones:** Yes, 1
Pump-Out: No **Fee:** n/a **Closed Heads:** Yes

Marina Operations
Owner/Manager: Jim Robertson **Dockmaster:** Shirley Command
In-Season: Year-Round, 8am-6pm **Off-Season:** n/a
After-Hours Arrival: Call ahead. Oct-May find slip and register in the morning
Reservations: Yes, Preferred **Credit Cards:** Visa/MC, Debit
Discounts: 4+ boats **Dockage:** 5-10% **Fuel:** n/a **Repair:** n/a
Pets: Welcome, Dog Walk Area **Handicap Access:** Yes, Heads, Docks

Marina Services and Boat Supplies
Services - Docking Assistance, Boaters' Lounge, Security *(video)*, Trash Pick-Up, Dock Carts **Communication -** Phone Messages, Fax in/out *($3/pg)*, Data Ports *(In Office or Wi-Fi - Broadband Xpress)*, FedEx, DHL, UPS, Express Mail *(Sat Del)* **Supplies - OnSite:** Ice *(Block, Cube)*, Ships' Store, Bait/Tackle **Under 1 mi:** Propane

Boatyard Services
Nearest Yard: Harbour's End Marine & Equip. (250) 537-4202

Restaurants and Accommodations
Near: Restaurant *(Golden Island 537-2535)*, *(House Piccolo 537-1844, D $25-33)*, *(Treehouse Café B $3-8, L $4-8, D $8-10)*, *(The Oystercatcher 537-5041, L $9-12, D $9-23)*, *(La Cucina Italian Grill 537-5747, B $6-10, L $10-18, D $12-25)*, *(Moby's Marine Pub 537-5559, L $10-14, D $10-17)*, *(Artists Bistro 537-1701, D $22-29)*, Pizzeria *(Luigi's 537-5660)*, Hotel *(Harbour House 537-5571, $59-295)* **Under 1 mi:** Motel *(Sea Breeze 537-4145, $59-119)*, Inn/B&B *(Wisteria Guest House 537-5899, $55-95)*

Recreation and Entertainment
OnSite: Picnic Area, Fishing Charter **Near:** Playground *(Centennial Park)*, Fitness Center *(North End Fitness 537-5217)*, Boat Rentals *(Sea Otter Kayaking 537-5678)*, Park *(Centennial)*, Cultural Attract, Galleries *(Pegasus 537-2421, Coastal Currents 537-0070)* **Under 1 mi:** Beach, Jogging Paths *(Mouat Park)*, Group Fishing Boat *(Salmon Charters 653-4902)*, Movie Theater *(Central Hall)*, Video Rental, Sightseeing *(Studio Tour, map at Visitor Info Centre 537-5075)*, Special Events *(Artcraft midMay-midSep 537-0899, Sea Capers: sand castles, boatbuilding & treasure hunts - Jun, Fall Fair 3rd*

wknd Sep) **1-3 mi:** Golf Course *(Salt Springs 537-2121)*, Horseback Riding *(Salt Spring Guided Rides 537-5761)*

Provisioning and General Services
OnSite: Fishmonger *(The Fishery)* **Near:** Supermarket *(Thrifty Foods 537-1522 free delivery, $25 min. order)*, Gourmet Shop *(Old Salty Shop 537-5593)*, Delicatessen *(Aunty Pesto's 537-4181)*, Health Food *(Nature Works 537-2325)*, Farmers' Market *(Sat in Centennial Park)*, Meat Market, Bank/ATM, Library *(537-4666 Mon-Sat 10-5, Internet access)*, Beauty Salon, Barber Shop, Pharmacy *(Pharmasave 537-5534)*, Clothing Store **Under 1 mi:** Market *(Ganges Village 537-4144)*, Wine/Beer, Liquor Store, Bakery *(Barb's Buns 537-4491)*, Catholic Church, Protestant Church, Laundry *(Mrs. Clean's 537-4133)*, Bookstore *(Volume II 537-9223)*, Newsstand, Hardware Store *(Mouat's Home 537-5551)*, Department Store *(Field's)*, Copies Etc. *(Apple Photo)* **1-3 mi:** Florist *(Flowers By Arrangement 537-9280)*

Transportation
OnSite: Rental Car *(reservations required; 4 hour min. $30-50/day, 537-5242)* **OnCall:** Water Taxi *(Gulf Islands Water Taxi 537-2510)*, Taxi *(Silver Shadow Taxi 537-3030)* **Near:** Local Bus **1-3 mi:** Ferry Service *(BC Ferries to Islands and Tsawwassen 888-223-3779)* **Airport:** Saltspring Air 537-9880 seaplanes only *(On-Site)*

Medical Services
911 Service **OnCall:** Ambulance **Under 1 mi:** Doctor, Dentist *(Hayden 537-1400)*, Holistic Services *(Skin Sensations 537-8807)*, Veterinarian *(Pickups 537-5334)*, Optician *(Gulf Islands 537-2648)* **1-3 mi:** Chiropractor *(Richardson 537-9399)* **Hospital:** Lady Minto 538-4800 *(0.5 mi.)*

Setting -- Around Grace Point and past the Kanaka Visitors Dock, Ganges Marina's new 500 foot floating breakwater protects the series of recently upgraded docks and provides more slips and alongside tie-ups. The shoreline is dominated by a bold cream-colored main office topped by a sage green roof punctuated with a string of dormers. A short distance across or around the harbor is the village of Ganges.

Marina Notes -- *15 & 30 amp power: $3/day up to 35', $4/day for 36-50', $5/day above 50'; 50 amp power $5/day for all vessels. Fuel on-site. Pump-out at Salt Spring HA Dock at Centennial Wharf. New owners in 2001 - expect more changes and upgrades. Caters to groups: discount of 5% for groups of 4-9 vessels, 10% for 10+ boats. Free coffee and muffins May 1-Oct 14 in the attractive and inviting office. The marina features two party docks complete with tables and chairs. A lovely pine panelled upstairs room is also available for groups. Free Internet service; just bring your laptop. Saltspring Air office is on-site. Clean, cedar-lined refurbished bathhouse and laundry with complimentary detergent. Note: The pool has been closed indefinitely.

Notable -- At 17 miles long and 9 miles wide, Salt Spring Island is the largest and most populated of the Gulf Islands, and its "metropolitan" area is Ganges, where most of the island's shops, stores, galleries and services can be found. The self-guided Studio Tour is a wonderful way to visit up to 40 artists' studios on the island - maps are available at the Ganges Info Center. The Saturday Farmers' Market at Centennial Park - famous for its organic produce - doubles as an arts and crafts fair too. ArtSpring is the Island's premier venue for performing arts, including concerts and theater.

PHOTOS ON DVD: 17

Navigational Information
Lat: 48°51.330' **Long:** 123°30.051' **Tide:** 11 ft. **Current:** n/a **Chart:** 3478
Rep. Depths (*MLW*): **Entry** 12 ft. **Fuel Dock** n/a **Max Slip/Moor** 20 ft./-
Access: East side of Salt Spring Island at the head of Ganges Harbour

Marina Facilities *(In Season/Off Season)*
Fuel: No
Slips: 65 Total, 35 Transient **Max LOA:** 150 ft. **Max Beam:** n/a
  Rate *(per ft.)*: **Day** $1.25/0.85* **Week** n/a **Month** n/a
  Power: 30 amp $3.75, **50 amp** n/a, **100 amp** n/a, **200 amp** n/a
  Cable TV: No **Dockside Phone:** No
  Dock Type: Floating, Long Fingers, Alongside, Wood
Moorings: 0 Total, 0 Transient **Launch:** n/a
  Rate: Day n/a **Week** n/a **Month** n/a
Heads: 6 Toilet(s), 6 Shower(s)
Laundry: 4 Washer(s), 4 Dryer(s) **Pay Phones:** Yes, 2
Pump-Out: No **Fee:** n/a **Closed Heads:** Yes

Marina Operations
Owner/Manager: Leslie Cheeseman **Dockmaster:** Amanda Sykes
In-Season: May-Sep, 8am-8pm **Off-Season:** Oct-Apr, 10am-2pm
After-Hours Arrival: Call ahead
Reservations: Yes, Preferred **Credit Cards:** Visa/MC, Debit
Discounts: 8+ boats **Dockage:** 10% **Fuel:** n/a **Repair:** n/a
Pets: Welcome **Handicap Access:** No

Salt Spring Marina

124 Upper Ganges Road; Saltspring Island, BC V8K2S2

Tel: (250) 537-5810; (800) 334-6629 **VHF: Monitor** Ch. 66A **Talk** n/a
Fax: (250) 537-5809 **Alternate Tel:** n/a
Email: info@saltspringmarina.com **Web:** www.saltspringmarina.com
Nearest Town: Ganges **Tourist Info:** (250) 537-5252

Marina Services and Boat Supplies
Services - Trash Pick-Up, Dock Carts **Communication -** Data Ports *(in the pub + Wi-Fi - Broadband Xpress)*, FedEx, UPS, Express Mail **Supplies -** **OnSite:** Ice *(Cube)*, Ships' Store *(Harbours End Marine and Equipment 537-4202)*, Bait/Tackle **Under 1 mi:** Propane

Boatyard Services
OnSite: Launching Ramp *($10)* **Nearest Yard:** Canoe Cove (250) 656-5566

Restaurants and Accommodations
OnSite: Restaurant *(Moby's Marine Pub 537-5559, L $8-11, D $8-13, Sun brunch 11am-3pm $7-12; late night menu 10-12pm $8-12)*, Lite Fare *(Rogue Café B, L soup & sandwiches)* **Near:** Restaurant *(Harbour House Hotel 537-5571, B, L, D)*, Lite Fare *(Tree House Café B $3-8, L $4-8, D $8-10)*, Motel *(Ganges Marina and Boatel 537-5810, $59-119)*, Hotel *(Harbour House Hotel 537-5571, $59-295)*, *(Hastings House $525-910)* **Under 1 mi:** Restaurant *(Golden Island 537-2535)*, *(The Oystercatcher 537-5041,L $9-12, D $9-23)*, *(La Cucina Ital. Grill 537-5747, B $6-10, L $10-18, D $12-25)*, Seafood Shack *(Shipstone's English Pub L $6-11, D $6-15)*, Inn/B&B *(Wisteria Guest House 537-5899, $55-95)*

Recreation and Entertainment
OnSite: Picnic Area, Boat Rentals *(SS Marine 537-5464)* **Near:** Fishing Charter *(Something Fishy 537-5810)* **Under 1 mi:** Beach, Movie Theater *(Central Hall)*, Video Rental, Park *(Centennial Park)*, Special Events *(Canada Day - Jul 1, Festival of the Arts - Jul)* **1-3 mi:** Golf Course *(SS Golf and Country Club 537-2121)*, Fitness Center *(Curves 538-5575)*, Horseback

Riding *(SS Guided Rides 537-5761)*, Museum *(Akerman Museum by appt. 653-4228)*, Galleries *(Serendipity Studio 537-4535, Pegasus 537-2421)*

Provisioning and General Services
OnSite: Bakery *(Rogue Café)* **Near:** Wine/Beer *(Harbour House)*, Fishmonger *(The Fishery)*, Bookstore *(Salt Spring Books 537-2812, Sabine's Fine Used Books 538-0025)* **Under 1 mi:** Market *(Ganges 537-4144)*, Supermarket *(Thrifty Foods 537-1522, free delivery with $25 min. order)*, Gourmet Shop *(Old Salt Shop 537-5593)*, Delicatessen *(Aunty Pesto's 537-4181)*, Health Food *(Nature Works 537-1919)*, Liquor Store *(Licensed Liquor Store 537-1919)*, Farmers' Market *(Saturdays in Centennial Park at Grace Point)*, Bank/ATM, Post Office, Catholic Church, Protestant Church, Library *(537-4666, Mon-Sat 10am-5pm, Internet access)*, Pharmacy *(Pharmasave 537-5534)*, Hardware Store *(Mouat's Home Hardware 537-5551)*

Transportation
OnSite: Bikes *(bikes & scooters)*, Rental Car *(Salt Spring Marine Rentals 537-5464)* **OnCall:** Water Taxi *(Gulf Islands Water Taxi 537-2510)*, Taxi *(Silver Shadow Taxi 537-3030)* **Near:** Local Bus *(The Ganges Mini Shuttle 537-6758)*, Ferry Service *(BC Ferries to Islands and Tsawwassen 888-223-3779)* **Airport:** Salt Spring Air to YVR *(On-Site)*

Medical Services
911 Service **OnCall:** Ambulance **Near:** Dentist *(Hayden 537-1400)* **Under1 mi:** Chiropractor *(Richardson 537-9399)*, Holistic Services *(Acupressure & Jin Shin Do 537-8525, 3mi. - Salt Spring Wellness Center for Ayurvedic Treatments 537-2326)*, Veterinarian *(Gulf Islands Vet 537-5334)* **Hospital:** Lady Minto 538-4800 *(0.5 mi.)*

Setting -- Salt Spring Marina sits at the head of protected Ganges Harbor - but watch the yellow barrels that warn boaters off the aptly named Moneymaker Reef. From the extensive network of floating docks, there is an expansive panorama of the harbor and of the village of Ganges across the water. The views directly landward are of the rustic marina building and the gray, contemporary Moby's Marine Pub. Pots of flowers lining the ramps step up the atmosphere.

Marina Notes -- *May-Sep $1.25/ft.; Apr $0.85/ft.; Oct-Mar $15/night flat fee. Up to 150 ft. boats on the outside of the breakwater. Garbage $2/bag; recycling free. Onsite are salmon fishing charters, sailing charters, coast and harbor tours, kayak rentals and aluminum power boat rentals. Pump-out at Salt Spring HA dock at Centennial Wharf. New owners as of July 2005 also own Moby's Marine Pub. Every August, the annual Street Dance, held in the parking lot, raises money to support a local youth project. Fresh new bathhouses are in the same building as the refurbished laundromat.

Notable -- Across the street is Harbour House Hotel and just up the road is the impeccable Hastings House Country Estate (a Relais & Chateaux and one of Canada's top rated resorts). Heading toward the village is the Stone Fish Sculpture Studio, The Fishery Seafood Market and the Treehouse Café. Ganges, the hub of Salt Spring Island, is a short ten-minute walk around the harbor. This fun village is the largest community in all the Gulf Islands. The streets and byways are lined with galleries and shops selling pottery, jewelry, stained glass, clothing, books, and more. Over 200 artists come together for ArtCraft, a summer-long exhibit at Mahon Hall. Nearby Mount Regional Park has picnic tables and easy hiking trails.

Navigational Information
Lat: 48°46.105' **Long:** 123°27.050' **Tide:** 11 ft. **Current:** n/a **Chart:** 3478
Rep. Depths *(MLW):* **Entry** 12 ft. **Fuel Dock** n/a **Max Slip/Moor** 25 ft./-
Access: Next to Roamer's fuel dock at the end of the Fulford Harbour

Marina Facilities *(In Season/Off Season)*
Fuel: *at Roamer's Landing* - Gasoline, Diesel
Slips: 6 Total, 6 Transient **Max LOA:** 100 ft. **Max Beam:** n/a
  Rate *(per ft.):* **Day** $0.43 **Week** n/a **Month** n/a
  Power: 30 amp Incl., **50 amp** n/a, **100 amp** n/a, **200 amp** n/a
  Cable TV: No **Dockside Phone:** No
  Dock Type: Floating, Alongside, Wood
Moorings: 0 Total, 0 Transient **Launch:** n/a
  Rate: Day n/a **Week** n/a **Month** n/a
Heads: None
Laundry: None **Pay Phones:** Yes, 1
Pump-Out: No **Fee:** n/a **Closed Heads:** Yes

Marina Operations
Owner/Manager: Salt Spring Island HA **Dockmaster:** Reid Collins
In-Season: Summer, 8am-8pm **Off-Season:** Winter, 9am-5pm*
After-Hours Arrival: Use the pay box
Reservations: No **Credit Cards:** No
Discounts: None
Pets: Welcome **Handicap Access:** No

Fulford Harbour Wharf

Saltspring Island, BC V8K 2T9

Tel: (250) 537-5711 **VHF: Monitor** n/a **Talk** n/a
Fax: (250) 537-5711 **Alternate Tel:** n/a
Email: ssha@saltspring.com **Web:** n/a
Nearest Town: Fulford Village **Tourist Info:** (250) 537-5252

Marina Services and Boat Supplies
Supplies - Near: Ice *(Cube)* **3+ mi:** Ships' Store *(Ganges Marina 537-4242, 9 mi.)*, Bait/Tackle *(9 mi.)*, Propane *(9 mi.)*

Boatyard Services
Nearest Yard: Canoe Cove Marina (250) 656-5566

Restaurants and Accommodations
OnSite: Coffee Shop *(Morningside Organic Bakery & Café, B & L)*, Lite Fare *(Tree House Café South 537-5379, B $3-8, L $4-8, D $8-10)* **Near:** Restaurant *(Fulford Inn 653-4432, L $6-10, D $7-15)* **Under 1 mi:** Inn/B&B *(Anchor Point B&B 538-0110, $160)* **1-3 mi:** Inn/B&B *(The Fulford Creek Guest House 653-0081, $175-250)*, *(Always Welcome 537-0785, $85-135)* **3+ mi:** Restaurant *(Artists Bistro 537-1701, D $22-29, 9 mi., D only, starting at 6pm; closed Sun)*, *(Salt Spring Roasting Company B $6-12, L $8-20, 9 mi., 800-332-8858)*, Pizzeria *(Luigi's 537-5660, 9 mi.)*

Recreation and Entertainment
Near: Beach *(Drummond Park)*, Picnic Area, Grills, Tennis Courts *(653-4467, $5)*, Special Events *(Fulford Days - Aug)* **Under 1 mi:** Video Rental *(Patterson Market and Video 653-4321)*, Park *(Drummond Park)*, Galleries *(Public Eye Gallery)* **1-3 mi:** Golf Course *(Blackburn Meadows Golf Course 537-1707, organically maintained)*, Horseback Riding *(Salt Island Guided Rides 537-5761)*, Boat Rentals *(Salt Spring Kayak 653-4222)*, Museum *(Akerman Museum by appt. 653-4228 Collection of First Nations and pioneer artifacts)* **3+ mi:** Movie Theater *(5 mi.)*

Provisioning and General Services
OnSite: Market *(Patterson Market and Video 653-4321)*, Gourmet Shop *(Jambalaya Jamaican Baskets, plus Salt Spring Island Cheese 653-2300, 1 mi.)*, Bakery *(Morningside Organic Bakery and Café 653-4414)*, Post Office, Clothing Store *(The Wardrobe, Stuff & Nonsense)* **Under 1 mi:** Wine/Beer, Liquor Store, Green Market, Bank/ATM, Catholic Church *(St. Paul's Catholic Church)*, Protestant Church, Beauty Salon, Barber Shop, Newsstand, Retail Shops **3+ mi:** Supermarket *(Thrifty Foods 537-1522, 9 mi.)*, Health Food *(Nature Works 537-2325, 9 mi.)*, Library *(537-4666, Mon-Sat 10am-5pm, Internet access, 9 mi.)*, Bookstore *(Salt Spring Books 537-2812, Sabine's Fine Used Books 538-0025, 9 mi.)*, Pharmacy *(Pharmasave 537-5534, 9 mi.)*, Hardware Store *(Mouat's Home Hardware 537-5551, 9 mi.)*

Transportation
OnSite: Ferry Service *(BC Ferries to Sidney, Swartz Bay 888-223-3779)* **OnCall:** Taxi *(Silver Shadow Taxi 537-3030)* **Near:** Local Bus *(The Ganges Faerie Mini Shuttle 537-6758)* **Airport:** Kenmore Air Seaplanes 425-486-1257 *(5 mi.)*

Medical Services
911 Service OnCall: Ambulance **3+ mi:** Dentist *(Hayden 537-1400, 9 mi.)*, Holistic Services *(Salt Spring Centre of Yoga and Wellness 537-2326, 6 mi.)*, Veterinarian *(Gulf Islands Vet 537-5334, 9 mi.)* **Hospital:** Lady Minto Hospital 538-4800 *(9 mi.)*

Setting -- Fulford Harbour is at the southern end of Salt Spring Island, and wide open to the southeast. The Wharf sits in the inner harbor, co-located with the Vancouver Island ferry terminal. Watch for the signature red ramps. St. Paul's Church, built in 1880 from materials transported to the island by ship, is a Fulford landmark that can be spotted from the harbor as you approach the dock.

Marina Notes -- *In season, dockmaster is available Wed-Sat 8am-8pm. The Fulford Harbour Wharf offers limited space on both sides of a 120 ft. wooden pier, so expect to raft. Power and water are available and there is a public phone on the shore. No other boater amenities are provided. Fuel is next door at Roamer's Landing.

Notable -- The area surrounding Fulford Harbour is quiet and rural, with little development interrupting expansive forested areas. Some delightful galleries, groceries and shops are nearby, but most services are 9 miles north in Ganges. The annual Fulford Days festival, held at Drummond Park in mid August has been the big local event since 1989. It offers music, food, and activities for kids and adults alike. Local specialties such as Salt Spring lamb, salmon, or beef barbecues entice visitors, while kids' races, a watermelon eating contest and a cake walk keep everyone entertained. On a more sophisticated level, Salt Spring Island is famous for its Studio Tour. Pick up a self-guided map at the Visitor Centre and head out to visit 40 local artists in their homes and studios.

Navigational Information

Lat: 48°46.243' **Long:** 123°27.135' **Tide:** 11 ft. **Current:** n/a **Chart:** 3478
Rep. Depths *(MLW)*: **Entry** 25 ft. **Fuel Dock** 12 ft. **Max Slip/Moor** 25 ft./-
Access: SE side of Salt Spring Island in Fulford Harbour off Satellite Channel

Marina Facilities *(In Season/Off Season)*

Fuel: *Roamer's Landing* - Gasoline, Diesel
Slips: 60 Total, 20 Transient **Max LOA:** 160 ft. **Max Beam:** n/a
  Rate *(per ft.)*: **Day** $0.85 **Week** $4.25 **Month** $8.75
  Power: 30 amp $3, **50 amp** n/a, **100 amp** n/a, **200 amp** n/a
  Cable TV: No **Dockside Phone:** No
  Dock Type: Floating, Long Fingers, Alongside, Concrete, Wood
Moorings: 0 Total, 0 Transient **Launch:** n/a
  Rate: Day n/a **Week** n/a **Month** n/a
Heads: 3 Toilet(s), 2 Shower(s)
Laundry: None, Book Exchange **Pay Phones:** Yes, 1
Pump-Out: No **Fee:** n/a **Closed Heads:** Yes

Marina Operations

Owner/Manager: Bill & Gay Perry **Dockmaster:** Same
In-Season: Apr-Sep, 9:30am-6pm **Off-Season:** Oct-Mar, Closed
After-Hours Arrival: Call ahead
Reservations: Yes, Preferred **Credit Cards:** Visa/MC, Debit
Discounts: None
Pets: Welcome, Dog Walk Area **Handicap Access:** Yes, Heads, Docks

Fulford Harbour Marina

2810 Fulford-Ganges Road; Saltspring Island, BC V8K1Z2

Tel: (250) 653-4467 **VHF: Monitor** Ch. 66A **Talk** n/a
Fax: (250) 653-4457 **Alternate Tel:** n/a
Email: fulfordmarina@saltspring.com **Web:** saltspring.com/fulfordmarina
Nearest Town: Fulford Village **Tourist Info:** (250) 537-5252

Marina Services and Boat Supplies

Services - Docking Assistance, Trash Pick-Up, Dock Carts
Communication - Mail & Package Hold, Phone Messages, FedEx, UPS,
Express Mail *(Sat Del)* **Supplies - OnSite:** Ice *(Block, Cube)* **3+ mi:** Ships'
Store *(Ganges Marina 537-4242, 9 mi.)*, Bait/Tackle *(9 mi.)*, Propane *(9 mi.)*

Boatyard Services

Nearest Yard: Canoe Cove Marina (250) 656-5566

Restaurants and Accommodations

Near: Restaurant *(Fulford Inn 653-4432, L $6-10, D $7-15)*, *(Tree House
Café South 537-5379, B $3-8, L $4-8, D $8-10)*, Coffee Shop *(Morningside
Organic Bakery 653-4414)* **Under 1 mi:** Inn/B&B *(Anchor Point B&B 538-
0110, $160)* **1-3 mi:** Inn/B&B *(The Fulford Creek Guest House 653-0081,
$175-250, 2 nt. stay)*, *(Always Welcome 537-0785, $85-135)* **3+ mi:** Rest-
aurant *(Artists Bistro 537-1701, D $22-29, 9 mi., D from 6pm; closed Sun)*,
(Salt Spring Roasting Company L $6-12, D $8-20, 9 mi.), Lite Fare *(Auntie
Pesto's 537-4181, 9 mi., deli)*, Pizzeria *(Luigi's 537-5660, 9 mi.)*

Recreation and Entertainment

OnSite: Picnic Area, Grills, Tennis Courts *(free to boaters)* **Near:** Beach
(Drummond Park), Playground, Video Rental *(Patterson Market & Video
653-4321)*, Special Events *(Fulford Days- Aug in Drummond Park)*
Under 1 mi: Boat Rentals *(Salt Spring Kayak 653-4222)*, Park *(Drummond)*,
Museum *(Akerman - by appt. 653-4228, collection of First Nations and
pioneer artifacts)*, Cultural Attract *(Artcraft)*, Galleries *(Public Eye)*

1-3 mi: Golf Course *(Blackburn Meadows Golf Course 537-1707 - only golf
course in Canada to be maintained entirely organically)*, Horseback Riding
(Salt Island Guided Rides 537-5761) **3+ mi:** Movie Theater *(9 mi.)*

Provisioning and General Services

Near: Market *(Patterson Market 653-4321)*, Wine/Beer, Liquor Store, Bakery
(Morningside Organic Bakery & Café 653-4414), Green Market, Protestant
Church, Beauty Salon, Barber Shop, Newsstand, Retail Shops
Under 1 mi: Gourmet Shop *(Salt Spring Island Cheese 653-2300)*, Bank/
ATM, Post Office, Catholic Church *(St. Paul's Catholic Church)*, Clothing
Store *(Phlying Phish Clothing 352-3844)* **3+ mi:** Supermarket *(Thrifty
Foods 537-1522, free delivery with $25 min., 9 mi.)*, Health Food *(Nature
Works 537-2325, 9 mi.)*, Library *(537-4666, Mon-Sat 10am-5pm, Internet
access, 9 mi.)*, Bookstore *(Salt Spring Books 537-2812, Sabine's Fine
Used Books 538-0025, 9 mi.)*, Pharmacy *(Pharmasave 537-5534, 9 mi.)*,
Hardware Store *(Mouat's Home Hardware 537-5551, 9 mi.)*

Transportation

OnCall: Taxi *(Silver Shadow Taxi 537-3030)* **Near:** Local Bus *(The Ganges
Faerie Mini-Shuttle 537-6758)*, Ferry Service *(BC Ferries to Sidney, Swartz
Bay 888-223-3779)* **Airport:** Kenmore Air Seaplanes 425-486-1257 *(4 mi.)*

Medical Services

911 Service **OnCall:** Ambulance **3+ mi:** Dentist *(Hayden 537-1400, 9 mi.)*,
Holistic Services *(Salt Spring Centre of Yoga and Wellness 537-2326, 6 mi.)*,
Veterinarian *(Gulf Islands Vet 537-5334, 9 mi.)* **Hospital:** Lady Minto 538-
4800 *(9 mi.)*

Setting -- Fulford Harbour opens off Satellite Channel at the south end of Salt Spring Island. At the head of the bay, the marina's facilities sit on two acres of tranquil, densely treed waterfront. A large two-story brown-stained building with a glass wind-break deck hosts the amenities and overlooks the concrete floating docks. It is surrounded by the peaceful hamlet of Fulford Harbour, which is largely hidden by foliage -- but offers a goodly number of useful services.

Marina Notes -- *15 amp power is available at $3/night. From Jun-Sep, pay for 5 consecutive nights and enjoy 2 nights free. The harbor is open to the southeast, but the ferry acts as a nighttime buffer when it overnights on the island. Built in 1988, the marina has been owned and operated by the service-oriented Perry family since 2002. As you approach the building from the docks, the marina office is on the right. Fuel (gasoline & diesel) is available next door at Roamer's Landing. A covered picnic area and picnic pavilion are perched way above the marina. Step back into childhood and ride one of the tire swings hanging in the trees. The nearby tennis court rents for $5/hr. but is free to boaters.

Notable -- Just across the harbor, Fulford Day is held in Drummond Park in August - it's a community event with music, food, games, and activities for the whole family. Stowell Lake, a good place to drown a few worms in the pursuit of trout, is about a half mile away in the other direction. Swimmers will also enjoy the lake and just past it, off Beaver Point Road, is Salt Spring Islands (handcrafted) Cheese. Some might prefer Fulford's relative peace and quiet and commute to the action in Ganges. Catch the mini-shuttle or a taxi into the village for shopping, restaurants, and the popular, summer-long Art/Craft exhibit at Mahon Hall.

Port Washington Public Wharf

1214 Bridges Road; North Pender Island, BC VON 2MI

Tel: (250) 629-6111 **VHF: Monitor** n/a **Talk** n/a
Fax: n/a **Alternate Tel:** n/a
Email: n/a **Web:** n/a
Nearest Town: Driftwood Centre *(6 mi.)* **Tourist Info:** (866) 468-7924

Navigational Information
Lat: 48°48.789' **Long:** 123°19.227' **Tide:** 12 ft. **Current:** n/a **Chart:** 3442
Rep. Depths *(MLW)*: **Entry** 15 ft. **Fuel Dock** n/a **Max Slip/Moor** 20 ft./-
Access: NW side of No. Pender Island north of Ferry landing in Brimmer Bay

Marina Facilities *(In Season/Off Season)*
Fuel: No
Slips: 6 Total, 3 Transient **Max LOA:** 100 ft. **Max Beam:** n/a
 Rate *(per ft.)*: **Day** $0.25* **Week** n/a **Month** n/a
 Power: 30 amp n/a, 50 amp n/a, 100 amp n/a, 200 amp n/a
 Cable TV: No **Dockside Phone:** No
 Dock Type: Floating, Alongside, Wood
Moorings: 0 Total, 0 Transient **Launch:** n/a
 Rate: Day n/a **Week** n/a **Month** n/a
Heads: None
Laundry: None **Pay Phones:** No
Pump-Out: No **Fee:** n/a **Closed Heads:** Yes

Marina Operations
Owner/Manager: S.G. Harbour Commission **Dockmaster:** Rod MacLean
In-Season: Year-Round, 24/7 **Off-Season:** n/a
After-Hours Arrival: Use the pay envelope
Reservations: No **Credit Cards:** No
Discounts: None
Pets: Welcome **Handicap Access:** No

Marina Services and Boat Supplies
Supplies - 1-3 mi: Ice *(Cube)*, Ships' Store *(Otter Bay Marina 629-3579)*, Bait/Tackle *(Otter Bay Marina)*, Propane *(Home Hardware 629-3455)*

Boatyard Services
Nearest Yard: Canoe Cove Marina (250) 656-5566

Restaurants and Accommodations
Near: Condo/Cottage *(Emily's Cottage 877-662-3414, 2-bedroom, $450/wknd, $950 wk)* **1-3 mi:** Restaurant *(Chippers at the Golf Course 629-6665)*, *(Islanders Restaurant 629-3929, D $15-30, steak & seafood, glassed-in deck with views)*, *(Hope Bay Café 629-6668)*, Lite Fare *(The Stand 629-3292, burgers, at ferry terminal)*, Inn/B&B *(Corbett House 629-6305, $110-125, 3-bedroom; will pick up guests)*, *(Sun Raven B&B & Wellness Center 629-6216, $85-110)* **3+ mi:** Restaurant *(Le Pistou Grill 629-3131, 6 mi.)*, Coffee Shop *(That Little Coffee Place 629-3080, 6 mi., 10am-4pm daily)*, Pizzeria *(Memories at the Inn 629-3353, 6 mi.)*, Hotel *(The Inn at Pender Island 629-3353, $79-139, 6 mi.)*

Recreation and Entertainment
Near: Beach **Under 1 mi:** Jogging Paths, Roller Blade/Bike Paths **1-3 mi:** Heated Pool, Picnic Area, Playground, Golf Course *(Pender Island G.C. 629-6614)*, Boat Rentals *(Kayak Pender Island Apr-Oct 629-6939)*, Video Rental *(Southridge Farms Country Store 629-2051)*, Park *(Pacific Marine Heritage Legacy Park)*, Cultural Attract *(concerts and plays at Community Hall 629-3669)*, Sightseeing, Special Events *(Fall Fair in Aug, Pender Sailing Race)*, Galleries *(Renaissance 629-3070, Red Tree)* **3+ mi:** Fitness Center *(Physiotherapy Clinic 929-9920, Pilates, 6 mi.)*

Provisioning and General Services
1-3 mi: Convenience Store *(Otter Bay Marina 629-3579)*, Market *(Southridge Farms Country Store 629-2051)*, Farmers' Market *(Saturday Morning Market at the Community Hall 629-3669)*, Library *(The Pender Lender 629-3722)*, Beauty Salon *(Hope Bay Hair Salon)*, Newsstand, Hardware Store *(Home Hardware 629-3455)* **3+ mi:** Supermarket *(True Value 629-8322, 9am-8:30pm, delivers, 6 mi.)*, Delicatessen *(at True Value, 6 mi.)*, Liquor Store *(Driftwood Centre 629-3413, 6 mi.)*, Bakery *(Pender Island Bakery 629-6453, 6 mi.)*, Bank/ATM *(629-6516, 6 mi.)*, Post Office *(629-3222, 6 mi.)*, Catholic Church *(St. Teresa 629-3141)*, Protestant Church *(Anglican, United Community)*, Laundry *(Driftood Centre 629-3005, 6 mi.)*, Bookstore *(Tallisman Books & Gallery 629-6944, 6 mi.)*, Pharmacy *(Pender Island Pharmacy 629-6555, 6 mi.)*, Florist *(Petals on Pender 629-3268, delivers, 6 mi.)*

Transportation
OnCall: Rental Car *(Local Motion 629-3366, pickup available)*, Taxi *(Pender Taxi 629-3555)* **1-3 mi:** Bikes *(Otter Bay Marina 629-3579)*, Ferry Service *(BC Ferries to Islands, Sidney & Vancouver 888-223-3779)* **Airport:** YVR/ daily flights with Seair Seaplanes 800-447-3247 *(20 mi./On-Site)*

Medical Services
911 Service **OnCall:** Ambulance **1-3 mi:** Chiropractor *(Chiro Clinic 629-9918)*, Holistic Services *(Sun Raven Wellness 629-6216: reflexology, Reiki, massage, herbal treatments)* **3+ mi:** Doctor *(Medical Clinic 639-3233, 7 mi.)*, Dentist *(Nord 629-6815, 6 mi.)*, Veterinarian *(Pender Island Vet 629-9909, 6 mi.)* **Hospital:** Saanich Peninsula Hospital 652-3911 *(20 mi.)*

Setting -- Port Washington Public Wharf is in North Pender Island's Grimmer Bay, flanked by James Point and Willey Point. Just offshore is Boat Islet with Port Washington Light on its western end. The brown "Shed" and black and white General Store edifices are at the head of the wharf. The land climbs steeply from the shore; contemporary homes are visible through the trees. This is a quiet residential area that promises solitude, beachcombing and lazy days.

Marina Notes -- *First 4 hours free; then flat rate for up to 12 hrs.: 30' $7.50, 40' $10, 50' $12.50; 13-24 hrs.: 30' $15, 40' $20, 50' $25. Commercial fishing vessels have priority. Seair Seaplanes operates 3 daily flights from the dock. At publication, the historic General Store at the wharf was boarded up and carried a For Sale sign, but the public phone outside and the public bulletin board were still active.

Notable -- If you have a kayak or dinghy, the Grimmer Bay shoreline offers lots of coves and crannies to duck into. Otter Bay Marina also rents kayaks, plus bikes and scooters. Picnic provisions can be found at Southridge Farms Country Store less than 3 miles away on Port Washington Road (Mon-Sat 9:30am-6pm, Sun 11am-5pm). Also within 3 miles are The Pender Lender Library and Community Center - which offers faxing, copying, and printing services, plus Internet access (run by volunteers Tue, Thu, Fri, and Sat from 10am to 4pm). In the summer, the Community Center is home to concerts, annual events and a Saturday farmers' market. A casual eatery and popular dinner restaurant are in the ferry landing area. The Driftwood Centre, 6 miles south, serves as "downtown" for the Penders with a pizza parlor, bank, bakery, laundromat, and more restaurants.

Navigational Information

Lat: 48°47.875' **Long:** 123°18.550' **Tide:** 12 ft. **Current:** n/a **Chart:** 3442
Rep. Depths (*MLW*): **Entry** 12 ft. **Fuel Dock** n/a **Max Slip/Moor** 9 ft./-
Access: On Swanson Channel on the northwest side of North Pender Island

Marina Facilities *(In Season/Off Season)*

Fuel: No
Slips: 65 Total, 45 Transient **Max LOA:** 150 ft. **Max Beam:** n/a
  Rate *(per ft.)*: **Day** $0.99/0.75 **Week** n/a **Month** $5.95*
  Power: 30 amp $5.50, **50 amp** n/a, **100 amp** n/a, **200 amp** n/a
  Cable TV: No **Dockside Phone:** No
  Dock Type: Floating, Long Fingers, Alongside, Concrete
Moorings: 0 Total, 0 Transient **Launch:** n/a, Dinghy Dock
  Rate: Day n/a **Week** n/a **Month** n/a
Heads: 4 Toilet(s), 4 Shower(s) *(with dressing rooms)*
Laundry: 2 Washer(s), 2 Dryer(s) **Pay Phones:** No
Pump-Out: No **Fee:** n/a **Closed Heads:** Yes

Marina Operations

Owner/Manager: Chuck Spense **Dockmaster:** Same
In-Season: Jun-Sep, 8am-8pm **Off-Season:** Oct-May
After-Hours Arrival: Call ahead
Reservations: Yes, Preferred **Credit Cards:** Visa/MC, Debit
Discounts: None
Pets: Welcome **Handicap Access:** Yes, Docks

Otter Bay Marina

2311 MacKinnon Road; RR 1; North Pender Island, BC V0N 2M0

Tel: (250) 629-3579 **VHF: Monitor** Ch. 66A **Talk** n/a
Fax: (250) 629-3589 **Alternate Tel:** n/a
Email: n/a **Web:** n/a
Nearest Town: Driftwood Centre *(5 mi.)* **Tourist Info:** (866) 468-7924

Marina Services and Boat Supplies

Services - Docking Assistance, Security *(owner onsite)*, Trash Pick-Up, Dock Carts **Communication -** Mail & Package Hold, Phone Messages, FedEx, UPS, Express Mail *(Sat Del)* **Supplies - OnSite:** Ice *(Cube)*, Ships' Store, Bait/Tackle

Boatyard Services

OnSite: Launching Ramp **Nearest Yard:** Canoe Cove Marina (250) 656-5566

Restaurants and Accommodations

Near: Restaurant *(Poplars 629-3955)*, *(Islanders Restaurant 629-3929, D $15-30, water views)*, Lite Fare *(The Stand 629-3292, burgers, hot dogs; at ferry teminal)*, Inn/B&B *(Sun Raven B&B & Wellness Center 629-6216, $85-110, onsite pool, balconies)* **Under 1 mi:** Restaurant *(Chippers at the Golf Course 629-6665, D $15-30)*, Inn/B&B *(Arcadia by the Sea 877-470-8439)*, Condo/Cottage *(The Otter Bay House 604-681-8959, $720-2120/week)* **1-3 mi:** Inn/B&B *(Corbett House B&B 629-6305, $110-125, guest pick up; 3 bedroom farmhouse)* **3+ mi:** Restaurant *(Le Pistou Grill 629-3131, 5 mi.)*

Recreation and Entertainment

OnSite: Heated Pool, Picnic Area, Grills, Playground, Boat Rentals *(Kayak Pender Island 629-6939, Apr-Oct)* **Near:** Jogging Paths, Roller Blade/Bike Paths **Under 1 mi:** Beach, Golf Course *(Pender Island Golf Course 629-6614)* **1-3 mi:** Video Rental *(Southridge Farms Country Store 629-2051)*, Park *(Pacific Marine Heritage Legacy Park)*, Cultural Attract *(concerts and plays at Community Hall 629-3669)*, Special Events *(Fall Fair - Aug, Art Shows)*, Galleries *(Renaissance Gallery 629-3070)*

Provisioning and General Services

OnSite: Convenience Store *(groceries, basic supplies, clothing)*
1-3 mi: Market *(Southridge Farms Country Store 629-2051)*, Farmers' Market *(Saturday Morning Market at the Community Hall, seasonal)*, Library *(The Pender Lender 629-3722)*, Beauty Salon *(Hope Bay Hair)*, Hardware Store *(Home 629-3455)* **3+ mi:** Supermarket *(True Value 628-8322 9am-8:30pm, delivers, 5 mi.)*, Delicatessen *(at True Value, 5 mi.)*, Liquor Store *(Driftwood Centre 629-3413, 5 mi.)*, Bakery *(Pender Island Bakery 629-6453, 5 mi.)*, Bank/ATM *(629-6516, 5 mi.)*, Post Office *(629-3222, 5 mi.)*, Catholic Church *(Chapel of Saint Teresa 629-3141)*, Protestant Church *(Pender Island United Community Church 629-3634)*, Laundry *(Driftood Centre 629-3005, 5 mi.)*, Bookstore *(Tallisman Books & Gallery 629-6944, 5 mi.)*, Pharmacy *(Pender Island Pharmacy 629-6555, 5 mi.)*, Florist *(Petals on Pender 629-3268, delivers, 5 mi.)*

Transportation

OnSite: Bikes *(Bikes and scooters)* **OnCall:** Rental Car *(Local Motion Car Rentals 629-3366)*, Taxi *(Pender Taxi 629-3555)* **Near:** Ferry Service *(BC Ferries to Islands, Swartz Bay & Tsawwassen 888-223-3779)* **Airport:** Seair Seaplanes 604-273-8900/ Victoria Int'l. *(On-Call/20 mi.)*

Medical Services

911 Service **OnCall:** Ambulance **Near:** Holistic Services *(Sun Raven Wellness 629-6216: reflexology, Reiki, massage, herbal treatments)* **1-3 mi:** Chiropractor *(Chiro Clinic 629-9918)* **3+ mi:** Doctor *(Medical Clinic 629-3233, 6 mi.)*, Dentist *(Nord 629-6815, 5 mi.)*, Veterinarian *(Pender Island Vet 629-9909, 5 mi.)* **Hospital:** Saanich Peninsula 652-3911 *(20 mi.)*

Setting -- Picturesque Otter Bay overlooks Swanson Channel with Salt Spring Island in the distance. Trees crowd the water's edge all around the bay. Nestled in a lovely, quiet cove on North Pender Island, the spacious grounds of Otter Bay Marina are attractive and well groomed; 31 brand new, upscale, contemporary west-coast cottages climb the hill above the docks and line the ridge overlooking the bay - landscaping underway.

Marina Notes -- *$5.95/ft. monthly rate for winter or annual (not summer). Service-oriented with immaculately kept facilities. Kayak, bike, and scooter rentals to see the island by land and water. The on-site store carries groceries, clothing, espresso, and gifts. The heated pool is for adults only from 4-6pm. A large group picnic area features a fireplace, grill, refrigerator, and sink. Bathhouse and laundry reside in a separate building. Those at anchor may dinghy in to use the facilities for a $5 charge. The new on-site "Currents" fully furnished cottage colony offers quarter fractional ownership and rental options.

Notable -- Picnics under the willow trees, golfing, swimming, kayaking, and bicycling are just a few of the activities available. Soak up the sun poolside or enjoy a Cappuccino on the deck. Islanders Restaurant offers romantic dinners with live piano on weekends. Or try The Stand next to the ferry terminal for casual lunches. Breakfast or lunch also available at the Golf Course restaurant less than a mile away (shuttle can be arranged). For relaxation, the Wellness Center at Sun Raven Retreat offers everything from herbal "potions" to limphatic drainage massages. Scenic hiking trails at the Pacific Marine Heritage Legacy Park lead to Roe Lake. A 35-minute ferry ride from the nearby terminal goes to Vancouver Island at Swartz Bay with easy connections to the mainland.

Hope Bay Public Wharf

Bedwell Harbor Road; North Pender Island, BC V0N 2M1

Tel: (250) 629-9990 **VHF: Monitor** Ch.66 **Talk** n/a
Fax: (250) 629-6751 **Alternate Tel:** (250) 539-3036
Email: pbinner@cablelan.net **Web:** n/a
Nearest Town: Driftwood Centre *(4 mi.)* **Tourist Info:** (866) 468-7924

Navigational Information
Lat: 48°48.210' **Long:** 123°16.484' **Tide:** 12 ft. **Current:** n/a **Chart:** 3442
Rep. Depths *(MLW)*: **Entry** 15 ft. **Fuel Dock** n/a **Max Slip/Moor** 10 ft./-
Access: Northeast corner of North Pender Island in Plumber Sound

Marina Facilities *(In Season/Off Season)*
Fuel: No
Slips: 6 Total, 6 Transient **Max LOA:** 220 ft. **Max Beam:** 12 ft.
 Rate *(per ft.)*: **Day** $0.55* **Week** n/a **Month** n/a
 Power: 30 amp n/a, **50 amp** n/a, **100 amp** n/a, **200 amp** n/a
 Cable TV: No **Dockside Phone:** No
 Dock Type: Floating, Alongside, Wood
Moorings: 2 Total, 0** Transient **Launch:** n/a, Dinghy Dock
 Rate: Day n/a **Week** n/a **Month** n/a
Heads: None
Laundry: None **Pay Phones:** No
Pump-Out: No **Fee:** n/a **Closed Heads:** Yes

Marina Operations
Owner/Manager: So. GI Harbour Commission **Dockmaster:** Peter Binner
In-Season: Year-Round, 24/7 **Off-Season:** n/a
After-Hours Arrival: Use pay envelope
Reservations: No **Credit Cards:** No
Discounts: None
Pets: Welcome **Handicap Access:** No

Marina Services and Boat Supplies
Communication - FedEx, DHL **Supplies - 1-3 mi:** Ice *(Cube)*, Ships' Store *(Otter Bay Marina 629-3579)*, Live Bait *(Otter Bay Marina)*

Boatyard Services
Nearest Yard: Canoe Cove Marina (250) 656-5566

Restaurants and Accommodations
OnSite: Restaurant *(Hope Bay Café 629-6668, L $6-12, D $9-28)*
Near: Condo/Cottage *(Morning Moon 537-0774, wkly only in summer, $750)* **Under 1 mi:** Inn/B&B *(Corbett House B&B 629-6305, $110-125, pick up available; 3 bedrooms)* **1-3 mi:** Restaurant *(Chippers at the Golf Course 629-6665)*, *(Islanders Restaurant 629-3929, D $15-30, seafood & steaks, fireplace, live piano on wknds)*, Lite Fare *(The Stand 629-3292, at ferry landing; burgers & more)*, Inn/B&B *(Arcadia by the Sea 877-470-8439)*, *(Sun Raven B&B & Wellness Center 629-6216, $85-110)* **3+ mi:** Pizzeria *(Memories at the Inn 629-3353, 4 mi.)*

Recreation and Entertainment
OnSite: Galleries *(The Goldsmith Shop - Peter Binner, the wharfinger, is the goldsmith; Red Tree Gallery 629-6800; Silver Grizzley)* **Under 1 mi:** Beach, Cultural Attract *(Community Hall concerts and plays in summer 629-3669)*, Special Events *(Fall Fair - Aug, Pender Sailing Race)* **1-3 mi:** Picnic Area, Playground, Golf Course *(Pender Island Golf Course 629-6614)*, Jogging Paths, Boat Rentals *(Kayak Pender Island, Apr-Oct 629-6939)*, Video Rental *(Southridge Farms 629-2051)*, Park *(Prior Centennial Park, Mount Norman Regional Park, Beaumont Provincial Marine Park)*

Provisioning and General Services
OnSite: Fishmonger *(in season)*, Beauty Salon *(Hope Bay Hair Salon)*, Retail Shops *(Pender Treasures, Sladen's home accessories)*
Under 1 mi: Convenience Store, Market *(Southridge Farms 629-2051 Country Store 629-2051)*, Wine/Beer, Library *(The Pender Lender 629-3722)* **1-3 mi:** Farmers' Market *(Saturday Morning Market at the Community Hall)*, Catholic Church *(St. Teresa 629-3141)*, Protestant Church *(United Community Church 629-3634, Anglican)*, Hardware Store *(Home Hardware 629-3455)* **3+ mi:** Supermarket *(True Value 9am-8:30pm, delivers 628-8322, 4 mi.)*, Delicatessen *(at True Value, 4 mi.)*, Liquor Store *(Driftwood Centre 629-3413, 4 mi.)*, Bakery *(Pender Island Bakery 629-6453, 4 mi.)*, Bank/ATM *(629-6516, 4 mi.)*, Post Office *(629-3222, 4 mi.)*, Laundry *(Driftood Centre 629-3005, 4 mi.)*, Bookstore *(Tallisman Books & Gallery 629-6944, 4 mi.)*, Pharmacy *(Pender Island Pharmacy 629-6555, 4 mi.)*, Florist *(Petals on Pender 629-3268, delivers, 4 mi.)*

Transportation
OnCall: Rental Car *(Local Motion 629-3366, pick-up available)*, Taxi *(Pender Taxi 629-3555)* **1-3 mi:** Bikes *(bikes and scooters at Otter Bay Marina 629-3579)*, Ferry Service *(BC Ferries to islands, Swartz Bay & Tsawwassen 888-223-3779)* **Airport:** Seair Seaplanes 604-273-8900/ YVR *(OnCall/20 mi.)*

Medical Services
911 Service **1-3 mi:** Chiropractor *(Chiro Clinic 629-9918)*, Holistic Services *(Sun Raven Wellness 629-6216, reflexology, Reiki, massage, herbal treatments)* **3+ mi:** Doctor *(Medical Clinic 629-3233, 5 mi.)*, Dentist *(Nord 629-6815, 4 mi.)* **Hospital:** Saanich Peninsula 652-3911 *(20 mi.)*

Setting -- On the east side of North Pender Island in Plumber Sound, Hope Bay looks out across Navy Channel to Mayne Island. A gangway, with signature bright blue railings, leads from a single wood float to the new shops and enticing café that populate the historic, elegantly restored Hope Bay General Store.

Marina Notes -- *First 4 hrs. free; 4-12 hrs: 30' $8.50, 40' $11, 50' $14; 13-24 hrs: 30' $16.50, 40' $22, 50' $27.50. **Moorings are for day use only. Leave dockage payment in the drop box and place back portion of envelope visible on the boat. Note: No garbage pick-up. In season, "Victoria" sells crabs from the dock. Note: Commercial fishing boats have priority at the public wharf.

Notable -- In the early 1900s, when the Hope Bay Wharf served ferries carrying supplies and passengers to and from Sidney, the Hope Bay General Store was one of the island's first commercial enterprises. It operated until '98 when it was destroyed by fire. Rebuilt by a group of local residents - in a corrugated aluminum, wood and clapboard contemporary style - it reopened in June '05. Sparkling, diminutive Hope Bay Café features a wall of windows that overlooks the Sound. A two-story addition wraps a stone patio with outdoor seating; it houses shops, intriguing galleries and offices. The wharfinger's goldsmith shop is upstairs along with a local Artist's Co-Op. Also in the complex are a Regional Parks Office, Island View Office Services, Silver Grizzley (Native items), West Central Computer Services, and a hair salon. The island Library, Community Center & well-stocked Country Market are nearby. Car/scooter rentals or taxis are available to explore the rest of the Pender Islands. Most other supplies and services, including a new spa, are about 4 miles away at the Drifwood Center.

Navigational Information
Lat: 48°46.680' **Long:** 123°16.070' **Tide:** 12 ft. **Current:** n/a **Chart:** 3477
Rep. Depths (*MLW*): Entry 6 ft. **Fuel Dock** n/a **Max Slip/Moor** 15 ft./-
Access: East side of North Pender Island

Marina Facilities *(In Season/Off Season)*
Fuel: No
Slips: 4 Total, 4 Transient **Max LOA:** 85 ft. **Max Beam:** n/a
  Rate *(per ft.)*: **Day** $0.25* **Week** n/a **Month** n/a
  Power: 30 amp n/a, **50 amp** n/a, **100 amp** n/a, **200 amp** n/a
  Cable TV: No **Dockside Phone:** No
  Dock Type: Floating, Alongside, Wood
Moorings: 0 Total, 0 Transient **Launch:** n/a
  Rate: **Day** n/a **Week** n/a **Month** n/a
Heads: None
Laundry: None **Pay Phones:** No
Pump-Out: No **Fee:** n/a **Closed Heads:** es

Marina Operations
Owner/Manager: S.G. Harbour Commission **Dockmaster:** Claude Kennedy
In-Season: Year-Round, 24/7 **Off-Season:** n/a
After-Hours Arrival: Use pay envelope
Reservations: No **Credit Cards:** None
Discounts: None
Pets: Welcome **Handicap Access:** No

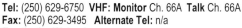

Browning Harbour Public Wharf
North Pender Island, BC V0N 2M0

Tel: (250) 629-6750 **VHF: Monitor** Ch. 66A **Talk** Ch. 66A
Fax: (250) 629-3495 **Alternate Tel:** n/a
Email: n/a **Web:** n/a
Nearest Town: Driftwood Centre *(0.75 mi.)* **Tourist Info:** (866) 468-7924

Marina Services and Boat Supplies
Supplies - Near: Ice *(Cube)*, Live Bait *(Port Browning Marina 629-3493)*, Propane

Boatyard Services
OnCall: Divers, Upholstery **Near:** Engine mechanic *(gas, diesel)*, Electrical Repairs. **Nearest Yard:** Canoe Cove Marina (250) 656-5566

Restaurants and Accommodations
Near: Restaurant *(The Dockside Café 629-3493, B $5-10, L $7-15, D $7-15)*, *(The A-License Pub 629-3493, L $7-15, D $7-15)*, Condo/Cottage *(Port Browning Guest Rooms 629-3493, $50-80)* **Under 1 mi:** Restaurant *(Le Pistou Grill 629-3131, L & D; French)*, Coffee Shop *(That Little Coffee Place 629-3080, Internet access)*, Pizzeria *(Memories at the Inn 629-3353, 5-8pm; pizza & entrees, takeout)*, Hotel *(The Inn at Pender 629-3353, $79-139)*, Inn/B&B *(Nosey Point Inn 629-3617, from $125)* **1-3 mi:** Restaurant *(Hope Bay Café 629-6668)*, Inn/B&B *(Delia's Shangri-la Oceanfront B&B 629-3808, $100-215)* **3+ mi:** Restaurant *(Islanders 629-3929, D $15-30, 5 mi., near ferry; water views)*

Recreation and Entertainment
Near: Heated Pool, Beach, Picnic Area *(Dinghy to Port Browning Marina)*, Grills, Tennis Courts, Group Fishing Boat *(Sound Passage Adventure 629-3920)* **Under 1 mi:** Fitness Center *(Physiotherapy Clinic 929-9920, Pilates)*, Jogging Paths, Video Rental, Sightseeing **1-3 mi:** Golf Course *(Pender Island GC 629-6614)*, Park *(Prior Centennial Park)*, Cultural Attract *(summer concerts & plays at Community Hall 629-3669)*, Special Events *(Fall Fair - Aug, Pender Sailing Race, Art Shows)*, Galleries *(Blood Star 629-6661, Galway 629-3176)* **3+ mi:** Roller Blade/Bike Paths *(Kayak Pender Isl. 629-6939, Apr-Oct, 5 mi.)*

Provisioning and General Services
Near: Wine/Beer *(Port Browning Marina)* **Under 1 mi:** Supermarket *(True Value 629-8322, 9am-8:30pm; delivers)*, Delicatessen *(at True Value)*, Liquor Store *(Driftwood Centre 629-3413)*, Bakery *(Pender Island Bakery 629-6453)*, Farmers' Market *(Driftwwod Centre, Sat May-Oct)*, Bank/ATM *(629-6516)*, Post Office *(629-3222)*, Beauty Salon *(Shear Delight 629-3582)*, Laundry *(Driftood Centre 629-3005)*, Bookstore *(Talisman Books & Gallery 629-6944)*, Pharmacy *(Pender Island 629-6555)*, Newsstand, Florist *(Petals on Pender 629-3268, delivers)* **1-3 mi:** Market *(Magic Lake Market)*, Catholic Church *(Chapel of St.Teresa 629-3141)*, Protestant Church *(Pender Island Community Church 629-3634)*, Library *(Pender Lender 629-3722)*, Hardware Store *(Home Hardware 629-3455)*

Transportation
OnCall: Water Taxi *(Sound Passage 629-3920)*, Rental Car *(Local Motion 629-3366, pickup from marinas)*, Taxi *(Pender Taxi 629-3555)* **3+ mi:** Bikes *(bikes & scooters at Otter Bay Marina 629-3579, 5 mi.)*, Ferry Service *(BC Ferries to Islands, Sidney & Vancouver 888-223-3779, 5 mi.)* **Airport:** Seair Seaplanes 604-273-8900/ Victoria Int'l. *(On-Call/20 mi.)*

Medical Services
911 Service **OnCall:** Ambulance **1-3 mi:** Doctor *(Medical Clinic 629-3233)*, Dentist *(Nord 629-6815)*, Veterinarian *(Pender Island Vet 629-9909)* **3+ mi:** Chiropractor *(Chiro Clinic 629-9918, 4 mi.)* **Hospital:** Saanich Peninsula 652-3911 *(20 mi.)*

Setting -- This small float on the north side of the bay offers about 90 feet of alongside tie-up with no other amenities. Commercial fishing boats have priority, but this provides an alternative for transient boaters when Port Browning Marina is full. The area immediately surrounding the wharf is rural and residential.

Marina Notes -- *First 4 hours are free; then flat rate 4-12 hrs: 30' $7.50, 40' $10, 50' $12.50; 13-24 hrs: 30' $15, 40' $20, 50' $25. NOTE: The Pender Canal between North and South Pender Islands is crossed by a fixed bridge (27 ft. of vertical clearance). The 89 foot side-tie dock offers no boater amenities, but Port Browning Marina is a short dinghy ride across the harbor where a restaurant, pub, and charter/tour operator await. Sound Passage Adventure offers fishing expeditions, scuba diving courses and charters, whale watching, ecotours, and yacht management.

Notable -- About 2.5 mi. south, the one-lane bridge that separates North and South Pender Islands affords an extensive view of Browning and Bedwell Harbours, and the wooded island shores. The canal was originally dredged to create a shorter passage for the ferry going to Hope Bay. From the Browning Harbour Wharf, it's less than 1 mi. to Driftwood Centre ("The Drift"). This area serves as the commercial center for the Penders; visitors can find a bank, laundromat, pharmacy, bookstore, bakery and restaurants here. There is over 5,000 square feet of shopping at the Tru Value supermarket, including a full service deli with an outside patio open in the summer. Also on the island, forests, fields and grazing cows meet visitors, with some galleries & craft shops scattered around. Hamilton Beach is around the harbor. Prior Centennial Park and a Disc Golf Park (629-6494) are farther east toward the Magic Lake area.

Port Browning Marina

P.O. Box 126; North Pender Island, BC V0N 2M0

Tel: (250) 629-3493 **VHF: Monitor** Ch. 66A **Talk** n/a
Fax: (250) 629-3495 **Alternate Tel:** n/a
Email: info@portbrowning.com **Web:** www.portbrowning.com
Nearest Town: Driftwood Centre *(0.2 mi.)* **Tourist Info:** (866) 468-7924

Navigational Information

Lat: 48°46.591' **Long:** 123°16.428' **Tide:** 12 ft **Current:** n/a **Chart:** 3477
Rep. Depths *(MLW)*: **Entry** 15 ft. **Fuel Dock** n/a **Max Slip/Moor** 100 ft./-
Access: East side of North Pender Island at the head of Port Browning Bay

Marina Facilities *(In Season/Off Season)*

Fuel: No
Slips: 90 Total, 75 Transient **Max LOA:** 100 ft. **Max Beam:** n/a
  Rate *(per ft.)*: **Day** $1.00/10.00* **Week** n/a **Month** $4
  Power: 30 amp $4** **50 amp** n/a, **100 amp** n/a, **200 amp** n/a
  Cable TV: No **Dockside Phone:** No
  Dock Type: Floating, Long Fingers, Pilings, Alongside, Wood
Moorings: 0 Total, 0 Transient **Launch:** n/a, Dinghy Dock
  Rate: Day n/a **Week** n/a **Month** n/a
Heads: 6 Toilet(s), 4 Shower(s)
Laundry: 2 Washer(s), 2 Dryer(s) **Pay Phones:** Yes, 2
Pump-Out: No **Fee:** n/a **Closed Heads:** Yes

Marina Operations

Owner/Manager: Lou Henshaw **Dockmaster:** Same
In-Season: Year-Round, 7am-Mid **Off-Season:** n/a
After-Hours Arrival: Take any slip not marked "Reserved"
Reservations: Yes, Preferred **Credit Cards:** Visa/MC, Debit
Discounts: None
Pets: Welcome **Handicap Access:** Yes, Heads, Docks

Marina Services and Boat Supplies

Services - Docking Assistance, Security *(staff onsite)* **Communication -** Mail & Package Hold, Phone Messages, Fax in/out *($1/pg)*, Data Ports *(Wi-Fi)*, FedEx, UPS, Express Mail *(Sat Del)* **Supplies - OnSite:** Ice *(Block, Cube)* **1-3 mi:** Propane

Boatyard Services

OnSite: Launching Ramp **OnCall:** Divers, Upholstery **Near:** Engine mechanic *(gas, diesel)*, Electrical Repairs. **Nearest Yard:** Canoe Cove Marina (250) 656-5566

Restaurants and Accommodations

OnSite: Restaurant *(The Dockside Café 629-3493, B $5-10, L $7-15, D $7-15, 7am-8:30pm daily)*, *(The A-License Pub 629-3493, L $7-15, D $7-15)*, Condo/Cottage *(Port Browning Guest Rooms 629-3493, $50-80)*
Near: Restaurant *(Le Pistou Grill 629-3131)*, Inn/B&B *(Nosey Point 629-3617, from $125)* **Under 1 mi:** Pizzeria *(Memories 629-3353)*, Hotel *(The Inn at Pender 629-3353, $79-139)* **1-3 mi:** Inn/B&B *(Sahhali Serenity 629-3664, $225-295)*, Condo/Cottage *(Magic Lake 598-0005, $325/wknd, up to 7 people)* **3+mi:** Restaurant *(Islanders 629-3929, D $15-30, 5 mi.)*

Recreation and Entertainment

OnSite: Heated Pool, Beach, Picnic Area, Grills, Tennis Courts **Near:** Jogging Paths, Video Rental, Sightseeing **Under 1 mi:** Horseback Riding **1-3 mi:** Golf Course *(Pender Island Golf Course 629-6614)*, Park *(Prior Centennial Park)*, Cultural Attract *(concerts and plays at Community Hall 629-3669)*, Special Events *(Fall Fair - Aug, Pender Sailing Race, Art Shows)*, Galleries *(Blood Star 629-6661, Galway 629-3176)* **3+ mi:** Boat Rentals *(Kayak Pender Island Apr-Oct 629-6939, 5 mi.)*

Provisioning and General Services

OnSite: Convenience Store *(groceries)*, Liquor Store, Fishmonger **Near:** Supermarket *(True Value 629-8322, 9am-8:30pm; delivers)*, Delicatessen *(at True Value)*, Bakery *(Pender Island Bakery 629-6453)*, Bank/ATM *(629-6516)*, Post Office *(629-3222)*, Beauty Salon *(Shear Delight 629-3582)*, Laundry *(Driftood Centre 629-3005)*, Bookstore *(Tallisman Books & Gallery 629-6944)*, Pharmacy *(Pender Island 629-6555)*, Newsstand, Florist *(Petals on Pender 629-3268, delivers)*, Clothing Store, Retail Shops **Under 1 mi:** Farmers' Market *(Driftwwod Centre Sat from May-Oct)* **1-3 mi:** Catholic Church *(St. Teresa 629-3141)*, Protestant Church *(United Community Church 629-3634)*, Library *(Pender Lender 629-3722)*, Hardware Store *(Home Hardware 629-3455)*

Transportation

OnCall: Water Taxi *(Sound Passage 629-3920)*, Taxi *(Pender Taxi 629-3555)* **3+ mi:** Bikes *(Bikes and scooters at Otter Bay Marina 629-3579, 5 mi.)*, Ferry Service *(BC Ferries to Islands, Sidney & Vancouver 888-223-3779, 5 mi.)* **Airport:** Seair Seaplanes 604-273-8900/ Victoria Int'l. *(On-Call/20 mi.)*

Medical Services

911 Service **OnCall:** Ambulance **1-3 mi:** Doctor *(Medical Clinic 629-3233)*, Dentist *(Nord 629-6815)*, Chiropractor *(Chiro Clinic 629-9918)*, Veterinarian *(Pender Island 629-9909)* **Hospital:** Saanich Peninsula 652-3911 *(20 mi.)*

Setting -- Port Browning Marina sits snugly at the head of Port Browning Bay and offers surprising amenities for a place with such a casual atmosphere. At the head of the main ramp, a tan contemporary with brick-red trim hosts the restaurant and pub - flanked by a deck and covered patio that overlook the pool and bay beyond. A designated seaplane float is in the northwest corner of the marina.

Marina Notes -- *Off-season rate is flat $10per day; taxes not included. **15 amp power at all slips, $4/night. Register at the bar or the café. Fresh drinking water is available upon request from a hose bib at the foot of the ramp. Trash can be dropped off for $2/small bag, $3/large bag. The on-site store sells groceries, convenience items, and tackle. Activities include a pool table, heated pool, tennis court, and a large open grassy area for get-togethers. This is one of the few British Columbia marinas that has a full-service liquor store on-site. New Wi-Fi available throughout the facility.

Notable -- The rustic wood-paneled marina café and pub offer good food at reasonable prices every day from 7am - with live music on Friday and Saturday nights during the summer. The on-site swimming pool and adjacent, popular Hamilton Beach offer a welcome choice of bathing options. A nicely maintained campground makes this a popular stop for RVers as well as boaters; self-contained cabins or rooms with shared facilities in the "mini-mansion" are available for a night ashore. Driftwood Centre with its wide variety of services and supplies is less than a quarter mile walk. Penderites take great pride in the sensitive ecology of their island and were among the first communities in Canada to institute recycling.

Navigational Information
Lat: 44°48.831' **Long:** 123°13.639' **Tide:** 12 ft. **Current:** n/a **Chart:** 3477
Rep. Depths *(MLW)*: **Entry** 30 ft. **Fuel Dock** 40 ft. **Max Slip/Moor** 50 ft./-
Access: Located on South Pender Island in Bedwell Harbour

Marina Facilities *(In Season/Off Season)*
Fuel: Gasoline, Diesel, High-Speed Pumps
Slips: 110 Total, 110 Transient **Max LOA:** 100 ft. **Max Beam:** n/a
  Rate *(per ft.)*: **Day** $1.40/0.80* **Week** n/a **Month** n/a
  Power: 30 amp $4-8** , **50 amp** n/a, **100 amp** n/a, **200 amp** n/a
  Cable TV: No **Dockside Phone:** No
  Dock Type: Floating, Long Fingers, Alongside, Wood
Moorings: 0 Total, 0 Transient **Launch:** n/a, Dinghy Dock
  Rate: Day n/a **Week** n/a **Month** n/a
Heads: 7 Toilet(s), 6 Shower(s) *(with dressing rooms)*
Laundry: 3 Washer(s), 3 Dryer(s) **Pay Phones:** Yes, 1
Pump-Out: No **Fee:** n/a **Closed Heads:** Yes

Marina Operations
Owner/Manager: Tara Hodgins **Dockmaster:** Emily Guenette
In-Season: Jun-Sep, 8am-8pm **Off-Season:** Oct-May, 8am-6pm
After-Hours Arrival: Call ahead
Reservations: Yes, Preferred **Credit Cards:** Visa/MC, Amex, Debit
Discounts: None
Pets: Welcome, Dog Walk Area **Handicap Access:** Yes, Heads, Docks

Marina Services and Boat Supplies
Services - Docking Assistance, Room Service to the Boat, Boaters' Lounge, Security, Trash Pick-Up, Dock Carts **Communication -** Mail & Package Hold, Phone Messages, Fax in/out, Data Ports *(Wi-Fi - Broadband Xpress)*, FedEx, UPS, Express Mail *(Sat Del)* **Supplies - OnSite:** Ice *(Block, Cube)*, Ships' Store, Bait/Tackle

Boatyard Services
OnSite: Bottom Cleaning, Brightwork, Divers, Compound, Wash & Wax, Interior Cleaning **OnCall:** Engine mechanic *(gas, diesel)* **3+ mi:** Launching Ramp *(3.5 mi.)*. **Nearest Yard:** Canoe Cove Marina (250) 656-5566

Restaurants and Accommodations
OnSite: Restaurant *(Aurora Restaurant 629-2115, D $27-36, reservations recommended)*, *(Syrens 629-2114, L $10-16, D $14-29)*, Condo/Cottage *(Poets Cove Seaside Resort 629-2100, $150-800, hotel rooms, lodges, cottages, and villas)* **3+ mi:** Restaurant *(Le Pistou Grill 629-3131, 7 mi.)*, Hotel *(The Inn at Pender Island 629-3353, $79-139, 7 mi.)*, Condo/Cottage *(Mount Norman Hideaway 4 mi., $650-850/wk; 2-bedroom cottage)*

Recreation and Entertainment
OnSite: Heated Pool *(free for guests; water workout $5)*, Spa, Beach, Picnic Area, Tennis Courts, Fitness Center *(daily 5am-10pm; yoga, Pilates, and fitness retreats available)*, Boat Rentals *(boats and kayaks)*, Group Fishing Boat *(dep sea salmon fishing)*, Sightseeing *(3 hr. guided Kayak Tours $50-60pp, 2p min.; Evening Island Tour around Penders $35pp, guided Cycling Tour on So. Pender $40/1.5 hrs.)* **Near:** Jogging Paths, Roller Blade/Bike Paths, Park *(Beaumont Marine Park, year-round - mooring buoys near*

Skull Islet; Mt. Norman Park - 30 min. hike to summit for panoramic views)
3+ mi: Golf Course *(Pender Island Golf Course 629-6659, 10 mi.)*, Cultural Attract *(festivals and concerts at No. Pender Community Hall in summer, 10 mi.)*, Galleries *(Blood Star 629-6661, Galway 629-3176, 7 mi.)*

Provisioning and General Services
OnSite: Market *(Moorings Market 9am-4pm: groceries, deli with sandwiches, drinks & ice cream, clothing)*, Wine/Beer *(Cold Beer & Wine Store 9:30am-3:30p)*, Bank/ATM, Newsstand, Copies Etc. **3+ mi:** Supermarket *(Tru Value 628-8322, 9am-8:30pm, delivers, 7 mi.)*, Liquor Store *(Driftwood Centre 629-3413, 7 mi.)*, Bakery *(Pender Island 629-6453, 7 mi.)*, Farmers' Market *(Sat at Driftwood Centre, 7 mi.)*, Post Office *(629-3222, 7 mi.)*, Library *(Pender Lender 629-3722, 10 mi.)*, Bookstore *(Tallisman 629-6944, 7 mi.)*, Pharmacy *(Pender Isl. 629-6555, 7 mi.)*, Hardware Store *(Home Hardware 629-3455, 10 mi.)*

Transportation
OnSite: Courtesy Car/Van *($10/pp each way)*, Bikes *(rentals $10/hr., $15/2 hrs., $30/4 hrs., $40/8 hrs.)*, Rental Car *(vans $25/half-day, $50/full day)* **OnCall:** Water Taxi, Taxi *(Pender Taxi 629-3555)* **3+ mi:** Ferry Service *(BC Ferries to Islands, Sidney & Vancouver 888-223-3779, 11 mi.)* **Airport:** Kenmore Air & Harbour Air/ YVR *(On-Site/20 mi.)*

Medical Services
911 Service **OnSite:** Holistic Services *(Susurrus Spa - classic services and signature treatments using clay, seaweed, mud and herbs)* **OnCall:** Ambulance **3+ mi:** Doctor *(Medical Clinic 629-3233, 8 mi.)*, Dentist *(Nord 629-6815, 9 mi.)* **Hospital:** Saanich Peninsula 652-3911 *(20 mi.)*

Poets Cove Resort & Spa

9801 Spalding Road RR 3; South Pender Island, BC V0N 2M3

Tel: (250) 629-2111; (866) 888-2683 **VHF: Monitor** Ch. 66A **Talk** n/a
Fax: (250) 629-2110 **Alternate Tel:** (250) 629-2100
Email: marina@poetscove.com **Web:** www.poetscove.com
Nearest Town: Driftwood Centre *(7 mi.)* **Tourist Info:** (866) 468-7924

Setting -- Rounding Tilly Point, the gorgeous new facilities of Poets Cove are a surprise. Sprawled on a perfectly manicured hillside above Bedwell Harbour, the 3-story brick-red shingle, stone and glass Arts & Craft style lodge is the centerpiece of an amenity-rich, impeccably turned out resort. Part of the new Gulf Islands National Park, the resort and the network of high-end docks are protected by the southern "arms" of North and South Pender Island.

Marina Notes -- *$1.40/ft. Jul-Sep 4, $1.25 May-Jun & Sep 5-30, $1 Mar-Apr, $0.80 Oct-Feb, Holida week-ends $1.45. Overflow dock $1/ft. - no power or water, but full access to all amenities. **15 & 30 amp power $4/40', $6/41-60', $8/60' +. Opened in May 2004. Reserve early. Floating docks have Ipe decking and high-end pedestals. Docks & power are being upgraded to megayacht needs. A guest of the marina is a guest of the resort. Book meals, spa services, and activities with marina staff; your "slip" account will be billed. On-site customs office. Well-supplied market, laundry and granite & tile bathhouse.

Notable -- This is clearly a destination marina. There's a dining room, pub, full-service spa, activity center, two pools, hot tub, tennis courts, playground, basketball, and more. An activities manager offers a comprehensive program: guided biking, kayaking, diving, hiking (to Mt. Norman), sailing, ecotours, plus kids' activities - sandcastle building, snorkeling, treasure hunts, and swim lessons. At Aurora, enjoy signature dishes inspired by fresh local ingredients. Sirens Lounge serves lunch and dinner from 11am-11pm Sun-Thu, 11am-12pm Fri & Sat. For a night ashore, choose one of 22 luxurious lodge rooms featuring decks with glass wind breaks or 15 intimate, fully-equipped 2 & 3 bedroom cottages or 9 villas. The Seaglass Ballroom accommodates up to 120 with fabulous views.

Boaters' Notes

_____
_____
_____
_____
_____
_____
_____
_____
_____
_____
_____
_____
_____
_____
_____
_____
_____
_____
_____
_____

Add Your Ratings and Reviews at www.AtlanticCruisingClub.com

A complimentary six-month Silver Membership is included with the purchase of this _Guide_. Select "Join Now," then "Silver," then follow the instructions.

The AtlanticCruisingClub website provides updated Marina Reports, Destination and Harbor Articles and much more — including an option within each online Marina Report for boaters to add their ratings and comments regarding that facility. Please log on frequently to share your experiences — and to read other boaters' comments.

On the website, boaters may rate marinas in one or more of the following categories — on a scale of 1 (basic) to 5 (world class) — and also enter additional commentary.

▸ **Facilities & Services** (Fuel, Reservations, Concierge Services and General Helpfulness)

▸ **Amenities** (Pool, Beach, Internet Access, including Wi-Fi, Picnic/Grill Area, Boaters' Lounge)

▸ **Setting** (Views, Design, Landscaping, Maintenance, Ship-Shapeness, Overall Ambiance)

▸ **Convenience** (Access — including delivery services — to Supermarkets, other Provisioning Sources, Shops, Services, Attractions, Entertainment, Sightseeing, Recreation, including Golf and Sport Fishing & Medical Services)

▸ **Restaurants/Eateries/Lodgings** (Availability of Fine Dining, Snack Bars, Lite Fare, OnCall food service, and Lodgings ashore)

▸ **Transportation** (Courtesy Car/Vans, Buses, Water Taxis, Bikes, Taxis, Rental Cars, Airports, Amtrak, Commuter Trains)

▸ **Please Add Any Additional Comments**

7. BC – Saanich Peninsula & Victoria

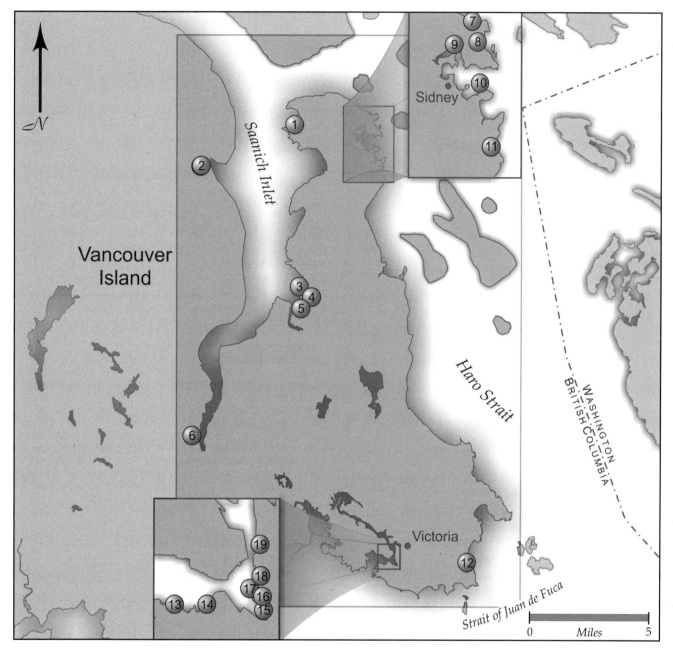

MAP	MARINA	HARBOR	PAGE	MAP	MARINA	HARBOR	PAGE
1	Deep Cove Marina	Saanich Inlet/Deep Cove	112	11	Port Sidney Marina	Sidney Harbour	122
2	Mill Bay Marina	Saanich Inlet/Mill Bay	113	12	Oak Bay Marina	Oak Bay	123
3	Compass Rose Cabins & Marina	Saanich Inlet/Brentwood Bay	114	13	Fisherman's Wharf	Victoria Middle Harbour	124
4	Brentwood Bay Lodge & Spa Marina	Saanich Inlet/Brentwood Bay	115	14	Coast Harbourside Hotel/Marina	Victoria Middle Harbour	125
5	Butchart Gardens	Saanich Inlet/Brentwood Bay	116	15	James Bay Causeway	Victoria Inner Harb./James Bay	126
6	Goldstream Boathouse Marina	Saanich Inlet/Finlayson Arm	117	16	Ship Point Wharf	Victoria Inner Harb./James Bay	127
7	Swartz Bay Public Wharf	Swartz Bay	118	17	Broughton Street Wharf	Victoria Inner Harbour	128
8	Canoe Cove Marina	Canoe Cove	119	18	Wharf Street Marina	Victoria Inner Harbour	129
9	Westport Marina	Tsehum Harbour	120	19	Johnson Street Public Wharf	Victoria Inner Harbour	130
10	Van Isle Marina	Tsehum Harbour	121				

Deep Cove Marina

PO Box ; 10990 Madrona Drive; North Saanich, BC V8L 5R7

Tel: (250) 656-0060 **VHF: Monitor** n/a **Talk** n/a
Fax: n/a **Alternate Tel:** n/a
Email: n/a **Web:** n/a
Nearest Town: Sidney *(6 mi.)* **Tourist Info:** (250) 656-0525

Navigational Information
Lat: 48°40.763' **Long:** 123°28.669' **Tide:** 10 ft. **Current:** n/a **Chart:** 3441
Rep. Depths *(MLW):* **Entry** 10 ft. **Fuel Dock** n/a **Max Slip/Moor** 10 ft./-
Access: In Deep Bay off Saanich Inlet

Marina Facilities *(In Season/Off Season)*
Fuel: No
Slips: 60 Total, 2 Transient **Max LOA:** 100 ft. **Max Beam:** n/a
  Rate *(per ft.):* **Day** $0.75/0.60 **Week** n/a **Month** n/a
  Power: 30 amp $5*, 50 amp $5, 100 amp n/a, 200 amp n/a
  Cable TV: No **Dockside Phone:** No
  Dock Type: Floating, Alongside, Wood
Moorings: 0 Total, 0 Transient **Launch:** n/a
  Rate: Day n/a **Week** n/a **Month** n/a
Heads: 1 Toilet(s), 1 Shower(s)
Laundry: None **Pay Phones:** Yes, 1
Pump-Out: OnCall **Fee:** Donations **Closed Heads:** Yes

Marina Operations
Owner/Manager: Bob Philipchalk **Dockmaster:** Michael Lang
In-Season: Year-Round **Off-Season:** n/a
After-Hours Arrival: Call ahead
Reservations: Yes, Preferred **Credit Cards:** No
Discounts: None
Pets: Welcome **Handicap Access:** No

Marina Services and Boat Supplies
Services - Security *(24/7, owner onsite)*, Dock Carts **Supplies - 1-3 mi:** Ice *(Cube)* **3+ mi:** Ships' Store *(Sidney Marine Supply, 6 mi.)*, West Marine *(654-0045, 6 mi.)*

Boatyard Services
3+ mi: Engine mechanic *(gas, diesel)*, Electrical Repairs *(4 mi.)*, Rigger *(4 mi.)*, Canvas Work *(4 mi.)*, Woodworking *(4 mi.)*, Upholstery *(4 mi.)*, Metal Fabrication *(4 mi.)*, Painting *(4 mi.)*. **Nearest Yard:** Philbrook's Boatyard (250) 656-1157

Restaurants and Accommodations
Under 1 mi: Restaurant *(Deep Cove Chalet 656-3541, L $15-35, D $26-55)*, Inn/B&B *(Deep Cove Chalet 656-3541, $325)* **1-3 mi:** Inn/B&B *(Alder Road Retreat 655-6378, $99-109)* **3+ mi:** Restaurant *(Odyssia Steakhouse 656-5596, 6 mi.)*, *(Taste of Tokyo 656-6862, 6 mi.)*, *(Homestead Café 655-1844, 4 mi.)*, Seafood Shack *(Fish on Fifth 656-4022, 6 mi.)*, Hotel *(Best Western 656-4441, 6 mi.)*, Inn/B&B *(Resthaven by the Sea 656-7510, 4 mi.)*

Recreation and Entertainment
1-3 mi: Picnic Area, Playground *(Toddler Park/Wain Park)*, Tennis Courts *(Wain Park)*, Park *(Denham Hill Park)* **3+ mi:** Golf Course *(Ardmore Golf Course 656-4621, 5 mi.)*, Fitness Center *(Body Barn 655-3393, 6 mi.)*, Movie Theater *(Star Cinema 655-1171, 6 mi.)*, Video Rental *(Video Express 656-3666, 6 mi.)*, Museum *(Marine Museum Sidney 656-1322, Historical Museum Sidney 655-6355, 6 mi.)*, Cultural Attract *(Charlie White Theatre 656-0271, 6 mi.)*, Sightseeing *(Chalet Estate Winery 656-2552, 1 mi.)*, Special Events

(Artisans at Mary Winspear Centre Jul-Aug, 6 mi.), Galleries *(Pottery Plus 656-7687, Village Gallery 656-3633, 6 mi.)*

Provisioning and General Services
1-3 mi: Convenience Store *(Deep Cove Store 656-2547)*, Wine/Beer, Bank/ATM, Protestant Church *(St. John's United Church 655-3043)* **3+ mi:** Market *(Thrifty Foods 656-0946, 6 mi.)*, Supermarket *(Safeway 656-1721, 6 mi.)*, Delicatessen *(Lunn's Pastries Deli & Coffee 656-1724, 6 mi.)*, Liquor Store *(BC Liquor 656-3041, 6 mi.)*, Bakery *(Big Rock Bakery Co. 655-3633, 6 mi.)*, Post Office *(6 mi.)*, Catholic Church *(5 mi.)*, Library *(North Saanich 656-0944, 5 mi.)*, Beauty Salon *(Gaea Salon & Spa 655-9377, 6 mi.)*, Dry Cleaners *(Sidney Professional 656-9555, 6 mi.)*, Bookstore *(Beacon 655-4447, Compass Rose Nautical 656-4674, 6 mi.)*, Pharmacy *(Pharma-save 656-1148, 6 mi.)*, Hardware Store *(Home Hardware 656-2712, 6 mi.)*, Clothing Store *(Hypersport Activewear 656-6161, 6 mi.)*, Retail Shops *(Sidney Sporting Goods 656-9255, Sidney Gift Shoppe 656-3232, 6 mi.)*

Transportation
OnCall: Taxi *(Peninsula Taxi 656-1111)* **Near:** InterCity Bus **1-3 mi:** Ferry Service *(BC Ferries 656-0757)* **3+ mi:** Rental Car *(Avis 656-6033, Hertz 656-2312, 5 mi.)* **Airport:** Victoria Int'l. *(5 mi.)*

Medical Services
911 Service **OnCall:** Ambulance **1-3 mi:** Dentist *(Novosad 656-2519)* **3+ mi:** Doctor *(Yam 655-2922, 4 mi.)*, Chiropractor *(Mill Bay 743-2170, 6 mi.)*, Holistic Services *(Blessed Organics 885-9097, 6 mi.)*, Veterinarian *(Sidney 656-3333, 6 mi.)* **Hospital:** Saanich Peninsula 652-3911 *(10 mi.)*

Setting -- Deep Cove is tucked in behind Moses Point at the entrance to Saanich Inlet. Look for Wain Rock when entering from Satellite Channel. Immediately to the left of the remnants of an old government dock is rustic Deep Cove Marina. A set of nicely maintained wooden docks offering alongside tie-ups radiates from one side of the main pier; the other side hosts side-tie larger vessels and smaller boats held away from the docks by whips. The surrounding area is rural and residential.

Marina Notes -- *15 amp power is available at $2/day. The new owner lives on-site. Quite busy during the summer months - call ahead; there's usually lots of room during the winter. Some boat repairs can be done here. Heads are basic at best. Pumpty Dumpty mobile pump-out serves the inlet: 250-480-9292.

Notable -- The beautiful area around Deep Cove is mostly residential and offers few boater amenities, but the calm waters of Deep Bay invite exploration by canoe, kayak, or dinghy. Denham Till Park, on Birch Road, has a playground and trails through the forest. Renowned Deep Cove Chalet Restaurant is around the bay, about one mile from the marina. Chef Pierre Koffel's superb dishes are matched by an amazing wine list; the extensive menu includes prix fixe meals for $60-95. Along the same road is the Chalet Estate Winery which welcomes visitors for tastings (656-2552). A First Nations reservation and the Victoria Int'l. Airport are a few miles south of the marina, separating Deep Cove from Sidney. Major rental car companies and some restaurants can be found at the airport. A short cab ride or local bus takes you to downtown Sidney for provisioning and entertainment.

Navigational Information
Lat: 48°39.100' **Long:** 123°33.100' **Tide:** 10 ft. **Current:** n/a **Chart:** 3441
Rep. Depths (*MLW*): Entry 20 ft. Fuel Dock 20 ft. Max Slip/Moor 18 ft./-
Access: On the western side of Saanich Inlet in Mill Bay

Marina Facilities (*In Season/Off Season*)
Fuel: Gasoline, Diesel
Slips: 150 Total, 15 Transient **Max LOA:** 100 ft. **Max Beam:** n/a
 Rate (*per ft.*): **Day** $1.00* **Week** n/a **Month** n/a
 Power: 30 amp $2** , **50 amp** n/a, **100 amp** n/a, **200 amp** n/a
 Cable TV: No **Dockside Phone:** No
 Dock Type: Floating, Long Fingers, Pilings, Alongside, Wood
Moorings: 0 Total, 0 Transient **Launch:** n/a, Dinghy Dock ($5)
 Rate: Day n/a **Week** n/a **Month** n/a
Heads: 4 Toilet(s), 2 Shower(s)
Laundry: 1 Washer(s), 1 Dryer(s) **Pay Phones:** Yes, 1
Pump-Out: OnCall **Fee:** Donation **Closed Heads:** Yes

Marina Operations
Owner/Manager: Fred & Marilyn Laba **Dockmaster:** Same
In-Season: May-Sep, 9am-6pm **Off-Season:** Oct-Apr, 9:30am-3pm
After-Hours Arrival: Call ahead
Reservations: Yes, Preferred **Credit Cards:** Visa/MC
Discounts: None
Pets: Welcome **Handicap Access:** No

Mill Bay Marina

PO Box 231; 740 Handy Road; Mill Bay, BC V0R 2P0

Tel: (250) 743-4112; (800) 253-4112 **VHF: Monitor** Ch. 66A **Talk** n/a
Fax: (250) 743-4122 **Alternate Tel:** n/a
Email: millbaymarina@shaw.ca **Web:** www.millbaymarina.com
Nearest Town: Mill Bay (0.5 mi.) **Tourist Info:** (250) 743-3566

Marina Services and Boat Supplies

Services - Docking Assistance, Security (*24 hr., video, owners onsite*), Trash Pick-Up, Dock Carts **Communication -** FedEx, UPS, Express Mail
Supplies - OnSite: Ice (*Cube*), Ships' Store, Bait/Tackle

Boatyard Services
OnSite: Launching Ramp ($7) **Nearest Yard:** Philbrook's Boatyard Ltd. (250) 656-1157

Restaurants and Accommodations
Near: Restaurant (*Catrina's Grill 743-7277, B, L, D*), Snack Bar (*Café Rusticana, mochas, coffee, and ice cream*), Pizzeria (*Mill Bay Pizza 743-8882, will deliver*) **1-3 mi:** Restaurant (*La Pommeraie Bistro 743-4293, L $11-15, D $20-40, at Merridale Estate Cidery; L Mon-Sat 11:30am-4pm, D Fri-Sat after 5pm, Sun brunch $11-15*), (*Friday's Restaurant 743-5533, pizza*), Fast Food (*McDonald's, Subway*), Motel (*Deer Lodge 743-2423*), Condo/Cottage (*Rosebank Oceanfront Cottages 743-5541, From $85*)
 3+ mi: Restaurant (*Steeples 743-1887, L $8-15, D $18-36, 4 mi., L Fri-Sun, brunch 10am-3pm Sat & Sun $8-13*)

Recreation and Entertainment
OnSite: Picnic Area **Near:** Tennis Courts, Video Rental (*Pioneer Video and Music 743-5522, Movie Gallery 743-4468*), Galleries **Under 1 mi:** Beach (*Mill Bay Beach or Bamberton Park Beach, over 700 ft. long*), Fitness Center (*Island Fitness 743-7600*) **1-3 mi:** Playground (*sports fields plus skating arena at Kerry Park Recreation Centre 743-5922*), Sightseeing (*Merridale Estate Cidery 800-998-9908; wineries*) **3+ mi:** Golf Course (*Arbutus Ridge

743-5000, 8 mi.*), Park (*Bamberton Provincial Park: picnic area, hiking, kayaking, 3 mi.*), Special Events (*annual Art-a-Fair - Aug*)

Provisioning and General Services
Near: Convenience Store (*Select Food Store*), Supermarket (*Thrifty Foods 743-3235*), Health Food (*Red Apple 743-7888*), Wine/Beer (*Valley Vines to Wines 743-4647*), Liquor Store (*BC Liquor 743-2161*), Bank/ATM, Post Office, Library (*Mill Bay 743-5436*), Dry Cleaners, Pharmacy (*Mill Bay Pharmacy 743-6679*), Clothing Store (*Needs and Desires 743-3531, Wear It's At 743-8914*) **Under 1 mi:** Beauty Salon (*Jada Hairstylists 743-4442*) **1-3 mi:** Delicatessen (*The Dutch Deli & Meats 743-4648/ Country Grocer 743-5600, 5 mi.*), Protestant Church (*Sylvan United 743-4659, St. John Anglican 743-3095*) **3+ mi:** Market (*Westside Market 743-1577, 4 mi.*), Bakery (*Island Bakery Limited 743-4244, 7 mi.*), Laundry (*Your Time Saved Laundromat 743-6902, 7 mi.*)

Transportation
OnCall: Taxi (*Cobble Hill Taxi 743-5555*), Airport Limo (*A-1 Airport Taxi 812-8900*) **3+ mi:** Ferry Service (*to Brentwood Bay, 3 mi.*) **Airport:** Victoria Int'l. (*27 mi.*)

Medical Services
911 Service **OnCall:** Ambulance **Near:** Doctor (*The Medical Centre 743-3211*), Dentist (*Sally 743-3112*), Optician (*Mill Bay Eye Care 743-8899*) **Under 1 mi:** Chiropractor (*Mill Bay Chiropractic 743-2170*) **3+ mi:** Holistic Services (*Choice Holistic Services 743-4243, 7 mi.*), Veterinarian (*Cobble Hill 743-4322, 8 mi.*) **Hospital:** Cowichan District 746-4141 (*12 mi.*)

Setting -- Dense woods isolate the marina from the nearby bustling community of Mill Bay. The peaceful setting and relaxed atmosphere are matched by a simple approach from Saanich Inlet. Seals take the sun on the floating breakwater, otters frequent the docks, a pair of Muscovy Ducks makes the marina home, and each spring baby seals are born on the floats. A single main pier hosts four docks with slips on one side and guest side-tie dockage on the other.

Marina Notes -- *Stay-aboard rates - based on length of slip: less than 30' $25 flat rate/night; 30' + $1/ft./night. Not staying on board: 20-24' $12 flat rate/night; over 24' $0.50/ft./night. **15 amp power $2/night. Mill Bay Marina is known for its prawning trip set-ups and advice - a fish cleaning station keeps the mess off the boat. The store at the end of the pier carries lots of fishing tackle, bait, groceries, ice and a wide variety of "other stuff." The coffee pot is always on. Say hello to Dock Duck, Daddy Duck and Mrs. Duck. A 14-site RV park is also on the property. Heads are up the hill from the docks.

Notable -- Mill Bay Centre is a short quarter-mile walk; it has banks, a supermarket, pharmacy, liquor store, restaurant, pizza takeout, clothing stores, a video rental and a dollar store. If you need help carrying all your purchases, talk to Fred or Marilyn about borrowing a cart. Café Rusticana is a great place to take time off & enjoy mochas or ice cream. Merridale Ciderworks, less than 2 miles away, offers self-guided or private estate tours, tastings, and reportedly fabulous meals at the new Pommerarie Bistro. Specials include a 3-course lunch with an orchard and facilities tour, plus a tasting, $30 (800-998-9908, reserve ahead). Picnic tables with views of the Saanich Peninsula & Mount Baker are at the Bamberton Provincial Park near the ferry dock.

Compass Rose Marina

799 Verdier Avenue; Brentwood Bay, BC V8M 1C5

Tel: (250) 544-1441 **VHF: Monitor** n/a **Talk** n/a
Fax: (250) 544-1015 **Alternate Tel:** n/a
Email: compassrosecabins@shaw.ca **Web:** www.compassrosecabins.com
Nearest Town: Brentwood Bay *(1 mi.)* **Tourist Info:** (250) 656-3260

Navigational Information

Lat: 48°34.661' **Long:** 123°28.012' **Tide:** 10 ft. **Current:** n/a **Chart:** 3441
Rep. Depths (*MLW*): Entry 10 ft. **Fuel Dock** n/a **Max Slip/Moor** 10 ft./-
Access: Located on the east side of Saanich Inlet

Marina Facilities *(In Season/Off Season)*

Fuel: No
Slips: 9 Total, 2 Transient **Max LOA:** 80 ft. **Max Beam:** n/a
  Rate *(per ft.)*: **Day** $1.00 **Week** n/a **Month** $5
  Power: 30 amp Incl., **50 amp** n/a, **100 amp** n/a, **200 amp** n/a
  Cable TV: No **Dockside Phone:** No
  Dock Type: Floating, Long Fingers, Alongside, Wood
Moorings: 0 Total, 0 Transient **Launch:** n/a
  Rate: Day n/a **Week** n/a **Month** n/a
Heads: 2 Toilet(s)
Laundry: None **Pay Phones:** Yes, 1
Pump-Out: OnCall *(250-480-9292)* **Fee:** Donation **Closed Heads:** Yes

Marina Operations

Owner/Manager: Pierre and Linda Picot **Dockmaster:** Same
In-Season: Year-Round, 24/7 **Off-Season:** n/a
After-Hours Arrival: Call ahead
Reservations: Yes, Preferred **Credit Cards:** Visa/MC, Debit
Discounts: None
Pets: Welcome **Handicap Access:** No

Marina Services and Boat Supplies

Services - Docking Assistance, Security *(owners onsite)*, Trash Pick-Up
Communication - Mail & Package Hold, Phone Messages, Fax in/out,
FedEx, UPS, Express Mail *(Sat Del)* **Supplies - Near:** Ice *(Shaved)*
3+ mi: West Marine *(654-0045, 10 mi.)*

Boatyard Services

Nearest Yard: Philbrook's Boatyard Ltd. (250) 656-1157

Restaurants and Accommodations

OnSite: Restaurant *(Sea Horses Café 544-1565, B $8-10, L $9-11,*
 D $15-22), Condo/Cottage *(Compass Rose Cabins 544-1441, $115-169)*
Near: Restaurant *(The Orient 652-2203, L $5-9, D $7-20, free del),*
(Aubutus Grille & Wine Bar 544-5100, D $15-40, Sun brunch), (Marine Pub
& Café 544-5102, B $5-17, L $10-18, D $10-25), (Jacky's Beach House
Caribbean 652-1554), Hotel *(Brentwood Bay Lodge and Marina 544-2079,*
$119-359) **Under 1 mi:** Restaurant *(Piccolo 652-5044, free del, $14*
specials)*, Lite Fare*(Zesto's Subs & Wraps 544-1833)*, Pizzeria *(Pizazz*
Gourmet Pizza 652-0109) **3+ mi:** Hotel *(Best Western 656-4441, 10 mi.)*

Recreation and Entertainment

OnSite: Boat Rentals *(Kayaks 544-1565)* **Near:** Picnic Area, Galleries
(quality artworks displayed in Brentwood Bay Lodge) **Under 1 mi:** Video
Rental *(Video Shop 652-5411)*, Park, Cultural Attract *(Central Saanich*
Cultural Center 605-6032) **1-3 mi:** Sightseeing *(Butchart Gardens 866-387-
8480)* **3+ mi:** Tennis Courts *(Panorama Recreation Centre 656-7271, 6*
mi.), Golf Course *(Glen Meadows Golf and Country Club 656-3921, 6 mi.)*,
Fitness Center *(Gold's Gym 652-5444, 4 mi.)*, Movie Theater *(Star Cinema*

655-3384, 10 mi.), Museum *(Sidney: Marine Ecology Centre 655-1555,*
Aviation Museum 655-3300, Marine Museum 656-1322, 10 mi.)

Provisioning and General Services

Near: Market *(Moodyville General Store 652-2081 - across the street)*,
Delicatessen *(Brentwood Lodge or Brentwood Bay Delicatessen and Bakery*
652-4623, 1 mi.), Bakery **Under 1 mi:** Supermarket *(True Value*
Food/Thrifty Foods 483-1600, 4 mi.), Health Food *(Grass Roots Health and*
Natural Foods 544-1718), Wine/Beer, Liquor Store *(BC Liquor 652-1212)*,
Bank/ATM, Post Office, Protestant Church *(Shady Creek United 652-2713,*
Baptist), Library *(Greater Victoria Public Library 652-2013)*, Beauty Salon
(Brentwood Coiffures 652-3333), Barber Shop *(Brentwood Barber 652-
6111)*, Dry Cleaners *(Fay's 652-5811)*, Laundry *(Brentwood Bay Laundromat*
652-8661), Bookstore *(Pages Used Books 652-4341/ Compass Rose*
Nautical Books 656-4674, 10 mi.), Pharmacy *(Rexall Drug Stores 652-8813)*,
Retail Shops **1-3 mi:** Green Market *(Peninsula Co-op 652-1188)*
3+ mi: Hardware Store *(Home Hardware 652-9119, 4 mi.)*

Transportation

OnCall: Taxi *(656-5588)* **Near:** Local Bus *(BC Transit 382-6161)*, Ferry
Service *(BC Ferries to Mill Bay 386-3431)* **Airport:** Victoria Int'l. *(10 mi.)*

Medical Services

911 Service **OnCall:** Ambulance **Near:** Holistic Services *(Brentwood Bay*
Lodge and Spa 544-2079) **Under 1 mi:** Doctor *(Family Medical 652-9191)*,
Dentist *(Brentwood Dental Clinic 652-3723)*, Chiropractor *(Brentwood Bay*
Chiropractic 652-5211), Veterinarian *(Brentwood Bay Veterinary Hospital*
652-3131) **Hospital:** Saanich Peninsula Hospital 652-3911 *(4 mi.)*

Setting -- Hard alongside the Mill Bay BC Ferry dock is diminutive Compass Rose Marina. The green-roofed Sea Horses Café and four cabins sit on pilings above the water - just up the ramp from the floatng docks. A deck populated with tables and chairs topped by colorful umbrellas rings the eatery. Flower boxes and hanging plants give the facilitiy a festive air. Note: The fjord-like Squally Reach and Finlayson Arm are a short distance to the south.

Marina Notes -- This is a small facility with limited guest moorage. The owners live on-site and they and their staff will help boaters get settled whether in the marina or in one of the charming cabins overlooking the water. Moorage is free for boaters staying in a cabin. The Picots also operate a large catamaran in Alaska during the summer and it takes up most of the dock during the winter. The simple heads are shared with the restaurant. There are no showers, laundry, fuel or pump-out on-site, but a laundromat is a short walk up the hill. For peace and quiet, this is a good place to target.

Notable -- The airy Sea Horses Café features West Coast cuisine and a garage-door-like wall that opens the pretty interior to the water. Adjacent to the docks, the four Compass Rose Cabins roost on pilinings above the water - contemporary interiors with kitchens, fireplaces, lofts, log-railed stairways, and a sundeck overlooking the docks make for a quiet respite ashore. Take your dinghy or rent a kayak from the Café and explore Butchart Gardens and Tod Inlet, or have a picnic on the beach (kayak rates: single 2 hrs. $20, 4 hrs. $35, 8 hrs. $49; double 2 hrs. $30, 4 hrs. $45, 8 hrs. $59). Right across the street, the acclaimed Brentwood Bay Lodge and Spa offers fine dining amid fine art. An easy walk up the hill is the community of Brentwood Bay.

Navigational Information
Lat: 48°34.623' **Long:** 123°27.940' **Tide:** 10 ft. **Current:** n/a **Chart:** 3441
Rep. Depths *(MLW)*: **Entry** 10 ft. **Fuel Dock** n/a **Max Slip/Moor** 8 ft./-
Access: Senanus Is. guards the entrance to Brentwood Bay off Saanich Inlet

Marina Facilities *(In Season/Off Season)*
Fuel: No
Slips: 50 Total, 25 Transient **Max LOA:** 100 ft. **Max Beam:** n/a
Rate *(per ft.)*: **Day** $1.50/1.00 **Week** n/a **Month** $10
Power: 30 amp $6*, **50 amp** $8, **100 amp** n/a, **200 amp** n/a
Cable TV: Yes **Dockside Phone:** No
Dock Type: Floating, Long Fingers, Alongside, Wood
Moorings: 0 Total, 0 Transient **Launch:** n/a, Dinghy Dock
Rate: Day n/a **Week** n/a **Month** n/a
Heads: 4 Toilet(s), 4 Shower(s)
Laundry: 1 Washer(s), 1 Dryer(s) **Pay Phones:** Yes
Pump-Out: OnCall **Fee:** Donation **Closed Heads:** Yes

Marina Operations
Owner/Manager: Matthew Smiley **Dockmaster:** Greg Duerksen
In-Season: Summer, 8am-8pm **Off-Season:** Fall & Spring, 9am-5pm**
After-Hours Arrival: Call ahead
Reservations: Yes, Preferred **Credit Cards:** Visa/MC, Amex
Discounts: None
Pets: Welcome, Dog Walk Area **Handicap Access:** Yes, Heads, Docks

Brentwood Bay Lodge Marina
849 Verdier Avenue; Brentwood Bay, BC V8M 1C5
Tel: (250) 652-3151; (888) 544-2079 **VHF: Monitor** Ch. 66A **Talk** n/a
Fax: (250) 544-2069 **Alternate Tel:** n/a
Email: marina@brentwoodbaylodge.com **Web:** brentwoodbaylodge.com
Nearest Town: Brentwood Bay *(1 mi.)* **Tourist Info:** (250) 656-3260

Marina Services and Boat Supplies
Services - Docking Assistance, Security, Trash Pick-Up, Dock Carts, Megayacht Facilities **Communication -** Data Ports *(Wi-Fi - Broadband Xpress)*, FedEx, UPS, Express Mail *(Sat Del)* **Supplies - Near:** Ice *(Block, Cube)*, Propane

Boatyard Services
Nearest Yard: Philbrooks' Boatyard Ltd. (250) 651-1157

Restaurants and Accommodations
OnSite: Restaurant *(Marine Pub & Café 544-5102, B $5-17, L $10-18, D $10-25, comp. wine tasting Sat 3-4pm; B from 7am)*, *(Arbutus Grille & Wine Bar 544-5100, D $15-40, 4-course tasting $39, 5-course seafood $54; Mon &Tue salmon BBQ & crab $39, Sun brunch)*, Hotel *(Brentwood Bay Lodge & Spa 544-2079, $149-509)* **Near:** Restaurant *(The Orient 652-2203, L $5-9, D $7-20, delivery)*, Seafood Shack *(Sea Horse Café 544-1565, B $8-10, L $9-11, D $15-22)*, Condo/Cottage *(Compass Rose Cabins 655-1441, $115-143)* **Under 1 mi:** Restaurant *(Brentwood Chinese 652-1812, L $6-18, D $6-18, delivery)*, *(Blue's Bayou 544-1194)*, Pizzeria *(Pizazz Gourmet 652-0109)*, Inn/B&B *(B.B. Heritage House 652-2012)*

Recreation and Entertainment
OnSite: Heated Pool, Spa, Picnic Area, Dive Shop *(classes, charters, rentals)*, Fitness Center, Boat Rentals *(Kayaks $25-35/2hrs., $40-60/day)*, Fishing Charter *(book at marina)*, Park, Sightseeing *(marine ecotours; shuttle to Butchart Gardens May 15-Oct 15, $40/28/Youth $19, under 12 free, 866-387-8480; Church & State Winery tours 652-2671)*, Galleries *(a revolving exhibit from Fran Willis Gallery is displayed throughout the lodge)*

Under 1 mi: Video Rental *(Video Shop 652-5411)* **3+ mi:** Golf Course *(Glen Meadows G & CC 656-3921, 6 mi.)*

Provisioning and General Services
OnSite: Convenience Store, Delicatessen, Wine/Beer, Liquor Store, Bakery, Bank/ATM *(ATM)* **Near:** Market *(Moodyville General Store 652-2081)*, Retail Shops **Under 1 mi:** Health Food *(Grass Roots Natural Foods 544-1718)*, Post Office, Protestant Church *(Shady Creek United 652-2013)*, Library *(Greater Victoria Public Library 652-2013)*, Beauty Salon *(Brentwood Coiffures 652-3333)*, Barber Shop *(Hair Flair 544-1197)*, Dry Cleaners *(Fay's 652-5811)*, Laundry *(Brentwood Bay 652-8661)*, Bookstore *(Pages Used Books 652-434)*, Pharmacy *(Rexall 652-8813)* **1-3 mi:** Supermarket *(True Value Food/Thrifty Foods 483-1600, 4 mi.)*, Green Market *(Peninsula Co-Op 652-1188)* **3+ mi:** Hardware Store *(Home 652-9119, 4 mi.)*

Transportation
OnSite: Water Taxi *(to Butchart Gardens $40/28/19 includes admission)* **OnCall:** Bikes *(Brentwood Cycle 652-4649)*, Taxi *(Express 381-2222)* **Near:** Local Bus *(382-6161)*, Ferry Service *(Mill Bay 386-3431)* **Airport:** YVR *(10 mi.)*

Medical Services
911 Service **OnSite:** Holistic Services *(full-service spa: massages $110-200, facials $110-175, body wraps & scrubs $105-130)* **OnCall:** Ambulance **Under 1 mi:** Doctor *(Family Medical 652-9191)*, Dentist *(Brentwood Dental 652-3723)*, Chiropractor *(Brentwood Bay 652-5211)*, Veterinarian *(Brentwood Bay 652-3131)* **Hospital:** Saanich Peninsula 652-3911 *(4 mi.)*

Setting -- South of Sluggett Point, the Mill Bay ferry landing lies adjacent to this destination marina. Moorage for fifty vessels - in slips and side-tie - nestle below the sweep of the luxurious three-story wood and glass Brentwood Bay Lodge. A gangway leads to the dockhouse and a natural-railed boardwalk spirals up from the docks to the pool, patio and dining levels. Art is everywhere - works by contemporary Canadian west coast artists hang throughout the resort.

Marina Notes -- *Also 15 amp power $4. **After Oct 15, reservations are made through the front desk. Opened in May 2004, this outstanding facility is a member of the prestigious Small Luxury Hotels of the World, and Victoria's only 5 Star oceanfront resort. The marina offers floating wooden piers with long slips and side-tie dockage, and accommodates a 100' yacht. Marina guests can enjoy the heated pool, a small fitness room, and beautiful parklike grounds. Picnic tables overlook the docks from intimate spots under the trees. The bathhouse facilities are new and inviting. Pumpty Dumpty boat: 250-480-9292.

Notable -- First-class amenities and quiet elegance are a hallmark of this small boutique resort. Dine on coastal cuisine at the Arbutus Grill - inside or on the heated patio - tasting menus daily and wine tastings Friday & Saturdays (5-6pm $15). The Pub & Café offers a more casual menu (seats 120 for special events). 33 oceanfront suites have private balconies and the spa entices with signature treatments - West Coast Granite Massage and Luscious Vino Scrub. The Eco-Adventure Centre offers kayak rentals, scuba courses - including PADI certification, and dive charters. A marine biologist conducts ecotours onboard a glass-domed vessel ($60/2 hrs., group rates available). Winery tours & tastings can also be arranged and a water taxi delivers guests to Butchart Gardens.

Navigational Information
Lat: 48°34.100' **Long:** 123°28.200' **Tide:** 10 ft. **Current:** n/a **Chart:** 3441
Rep. Depths (*MLW*): Entry 15 ft. **Fuel Dock** n/a **Max Slip/Moor** -/8 ft.
Access: Tod Inlet off Brentwood Bay to Butchart Cove

Marina Facilities *(In Season/Off Season)*
Fuel: No
Slips: 0 Total, 0 Transient **Max LOA:** 50 ft. **Max Beam:** n/a
  Rate *(per ft.)*: **Day** n/a **Week** n/a **Month** n/a
  Power: 30 amp n/a, 50 amp n/a, 100 amp n/a, 200 amp n/a
  Cable TV: No **Dockside Phone:** No
  Dock Type: n/a
Moorings: 4 Total, 4 Transient **Launch:** n/a, Dinghy Dock
  Rate: Day Free* **Week** n/a **Month** n/a
Heads: None
Laundry: None **Pay Phones:** No
Pump-Out: OnCall *(250-480-9292)* **Fee:** Donation **Closed Heads:** Yes

Marina Operations
Owner/Manager: Butchart Family **Dockmaster:** Same
In-Season: Jun-Sep, 9am-10:30pm **Off-Season:** Oct-May, Varies
After-Hours Arrival: Moorings only*
Reservations: First-come first-served **Credit Cards:** Visa/MC, Amex
Discounts: None
Pets: Welcome **Handicap Access:** No

Butchart Gardens

PO Box; 800 Benvenuto Avenue; Brentwood Bay, BC V8M 1B7

Tel: (250) 652-4422; (866) 652-4422 **VHF: Monitor** n/a **Talk** n/a
Fax: (250) 652-7751 **Alternate Tel:** n/a
Email: n/a **Web:** www.butchartgardens.com
Nearest Town: Brentwood Bay *(2 mi.)* **Tourist Info:** (250) 656-3616

Marina Services and Boat Supplies
Services - Security *(onsite attendant)*

Boatyard Services
Nearest Yard: Westport Marina (250) 656-2832

Restaurants and Accommodations
OnSite: Restaurant *(The Dining Room Restaurant 652-8222, L $14, D $29, L 11:30am-3pm Apr-Oct 8; D 5-8/9pm Jun 15-Dec 15; afternoon tea $29, Noon-4pm)*, *(The Blue Poppy L $6-13, D $6-13, cafeteria-style, L Apr-Dec 11am-4pm, D Jun 15-Sep 2, 4-9pm; group rates)*, Snack Bar *(hot dog cart L $5)*, Lite Fare *(The Coffee Shop - coffee, baked goods, boxed lunches to go; 9am-park closing)* **1-3 mi:** Restaurant *(Blue's Bayou 652-1194)*, *(The Orient 652-2203, L $5-9, D $7-20)*, *(Keating Cross Pizza & Pasta 652-0772)*, Lite Fare *(Just Joey's 652-9855)*, Motel *(Best Value Motel 652-1551, $50-75)*, Inn/B&B *(Brentwood Bay Lodge and Spa 544-2079, $120-400)*, *(Clinker's B&B 652-2110, $105-205)*

Recreation and Entertainment
OnSite: Picnic Area, Park, Museum, Special Events *(Sat fireworks; Night Illuminations, afternoon & evening entertainment)* **1-3 mi:** Beach, Playground, Fitness Center *(Gold's Gym 652-5444)*, Jogging Paths *(Gowlland Tod Provincial Park)*, Roller Blade/Bike Paths, Video Rental *(Video Shop 652-5411)*, Cultural Attract *(Central Saanich Cultural Center 605-6032)*, Sightseeing *(Victoria Butterfly Gardens 877-722-0272 via BC Transit)* **3+ mi:** Tennis Courts *(Panorama Recreation Centre 656-7271, 7 mi.)*, Golf Course *(Glen Meadows Golf and Country Club 656-3921, 6 mi.)*, Bowling *(Central Saanich Lawn Bowling Club 652-4774, 5 mi.)*

Provisioning and General Services
OnSite: Bookstore, Retail Shops *(Seed & Gift Store - treasure-trove of unique souvenirs)* **Under 1 mi:** Protestant Church *(Brentwood Chapel 652-3860)* **1-3 mi:** Market *(Peninsula Co-Op 652-1822)*, Supermarket *(Thrifty Foods 483-1600)*, Gourmet Shop *(Joey's Gourmet Essentials 544-1554)*, Delicatessen *(Brentwood Bay Delicatessen and Bakery 652-4623)*, Health Food *(Grass Roots Health & Natural Foods 644-1718)*, Liquor Store *(BC Liquor 652-1212)*, Bakery *(Brentwood Bay Delicatessen and Bakery 652-4623)*, Bank/ATM *(Royal Bank of Canada 356-3337)*, Catholic Church *(Our Lady of the Assumption 652-1909)*, Library *(Greater Victoria Public Library 652-2013)*, Beauty Salon *(Brentwood Coiffures 652-3333)*, Barber Shop *(Brentwood Barber Shop 652-6111)*, Dry Cleaners *(Fay's 652-5811)*, Laundry *(Brentwood Bay Laundromat 652-8661)*, Pharmacy *(Pharmasave 652-1235)*, Hardware Store *(Home Building Centre 652-1121)*

Transportation
OnSite: Local Bus *(BC Transit 382-6161, Rte. 75: N to Brentwood Bay or S to Keating Cross for shopping; continues to downtown Victoria)* **OnCall:** Taxi *(656-5588)* **1-3 mi:** Ferry Service *(Bentwwod Bay to Mill Bay 386-3431)* **3+ mi:** Rental Car *(Enterprise 475-6900, 14 mi.)* **Airport:** Victoria Int'l. *(10 mi.)*

Medical Services
911 Service OnCall: Ambulance **1-3 mi:** Doctor *(Family Medical Centre 652-9191)*, Dentist *(Brentwood Dental Clinic 652-3723)*, Chiropractor *(Brentwood Bay Chiropractic Office 652-5211)*, Holistic Services *(Spa at Brentwood Bay Lodge 652-3151)*, Veterinarian *(Brentwood Bay Veterinary Hospital 652-3131)* **Hospital:** Saanich Peninsula Hospital 652-3911 *(6 mi.)*

Setting -- Off Tod Inlet, four breathtaking gardens sprawl across fifty-five perfectly manicured and cultivated acres - creating what is arguably the premier tourist draw on Vancouver Island. The Sunken Garden replaced a limestone quarry; it was followed by the Japanese, Rose and Italian Gardens. A small mooring field and floating concrete dinghy landing make it accessible to cruisers. The ramp from the float ends at a short trail that leads to the Gardens.

Marina Notes -- *The Gardens provide 4 free buoys in Butchart Cove. Purchase admission from the dock attendant in the house at the top of the ramp: Adult rates from $13 in winter to $25 in summer; Youth $6.50-12.50; Kids (5-12 years) $2-3. A re-admission rate of $3/2 encourages coffee in the park the next morning. The dock is for dinghies and seaplanes only, and must be vacated when the gardens are closed. Hours: 9am -10:30pm during the peak season (Jun15-Sep15); call ahead for other hours. Closed Dec 26-Jan 6. Anchoring in Butchart Cove is permitted, but beware - the west side gets very shallow at low tide. Alternatively anchor at Tod Inlet and dinghy back. To provision, take the bus from outside the gate to Brentwood Bay or Keating Cross (under 3 miles).

Notable -- At the turn-of-the-last century, this cove bustled with activity from Butchart's Portland cement plant. Eventually, the limestone ran out and in 1904 Jennie Butchart began to convert the quarry into a sunken garden (a museum shows pictures of the Gardens' beginnings and its transformation). During the summer, there's musical entertainment every afternoon and evening. Extraordinary Night Illuminations transform the gardens and Saturday evenings bring truly sensational fireworks. One of four eateries, the Dining Room, in the original Butchart residence, serves lunch, tea, dinner, plus gourmet picnic baskets to go.

Navigational Information
Lat: 48°29.775' **Long:** 123°33.164' **Tide:** 10 ft. **Current:** n/a **Chart:** 3441
Rep. Depths (MLW): Entry 20 ft. **Fuel Dock** 10 ft. **Max Slip/Moor** 12 ft./-
Access: Off the south end of Saanich Inlet at the head of Finlayson Arm

Marina Facilities (In Season/Off Season)
Fuel: Gasoline, Diesel
Slips: 200 Total, 8 Transient **Max LOA:** 110 ft. **Max Beam:** n/a
Rate (per ft.): **Day** $1.00* **Week** n/a **Month** $225
Power: 30 amp $5, **50 amp** n/a, **100 amp** n/a, **200 amp** n/a
Cable TV: No **Dockside Phone:** No
Dock Type: Floating, Long Fingers, Alongside, Wood
Moorings: 0 Total, 0 Transient **Launch:** 2 lane ramp, Dinghy Dock
Rate: Day n/a **Week** n/a **Month** n/a
Heads: 2 Toilet(s)
Laundry: None **Pay Phones:** Yes
Pump-Out: OnCall (250-480-9292) **Fee:** Donations **Closed Heads:** Yes

Marina Operations
Owner/Manager: Lida Seymonsbergen **Dockmaster:** Mark Aitken
In-Season: Jun-Sep, 7:30am-Dark **Off-Season:** Oct-May, 8am-5pm
After-Hours Arrival: Tie up and register in the morning
Reservations: Yes, Preferred **Credit Cards:** Visa/MC, Debit
Discounts: None
Pets: Welcome **Handicap Access:** No

Goldstream Boathouse Marina
3540 Trans-Canada Highway; Victoria, BC V9B 6H6

Tel: (250) 478-4407 **VHF: Monitor** n/a **Talk** n/a
Fax: (250) 478-6882 **Alternate Tel:** (250) 478-7849
Email: See Marina Notes **Web:** www.goldstreamboathousemarina.com
Nearest Town: Langford (5 mi.) **Tourist Info:** (250) 478-5541

Marina Services and Boat Supplies
Services - Docking Assistance, Security (24, owners onsite), Trash Pick-Up, Dock Carts **Communication -** FedEx, UPS, Express Mail (Sat Del) **Supplies -** OnSite: Ice (Cube), Ships' Store, Bait/Tackle

Boatyard Services
OnSite: Travelift (40T to 50 ft.), Engine mechanic (gas, diesel), Launching Ramp (2-lane, with dock), Electrical Repairs, Electronics Repairs, Hull Repairs, Total Refits **Yard Rates:** $55/hr., Haul & Launch $6/ft., Power Wash $3/ft.

Restaurants and Accommodations
Under 1 mi: Pizzeria (Hot Spot Pizza 743-2811) **3+ mi:** Restaurant (Asia Express 474-7693, 5 mi.), (Ma Miller's Goldstream Inn 478-3773, 3 mi.), Seafood Shack (Salty's Fish & Chips 478-2277, 5 mi.), Fast Food (Subway, A&W, 5 mi.), Lite Fare (Water Wheel Pub 474-7004, 5 mi.), Pizzeria (Pizza Pieman 474-7383, 5 mi.), Hotel (Westwind Plaza 478-8334, 5 mi.), Inn/B&B (Wayward Navigator B&B 478-6836, $85-95, 10 mi.), (Glen Meadows 478-5340, 7 mi.)

Recreation and Entertainment
OnSite: Picnic Area **1-3 mi:** Park (Goldstream Park) **3+ mi:** Golf Course (Island Golf Centre 474-1275, 5 mi.), Fitness Center (Juan de Fuca Parks and Recreation 478-8384, 5 mi.), Movie Theater (Caprice Triplex Theatre 474-2700, 5 mi.), Video Rental (Rogers Video 478-8556, 5 mi.), Galleries (Reflections Gallery 474-7800, 7 mi.)

Provisioning and General Services
Under 1 mi: Convenience Store (Streamside Corner Store and Café 478-7751) **3+ mi:** Market (Peninsula Co-Op 478-7726, 5 mi.), Delicatessen (Admiral's Fleet Subs 474-3796, 5 mi.), Liquor Store (BC Liquor 478-9726, 7 mi.), Bakery (Lighthouse Cake Co. 478-4499, 6 mi.), Bank/ATM (Bank of Montreal 391-3450, 6 mi.), Post Office (474-4611, 5 mi.), Catholic Church (Our Lady of the Rosary 478-3482, 5 mi.), Protestant Church (Gordon United Church 478-6632, 5 mi.), Beauty Salon (Mystik Hair Designs 478-1539, 5 mi.), Barber Shop (Island Haircutting Co. 474-3241, 5 mi.), Dry Cleaners (Classic Drycleaners 474-4525, 5 mi.), Laundry (Westbrook Home Style Laundromat 478-1249, 6 mi.), Pharmacy (Drugstore Pharmacy 391-3135, 5 mi.), Hardware Store (Westbrook Home Hardware 474-2241, 5 mi.), Florist (Basket Boutique 391-9679, 5 mi.), Clothing Store (Mariposa 478-3911, 5 mi.), Retail Shops (Radio Shack 478-3822, Treasure Gifts 474-6559, 5 mi.), Department Store (Zellers 474-3148, 5 mi.), Buying Club (Costco 391-1181, 6 mi.), Copies Etc. (Staples 391-3070, 6 mi.)

Transportation
OnCall: Taxi (Westshore Taxi 474-1369) **3+ mi:** Rental Car (Budget 952-5300, 8 mi.) **Airport:** Victoria Int'l. (8 mi.)

Medical Services
911 Service **OnCall:** Ambulance **3+ mi:** Doctor (Losie, K Dr. 478-4421, 5 mi.), Dentist (Can West Dental 474-2296, 5 mi.), Chiropractor (All Care Chiropractic 478-2225, 6 mi.), Veterinarian (Belmont Langford Veterinary Hospital 478-0711, 5 mi.), Optician (Iris 478-0213, 6 mi.) **Hospital:** Victoria General Hospital 727-4212 (10 mi.)

Setting -- Although the Trans-Canada Highway is at the end of the driveway, Goldstream Boathouse Marina feels very isolated and remote. It's at the tip of Finlayson Arm, a true fjord at the end of Saanich Inlet surrounded by high mountains. A grid of docks fills a corner of the fjord and several aquaculture structures line the shore. Dominating the view are two two-story shingled buildings edged with white railings. Separating them is a deck furnished with umbrella-topped picnic tables. The farthest from the docks is the office & store. Potted flowers step up the atmosphere.

Marina Notes -- *$15 min. Owners live on-site. The guest dock is the first dock encountered and the fuel dock is the last. The marina office & store are a bit of a walk, but you may use the office dinght dock if you have supplies to carry to the boat. The small store has ice and fishing tackle. The outside picnic tables provide a pleasant place for a morning cup of coffee or a snack. Victoria General Marine provides a complete repair facility. Very basic heads, no showers. Beware of encroaching shallows near the fuel dock. Email: seymonsbergen@lincsat.com or admin@goldstreamboathouse.com

Notable -- Beyond Brentwood Bay, most of the eastern shore is protected by Gowlland Tod Provincial Park and the western shore is too steep for development. Goldstream Provincial Park, which surrounds the marina, offers pleasant trails among some of the most accessible temperate rainforests in the area. It boasts amazing old growth trees, including 300-year-old Black Cottonwoods and 100-year old Broad-leafed Maples in the lower areas, and 600-year-old Douglas Firs at higher elevations. The Freeman King Visitor Centre offers naturalist programs for all ages. The nearest services are in Langford.

7. BC - SAANICH PENINSULA & VICTORIA
PHOTOS ON DVD: 15

Swartz Bay Public Wharf

Barnacle Road; Sidney, BC V8L 3X9

Tel: (250) 655-3256 **VHF: Monitor** n/a **Talk** n/a
Fax: (250) 246-1595 **Alternate Tel:** n/a
Email: n/a **Web:** n/a
Nearest Town: Sidney *(4 mi.)* **Tourist Info:** (250) 656-3616

Navigational Information
Lat: 48°41.290' **Long:** 123°24.475' **Tide:** 10 ft. **Current:** n/a **Chart:** 3476
Rep. Depths *(MLW)*: **Entry** 8 ft. **Fuel Dock** n/a **Max Slip/Moor** 8 ft./-
Access: On the south side of Colburne Pass

Marina Facilities *(In Season/Off Season)*
Fuel: No
Slips: 3 Total, 3 Transient **Max LOA:** 50 ft. **Max Beam:** n/a
  Rate *(per ft.)*: **Day** $0.43* **Week** n/a **Month** n/a
  Power: 30 amp n/a, 50 amp n/a, 100 amp n/a, 200 amp n/a
  Cable TV: No **Dockside Phone:** No
  Dock Type: Floating, Alongside, Wood
Moorings: 0 Total, 0 Transient **Launch:** n/a
  Rate: Day n/a **Week** n/a **Month** n/a
Heads: None
Laundry: None **Pay Phones:** No
Pump-Out: No **Fee:** n/a **Closed Heads:** No

Marina Operations
Owner/Manager: Swartz Bay H. A. **Dockmaster:** Mike Smart
In-Season: Year-Round, 24/7 **Off-Season:** n/a
After-Hours Arrival: Drop box
Reservations: No **Credit Cards:** No
Discounts: None
Pets: Welcome **Handicap Access:** Yes, Docks

Marina Services and Boat Supplies
Supplies - Near: Ice *(Cube)* **1-3 mi:** Ships' Store *(Canoe Cove Marina 656-5566)* **3+ mi:** West Marine *(654-0045, 5 mi.)*, Bait/Tackle *(Sidney Sporting Goods 656-9255, 5 mi.)*

Boatyard Services
Near: Launching Ramp. **1-3 mi:** Travelift *(50T)*, Forklift, Hydraulic Trailer *(20T)*, Engine mechanic *(gas, diesel)*, Electrical Repairs, Electronic Sales, Hull Repairs, Rigger *(Blanchard Rigging Ltd. 656-6199)*, Sail Loft, Canvas Work *(Bea's Sail and Canvas 656-7557)*, Bottom Cleaning, Brightwork, Air Conditioning, Refrigeration, Divers, Compound, Wash & Wax, Woodworking, Metal Fabrication, Painting, Awlgrip, Yacht Broker *(Thunderbird Yacht Sales 656-5832)*, Total Refits. **Nearest Yard:** Westport Marina (250) 656-2832

Restaurants and Accommodations
Near: Snack Bar *(Land's End Café at ferry terminal: espresso, snacks, pizza $3.70/slice, 6am-8:30pm)* **1-3 mi:** Restaurant *(Stonehouse Pub & Restaurant 656-3498, L $9-15, D $9-22)*, *(Seaside Café 656-5557, B & L 6:30am-4pm)*, *(Dock 503 656-0828, L $9-12, D $19-29, Sun brunch 10:30am-2:30pm $8-11)*, Inn/B&B *(Shoal Harbour Inn 656-6622, $119-329)* **3+ mi:** Fast Food *(KFC, McDonald's 655-3511, 4 mi.)*, Motel *(TraveLodge 656-1176, $71-100, 4 mi.)*

Recreation and Entertainment
Near: Picnic Area **1-3 mi:** Park *(North Hill Regional)* **3+ mi:** Tennis Courts *(Panorama Recreation Centre 656-7271, 4 mi.)*, Golf Course *(Ardmore 656-4621, 6 mi.)*, Fitness Center *(50 Plus Fitness 655-1946, 4 mi.)*, Horseback Riding *(Woodgate Stables 652-0287, 6 mi.)*, Boat Rentals *(SeaQuest Adventures 656-7599, 5 mi.)*, Movie Theater *(Star Cinema 655-3384, 3 mi.)*, Video Rental *(Video Express 656-3666, 4 mi.)*, Museum *(Maritime Museum 385-4222; Sidney Historical Museum 652-5522, donation; Aviation Museum 655-3300, The Marine Ecology Centre 655-1555, 4 mi.)*, Cultural Attract *(Theatre at Mary Winspear Centre 656-0271, 4 mi.)*, Galleries *(Peninsula 655-1722, Village 656-3633, 4 mi.)*

Provisioning and General Services
OnSite: Convenience Store *(The Market)* **Near:** Bank/ATM, Newsstand **1-3 mi:** Delicatessen *(Harbour Road 655-0005)*, Wine/Beer, Protestant Church **3+ mi:** Market *(Queens Grocery 656-1912, 3 mi.)*, Supermarket *(Safeway 656-1721, 4 mi.)*, Liquor Store *(BC Liquor 656-3041, 4 mi.)*, Farmers' Market *(Beacon Ave. Thu 5:30-8:30pm Jun-Aug, 4 mi.)*, Catholic Church *(4 mi.)*, Library *(656-0944, 4 mi.)*, Beauty Salon *(Beechwood Salon 655-5061, 4 mi.)*, Barber Shop *(Beacon 656-5244, 4 mi.)*, Bookstore *(Tanner's 656-2345, 4 mi.)*, Pharmacy *(Sidney 656-1168, 4 mi.)*, Hardware Store *(Home 656-2712, 4 mi.)*

Transportation
OnCall: Taxi *(Empress 381-2222)* **Near:** Ferry Service *(BC 888-223-3779)* **1-3 mi:** Local Bus *(BC Transit 382-6161, Rte 70)* **3+ mi:** Rental Car *(Avis 656-6033, Hertz 656-2312, 4 mi.)* **Airport:** YVR *(4 mi.)*

Medical Services
911 Service **OnCall:** Ambulance *(656-1001)* **3+ mi:** Doctor *(Green 656-1840, 4 mi.)*, Dentist *(Scott 656-0701, 4 mi.)*, Chiropractor *(Roper 656-4611, 4 mi.)* **Hospital:** Saanich Peninsula 652-3911 *(8 mi.)*

Setting -- Swartz Bay Public Wharf, like many of British Columbia's public docks, lies next to a ferry terminal; in this case, Vancouver Island's busiest one. Fortunately, the terminal dock is in an adjacent cove -- close enough to provide boaters with a light meal & a ride, but not so close as to be a nuisance.

Marina Notes -- *First 4 hours free; flat rate 4-12 hrs: 30' $7.50, 40' $10, 50' $12.50; 13-24 hrs: 30' $15, 40' $20, 50' $25. Rafting is permitted at this small dock. Leave payment in the drop box and place the back portion of the envelope on the boat so it is visible. The are no amenities on-site, but there are 4 pay phones at the ferry terminal as well as at The Market, which sells arts, crafts, and food. The terminal also has an ATM, newsstand, information center, frozen lemonade, and restrooms.

Notable -- This is an ideal place from which to explore the remarkable BC Ferries system, since the Swartz Bay terminal has service to Vancouver & numerous Gulf Islands. The 1.5 hour crossing to Tsawwassen ($10.25/5.25) runs on odd hours. A trip to Salt Spring Island takes about 35 minutes ($6.50/3.25). The Pier Island ferry & taxi schedule is posted at the top of the ramp (655-5211). Or stay around and visit Sidney's unique museums. The Historical Museum is devoted to the life & times of the early Saanich settlers. The Sidney Marine Museum is currently undergoing renovation and will reopen in the spring of '07 with new exhibits. Meanwhile, visitors can still see the whales (655-6355). The BC Aviation Museum near the airport displays and restores small aircraft - learn about float planes, old engines, and visit the gift shop year-round 10am-4pm in summer, 11am-3pm in winter (655-3300, $7/under 12 free).

Navigational Information
Lat: 48°41.020' **Long:** 123°24.353' **Tide:** 10 **Current:** n/a **Chart:** 3476
Rep. Depths (*MLW*): **Entry** 22 ft. **Fuel Dock** 8 ft. **Max Slip/Moor** 8 ft./-
Access: In Canoe Cove off Sidney Channel

Marina Facilities (*In Season/Off Season*)
Fuel: Gasoline, Diesel
Slips: 460 Total, 1 Transient **Max LOA:** 75 ft. **Max Beam:** n/a
  Rate (*per ft.*): **Day** $1.00 **Week** n/a **Month** $11.75
  Power: 30 amp $6.50, **50 amp** n/a, **100 amp** n/a, **200 amp** n/a
Cable TV: No **Dockside Phone:** No
Dock Type: Floating, Long Fingers, Alongside, Wood
Moorings: 0 Total, 0 Transient **Launch:** n/a
  Rate: Day n/a **Week** n/a **Month** n/a
Heads: 2 Toilet(s), 2 Shower(s) (*with dressing rooms*)
Laundry: 2 Washer(s), 1 Dryer(s) **Pay Phones:** Yes, 2
Pump-Out: No **Fee:** n/a **Closed Heads:** Yes

Marina Operations
Owner/Manager: Don Prettie **Dockmaster:** Terry Crawford
In-Season: Summer, 8am-6pm **Off-Season:** Fall-Spring, 8am-4:30pm
After-Hours Arrival: Call the fuel dock
Reservations: Yes, Required **Credit Cards:** Visa/MC, Interac
Discounts: None
Pets: Welcome **Handicap Access:** No

Canoe Cove Marina

2300 Canoe Cove Road; Sidney, BC V8L 3X9

Tel: (250) 656-5566 **VHF: Monitor** Ch. 66A **Talk** n/a
Fax: (250) 655-7197 **Alternate Tel:** (250) 656-5515
Email: wharfage@canoecovemarina.com **Web:** canoecovemarina.com
Nearest Town: Sidney (*4.5 mi.*) **Tourist Info:** (250) 656-3616

Marina Services and Boat Supplies
Services - Security (*video*), Trash Pick-Up, Dock Carts **Communication -** Data Ports (*Wi-Fi - Broadband Xpress*), FedEx, DHL, UPS, Express Mail
Supplies - OnSite: Ice (*Cube*), Ships' Store, Bait/Tackle, Propane
3+ mi: West Marine (*654-0045, 4 mi.*)

Boatyard Services
OnSite: Travelift (*35T*), Hydraulic Trailer (*40T*), Electronics Repairs, Hull Repairs, Rigger, Brightwork, Air Conditioning, Refrigeration, Compound, Wash & Wax, Propeller Repairs, Woodworking, Yacht Interiors, Metal Fabrication, Painting, Awlgrip, Yacht Broker (*Beacon Yacht Sales*) **Dealer for:** Yanmar, Detroit, Cummins, Volvo, Caterpillar, Mann, Wesmar, Espar, Hurricane, Onan, Westerbeke, Northern Lights, Kohler, Bluestar, Trace, Heart, Statpower, Furuno, Raytheon, Autohelm, Standard/Horizon, Icom.

Restaurants and Accommodations
OnSite: Restaurant (*Sofie's Café 656-5557, B & L 6:30am-4pm, patio seating*), (*Stonehouse Pub and Restaurant 656-3498, L $9-15, D $9-19*)
1-3 mi: Restaurant (*Dock 503 656-0828, D $20-32, L, D, Brunch*), (*Blue Peter Pub & Restaurant 656-4551, D $16-30, L, D*), Motel (*Resthaven by the Sea 656-7510*) **3+ mi:** Pizzeria (*Odyssia Pizza and Steak House 656-5596, 4 mi., free del. after 5pm*), Hotel (*TraveLodge 656-1176, $71-100, 4 mi.*)

Recreation and Entertainment
OnSite: Sightseeing (*Sidney Harbor Cruise 655-1562;*), Galleries (*Morgan Warren 655-1081*) **1-3 mi:** Picnic Area, Playground, Movie Theater (*Star Cinema 655-3384*), Video Rental, Park (*McDonald Provincial*) **3+ mi:** Tennis Courts (*Panorama Recreation Centre 656-7271, 4 mi.*), Golf Course

(*Ardmore GC 656-4621, 6 mi.*), Fitness Center (*Body Barn 655-3393, 4 mi.*), Bowling (*Miracle Lanes 656-2431, 4 mi.*), Museum (*Historical Museum 655-6355, donation; Aviation Museum 655-3300, $7; Marine Ecology Centre 655-1555, 4 mi.*), Cultural Attract (*Charlie White Theatre 656-0271, 4 mi.*), Special Events (*Sidney Days - Jul, Saanich Fall Fair - Sep, 5 mi.*)

Provisioning and General Services
1-3 mi: Delicatessen (*Harbour Road 655-0005*), Bank/ATM, Protestant Church, Library (*656-0944*) **3+ mi:** Market (*Thrifty Foods 656-0946, 4 mi.*), Supermarket (*Safeway 656-1721, 4 mi.*), Liquor Store (*BC Liquor 656-3041, 4 mi.*), Bakery (*Big Rock 655-3633, 4 mi.*), Farmers' Market (*Jun-Aug 5:30-8:30pm on Beacon Ave., 4 mi.*), Catholic Church (*4 mi.*), Beauty Salon (*Marinello 656-7488, 4 mi.*), Laundry (*Wash-Rite 656-0025, 4 mi.*), Bookstore (*Compass Rose Nautical 656-4674, 4 mi.*), Pharmacy (*Sidney 656-1168, 4 mi.*), Hardware Store (*Home 656-2712, 4 mi.*)

Transportation
OnSite: Water Taxi (*Sidney Harbour Cruise*) **OnCall:** Taxi (*Peninsula 656-1111*), Airport Limo (*Royal 389-0004*) **Under 1 mi:** Ferry Service (*BC Ferries 656-0757*) **1-3 mi:** Local Bus (*BC Transit 382-6161*) **3+ mi:** Rental Car (*Thrifty 656-8804, Budget 953-5300, 4 mi.*) **Airport:** Victoria Int'l. (*5 mi.*)

Medical Services
911 Service **OnCall:** Ambulance (*656-1001*) **3+ mi:** Doctor (*Bevan Ave. Family Medical 656-4143, 4 mi.*), Dentist (*Sidney Dental 656-4848, 4 mi.*), Chiropractor (*Laidley 656-6643, 4 mi.*), Veterinarian (*Sidney Animal Hosp. 656-3333, 4 mi.*), Optician (*Iris 656-1413, 4 mi.*) **Hospital:** Saanich Peninsula 652-3911 (*8 mi.*)

Setting -- Canoe Bay, better known as Canoe Cove, lies behind a clutter of offshore islands that guard the large web of floating docks but complicate navigation. Five sets of docks curve into a protected basin - the outer slips open and the inner ones tucked under boat houses. Landside is a serious boatyard, two eateries and a full complement of marine and boater services.

Marina Notes -- This is truly a full-service marina with a fuel dock, extensive boatyard operation, restaurants, and a chandlery. Canoe Cove Marina has a long history, starting in the '30s as a wooden boat building facility - a tradition that continued until the '60s. The current owners bought the marina in '74 and have been expanding ever since. There are 450 in-water slips, 100 covered slips, 140 boathouses, and dry storage for 120 boats. The well-supplied chandlery offers items from general hardware to electrical supplies, varnishes, and batteries. Fuel, propane, stove oil, and 50:1 available. Ice is on the fuel dock. Home to Gulf Islands Cruising School (656-2628). The Morgan Warren Art Gallery rounds out the amenities. Note: Navigate the passes between the islands carefully.

Notable -- Just over a mile away, McDonald Park, part of the Gulf Islands National Reserve, has picnic areas and trails. Or catch a taxi to Sidney for big city fun in a small town atmosphere. Sidney Days in early July features a pancake breakfast, sidewalk sale, beer garden, fireworks, and a Build-a-Quick-Boat Race (656-4365). Sunday concerts are held July through September at Bandstand Park (656-4365) and the annual Saanich Fair is in early September (652-3314). The Marine Ecology Centre at Port Sidney Marina offers summer science camps for kids and marine ecotours ($9 tourist pass for access to several attractions).

Westport Marina

2075 Tryon Road; Sidney, BC V8L 3X9

Tel: (250) 656-2832 **VHF: Monitor** n/a **Talk** n/a
Fax: (250) 655-1981 **Alternate Tel:** n/a
Email: westport@thunderbirdmarine.com **Web:** thunderbirdmarine.com
Nearest Town: Sidney (4 mi.) **Tourist Info:** (250) 656-3616

Navigational Information
Lat: 48°40.773' **Long:** 123°24.836' **Tide:** 10 ft. **Current:** n/a **Chart:** 3415
Rep. Depths (MLW): Entry 6 ft. **Fuel Dock** n/a **Max Slip/Moor** 7 ft./-
Access: Northern side of Tsehum Harbour

Marina Facilities (In Season/Off Season)
Fuel: No
Slips: 600 Total, 10 Transient **Max LOA:** 70 ft. **Max Beam:** n/a
 Rate (per ft.): **Day** $1.10 **Week** $5.50 **Month** $135-730
 Power: 30 amp $0.10/kwh*, **50 amp** $0.10/kwh, **100 amp** n/a, **200 amp** n/a
 Cable TV: No **Dockside Phone:** No
 Dock Type: Floating, Long Fingers, Alongside, Concrete, Wood
Moorings: 0 Total, 0 Transient **Launch:** n/a, Dinghy Dock
 Rate: Day n/a **Week** n/a **Month** n/a
Heads: 5 Toilet(s), 2 Shower(s) (with dressing rooms)
Laundry: None **Pay Phones:** Yes, 2
Pump-Out: No **Fee:** n/a **Closed Heads:** No

Marina Operations
Owner/Manager: Thunderbird Marine Corp. **Dockmaster:** Ken Gowan
In-Season: Apr-Sep, 7am-7pm **Off-Season:** Oct-Mar, 8am-5pm
After-Hours Arrival: Call ahead
Reservations: Yes, Required **Credit Cards:** Visa/MC, Debit
Discounts: None
Pets: Welcome **Handicap Access:** Yes, Heads, Docks

Marina Services and Boat Supplies
Services - Security (locked gates), Trash Pick-Up, Dock Carts
Communication - Fax in/out, FedEx, UPS, Express Mail **Supplies -**
OnSite: Ice (Cube), Ships' Store (Jensen's 656-1114)
Under 1 mi: Bait/Tackle (Westport Seafood 268-0133, Hungry Whale 268-0136, Angler Charters 268-1030) **3+ mi:** West Marine (654-0045, 4 mi.)

Boatyard Services
OnSite: Travelift (50T), Forklift, Crane, Hydraulic Trailer (20T), Engine mechanic (gas, diesel), Electrical Repairs, Electronic Sales, Hull Repairs, Rigger (Blanchard Rigging 656-6199), Canvas Work (Bea's Sail & Canvas 656-7557), Bottom Cleaning, Brightwork, Air Conditioning, Refrigeration, Divers, Compound, Wash & Wax, Woodworking, Metal Fabrication, Painting, Awlgrip, Yacht Broker (Thunderbird Yacht Sales 656-5832), Total Refits
OnCall: Propeller Repairs, Inflatable Repairs, Yacht Interiors
Under 1 mi: Launching Ramp, Sail Loft **Yard Rates:** Haul & Launch $400**
Power Wash $2.40/ft.** **Storage:** On-Land $312 Oct-Feb

Restaurants and Accommodations
Under 1 mi: Restaurant (Sofie's Café 656-5557, 6:30am-4pm), (Stonehouse 656-3498, L $9-15, D $9-22) **1-3 mi:** Restaurant (Blue Peter 656-4551), (Dock 503 656-0828, L $9-12, D $19-29, Sun brunch 10:30am-2:30pm $8-11), Seafood Shack (Hyland's 656-4435), Inn/B&B (Resthaven by the Sea 655-7510), (Shoal Harbour 656-6622, $119-329) **3+ mi:** Motel (TraveLodge 656-1176, $71-100, 4 mi.)

Recreation and Entertainment
1-3 mi: Picnic Area, Jogging Paths, Park **3+ mi:** Golf Course

(Ardmore 656-4621, 6 mi.), Fitness Center (Body Barn 655-3393, 4 mi.), Movie Theater (Star 655-3384, 4 mi.), Video Rental (Video Express 655-0819, 4 mi.), Museum (Historical 655-6355, Marine Ecology Centre 655-1555, 4 mi.), Cultural Attract (Charlie White Theatre 656-0271, 4 mi.)

Provisioning and General Services
Near: Laundry (RV Park Laundry & Showers 268-0025) **1-3 mi:** Convenience Store (Resthaven 656-1140), Market (Queens 656-1912), Delicatessen (Harbour Road 655-0005), Wine/Beer, Bank/ATM, Catholic Church, Protestant Church, Synagogue, Library (656-0944) **3+ mi:** Supermarket (Safeway 656-1721, 4 mi.), Health Food (Lifestyle Select 656-2326, 4 mi.), Liquor Store (BC 656-3041, 4 mi.), Bakery (Big Rock 655-3633, 4 mi.), Farmers' Market (Thu 5:30-8:30pm Beacon Ave., 4 mi.), Beauty Salon (Cameo 656-1713, 4 mi.), Barber Shop (Bell Buoy 656-4111, 4 mi.), Bookstore (Compass Rose 656-4674, 4 mi.), Pharmacy (Sidney 656-1168, 4 mi.), Hardware Store (True Value 656-8611, 4 mi.)

Transportation
OnCall: Rental Car (Enterprise 475-6900), Taxi (Empress 381-2222)
Under 1 mi: Local Bus (BC Transit 382-6161), Ferry Service (BC 656-0757) **Airport:** YVR (4 mi.)

Medical Services
911 Service **OnCall:** Ambulance (656-1001) **1-3 mi:** Doctor (Associated Physicians 656-1164), Dentist (Sidney 656-4848), Chiropractor (Sidney 655-0543), Holistic Services (Sidney Acupuncture 655-6409; Bella 267-0843, massages, facials, nail care), Veterinarian (Sidney 656-3333)
Hospital: Saanich Peninsula 652-3911 (8 mi.)

Setting -- Westport Marina's seven main floating concrete docks, hosting over 600 slips, are safely tucked into a basin off the north corner of Tsehum Harbour. Onshore, a relatively new boatyard operation informs the ambiance and houses a small mall of marine service contractors.

Marina Notes -- *15 amp included in the slip fee, 30 amp is by the kwh and varies according to size. **Boatyard rates are for a sample 40 ft. vessel. Westport offers limited guest moorage, so call ahead. The boatyard, completed in '03, sports a 50 ton lift for haul outs and a wide range of shops to meet most boaters' needs - do the work yourself or hire a contractor. A new bathhouse was completed in 2005. Note: Numerous hazards litter the approach to the marina so deliberate, careful navigation carries the day. In the summertime, at least, most of the problems can be avoided by staying out of the kelp.

Notable -- The nearest restaurants are Sophie's Café and Stonehouse Pub at Canoe Cove Marina plus the BC Ferry terminal is less than a mile away. Avid shoppers will find that the Saanich Peninsula offers a wide range of specialty shops featuring both local and international wares. Beacon Avenue, Sidney's main street, is lined with shops, galleries, bookstores, and cafés. Follow it right down to the waterfront where restaurants and museums await. The Marine Museum, currently undergoing renovation, has an outstanding exhibit on the evolution of whales. Sidney Historical Museum tells the story of the peninsula and its settlers. Special events run throughout the summer, and include outdoor concerts, a weekly open air market, and maritime-themed events. An annual antique toy show at the Mary Winspear Centre draws crowds in October; the annual Artisans Show & Sale from mid-June to mid-August showcases over 60 artists.

Navigational Information
Lat: 48°40.310' **Long:** 123°24.300' **Tide:** 10 ft. **Current:** n/a **Chart:** 3476
Rep. Depths (MLW): Entry 12 ft. **Fuel Dock** 18 ft. **Max Slip/Moor** 18 ft./-
Access: Southern side of Tsehum Harbour

Marina Facilities *(In Season/Off Season)*
Fuel: Gasoline, Diesel
Slips: 538 Total, 20 Transient **Max LOA:** 300 ft. **Max Beam:** n/a
Rate *(per ft.):* **Day** $1.35/1.15 **Week** n/a **Month** $14.90*
Power: 30 amp $6** **50 amp** $9-10, **100 amp** $30, **200 amp** $50
Cable TV: Yes **Dockside Phone:** Yes
Dock Type: Floating, Long Fingers, Alongside, Wood
Moorings: 0 Total, 0 Transient **Launch:** n/a
Rate: Day n/a **Week** n/a **Month** n/a
Heads: 7 Toilet(s), 3 Shower(s) *(with dressing rooms)*
Laundry: 6 Washer(s), 6 Dryer(s), Iron, Iron Board **Pay Phones:** Yes
Pump-Out: OnSite, Self Service, 1 Central **Fee:** $10 **Closed Heads:** No

Marina Operations
Owner/Manager: Greg Dickinson **Dockmaster:** Same
In-Season: May-Oct, 7:30am-8:30pm **Off-Season:** Nov-Apr, 8am-5pm
After-Hours Arrival: Tie to check-in dock
Reservations: Yes, Preferred **Credit Cards:** Visa/MC, Amex
Discounts: None
Pets: Welcome **Handicap Access:** Yes, Heads, Docks

Van Isle Marina

2320 Harbour Road; Sidney, BC V8L 2P6

Tel: (250) 656-1138 **VHF: Monitor** Ch. 66A **Talk** n/a
Fax: (250) 656-0182 **Alternate Tel:** n/a
Email: info@vanislemarina.com **Web:** www.vanislemarina.com
Nearest Town: Sidney *(1.5 mi.)* **Tourist Info:** (250) 656-3616

Marina Services and Boat Supplies
Services - Docking Assistance, Trash Pick-Up, Dock Carts, Megayacht
Facilities **Communication -** Data Ports *(Wi-Fi - Broadband Xpress)*, FedEx,
DHL, UPS *(Sat Del)* **Supplies - OnSite:** Ice *(Block, Cube)*, Ships' Store,
Bait/Tackle **Under 1 mi:** CNG **1-3 mi:** West Marine *(654-0045)*, Propane

Boatyard Services
OnSite: Railway *(2 150T to 130 ft.***)*, Engine mechanic *(gas, diesel)*,
Launching Ramp *($10)*, Electrical Repairs, Electronics Repairs, Hull Repairs,
Canvas Work, Brightwork, Air Conditioning, Compound, Wash & Wax,
Propeller Repairs, Woodworking, Metal Fabrication, Painting, Awlgrip
Yard Rates: $39-59/hr. **Storage:** On-Land $0.33/sq.ft./mo.

Restaurants and Accommodations
OnSite: Restaurant *(Dock 503, 656-0828, L $9-12, D $19-29, Sun brunch
10:30am-2:30pm $8-11; three-course specials $30)* **Near:** Restaurant *(Blue
Peter Pub 656-4551, dock)*, Inn/B&B *(Shoal Harbour Inn 656-6622, $119-
329)* **1-3 mi:** Restaurant *(Taste of Tokyo 656-6862)*, *(Thyme Out 656-
4115)*, *(Sofie's Café 656-5557, B & L)*, *(Stonehouse Pub 656-3498, L $9-15,
D $9-19)*, Fast Food *(McDonald's, Dairy Queen)*, Lite Fare *(Lunn's Pastries
Deli 656-1724)*, Pizzeria *(Pizzability 656-5222)*, Motel *(TraveLodge 656-
1176, $71-100)*, Inn/B&B *(Beacon Inn 655-3288, $119-259)*

Recreation and Entertainment
OnCall: Fishing Charter *(Tailour 361-7119)* **Under 1 mi:** Picnic Area,
Playground, Tennis Courts **1-3 mi:** Fitness Center *(Body Barn 655-3393)*,
Jogging Paths, Bowling *(Miracle Lanes 656-2431)*, Movie Theater *(Star
Cinema 655-1171)*, Video Rental *(Video Vault 656-1215)*, Park, Museum

(Sidney Historical 655-6355, Sidney Marine 656-1322,), Cultural Attract
(Charlie White Theatre 656-0271), Special Events *(Arts & Crafts Show,
Sidney Days, bandstand summer concerts)*, Galleries *(Peninsula 655-1282)*
3+ mi: Golf Course *(Ardmore GC 656-4621, 7 mi.)*

Provisioning and General Services
Near: Convenience Store *(Resthaven 656-1140)*, Fishmonger *(Blue Peter
Pub - fresh salmon)*, Bank/ATM *(ATM)* **Under 1 mi:** Delicatessen *(Harbour
Road 655-0005)*, Protestant Church *(St. Paul)* **1-3 mi:** Market *(Thrifty 656-
0946)*, Supermarket *(Safeway 656-1721)*, Health Food *(Sidney Natural 656-
4634)*, Liquor Store *(BC 656-3041)*, Bakery *(Big Rock 655-3633)*, Farmers'
Market *(Beacon Ave. Thu 5:30-8:30pm Jun-Aug)*, Catholic Church, Library
(656-0944), Beauty Salon *(Gaea 655-9377)*, Barber Shop *(Beacon 656-
5244)*, Dry Cleaners *(Classic 656-7501)*, Laundry *(Wash-Rite 656-0025)*,
Bookstore *(Beacon 655-4447, Compass Rose Nautical 656-4674)*,
Pharmacy *(Pharmasave 656-1148)*, Hardware Store *(True Value 656-8611)*,
Copies Etc. *(Sidney Copyprint 656-1233)*

Transportation
OnCall: Rental Car *(Discount 310-2277, Hertz 656-2312)*, Taxi *(Empress
381-2222)* **Near:** Local Bus *(BC Transit 382-6161)* **1-3 mi:** Ferry Service
(BC Ferries 656-0757) **Airport:** YVR *(3 mi.)*

Medical Services
911 Service **OnCall:** Ambulance **1-3 mi:** Doctor *(Associated Physicians
656-1164)*, Dentist *(Sidney Dental 656-4848)*, Chiropractor *(Galloway 655-
3233)*, Holistic Services *(Sidney Acupuncture 655-6409)*, Veterinarian
(Sidney 656-3333) **Hospital:** Saanich Peninsula 652-3911 *(7 mi.)*

Setting -- Van Isle is the first marina along Tsehum Harbour's south shore. 538 slips on nine open and covered docks are protected by a 750 foot breakwater. At the entrance to the basin, a long, multi-pump sheltered fuel dock is followed by customs and visitor check-in. At the head of the docks a recent, attractive two-story blue and gray contemporary hosts the offices and a matching building, with a deck overlooking the docks, houses "Dock 503" and the bathhouse.

Marina Notes -- *$10.40/ft./mo. for 3 mo., $7.75/ft./mo. for 12 mo.; liveaboard fee $160/mo. **15 amp power $5/nt.; also 120 & 208 volt power. ***Onsite Philbrook's BY (656-1157), one of the area's largest, has two 150 ton railways. Owned & operated by the Dickinson family since '55, Van Isle accommodates 300 ft. yachts and provides Wi-Fi, cable TV and phone service. The fuel dock, near the harbor entrance, supplies gas, diesel, stove oil, 50:1 outboard mix, oils, and ice. The store sells many other supplies, from tackle to charts, mooring lines to life jackets. Pump-out stations for sewage & oil are also on the fuel dock. Marina staff can arrange fishing and diving charters or whale watching trips. Visitors can clear Canadian customs by phone. There is a well-maintained all-tile bathhouse and a large laundry room. Note: The Tsehum Harbour Public Wharf next door has three wooden piers intended only for commercial fishermen.

Notable -- Mostly glass Dock 503 serves innovative, local, seasonal West Coast cuisine - with wine pairings. Fine dining inside or on the deck right at the waterfront. The relatively quiet neighborhood sports a small market, a deli, and a couple of eateries along Harbour Road. Local buses, taxis, and rental cars are available for exploring downtown Sidney, the Saanich Peninsula, or Victoria. With the airport and ferry landing nearby, this is a great spot to leave the boat.

Port Sidney Marina

9835 Seaport Place; Sidney, BC V8L 4X3

Tel: (250) 655-3711 **VHF: Monitor** Ch. 66A **Talk** n/a
Fax: (250) 655-3771 **Alternate Tel:** n/a
Email: info@portsidney.com **Web:** www.portsidney.com
Nearest Town: Sidney **Tourist Info:** (250) 656-3616

Navigational Information
Lat: 48°39.100' **Long:** 123°23.500' **Tide:** 10 ft. **Current:** n/a **Chart:** 3476
Rep. Depths (MLW): Entry 14 ft. **Fuel Dock** 9 ft. **Max Slip/Moor** 12 ft./-
Access: Located on the east side of the Saanich Peninsula

Marina Facilities (In Season/Off Season)
Fuel: No
Slips: 320 Total, 120 Transient **Max LOA:** 110 ft. **Max Beam:** n/a
 Rate (per ft.): **Day** $1.35/0.85 **Week** n/a **Month** $10.65
 Power: 30 amp $4-6*, **50 amp** $4-6*, **100 amp** Yes*, **200 amp** n/a
 Cable TV: Yes, $3/night, $20/mo. **Dockside Phone:** Yes
 Dock Type: Floating, Long Fingers, Alongside, Concrete
 Moorings: 0 Total, 0 Transient **Launch:** n/a, Dinghy Dock
 Rate: Day n/a **Week** n/a **Month** n/a
Heads: 5 Toilet(s), 5 Shower(s)
Laundry: 5 Washer(s), 5 Dryer(s) **Pay Phones:** Yes
Pump-Out: OnSite **Fee:** $10 **Closed Heads:** Yes

Marina Operations
Owner/Manager: Lyndell Curry, GM **Dockmaster:** Wayne Pullen, HM
In-Season: Jun-Sep, 8am-8pm **Off-Season:** Oct-May, 8am-5pm
After-Hours Arrival: Reserve and slip will be assigned
Reservations: Yes, Preferred **Credit Cards:** Visa/MC, Amex, Debit
Discounts: None
Pets: Welcome, Dog Walk Area **Handicap Access:** Yes, Heads, Docks

Marina Services and Boat Supplies
Services - Docking Assistance, Concierge, Boaters' Lounge, Dock Carts, Megayacht Facilities **Communication** - Mail & Package Hold, Fax in/out - ($0.25), Data Ports (WiFi - Broadband Xpress), FedEx, UPS, Express Mail (Sat Del) **Supplies** - OnSite: Ice (Block, Cube) **Near:** Bait/Tackle, Live Bait **Under 1 mi:** WestMarine (654-0045)

Boatyard Services
Nearest Yard: Philbrook's Boatyard Ltd. (250) 656-1157

Restaurants and Accommodations
OnSite: Restaurant (The Rumrunner 656-5643 - patio), (Captain's Table 656-3320, B $6-10, L $8-14, D $10-20, waterfront dining), Lite Fare (Breakwater Café 655-3757,B $5-8, L $6-8, Internet access) **Near:** Restaurant (Pier 1 656-1224), (Beacon Landing 656-6690), (Japanese Garden 655-1833, L $7-12,D $7-12, delivery min. $25), Lite Fare (Pelicanos Café & Bakery 656-4116), Inn/B&B (Waterfront Inn & Suites 655-1131) **Under 1 mi:** Fast Food, Pizzeria (Mariner Pizza 656-9606), Motel (TraveLodge 656-1176, $71-100), Hotel (Best Western 656-4441, $125-249), (Beacon Inn 656-3288)

Recreation and Entertainment
OnSite: Picnic Area, Museum (Marine Ecology Centre 655-1555, Sidney Marine Museum 656-1322, near/ BC Aviation Museum 655-3300, 1 mi.), Sightseeing (Sea Quest Whale Watching 655-9256) **Near:** Video Rental (Video Vault 656-1215), Park (Bandstand Park), Cultural Attract (Charlie White Theatre 656-0271), Special Events (summer concerts, crafts sale, Sidney Days), Galleries (Peninsula 655-1722, Village Gallery 656-3633) **Under 1 mi:** Fitness Center (Body Barn 655-3393), Bowling (Miracle Lanes

656-2431), Group Fishing Boat (Sea Esta Boat Charters 656-9919), Movie Theater (Star Cinema 655-1171) **1-3 mi:** Tennis Courts, Jogging Paths, Roller Blade/Bike Paths **3+ mi:** Golf Course (Ardmore 656-4621, 4 mi.)

Provisioning and General Services
OnSite: Convenience Store, Gourmet Shop, Delicatessen (Corner Deli 656-0435) **Near:** Market (Super Foods 656-0727), Health Food (Sidney Natural Foods 656-4634), Bakery (Pelicanos 655-4116), Farmers' Market (Thu 5:30-8:30pm Jul-Aug), Bank/ATM, Post Office, Beauty Salon (Cameo 656-1713), Barber Shop (Bell Buoy 656-4111), Dry Cleaners (Classic Drycleaners 656-7501), Bookstore (Time Enough for Books 655-1964 , Compass Rose Nautical 656-4674), Hardware Store (True Value 656-8611), Florist, Clothing Store (Cottons and Blues 656-6933), Retail Shops **Under 1 mi:** Supermarket (Safeway 656-1721, Thrifty Foods 656-0946), Wine/Beer, Liquor Store (BC Liquor 656-3041), Fishmonger (Satillite Fish Co. 656-2642), Catholic Church, Protestant Church, Library (656-0944), Pharmacy (Pharmasave 656-1148)

Transportation
OnCall: Taxi (Peninsula Taxi 656-1111) **Near:** Local Bus (BC Transit 382-6161) **Under 1 mi:** Rental Car (Budget 953-5300, Discount 310-2277) **1-3 mi:** Ferry Service (BC Ferries 656-0757) **Airport:** YVR (3 mi.)

Medical Services
911 Service **OnCall:** Ambulance (656-1001) **Under 1 mi:** Doctor (Family Medical 656-4143), Dentist (Sidney Dental 656-4848), Chiropractor (Sidney Chiro 655-0543), Holistic Services (Gaea Spa 655-9377), Veterinarian (Sidney Animal H. 656-3333) **Hospital:** Saanich Peninsla 652-3911 (5 mi.)

Setting -- Sidney-by-the-Sea presents its best face to the harbor. The delightful Port is easily recognizable by its floating white administration building with a dramatic, angled red roof. Enter through the overlapping arms of the protective breakwater to find wide, floating concrete docks, over 300 slips with plenty of visitor space, well-tended grounds, and numerous shipshape boater amenities. Classic lampposts and magnificent hanging baskets enhance the promenades along the basin, and from the streetside, the entrance to the docks is through a gourmet/gift shop.

Marina Notes -- Up to 110' on the "T" dock. *Power rates based on boat size: under 31' $3/night, 31-50' $4, 51-80' $5, over 80' $6. Monthly liveaboard fee $13.20/ft. Privately owned since it was built in 1990, the marina has fresh potable water, telephone, and cable TV. Radio ahead on VHF, or go to the check-in dock for berth allocation. A customs dock is on-site. Sidney Express Courier offers pick up & delivery to your boat and will handle groceries, liquor, parts, mail, and more (656-5126). Floating tent available for groups and dinners. The newly redone tile bathhouse and laundry are very roomy with good natural lighting.

Notable -- The Marine Ecology Centre, which floats next to the admin building, offers programs for all ages and public walk-in hours ($4/3, $10 for families). Sea Quest Whale Watching is always on the lookout for marine mammals. On-site or just steps from the marina are restaurants, delis, bakeries, museums. Downtown Sidney, with more museums, attractions, shops, galleries, bookshops and eateries is one block; supermarkets, marine stores, attractions and major car rentals are easily within a mile. Kids can pan for gold at nearby Mineral World. Those with cars can take a day trip to the gem mine for $100 (655-4367).

Navigational Information
Lat: 48°25.500' **Long:** 123°18.150' **Tide:** 9 ft. **Current:** n/a **Chart:** 3415
Rep. Depths (*MLW*): **Entry** 6 ft. **Fuel Dock** 6 ft. **Max Slip/Moor** 9 ft./-
Access: Located on the southeastern side of Vancouver Island

Marina Facilities *(In Season/Off Season)*
Fuel: Gasoline, Diesel
Slips: 420 Total, 20 Transient **Max LOA:** 80 ft. **Max Beam:** n/a
 Rate *(per ft.)*: **Day** $1.20/1.00 **Week** n/a **Month** n/a
 Power: 30 amp $5* **, 50 amp** n/a **, 100 amp** n/a **, 200 amp** n/a
 Cable TV: No **Dockside Phone:** No
 Dock Type: Floating, Long Fingers, Wood
Moorings: 0 Total, 0 Transient **Launch:** n/a
 Rate: Day n/a **Week** n/a **Month** n/a
Heads: 4 Toilet(s), 4 Shower(s)
Laundry: 1 Washer(s), 1 Dryer(s) **Pay Phones:** Yes, 2
Pump-Out: No **Fee:** n/a **Closed Heads:** Yes

Marina Operations
Owner/Manager: Dave Gosnell **Dockmaster:** Same
In-Season: Jun-Sep, 7am-7pm **Off-Season:** Oct-May, 9am-5pm
After-Hours Arrival: Call ahead
Reservations: Yes, Preferred **Credit Cards:** Visa/MC, Amex
Discounts: None
Pets: Welcome **Handicap Access:** Yes, Heads, Docks

Oak Bay Marina

1327 Beach Drive; Victoria, BC V8S 2N4

Tel: (250) 598-3369; (800) 663-7090 **VHF: Monitor** Ch. 66A **Talk** n/a
Fax: (250) 598-1361 **Alternate Tel:** n/a
Email: obmg@obmg.com **Web:** www.oakbaymarina.com
Nearest Town: Oak Bay Village *(0.5 mi.)* **Tourist Info:** (250) 383-7191

Marina Services and Boat Supplies
Services - Docking Assistance, Security *(locked gates at night)*, Trash Pick-Up, Dock Carts **Supplies - OnSite:** Ice *(Block, Cube)*, Ships' Store, Bait/Tackle **1-3 mi:** Propane

Boatyard Services
OnSite: Engine mechanic *(gas, diesel)*, Electrical Repairs, Propeller Repairs, Yacht Broker *(Vela)* **Dealer for:** Perkins, Sole, Yanmar, Volvo, Universal, Westerbeke, Lehman, Hino, Cummins, Mercruiser, B.M.C., Vetus.

Restaurants and Accommodations
OnSite: Restaurant *(Marina Restaurant 598-8555, L $7-18, D $20-35, sushi)*, Coffee Shop *(Marina Coffee House 7am-9pm Year-Round, outside deck)* **Near:** Lite Fare *(Windsor House Tea Room 595-3135)*, Hotel *(Oak Bay Beach Resort 598-4556, $175-418)*, Inn/B&B *(Oak Bay Guest House 598-3812, $89-180)* **Under 1 mi:** Restaurant *(Side Street Bistro 370-6030, L, D)*, *(Splendid Chinese 592-3318)*, *(Palmer's 598-1100)*, Lite Fare *(Ottavio Italian Bakery & Deli 592-4080)*, Pizzeria *(Village Pizza 598-3623)* **3+ mi:** Motel *(Quality Inn 385-6787, 4 mi.)*, Hotel *(Travellers Inn 953-1000, $50-110, 4 mi.)*

Recreation and Entertainment
OnSite: Picnic Area, Fishing Charter *(Sailing Charters, Kayak Tours)* **Near:** Beach, Jogging Paths *(Turkey Head Walkway)*, Boat Rentals *(Oak Bay Resort 592-3474)*, Special Events *(Oak Bay Tea Party - Jun 382-6161)* **Under 1 mi:** Playground, Fitness Center *(Lady Fitness 595-3354)*, Park *(Windsor Park)*, Galleries *(Oak Bay Village 598-9890)* **1-3 mi:** Pool *(Oak Bay Recreation 595-7946: classes, aquatics, sauna, sports fields)*, Spa,

Tennis Courts, Video Rental *(Yo Video 592-5678)* **3+ mi:** Golf Course *(Cedar Hill 595-3103, 5 mi.)*, Movie Theater *(National Geographic IMAX 480-4887, 4 mi.)*, Museum *(BC Maritime 385-4222, Craigdarroch Castle 592-5323, 4 mi.)*, Cultural Attract *(Royal Theatre 386-6121, McPherson Playhouse 386-6121, 4 mi.)*

Provisioning and General Services
OnSite: Retail Shops *(Giftshop)* **Under 1 mi:** Market *(Fairway 592-2433)*, Bank/ATM, Library *(592-2489)*, Beauty Salon *(Reflections 598-5421)*, Barber Shop *(Monterey Mews 595-0201)*, Hardware Store *(Oak Bay 598-4222)* **1-3 mi:** Supermarket *(Safeway 370-1779)*, Gourmet Shop *(Casey's 598-3432)*, Liquor Store *(BC Liquor 952-4220)*, Bakery *(Demitasse 598-6668)*, Meat Market *(Slater's 592-0823)*, Post Office, Catholic Church, Protestant Church, Dry Cleaners *(Individual 598-2950)*, Laundry *(598-7977)*, Bookstore *(Books & More 592-2933)*, Pharmacy *(Pharmasave 598-3380)*, Clothing Store *(Ellswear 598-5393, Country Life 598-3388)* **3+ mi:** Health Food *(Logical Health 595-6690, 4 mi.)*, Copies Etc. *(Staples 383-8178, 6 mi.)*

Transportation
OnCall: Rental Car *(Enterprise 475-6900)*, Taxi *(Blue Bird Cabs 382-4235; marina has a direct line to the company)* **Near:** Local Bus *(BC Transit 382-6161, #1 & 2/2N at night to downtown)* **Airport:** YVR *(19 mi.)*

Medical Services
911 Service **OnCall:** Ambulance *(388-5505)* **Under 1 mi:** Doctor *(Stubbs 595-4131)*, Dentist *(Oak Bay Dental 595-3833)* **1-3 mi:** Chiropractor *(Russell 595-2252)*, Holistic Services, Veterinarian *(Oak Bay Pet Clinic 598-4595)*, Optician *(Iris 953-8008)* **Hospital:** Royal Jubilee 370-8000 *(2 mi.)*

Setting -- Oak Bay is located at the southeastern tip of Vancouver Island - at Victoria's "back door" - overlooking both Maro and Juan de Fuca Straits. Protected by a stone breakwater, the 420 floating docks look out on a jumble of islands, most notably the Chatham Islands group - Mary Todd Island guards the entrance. Looking landward, the zig-zag roofed two-story Marina Restaurant building dominates, surrounded by the quiet residential village of Oak Bay.

Marina Notes -- *15 amp power also available. TThe service-oriented staff enhances the marina experience. Well maintained wooden docks with long finger piers have power and water at every slip - accented with potted plants. A well stocked store offers boat supplies, fishing tackle, clothing, and gifts. On-site Gartside Marine Engines (598-3378) provides many boatyard services. A fish cleaning table awaits your catch of the day. Clear Canadian customs here by telephone.

Notable -- Great dining is on-site: the mostly glass, dramatic Marina Restaurant serves seafood, pasta, steaks, and sushi in an elegant setting wtih fabulous views of Mt. Baker and the Straits. On the level below, the Coffee House and deck is a popular place to enjoy light fare and watch the boat traffic. Oak Bay Avenue starts about 0.3 mi. north of the marina and offers most services plus additional eateries. Around the Bay is Willows Beach, popular for family outings - it has a playground, sandy beach, picnic tables, restrooms and concession stand. Four beautiful gardens are open to visitors: Ada Beavan, Abkhazi Gardens, Lokier Gardens, and the Oak Bay Beach Hotel Native Plant Garden. For sightseeing & nightlife, the bus to Victoria stops just outside the marina. Craigdarroch Castle, the former home of coal magnate Robert Dunsmuir, offers a glimpse into the life of the 19th century very rich (9am-7pm in summer, $10/$3.50).

Fisherman's Wharf

12 Erie Street; Victoria, BC V8V 1V4

Tel: (250) 383-8326; (877) 783-8300 **VHF: Monitor** Ch. 66A **Talk** n/a
Fax: (250) 383-8306 **Alternate Tel:** n/a
Email: moorage@victoriaharbour.org **Web:** www.victoriaharbour.org
Nearest Town: Victoria **Tourist Info:** (250) 385-5711

Navigational Information
Lat: 48°25.420' **Long:** 123°23.007' **Tide:** 8 ft. **Current:** n/a **Chart:** 3415
Rep. Depths (*MLW*): **Entry** 17 ft. **Fuel Dock** n/a **Max Slip/Moor** 17 ft./-
Access: Around Laurel Point, adjacent to the Coast Harbourside Hotel

Marina Facilities *(In Season/Off Season)*
Fuel: No
Slips: 80 Total, 80 Transient **Max LOA:** 100 ft. **Max Beam:** n/a
Rate *(per ft.)*: **Day** $1.00/0.87 **Week** $6.09/3.15 **Month** $9.65*
Power: 30 amp $4.60** 50 amp n/a, 100 amp n/a, 200 amp n/a
Cable TV: No **Dockside Phone:** No
Dock Type: Floating, Alongside, Wood
Moorings: 0 Total, 0 Transient **Launch:** n/a
Rate: Day n/a **Week** n/a **Month** n/a
Heads: 4 Toilet(s), 2 Shower(s)
Laundry: Yes **Pay Phones:** Yes, 2
Pump-Out: No **Fee:** n/a **Closed Heads:** Yes

Marina Operations
Owner/Manager: Greater Victoria HA **Dockmaster:** Ian Crocker
In-Season: Jun-Sep, 8am-8pm **Off-Season:** Oct-May, 9am-4pm
After-Hours Arrival: Safe/drop box located at bottom of West gangway
Reservations: Yes*** **Credit Cards:** Visa/MC, Amex, Debit
Discounts: None
Pets: Welcome **Handicap Access:** No

Marina Services and Boat Supplies
Communication - Data Ports *(Wi-Fi - BB Xpress)* **Supplies - Near:** Ice
(Cube), Bait/Tackle **1-3 mi:** Ships' Store *(Bosuns 386-1308)*, West Marine

Boatyard Services
Nearest Yard: Point Hope Maritime (250) 385-3623

Restaurants and Accommodations
OnSite: Lite Fare *(Moka House 386-1313)*, *(Barb's Place 384-6515, L $8-10,
D $8-10, 10am-Dark, kids' menu $6)* **Near:** Restaurant *(Blue Crab Bar
and Grill 480-1999)*, *(Santiago's Café 388-7376)*, *(Victoria Harbour House
386-1244, D $19-36, from 4:30pm)*, Hotel *(Coast Harbourside 360-1211,
$149-237)* **Under 1 mi:** Hotel *(Grand Pacific 386-0450, $249-1000)*,
(Best Western 384-5122), Inn/B&B *(Robert Porter 385-8787, $145-215)*
1-3 mi: Restaurant *(Canoe Brewpub 361-1940, L $10-15, D $12-26)*,
(Azuma Sushi 382-8768, L & D $11-20), Pizzeria *(Pizza Factory 386-7888)*

Recreation and Entertainment
OnSite: Picnic Area, Fishing Charter *(Marine Adventure Gallery 361-3684:
kayaking, whale watching, fishing charters)*, Park *(Fisherman's Wharf Park)*,
Sightseeing *(SeaKing Whale Watching 361-8504)*, Galleries *(Marine
Adventure 361-3684)* **OnCall:** Boat Rentals *(Great Pacific Adventures 386-
2277, free pick-up)* **Near:** Museum *(Royal BC Museum 356-7226 9am-
6:30pm, Wax Museum 388-4461, Maritime Museum 385-4222; Parliament
Buildings)* **Under 1 mi:** Movie Theater *(National Geographic IMAX 480-
4887)* **1-3 mi:** Pool *(YMCA 386-7511)*, Fitness Center *(Grand Pacific 380-
4460)*, Cultural Attract *(Royal Theatre 386-6121, Music Conservatory 386-
5311)*, Special Events **3+ mi:** Golf Course *(Ceder Hill 595-3103, 5 mi.)*

Provisioning and General Services
OnSite: Fishmonger *(The Fish Store 383-6462, fresh fish on the dock)*
Near: Convenience Store *(Little Gem 386-3632)* **Under 1 mi:** Super-
market *(Thrifty Foods 380-2867)*, Delicatessen *(Jeffrey's 385-0077)*,
Health Food *(Seed of Life 382-4343)*, Liquor Store *(James Bay 356-2883)*,
Bakery *(Fort St. Bakery 383-0098)*, Bank/ATM, Protestant Church, Beauty
Salon *(Hairtrends 383-3245)*, Barber Shop *(Gary's Hair Place 385-3061)*, Dry
Cleaners *(Individual 382-5533)*, Laundry *(Pristine Cleaners & Laundry 386-
2220)*, Bookstore *(Snowden's 383-8131)*, Pharmacy *(Pharmasave 383-
7196)*, Florist *(Blossoms Cottage 386-1700)*, Clothing Store *(Oxfords
Sportswear 381-1532, Designers International 383-2115)* **1-3 mi:** Post
Office, Catholic Church, Synagogue, Library *(382-7241)*, Department Store
(Sears 385-8171), Copies Etc. *(Copy Copy 383-2679)*

Transportation
OnSite: Water Taxi *(Victoria Harbour 708-0201 $4/2)* **OnCall:** Rental Car
(Enterprise 475-5900), Taxi *(Victoria Taxi 383-7111)* **Near:** Local Bus
Under 1 mi: Ferry Service *(to Port Angeles 386-2202)* **1-3 mi:** Bikes *(Cycle
BC Rentals 380-2453)*, InterCity Bus *(Pacific Coach Lines 385-4411)*, Airport
Limo *(AKAL Airport Shuttle Bus 386-2525)* **Airport:** YVR *(17 mi.)*

Medical Services
911 Service **OnCall:** Ambulance *(388-5505)* **Under 1 mi:** Doctor *(Cool Aid
Community Health Centre 385-1466)*, Dentist *(James Bay Dental 380-6655)*,
Chiropractor *(James Bay Chiropractic & Massage 385-5583)*, Optician *(RX
Eyewear 385-5530)* **1-3 mi:** Holistic Services *(Heartstone Holistic Health
389-0858)*, Veterinarian *(Napier Lane Animal Clinic 381-7729)*
Hospital: Royal Jubilee 370-8000 *(3.5 mi.)*

Setting -- The first facility at the entrance to Victoria's Inner Harbour, Fisherman's Wharf is also known as Erie Street Wharf. A fun, funky ambiance is created by the mix of well-maintained floating homes, the fishing boats and pleasure vessels, and the businesses strung along the waterfront. Curved "greenhouse" windows signal the perched-on-pilings Wharf Office, which is backed by a park and high-rise buildings. Around Laurel Point is downtown.

Marina Notes -- *Monthly rate for pleasure boats $9.65/ft./mo. plus $12.05/mo. facility charge. **20 amp power. ***Moorage is available on a first-come, first-served basis, with reservations possible for vessels over 65', for groups, and during holidays/special events. ($20/night plus the moorage fee). This is one of several marinas managed by the Greater Victoria Harbour Authority (GVHA), which is curently in the process of upgrading facilities and services at all of its locations - call for updates. Rafting is mandatory during the busy summer season. The tiled bathhouse and laundry are across the street.

Notable -- Many businesses front the wharf, including: Moka House - with patio tables and great coffee, Marine Adventure Gallery - for art, gifts, and marine tours, SeaKing Whale Watching, and The Fish Store. Fisherman's Wharf Park is across the street with grass, a baseball diamond, picnic tables, and a bus stop. Stroll the docks and admire the colorful floating homes, then stop at Barb's Place on the dock for some fish and chips before heading into Victoria to savor the nightlife. The 12-person, enclosed Harbour Ferries run from Mar-Oct, every 12-20 min. depending on season; you can hop off downtown near the Visitors' Centre ($4/2) or choose a tour of the harbor or gorge ($16/$14 seniors/$12 under 8 yrs.).

Navigational Information
Lat: 48°25.420' **Long:** 123°22.788' **Tide:** 8 ft. **Current:** n/a **Chart:** 3415
Rep. Depths (*MLW*): **Entry** 16 ft. **Fuel Dock** n/a **Max Slip/Moor** 16 ft./-
Access: Between Raymour & Laurel Points, adjacent to Erie St. Public Wharf

Marina Facilities *(In Season/Off Season)*
Fuel: No
Slips: 50 Total, 40 Transient **Max LOA:** 120 ft. **Max Beam:** n/a
  Rate (*per ft.*): **Day** $1.65/1.00 **Week** n/a **Month** n/a
  Power: 30 amp Incl., **50 amp** Incl., **100 amp** n/a, **200 amp** n/a
  Cable TV: No **Dockside Phone:** No
  Dock Type: Floating, Long Fingers, Alongside, Concrete, Wood
Moorings: 0 Total, 0 Transient **Launch:** n/a
  Rate: Day n/a **Week** n/a **Month** n/a
Heads: 4 Toilet(s), 6 Shower(s) *(with dressing rooms)*, Hair Dryers
Laundry: None **Pay Phones:** Yes, 2
Pump-Out: OnSite, Self Service **Fee:** $10* **Closed Heads:** Yes

Marina Operations
Owner/Manager: Coast Hotels & Resorts **Dockmaster:** Gail Windle
In-Season: May-Sep, 24/7 **Off-Season:** Oct-Apr
After-Hours Arrival: Register at the front desk
Reservations: Yes, Preferred **Credit Cards:** Visa/MC, Dscvr, Din, Amex, Debit
Discounts: None
Pets: Welcome, Dog Walk Area **Handicap Access:** Yes, Heads, Docks

Coast Harbourside Hotel/Marina

146 Kingston Street; Victoria, BC V8V 1V4

Tel: (250) 360-1211 **VHF: Monitor** Ch. 66A **Talk** n/a
Fax: (250) 360-1418 **Alternate Tel:** n/a
Email: victoriamarina@coasthotels.com **Web:** www.coasthotels.com
Nearest Town: Victoria **Tourist Info:** (250) 385-5711

Marina Services and Boat Supplies
Services - Docking Assistance, Concierge, Security *(24 hrs.)*, Trash Pick-Up, Megayacht Facilities **Communication -** Mail & Package Hold, Phone Messages, Fax in/out *($1/pg)*, Data Ports *(Wi-Fi - Broadband Xpress)*, FedEx, UPS, Express Mail *(Sat Del)* **Supplies - Near:** Bait/Tackle **1-3 mi:** Ships' Store *(Bosun's Locker 386-1308)*, West Marine *(380-4097)*

Boatyard Services
Nearest Yard: Point Hope Maritime (250) 385-3623

Restaurants and Accommodations
OnSite: Restaurant *(Blue Crab Bar & Grill 480-1999, B $11-19, L $10-15, D $27-68, daily specials)*, Hotel *(Coast Harbourside 360-1211, $149-237)* **Near:** Restaurant *(The Cooks Landing 386-8712)*, *(Victoria Harbour House 386-1244, D $19-36)* **Under 1 mi:** Restaurant *(The Bent Mast 383-6000, L $7-12, D $10-18)*, Coffee Shop *(Liberty Java & Juice 383-2213)*, Hotel *(Hotel Grand Pacific 386-0450, $249-1000)*, Inn/B&B *(Robert Porter House 385-8787, $145-215)* **1-3 mi:** Restaurant *(Crown Palace 383-5322, L $8-15, D $8-15, delivery after 4pm w/$16 order)*, Fast Food *(Chicken on the Run 385-3468, D $8-23, delivery w/$16 order)*

Recreation and Entertainment
OnSite: Heated Pool, Spa, Fitness Center, Sightseeing *(Seacoast Expeditions 383-2254, Orca Spirit Adventures 383-8411, 3 hr. trips: summer $89/$69 Youth/$59 under 12, $20 less off-season; 9am, 1pm, and 5pm in spring & summer, and 1pm in fall & winter)* **Near:** Boat Rentals *(kayaks at Fisherman's Wharf)* **Under 1 mi:** Movie Theater *(National Geographic IMAX 480-4887)*, Video Rental *(Oswego 384-9982)*, Park *(Beacon Hill)*,

Museum *(Royal London Wax 388-4461, BC Maritime 385-4222, Undersea Gardens 382-5717)*, Cultural Attract *(McPherson Playhouse 386-6121, concerts at Music Conservatory 386-5311)*, Galleries *(Art of Man 383-3800)* **3+ mi:** Golf Course *(Henderson Park 370-7200, 6 mi.)*

Provisioning and General Services
OnSite: Dry Cleaners, Copies Etc. **Near:** Convenience Store *(Little Gem 386-3632)*, Retail Shops **Under 1 mi:** Supermarket *(Thrifty Foods 380-2867)*, Health Food *(Seed of Life 382-4343)*, Liquor Store *(James Bay 356-2883)*, Bakery *(Fort St. Bakery 383-0098)*, Fishmonger *(The Fish Store 383-6462)*, Bank/ATM, Protestant Church, Library *(382-7241)*, Beauty Salon *(Hairtrends 383-3245)*, Barber Shop *(Gary's 385-3061)*, Bookstore *(Avalon 380-1721)*, Pharmacy *(Pharmasave 383-7196)*, Clothing Store *(Sportex 382-2521)* **1-3 mi:** Delicatessen *(Sam's "Home of the Extraordinary Sandwich" 382-8424)*, Catholic Church, Hardware Store *(Do-It Centre 384-8181)*

Transportation
OnCall: Taxi *(Empress 381-2222)* **Near:** Water Taxi *(Victoria Harbour 708-0201, 9am-8:15pm in summer $4/2)*, Rental Car *(Hertz 360-2822)*, Ferry Service *(Port Angeles Ferry 386-2202)* **Under 1 mi:** Local Bus, InterCity Bus *(Pacific Castlines 385-4411)* **1-3 mi:** Bikes *(Cycle Rentals 380-2453)*, Airport Limo *(AKAL Shuttle 386-2525)* **Airport:** YVR *(17 mi.)*

Medical Services
911 Service **OnCall:** Ambulance **Under 1 mi:** Doctor *(James Bay 388-9934)*, Dentist *(James Bay 380-6655)*, Chiropractor *(James Bay 385-5583)* **1-3 mi:** Holistic Services *(Hearthstone 389-0858)*, Veterinarian *(Napier Lane 381-7729)* **Hospital:** Royal Jubilee 370-8000 *(5 mi.)*

Setting -- The Coast Harbourside Hotel & Marina is tucked into a small bay between Raymour Point and Laurel Point, past Fisherman's Wharf. It's a short walk to the ferry terminal, and about ten minutes to downtown. The view across the bay is of upscale condominiums and landward of the ten-story brick hotel - topped by its signature blue roof. All the passing traffic in and out of Victoria provides plenty of interest.

Marina Notes -- *Pump-out free with moorage; this is the only pump-out station in the Inner Harbour. Excellent concrete docks with long finger piers are kept shipshape, and alongside space is available for larger yachts. Guests of the marina are guests of the hotel, which boasts the only indoor-outdoor swimming pool on Vancouver Island. Other amenities include an exercise room, sauna and whirlpool, and all-tile heads and showers. Liveaboards permitted during the off-season. Note: Sailing is not allowed inside Victoria's breakwater and the float plane area is designated by yellow buoys.

Notable -- The elegant Coast Harbourside is one of Victoria's premier waterfront hotels, and The Blue Crab restaurant is highly regarded for its seafood dishes. Try chef Phil Lavoie's signature crab cakes, paella, or indulge with a "Sea & Pasture" entrée for 2 (Black Angus steaks, prawns, scallops, and your choice of crabs or lobsters) for $80-95. Lighter fare at the Lounge from 5-10pm. The hotel rooms feature balconies to enjoy the harbor scene, 24-hr. room service. Golf packages available. A stroll along the boardwalk or a ride in a miniature tug will quickly take you downtown. Don't miss high tea at the Empress Hotel ($49) or the famous Butchart Gardens (652-5256). Buses make the trip several times a day, or rent a car.

James Bay Causeway

Government Street; Victoria, BC V8V 1W9

Tel: (250) 383-8326; (844) 783-8300 **VHF: Monitor** Ch. 66A **Talk** n/a
Fax: (250) 383-8306 **Alternate Tel:** n/a
Email: moorage@victoriaharbour.org **Web:** www.victoriaharbour.org
Nearest Town: Victoria **Tourist Info:** (250) 385-5711

Navigational Information
Lat: 48°25.320' **Long:** 123°22.165' **Tide:** 8 ft. **Current:** n/a **Chart:** 3415
Rep. Depths (*MLW*): **Entry** 10 ft. **Fuel Dock** n/a **Max Slip/Moor** 10 ft./-
Access: In James Bay, in front of the Empress Hotel on the Inner Harbour

Marina Facilities *(In Season/Off Season)*
Fuel: No
Slips: 33 Total, 33 Transient **Max LOA:** 25057 ft. **Max Beam:** n/a
 Rate *(per ft.)*: **Day** $1.25/0.52 **Week** $7.50 **Month** 5.00
 Power: 30 amp $4.60*, **50 amp** $6.90, **100 amp** n/a, **200 amp** n/a
 Cable TV: No **Dockside Phone:** No
 Dock Type: Floating, Alongside, Wood
Moorings: 0 Total, 0 Transient **Launch:** n/a
 Rate: Day n/a **Week** n/a **Month** n/a
Heads: 6 Toilet(s), 6 Shower(s)
Laundry: 3 Washer(s), 3 Dryer(s) **Pay Phones:** Yes, 4
Pump-Out: OnSite **Fee:** n/a **Closed Heads:** Yes

Marina Operations
Owner/Manager: Greater Victoria HA **Dockmaster:** Ian Crocker
In-Season: Jun-Sep **Off-Season:** Oct-May
After-Hours Arrival: Safe/drop box located at bottom of gangway
Reservations: Yes, $25 fee** **Credit Cards:** Visa/MC, Amex, Debit
Discounts: Groups **Dockage:** n/a **Fuel:** n/a **Repair:** n/a
Pets: Welcome **Handicap Access:** Yes

Marina Services and Boat Supplies
Communication - Data Ports *(Wi-Fi - Broadband Xpress)*, FedEx, UPS
Supplies - Near: Ice *(Cube)*, Ships' Store *(Bosun's Locker 386-1308)*,
Bait/Tackle **1-3 mi:** West Marine *(380-4097)*

Boatyard Services
Nearest Yard: Point Hope Maritime (250) 385-3623

Restaurants and Accommodations
Near: Restaurant *(Milestone's 388-7313, L $9-18, D $15-26)*, *(Blackfish Café 385-9996, L $8-13, D $8-13)*, *(Panda Szechuan 388-0080)*, *(Pescatore's 385-4512, L $12-22, D $19-30)*, *(Old Spaghetti Factory 381-8444)*, *(Barkley's Steak & Seafood 382-7111, D $23-55)*, *(Ebizo Japanese 383-3234)*, *(Empress Room 389-2727)*, Coffee Shop *(Timothy's 388-3770)*, Motel *(Crystal Court 384-0551, $70-135, kitchenettes avail., seasonal rates)*, Hotel *(Fairmont Empress 384-8111, $175-749, seasonal rates)*, *(Bedford Regency 384-6835, $99-219)*, Inn/B&B *(Helm's Inn 385-5767, $85-205, kitchenettes)*

Recreation and Entertainment
Near: Boat Rentals *(Inner Harbour Centre 995-2211)*, Fishing Charter *(Inner Harbour)*, Group Fishing Boat *(Beasley's 381-8000)*, Movie Theater *(IMAX 480-4887)*, Museum *(Royal London Wax Museum 388-4461, Royal BC Museum 356-7226, Maritime Museum 385-4222)*, Cultural Attract *(Royal Theatre 386-6121, Victoria Symphony 385-6515)*, Sightseeing *(Undersea Gardens 382-5717, BC Legislative Bldgs. 387-3046, Prince of Whales 383-4884)*, Special Events, Galleries **Under 1 mi:** Pool *(YMCA 386-7511)*, Fitness Center *(YMCA 386-7511)*, Jogging Paths, Video Rental *(Prime Time 384-2254)* **3+ mi:** Golf Course *(Henderson Park 370-7200, 5 mi.)*

Provisioning and General Services
Near: Convenience Store *(7-Eleven 873-0516)*, Delicatessen *(Jeffrey's 385-0077)*, Health Food *(Seed of Life 382-4343)*, Bakery *(Fort St. 383-0098)*, Bank/ATM, Protestant Church *(St. Andrew's Presb.)*, Library *(382-7241)*, Beauty Salon *(Eclectic Hair & Day Spa 382-4641)*, Pharmacy *(Rexall 384-1195)*, Clothing Store *(Collections 380-0697, Oxfords 381-1532)*, Retail Shops **Under 1 mi:** Market *(GPL 381-6522)*, Liquor Store *(Swiftsure 384-0212)*, Post Office *(386-1214)*, Catholic Church, Barber Shop, Dry Cleaners *(Individual 382-5533)*, Laundry, Bookstore *(Munro's 382-2464, Annabelle's 380-7708)* **1-3 mi:** Supermarket *(Thrifty Foods 380-2867)*, Farmers' Market *(Moss St. Sat 10am-2pm, 361-1747)*, Fishmonger *(The Fish Store 383-6462)*, Synagogue, Hardware Store *(Pacific Brass 388-5311)*, Department Store *(Hudson's Bay 385-1311)*, Copies Etc. *(Copy Copy 383-2679)*

Transportation
OnSite: Water Taxi *(Victoria Harbour 708-0201 $4/2)* **OnCall:** Taxi *(Victoria Taxi 383-7111)* **Near:** Bikes *(Cycle BC 885-2453 bikes $6-18/hr., scooters $16/hr., motorcycles $25-35/hr.)*, Rental Car *(National 386-1213, Budget 953-5300)*, Local Bus *(BC Transit 382-6161)* **Under 1 mi:** Ferry Service *(Port Angeles Ferry 386-2202)* **Airport:** Victoria Int'l. *(16 mi.)*

Medical Services
911 Service **OnCall:** Ambulance *(388-5505)* **Near:** Doctor *(Broad St. 380-2620)*, Dentist *(Geddo 389-0669)*, Chiropractor *(City Chiropractic 382-4476)*, Holistic Services *(Willow Stream Spa at Empress Hotel 995-4650)* **Under 1 mi:** Optician *(EyeMax 383-3937)* **1-3 mi:** Veterinarian *(Quadra Animal Clinic 383-7124)* **Hospital:** Royal Jubilee 370-8000 *(2.5 mi.)*

Setting -- "The Causeway" is the main destination for pleasure craft visiting Victoria. Tucked into James Bay, it sits along the beautifully tended waterfront promenade at the foot of the majestic Empress Hotel - and provides easy foot access to a host of attractions. The wharf is not fenced or secured, so the level of activity on a summer weekend could be a bit frenetic. The boats are a "tourist attraction" along with the buskers, sidewalk vendors, tours, and carriages.

Marina Notes -- *20 amp $3.40. **Reservations during holidays & special events - under 100 ft. $25 non-refundable fee. Reservations all times 65-100 ft. $25, 100-200 $50, 200+ ft. $100. 48-hour max. stay in summer. Causeway floats expanded and upgraded in '06. Alongside moorage on long piers, or stern-to on dock section closer to the boardwalk. Rafting mandatory at all GVHA facilities; payment due upon arrival. Customs office nearby at Government & Wharf Streets. GVHA's new home is around the corner on Bellville St. Bathhouse and laundry share a building with Milestone's Restaurant.

Notable -- In downtown Victoria, you are more likely to run out of time than of things to do. Endless shopping opportunities are within walking distance - Native crafts to Callebaut chocolates and Irish linens. Major BC museums, an IMAX and several live theaters are close at hand. The seasonal Government St. Market features gifts, crafts, and food on Saturdays from 11am to 5pm. Dining options range from to-go lunches at quaint coffee shops to sophisticated meals and high tea at the Empress - a do-not miss for visitors (reservations required). Tours by car, boat, horse drawn carriage, double decker bus, and float plane are available. To explore the city on your own, hourly or daily rentals are available at Cycle BC on Wharf St., and several rental car offices are also nearby.

Navigational Information
Lat: 48°25.380' **Long:** 123°22.249' **Tide:** n/a **Current:** n/a **Chart:** 3415
Rep. Depths *(MLW):* **Entry** n/a **Fuel Dock** n/a **Max Slip/Moor** -/-
Access: Between James Bay floats and Wharf Street Public Wharf

Marina Facilities *(In Season/Off Season)*
Fuel: No
Slips: 20 Total, 20 Transient **Max LOA:** 250 ft. **Max Beam:** n/a
  Rate *(per ft.):* **Day** $1.00/0.87 **Week** $6.09/3.15 **Month** n/a
  Power: 30 amp $4.60** , **50 amp** $6.90, **100 amp** n/a, **200 amp** n/a
  Cable TV: No **Dockside Phone:** No
  Dock Type: Floating, Alongside, Wood
Moorings: 0 Total, 0 Transient **Launch:** n/a
  Rate: Day n/a **Week** n/a **Month** n/a
Heads: 6 Toilet(s), 6 Shower(s)
Laundry: 3 Washer(s), 3 Dryer(s) **Pay Phones:** No
Pump-Out: No **Fee:** n/a **Closed Heads:** Yes

Marina Operations
Owner/Manager: Greater Victoria HA **Dockmaster:** Ian Crocker
In-Season: Jun-Sep, 8:30am-4:30pm **Off-Season:** Oct-May
After-Hours Arrival: Call ahead
Reservations: Yes, $25 fee* **Credit Cards:** Visa/MC, Amex, Debit
Discounts: None
Pets: Welcome **Handicap Access:** No

Ship Point Wharf

814 Wharf Street; Victoria, BC V8W 1T3

Tel: (250) 383-8326; (877) 783-8300 **VHF: Monitor** Ch. 66A **Talk** n/a
Fax: (250) 383-8306 **Alternate Tel:** n/a
Email: moorage@victoriaharbour.org **Web:** www.victoriaharbour.org
Nearest Town: Victoria **Tourist Info:** (250) 385-5711

Marina Services and Boat Supplies
Communication - Data Ports *(Wi-Fi - Broadband Xpress)* **Supplies -**
Near: Ice *(Cube)*, Ships' Store *(Bosun's Locker 386-1308)* **1-3 mi:** West Marine

Boatyard Services
Nearest Yard: Point Hope Maritime (250) 385-3623

Restaurants and Accommodations
OnSite: Restaurant *(Milestone's 388-7313, L $9-18, D $15-26)* **Near:** Restaurant *(The Blackfish Café 385-9996, L $8-13, D $8-13)*, *(Pescatore's 385-4512, L $12-22, D $19-30)*, *(Siam Thai 383-9911)*, *(Empress Room 389-2727, 7am-9pm daily)*, *(Nautical Nellies 380-2260, L $8-16, D $15-30, brunch Sat & Sun 11:30am-3pm $10-13)*, Snack Bar *(hot dogs & ice cream)*, Pizzeria *(Pacific Rim 385-1167)*, Hotel *(Fairmont Empress 384-8111, $175-749)*, *(Marriott 480-3800)*, *(Grand Pacific 386-0450)* **Under 1 mi:** Inn/B&B *(Rosewood Inn 384-6644, $150-275)*

Recreation and Entertainment
OnSite: Special Events *(Dragon Boat Fest - Aug)* **Near:** Jogging Paths, Boat Rentals *(Inner Harbour Centre 995-2211)*, Fishing Charter *(Inner Harbour)*, Group Fishing Boat *(Beasley's 381-8000)*, Movie Theater *(National Geographic IMAX 480-4887)*, Museum *(Royal BC Museum 356-7226, Maritime Museum of BC 385-4222, Art Gallery of Greater Victoria 384-4101)*, Cultural Attract *(McPherson Playhouse 386-6121, Music Conservatory 386-5311)*, Sightseeing *(Great Pacific Adventures 386-2277)*, Galleries *(Out of Hand 384-5221, Indian Crafts 386-2529)* **Under 1 mi:** Pool *(YMCA 386-7511)*, Golf Course *(Henderson Park 370-7200)*, Fitness Center *(Grand Pacific Athletic Club 380-4460)*, Video Rental *(Prime Time 384-2254)*, Park

Provisioning and General Services
Near: Convenience Store *(7-Eleven 873-0516)*, Delicatessen *(Jeffrey's 385-0077)*, Health Food *(Seed of Life Natural Foods 382-4343)*, Wine/Beer *(Wine Barrel 388-0606)*, Bakery *(Fort St. Bakery 383-0098)*, Bank/ATM, Protestant Church *(St. Andrew's Presbyterian 384-5734)*, Library *(382-7241)*, Beauty Salon *(Crescendo for Hair 386-7550)*, Pharmacy *(Rexall 384-1195)*, Clothing Store *(Adventure Clothing 384-3337, Breze 383-8871)* **Under 1 mi:** Market *(GPL Grocery 381-6522)*, Liquor Store *(Swiftsure 384-0212)*, Catholic Church *(St. Andrew's Cathedral 388-5571)*, Barber Shop *(Jimmy's 384-2629)*, Dry Cleaners *(Viking Cleaners & Laundry 384-4600)*, Laundry, Bookstore *(Avalon 380-1721, Earth Quest Books 361-4533)*, Florist *(Flowers on Top 383-5262)* **1-3 mi:** Supermarket *(Thrifty Foods 380-2867)*, Farmers' Market *(Fairfield & Moss St. Sat 10am-2pm 361-1747)*, Synagogue *(Temple Emanuel 382-5222)*, Copies Etc. *(Copy Copy 383-2679)*

Transportation
OnSite: Bikes *(Cycle BC 885-2453 bikes $6-18/hr., scooters $16/hr., motorcycles $25-35/hr.)* **OnCall:** Rental Car *(Enterprise 476-5900/Avis 386-8468, nearby)*, Taxi *(Victoria Taxi 383-7111)* **Near:** Water Taxi *(Victoria Harbour 708-0201 $4/2)*, Local Bus *(BC Transit 382-6161)* **Under 1 mi:** Ferry Service *(Port Angeles Ferry 386-2202)* **Airport:** Victoria Int'l. *(16 mi.)*

Medical Services
911 Service **OnCall:** Ambulance *(388-5505)* **Near:** Doctor *(Downtown Medical Centre 380-2210)*, Dentist *(Bayshore Dental 381-6433)*, Chiropractor *(City Chiropractic 382-4476)*, Holistic Services *(European Skin Care 383-2818)* **Under 1 mi:** Optician *(EyeMax 383-3937)* **1-3 mi:** Veterinarian *(Quadra 383-7124)* **Hospital:** Royal Jubilee 370-8000 *(2.5 mi.)*

7. BC - SAANICH PENINSULA & VICTORIA

PHOTOS ON DVD: 12

Setting -- Wide, public Ship Point Wharf juts out into the harbor in the heart of downtown Victoria - between the James Bay Causeway and the Wharf Street Marina. A 500 foot side-tie floating dock is attached to Ship Point Wharf's south side. The pier to the north is home to seaplane services plus several diving and whale-watching operators. The waterfront promenade, with antique lamp posts dripping with flowers, reminds one why this is called the Garden City.

Marina Notes -- *Reservations - boats over 65 ft., groups, and special events & holidays (see James Bay). **20 amp $3.40. Max. 48 hr. stay in summer. Dockage primarily meant for larger yachts and fishing vessels - the latter are usually out in the summer, making room for more pleasure craft. The GVHA is replacing some of the floats at this facility, and added a customer service kiosk in 2006. Vehicle access requires advance permission. A white building just behind the wharf houses Milestone Restaurant, several shops, and the bathhouse and laundry. On top is Tourism Victoria's downtown information center.

Notable -- Ship Point is a great place to relax and enjoy the perpetual harbor activity; watch seaplanes maneuver nearby, slicker-clad tourists heading out for whale watching trips, visiting boats arriving at nearby docks, and Harbour Ferries carrying tourists to their destinations. Street vendors and entertainers line the waterfront promenade to intrigue and entertain visitors. Milestone's, a very popular Canadian eatery with about 30 locations, serves an impressive menu plus an innovative wine list. Enjoy the casual atmosphere and harbor views, or carry out to your boat. For a unique way to see the city, try a horse drawn carriage: Tally Ho (514-9257), Black Beauty (361-1220), and Victoria Carriage Tours (383-2207) are nearby.

Navigational Information

Lat: 48°25.450' **Long:** 123°22.280' **Tide:** 8 ft. **Current:** n/a **Chart:** 3415
Rep. Depths (MLW): Entry 9 ft. **Fuel Dock** n/a **Max Slip/Moor** -/9 ft.
Access: Located next to the Customs Office on the Inner Harbour

Marina Facilities *(In Season/Off Season)*

Fuel: No
Slips: 4 Total, 4 Transient **Max LOA:** 100 ft. **Max Beam:** n/a
  Rate *(per ft.):* **Day** $1.25/0.87 **Week** n/a **Month** n/a
  Power: 30 amp $4.60, **50 amp** n/a, **100 amp** n/a, **200 amp** n/a
  Cable TV: No **Dockside Phone:** No
  Dock Type: Floating, Alongside, Wood
Moorings: 0 Total, 0 Transient **Launch:** n/a
  Rate: Day n/a **Week** n/a **Month** n/a
Heads: 2 Toilet(s), 2 Shower(s)
Laundry: 3 Washer(s), 3 Dryer(s) **Pay Phones:** Yes, 2
Pump-Out: No **Fee:** n/a **Closed Heads:** Yes

Marina Operations

Owner/Manager: GVHA/Louise Carlow **Dockmaster:** Ian Crocker
In-Season: Jun-Sep, 8:30am-4:30pm **Off-Season:** Oct-May
After-Hours Arrival: Tie up and pay in the morning
Reservations: Yes* **Credit Cards:** Visa/MC, Amex, Debit
Discounts: None
Pets: Welcome **Handicap Access:** No

Broughton Street Wharf

PO Box; 1002A Wharf Street; Victoria, BC V8V 1W9

Tel: (250) 383-8326 **VHF: Monitor** Ch. 66A **Talk** n/a
Fax: n/a **Alternate Tel:** n/a
Email: moorage@victoriaharbour.org **Web:** n/a
Nearest Town: Victoria **Tourist Info:** (250) 385-5711

Marina Services and Boat Supplies

Services - Security *(locked gate)* **Supplies - Near:** Ice *(Cube)*, Ships' Store *(Bosun's Locker 386-1308)* **1-3 mi:** West Marine *(380-4097)*

Boatyard Services

Nearest Yard: Point Hope Maritime (250) 385-3623

Restaurants and Accommodations

OnSite: Restaurant *(Blackfish Café 385-9996, L & D $8-13)* **Near:** Restaurant *(Milestone's 388-7313, L $9-18, D $15-26)*, *(Camille's 381-3433, D $20-33)*, *(Nautical Nellie's 380-2260, L $8-16, D $15-30)*, *(The Keg Steakhouse 386-7789)*, *(Pagliaci's 386-1662, L $8-11, D $11-20, live music Sun-Wed)*, *(Rebecca's 380-6999, seafood)*, *(JR's Curry House 384-5622)*, *(Ebizo 383-3234, sushi)*, Coffee Shop *(Bean Bandits 384-6635)*, Pizzeria *(Pacific Rim 385-1167)*, Hotel *(Fairmont Empress 384-8111, $175-749)*, *(Magnolia 381-0999)*, *(Quality Inn 385-6787)*, *(Bedford Regency 384-6835)*

Recreation and Entertainment

OnSite: Sightseeing *(Whale Watching Adventures; Old Town walking tours)* **Near:** Jogging Paths, Boat Rentals *(Inner Harbour Centre 995-2211)*, Fishing Charter *(Inner Harbour Centre)*, Group Fishing Boat *(Beasley's 381-8000)*, Movie Theater *(Nat'l Geographic IMAX 480-4887)*, Museum *(Maritime Museum of BC 385-4222, Royal BC 356-7226, Art Gallery of Greater Victoria 384-4101)*, Cultural Attract *(Royal Theatre 386-6121, McPherson Playhouse 386-6121, Victoria Symphony 385-6515)*, Galleries *(Eagle's Moon 361-4174, Scarmouche 386-2215)* **Under 1 mi:** Pool *(YMCA 386-7511)*, Fitness Center *(YMCA 386-7511)*, Video Rental *(Hollywood 385-4525)* **1-3 mi:** Park *(Beacon Hill)* **3+ mi:** Golf Course *(Henderson Park 370-7200, 5 mi.)*

Provisioning and General Services

Near: Convenience Store *(7-Eleven 873-0516)*, Health Food *(Seed of Life 382-4343)*, Wine/Beer *(Wine Barrel 388-0606)*, Liquor Store *(Strath Spirits 370-9463)*, Bakery *(Fort St. Bakery 383-0098)*, Fishmonger *(Cross Meats & Seafood 384-2631)*, Bank/ATM, Post Office *(386-1214)*, Protestant Church *(St. Andrew's Presbyterian)*, Beauty Salon *(Eclectic Hair Studio & Day Spa 382-4641)*, Barber Shop *(Martini 's Hair 386-4247)*, Dry Cleaners *(Individual Dry Cleaners 382-5533)*, Bookstore *(Russell Books 361-4447)*, Pharmacy *(Rexall 384-1195)*, Florist *(Flowers On Top 383-5262)*, Clothing Store *(Adventure Clothing 384-3337)*, Copies Etc. **Under 1 mi:** Market *(Moon Key 380-9681, GPL Grocery 381-6522)*, Delicatessen *(Souped Up 380-0856)*, Catholic Church, Library *(382-7241)*, Laundry *(Viking 384-4600)*, Hardware Store *(Pacific Brass 388-5311)* **1-3 mi:** Supermarket *(Save On Foods 389-6115)*, Farmers' Market *(Moss St. 361-1747, 10am-2pm Sat)*, Synagogue

Transportation

OnCall: Rental Car *(Enterprise 476-5900)*, Taxi *(Empress 381-2222)* **Near:** Bikes *(Cycle BC 885-2453, bikes $6-18/hr., scooters $16/hr., motorcycles $25-35/hr.)*, Water Taxi *(Harbour Ferry 708-0201, $4/2)*, LocalBus *(BC Transit 382-6161)* **Under 1 mi:** Ferry Service *(Port Angeles 386-2202)* **Airport:** YVR *(16 mi.)*

Medical Services

911 Service **OnCall:** Ambulance *(388-5505)* **Near:** Dentist *(Bayshore 381-6433)*, Chiropractor *(City Chiropractic 382-4476)*, Holistic Services *(Le Spa Sereine 388-4419)* **Under 1 mi:** Doctor *(Downtown Medical Centre 380-2210)*, Optician *(EyeMax 383-3937)* **1-3 mi:** Veterinarian *(Quadra Animal Clinic 383-7124)* **Hospital:** Royal Jubilee 370-8000 *(2.5 mi.)*

Setting -- The Broughton Street Wharf is just north of the commercial dock used by float planes and adventure tour businesses. Two piers provide side-tie moorage and a small gray building houses the Greater Victoria Harbour Authority's Services. The historic Customs House building rises above the docks.

Marina Notes -- *Over 65' & holidays. 48 hr. max stay during summer. Mandatory rafting. Limited space is available at the wharf which is often used as temporary mooring while clearing customs. Customs office is open 24/7, and all vessels arriving from the U.S. must check in before heading for other berths. Once tied up, use the phone on the dock to contact the office. For general information, the Canada Border Information Service has an automated line, or call Mon-Fri 8am-4pm for specific inquiries: 800-461-9999 or 204-983-3500 in Canada. Public restrooms, showers and laundry are down the street by the Wharf Street Marina in the white Milestone's Restaurant building. Note: The GVHA main office is located at 202-468 Belleville Street.

Notable -- The quiet, slightly less commerical surrounding area is home to some of the oldest buildings in Victoria - lovingly preserved and restored. The old Customs House, known as the Malahat Building, initially served gold miners arriving to the region in the 1870s. Behind Bastion Square, the old Courthouse designed by the same German architect houses the Maritime Museum of BC, one of Victoria's main attractions (daily 9:30am-4pm, 'til 5pm Jun15-Sep15, $8/$5 seniors/$3 kids). Look for the Mooring Rings dating back to the Fort Victoria days - this is where ships tied up before the wharfs were built. One may even encounter a mother otter and her two young on the wharf soaking up some sun. The downtown action is an easy walk.

Navigational Information
Lat: 48°25.502' **Long:** 123°22.309' **Tide:** 8 ft. **Current:** n/a **Chart:** 3415
Rep. Depths (*MLW*): **Entry** 22 ft. **Fuel Dock** n/a **Max Slip/Moor** 22 ft./-
Access: East side of Inner Harbour after leaving Middle Harbour

Marina Facilities (*In Season/Off Season*)
Fuel: No
Slips: 45 Total, 45 Transient **Max LOA:** 65 ft. **Max Beam:** n/a
Rate (*per ft.*): **Day** $1.25/0.52 **Week** $6.09/3.15 **Month** $9.65*
Power: 30 amp $4.60** , **50 amp** $6.90, **100 amp** n/a, **200 amp** n/a
Cable TV: No **Dockside Phone:** No
Dock Type: Floating, Alongside, Wood
Moorings: 0 Total, 0 Transient **Launch:** n/a
Rate: Day n/a **Week** n/a **Month** n/a
Heads: 2 Toilet(s), 2 Shower(s)
Laundry: 3 Washer(s), 3 Dryer(s) **Pay Phones:** Yes, 2
Pump-Out: No **Fee:** n/a **Closed Heads:** Yes

Marina Operations
Owner/Manager: Greater Victoria HA **Dockmaster:** Ian Crocker
In-Season: Jun-Sep, 8:30am-4:30pm **Off-Season:** Oct-May
After-Hours Arrival: Tie up and pay in the morning
Reservations: Yes*** **Credit Cards:** Visa/MC, Amex, Cash, Check, Debit
Discounts: 15% for Yacht Clubs **Dockage:** n/a **Fuel:** n/a **Repair:** n/a
Pets: Welcome **Handicap Access:** No

Wharf Street Marina

1002A Wharf Street; Victoria, BC V8V 1W9

Tel: (250) 383-8326; (877) 783-8300 **VHF: Monitor** Ch. 66A **Talk** n/a
Fax: (250) 383-8306 **Alternate Tel:** n/a
Email: moorage@victoriaharbour.org **Web:** www.victoriaharbour.org
Nearest Town: Victoria **Tourist Info:** (250) 385-5711

Marina Services and Boat Supplies
Services - Docking Assistance, Concierge, Security (*locked gate*), Trash Pick-Up **Communication -** Fax in/out (*$1/pg*) **Supplies - Near:** Ice (*Cube*), Ships' Store (*Bosun's 386-1308*) **1-3 mi:** West Marine (*380-4097*)

Boatyard Services
Nearest Yard: Point Hope Maritime (250) 385-3623

Restaurants and Accommodations
OnSite: Hotel (*Victoria Regent 386-2211, $129-730*) **Near:** Restaurant (*Blackfish Café 385-9996, L $8-13, D $8-13*), (*Ferris Oyster Bar & Grill 360-1824, L $7-11, D $12-19, Sun brunch 9am-2:30pm*), (*Green Cuisine 385-1809, vegan buffet, $7/lb.*), (*Siam Thai 383-9911*), (*Koto Japanese 382-1514*), (*Golden Chopsticks 388-3148*), (*Camille's 381-3433, D $20-33*), (*The Keg Steakhouse 386-7789*), (*Chandlers Seafood 385-3474*), Snack Bar (*Paradiso Di Stelle 920-7266, espresso & gelato*), Hotel (*Magnolia Hotel & Spa 381-0999*), (*Best Western Carlton Plaza 388-5513*)

Recreation and Entertainment
Near: Jogging Paths, Boat Rentals (*Inner Harbour 995-2211*), Group Fishing Boat (*Beasley's 381-8000*), Movie Theater (*National Geographic IMAX 480-4887*), Museum (*Maritime Museum of BC 385-4222, Royal BC Museum 356-7226, Wax Museum 388-4461*), Cultural Attract (*Victoria Symphony 385-6515, Royal Theatre 386-6121*), Special Events (*Swiftsure Yacht Race - Mem Day wknd, Jazz Festival - Jun, Folk Fest - Jul, First Peoples Festival Jul/Aug, Symphony Splash - Aug, Dragon Boat Festival - Aug, Fringe Festival - Aug*), Galleries (*Winchester Gallery 386-2773*) **Under 1 mi:** Pool (*YMCA 386-7511*), Fitness Center (*YMCA*), Video Rental (*Hollywood

Video 385-4525*) **3+ mi:** Golf Course (*Henderson Park 370-7200, 5 mi.*)

Provisioning and General Services
Near: Convenience Store (*7-Eleven 873-0516*), Delicatessen (*Rising Star 386-6660*), Health Food (*Seed of Life 382-4343*), Wine/Beer (*Wine Barrel 388-0606*), Liquor Store (*Strath 370-9463*), Bakery (*Dutch 385-1012*), Bank/ATM, Post Office (*386-1214*), Catholic Church, Protestant Church, Library (*382-7241*), Beauty Salon (*Eclectic Hair & Day Spa 382-4641*), Dry Cleaners (*Individual Dry Cleaners 382-5533*), Bookstore (*Munro's Books 382-2464*), Pharmacy (*Rexall 384-1195*), Hardware Store (*Pacific Brass 388-5311*), Florist (*Flowers On Top 383-5262*), Clothing Store (*Pacific Trekking 388-3976*), Copies Etc. (*L&J Copy 386-3333*) **Under 1 mi:** Market (*GPL Grocery 381-6522*), Laundry (*Viking 384-4600*) **1-3 mi:** Supermarket (*Save On Foods 389-6115*), Farmers' Market (*Moss St. Sat 10am-2pm*)

Transportation
OnCall: Rental Car (*Enterprise 476-5900/Avis 386-8468, near*), Taxi (*Empress 381-2222*) **Near:** Bikes (*Cycle BC 885-2453, bikes $6-18/hr., scooters $16/hr., motorcycles $25-35/hr.*), Water Taxi (*Victoria Harbour Ferry 708-0201, $4/2*), Local Bus (*BC Transit 382-6161*), Ferry Service (*Port Angeles 386-2202*) **Airport:** Victoria Int'l. (*16 mi.*)

Medical Services
911 Service **OnCall:** Ambulance **Near:** Doctor (*Downtown Medical Centre 380-2210*), Dentist (*Bayshore Dental 381-6433*), Chiropractor (*City Chiropractic 382-4476*), Holistic Services (*Le Spa Sereine 388-4419*) **Under 1 mi:** Optician (*EyeMax EyeCare 383-3937*) **1-3 mi:** Veterinarian (*Quadra Animal Clinic 383-7124*) **Hospital:** Royal Jubilee 370-8000 (*2.5 mi.*)

Setting -- Following the channel into Victoria's Inner Harbour, the Wharf Street Marina is straight ahead as the harbor narrows - just north of the orange-signed "Customs" float. 1,500 feet of slip and linear dockage lie in the shadow of the Victoria Regent Hotel - a bit away from the hustle and bustle of the main waterfront. A large parking area occupies the area immediately behind the docks, and Old Town with its historic buildings is just across Wharf Street.

Marina Notes -- *Limited monthly space; fee $9.65/ft./mo., plus $12.05/mo. facilities charge. **50 amp on float "D", also 20 amp service. ***Reservations for groups & holidays/special events ($20 fee). Transients use floats "B" through "D," in the southern part of the marina. Mandatory rafting & 48 hr. max stay during summer. Hyack, the Kenmore Air operator occupying the dock at the foot of the hotel, has expanded to create a new Adventure Centre, with scenic flights and whale watching tours. Public heads, showers and laundry. Because it is more likely than James Bay Causeway to have space available, this is the primary transient facility in Victoria. Note: The GVHA Office has moved to 202-468 Belleville St., Victoria, BC, V8V 1W9.

Notable -- The Wharf Street floats are situated midway between downtown Victoria and Chinatown - each just a couple of blocks away. Excellent restaurants are nearby, as are several attractions. The Royal BC Museum explores the social and natural history of British Columbia, with a recent addition on the region's climate and its future (daily 9am-5pm, $12.50/$8.70 kids & seniors/under 5 free); the IMAX theater has shows daily 10am-8pm. Victoria's mild climate allows for an amazing array of special events year-round. Symphony Splash, a concert performed on a barge in the Inner Harbour, is the highlight of the music season.

Johnson Street Public Wharf

1314 Wharf Street; Victoria, BC

Tel: (250) 383-8326; (877) 783-8300 **VHF:** Monitor Ch. 66A **Talk** n/a
Fax: (250) 383-8306 **Alternate Tel:** n/a
Email: moorage@victoriaharbour.org **Web:** www.victoriaharbour.org
Nearest Town: Victoria **Tourist Info:** (250) 385-5711

Navigational Information
Lat: 48°25.750' **Long:** 123°22.281' **Tide:** 8 ft. **Current:** n/a **Chart:** 3415
Rep. Depths *(MLW)*: **Entry** 9 ft. **Fuel Dock** n/a **Max Slip/Moor** -/9 ft.
Access: North end of Inner Harbour just before Upper Harbour

Marina Facilities *(In Season/Off Season)*
Fuel: No
Slips: 6 Total, 6 Transient **Max LOA:** 42 ft. **Max Beam:** n/a
  Rate *(per ft.)*: **Day** $1.25/0.52 **Week** n/a **Month** n/a
  Power: 30 amp n/a, **50 amp** n/a, **100 amp** n/a, **200 amp** n/a
  Cable TV: No **Dockside Phone:** No
  Dock Type: Floating, Alongside, Wood
Moorings: 0 Total, 0 Transient **Launch:** n/a
  Rate: Day n/a **Week** n/a **Month** n/a
Heads: None
Laundry: None **Pay Phones:** No
Pump-Out: No **Fee:** n/a **Closed Heads:** Yes

Marina Operations
Owner/Manager: Greater Victoria HA **Dockmaster:** Ian Crocker
In-Season: Jun-Sep, 8:30am-4:30pm **Off-Season:** Oct-May
After-Hours Arrival: Tie up and pay at the office in the morning
Reservations: Yes* **Credit Cards:** Visa/MC, Amex, Debit
Discounts: None
Pets: Welcome **Handicap Access:** Yes, Docks

Marina Services and Boat Supplies
Supplies - Near: Ice *(Cube)*, Ships' Store *(Bosun's Locker 386-1308)*
1-3 mi: West Marine *(380-4097)*

Boatyard Services
Nearest Yard: Point Hope Maritime (250) 385-3623

Restaurants and Accommodations
Near: Restaurant *(Wild Saffron 361-3310, D $15-32, fondue for 2 $25)*, *(Swans Brew Pub 361-3310, L $9-16, D $9-16, Buckerfield Brewery onsite)*, *(Il Terrazzo 361-0028, L $8-14, D $14-35, highly acclaimed)*, *(Tamami Sushi 382-3529)*, *(Green Cuisine 385-1809, vegan buffet, $7/lb.)*, *(Tommy Bahamas' 386-1140, live jazz Thu)*, *(Sour Pickle 384-9390)*, Coffee Shop *(Bean Around the World 386-7115)*, Fast Food *(Burrito Express 381-2333)*, Hotel *(Victoria Regent 386-2211, $129-730, water views)*, *(Best Western Carlton Plaza 388-5513, $89-200, kitchenettes)*, *(Swans Hotel 361-3310, $179-359)* **Under 1 mi:** Hotel *(Embassy Inn 382-8161, $119-239)*

Recreation and Entertainment
Near: Fitness Center *(Club Phoenix 920-0300)*, Boat Rentals *(Inner Harbour Centre 995-2211)*, Museum *(Maritime Museum 385-4222, Royal BC Museum 356-7226, Art Gallery of Greater Victoria 384-4101)*, Special Events *(Classic Boat Festival - Sep)* **Under 1 mi:** Pool *(YMCA 386-7511)*, Group Fishing Boat *(Beasley's 381-8000)*, Movie Theater *(IMAX 480-4887)*, Video Rental, Cultural Attract *(Pacific Opera 385-0222, Belfry Thearte 385-6815, Victoria Symphony 385-6515)*, Sightseeing *(Miniature World 385-9731, Parliament Bldg 387-3046)*, Galleries *(Store Street 480-7505)*
3+ mi: Golf Course *(Henderson Park 370-7200, 4 mi.)*

Provisioning and General Services
Near: Market *(Fisgard 383-6969, The Bay 385-1311)*, Delicatessen *(Pita Pocket 382-4432)*, Health Food *(Seed of Life 382-4343)*, Wine/Beer *(Wine Barrel 388-0606)*, Liquor Store *(Swams 361-3365)*, Bakery *(Dutch 385-1012)*, Bank/ATM, Beauty Salon *(Vivre Hair 920-7725)*, Barber Shop *(Riviera Barber 385-3931)*, Dry Cleaners *(China 384-6123)*, Laundry, Bookstore *(Annabelle's 380-7708)*, Florist, Clothing Store *(Funky Town 382-2626, Urban Apparel 386-8778)* **Under 1 mi:** Post Office *(953-1351)*, Catholic Church *(St. Andrew's 388-5571)*, Protestant Church, Pharmacy *(Rexall 384-1195)*, Hardware Store *(Pacific Brass 388-5311)*, Copies Etc. *(Copy Copy 383-2679)* **1-3 mi:** Supermarket *(Save On Foods 475-3302)*, Farmers' Market *(Moss St. Sat 10am-2pm 361-1747)*, Synagogue *(Emanuel 382-5222)*, Library *(382-7241)*

Transportation
OnSite: Water Taxi *(Victoria Harbour 708-0201 $4/2)* **OnCall:** Rental Car *(Enterprise 476-5900)*, Taxi *(Victoria Taxi 383-7111)* **Near:** Local Bus *(BC Transit 382-6161)* **1-3 mi:** Ferry Service *(Port Angeles 386-2202)*
Airport: YVR *(16 mi.)*

Medical Services
911 Service **OnCall:** Ambulance *(388-5505)* **Near:** Doctor *(Downtown Medical Centre 380-2210)*, Dentist *(Bayshore Dental 381-6433)*, Chiropractor *(City Chiropractic 382-4476)*, Holistic Services *(Le Spa Sereine 388-4419, Metro Massage Therapy 383-9775)* **Under 1 mi:** Optician *(EyeMax 383-3937)* **1-3 mi:** Veterinarian *(Quadra Animal Clinic 383-7124)*
Hospital: Royal Jubilee 370-8000 *(2.5 mi.)*

Setting -- The Johnson Street Wharf is a relatively small wooden float immediately before the blue Johnson Street Bridge - which marks the northern limit of Victoria's Inner Harbour and the beginning of Upper Harbour. The surrounding streets are part of Victoria's Chinatown; downtown is less than a mile south.

Marina Notes -- *Greater Victoria Harbour Authority facilities are available on a first-come, first-served basis, with reservations possible for groups, and during special events or holidays ($20 fee). Mandatory rafting applies to all GVHA docks. The Johnson Street facility consists of one wooden pier which provides 200 ft. of open dock space. No power or water are available at this location, and no boater amenities. The facilities at the Wharf St. Marina are the closest, with additional restrooms, showers, and laundry near the Ship Point Wharf.

Notable -- Chinatown, with its traditional gate and narrow alleys, is a great place to provision with fresh produce or to shop for Asian treasures. Stores and restaurants line Fisgard Street near the bridge, and tiny Fan Tan Alley attracts tourists with charming boutiques. Just a stone's throw away is Market Square surrounded by more shops. Follow Wharf St. or Government St. downtown, past the old buildings of Fort Victoria, and a miriad restaurants offering everything from steaks and seafood to vegan or Vietnamese dishes. The Maritime Museum of BC is nearby at Bastion Square; see Tilikum, a dugout canoe sailed around the world by Captain Voss in 1901-04. The Royal BC Museum is close to the Fairmount Empress Hotel, about 0.5 miles south. The Art Gallery is one of the best fine art museums in Canada; it's about 2 miles away, and open daily 10am-5pm, until 9pm on Thursdays ($8/$6 seniors/$2 kids/under 6 free).

8. WA – Point Roberts to Bellingham

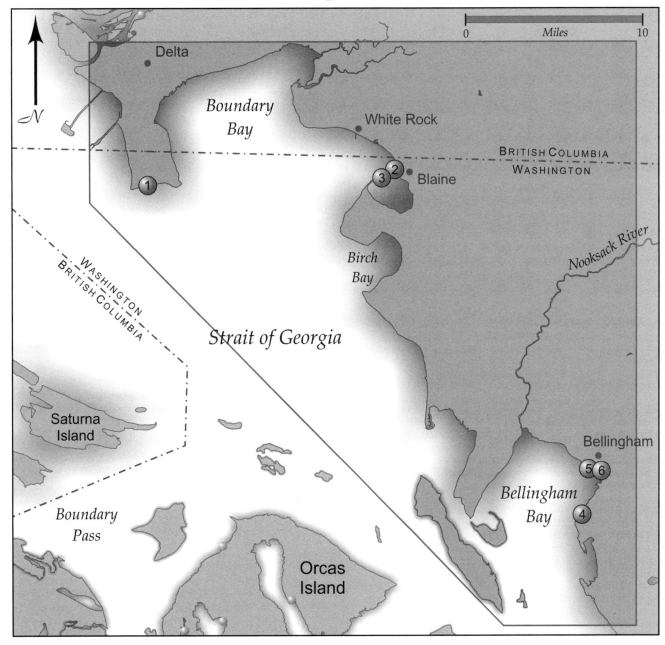

MAP	MARINA	HARBOR	PAGE	MAP	MARINA	HARBOR	PAGE
1	**Point Roberts Marina Resort**	*Point Roberts*	132	4	**Fairhaven Moorage**	*Bellingham Bay*	135
2	**Blaine Harbor**	*Drayton Harbor*	133	5	**Squalicum Harbor**	*Bellingham Bay/Squalicum Harbor*	136
3	Semiahmoo Marina	*Drayton Harbor*	134	6	**Hotel Bellwether**	*Bellingham Bay/Squalicum Harbor*	137

▸ **Currency** — In Canadian Marina Reports, all prices are in Canadian dollars. In U.S. Marina Reports, all prices are in U.S. dollars.

▸ **"CCM"** — Denotes a Certified Clean Marina, a state/provincial award for environmental excellence. See page 298 for an explanation and page 299 for a list of Pump-Out facilities.

▸ **Ratings & Reviews** — An overview of the Atlantic Cruising Club's rating system is on page 6 and details on the content of each Marina Report are on pages 7 – 11.

▸ **Marina Report Updates** — Comments from boaters and new information from ACC reviewers and marinas are posted regularly on *www.AtlanticCruisingClub.com*.

Navigational Information
Lat: 48°58.333' **Long:** 123°03.817' **Tide:** 14 ft. **Current:** 0 kt. **Chart:** 18400
Rep. Depths *(MLW)*: **Entry** 12 ft. **Fuel Dock** 14 ft. **Max Slip/Moor** 14 ft./-
Access: 10 mi. east of Active Pass, just south of the Canadian border

Marina Facilities *(In Season/Off Season)*
Fuel: *Shell* - Gasoline, Diesel, High-Speed Pumps
Slips: 1023 Total, 200 Transient **Max LOA:** 120 ft. **Max Beam:** n/a
  Rate *(per ft.)*: **Day** $1.00 **Week** n/a **Month** $6.75
  Power: 30 amp Incl., **50 amp** Incl., **100 amp** Incl., **200 amp** n/a
  Cable TV: No **Dockside Phone:** No
  Dock Type: Floating, Long Fingers, Concrete
Moorings: 0 Total, 0 Transient **Launch:** n/a
  Rate: **Day** n/a **Week** n/a **Month** n/a
Heads: 10 Toilet(s), 8 Shower(s) *(with dressing rooms)*
Laundry: 4 Washer(s), 4 Dryer(s), Book Exchange **Pay Phones:** Yes, 4
Pump-Out: OnSite, Self Service, 2 Central **Fee:** Free **Closed Heads:** Yes

Point Roberts Marina Resort

713 Simundson Drive; Point Roberts, WA 98281

Tel: (360) 945-2255 **VHF:** **Monitor** Ch. 16 **Talk** Ch. 66A
Fax: (360) 945-0927 **Alternate Tel:** (360) 945-2166
Email: prmarina@pointrobertsmarina.com **Web:** pointrobertsmarina.com
Nearest Town: Point Roberts *(1 mi.)* **Tourist Info:** (360) 945-2313

Marina Operations
Owner/Manager: Allan Sharp **Dockmaster:** Jacquelyne Everett, HM
In-Season: Year-Round, 9am-4pm* **Off-Season:** n/a
After-Hours Arrival: Call security 360-945-2166
Reservations: No **Credit Cards:** Visa/MC
Discounts: MarinaLife **Dockage:** 20% **Fuel:** n/a **Repair:** n/a
Pets: Welcome, Dog Walk Area **Handicap Access:** Yes, Heads, Docks

Marina Services and Boat Supplies
Services - Docking Assistance, Security, Trash Pick-Up, Dock Carts, Megayacht Facilities **Communication -** Phone Messages, Fax in/out *(Free)*, Data Ports *(Wi-Fi - Broadband Xpress)*, FedEx, DHL, UPS
Supplies - OnSite: Ice *(Block, Cube)*, Ships' Store *(Westwind Marine 945-5523)*, Bait/Tackle, Propane

Boatyard Services
OnSite: Travelift *(35T)*, Engine mechanic *(gas, diesel)*, Launching Ramp *($20 two-way)*, Electrical Repairs, Electronic Sales, Hull Repairs, Rigger, Bottom Cleaning, Brightwork, Compound, Wash & Wax, Woodworking, Painting, Yacht Broker, Total Refits **Dealer for:** Mercruiser, Mercury Outboard, Johnson/Evinrude, Volvo, Yanmar, Interlux Paints, and Vacuflush. **Yard Rates:** $74/hr., Haul & Launch $8/ft., Power Wash $2/ft.
Storage: In-Water $5-8.50/ft./mo.

Restaurants and Accommodations
OnSite: Restaurant *(Dockside Café & Pub 945-1208, B $7-8, L $4-10, D $9-15, Wed-Sun, live music Fri & Sat)* **Near:** Restaurant *(South Beach House 945-0717, D $7-20)* **Under 1 mi:** Restaurant *(Brewsters 945-4545, L $5-12, D $7-20)*, *(TJ's 945-3663, L $5-11, D $8-17)*, Lite Fare *(Caffé Capanna 945-0234)*, *(Dylan's at Shell 945-2454)*, Inn/B&B *(Maple Meadows 945-5536, $100)* **1-3 mi:** Restaurant *(Eagle's Roost at Point Roberts Golf & Country 945-4653)*

Recreation and Entertainment
OnSite: Beach, Picnic Area, Playground **Near:** Jogging Paths, Roller Blade/Bike Paths **Under 1 mi:** Video Rental *(International Marketplace)*, Park *(Lighthouse Marine Park, Monument Park)*, Sightseeing *(whale watching)*, Galleries *(Blue Heron Gallery 945-2747)* **1-3 mi:** Golf Course *(Point Roberts Golf and Country Club 945-4653)*

Provisioning and General Services
OnSite: Convenience Store *(Fuel Dock, or Starvin' Sam's 945-7611, nearby)*, Wine/Beer **Near:** Newsstand, Hardware Store *(Nielson's Building Center 945-3116)*, Retail Shops **Under 1 mi:** Supermarket *(International Marketplace 945-0237)*, Bank/ATM, Post Office *(945-7770)*, Protestant Church, Library *(945-6545)*, Beauty Salon, Barber Shop, Florist *(Country Farm 945-0621)* **1-3 mi:** Gourmet Shop, Delicatessen *(Select Deli 250-563-3115)*, Health Food, Liquor Store, Dry Cleaners, Pharmacy **3+ mi:** Catholic Church *(5 mi.)*, Bookstore *(Albany Books 604-943-2293, 5 mi.)*, Department Store *(Richmond, 17 mi./ Vancouver, 22 mi.)*, Buying Club *(Vancouver, BC, 22 mi.)*

Transportation
OnCall: Water Taxi *(Pacific Sea 393-7123, $20-110 dependng on number of passengers & distance)*, Taxi *(Point Roberts 945-2757)* **Near:** Local Bus *(WTA Van Pool 945-2844)* **3+ mi:** Ferry Service *(Delta, BC to Vancouver Island, 5 mi.)* **Airport:** YVR *(18 mi.)*

Medical Services
911 Service **OnCall:** Ambulance **Under 1 mi:** Doctor *(Nurse Practioner)*, Holistic Services *(Massage therapist)* **3+ mi:** Dentist *(5 mi.)*, Chiropractor *(5 mi.)*, Veterinarian *(5 mi.)*, Optician *(5 mi.)* **Hospital:** Delta Hospital 604-946-1121 *(7 mi.)*

Setting -- At the tip of Tsawwassen Peninsula, a large man-made circular basin hosts 17 floating concrete docks with 1,000 slips. The customs dock, with pump-out, is at the mouth of the harbor, followed by the fuel dock.with a second pump-out. Landside, a red roofed cuploa tops a large, white two-story contemporary that houses the office, eatery, club and chandlery. Guest moorage is at the back of the fairway on "T" dock. Beautiful hanging plants are everwhere.

Marina Notes -- *10am-4pm on Sun. Amenities include Wi-Fi, online weather station, and several security gates for convenient slip access. Sandwiches and drinks (including wine and beer) are available at the fuel dock, along with ice, bait, and propane. Westwind Marine provides boatyard services and runs the chandlery (Mon-Sat 9am-5pm, Sun 10am-3pm). Excellent crabbing and fishing is reported at the marina. On both sides of the basin, well-maintained bathhouses have heads, showers, and laundry facilities. NOTE: Marina is an entry point into the U.S. -- Point Roberts Customs & Border Patrol 945-2314.

Notable -- Extending south from Canada and separated from the state of Washington by Boundary Bay, 5-square-mile Point Roberts has been part of the U.S. since the 1846 Treaty of Washington. In the "Banana Belt," Point Roberts gets about half the rainfall that dampens Seattle and Vancouver every year. The International Marketplace grocery - with video rentals, film processing, photocopying services, bank, and liquor store - and the Visitors' Center are on Tyee Drive. West of the marina, a nature trail leads to Lighthouse Marine Park, which offers a full-service campground, picnic tables, playground, boat launch, boardwalk, and a good vantage point for bird and whale watching. Monument Park farther north has a walking trail and hosts the U.S. Border Marker No. 1.

Navigational Information
Lat: 48°59.556' **Long:** 122°45.626' **Tide:** 10 ft. **Current:** 0 ft. **Chart:** 18424
Rep. Depths (MLW): Entry 30 ft. **Fuel Dock** 25 ft. **Max Slip/Moor** 12 ft./-
Access: First marina to port after entering Drayton Harbor

Marina Facilities *(In Season/Off Season)*
Fuel: Gasoline, Diesel
Slips: 629 Total, 25* Transient **Max LOA:** 80 ft. **Max Beam:** n/a
  Rate *(per ft.):* **Day** $0.50 **Week** $2 **Month** $4.90
  Power: 30 amp $0.06/kwh, **50 amp** $0.06/kwh., **100 amp** n/a, **200 amp** n/a
  Cable TV: No **Dockside Phone:** Yes
  Dock Type: Fixed, Long Fingers, Alongside, Concrete
Moorings: 0 Total, 0 Transient **Launch:** n/a, Dinghy Dock
  Rate: Day n/a **Week** n/a **Month** n/a
Heads: 6 Toilet(s), 4 Shower(s)
Laundry: 2 Washer(s), 2 Dryer(s) **Pay Phones:** Yes
Pump-Out: OnSite, 4 Central, 4 Port **Fee:** n/a **Closed Heads:** Yes

Marina Operations
Owner/Manager: Port of Bellingham **Dockmaster:** Pam Taft
In-Season: Year-Round, 8am-5pm **Off-Season:** n/a
After-Hours Arrival: See marina staff
Reservations: No **Credit Cards:** Visa/MC, Dscvr
Discounts: None
Pets: Welcome **Handicap Access:** Yes, Heads, Docks

Blaine Harbor

PO Box 1245; 235 Marine Drive; Blaine, WA 98230

Tel: (360) 647-6176 **VHF: Monitor** Ch. 16 **Talk** Ch. 68
Fax: (360) 332-1043 **Alternate Tel:** n/a
Email: blaineharbor@portofbellingham.com **Web:** portofbellingham.com
Nearest Town: Blaine *(0.25 mi.)* **Tourist Info:** (360) 332-8311

Marina Services and Boat Supplies
Services - Docking Assistance, Security *(after hrs.)*, Trash Pick-Up, Dock Carts **Communication -** Fax in/out *($1/pg)*, Data Ports *(Wi-Fi - Broadband Xpress)* **Supplies - OnSite:** Ice *(Block)* **Near:** Ships' Store, West Marine *(332-1918)*, Bait/Tackle, Live Bait, Propane *(Chevron 332-8412)*

Boatyard Services
OnSite: Engine mechanic *(gas)*, Launching Ramp, Electrical Repairs, Electronic Sales, Hull Repairs, Rigger, Yacht Broker *(Diamond Yachts)*, Total Refits **Near:** Engine mechanic *(diesel)*, Electronics Repairs, Canvas Work, Bottom Cleaning, Brightwork, Air Conditioning, Refrigeration, Divers, Propeller Repairs, Woodworking, Inflatable Repairs, Metal Fabrication, Painting. **Nearest Yard:** Blaine Marine Services (360) 332-3324

Restaurants and Accommodations
Near: Restaurant *(Portal Café 332-7274, B $4-10, L $7-12, D $8-15)*, *(Ocean Bay 332-6311, L $8-16, D $10-20)*, *(Wheel House Tavern 332-3512)*, Fast Food *(Subway)*, Pizzeria *(Pizza Factory 332-3636)*, Motel *(Motel International 332-8222, $46-75)*, *(Northwoods 332-5603)* **Under 1 mi:** Restaurant *(Kira's Grill 332-2040)*, Motel *(Anchor Inn 332-5539, $60-125)* **1-3 mi:** Inn/ B&B *(Smugglers Inn 332-1749)* **3+ mi:** Hotel *(Semiahmoo Resort 318-2010 , $139-409, 8 mi., less than 1 mi. by dinghy or foot ferry)*

Recreation and Entertainment
OnCall: Dive Shop **Near:** Picnic Area, Jogging Paths, Park *(Marine Park)* **Under 1 mi:** Fitness Center *(Blaine Athletic Club 332-2090)*, Fishing Charter *(Jim's Salmon Charter 332-6724)* **1-3 mi:** Golf Course *(Birch Bay 371-2026/*

Peace Portal 604-538-4737)*, Video Rental *(Blockbuster 332-2441)*, Museum *(Drayton Harbor Maritime at Semiahmoo Park, dinghy)*, Cultural Attract *(Art by the Bay self-guided tour)*, Sightseeing *(historic home tour 332-4544)*, Special Events *(Hands Across the Border - Jun, Jazz Fest - Jul, Summer Aire Art - Jul, Plover Days - Aug)*, Galleries *(Art Center 332-1790)*

Provisioning and General Services
Near: Convenience Store *(USA Mini Mart 332-7730)*, Wine/Beer, Bank/ATM, Protestant Church, Library *(Blaine 332-8146)*, Beauty Salon *(Your Style 332-4343)*, Dry Cleaners *(Corry's 332-6800)*, Hardware Store *(True Value 332-4077)*, Florist *(Blaine Bouquets 332-6700)*, Department Store *(Goff's Department Store 332-6663)* **Under 1 mi:** Post Office *(332-7184)*, Catholic Church, Barber Shop *(Guys 'n Gals 332-3109)*, Retail Shops *(Bayside Treasures 332-2233)* **1-3 mi:** Supermarket *(Cost Cutters 332-5909)*, Liquor Store *(Blaine Liquor 332-5671)*

Transportation
OnSite: Courtesy Car/Van *(seasonal, to downtown Blaine)*, Ferry Service *(Plover Foot Ferry to Semiahoo - summer wknds, free/donations)* **OnCall:** Rental Car *(Enterprise 714-0243)*, Taxi *(Yellow Cab 332-8294)*, Local Bus, Airport Limo *(Classic Limousine 332-1749)* **Airport:** Blaine Municipal/ Bellingham Int'l. *(2 mi./16 mi.)*

Medical Services
911 Service **OnCall:** Ambulance **Near:** Doctor *(Bay Medical Center 332-6327)*, Dentist *(Blaine Harbor Dental 332-2400)* **1-3 mi:** Veterinarian *(Cat & Dog Clinic 332-280)* **Hospital:** St. Joseph 734-5400 *(20 mi.)*

Setting -- Blaine and Semiahmoo marinas share the same Drayton Harbor entry channel. To port, a commanding two-story blue-roofed West Coast-style building, topped by a cuploa, welcomes guests to Blaine Harbor. The floating pier directly in front offers 850 feet of side-tie dockage for visitors, sheltered by a stone breakwater and served by a wide waterway. Colorful hanging baskets adorn the dockside period lamp posts, waterfront deck and shaded picnic tables.

Marina Notes -- CCM. *850 ft. alongside at Gate 2. Three-day limit at the visitor dock; slips may be available for longer stays through seasonal tenants' "loan- a-slip" program. Proof of vessel insurance required. Owned by Port of Bellingham, the marina is staffed 24-hours, and offers a chandlery, boat ramp, on-site restaurant, a wonderful waterfront boardwalk with shaded picnic spots, and seasonal shuttle into town. New tile bathhouse and laundry.

Notable -- The foot ferry Plover, built over 60 years ago in Seattle, docks here and connects to Semiahmoo on summer weekends. Along Marine Drive is a small park with walking trails and great bird watching spots. At the end of the road, the town fishing pier affords expansive views of the harbor and the San Juans. The town of Blaine, the busiest border crossing point between BC and Washington, is an easy walk in the opposite direction with restaurants, bars and shops. It's hard to miss the flags atop the 67-foot Peace Arch; it straddles the border and is surrounded by a park. Semiahmoo Resort features a spa & salon, pool, exercise equipment, shuttle to the golf courses, racquetball, tennis courts, and video arcade and is just a dinghy/ferry ride away; all of its amenities are available for $15/day. Note: Watch for the new Whatcom Redevelopment Downtown Marina.

Semiahmoo Marina

9450 Semiahmoo Parkway; Blaine, WA 98230

Tel: (360) 371-0440 **VHF: Monitor** Ch. 68 **Talk** n/a
Fax: (360) 371-0200 **Alternate Tel:** n/a
Email: semimarina@bbxmail.net **Web:** www.semiahmoomarina.com
Nearest Town: Blaine, WA *(8 mi.)* **Tourist Info:** (360) 671-3990

Navigational Information
Lat: 48°59.196' **Long:** 122°46.366' **Tide:** 10 ft. **Current:** 0 kt. **Chart:** 18424
Rep. Depths (*MLW*): Entry 30 ft. **Fuel Dock** 10 ft. **Max Slip/Moor** 12 ft./-
Access: First marina to starboard on entering Drayton Harbor

Marina Facilities *(In Season/Off Season)*
Fuel: Gasoline, Diesel
Slips: 294 Total, 10 Transient **Max LOA:** 120 ft. **Max Beam:** n/a
 Rate *(per ft.)*: **Day** $0.80 **Week** n/a **Month** $4.85
 Power: 30 amp $3, 50 amp $5, **100 amp** n/a, **200 amp** n/a
 Cable TV: No **Dockside Phone:** Yes
 Dock Type: Floating, Alongside, Concrete
Moorings: 0 Total, 0 Transient **Launch:** n/a
 Rate: Day n/a **Week** n/a **Month** n/a
Heads: 10 Toilet(s), 8 Shower(s)
Laundry: 4 Washer(s), 4 Dryer(s), Iron, Iron Board **Pay Phones:** Yes
Pump-Out: OnSite, 1 Central, 1 Port **Fee:** n/a **Closed Heads:** Yes

Marina Operations
Owner/Manager: Lou Herrick **Dockmaster:** Same
In-Season: Jun-Sep, 8:30am-6pm **Off-Season:** Oct-May, 8:30am-5pm
After-Hours Arrival: Call ahead
Reservations: Yes, Preferred **Credit Cards:** Visa/MC, Amex
Discounts: None
Pets: Welcome, Dog Walk Area **Handicap Access:** Yes, Heads, Docks

Marina Services and Boat Supplies
Services - Docking Assistance, Room Service to the Boat, Dock Carts
Communication - Mail & Package Hold, Data Ports *(Wi-Fi - Broadband Xpress)*, FedEx, DHL, UPS, Express Mail *(Sat Del)* **Supplies - OnSite:** Ice *(Block, Cube)*, Propane **3+ mi:** West Marine *(332-1918, 8 mi.)*

Boatyard Services
OnSite: Travelift *(35T)*, Engine mechanic *(gas, diesel)*, Electrical Repairs, Hull Repairs, Rigger, Bottom Cleaning, Brightwork, Compound, Wash & Wax, Yacht Interiors

Restaurants and Accommodations
Near: Restaurant *(Packers Lounge 318-2000, L & D Sun-Thu 11am-Mid, Fri-Sat 11am-1am, B Sat 9am-1pm)*, *(Stars 318-2000, B Mon-Fri 6:30-11:30am, till Noon wknds, D daily 5-10pm)*, *(Pierside 318-2000)*, *(Blue Heron 371-5745, at Semiahmoo Golf; casual and to go)*, Hotel *(Semiahmoo Resort 318-2010, $139-409)* **3+ mi:** Restaurant *(Stephanie's 371-7033, B $5-9, L $7-28, D $10-24, 5 mi.)*, Motel *(Anchor Inn 332-5539, $60-125, 8 mi.)*

Recreation and Entertainment
OnSite: Picnic Area, Grills, Boat Rentals *(kayaks, paddle boats)*, Roller Blade/Bike Paths *(rentals too)* **Near:** Heated Pool *(at resort)*, Spa, Beach, Tennis Courts, Golf Course *(shuttle to Semiahmoo Golf 371-7005 & Loomis Trail 332-1725)*, Fitness Center, Jogging Paths, Video Arcade *(at resort)*
1-3 mi: Park *(Semiahmoo Park 733-2900)*, Museum *(Drayton Harbor Museum at Semiahmoo Park - local history)* **3+ mi:** Fishing Charter *(Eagle Point - Bellingham 966-3334, 25 mi.)*, Video Rental *(Blockbuster 332-2441, 9 mi.)*, Cultural Attract *(Peace Arch Playhouse 332-4678, 8 mi.)*, Special Events

(Blaine: Ski to Sea Fest - May, Hands Across the Border - Jun, Jazz Fest - Jul, Summer Aire Art - Jul, 8 mi.), Galleries *(Peace Arch Park International Sculpture Exhibition 322-7165, 8 mi.)*

Provisioning and General Services
OnSite: Convenience Store *(snacks, deli items, sundries, gifts)*
Near: Beauty Salon *(at resort 318-2009)*, Barber Shop **1-3 mi:** Retail Shops *(gifts, antiques)* **3+ mi:** Market *(Linda's Maxi Market 371-2804, 6 mi.)*, Supermarket *(Cost Cutters 332-5909, 8 mi.)*, Liquor Store *(8 mi.)*, Green Market *(Grace Harbor Farms 371-9060, 4 mi.)*, Bank/ATM *(4 mi.)*, Post Office *(332-7184, 8 mi.)*, Catholic Church *(8 mi.)*, Protestant Church *(8 mi.)*, Library *(Blaine 332-8146, 8 mi.)*, Dry Cleaners *(Biz Center Plus 332-2030, 8 mi.)*, Bookstore *(Book Warehouse 366-5354, 10 mi.)*, Hardware Store *(True Value 332-4077, 8 mi.)*, Clothing Store *(Bellis Fair Mall, BC 360-734-5022; by bus, 22 mi.)*, Department Store *(Goff's 332-6663, 8 mi.)*

Transportation
OnSite: Bikes *($12 at store)*, Ferry Service *(Plover Foot Ferry to Blaine, Fri-Sun May-Sep)* **OnCall:** Rental Car *(Enterprise 714-0243)*, Taxi *(City Cab 332-8294)* **Near:** Courtesy Car/Van *(to golf & casino)* **3+ mi:** Local Bus *(Wilcom Transportation Authority 676-7433, 8 mi.)* **Airport:** Blaine Municipal/ Bellingham Int'l. *(9 mi /18 mi.)*

Medical Services
911 Service **OnCall:** Ambulance **Near:** Holistic Services *(full-service European spa)* **3+ mi:** Doctor *(Bay Medical Clinic 332-6327, 8 mi.)*, Dentist *(Blaine Harbor Dental 332-2400, 8 mi.)*, Veterinarian *(Cat & Dog Clinic 332-280, 9 mi.)* **Hospital:** St. Joseph 734-5400 *(28 mi.)*

Setting -- Located on the west side of the entrance to Drayton Harbor, quiet, clean, and well-groomed Semiahmoo Marina is situated at the end of a long spit of undeveloped land. Approaching the marina, to starboard, the facilities of the world-class Semiahmoo Resort sprawl along the shoreline. There are stunning harbor views with snowcapped Mount Baker lying to the south and the Peace Arch - marking the border between Canada and the U.S. - to the north.

Marina Notes -- Stay in the channel, as a shoal parallels the fuel dock. Call ahead for slips or check in at the fuel dock. Overlooking the docks are the marina office, chandlery, picnic tables (some covered) and a large charcoal grill. A lift is on-site and Blaine Marine provides full boatyard services (371-5700). Kayak, bike, and roller blade rentals are available, as well as nature hikes and beachcombing. Modern tile bathhouse and laundry. Semiahmoo Yacht Club also makes its home here (reciprocal, 371-0440). The 32-foot wooden ferry "Plover" runs in season between the marina and Blaine Harbor.

Notable -- The adjacent Inn at Semiahmoo, on the site of an old salmon cannery, offers three outstanding restaurants, a European-style spa & salon, and entry to the two top-rated golf courses in Washington. A fee of $15/day per adult ($5 per child) provides access to the Inn's amenities: heated pool, fitness center, tennis courts, indoor track, racquetball court, and aerobics classes. Dining options include informal Packers Lounge, fine dining at Stars, and Sunday brunch buffet at Pierside - with views out to the bay (9am-1pm, under 5 free). Shuttle service to golf courses, the Skagit Casino, and room service to the boats are provided by the Inn. Several antique/gift shops and casual eateries are on this side of town. Blaine is a short dinghy/ferry ride, or about 8 miles by road.

Navigational Information
Lat: 48°43.400' **Long:** 122°30.600' **Tide:** 9 ft. **Current:** n/a **Chart:** 18424
Rep. Depths (*MLW*): **Entry** 12 ft. **Fuel Dock** n/a **Max Slip/Moor** 12 ft./20 ft.
Access: South Bellingham Bay past the ferry terminal

Marina Facilities (*In Season/Off Season*)
Fuel: No
Slips: 8 Total, 8 Transient **Max LOA:** 40 ft. **Max Beam:** n/a
  Rate (*per ft.*): **Day** $0.50 **Week** n/a **Month** n/a
  Power: 30 amp n/a, **50 amp** n/a, **100 amp** n/a, **200 amp** n/a
  Cable TV: No **Dockside Phone:** No
  Dock Type: n/a
Moorings: 10 Total, 10 Transient **Launch:** n/a
  Rate: Day 0.50/ft. **Week** n/a **Month** n/a
Heads: 6 Toilet(s)
Laundry: None **Pay Phones:** No
Pump-Out: No **Fee:** n/a **Closed Heads:** Yes

Marina Operations
Owner/Manager: Port of Bellingham **Dockmaster:** Dave Walter
In-Season: May - Oct, 24/7 **Off-Season:** n/a
After-Hours Arrival: Call ahead
Reservations: No **Credit Cards:** No; Cash only
Discounts: None
Pets: Welcome **Handicap Access:** No

Fairhaven Moorage

Harris Street; Bellingham, WA 98225

Tel: (360) 676-7567 **VHF: Monitor** n/a **Talk** n/a
Fax: (360) 671-6411 **Alternate Tel:** n/a
Email: info@portofbellingham.com **Web:** n/a
Nearest Town: Fairhaven **Tourist Info:** (360) 734-1330

Marina Services and Boat Supplies
Supplies - Under 1 mi: Ice (*Cube*) **1-3 mi:** Bait/Tackle (*H & H Outdoor 733-2050*) **3+ mi:** West Marine (*650-1100, 4 mi.*), Boater's World (*715-7304, 5 mi.*)

Boatyard Services
1-3 mi: Travelift (*35T*). **Nearest Yard:** Seaview North at Squalicum (360) 676-8282

Restaurants and Accommodations
Near: Restaurant (*Skylark's Hidden Café 715-3642, B $4-9, L $6-11, D $6-12*), (*Harris Ave. Café 738-0802*), Seafood Shack (*Jacci's Fish and Chips 733-5021, L $7-13, salmon & halibut chips*), Coffee Shop (*Shipwrecked Coffee, coffee & snacks*), Lite Fare (*Archer Ale House 647-7002*)
Under 1 mi: Restaurant (*The Black Cat 733-6136, L & D daily*), (*Mambo Italiano 734-7677*), (*Coppa Mediteranean Bistro 312-5050*), (*Flat Tapas Bar 738-6001*), (*Stanello's 676-1304*), Pizzeria (*Fairhaven Pizza 756-7561*), Inn/ B&B (*Fiarhaven B&B 734-7243*), (*Chrysalis Inn & Spa 756-1005, $170-270, waterfront*) **1-3 mi:** Motel (*Travelodge 733-8280, $39-89*), Hotel (*Ramada 734-8830, $61-95*)

Recreation and Entertainment
OnSite: Picnic Area, Playground **Near:** Jogging Paths (*scenic sidewalks & trails along the bay, plus waterfront Boulevard Park and Chuckanut Dr.*), Park, Sightseeing (*Fairhaven Historic District*) **Under 1 mi:** Boat Rentals (*Fairhaven Boatworks 714-8891- rowboats and kayaks $7-18/hr., $30-70/day - Tue-Sun in summer, wknds only off-season*), Video Rental (*Trek Video 671-1478*) **1-3 mi:** Golf Course (*Lake Padden 738-7400*), Fitness Center (*YMCA 733-8630*), Bowling (*20th Century 734-5250*), Movie Theater (*Pickford 738-0735*), Museum (*History & Art 676-6981, Children's Museum 733-8769, Museum of Radio 738-3886*), Galleries (*Gallery West 734-8414*)

Provisioning and General Services
Near: Beauty Salon (*Fairhaven Salon 733-2733*), Barber Shop (*Barber Shop at Fairhaven 738-8081*), Hardware Store (*Fairhaven Hardware 756-5033*) **Under 1 mi:** Convenience Store (*Starvin' Sam's 671-0455*), Market (*Fairhaven Red Apple 733-4370*), Supermarket (*Super Store 676-1962*), Gourmet Shop (*Gourmet House 756-1611*), Delicatessen (*Avenue Bread 676-1809*), Health Food, Liquor Store, Farmers' Market (*Village Green, Jun-Sep, Wed 3-7 pm*), Bank/ATM, Library (*Fairhaven 676-6877*), Dry Cleaners, Laundry (*Fairhaven Laundry & Cleaners 734-9647*), Bookstore (*Village Books 671-2626 has coffee, sandwiches, and wine*), Pharmacy (*Fairhaven 734-3340*), Florist (*Rebecca's 715-3066*) **1-3 mi:** Catholic Church, Protestant Church, Copies Etc.

Transportation
OnCall: Taxi (*Evergreen 714-0502*) **Near:** Bikes (*Fairhaven Bike 733-4433*), Local Bus, Rail (*Amtrak 374-8851*), Ferry Service (*Alaska Ferry Terminal 676-8445 - Victoria, San Juans, Alaska & whale watching*) **Airport:** Bellingham Int'l. (*5 mi.*)

Medical Services
911 Service **OnCall:** Ambulance **Near:** Dentist (*Fairhaven 733-7988*), Chiropractor (*Back in Motion 647-1970*) **Under 1 mi:** Holistic Services (*Acupuncture and Herbal 715-1824*) **1-3 mi:** Doctor (*Bunks Medical 752-2865*) **Hospital:** St. Joseph 734-5400 (*4 mi.*)

Setting -- The historic neighborhood of Fairhaven is approximately two miles south of Bellingham - by both land and water. The public Fairhaven Moorage is tucked past the commercial docks where it is convenient to the beautifully preserved 1890s district. Both a linear side-tie dock and seasonal buoys are available. The south end of Bellingham Bay is also home to the venerable Alaska Ferry Terminal, with connections to Alaska, Victoria, and the San Juans.

Marina Notes -- Port of Bellingham owns the moorage. 10 buoys, available from May-Oct 31, hold up to a 40' boat. The linear moorage is available year-round. Boats tie up along a 300-ft. line strung between several buoys. The buoys are available on a first-come, first-served basis. On the linear moorage, the Port also accepts seasonal tenants - so space may be harder to find. To go ashore from the moorings, guests can tie up at the launch ramp - there is no dinghy dock. A payment box is located at the ramp. Heads are located at the nearby boat launch parking lot. Note: The dock is exposed to the ferry wake.

Notable -- Restaurants, shops, galleries are housed in restored Romanesque brick structures built during the 1890 boom days, when the regions best deep water wharves attracted serious ship traffic. Bike rentals, parks, and a seasonal farmers' market are also nearby. Fairhaven Boatworks has kayak rentals. Dinghy to their dock and enjoy a muffin and a latte on the small deck outside Shipwreck Coffee - the views of Bellingham Bay and the Ferry Terminal are awesome. During the summer, the Whatcom Film Association shows movies in the park at 10th and Mill Streets. Scenic sidewalks and trails run along the bay, including waterfront Boulevard Park. About two miles away, at 165-acre Sehome Hill Arboretum, trails lead to a hilltop observation tower with panoramic views.

Squalicum Harbor

PO Box 1677; 722 Coho Way; Bellingham, WA 98227

Tel: (360) 676-2542 **VHF: Monitor** Ch. 16 **Talk** Ch. 68
Fax: (360) 671-6149 **Alternate Tel:** (360) 739-8131
Email: squalicum@portofbellingham.com **Web:** www.portofbellingham.com
Nearest Town: Bellingham **Tourist Info:** (360) 734-1330

Navigational Information
Lat: 48°45.300' **Long:** 122°30.195' **Tide:** 8.5 ft. **Current:** n/a **Chart:** 18424
Rep. Depths (MLW): Entry 12 ft. **Fuel Dock** n/a **Max Slip/Moor** 12 ft./-
Access: Located in Bellingham Bay north of Chuckanut Bay

Marina Facilities *(In Season/Off Season)*
Fuel: No
Slips: 1417 Total, 35 Transient **Max LOA:** 100 ft. **Max Beam:** n/a
  Rate *(per ft.)*: **Day** $0.50* **Week** $2 **Month** $4.90
  Power: 30 amp Incl., **50 amp** Incl., **100 amp** n/a, **200 amp** n/a
  Cable TV: No **Dockside Phone:** No
  Dock Type: Floating, Long Fingers, Alongside, Concrete, Wood
Moorings: 0 Total, 0 Transient **Launch:** n/a
  Rate: Day n/a **Week** n/a **Month** n/a
Heads: 6 Toilet(s), 5 Shower(s)
Laundry: 6 Washer(s), 6 Dryer(s) **Pay Phones:** Yes, 2
Pump-Out: OnSite, 3 Central, 10 Port **Fee:** n/a **Closed Heads:** Yes

Marina Operations
Owner/Manager: Port of Bellingham/Dan Stahl **Dockmaster:** Mike Endsley
In-Season: May-Sep, 8am-7pm **Off-Season:** Oct-Apr, 8am-5pm
After-Hours Arrival: Call ahead
Reservations: No, First-come first-served **Credit Cards:** Visa/MC, Dscvr
Discounts: None
Pets: Welcome **Handicap Access:** Yes

Marina Services and Boat Supplies
Services - Security *(after hours)*, Megayacht Facilities **Communication -** Data Ports *(Wi-Fi - Broadband Xpress)* **Supplies - Near:** Ice *(Block, Cube)*, Ships' Store, Bait/Tackle *(LFS 734-3336)* **Under 1 mi:** West Marine *(686-8020)* **1-3 mi:** Boater's World *(715-8304)*

Boatyard Services
OnSite: Travelift *(150T)*, Railway, Crane, Hydraulic Trailer, Engine mechanic *(gas, diesel)*, Launching Ramp, Electrical Repairs, Electronics Repairs, Hull Repairs, Rigger, Propeller Repairs, Inflatable Repairs, Yacht Broker **OnCall:** Divers **Near:** Sail Loft, Canvas Work, Bottom Cleaning, Brightwork.

Restaurants and Accommodations
OnSite: Restaurant *(Anthony's Homeport 647-5588, B $8-12, L $8-17, D $17-30, B wknds only, fresh seafood specialties)* **Near:** Restaurant *(Bayside Café 715-0975, B $4-10, L $6-12, D $13-21)*, Coffee Shop *(Web Locker 734-5163, B $2-7, L $3-11)*, Inn/B&B *(Decann House B&B 734-9172)* **Under 1 mi:** Restaurant *(Harborside Bistro 392-3200, L $11-15, D $16-26)*, Hotel *(Hotel Bellwether 392-3100, $124-699)* **1-3 mi:** Restaurant *(Jalepenos 671-3099)*, *(Wild Garlic 671-1955, L $6-11, D $13-22)*, Hotel *(Best Western 333-2080, $74-84, comp. pick-up)*

Recreation and Entertainment
OnSite: Picnic Area, Fishing Charter **Near:** Playground *(Zuranich Point Park)*, Tennis Courts *(Elizabeth Park)*, Jogging Paths, Video Rental, Park **Under 1 mi:** Galleries *(Fish Boy Gallery 714-0815)* **1-3 mi:** Golf Course *(Bellingham G & C.C. 733-3450)*, Fitness Center *(Bellingham Athletic Club 734-1616)*, Movie Theater *(Regal 676-2280)*, Museum *(American Museum of Radio 738-3886, Mindport 647-5614)*, Cultural Attract *(Mt. Baker Theatre 734-6080 - a National Historic Monument)*, Sightseeing *(Historic Fairhaven)*, Special Events *(Bellingham Festival of Music - Aug 800-335-5550)*

Provisioning and General Services
Near: Fishmonger *(right off the boat)*, Newsstand **Under 1 mi:** Market *(Haggen 671-3300)*, Bakery, Bank/ATM, Post Office, Beauty Salon *(CJ's Harbor Salon and Barber 734-4843)*, Pharmacy *(Fountain Drug 733-6200)* **1-3 mi:** Supermarket *(Albertson's 733-9244)*, Delicatessen *(Avenue Bread & Deli 676-9274)*, Health Food *(Terra Organica 715-8020)*, Farmers' Market *(Sat 10am-3pm, Apr-Oct Railroad & Chestnut St.)*, Library *(Bellingham 676-6860)*, Dry Cleaners *(Vienna 734-3333)*, Bookstore *(Village Books 671-2626)*, Hardware Store *(Lowes 734-2659)*, Department Store *(Bellis Fair Mall 734-5022)*, Buying Club *(Costco 671-6947)*

Transportation
OnCall: Rental Car *(Enterprise 733-4363/ Budget 671-3800, near)*, Taxi *(Yellow Cab 734-8294)*, Airport Limo *(Bellair Chargers 380-8800)* **1-3 mi:** Rail *(Amtrak 734-8851)*, Ferry Service *(Alaska Ferry 676-8445 to Victoria, San Juan, Alaska)* **Airport:** Bellingham Int'l. *(3 mi.)*

Medical Services
911 Service **OnCall:** Ambulance **Near:** Holistic Services *(Zazen Spa 715-1050)* **Under 1 mi:** Doctor, Dentist *(Bayside Dental 738-9791)*, Chiropractor *(Bedry 676-8227)*, Veterinarian **Hospital:** St. Joseph 734-5400 *(2 mi.)*

Setting -- Located off Bellingham Bay are two large moorage basins with over 1400 slips - the outer one protected by jetties and the inner one virtually enclosed. Each has a complete set of amentities and facilities - potted flowering plants add a touch of color. Shops and restaurants line the shore and Mount Baker rises majestically in the background. The waterside view stretches to the San Juan Islands.

Marina Notes -- Squalicum Harbor is owned by the Port of Bellingham. Open all year, this clean, beautifully maintained marina has three guest moorage docks. There is a 3 day limit on transient docks but ask about the "loan-a-slip" program for stays up to 14 days. Harbor office is located across the parking lot. Pay envelopes and drop boxes are located at the Harbor Center Building, top of Gate 3, and at Gate 12. Security gates are locked at dusk and opened at dawn. Seaview North (676-8282) provides onsite boatyard services. Bathhouses and the laundry rooms are locked at 10pm and opened at 6:30am. Customs clearance is available. Bellingham is a frequent kickoff site for island cruising and the Squalicum Boathouse works for group cruise and flotilla gatherings.

Notable -- Downtown Bellingham is filled with shops, restaurants, galleries, bookstores, antique shops, and coffeehouses. The area also boasts unique cultural destinations - The American Museum of Radio and Electricity (738-3886, Wed-Sat 11am-4pm) explores the history of broadcasting. Mindport (647-5614, Wed-Sun) offers interactive exhibits for all ages. The Whatcom Museum (767-6981, Tue-Sun) features local history and art. Restored Mt. Baker Theatre (733-5793) offers music, dance, and theatre. In nearby Fairhaven, restaurants, galleries and ateliers occupy many of the original 1890s boom-days buildings.

Navigational Information

Lat: 48°45.304' **Long:** 122°29.801' **Tide:** 10 ft. **Current:** 0 ft. **Chart:** 18424
Rep. Depths (*MLW*): **Entry** 10 ft. **Fuel Dock** n/a **Max Slip/Moor** 12 ft./-
Access: Enter from Bellingham Bay to first dock to starboard

Marina Facilities *(In Season/Off Season)*

Fuel: No
Slips: 10 Total, 10 Transient **Max LOA:** 220 ft. **Max Beam:** n/a
  Rate *(per ft.)*: **Day** $1.00* **Week** n/a **Month** n/a
  Power: 30 amp Incl., **50 amp** Incl., **100 amp** Incl., **200 amp** n/a
  Cable TV: No **Dockside Phone:** No
  Dock Type: Fixed, Concrete
Moorings: 0 Total, 0 Transient **Launch:** n/a
  Rate: Day n/a **Week** n/a **Month** n/a
Heads: 4 Toilet(s)
Laundry: None **Pay Phones:** Yes
Pump-Out: No **Fee:** n/a **Closed Heads:** Yes

Marina Operations

Owner/Manager: Peter Paulsen **Dockmaster:** Same
In-Season: Year-Round, 24/7 **Off-Season:** n/a
After-Hours Arrival: Open 24 hrs.
Reservations: Yes, Required **Credit Cards:** Visa/MC, Dscvr, Din, Amex
Discounts: None
Pets: Welcome, Dog Walk Area **Handicap Access:** Yes, Heads, Docks

Hotel Bellwether

One Bellwether Way; Bellingham, WA 98225

Tel: (360) 392-3100; (877) 411-1200 **VHF: Monitor** n/a **Talk** n/a
Fax: (360) 392-3101 **Alternate Tel:** (360) 392-3100
Email: reservations@hotelbellwether.com **Web:** www.hotelbellwether.com
Nearest Town: Bellingham *(1 mi.)* **Tourist Info:** (360) 734-1330

Marina Services and Boat Supplies

Services - Concierge, Room Service to the Boat, Boaters' Lounge, Megayacht Facilities **Communication -** Mail & Package Hold, Phone Messages, Fax in/out *($0.50/pg)*, Data Ports *(Hotel)*, FedEx, DHL, UPS, Express Mail **Supplies - Near:** Ice *(Block, Cube, Shaved)*, Ships' Store *(LFS 734-3336)*, Bait/Tackle, Propane **1-3 mi:** West Marine *(676-8020)*

Boatyard Services

Under 1 mi: Forklift, Crane, Engine mechanic *(gas, diesel)*, Launching Ramp, Electrical Repairs, Hull Repairs, Rigger, Sail Loft, Canvas Work, Bottom Cleaning, Brightwork, Compound, Wash & Wax, Interior Cleaning, Propeller Repairs, Woodworking, Inflatable Repairs, Life Raft Service, Upholstery, Metal Fabrication, Painting, Awlgrip, Yacht Broker, Total Refits. **1-3 mi:** Railway, Divers. **Nearest Yard:** Seaview North (360) 676-8282

Restaurants and Accommodations

OnSite: Restaurant *(Harborside Bistro 392-3200, B $8, L $11-15, D $16-26, Cont B Mon-Sat, L 11:30am-2:30 pm, D Sun-Thu 5:30-9pm, Fri-Sat 5:30-9:30pm, Sun brunch 10am-1pm)*, Lite Fare *(Sunset Lounge 11am-11pm)*, Hotel *(Hotel Bellwether 392-3100, $124-699)* **Near:** Restaurant *(The Marina Restaurant 733-8292, L $7-13, D $18-25)*, *(Anthony's Homeport 647-5588, L $6-12, D $11-25)*, *(Bayside Café 715-0975, B $4-10, L $6-12, D $13-21)*, Pizzeria *(Mercata Italiano 527-3373, B $6-11, L $7-16)*, Motel *(Maple Alley Inn 738-0199)* **1-3 mi:** Motel *(Best Western 671-1011, $74-86)*

Recreation and Entertainment

OnSite: Golf Course *(5-hole putting green or Lake Padden 738-7400, 6 mi.)*, Fitness Center *(Fitness Alliance 738-4575)*, Roller Blade/Bike Paths,

Galleries *(Mark Bergsma 671-6818)* **Near:** Picnic Area, Grills, Playground, Tennis Courts *(Elizabeth Park)*, Boat Rentals, Fishing Charter, Park **Under 1 mi:** Video Rental *(Video Extreme 647-7811)*, Museum *(Childrens Museum 671-6818, History and Art 676-6981)*, Cultural Attract *(Mt. Baker Theatre 734-6080)*, Sightseeing *(Fish Boy 714-0815)*, Special Events *(Ski to Sea Grand Parade - May)* **1-3 mi:** Heated Pool, Beach, Dive Shop, Movie Theater *(Regal 676-2280)*

Provisioning and General Services

OnSite: Gourmet Shop, Wine/Beer, Bank/ATM *(ATM)*, Beauty Salon, Dry Cleaners, Clothing Store *(Blue Willi's 756-1998)*, Copies Etc. **Near:** Convenience Store *(Kwik Stop 733-8982)*, Pharmacy *(Fountain 733-6200)* **Under 1 mi:** Market *(Associated Grocers 647-9677)*, Farmers' Market *(Sat 10am-3pm, Apr-Oct Railroad & Chestnut St.)*, Fishmonger, Post Office, Catholic Church, Protestant Church, Library *(Bellingham 676-6860)*, Laundry *(Cascade 734-4200)*, Bookstore *(Henderson 734-6855)*, Florist **1-3 mi:** Supermarket *(Albertsons 733-9244)*, Hardware Store *(Lowes 734-2659)*

Transportation

OnSite: Bikes **OnCall:** Rental Car *(Enterprise 733-4363/ Budget 671-3800, near)*, Taxi *(Yellow Cab 332-8294)* **Near:** Local Bus, InterCity Bus **1-3 mi:** Rail *(Amtrak 734-8851)*, Ferry Service *(Alaska Ferry 676-8445 to San Juan Islands, Vistoria, Alaska)* **Airport:** Bellingham Int'l. *(3 mi.)*

Medical Services

911 Service **OnSite:** Holistic Services *(Zazen Spa 715-1050)* **OnCall:** Ambulance **Under 1 mi:** Dentist, Veterinarian *(Fountain 733-2660)* **1-3 mi:** Doctor *(Family Health 671-635)* **Hospital:** St. Joseph 734-5400 *(1.5 mi.)*

Setting -- Located to starboard just inside the entry to Bellingham Bay, this luxurious waterfront European-style inn features a 220-foot side-tie concrete floating transient dock. This is a wonderful oasis - unwind, be pampered and enjoy the views of the pretty hotel, Squalicum Harbor, and the San Juan Islands.

Marina Notes -- *$1/ft. if over 30'/$30 flat fee under 30'. Total 400 feet of linear dockage. Reservations are strongly recommended. Welcomes megayachts to 220' (side-tie to the outside of the dock) - 100 amp available. Built in 2000. A guest of the dock is a guest of the hotel, with access to the fitness center and beautiful grounds. A variety of meeting/event rooms are available for private functions - the largest accommodates 700. Hotel rooms offer fireplaces, fully stocked bars, original artwork, and private balconies. For the ultimate retreat, reserve the 3-story lighthouse suite with its own observation deck.

Notable -- The onsite Bistro offers a seasonal menu (lunch, dinner, and all-day options), an award-winning wine list and harbor views; lite fare is at the Sunset Lounge, and British-style afternoon tea is served in the Compass Room (a breakfast buffet is available to guests for $8). Practice your golf game on the 5-hole waterfront putting green, or visit the Zazen Spa (715-1050, spa/room packages $245-265). Just outside the hotel gates, Marine Life Center, with observation tanks and touch pools, is a great destination for kids (671-2431, daily 8am-Sunset, donations). A wide range of restaurants, coffee shops, boutiques, and galleries surround the hotel. Enjoy a casual microbrew, a pizza cooked outside in a wood-fired oven, or a full-course dinner by candlelight. Two wonderful parks are nearby, and a short drive will land a shopping mall, movie theaters, and nightlife.

Boaters' Notes

Add Your Ratings and Reviews at www.AtlanticCruisingClub.com

A complimentary six-month Silver Membership is included with the purchase of this *Guide*. Select "Join Now," then "Silver," then follow the instructions.

The AtlanticCruisingClub website provides updated Marina Reports, Destination and Harbor Articles and much more — including an option within each online Marina Report for boaters to add their ratings and comments regarding that facility. Please log on frequently to share your experiences — and to read other boaters' comments.

On the website, boaters may rate marinas in one or more of the following categories — on a scale of 1 (basic) to 5 (world class) — and also enter additional commentary.

▸ **Facilities & Services** (Fuel, Reservations, Concierge Services and General Helpfulness)

▸ **Amenities** (Pool, Beach, Internet Access, including Wi-Fi, Picnic/Grill Area, Boaters' Lounge)

▸ **Setting** (Views, Design, Landscaping, Maintenance, Ship-Shapeness, Overall Ambiance)

▸ **Convenience** (Access — including delivery services — to Supermarkets, other Provisioning Sources, Shops, Services, Attractions, Entertainment, Sightseeing, Recreation, including Golf and Sport Fishing & Medical Services)

▸ **Restaurants/Eateries/Lodgings** (Availability of Fine Dining, Snack Bars, Lite Fare, OnCall food service, and Lodgings ashore)

▸ **Transportation** (Courtesy Car/Vans, Buses, Water Taxis, Bikes, Taxis, Rental Cars, Airports, Amtrak, Commuter Trains)

▸ **Please Add Any Additional Comments**

9. WA – San Juan Islands

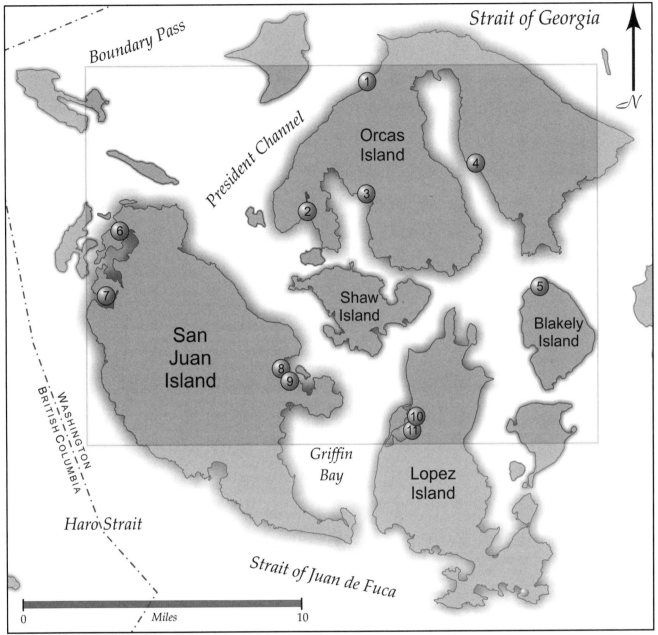

MAP	MARINA	HARBOR	PAGE		MAP	MARINA	HARBOR	PAGE
1	West Beach Resort	President Channel	140		7	Snug Harbor Marina Resort	Mitchell Bay	146
2	Deer Harbor Marina	Deer Harbor	141		8	Port of Friday Harbor	Friday Harbor	147
3	West Sound Marina	West Sound	142		9	Charters Northwest	Friday Harbor	148
4	Rosario Resort Hotel	East Sound/Cascade Bay	143		10	Islands Marine Center	Fisherman Bay	149
5	Blakely Island Marina	Peavine Pass	144		11	Lopez Islander Resort & Marina	Fisherman Bay	150
6	Roche Harbor Resort	Roche Harbor	145					

▸ **Currency** — In Canadian Marina Reports, all prices are in Canadian dollars. In U.S. Marina Reports, all prices are in U.S. dollars.

▸ **"CCM"** — Denotes a Certified Clean Marina, a state/provincial award for environmental excellence. See page 298 for an explanation and page 299 for a list of Pump-Out facilities.

▸ **Ratings & Reviews** — An overview of the Atlantic Cruising Club's rating system is on page 6 and details on the content of each Marina Report are on pages 7 – 11.

▸ **Marina Report Updates** — Comments from boaters and new information from ACC reviewers and marinas are posted regularly on *www.AtlanticCruisingClub.com*.

Navigational Information

Lat: 48°41.291' **Long:** 122°57.483' **Tide:** 8 ft. **Current:** n/a **Chart:** 18421
Rep. Depths *(MLW):* **Entry** 25 ft. **Fuel Dock** n/a **Max Slip/Moor** 40 ft./40 ft.
Access: Off President Channel between Orcas and Waldron Islands

Marina Facilities *(In Season/Off Season)*

Fuel: *Texaco* - Gasoline
Slips: 20 Total, 4 Transient **Max LOA:** 30 ft. **Max Beam:** n/a
  Rate *(per ft.):* **Day** $0.60* **Week** $4.20 **Month** n/a
  Power: 30 amp n/a, 50 amp n/a, 100 amp n/a, 200 amp n/a
  Cable TV: No **Dockside Phone:** No
  Dock Type: Floating, Wood
Moorings: 11 Total, 2 Transient **Launch:** n/a, Dinghy Dock
  Rate: Day $25 **Week** $175 **Month** n/a
Heads: 6 Toilet(s), 4 Shower(s)
Laundry: 2 Washer(s), 2 Dryer(s), Book Exchange **Pay Phones:** Yes, 1
Pump-Out: No **Fee:** n/a **Closed Heads:** Yes

Marina Operations

Owner/Manager: Jamey Hance **Dockmaster:** Same
In-Season: MemDay-Sep15, 8-8 **Off-Season:** Sep16-MemDay, 9-4
After-Hours Arrival: See front door posting for slip asignment
Reservations: Yes, Preferred **Credit Cards:** Visa/MC, Tex
Discounts: None
Pets: Welcome, Dog Walk Area **Handicap Access:** No

West Beach Resort

190 Waterfront Way; Eastsound, WA 98245

Tel: (360) 376-2240; (877) 937-8224 **VHF: Monitor** n/a **Talk** n/a
Fax: (360) 376-4746 **Alternate Tel:** n/a
Email: vacation@westbeachresort.com **Web:** www.westbeachresort.com
Nearest Town: Eastsound *(2.5 mi.)* **Tourist Info:** (360) 376-8888

Marina Services and Boat Supplies

Services - Boaters' Lounge, Dock Carts **Communication -** Phone
Messages, Fax in/out *($1/pg)*, Data Ports *(DSL in lobby)*, FedEx, DHL, UPS,
Express Mail *(Sat Del)* **Supplies - OnSite:** Ice *(Block, Cube)*, Bait/Tackle
(Herring, crab), Propane

Boatyard Services

OnSite: Launching Ramp *(Free)* **OnCall:** Engine mechanic *(gas, diesel)*,
Divers **1-3 mi:** Electrical Repairs, Electronic Sales, Electronics Repairs,
Compound, Wash & Wax, Interior Cleaning. **Nearest Yard:** Deer Harbor
Boatworks (360) 376-4056

Restaurants and Accommodations

OnSite: Snack Bar *(West Beach B $5-10, L $5-10)*, Condo/Cottage *(West
Beach Resort $100-250, fully equiped oceanfront cottages, up to 6 people;
pets welcome)* **1-3 mi:** Restaurant *(Vern's Bayside 376-2231, B $6-13,
L $7-13, D $19-30)*, *(Bilbo's Festivo 376-4728, L $8-20, D $15-30)*, *(Sun-
flower Café 376-2335, B $10-20, L $10-15, D $18-23, brunch also)*,
(Christina's 376-4904, D $15-30), Lite Fare *(Country Corner 376-6900,
sandwiches, salads, pizza)*, Pizzeria *(Portofino 376-2085)*, Inn/B&B
(Barwood Lodge 376-2242, $100-200), *(Outlook Inn 376-2200, $129-300)*

Recreation and Entertainment

OnSite: Spa *(Hot Tub)*, Beach, Picnic Area, Playground, Volleyball *(plus
badminton, soccer, horseshoes, ping-pong)*, Boat Rentals *(kayaks, canoes,
16' power boats)* **OnCall:** Group Fishing Boat **1-3 mi:** Golf Course *(Orcas
Island 376-4400)*, Fitness Center *(Fitness Quest 376-9568)*, Jogging Paths,
MovieTheater *(Sea View Theatre 376-5724)*, Video Rental *(Sea View Theater
a

and Video Center 376-2864)*, Museum *(Orcas Island Historical Museum
376-4849, Tue-Sun)*, Sightseeing *(Fun House 376-7177)* **3+ mi:** Park
(5,000-acre Moran State Park, 7 mi.)

Provisioning and General Services

OnSite: Convenience Store *(Village Stop 376-2093)* **1-3 mi:** Market,
Supermarket *(Island Market 376-6000)*, Gourmet Shop, Delicatessen *(Orcas
Home Grown 376-2009)*, Health Food *(Orcas Herbal Apothecary 376-8272)*,
Wine/Beer, Liquor Store *(East Sound 376-2616)*, Bakery *(Roses Bread &
Specialties 376-5805)*, Farmers' Market, Green Market, Meat Market,
Bank/ATM, Post Office, Catholic Church, Protestant Church, Library *(376-
4985)*, Beauty Salon *(Tangle's 376-5077)*, Barber Shop *(Susie's Barber Shop
376-1911)*, Dry Cleaners, Laundry *(Country Corner 376-6900)*, Bookstore
(Darvill's 376-2135), Pharmacy *(Ray's 376-2230)*, Newsstand, Hardware
Store *(Island Hardware 376-4200)*, Florist *(Orcas Island 376-3800)*, Clothing
Store *(Monkey Puzzle 376-2275, Eastsound Sporting 376-5588)*

Transportation

OnSite: Bikes **OnCall:** Taxi *(Orcas Island 376-8294)* **1-3 mi:** Rental Car,
Local Bus *(Island Shuttle 376-7433, May-Sep)* **3+ mi:** Ferry Service *(376-
6253, 8 mi.)* **Airport:** Orcas Airport *(3 mi.)*

Medical Services

911 Service **OnCall:** Ambulance **1-3 mi:** Doctor *(O.I. Medical 376-6604)*,
Dentist *(O.I. Dentistry 376-4301)*, Chiropractor *(Inter Island Chiropractic 376-
2100)*, Holistic Services *(Clincicof Tibetan Medicine 376-8272, Orcas Spa
376-6361)* **Hospital:** St. Joseph 734-5400 *(21 mi.)*

Setting -- Located on President Channel on the northwest side of Orcas Island, West Beach Resort's pretty, rustic cottages edge a deep curve of beach and a wooded wildlife pond. Facing due west, the resort enjoys spectacular sunsets over the water. 750 feet of dockage and 11 mooring buoys welcome smaller vessels. At the foot of the dock, the natural-sided L-shaped store includes a snack bar, wide-screen TV, comfortable seating, fireplace and Internet desk.

Marina Notes -- *$25 flat rate under 26'. Free moorage for cabin guests. Transients are welcome when dock space is available - there is no power. If on a mooring buoy, guests may use the resort's dinghy. The store stocks some groceries, wine and beer, snacks, ice cream, plus bait and tackle; an inviting corner with satelliet TV serves as a boaters/guest lounge. Old photos show the beginnings of the resort, which was established in the 1930s. The rustic, wood-sided bathhouse (coin-op showers) and laundry are in the campground area. Note: Very exposed to SW weather.

Notable -- West Beach Resort is a great destination for families; the staff organizes daily activities for kids. Nightly beach bonfires are a family favorite. A hot tub is close to the general store. Kayak rentals and tours are available - including dry land instruction, snacks, and an experienced guide. Orca whales are frequently sighted during the summer months. The resort's guided fishing trips guarantee a catch, or try fishing and crabbing from the dock; a cleaning station and smokehouse are nearby. Scuba diving in the area is second only to the Great Barrier Reef, according to Jacques Cousteau (air is available at West Sound Marina). Two pottery studios are nearby, everything else is three miles east in the village - call a cab or the Island Shuttle (which also has rentals).

Navigational Information
Lat: 48°37.210' **Long:** 123°00.073' **Tide:** 12 ft. **Current:** n/a **Chart:** 18421
Rep. Depths (*MLW*): **Entry** 45 ft. **Fuel Dock** 20 ft. **Max Slip/Moor** 20 ft./45 ft.
Access: 1 mi. east of San Juan Channel, through marked passages

Marina Facilities (*In Season/Off Season*)
Fuel: Gasoline, Diesel
Slips: 125 Total, 75 Transient **Max LOA:** 100 ft. **Max Beam:** n/a
  Rate (*per ft.*): **Day** $1.30/0.65 **Week** $4.00 **Month** $5.50/4.00
  Power: 30 amp Incl., **50 amp** n/a, **100 amp** n/a, **200 amp** n/a
  Cable TV: No **Dockside Phone:** No
  Dock Type: Floating, Long Fingers, Aluminum
Moorings: 3 Total, 3 Transient **Launch:** Yes ($5/p), Dinghy Dock (Free)
  Rate: Day $20 **Week** $140 **Month** n/a
Heads: 4 Toilet(s), 4 Shower(s), Hair Dryers
Laundry: 2 Washer(s), 2 Dryer(s) **Pay Phones:** Yes, 2
Pump-Out: OnSite, 1 Central **Fee:** $8 **Closed Heads:** Yes

Marina Operations
Owner/Manager: Bellport Inc. **Dockmaster:** Marc Broman
In-Season: Mem-LabDay, 7:30am-9:30pm **Off-Season:** Sep-May, 8am-6pm
After-Hours Arrival: Check in in the morning
Reservations: Yes, Preferred **Credit Cards:** Visa/MC, Dscvr, Amex
Discounts: None
Pets: Welcome, Dog Walk Area **Handicap Access:** Yes, Heads, Docks

Deer Harbor Marina
PO Box 344; Deer Harbor, WA 98243

Tel: (360) 376-3037 **VHF: Monitor** Ch. 78A **Talk** n/a
Fax: (360) 376-6091 **Alternate Tel:** n/a
Email: deerharbor@rockisland.com **Web:** www.deerharbormarina.com
Nearest Town: Eastsound (*9.5 mi.*) **Tourist Info:** (360) 376-8888

Marina Services and Boat Supplies
Services - Docking Assistance, Concierge, Trash Pick-Up, Dock Carts, Megayacht Facilities **Communication -** Mail & Package Hold, Phone Messages, Fax in/out (*$1/pg*), Data Ports (*Wi-Fi - Broadband Xpress*), FedEx, DHL, UPS, Express Mail (*Sat Del*) **Supplies - OnSite:** Ice (*Block, Cube*), Bait/Tackle, Live Bait **OnCall:** Boat/US **1-3 mi:** Propane

Boatyard Services
OnSite: Crane, Launching Ramp **OnCall:** Divers (*Jen-Jay Diving 376-4664*) **Near:** Travelift, Engine mechanic (*gas, diesel*), Electrical Repairs, Electronic Sales, Electronics Repairs, Hull Repairs, Rigger, Sail Loft, Canvas Work, Bottom Cleaning, Brightwork, Air Conditioning, Refrigeration, Propeller Repairs, Inflatable Repairs, Metal Fabrication, Total Refits.
Nearest Yard: Deer Harbor Boatworks (360) 376-4056

Restaurants and Accommodations
OnSite: Lite Fare (*Dockside Galley 376-3037, B $2-9, L $4-10*), Condo/Cottage (*Resort at Deer Harbor 376-4420, now available only to Trendwest timeshare owners*) **Near:** Restaurant (*Deer Harbor Inn 376-4110, D $17-32*), Inn/B&B (*Deer Harbor 376-4110, $119-219*)
1-3 mi: Inn/B&B (*Inn on Orcas Island 376-5531*), (*The Place at Cayou Cove 376-3199, $195-425*) **3+ mi:** Restaurant (*Vern's 376-2231, B $6-13, L $7-13, D $19-30, 9 mi.*)

Recreation and Entertainment
OnSite: Heated Pool, Beach, Picnic Area, Grills, Playground, Boat Rentals (*Emerald Isle 376-3472 & Orcas Boat - power and sail 376-7616, Shearwater Kayaks 376-4699*), Fishing Charter (*Deer Harbor Charters 376-5989, whale watching & nature cruises*), Group Fishing Boat, Video Rental, Park, Galleries (*Rutabaga - art, gifts, books*) **Near:** Jogging Paths, Horseback Riding, Special Events (*Arts Afloat - May, Orcas Fly-In - Aug, Wooden Boat Fest - Sep*) **1-3 mi:** Volleyball, Tennis Courts, Roller Blade/Bike Paths
3+ mi: Golf Course (*Orcas Island Golf Course 376-4400, 6 mi.*), Movie Theater (*Sea View 376-5724, 9 mi.*), Museum (*Heritage Flight Museum 376-7654/ O.I. Historical Museum 376-4849, 4 mi./9 mi.*), Sightseeing (*Fun House 376-7177, 9 mi.*)

Provisioning and General Services
OnSite: Convenience Store, Delicatessen, Wine/Beer, Bank/ATM, Post Office, Retail Shops (*The Barge*) **Near:** Gourmet Shop **1-3 mi:** Market (*Orcas Market 376-8860*), Liquor Store **3+ mi:** Bakery (*Roses 376-5805, 9 mi.*), Farmers' Market (*Sat 10am-3pm, May-Sep next to museum, 9 mi.*), Green Market (*Orcas Home Grown Market & Deli 376-2009, 9 mi.*), Catholic Church (*9 mi.*), Protestant Church (*7 mi.*), Library (*376-4985, 9 mi.*), Laundry (*Lowrey's 376-2102, 5 mi.*), Pharmacy (*Ray's 376-2230, 9 mi.*), Hardware Store (*Island Hardware 376-4200, 9 mi.*)

Transportation
OnSite: Courtesy Car/Van (*$10 RT*), Bikes (*$8/hr., $35/day*), Rental Car (*376-4420*), Local Bus (*O.I. Shuttle 376-7433, seasonal*) **OnCall:** Taxi (*376-8294*) **3+ mi:** Ferry Service (*376-6253, 7 mi.*) **Airport:** Orcas (*12 mi.*)

Medical Services
911 Service **Near:** Ambulance **1-3 mi:** Veterinarian **3+ mi:** Doctor (*Orcas Island Medical 376-2561, 9 mi.*), Dentist (*Nordquist 376-5088, 4 mi.*)
Hospital: St. Joseph 374-5400 (*26 mi.*)

Setting -- From the San Juan Channel, wiggle through the Wasp Islands past Fawn Island, to the entrance of Deer Harbor. Tucked into the northeast corner, the marina's four main piers sit at the foot of a small hill, with the resort's craftsman-style shingled cottages scattered behind it. Hunter-green roofs top natural clapboard structures that house a store, snack bar, offices, gift shop, and bathhouse. Views are of the wooded peninsula across the bay - gorgeous at sunset.

Marina Notes -- The floating fuel dock (with intercom to the store and Gasboy CFN payment machine) and shop are open daily from 7:30am-9:30pm in summer, and 8am-6pm off-season. Welcoming touches include beautifully maintained docks with resident picnic tables and a well-stocked store and snack bar with sundries, some produce, wine and beer, Starbucks coffee and ice cream. Groups can rent BBQs - $25 includes the starter and charcoal. A few charters, kayak and whale watching tours are on-site. Sparkling tile bathhouse, at the head of the wharf, is open 24 hrs. The laundry is open during store hours (get $2 tokens). Note: The lovely cottages on the hill are now Trendwest timeshares.

Notable -- Relax at the beach, fish from the dock, enjoy the resort's heated pool, or get videos at the store. Elegant dinner and breakfast in season are served at the nearby Deer Harbor Inn's main dining room, and light fare is at the Dockside Galley in the store. Rental bicycles at the marina are a great way to explore the island. Daily shuttle rides to Eastsound, the golf course, and other points are provided by the marina in season. Alternatives are rental cars or the Island Shuttle which circles the island ($5 one way/$10 all day). The shuttle also offers a 3-hour island tour ($40) - stops include pottery studios and Moran State Park.

West Sound Marina

West Sound Marina

PO Box 119; 525 Deer Harbor Road; Orcas, WA 98280

Tel: (360) 376-2314 **VHF: Monitor** Ch. 16 **Talk** Ch. 9
Fax: (360) 376-4634 **Alternate Tel:** n/a
Email: betsy@westsoundmarina.com **Web:** www.westsoundmarina.com
Nearest Town: Orcas *(3 mi.)* **Tourist Info:** (360) 376-8888

Navigational Information
Lat: 48°37.761' **Long:** 122°57.400' **Tide:** 12 ft. **Current:** n/a **Chart:** 18434
Rep. Depths (*MLW***): Entry** 30 ft. **Fuel Dock** 6 ft. **Max Slip/Moor** 40 ft./-
Access: NE side of West Sound behind Picnic Island

Marina Facilities *(In Season/Off Season)*
Fuel: Gasoline, Diesel
Slips: 180 Total, 6 Transient **Max LOA:** 140 ft. **Max Beam:** n/a
  Rate *(per ft.)*: **Day** $0.75/0.60 **Week** $4.50 **Month** $9
  Power: 30 amp $2, 50 amp n/a, 100 amp n/a, 200 amp n/a
  Cable TV: No **Dockside Phone:** No
  Dock Type: Floating, Concrete, Wood
Moorings: 0 Total, 0 Transient **Launch:** n/a
  Rate: Day n/a **Week** n/a **Month** n/a
Heads: 2 Toilet(s)
Laundry: None **Pay Phones:** Yes, 1
Pump-Out: OnSite **Fee:** $15 **Closed Heads:** Yes

Marina Operations
Owner/Manager: Betsy Wareham **Dockmaster:** Same
In-Season: Year-Round*, 8am-4:45pm **Off-Season:** n/a
After-Hours Arrival: Tie up to guest float
Reservations: Yes, Preferred **Credit Cards:** Visa/MC
Discounts: None
Pets: Welcome **Handicap Access:** Yes, Docks

Marina Services and Boat Supplies
Services - Security *(owner lives onsite)*, Dock Carts **Supplies - OnSite:** Ice *(Block, Cube)*, Ships' Store, Bait/Tackle, Propane

Boatyard Services
OnSite: Travelift *(30T- 18ft. beam)*, Crane *(1T)*, Engine mechanic *(gas, diesel)*, Electrical Repairs, Electronic Sales, Electronics Repairs, Bottom Cleaning, Brightwork, Divers, Compound, Wash & Wax, Interior Cleaning, Woodworking, Inflatable Repairs, Painting, Awlgrip **OnCall:** Launching Ramp **Dealer for:** Volvo, Yanmar, Mercruiser, Johnson Evinrude. Other Certifications: NWMTA **Yard Rates:** $70/hr., Haul & Launch Inq.

Restaurants and Accommodations
Near: Lite Fare *(West Sound Café 376-4440, B only)*, Inn/B&B *(Kingfish Inn 376-4440, $100-160)* **1-3 mi:** Restaurant *(Mamie's Boardwalk 376-2971, B $7-15, L $8-16, D $15-22)*, *(Octavia's Bistro 376-4300, B $7-11, D $15-25, steaks & seafood)*, Lite Fare *(Orcas Village Store 376-8860, deli)*, *(Orcas Hotel Café 376-4300, coffee, baked goods, lunch)*, Inn/B&B *(Orcas Hotel 376-4300, $85-198, rent the entire hotel, 12 rooms, for $1,895 a night)* **3+ mi:** Restaurant *(Country Corner 7 mi.)*, Lite Fare *(Dockside Galley 376-2482, 4 mi.)*, Inn/B&B *(Deer Harbor Resort 376-4420, $129-169, 4 mi.)*

Recreation and Entertainment
1-3 mi: Picnic Area, Playground, Video Rental *(Orcas Store 376-8860)* **3+ mi:** Golf Course *(Orcas Island Golf Course 376-4400, 3 mi.)*, Fitness Center *(Fitness Quest 376-9568, 6 mi.)*, Boat Rentals *(Orcas Boat Rentals 376-7616, 4 mi.)*, Movie Theater *(Sea View 376-5724, 6 mi.)*, Park *(Moran St. Park 376-2326, Rosario Resort, 10 mi.)*, Museum *(Lambiel Museum*

376-4555, Orcas Island Historical Museum 376-4849, Tue-Sun, 6 mi.)*, Cultural Attract *(Dolphin Bay Woodworking and Star 376-5884, 6 mi.)*, Sightseeing *(Funhouse 376-4010, 6 mi.)*, Special Events *(Library Fair - Aug, Orcas Fly-In - Aug, Chamber Music - Aug, Wooden Boat - Sep, 3-6 mi.)*, Galleries *(Howe Art 376-2945, 6 mi.)*

Provisioning and General Services
1-3 mi: Market *(Orcas Market 376-8860)*, Delicatessen, Liquor Store *(State Liquor Store 376-4389)*, Bank/ATM, Post Office, Retail Shops *(Orcas Landing)* **3+ mi:** Health Food *(Orcas Home Grown Market & Deli 376-2009, 6 mi.)*, Bakery *(Roses Bread and Specialties 376-5805, 6 mi.)*, Farmers' Market *(Sat 10am-3pm, May-Sep next to museum, 6 mi.)*, Catholic Church *(7 mi.)*, Protestant Church *(4 mi.)*, Library *(Orcas 76-4985, 6 mi.)*, Beauty Salon *(Altier Salon 378-8835, 6 mi.)*, Laundry *(Deer Harbor, 4 mi.)*, Bookstore *(Darvill's Book Store 376-2135, 6 mi.)*, Pharmacy *(Ray's 376-2230, 6 mi.)*, Hardware Store *(Island Hardware 376-4200, 6 mi.)*

Transportation
OnCall: Taxi *(O.I. Taxi 376-8294)* **Near:** Local Bus *(O.I. Shuttle 376-7433, May-Sep)* **1-3 mi:** Ferry Service *(376-6253)* **3+ mi:** Bikes *(Orcas Mopeds at ferry landing 376-5266, 3 mi.)*, Rental Car *(Deer Harbor Resort 376-4420, 4 mi.)* **Airport:** Orcas Airport *(9 mi.)*

Medical Services
911 Service **OnCall:** Ambulance **3+ mi:** Doctor *(Orcas Island Medical 376-2561, 6 mi.)*, Dentist *(Ivans 376-4774, 6 mi.)*, Chiropractor *(7 mi.)*, Holistic Services *(Orcas Spa 376-6361, 6 mi.)*, Veterinarian *(Animal Clinic 376-7838, 6 mi.)* **Hospital:** St. Joseph 374-5400 *(26 mi.)*

Setting -- West Sound is located in a quiet bay of the same name on the south side of Orcas Island. Four main docks host 180 slips plus a 250-foot side-tie guest float that also houses the seaplane landing and fueling station. Onshore an imposing gray shed towers over a two-story, gray barn-roofed office trimmed in green. Above the marina a tiny community is perched on the wooded hillside. Views across the sound are pristine and untouched.

Marina Notes -- *10am-3pm Sat. Closed off-season Sun, Thanksgiving, Christmas, New Year's. Established in 1950, expanded in 1999 - the Warehams have been at the helm since 1974. Focus is on seasonal slips and marine services; a 30T travelift is one of the largest in the area. The well-stocked ships' store is also one of the best in the islands. Amenities for the visiting boater are sparse. Kenmore Air Seaplane is on-site. Simple heads are at the top of the driveway.

Notable -- The Kingfish Inn and West Sound Café are a few steps away. The area around the ferry landing, about 3 miles south the marina, offers additional restaurants, a gift shop, kayaking and whale watching tours, and scooter rentals at Suzi's Mopeds. Orcas Hotel, a restored waterfront Victorian inn, serves light fare and houses Octavia's Bistro Restaurant, open for breakfast and dinner. Orcas Village Store has a deli, and sells produce and fresh meat and seafood. The largest island in the San Juan group, Orcas also owns the highest point in the islands, 2,409-ft. Mount Constitution in Moran State Park - the Island Shuttle tour stops here. The park sports 30 miles of trails, picnic shelters, hike-in lakes, bridges and waterfalls, and 150 campsites. The community of Eastsound, some 6 miles away, the commercial and cultural center of the island, offers many attractions. Orcas is also known as the Emerald Isle and the Horseshoe Island.

Navigational Information
Lat: 48°38.800' **Long:** 122°52.230' **Tide:** 10 ft. **Current:** 0.5 kt. **Chart:** 1843
Rep. Depths *(MLW):* **Entry** 12 ft. **Fuel Dock** 9 ft. **Max Slip/Moor** 12 ft./75 ft.
Access: Starboard side of East Sound, about 3.5 mi. from entrance

Marina Facilities *(In Season/Off Season)*
Fuel: Gasoline, Diesel
Slips: 32 Total, 32 Transient **Max LOA:** 105 ft. **Max Beam:** n/a
  Rate *(per ft.):* **Day** $1.62/$32* **Week** n/a **Month** n/a
  Power: 30 amp Incl., **50 amp** n/a, **100 amp** n/a, **200 amp** n/a
  Cable TV: No **Dockside Phone:** No
  Dock Type: Floating, Long Fingers, Short Fingers, Pilings, Wood
Moorings: 20 Total, 20 Transient **Launch:** n/a, Dinghy Dock
  Rate: Day $25 **Week** $175 **Month** n/a
Heads: 5 Toilet(s) 8 Shower(s) *(with dressing rooms)*
Laundry: 4 Washer(s) 6 Dryer(s) **Pay Phones:** Yes, 2
Pump-Out: No **Fee:** n/a **Closed Heads:** Yes

Marina Operations
Owner/Manager: Rock Resorts/Nelson Moulton **Dockmaster:** Gary Joseph
In-Season: May 15-Sep 15, 8am-8pm **Off-Season:** Oct 1-May 1
After-Hours Arrival: Call ahead
Reservations: Yes, Preferred **Credit Cards:** Visa/MC, Dscvr, Din, Amex
Discounts: None
Pets: Welcome, Dog Walk Area **Handicap Access:** No

Rosario Resort Hotel
1400 Rosario Road; Eastsound, WA 98245

Tel: (360) 376-2222; (800) 562-8820 **VHF: Monitor** Ch. 78A **Talk** n/a
Fax: (360) 376-3680 **Alternate Tel:** n/a
Email: harbormaster@rosarioresort.com **Web:** www.rosarioresort.com
Nearest Town: Eastsound *(4.5 mi.)* **Tourist Info:** (360) 376-8888

Marina Services and Boat Supplies
Services - Docking Assistance, Concierge, Room Service to the Boat, Boaters' Lounge, Security *(24 hrs., staff)*, Trash Pick-Up, Dock Carts
Communication - Mail & Package Hold, Phone Messages, Fax in/out, Data Ports *(Front desk)*, FedEx, UPS **Supplies - OnSite:** Ice *(Block, Cube)*
3+ mi: Propane *(Crescent Service 376-4076, 4 mi.)*

Boatyard Services
OnCall: Divers *(Jen-Jay Diving 376-4664)* **Nearest Yard:** West Sound Marina (360) 376-2314

Restaurants and Accommodations
OnSite: Restaurant *(Mansion Dining Room 376-2222, B $10-15, D $11-35, B 7:30 -11am, D 5:30-9pm)*, *(Moran Lounge 376-2222, L $10-15, D $11-35, L 11am-3pm, all-day menu 3:30-10pm Sun-Thu, 'til 11pm Fri-Sat)*, *(Cascade Bay Grill 376-2222, L $8-14, D $12-30)*, Hotel *(Rosario Resort 376-2222, $129-349)* **Near:** Hotel *(Cascade Harbor Inn 376-6350, $75-299)*
1-3 mi: Restaurant *(Olga Café 376-5098, B $7-12, L $9-15, D $14-25)*
3+ mi: Hotel *(Resort at Deer Harbor 376-4420, $129-399, 6 mi.)*, Inn/B&B *(Palmers Chart House 376-4231, $60-80, 3 mi.)*

Recreation and Entertainment
OnSite: Heated Pool, Spa, Beach, Grills, Playground, Fitness Center, Boat Rentals *(kayaks)*, Fishing Charter, Museum *(Moran Museum, plus O.I. Historical Museum 376-4849, Tue-Sun, 4 mi.)*, Cultural Attract *(organ concerts Sat in season)* **OnCall:** Dive Shop **Near:** Jogging Paths, Park *(Moran State Park - observation tower at top of Mt. Constitution)* **Under 1 mi:** Picnic Area **1-3 mi:** Video Rental *(Island Market 376-6000)*, Sightseeing *(Heritage Flight Museum 376-7654)* **3+ mi:** Golf Course *(Orcas Island Golf Course 376-4400, 7 mi.)*, Movie Theater *(Sea View 376-5724, 5 mi.)*

Provisioning and General Services
OnSite: Convenience Store *(Dockside General Store)*, Wine/Beer, Bank/ATM *(ATM)*, Newsstand, Copies Etc. **1-3 mi:** Bakery *(Roses Bread and Specialties 376-5805)*, Post Office *(800-275-8777)*, Protestant Church, Barber Shop *(Suzie's 376-1911)*, Pharmacy *(Ray's 376-2230)*
3+ mi: Market *(Island 376-6000, 4 mi.)*, Delicatessen *(Country Corner 376-6900, 4 mi.)*, Health Food *(Orcas Home Grown Market and Deli 376-2009, 4 mi.)*, Liquor Store *(Eastsound Liquor 376-2616, 4 mi.)*, Farmers' Market *(Eastsound, Sat 10am-3pm, May-Sep, 4 mi.)*, Green Market *(Maple Rock Farm 376-5994, 6mi.)*, Library *(Orcas 376-4985, 4 mi.)*, Beauty Salon *(Tangle's 376-5077, 4 mi.)*, Laundry *(Country Corner 376-6900, 5 mi.)*, Bookstore *(Darvill's 376-2135, 4 mi.)*, Hardware Store *(Island Hardware 376-4200, 4 mi.)*, Florist *(Blossoms and Branches 376-5075, 5 mi.)*

Transportation
OnSite: Courtesy Car/Van, Rental Car, Local Bus *(seasonal shuttle 376-7433)* **OnCall:** Water Taxi *(Paraclete Charters 800-808-2999)*, Taxi *(376-8294)* **3+ mi:** Bikes *(Marina Bicycle Rentals 376-3037, 6 mi.)*, Ferry Service *(Orcas 376-6253, 12 mi.)* **Airport:** Port of Orcas 376-5285 *(5 mi.)*

Medical Services
911 Service **OnSite:** Holistic Services *(Avanyu Spa)* **OnCall:** Ambulance
1-3 mi: Doctor *(Orcas Island Medical Center 376-6604)*, Dentist *(Nordquist 376-5088)*, Veterinarian *(Orcas Animal Clinic 376-7838)* **3+ mi:** Optician *(San Juan Vision, 5 mi.)* **Hospital:** St. Joseph 374-5400 *(27 mi.)*

Setting -- Overlooking Cascade Bay, the beautifully landscaped 30-acre Rosario Resort sits prominently on the east side of Orcas Island's East Sound. From the sea, the resort's centerpiece - the magnificent, off-white stucco, five-story 1906 Moran Mansion - creates a sharp contrast with the surrounding lush evergreen hills. The protected 30-slip destination marina (soon to be 165 slips) is tucked into a cove with convenient access to all the facilities and amenities.

Marina Notes -- *Flat fees in summer for dockage: 28' $55, 40' $65, 50' $75, T-dock $110/night. Moorings max 45 ft. LOA. Fees include access for up to 12 guests to the spa, indoor therapeutic pool, hot tub, exercise room, fitness classes, 2 pools, bathhouse, laundry and scheduled van service. 25% discount if boater is staying in hotel (excl. Aug & hols) Off-season dockage $32/night (or whatever a boater considers fair - less than $10 - stay limited to 4 nights). Facilities available to anchored boats - $25/day. A guest of the marina is a guest of the resort. Day use: free up to 4 hrs. on dock or 8 hrs. on buoy.

Notable -- On the Nat'l. Register of Historic Places, the "Northwest's San Simeon," was built by shipbuilding magnate and former Seattle Mayor Robert Moran - mahogany ceiling beams, intricate parquet floors, and an Arts & Crafts style informed by nautical sensibilities. It houses three eateries, lounges, library, music room, fitness center, two pools, a boutique, an art gallery and a full service spa - everything from manicures to therapeutic massages to body wraps and Reiki. The 116 guest rooms, suites and cottages are spread out across the acreage. Dockside General Store, a pool, and the marine center are at the docks (concierge and room service to the boat). Don't miss a silent film showing accompanied by curator Christopher Peacock on the 26-rank, Aeolian pipe organ.

Blakely Island Marina

Navigational Information
Lat: 48°35.120' **Long:** 122°48.950' **Tide:** 10 ft. **Current:** n/a **Chart:** 18430
Rep. Depths (*MLW*): Entry 8 ft. **Fuel Dock** 20 ft. **Max Slip/Moor** 20 ft./-
Access: Enter Peavine Pass off Rosario Strait; marina is at the north end

Marina Facilities *(In Season/Off Season)*
Fuel: *Chevron* - Gasoline, Diesel
Slips: 68 Total, 30 Transient **Max LOA:** 80 ft. **Max Beam:** 16 ft.
  Rate *(per ft.)*: Day $0.95/0.75 **Week** n/a **Month** n/a
  Power: 30 amp Incl., **50 amp** n/a, **100 amp** n/a, **200 amp** n/a
  Cable TV: No **Dockside Phone:** No
  Dock Type: Floating, Long Fingers, Concrete
Moorings: 0 Total, 0 Transient **Launch:** n/a
  Rate: Day n/a **Week** n/a **Month** n/a
Heads: 5 Toilet(s), 4 Shower(s) *(with dressing rooms)*
Laundry: 2 Washer(s), 3 Dryer(s), Book Exchange **Pay Phones:** Yes, 1
Pump-Out: No **Fee:** n/a **Closed Heads:** Yes

Marina Operations
Owner/Manager: Richard and Norma Reed **Dockmaster:** Same
In-Season: Jul-Aug, 8am-8pm **Off-Season:** Sep-Dec, Varies
After-Hours Arrival: Sign in at store and pay through the envelope door slot
Reservations: Yes, Preferred **Credit Cards:** Visa/MC
Discounts: Groups* **Dockage:** Inq. **Fuel:** n/a **Repair:** n/a
Pets: Welcome, Dog Walk Area **Handicap Access:** No

Blakely Island Marina

One Marina Drive; Blakely Island, WA 98222

Tel: (360) 375-6121 **VHF: Monitor** Ch. 66A **Talk** n/a
Fax: n/a **Alternate Tel:** n/a
Email: info@blakelymarina.com **Web:** www.rockisland.com/~blakely
Nearest Town: Anacortes *(8 mi.)* **Tourist Info:** (360) 293-7911

Marina Services and Boat Supplies
Services - Trash Pick-Up, Dock Carts **Supplies - OnSite:** Ice *(Cube)*,
Bait/Tackle **3+ mi:** West Marine *(Friday Harbor, by boat, 7 mi.)*

Boatyard Services
3+ mi: Travelift *(30T, 15 mi.)*, Engine mechanic *(gas, diesel)*, Launching
Ramp *(15 mi.)*, Electrical Repairs *(15 mi.)*, Electronics Repairs *(15 mi.)*, Hull
Repairs *(15 mi.)*, Rigger *(15 mi.)*, Bottom Cleaning *(15 mi.)*, Brightwork *(15
mi.)*, Divers *(15 mi.)*, Compound, Wash & Wax *(15 mi.)*, Interior Cleaning
(15 mi.), Woodworking *(15 mi.)*, Inflatable Repairs *(15 mi.)*, Painting *(15 mi.)*,
Awlgrip *(15 mi.)*. **Nearest Yard:** West Sound Marina (360) 376-2314

Restaurants and Accommodations
Under 1 mi: Coffee Shop *(Olga Café 375-5098, B $7-12, L $9-15, D $14-25,
by dinghy)* **1-3 mi:** Restaurant *(Rosario Resort 376-2222, B $10-15, L $10-
15, D $11-35, by large dinghy or boat)*, Condo/Cottage *(cottages and cabins
at Olga on Orcas)* **3+ mi:** Restaurant *(Vern's Bayside 376-2231, B $6-13, L
$7-13, D $19-30, 10 mi., by dinghy, then taxi)*, Hotel *(BW Inn 794-3111, $59-
99, 15 mi., Anacortes)*

Recreation and Entertainment
OnSite: Beach, Picnic Area, Grills, Video Rental, Park **Near:** Boat Rentals
(Kayaks) **1-3 mi:** Fishing Charter *(Salmon charters@earthlink.net)*
3+ mi: Golf Course *(Lopez Island Golf Course 468-2679 by boat, 12 mi.)*

Provisioning and General Services
OnSite: Convenience Store, Wine/Beer, Clothing Store **3+ mi:** Market
(Store Grocery 293-2851, Anacortes), Supermarket *(Safeway 293-5393,
Anacortes)*, Delicatessen *(Jackpot Food Mart 293-6222, Anacortes)*, Catholic
Church *(Anacortes)*, Protestant Church *(Anacortes)*, Synagogue *(Anacortes)*,
Library *(Orcas 376-4985/Anacortes 293-1910, 10 mi./16 mi.)*, Beauty Salon
(Anacortes), Florist *(Anacortes)*, Department Store *(By boat to Anacortes,
3+ mi.)*

Transportation
3+ mi: Taxi *(Orcas Island Taxi 376-TAXI, on Orcas Island - by dinghy)*
Airport: Anacortes 299-1803 *(15 mi.)*

Medical Services
911 Service **3+ mi:** Doctor *(Anacortes Family Medicine 293-5059, 15 mi.)*,
Dentist *(Sound Health 299-9347, Anacortes, 15 mi.)* **Hospital:** Skagit Valley
424-4111 *(15 mi.)*

Setting -- Located in Peavine Pass at the north end of Blakely Island, the Blakely Island Marina and General Store provides a beautiful sheltered yacht harbor
in the San Juans. Flowers are everywhere - even in the restrooms. Rosebushes line the front of the store. Patio tables and chairs are scattered about
encouraging visitors to relax and enjoy the ambiance. A picnic shelter provides the perfect setting for a group to gather for a BBQ. The vista across the water,
from the front of the store, delivers six islands in a single sweeping glance. Turn around and Cypress Island and Mt. Baker come into view.

Marina Notes -- *Yacht club and group discounts. The island is owned by private residents - the marina is the only commercial enterprise. Family owned
since 1955. Richard and Norma Reed enthusiastically welcome visitors and the marina reflects their pride in the facility. Club cruises are encouraged. This is a
good spot for a family reunion of boaters. The general store serves as the downtown "mall" for island residents. It's well-stocked with groceries, wine, sundries,
gifts, fishing supplies, nautical books, and rental videos. Simple but sparkling bathhouse (showers are $1 for 5 minutes).

Notable -- Blakely Island is accessible only by boat or plane (no ferries), and protected by conservation easements designed to preserve its beauty and limit
development. Blakely Island is a perfect place to slip away for a few days, kick back and tune in to nature. The western tip of Orcas Island is a dinghy or small
boat ride away; there is a public dock in the hamlet of Olga - a café and O.I. Artworks are nearby. To get from there to the O.I. ferry terminal requires a larger
boat or taxi - foot passengers can ride from one island to another free. Friday Harbor (an hour by ferry) offers restaurants, whale watching trips, and lodgings.

Navigational Information
Lat: 48°36.589' **Long:** 123°09.200' **Tide:** 14 ft. **Current:** 1 kt. **Chart:** 18433
Rep. Depths *(MLW)*: **Entry** 30 ft. **Fuel Dock** 30 ft. **Max Slip/Moor** 30 ft./-
Access: Enter either by way of Spieden Channel or Mitchell Bay

Marina Facilities *(In Season/Off Season)*
Fuel: Slip-Side Fueling, Gasoline, Diesel, High-Speed Pumps
Slips: 377 Total, 250 Transient **Max LOA:** 200 ft. **Max Beam:** n/a
Rate *(per ft.)*: **Day** $1.40/0.70* **Week** n/a **Month** n/a
Power: 30 amp Incl., **50 amp** $5, **100 amp** Incl., **200 amp** n/a
Cable TV: Yes, Free Satellite **Dockside Phone:** Yes
Dock Type: Floating, Long Fingers, Concrete, Wood
Moorings: 0 Total, 0 Transient **Launch:** n/a, Dinghy Dock (Free)
Rate: Day n/a **Week** n/a **Month** n/a
Heads: 10 Toilet(s), 10 Shower(s) *(with dressing rooms)*
Laundry: 7 Washer(s), 6 Dryer(s) **Pay Phones:** Yes, 4
Pump-Out: OnSite, 1 Central, 1 Port **Fee:** Free **Closed Heads:** Yes

Marina Operations
Owner/Manager: Brent Snow **Dockmaster:** Kevin Carlton/Troy Buck
In-Season: May-Sep, 8am-Sunset **Off-Season:** Oct-May, 9am-5pm
After-Hours Arrival: Check in at the hotel
Reservations: Yes, Preferred **Credit Cards:** Visa/MC, Amex
Discounts: None
Pets: Welcome, Dog Walk Area **Handicap Access:** Yes, Heads, Docks

Roche Harbor Resort

PO Box 4001; 248 Reuben Mem. Dr.; Roche Harbor, WA 98250

Tel: (360) 378-9800; (800) 451-8910 **VHF: Monitor** Ch. 78A **Talk** n/a
Fax: (360) 378-9800 **Alternate Tel:** n/a
Email: rhmarina@rockisland.com **Web:** www.rocheharbor.com
Nearest Town: Friday Harbor *(9 mi.)* **Tourist Info:** (360) 378-5240

Marina Services and Boat Supplies
Services - Docking Assistance, Concierge, Security *(24 hrs.)*, Trash Pick-Up, Dock Carts, Megayacht Facilities **Communication -** Mail & Package Hold, Phone Messages, Fax in/out *(Free)*, Data Ports *(Wi-Fi - Broadband Xpress)*, FedEx, DHL, UPS, Express Mail *(Sat Del)* **Supplies - OnSite:** Ice *(Block, Cube)*, Ships' Store *(Roche Harbor Market)*, Bait/Tackle, Propane

Boatyard Services
OnSite: Engine mechanic *(gas, diesel)* **OnCall:** Electrical Repairs, Divers, Compound, Wash & Wax, Interior Cleaning, Metal Fabrication
3+ mi: Travelift *(35T, 10 mi.)* **Nearest Yard:** Albert Jensen (360) 378-4343

Restaurants and Accommodations
OnSite: Restaurant *(McMillin's 378-5757, D $16-25)*, (Madrone Bar & Grill 378-2155, L $8-16, D $10-18, harborside deck)*, Snack Bar *(Beechtree Expresso & Ice Creamery)*, Lite Fare *(Lime Kiln Café B $7-13, L $7-10)*, Hotel *(Hotel de Haro 378-2155, $70-183, MacMillan Suites $205-325)*, Condo/Cottage *(Harbor View $143-310)*, *(Company Town Cottages $130-250)*, *(Carriage Houses & Cottages on the Green $375-475)*

Recreation and Entertainment
OnSite: Heated Pool, Spa, Beach, Picnic Area, Grills, Playground, Dive Shop *(Island Dive & Water Sports 378-2772)*, Tennis Courts, Jogging Paths, Boat Rentals *(Kayaks 378-1323 $15-18/hr., $90-108/day & 3-5 hr. tours $59-75)*, Video Rental, Park *(19-acre Wescott Bay Reserve - displays over 85 sculptures)*, Cultural Attract *(Island Stage Left at Roche Harbor Outdoor Stage - open air Shakespeare)*, Sightseeing *(Whale Watching 378-1323 $59/39 11am & 3pm; self-guided walking and hiking tours)*, Special Events

(at sunset U.S.and Canadian flags are lowered to their respective national anthems; Salmon Classic Invitational - Feb) **1-3 mi:** Bowling *(Paradise Lanes 370-5667)* **3+ mi:** Golf Course *(San Juan G & CC 378-5158, 4 mi.)*, Movie Theater *(The Palace 370-5666, 5 mi.)*

Provisioning and General Services
OnSite: Market *(Roche Harbor Village Market)*, Wine/Beer, Post Office, Protestant Church *(Our Lady of Good Voyage Chapel)*, Newsstand, Retail Shops *(R.H. Gift Shop; Shirt Shoppe; R.H. Sportswear; "rock n' roche" Children's)*, Copies Etc. **Under 1 mi:** Fishmonger *(Westcott Bay Sea Farms 378-2489 oysters, clams & mussels)* **3+ mi:** Supermarket *(King's Market 378-4505, 8 mi.)*, Liquor Store *(Island Wine 378-3229, 7 mi.)*, Farmers' Market *(Sat 10am-1pm, midApr-midOct, courthouse parking, 8 mi.)*, Catholic Church *(7 mi.)*, Library *(378-2798, 7 mi.)*, Pharmacy *(Friday Harbor Drug 378-4421, 7 mi.)*, Hardware Store *(Friday Harbor 378-4622, 7 mi.)*

Transportation
OnSite: Local Bus *(San Juan Transit 378-8887 $7/3 1-way, $10 RT May-LabDay)* **OnCall:** Taxi *(Bob's 378-6777)* **Near:** Rental Car *(Susie's 378-5244: GeoTracker cars $96/day, Scootcars $125/day, Mopeds $62-125; West Isle 378-2440)* **3+ mi:** Ferry Service *(State Ferries 468-2142, Foot Ferry to Victoria 452-8088, 8 mi.)* **Airport:** Roche Harbor/Friday Harbor 378-4724/ Kenmore Air & Northwest Seaplane *(0.5mi./8 mi./onsite)*

Medical Services
911 Service **3+ mi:** Doctor *(Inter Island Medical Center 378-1338, 7 mi.)*, Dentist *(Tooth Ferry 378-5300, 8 mi.)*, Chiropractor *(7 mi.)*, Veterinarian *(7 mi.)* **Hospital:** Whidbey General 678-5151 *(29 mi.)*

Setting -- Off Spieden Channel through Mosquito Pass, a glorious, historic resort and first-class marina fill most of protected Roche Harbor. A profusion of blooms highlight repurposed and restored white clapboard buildings and the landmarked 1886 Hotel de Haro - all surrounded by magnificent, carefully tended grounds and cobblestone walkways. A white-railed boardwalk leads from the docks through exquisite Victorian gardens to the "village."

Marina Notes -- *Holiday weekends - 3 day min. + add'l. $10-20 nightly (reserve far in advance). 40 summertime dockhands greet boats, handle lines and describe the activities and state-of-the-art amenities. Transients use the Guest Dock in front of the hotel; new guest docks and walkway were installed in 2006. Registration is on the dock as is U.S. Customs. A guest of the marina is a guest of the resort. 40' x 40' barge with 2 propane BBQs for group events. Stationary pump-out or "Phecal Phreak" travels to the slips. Laundry has heavy-duty machines; 2 bathhouses feature varnished door stalls or lovely tiled full bathrooms.

Notable -- The village, listed on the National Register of Historic Sites, has come a long way from a 19th century lime kiln operation to a gorgeous family-oriented resort and one of the top boating destinations in the northwest. Marine Adventures features kayak classes, whale watching and scuba diving. Within steps are enticing boutiques, three eateries, the Hotel de Haro, and a well-supplied grocery. All summer, artists display their work in booths under the trees. Hiking trails abound and around the harbor the Company Town Cottages overlook the pool, tennis and playground complex. Above the hotel are neo-Victorian Cottages and up on the hill are new de Haro townhomes. Coming in '07: a 3-story "Commercial Building" - 12 deluxe hotel suites, a full-service spa, and shops.

Snug Harbor Marina Resort

1997 Mitchell Bay Road; Friday Harbor, WA 98250

Tel: (360) 378-4762 **VHF: Monitor** n/a **Talk** n/a
Fax: (360) 378-8859 **Alternate Tel:** n/a
Email: sneakaway@snugharbor.com **Web:** www.snugresort.com
Nearest Town: Friday Harbor *(9 mi.)* **Tourist Info:** (360) 378-5240

Navigational Information
Lat: 48°34.298' **Long:** 123°10.141' **Tide:** 10 ft. **Current:** n/a **Chart:** 18433
Rep. Depths *(MLW)*: **Entry** 6 ft. **Fuel Dock** 6 ft. **Max Slip/Moor** 6 ft./-
Access: At the head of Mitchell Bay off the Haro Strait

Marina Facilities *(In Season/Off Season)*
Fuel: Gasoline
Slips: 70 Total, 25 Transient **Max LOA:** 40 ft. **Max Beam:** n/a
 Rate *(per ft.)*: **Day** $0.75/0.50 **Week** n/a **Month** $12/6
 Power: 30 amp $2, **50 amp** n/a, **100 amp** n/a, **200 amp** n/a
 Cable TV: No **Dockside Phone:** No
 Dock Type: Floating, Long Fingers, Wood
Moorings: 0 Total, 0 Transient **Launch:** n/a
 Rate: Day n/a **Week** n/a **Month** n/a
Heads: 2 Toilet(s), 2 Shower(s)
Laundry: None **Pay Phones:** No
Pump-Out: No **Fee:** n/a **Closed Heads:** Yes

Marina Operations
Owner/Manager: Glenn Kalmus **Dockmaster:** Same
In-Season: May-Sep **Off-Season:** Oct-Apr
After-Hours Arrival: Take a slip, check in next morning
Reservations: Yes, Preferred **Credit Cards:** Visa/MC, Dscvr, Amex
Discounts: None
Pets: Welcome **Handicap Access:** No

Marina Services and Boat Supplies
Services - Dock Carts **Supplies - OnSite:** Ice *(Block, Cube)*, Bait/Tackle
3+ mi: West Marine *(378-1086, 9 mi.)*

Boatyard Services
OnCall: Electrical Repairs, Divers, Compound, Wash & Wax, Interior
Cleaning, Metal Fabrication **3+ mi:** Travelift *(35T, 9 mi.)*, Yacht Broker
(Friday Harbor Yacht Sales 378-4047, 8 mi.). **Nearest Yard:** Albert Jensen
and Sons Inc. (360) 378-4343

Restaurants and Accommodations
OnSite: Condo/Cottage *(Snug Harbor Resort 378-4762, $95-210)* **Near:**
Inn/B&B *(Longhouse B&B 378-2568, $110-150)* **1-3 mi:** Inn/B&B *(States
Inn 378-6240, $80-200)* **3+ mi:** Restaurant *(China Pearl 378-5254, B $6-10,
L $9-22, D $9-22, 8 mi.)*, *(Bella Luna 378-4118, B $4-8, L $7-20, D $12-24, 8
mi.)*, *(Madrone Bar & Grill 378-2155, L $8-16, D $10-18, 6 mi.)*, *(McMillin's
378-5757, D $16-25, 6 mi.)* *(Fat Cat 378-8646, 8 mi.)*, *(Hungry Clam 378-
3474, 8 mi.)*, Pizzeria *(Cousin Vinnie's 378-8308, 8 mi., no del.)*

Recreation and Entertainment
OnSite: Picnic Area, Grills, Playground, Boat Rentals *(power boats and
kayaks)*, Fishing Charter *(Maya's Charters or Salish Sea for whale watching)*,
Sightseeing *(Center for Whale Watching 378-5835)* **1-3 mi:** Park *(Lime Kiln
St. Park 378-2040, Stuart Island St. Park 378-2044)* **3+ mi:** Dive Shop
(Island Dive and Water Sports 800-303-8386, 9 mi.), Golf Course *(San Juan
Golf 378-5158, 8 mi.)*, Fitness Center *(Island Fitness 378-4449, 8 mi.)*,
Bowling *(San Juan Bowling 370-5667, 8 mi.)*, Group Fishing Boat *(Buffalo
Works 378-4612, 8 mi.)*, Museum *(Whale Museum 378-4710, 8 mi.)*,

Cultural Attract *(Community Theatre 378-3210, 8 mi.)*, Galleries *(Island
Studios 378-6550, Arctic Raven 378-3433, 5 mi.)*

Provisioning and General Services
OnSite: Convenience Store **3+ mi:** Market *(King's Market 378-4505, 7 mi.)*,
Gourmet Shop *(Gourmet's Galley 378-2251, 8 mi.)*, Delicatessen *(Little Store
378-4422, 8 mi.)*, Health Food *(Sound Nutrition 378-2939, 7 mi.)*, Liquor
Store *(State Liquor Store 378-2956, 7 mi.)*, Bakery *(Kneadful Things 378-
7089, 7 mi.)*, Farmers' Market *(Sat 10am-1pm, midApr - midOct, Courthouse
parking, 8 mi.)*, Meat Market *(Jim's Meat Market 378-2373, 7 mi.)*, Bank/ATM
(Key Bank 378-2111, 5 mi.), Library *(378-2798, 7 mi.)*, Beauty Salon *(Studio
55 378-6899, 7 mi.)*, Barber Shop *(Clipper Ship 378-2636, 7 mi.)*, Laundry
(Sunshine Laundry and Dry Clean 378-7223, 8 mi.), Pharmacy *(Friday
Harbor Drug 378-4451, 7 mi.)*, Hardware Store *(Friday Harbor Hardware
378-4622, 7 mi.)*

Transportation
OnCall: Taxi *(378-3550)* **Near:** Local Bus *(San Juan Shuttle 378-8887,
seasonal)* **3+ mi:** Bikes *(Susie's Mopeds 800-532-0087, 8 mi.)*, Rental Car
(M & W Auto 800-323-6037, 8 mi.), Ferry Service *(WA State 468-2142,
Victoria 452-8088, 8 mi.)* **Airport:** Friday Harbor *(7 mi.)*

Medical Services
911 Service **3+ mi:** Doctor *(Inter Island Medical Center 378-1338, 7 mi.)*,
Dentist *(Friday Harbor Dentistry 378-4944, 7 mi.)*, Chiropractor *(Infinity
Chiropractic 378-8699, 7 mi.)*, Holistic Services *(Healing Touch 378-4275, 7
mi.)*, Veterinarian *(Harbor Vet 378-3959, 7 mi.)* **Hospital:** Whidbey General
678-5151 *(33 mi.)*

Setting -- Located on the sunny, west side of San Juan Island, the aptly named marina is snugged up to the east end of Mitchell Bay off Haro Strait. Nestled among evergreens, a string of attractively landscaped, green-trimmed cedar cottages follow the curve of the shoreline. Scattered along the stone bulkhead are green picnic tables, teak chairs, swings, and a fire pit surrounded by benches - all overlooking the network of quality docks and large homes across the bay.

Marina Notes -- The resort caters to families, couples, and active vacationers and encourages group retreats and weddings. A guest of the marina is a guest of the resort. The floating wood docks are well-maintained, with lighthouse pedestals. A caretaker is on-site 24 hours and the energetic staff keeps many activities happening. Several power boats, from 14 to 22 ft., are available for rent ($75-300/half-day, $110-380/day). Crystal Sea Kayaking offers guided tours lasting from 3 hours to multiday trips. Summer kayaking is made easy by the protected bay waters. All of the cottages face the water, as do several swings designed for two. The well-supplied store features a comfortable boaters' lounge and, behind the store, is a bathhouse with a tub/shower combinations.

Notable -- Reportedly good crabbing and shrimping right from the dock - just toss out a pot - and excellent salmon and bottom fishing in the surrounding waters. The resort partners with Buffalo Works for guided fishing trips and sponsors a monthly salmon derby called "Snug Jug." Explore the tide pools or investigate the great scuba diving just outside the bay. Two state parks are on the west side of the island: Lime Kiln State Park (378-2040) & Stuart Island State Park (378-2044). Take a taxi or the seasonal Island Shuttle 6 miles north to Roche Harbor or 8 miles to Friday Harbor - enjoy dinner or explore the galleries.

Navigational Information
Lat: 48°32.180' **Long:** 123°00.980' **Tide:** 14 ft. **Current:** 2 kt. **Chart:** 18424
Rep. Depths *(MLW)*: **Entry** 40 ft. **Fuel Dock** 20 ft. **Max Slip/Moor** 40 ft./-
Access: Off the San Juan Channel between Shaw and San Juan Islands

Marina Facilities *(In Season/Off Season)*
Fuel: *Texaco* - Gasoline, Diesel
Slips: 500 Total, 150 Transient **Max LOA:** 250 ft. **Max Beam:** n/a
  Rate *(per ft.)*: **Day** $0.90/0.65-0.75* **Week** n/a **Month** $7.34
  Power: 30 amp $0.10/ft., 50 amp $0.20/ft., 100 amp $0.30/ft., 200 amp n/a
  Cable TV: No **Dockside Phone:** No
  Dock Type: Floating, Long Fingers, Concrete, Wood
Moorings: 0 Total, 0 Transient **Launch:** n/a, Dinghy Dock (Free)
  Rate: Day n/a **Week** n/a **Month** n/a
Heads: 10 Toilet(s), 14 Shower(s), Hair Dryers
Laundry: None, Book Exchange **Pay Phones:** Yes, 3
Pump-Out: OnSite, OnCall, 1 Central, 2 Port **Fee:** $10 **Closed Heads:** Yes

Marina Operations
Owner/Manager: Steve Simpson **Dockmaster:** Tami Hayes
In-Season: May-Sep, 7am-7pm **Off-Season:** Oct-Apr, 8am-5pm
After-Hours Arrival: Night drop at office
Reservations: Yes, Preferred** **Credit Cards:** Visa/MC
Discounts: None
Pets: Welcome, Dog Walk Area **Handicap Access:** Yes, Docks

Port of Friday Harbor

PO Box 889; 204 Front Street; Friday Harbor, WA 98250

Tel: (360) 378-2688 **VHF: Monitor** Ch. 66A **Talk** n/a
Fax: (360) 378-6114 **Alternate Tel:** (360) 317-8672
Email: tamih@portfridayharbor.org **Web:** www.portfridayharbor.org
Nearest Town: Friday Harbor **Tourist Info:** (360) 378-5240

Marina Services and Boat Supplies
Services - Docking Assistance, Security (8 hrs., night security), Trash Pick-Up **Communication -** Data Ports *(Wi-Fi - Broadband Xpress)*, FedEx, DHL, UPS **Supplies -** OnSite: Ice *(Block, Cube)*, Propane *(at fuel pier)* **Near:** Ships' Store, West Marine *(378-1086)*, Bait/Tackle, Live Bait

Boatyard Services
OnSite: Forklift, Crane, Sail Loft, Canvas Work, Yacht Broker *(Friday Harbor Yacht Sales 378-4047)* **OnCall:** Engine mechanic *(gas, diesel)*, Electrical Repairs, Electronics Repairs **Near:** Launching Ramp, Divers, Compound, Wash & Wax, Interior Cleaning, Metal Fabrication. **Under 1 mi:** Travelift *(35T)*, Hull Repairs, Propeller Repairs, Inflatable Repairs, Upholstery. **Nearest Yard:** Albert Jensen and Sons, Inc. (360) 378-4343

Restaurants and Accommodations
OnSite: Restaurant *(Downriggers 378-2700, B $5-13, L $7-24, D $8-33)* **Near:** Restaurant *(Vinnie's 378-1934, D $15-35, upscale Italian)*, *(China Pearl 378-5254, B $6-10, L $9-22, D $9-22)*, *(Amigo 378-TACO, L $6-15, D $6-15)*, *(Thai Kitchen 378-1917)*, *(Front St. Ale House 378-2337)*, Seafood Shack *(Friday's Crabhouse 378-8801, L $8-16, D $13-24)*, Lite Fare *(San Juan Brewing Co. 378-2337, kids' menu)*, Pizzeria *(Tommy's Pizza & Subs 378-8440, delivers)*, Inn/B&B *(Trumpeter Inn 378-3884, $100-200)*, *(Friday's Historic Inn 378-5848, $59-230)* **Under 1 mi:** Motel *(Best Western 378-3031)*, Inn/B&B *(Hillside House 378-4730, $90-200)*

Recreation and Entertainment
OnSite: Dive Shop **Near:** Tennis Courts, Fitness Center, Boat Rentals *(Charters Northwest 378-7196)*, Bowling *(Paradise 370-5667)*, Movie Theater *(The Palace 378-3578)*, Video Rental, Park, Museum *(Whale 800-946-7227 $6/3, San Juan Historical 378-3949)*, Cultural Attract *(Garuda & I 378-3733, Community Theatre 378-3210)*, Sightseeing *(San Juan Safaris whale watching 378-1323 $49/39)*, Galleries *(Artic Raven 378-5788)* **Under 1 mi:** Picnic Area, Grills, Playground **1-3 mi:** Beach *(Jackson's - 2 mi.)*, Golf Course *(San Juan 378-5158)*

Provisioning and General Services
OnSite: Fishmonger *(Friday Harbor Seafoods)*, Newsstand **Near:** Convenience Store *(Little Store 378-4422)*, Market *(King's 378-4505)*, Gourmet Shop *(Gourmet's Galley 378-2251)*, Delicatessen *(Market Chef 378-4546)*, Liquor Store *(378-2956)*, Bank/ATM, Post Office, Protestant Church, Library *(378-2798)*, Beauty Salon *(Studio 55 378-8835)*, Laundry *(Sunshine Laundry and Dry Clean 378-7223)*, Bookstore *(Boardwalk 378-2787)*, Pharmacy *(Friday Harbor Drug 378-4421)*, Hardware Store *(Ace 378-4622)*, Copies Etc. **Under 1 mi:** Health Food, Bakery, Farmers' Market *(at courthouse parking, Sat 10am-1pm, midApr-midOct)*, Catholic Church

Transportation
OnCall: Taxi *(Bob's 378-6777)* **Near:** Bikes *(Island Bicycles 378-4941)*, Local Bus *(seasonal shuttle 378-8887)*, Ferry Service *(468-2142)* **Under 1 mi:** Rental Car *(West Isle 378-2440)* **Airport:** Friday Harbor *(2 mi.)*

Medical Services
911 Service **OnCall:** Ambulance **Near:** Dentist *(Tooth Ferry 378-5300)*, Chiropractor *(Infinity Chiropractic 378-8699)*, Holistic Services **Under 1 mi:** Doctor *(Inter Island Medical Center 378-1338)*, Veterinarian *(Harbor Vet 378-3959)* **Hospital:** St. Joseph Hospital (Bellingham) *(20 mi.)*

Setting -- On San Juan's east side, Port of Friday Harbor is protected by a unique floating breakwater designed by the U.S. Army Corps of Engineers. The 500 slip marina sprawls along Front Street - with a view of the ebullient, charming Friday Harbor downtown that marches up the hill. Adjacent to the breakwater is the Washington State Ferry terminal (beware the wakes) and just behind the marina listen to the whales at the only museum dedicated to those in the wild.

Marina Notes -- *Jul-Aug, $0.90 up to 74' LOA, $1.20 over 75'. May-Jun & Sep, $0.75 all boats; Jan-Apr & Oct-Dec, $0.65 all boats. **$5 res. fee for boats 20-44 ft. Pump-out boat $10. Check in on the "G" float dock, the main pier, or at the port office. Self-register after hours outside the door at the port office. Owned by the Port of Friday Harbor; operating for over 50 yrs. Wide, well kept wooden or cement docks with over 150 transient slips, and a seaplane base. Southern breakwater reserved for boats over 45 ft. LOA. Customs clearance is on-site. Well-maintained cinderblock & tile bathhouse (showers $1.25/5 min.).

Notable -- Friday Harbor is the San Juan county seat and largest community in the islands. It's simply delightful, especially in the spring & fall, when the island is unhurried and not so crowded. Downriggers Restaurant, located on the dock, has a friendly, casual atmosphere (L Mon-Sat 11am-5pm, Sun to 4pm; D Mon-Sat 5-10pm, Sun 4-10pm; B wknds only, 9am-Noon/2pm). Everything is within walking distance - shopping, theater, bowling, whale watching tours, and scenic flights to see it all from above. The Menu Bar (378-1987) has Internet access. Festivals include Silver Tea & Studios Open House in May, and Artists in Action in September. Explore the far corners of the San Juans via shuttle buses, taxis, rental cars, bicycles, mopeds and ferries - free for foot passengers.

Navigational Information

Lat: 48°32.120' **Long:** 123°00.915' **Tide:** 14 ft. **Current:** 2 kt. **Chart:** 18434
Rep. Depths *(MLW):* **Entry** 40 ft. **Fuel Dock** n/a **Max Slip/Moor** 40 ft./-
Access: West side of San Juan Channel

Marina Facilities *(In Season/Off Season)*

Fuel: No
Slips: 16 Total, 3 Transient **Max LOA:** 55 ft. **Max Beam:** n/a
  Rate *(per ft.):* **Day** $1.25/0.65 **Week** n/a **Month** $9.90
  Power: 30 amp $0.10/ft., 50 amp $0.20/ft., 100 amp n/a, 200 amp n/a
  Cable TV: No **Dockside Phone:** No
  Dock Type: Floating, Long Fingers, Concrete, Wood
Moorings: 0 Total, 0 Transient **Launch:** n/a
  Rate: Day n/a **Week** n/a **Month** n/a
Heads: None
Laundry: None **Pay Phones:** No
Pump-Out: 1 Central, 2 Port **Fee:** $10 **Closed Heads:** Yes

Charters Northwest

2 Spring Street; Friday Harbor, WA 98250*

Tel: (360) 378-7196 **VHF: Monitor** n/a **Talk** n/a
Fax: (360) 378-7197 **Alternate Tel:** n/a
Email: info@chartersnorthwest.com **Web:** www.chartersnorthwest.com
Nearest Town: Friday Harbor **Tourist Info:** (360) 378-5240

Marina Operations

Owner/Manager: Bill Jenkins **Dockmaster:** Same
In-Season: Oct-Apr, 9am-6pm **Off-Season:** Sep-May, 9am-4pm
After-Hours Arrival: Call ahead
Reservations: Not Required **Credit Cards:** Visa/MC
Discounts: None
Pets: No **Handicap Access:** No

Marina Services and Boat Supplies

Services - Security (8 hrs., night security), Dock Carts **Communication -** Data Ports (In town) **Supplies - OnSite:** Ice (Block, Cube) **Near:** Ships' Store, West Marine (378-1086), Bait/Tackle

Boatyard Services

OnSite: Yacht Broker (Friday Harbor Yacht Sales 378-4047)
OnCall: Electrical Repairs **Near:** Launching Ramp, Divers, Compound, Wash & Wax, Interior Cleaning, Metal Fabrication **Under 1 mi:** Travelift (35T), Engine mechanic (gas, diesel), Hull Repairs, Bottom Cleaning, Propeller Repairs, Woodworking, Inflatable Repairs, Upholstery.
Nearest Yard: Albert Jensen and Sons, Inc. (360) 378-4343

Restaurants and Accommodations

OnSite: Restaurant (Downriggers 378-2700, B $7-17, L $7-20, D $12-24, B wknds only), Seafood Shack (Friday Harbor Seafood 378-5779, takeout)
Near: Restaurant (Backdoor Kitchen 378-9540), (Place Bar & Grill 378-8707, D $10-18, D only), (The Flying Burito 378-1077, L $6-14, D $10-19, closed Sun), Lite Fare (Front St. Ale House 378-2337, L $5-12, D $9-16), Pizzeria (Tommy's Pizza and Subs 378-8440, L, D), Inn/B&B (Friday Harbor House $15-0325), (Friday's Historic Inn 378-5848, $59-230) **Under 1 mi:** Motel (Best Western 378-3031, $90-285)

Recreation and Entertainment

OnSite: Dive Shop (Island Dive and Water Sports 378-2772), Boat Rentals (Charters Northwest 378-7196, Capt'n Howards Sailing Charters 378-3959), Group Fishing Boat **Near:** Heated Pool, Spa, Tennis Courts, Fitness Center (Island Fitness 378-4449), Jogging Paths, Movie Theater (The Palace 378-3578), Video Rental, Park, Museum (Pig War Museum 378-6495, San Juan Historical Museum 378-3949), Cultural Attract (Island Stage Left 378-5649), Sightseeing (San Juan Safaris whale watching 378-1323 $49/39), Galleries (Friday Harbor Art Studio 378-5788) **Under 1 mi:** Beach, Picnic Area, Grills, Playground **1-3 mi:** Golf Course (San Juan 378-5158), Fishing Charter (A Trophy 378-4612), Special Events (San Juan County Fair - Aug)

Provisioning and General Services

OnSite: Fishmonger, Newsstand **Near:** Convenience Store (Entertainment Tonite 378-2424), Market (King's Market 378-4505), Gourmet Shop (Gourmet's Galley 378-2251), Delicatessen (Market Chef 378-4546), Liquor Store (378-2956), Meat Market, Bank/ATM, Post Office, Protestant Church, Library (378-2798), Laundry (Sunshine 378-7223), Bookstore (Boardwalk 378-2787), Pharmacy (Friday Harbor Drug 378-4421), Hardware Store (Ace 378-4622), Retail Shops, Copies Etc. **Under 1 mi:** Bakery (Kneadful Things 378-7089), Farmers' Market (Sat 10am-1pm in season), Catholic Church

Transportation

OnCall: Taxi (Bob's 378-6777) **Near:** Bikes (Island Bicycles 378-4941, Suzie's Mopeds 378-5244), Local Bus (seasonal shuttle 378-8887), Ferry Service (468-2142, Foot Ferry to Victoria 452-8088) **Under 1 mi:** Rental Car (West Isle 378-2440) **Airport:** Friday Harbor (2 mi.)

Medical Services

911 Service **Near:** Dentist (Tooth Ferry 378-5300), Chiropractor, Holistic Services **Under 1 mi:** Doctor (Inter Island Medical 378-1338), Veterinarian (Harbor Vet 378-3959) **Hospital:** Whidbey General 678-5151 (27 mi.)

Setting -- Charters Northwest sits right at the foot of Friday Harbor's main artery, Spring Street, with the ferry dock to port and Port of Friday Harbor to starboard. The town is visible just beyond Charters' blue two-story building. Charters' main dock parallels the Spring Street Aquarium Wharf, which has a large pavilion with a saltwater tank and heads open to the public. Colorful marine and adventure business along with eateries line the waterfront.

Marina Notes -- *Mailing address: PO Box 996, Anacortes, WA 98221. This is primarily a charter and brokerage business, but the docks, with deep slips and some side-tie space, are open to visitors when available. The dinghy dock is also available to those anchored in the harbor. Rates are the same as for the Port docks. Classic trawlers, and a sailboat fleet are available for charter. Some fishing and whale watching tours also base here. The inviting cinderblock bathhouse is next door at Port of Friday (showers $1.25/5 min.) or use the municipal heads in the Wharf building, open same hours as the marina.

Notable -- Step off your boat and you are literally in the heart of Friday Harbor - which caters to visitors with a great selection of dining and entertainment options. Visit the Pig War Museum (378-6495) to learn about the "Pig War" between the American Camp and the British Camp, situated at opposite ends of the island. The Friday Harbor Art Studio (378-5788) is where the Free Willy II amulet was created. Rent a car or moped and drive around the island. You will be amazed at the diversity of the terrain. Roche Harbor (800-451-8910), Pelindaba Lavender Farm and Spa (378-4248), San Juan Vineyards (378-9463), Kristal Acres Alpaca Farm (378-6125), and Shepherd's Croft (378-6372) are some of the unique island destinations.

Navigational Information

Lat: 48°30.950' **Long:** 122°54.790' **Tide:** 13 ft. **Current:** 4 kt. **Chart:** 18424
Rep. Depths *(MLW)*: **Entry** 4 ft. **Fuel Dock** n/a **Max Slip/Moor** 20 ft./-
Access: Enter from San Juan Channel along west shore of Lopez Island

Marina Facilities *(In Season/Off Season)*

Fuel: No
Slips: 100 Total, 30 Transient **Max LOA:** 150 ft. **Max Beam:** 10 ft.
  Rate *(per ft.)*: **Day** $0.88* **Week** n/a **Month** n/a
  Power: 30 amp $4, **50 amp** n/a, **100 amp** n/a, **200 amp** n/a
  Cable TV: No **Dockside Phone:** No
Dock Type: Floating, Concrete
Moorings: 0 Total, 0 Transient **Launch:** n/a, Dinghy Dock
  Rate: **Day** n/a **Week** n/a **Month** n/a
Heads: 2 Toilet(s), 3 Shower(s)
Laundry: None **Pay Phones:** No
Pump-Out: OnSite, Full Service, 5 Central **Fee:** $5 **Closed Heads:** Yes

Marina Operations

Owner/Manager: Ron Meng **Dockmaster:** Chela King
In-Season: Mon-Sat, 8:30am-6pm **Off-Season:** Mon-Sat, 8:30am-5pm
After-Hours Arrival: Lock box at door of store
Reservations: Yes, Preferred **Credit Cards:** Visa/MC, Dscvr, Amex
Discounts: Boat/US **Dockage:** 25% off trans **Fuel:** n/a **Repair:** n/a
Pets: Welcome, Dog Walk Area **Handicap Access:** Yes, Heads

Islands Marine Center

PO Box 88; 2790 Fisherman Bay Road; Lopez Island, WA 98261

Tel: (360) 468-3377 **VHF: Monitor** Ch. 69 **Talk** Ch. 69
Fax: (360) 468-2283 **Alternate Tel:** n/a
Email: imc@rockisland.com **Web:** www.islandsmarinecenter.com
Nearest Town: Lopez Village *(0.75 mi.)* **Tourist Info:** (360) 468-4664

Marina Services and Boat Supplies

Services - Docking Assistance, Dock Carts **Communication -** Data Ports *(Wi-Fi - Broadband Xpress)* **Supplies - OnSite:** Ice *(Block, Cube)*, Ships' Store, Bait/Tackle *(frozen herring & crab bait)*

Boatyard Services

OnSite: Travelift *(15T)*, Engine mechanic *(gas, diesel)*, Launching Ramp *($5/ft.)*, Electrical Repairs, Bottom Cleaning, Compound, Wash & Wax, Interior Cleaning, Inflatable Repairs, Painting, Awlgrip, Yacht Broker
OnCall: Electronics Repairs, Divers, Propeller Repairs, Woodworking, Metal Fabrication **Near:** Electronic Sales. **Dealer for:** Pursuit, Parker, Ocean Sport, Avon, Duraboat;. Other Certifications: OMC, Volvo, Yamaha, Mercruiser **Yard Rates:** $40/hr., Haul & Launch $5/ft. *(blocking incl.)*, Power Wash $1.75/ft., Bottom Paint $5.50-6/ft.

Restaurants and Accommodations

OnSite: Condo/Cottage *(Islands Marine Center 468-3377, $85-130, 2 apartments)* **Near:** Restaurant *(The Galley 468-2713, B $5-7, L $8-15, D $9-29)*, *(Lopez Islander Resort 468-2233, L $6-9, D $8-19)*, Hotel *(Lopez Islander Resort 468-2233, $79-142)* **Under 1 mi:** Restaurant *(The Bay 468-3700, D $20-50)*, *(Love Dog Café 468-2150)*, Coffee Shop *(Isabel's Espresso 468-4114)*, Inn/B&B *(Lopez Lodge 468-2777, $65-135)*

Recreation and Entertainment

OnSite: Beach, Picnic Area, Grills, Special Events *(4th of July Parade, BBQ, and fireworks)* **Near:** Pool, Heated Pool, Jogging Paths
Under 1 mi: Cultural Attract *(Lopez Community Center 468-2203)*

1-3 mi: Golf Course *(Lopez Island Golf Club 468-2679)*, Fitness Center *(Lopez Island Fitness 468-4911)*, Boat Rentals *(Harmony Charters 468-3310)*, Park, Museum *(Lopez Island Historical Society Museum 468-2049)*, Galleries *(Chimera Gallery 468-3265)*

Provisioning and General Services

Under 1 mi: Supermarket *(Red Apple 468-2266)*, Delicatessen *(Vita's 468-4268)*, Wine/Beer, Liquor Store *(Washington Liquor Store 468-2407)*, Bakery *(Holly B's Bakery 468-2133)*, Farmers' Market *(seasonal next to Community Center, Sat 10am-2pm)*, Protestant Church, Library *(Lopez 468-2265)*, Beauty Salon *(A Touch of Eden 468-2324)*, Bookstore *(Isle Haven Books 468-2132)*, Clothing Store, Retail Shops **1-3 mi:** Health Food *(Blossom Natural Foods 468-2204)*, Bank/ATM, Post Office *(468-2282)*, Catholic Church, Laundry *(Keep It Clean 468-3466)*, Pharmacy *(Lopez Island Pharmacy 468-2616)*, Hardware Store *(Sunset Builders Supply 468-2241)*, Florist *(Sticks & Stems Floral and Gifts 468-4377)*

Transportation

OnCall: Water Taxi *(A Lopez Cab Courier 468-2227)* **Near:** Bikes *(Lopez Bicycle 468-2847)* **3+ mi:** Ferry Service *(Washington State Ferries 468-2142, 5 mi.)* **Airport:** Lopez Airport *(3 mi.)*

Medical Services

911 Service **Near:** Dentist *(Bayview Dental 468-2551)* **Under 1 mi:** Doctor *(Lopez Island Medical Clinic 468-2245)* **1-3 mi:** Chiropractor *(Inter Island Chiropractic 468-2415)*, Holistic Services *(Lopez Massage Works 468-3239)*, Veterinarian **Hospital:** Whidby General 648-5151 *(23 mi.)*

Setting -- Fisherman Bay on Lopez Island is one of the most protected spots in the San Juans - a secure harbor in a delightful setting. A meandering set of quality floating concrete docks, hosting 100 slips, wend their way to the main wharf. On shore the large boat yard operation is dominated by a gray two-story main office, chandlery and Napa store. Look the other way for serene vistas across the bay and spectacular sunsets. ISM delivers the best of both worlds.

Marina Notes -- *Rates vary by size: 28-30' $25, 31-33' $28, 34-36' $30, 37-39' $33, 40-42' $35, 43-45' $38, 46-48' $40, 49-52' $43, 53-54' $45, 55-57' $48, 58-60' $50, 61' and over $0.85/ft. There are five 80 ft. piers and a "T" dock for guest moorage. The same owner has been managing the marina since 1972, and the accommodating staff go the extra mile to create a pleasant visit. Islands Marine Center is far more than a boatyard. It offers a well-stocked chandlery, a NAPA auto parts store, and two apartments upstairs. Often the apartments are rented to people who are having work done in the yard. The basic bathhouse is located on the north side of the main building (showers $0.25 per minute).

Notable -- Lopez Village is an easy 3/4-mile stroll to the north. Next door, on the south side, is the Lopez Islander Resort with a restaurant and full-service lounge. The Galley restaurant (468-2713), south of the marina, has a dinghy dock for guests. Playing tourist in the Village has its benefits. The flat, pastoral island lends itself to biking (rentals nearby). Chimera Gallery (468-3265) displays the work of very talented local artists. Visit the Lopez Island Historical Museum (468-2049) or the excellent library in an historic schoolhouse. Try some gourmet ice cream, visit the bakery, or one of the great delis in town.

Lopez Islander Resort & Marina

Lopez Islander Resort & Marina

PO Box 459; 2864 Fisherman Bay Road; Lopez Island, WA 98261

Tel: (360) 468-2233; (800) 736-3434 **VHF: Monitor** Ch. 78A **Talk** Ch. 78
Fax: (360) 468-3382 **Alternate Tel:** n/a
Email: li@rockisland.com **Web:** www.lopezislander.com
Nearest Town: Lopez Village *(1 mi.)* **Tourist Info:** (360) 468-4664

Navigational Information
Lat: 48°30.781' **Long:** 122°54.992' **Tide:** 10 ft. **Current:** 4 kt. **Chart:** 18434
Rep. Depths (*MLW*): Entry 4.5 ft. **Fuel Dock** 4 ft. **Max Slip/Moor** 7 ft./-
Access: Enter from San Juan Channel along the west side of Lopez Island

Marina Facilities *(In Season/Off Season)*
Fuel: Gasoline, Diesel
Slips: 64 Total, 4 Transient **Max LOA:** 80 ft. **Max Beam:** n/a
  Rate *(per ft.):* **Day** $1.00/0.87* **Week** n/a **Month** n/a
  Power: 30 amp Incl., **50 amp** Incl., **100 amp** n/a, **200 amp** n/a
  Cable TV: No **Dockside Phone:** Yes
  Dock Type: Fixed, Floating, Long Fingers, Short Fingers, Concrete, Wood
Moorings: 0 Total, 0 Transient **Launch:** n/a, Dinghy Dock ($1/day)
  Rate: Day n/a **Week** n/a **Month** n/a
Heads: 4 Toilet(s), 4 Shower(s) *(with dressing rooms)*
Laundry: 4 Washer(s), 4 Dryer(s), Iron, Iron Board **Pay Phones:** Yes, 1
Pump-Out: No **Fee:** n/a **Closed Heads:** Yes

Marina Operations
Owner/Manager: Kathy Casey **Dockmaster:** Same
In-Season: Year-Round, 8am-8pm **Off-Season:** Oct-May
After-Hours Arrival: Phone on desk
Reservations: Yes, Preferred **Credit Cards:** Visa/MC, Dscvr, Amex
Discounts: None
Pets: Welcome, Dog Walk Area **Handicap Access:** No

Marina Services and Boat Supplies
Services - Docking Assistance, Security *(24 hrs.)*, Dock Carts
Communication - Mail & Package Hold, Phone Messages, Fax in/out
($1/pg), Data Ports *(Wi-Fi - Broadband Xpress or Lobby & Espresso Bar ,
Free)*, FedEx, DHL, UPS, Express Mail **Supplies - OnSite:** Ice *(Cube)*,
Bait/Tackle *(Crab and herring)*, Propane **Near:** Ice *(Block)*, Ships' Store

Boatyard Services
OnCall: Divers **Near:** Travelift *(15T)*, Engine mechanic *(gas, diesel)*,
Launching Ramp, Electrical Repairs, Electronic Sales, Electronics Repairs,
Bottom Cleaning, Compound, Wash & Wax, Interior Cleaning, Propeller
Repairs, Woodworking, Metal Fabrication, Painting. **Nearest Yard:** Islands
Marine Center (360) 468-3377

Restaurants and Accommodations
OnSite: Restaurant *(Islander 468-2233, L $6-9, D $8-19, coffee and juice in
the morning, plus tiki bar)*, Hotel *(Lopez Islander Resort 468-2233, $79-142)*
Under 1 mi: Restaurant *(Galley Restaurante 468-2713, B $5-7, L $8-15, D
$9-29, waterfront)*, *(Bay Café 468-3700, D $20-50)*, *(Buckey's Grill 468-2595,
L $7-15, D $9-29)*, *(Love Dog Café 468-2150)*, Lite Fare *(Vortex 468-4740,
wraps, burritos, salads, smoothies)*, Motel *(Lopez Lodge 468-2500, $65-
135)*, Inn/B&B *(Edenwild 468-3238, $60-180)* **1-3 mi:** Inn/B&B *(Inn at Swifts
Bay 468-3636)*

Recreation and Entertainment
OnSite: Heated Pool, Spa, Beach, Picnic Area, Grills, Playground,
Volleyball, Fitness Center *(small fee)*, Jogging Paths, Boat Rentals *(kayaks,
incl. tours)*, Roller Blade/Bike Paths **OnCall:** Group Fishing Boat

Under 1 mi: Tennis Courts, Video Rental, Museum *(Lopez Island Historical
Museum 468-2049)*, Cultural Attract *(Art in Sumas 988-9056)*, Galleries
(Chimera Gallery 468-3265) **1-3 mi:** Golf Course *(Lopez Island Golf Course
468-2679)*, Park

Provisioning and General Services
OnSite: Wine/Beer **Near:** Convenience Store, Bank/ATM, Newsstand
Under 1 mi: Gourmet Shop *(Vita's 468-4268)*, Delicatessen, Liquor Store
(Washington Liquor 468-2407), Bakery *(Holly B's 468-2133)*, Farmers'
Market *(in summer, Sun 10am-2pm next to Community Center)*, Fishmonger,
Post Office *(468-2282)*, Library *(Lopez 468-2265)*, Beauty Salon *(Fiddlehead
468-2324)*, Clothing Store *(Village Apparel 468-2022)*, Retail Shops, Copies
Etc. **1-3 mi:** Supermarket *(Red Apple 468-2266)*, Health Food *(Blossom
Natural Foods 468-2204)*, Catholic Church, Protestant Church, Laundry
(Keep It Clean 468-3466), Bookstore *(Islehaven Books 468-2132)*, Pharmacy
(Lopez Island Pharmacy 468-2616), Hardware Store *(Sunset Builders Supply
468-2241)*, Florist *(Sticks & Stems 468-4377)*

Transportation
OnCall: Taxi *(A Lopez Cab 468-2227)* **Under 1 mi:** Bikes *(Bike Shop 468-
3497)* **3+ mi:** Ferry Service *(Washington State Ferries 468-2142, 5 mi.)*
Airport: Kenmore Seaplanes/Lopez Airport *(On-Call/3 mi.)*

Medical Services
911 Service **OnCall:** Ambulance **Under 1 mi:** Doctor *(Lopez Medical Clinic
468-2245)*, Dentist *(Bayview Dental Clinic 468-2551)*, Chiropractor *(Inter
Islands Chiropractic 468-2415)* **1-3 mi:** Veterinarian *(Ark Veterinary Clinic
468-2477)* **Hospital:** Whidbey General 648-5151 *(23 mi.)*

Setting -- Lopez Islander Resort & Marina is located on the west side of the island in protected Fisherman Bay. Views from the marina are of the bay with
eagles flying overhead and calm seas. From the water, the visitor sees well manicured lawns and the resort built in low-slung 60's style architecture - pale
yellow with sage green trim. The hotel rooms rise behind the restaurant and lounge overlooking the bay and its spectacular sunsets.

Marina Notes -- *Price varies by size & time of year. A 40' boat is $39.95 between May 15 & Sep 30 and $34.95 between Oct 1 & May 14. Average is $1/ft.
A marina guest is considered a hotel guest with access to the heated indoor pool, Jacuzzi, full gym, massage & tanning bed, laundry, book exchange, volleyball
court, horseshoes, children's playroom, plus restaurant, lounge and sports bar with pool table (manager boasts this is the best bar in the San Juans). A store on
the dock provides limited groceries, fuel, and propane. A seaplane float and bicycle & kayak rentals are also available. Bathhouses are at the resort.

Notable -- The resort sprawls across 14 green acres, and makes its facilities available for meetings, weddings, and boaters' rendezvous. Picnic tables are on
the lawn at the foot of the dock. The waterfront restaurant serves lunch & dinner daily from 11:30am to 10pm - the tiki bar stays open late. Because of its affable
people and gentle, rolling farmland, Lopez Island is known as "The Friendly Isle." Rent a bike or walk 15 min. to town for a supermarket, a clinic, shops,
community theater, and galleries. The Historical Museum details the island's maritime & farming history (468-2049 - call for hours). Around the bay & off the
beaten path, quiet Otis Perkins Park, about 2 mi. from the resort, has one of the longest beaches in the county, with breathtaking views out to San Juan Island.

PHOTOS ON DVD: 12

10. WA – Rosario Strait / Skagit Bay

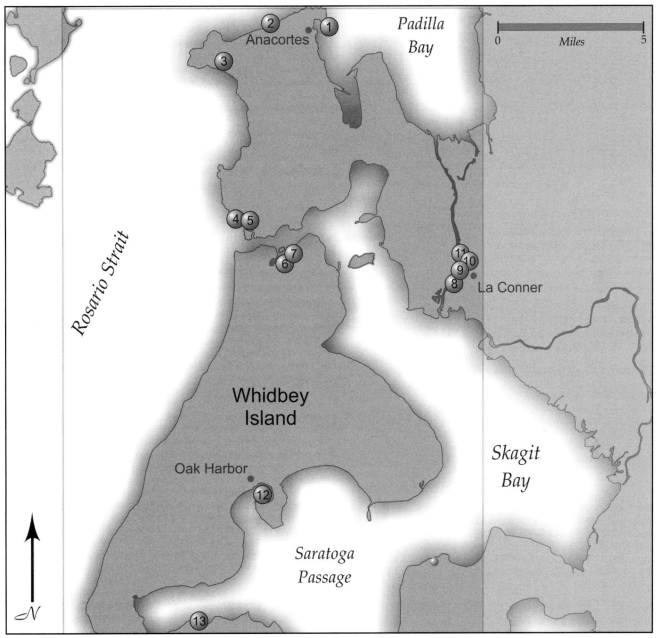

MAP	MARINA	HARBOR	PAGE	MAP	MARINA	HARBOR	PAGE
1	Cap Sante Boat Haven	Fidalgo Bay/Cap Sante	152	8	La Conner Town Docks	Swinomish Channel	159
2	Lovric's Sea-Craft	Guemes Channel	153	9	La Conner Marina	Swinomish Channel	160
3	Skyline Marina	Flounder Bay	154	10	Boater's Discount Center	Swinomish Channel	161
4	Rosario Beach State Park	Rosario Bay	155	11	La Conner Maritime Service	Swinomish Channel	162
5	Bowman Bay State Park	Bowman Bay	156	12	Oak Harbor Marina	Oak Harbor	163
6	Deception Pass Marina	Cornet Bay	157	13	Coupeville Wharf	Penn Cove	164
7	Deception Pass State Park	Cornet Bay	158				

▶ **Currency** — In Canadian Marina Reports, all prices are in Canadian dollars. In U.S. Marina Reports, all prices are in U.S. dollars.

▶ **"CCM"** — Denotes a Certified Clean Marina, a state/provincial award for environmental excellence. See page 298 for an explanation and page 299 for a list of Pump-Out facilities.

▶ **Ratings & Reviews** — An overview of the Atlantic Cruising Club's rating system is on page 6 and details on the content of each Marina Report are on pages 7 – 11.

▶ **Marina Report Updates** — Comments from boaters and new information from ACC reviewers and marinas are posted regularly on *www.AtlanticCruisingClub.com*.

Cap Sante Boat Haven

PO Box 297; 1019 Q Avenue; Anacortes, WA 98221

Tel: (360) 293-0694 **VHF: Monitor** Ch. 66A **Talk** Ch. 66A
Fax: (360) 299-0998 **Alternate Tel:** n/a
Email: marina@portofanacortes.com **Web:** www.portofanacortes.com
Nearest Town: Anacortes **Tourist Info:** (360) 293-3832

Navigational Information
Lat: 48°30.682' **Long:** 122°35.670' **Tide:** 12 ft **Current:** 0 kt. **Chart:** 18421
Rep. Depths (*MLW*): **Entry** 12 ft. **Fuel Dock** 12 ft. **Max Slip/Moor** 12 ft./-
Access: East side of Fidalgo Bay

Marina Facilities *(In Season/Off Season)*
Fuel: Gasoline, Diesel, High-Speed Pumps, On Call Delivery
Slips: 1000 Total, 150 Transient **Max LOA:** 120 ft. **Max Beam:** 40 ft.
  Rate *(per ft.):* **Day** $0.75/0.60 **Week** $4.00 **Month** Inq.
  Power: 30 amp $3, **50 amp** $6, **100 amp** $12, **200 amp** n/a
  Cable TV: No **Dockside Phone:** No
  Dock Type: Floating, Long Fingers, Alongside, Concrete, Wood
Moorings: 0 Total, 0 Transient **Launch:** n/a, Dinghy Dock (Free)
  Rate: Day n/a **Week** n/a **Month** n/a
Heads: 19 Toilet(s), 7 Shower(s) *(with dressing rooms)*
Laundry: 5 Washer(s), 6 Dryer(s) **Pay Phones:** Yes, 5
Pump-Out: OnSite, Self Service, 4 Port **Fee:** Free **Closed Heads:** Yes

Marina Operations
Owner/Manager: Port of Anacortes **Dockmaster:** Dale Fowler
In-Season: Year-Round, 7am-5pm **Off-Season:** n/a
After-Hours Arrival: Call Port Security 661-5000
Reservations: Yes, Preferred **Credit Cards:** Visa/MC
Discounts: None
Pets: Welcome **Handicap Access:** Yes, Heads, Docks

Marina Services and Boat Supplies
Services - Docking Assistance, Security *(night)*, Trash Pick-Up, Dock Carts **Communication -** Fax in/out, Data Ports *(Wi-Fi - Broadband Xpress)*, FedEx, UPS *(Sat Del)* **Supplies - OnSite:** Ice *(Block, Cube)*, Ships' Store, Bait/Tackle, Live Bait *(herring, squid, crab)*, Propane **Near:** West Marine *(293-4262)* **Under 1 mi:** Marine Discount Store *(Marine Supply 293-4014)*

Boatyard Services
OnSite: Travelift *(60T)*, Forklift, Engine mechanic *(gas, diesel)*, Electrical Repairs, Electronic Sales, Electronics Repairs, Hull Repairs, Rigger, Bottom Cleaning, Brightwork, Divers, Propeller Repairs, Woodworking, Inflatable Repairs, Life Raft Service, Metal Fabrication, Painting, Awlgrip, Yacht Broker, Yacht Building **Near:** Railway, Crane. **Under 1 mi:** Hydraulic Trailer, Sail Loft, Canvas Work, Air Conditioning, Refrigeration, Compound, Wash & Wax, Upholstery **1-3 mi:** Launching Ramp

Restaurants and Accommodations
OnSite: Restaurant *(Captain's Place 293-3196, L $5-12, D $8-18, 7 days)* **Near:** Restaurant *(Randy's Pier 61 293-5108, L $8-16, D $10-20)*, *(Brown Lantern Ale House 293-2544, B $6-11, L $6-8, D $9-13, best burgers in town)*, *(Teriyaki 588-8025)*, Fast Food *(Subway, McDonald's)*, Pizzeria *(Domino's 299-2000)*, Inn/B&B *(Lowman House 293-0590, $75-85)*, *(Cap Sante Inn 293-0602)* **Under 1 mi:** Restaurant *(Majestic L $6-9, D $16-24)*, Hotel *(Majestic Inn & Spa $130-170)*

Recreation and Entertainment
OnSite: Beach, Picnic Area, Grills, Fishing Charter *(Catchmore 293-7093)*, Group Fishing Boat, Park, Sightseeing *(Mystic Charters whale watching &* island tours)*, Special Events *(Arts Fest - Aug, Oyster Run, Waterfront Festival, Jazz Fest, Antique Engine & Machine Show - Sep)* **Near:** Playground, Movie Theater *(Anacortes 293-7000)*, Video Rental *(Blockbuster 588-9113)*, Museum *(Anacortes Historical 293-1915 - donation)* **Under 1 mi:** Heated Pool, Dive Shop *(Anacortes Dive 293-2070)*, Tennis Courts, Bowling *(San Juan Lanes 293-5185)*, Cultural Attract *(Keystone Center Forthe Arts 293-5908)* **3+ mi:** Golf Course *(Similk Beach 293-3444, 4 mi.)*

Provisioning and General Services
OnSite: Newsstand **Near:** Convenience Store *(7-Eleven 293-6780)*, Market *(Food Pavilion 588-8181)*, Supermarket *(Safeway 293-5393)*, Bakery *(Calico Cupboard 293-7315)*, Farmers' Market *(Depots Arts Center Sat 9am-2pm)*, Bank/ATM, Post Office *(299-6689)*, Protestant Church, Library *(293-9052)*, Beauty Salon *(Profiles 293-5901)*, Dry Cleaners *(Thrifty 293-8783)*, Pharmacy *(Rite Aid 293-2119)*, Hardware Store *(Sebo's 293-4575)* **Under 1 mi:** Liquor Store *(293-2715)*, Laundry *(Econo Wash 293-2974)*, Bookstore *(Watermark 293-4277)* **3+ mi:** Fishmonger *(Thibert's 293-2525, 4 mi.)*

Transportation
OnSite: Water Taxi *(Island Express 877-473-9777)*, Rental Car **OnCall:** Taxi *(Anacortes 293-3979)* **Near:** Local Bus **3+ mi:** Ferry Service *(IslandTrans. 293-6060, 4 mi.)* **Airport:** Anacortes/Bellingham *(3 mi./45 mi.)*

Medical Services
911 Service **OnCall:** Ambulance **Near:** Dentist *(Damon 293-2888)*, Chiropractor *(Jennings 299-8999)*, Veterinarian *(Animal Hosp. 293-3431)* **Under 1 mi:** Doctor *(Family Medicine 299-4211)*, Holistic Services *(Anacortes Wellness 293-0927)* **Hospital:** Island Hospital 299-1300 *(1 mi.)*

Setting -- This popular, port-operated marina is located on the east side of Fidalgo Bay in the heart of Anacortes. Breakwaters of rock pile and wood keep the quality moorage very sheltered. The sleek, blue-roofed marina and U.S. Customs offices are at the head of the docks. A spacious, well-lit esplanade and jogging trail, edged with mahogany-railings, surrounds the large web of docks and connects the two parks that are situated at each end of the facility.

Marina Notes -- Slips range from 24-65 ft. Call for a slip assignment on arrival day. The helpful staff always makes room for guests, and the marina provides most amenities a boater might need. A large chandlery offers marine supplies & hardware, ice, tackle, and live bait. Temporary winter moorage is available for about 100 boats, Sep 9-May 31 ($0.60/day plus electric). The marina is also a U.S. Port of Entry (360-293-2331). Inviting tile bathhouses (showers $0.25 for 1-3/4 mins.) are well-maintained.

Notable -- On-site Captain's Place boasts "the Best Fish & Chips in Town." Cap Sante is in the heart of old-fashioned, flower-bedecked Anacortes - with beautiful water views, an active arts community, and small town charm. Blooms even cascade from lamp posts. Everything is an easy walk, including a shopping mall, well-stocked Safeway supermarket (across the street), West Marine, fast-food, anqique shops and many unique restaurants. Marine Supply & Hardware, founded in 1911, claims to be the oldest marine store on the West Coast. The town's fishing, canning, and milling history is depicted at nearby Anacortes Museum. Annual events include Waterfront Festival (3rd weekend in May), Shipwreck Day & Flea Market in July, and Arts Fest in early August.

Navigational Information
Lat: 48°30.790' **Long:** 122°38.450' **Tide:** 11 ft. **Current:** n/a **Chart:** 18421
Rep. Depths (MLW): Entry 7 ft. **Fuel Dock** n/a **Max Slip/Moor** 12 ft./-
Access: Guemes Channel crosses Bellingham Channel

Marina Facilities *(In Season/Off Season)*
Fuel: Gasoline, Diesel, On Call Delivery
Slips: 30 Total, 2 Transient **Max LOA:** 75 ft. **Max Beam:** n/a
  Rate *(per ft.):* **Day** $0.75* **Week** n/a **Month** $6*
  Power: 30 amp 0.10/kw., 50 amp 0.10/kw, 100 amp n/a, 200 amp n/a
  Cable TV: No **Dockside Phone:** No
  Dock Type: Floating, Long Fingers, Alongside, Wood
Moorings: 0 Total, 0 Transient **Launch:** n/a
  Rate: Day n/a **Week** n/a **Month** n/a
Heads: 1 Toilet(s)
Laundry: None **Pay Phones:** No
Pump-Out: No **Fee:** n/a **Closed Heads:** Yes

Marina Operations
Owner/Manager: Florence Lovric **Dockmaster:** Same
In-Season: Year-Round, Mon-Fri 8am-4:30pm **Off-Season:** n/a
After-Hours Arrival: Call ahead
Reservations: Yes, Preferred **Credit Cards:** No
Discounts: None
Pets: Welcome **Handicap Access:** Yes, Docks

Lovric's Sea-Craft

3022 Oakes Avenue; Anacortes, WA 98221

Tel: (360) 293-2042 **VHF: Monitor** n/a **Talk** n/a
Fax: (360) 293-2042 **Alternate Tel:** n/a
Email: n/a **Web:** n/a
Nearest Town: Anacortes *(3 mi.)* **Tourist Info:** (360) 293-7911

Marina Services and Boat Supplies
Services - Security *(24/7, owners live onsite)* **Supplies - Under 1 mi:** Ice *(Block)* **1-3 mi:** Ships' Store *(Marine Supply 293-4014)*, West Marine *(293-4262)*, Bait/Tackle, Live Bait, Propane

Boatyard Services
OnSite: Railway *(1000T and 500T)*, Forklift *(5T)*, Crane *(20T)*, Engine mechanic *(gas, diesel)*, Electrical Repairs, Hull Repairs, Bottom Cleaning, Brightwork, Divers, Propeller Repairs, Woodworking, Metal Fabrication, Painting, Total Refits **1-3 mi:** Electronics Repairs *(Anacortes Marine Electronics 293-6100)*, Rigger, Sail Loft, Canvas Work, Yacht Broker *(Cannon Yacht Sales 293-0491)*. **Yard Rates:** $49/hr., Haul & Launch $6-12/ft. *(blocking $49/hr.)*, Power Wash $49/hr., Bottom Paint $49/hr.

Restaurants and Accommodations
Near: Motel *(Ship Harbor Inn 293-5177)* **Under 1 mi:** Inn/B&B *(Autumn Leaves 293-4920)* **1-3 mi:** Restaurant *(Captain's Place 293-3196, L $5-12, D $8-18)*, *(Randy's Pier 61 293-5108, L $8-16, D $10-20)*, *(Lucky Chopsticks 588-8899)*, *(Esteban's Mexican 299-1060)*, *(Tokyo 293-9898)*, Fast Food *(McD's, Burger King, Subway)*, Lite Fare *(Calico Cupboard Café & Bakery 293-7315)*, Pizzeria *(Village 293-7847)*, *(Papa Murphy's 299-2299)*, Motel *(San Juan 293-5105, $50-60)*, Hotel *(Anaco Bay 299-3320, $59-119)*

Recreation and Entertainment
1-3 mi: Pool *(Fidalgo Pool and Fitness 293-0673)*, Beach, Picnic Area, Grills, Playground, Dive Shop *(Anacortes Dive 293-2070)*, Fitness Center, Jogging Paths, Bowling *(San Juan Lanes 293-5185)*, Fishing Charter *(Catchmore Charters 293-7093)*, Movie Theater *(Anacortes 293-7000)*, Video Rental

(Blockbuster 588-9133), Park *(Cap Sante Park)*, Museum *(Anacortes 293-1915)*, Sightseeing *(Mystic Sea Charters 800-308-9387, Deception Pass Sunset Cruises 465-4604, Bill Mitchell tour of town murals 293-3328)*, Special Events *(Shipwreck Day - Jul, Arts Fest - Aug, Oyster Run last wknd Sep)* **3+ mi:** Golf Course *(Similk Beach 293-3444, 5 mi.)*

Provisioning and General Services
Under 1 mi: Market *(Steiman's Grocery 293-6456)* **1-3 mi:** Supermarket *(Safeway 293-5393)*, Delicatessen *(Easy Street Grocery & Deli 588-9328)*, Wine/Beer *(Compass Wines 293-6500)*, Liquor Store *(WA Liquor 293-2715)*, Bakery *(La Vie El Rose 299-9546)*, Farmers' Market *(Depots Arts Center, Sat 9am-2pm)*, Bank/ATM, Post Office, Catholic Church, Protestant Church, Synagogue, Library *(293-9052)*, Beauty Salon *(Studio One 424-2211)*, Barber Shop *(15th St. Barber 293-2288)*, Dry Cleaners, Laundry *(Econo-Wash 293-2974)*, Bookstore *(Living World 293-6321)*, Pharmacy *(Rite Aid 293-2119)*, Hardware Store *(Ace Hardware 293-3535)*, Copies Etc. *(Old Fashioned Print Shop 293-2273)*

Transportation
OnSite: Local Bus **OnCall:** Rental Car *(Enterprise 293-4325)*, Taxi *(Anacortes Van and Taxi 588-TAXI)* **1-3 mi:** Water Taxi *(Island Express to all islands, 877-473-9777)*, Ferry Service *(Island Transport 293-6060)* **Airport:** Anacortes/Bellingham Int'l. *(1 mi./47 mi.)*

Medical Services
911 Service **OnCall:** Ambulance **1-3 mi:** Doctor *(Fidalgo Medical 293-3101)*, Dentist *(Anacortes Dental 293-2000)*, Veterinarian *(Animal Hospital 293-3431)* **Hospital:** Island Hospital 299-1300 *(1.5 mi.)*

PHOTOS ON DVD: 13

Setting -- Lovric's Sea-Craft rests along the Guemes Channel at the north end of Fidalgo Island. Shoreside views are dominated by the boatyard facilities. Part of the breakwater, which protects 600 feet of alongside moorage, consists of an old vessel embedded in the wall. Boatyard work surrounds the docks, which are occupied mostly by large power and sail vessels in for repairs. Seaward, Guemes and Cypress Islands can be seen to the north.

Marina Notes -- *Flat rates: monthly 30-69' $6/ft. **30 & 50 amp power available at $0.10/kwt. Fuel is by appointment. Family owned since 1965, Lovric's Sea-Craft is primarily a boatyard and boatbuilding facility. A floating lift, crane, and two marine railways provide haulout. If you need work done, call ahead and grab a side-tie or an available slip. Doesn't specifically cater to maegayachts, but have moored, serviced and repaired them. The owner lives on-site in the house at the top of the hill overlooking the marina. Water and power are on the dock, and only one head is available for the boat owner or crew.

Notable -- Steiman's Grocery, a mini-mart, is within a mile and provides basic provisions. The town of Anacortes is approximately three miles east, and the bus conveniently stops at the top of the hill next to the marina - it's well worth the short bus ride. Restaurants, shops, and chandleries stretch along 2 miles on Commercial Avenue. The island's maritime spirit is reflected in several annual events, and a museum features exhibits on local history. The downtown area can be explored on foot, but public transportation, rental cars and water taxis are also available. The Washington State Ferry links Anacortes with four of the San Juan Islands, as well as Sidney on Canada's Vancouver Island. Passengers pay only to travel west; those on foot travel between the islands for free.

Skyline Marina

2011 Skyline Way; Anacortes, WA 98221

Tel: (360) 293-5134 **VHF: Monitor** Ch. 16 **Talk** Ch. 68
Fax: (360) 293-2427 **Alternate Tel:** n/a
Email: n/a **Web:** n/a
Nearest Town: Anacortes *(4 mi.)* **Tourist Info:** (360) 293-7911

Navigational Information
Lat: 48°29.410' **Long:** 122°40.670' **Tide:** 8 ft. **Current:** n/a **Chart:** 18421
Rep. Depths *(MLW)*: **Entry** 10 ft. **Fuel Dock** 15 ft. **Max Slip/Moor** 35 ft./-
Access: North end of Burrows Bay

Marina Facilities *(In Season/Off Season)*
Fuel: Gasoline, Diesel, High-Speed Pumps
Slips: 660 Total, 20 Transient **Max LOA:** 120 ft. **Max Beam:** 24 ft.
 Rate *(per ft.)*: **Day** $0.90 **Week** $12 **Month** $9.50
 Power: 30 amp $3, **50 amp** n/a, **100 amp** n/a, **200 amp** n/a
 Cable TV: No **Dockside Phone:** No
 Dock Type: Floating, Long Fingers, Alongside, Concrete
Moorings: 0 Total, 0 Transient **Launch:** n/a
 Rate: Day n/a **Week** n/a **Month** n/a
Heads: 2 Toilet(s), 2 Shower(s)
Laundry: 4 Washer(s), 4 Dryer(s) **Pay Phones:** No
Pump-Out: OnSite, Full Service **Fee:** $5 **Closed Heads:** Yes

Marina Operations
Owner/Manager: Dick Britton **Dockmaster:** Bethany Britton
In-Season: Year-Round, 8am-5pm **Off-Season:** n/a
After-Hours Arrival: Tie up, pay in the morning
Reservations: Yes, Preferred **Credit Cards:** Visa/MC, Dscvr
Discounts: None **Dockage:** n/a **Fuel:** Call **Repair:** n/a
Pets: Welcome, Dog Walk Area **Handicap Access:** Yes, Heads, Docks

Marina Services and Boat Supplies
Services - Security *(24 hrs.)*, Trash Pick-Up, Dock Carts **Communication -** Mail & Package Hold, Phone Messages, Fax in/out, Data Ports *(Wi-Fi - Broadband Xpress)*, FedEx, UPS **Supplies - OnSite:** Ice *(Block, Cube)*, Ships' Store, Bait/Tackle, Propane, CNG **3+ mi:** West Marine *(293-4262)*

Boatyard Services
OnSite: Travelift *(55T)*, Forklift, Crane, Hydraulic Trailer *(55T)*, Engine mechanic *(gas, diesel)*, Electrical Repairs, Electronic Sales, Electronics Repairs, Hull Repairs, Rigger, Bottom Cleaning, Brightwork, Air Conditioning, Refrigeration, Compound, Wash & Wax, Interior Cleaning, Propeller Repairs, Woodworking, Inflatable Repairs, Upholstery, Metal Fabrication, Painting, Awlgrip, Yacht Broker *(Sea Breeze 202-9445)* **Member:** ABYC - 2 Certified Tech(s) **Yard Rates:** $58/hr., Haul & Launch $5/ft. *(blocking incl.)*, Power Wash $1.75/ft., Bottom Paint $58/hr. **Storage:** On-Land $5.65/ft./mo.

Restaurants and Accommodations
OnSite: Restaurant *(Flounder Bay Café 293-3680, L $7-11, D $12-28, Sun brunch, kids' menu, 1st & 3rd Thu live music + martinis)* **Under 1 mi:** Restaurant *(Charlie's Café 293-3680, L $5-10, D $8-20)*, *(Billy Ray's 588-0491, L $7-12, D $12-20)*, *(Randy's Pier 61 293-5108)*, Inn/B&B *(Ship Harbor Inn 293-5177)* **1-3 mi:** Restaurant *(Captain's Place 293-3196, L $8-13, D $11-18)*, Inn/B&B *(Autumn Leaves B&B 293-4920, $145)* **3+ mi:** Restaurant *(Hong Kong 293-9595, 4 mi.)*, *(Greek Islands 293-6911, 4 mi.)*, Fast Food *(Subway, Taco Bell, 4 mi.)*

Recreation and Entertainment
OnSite: Beach, Jogging Paths, Boat Rentals *(Adventure Charters)*, Fishing Charter **Near:** Tennis Courts *(Beach Club 293-0277)*, Park **1-3 mi:** Heated Pool, Picnic Area, Grills **3+ mi:** Dive Shop *(Anacortes 293-2070, 4mi.)*, Fitness Center *(Water's Edge 299-2180, 4 mi.)*, Bowling *(San Juan Lanes 293-5185, 4 mi.)*, Movie Theater *(Anacortes 293-7000, 4 mi.)*, Video Rental *(Island 293-8611, 4 mi.)*, Museum *(Anacortes Historical 293-1915, 4mi.)*, Sightseeing *(Anacortes Walking Tour, 4 mi.)*, Special Events *(Waterfront Fest - May, Arts Fest - Aug, Oyster Run - Sep 293-6211, 4 mi.)*

Provisioning and General Services
OnSite: Barber Shop **Near:** Market *(Old Salt's Deli & Market 293-0618)*, Wine/Beer, Bank/ATM, Newsstand **Under 1 mi:** Beauty Salon *(Skyline Styling 293-5623)*, Retail Shops *(San Juan Souvenir 588-8008)* **1-3 mi:** Supermarket *(Safeway 293-5393)* **3+ mi:** Gourmet Shop *(Sea Bear Smokehouse 293-4661, 4 mi.)*, Liquor Store *(WA Liquor 293-2715, 4 mi.)*, Farmers' Market *(Sat 9am-2pm at Depots Arts Center, 4 mi.)*, Fishmonger *(Knudson Crab 293-3696, 4 mi.)*, Library *(293-9052, 4 mi.)*, Dry Cleaners *(Econo Wash & Dry Cleaning 293-2974, 4 mi.)*, Bookstore *(Living World 293-6321, 4 mi.)*, Pharmacy *(Anacortes 299-8560, 4 mi.)*, Hardware Store *(Marine Supply 293-4014, Ace 293-3535, 4 mi.)*

Transportation
OnSite: Airport Limo **OnCall:** Rental Car *(Enterprise 293-4325)*, Taxi *(Triangle 293-3979)* **Near:** Local Bus **Under 1 mi:** Ferry Service *(Sindey & San Juans 293-6060)* **Airport:** Anacortes/Bellingham Int'l. *(2 mi./50 mi.)*

Medical Services
911 Service **OnCall:** Ambulance **3+ mi:** Doctor *(4 mi.)*, Dentist *(4 mi.)*, Veterinarian *(4 mi.)* **Hospital:** Island Hospital 299-1300 *(4 mi.)*

Setting -- On Fidalgo Island's northwest side is Flounder Bay, a fully protected harbor off Burrows Bay. Skyline Marina is tucked into a residential area - its upland filled with a series of two-story white buildings topped by green roofs, where boatyard services are managed. The innermost docks are tucked under sheds. The view from the docks is a panorama of attractive homes perched along the hillside.

Marina Notes -- This large marina, built in 1965, is able to accommodate smaller megayachts with advance notice. Facility is a dockominium, so access is based on availability. The neat, uncluttered concrete docks are protected by locked gates and 24-hour security surveillance. A marine store, large boatyard that can deal with most repair and maintenance needs, and a yacht broker are all on-site. Call for special discounts, especially during the off-season. Guests must use the public restrooms - head & shower combinations. Note: Favor the west side of the channel at the entrance.

Notable -- The on-site Flounder Bay Café has a wall of glass overlooking the docks, plus outdoor seating. Lunch is served Monday-Saturday 11am-4pm, dinner daily after 4pm with Sunset Supper specials from 4-6pm (a house special is Halibut Macadamia). Flounder Bay, a good jumping-off point for the Islands, is peaceful and quiet. Nearby Washington Park, with over 200 wooded acres, has a two-mile loop that can be enjoyed on foot or by car, and offers great views of Rosario Straight and the San Juans. It's a short taxi ride to Anacortes which is filled with charming shops, art and antique galleries, and eateries. An enchanting series of murals, depicting scenes of early Anacortes, sets this town apart; a self-guided tour is available at the marina office.

Navigational Information
Lat: 48°24.970' **Long:** 122°39.800' **Tide:** 11 ft. **Current:** n/a **Chart:** 18427
Rep. Depths *(MLW)*: **Entry** 6 ft. **Fuel Dock** n/a **Max Slip/Moor** 6 ft./6 ft.
Access: South end of Fidalgo Island, east side of Rosario Strait

Marina Facilities *(In Season/Off Season)*
Fuel: No
Slips: 4 Total, 4 Transient **Max LOA:** 120 ft. **Max Beam:** n/a
 Rate *(per ft.)*: **Day** $0.40 **Week** n/a **Month** n/a
 Power: 30 amp n/a, **50 amp** n/a, **100 amp** n/a, **200 amp** n/a
 Cable TV: No **Dockside Phone:** No
 Dock Type: n/a
Moorings: 5 Total, 2 Transient **Launch:** n/a, Dinghy Dock
 Rate: Day $10 **Week** n/a **Month** n/a
Heads: Toilet(s)
Laundry: None **Pay Phones:** No
Pump-Out: No **Fee:** n/a **Closed Heads:** Yes

Marina Operations
Owner/Manager: Washington State Parks Department **Dockmaster:** n/a
In-Season: May-Oct, 6:30am-Dusk **Off-Season:** Nov-Apr, 8am-Dusk
After-Hours Arrival: Pay envelope
Reservations: None Required **Credit Cards:** No; Cash only
Discounts: None
Pets: Welcome **Handicap Access:** No

Rosario Beach State Park

N.S.H. 20; Oak Harbor, WA 98277

Tel: (360) 675-2417 **VHF: Monitor** n/a **Talk** n/a
Fax: n/a **Alternate Tel:** n/a
Email: n/a **Web:** www.parks.wa.gov/moorage/parks
Nearest Town: Anacortes *(8 mi.)* **Tourist Info:** (360) 675-9438

Marina Services and Boat Supplies
Services - Trash Pick-Up **Supplies - 3+ mi:** Ice *(Cube)*, Ships' Store *(Marine Supply & Hardware 293-4014, 8 mi.)*, West Marine *(675-1976, 8 mi.)*

Boatyard Services
3+ mi: Travelift *(45T, 12 mi.)*, Engine mechanic *(gas, diesel)*, Hull Repairs *(12 mi.)*, Canvas Work *(The Canvas Riggers 672-4417, 12 mi.)*, Air Conditioning *(12 mi.)*, Refrigeration *(12 mi.)*, Divers *(12 mi.)*, Compound, Wash & Wax *(12 mi.)*, Propeller Repairs *(12 mi.)*, Woodworking *(12 mi.)*, Metal Fabrication *(12 mi.)*, Painting *(12 mi.)*, Awlgrip *(12 mi.)*, Yacht Broker *(12 mi)*. **Nearest Yard:** Oak Harbor Boat Works (360) 675-2659

Restaurants and Accommodations
1-3 mi: Restaurant *(Island Grill 679-3914, B $5-10, L $6-15)*, *(Deception Café & Grill 293-9250)*, *(Seabolt's Smokehouse 675-6485)* **3+ mi:** Motel *(Lake Campbell Lodge 293-5314, 5 mi.)*, Inn/B&B *(Anacortes Inn 293-3153, 8 mi.)*, *(Autumn Leaves B&B 293-4920, 7 mi.)*

Recreation and Entertainment
OnSite: Picnic Area, Grills, Park **3+ mi:** Golf Course *(Lam's Links 9-hole, 675-3412, 5 mi.)*

Provisioning and General Services
1-3 mi: Market *(Whidbey Market 679-1384/Soundview Shopper 675-6132)*
3+ mi: Wine/Beer *(Nick's Deception Pass Saloon 675-6916, 4 mi.)*, Bank/ATM *(3 mi.)*, Pharmacy *(Anacortes Pharmacy 299-8560, 8 mi.)*, Hardware Store *(Ace 293-3535, 8 mi.)*, Retail Shops *(Anacortes, 8 mi.)*

Transportation
OnCall: Rental Car *(Enterprise 293-4325)*, Taxi *(Anacortes City Cab 299-8294/ Oak Harbor 675-1244)* **3+ mi:** Ferry Service *(to San Juan Island & Sidney BC 293-6060, 8 mi.)* **Airport:** Anacortes *(7 mi.)*

Medical Services
911 Service **OnCall:** Ambulance **3+ mi:** Doctor *(Family Medicine 299-4211, 8 mi.)*, Dentist *(Huntsinger 299-9347, 7 mi.)*, Chiropractor *(Henning Chiropractic 293-972, 8 mi.)*, Veterinarian *(Bjerk 293-2468, 7 mi.)*
Hospital: Island Hospital 299-1300 *(8 mi.)*

Setting -- Rosario Beach State Park is located in Sharpe Cove, which is just west of Bowman Bay. The small, relatively unprotected dock lies on a rocky beach and looks out on Rosario Strait. The park offers a peaceful overnight stay among old-growth forests and abundant wildlife. A community college marine studies campus backs up on the cove as well. Many rocks lie close to shore, and there are strong tidal currents in the bay.

Marina Notes -- This state park offers little in the way of protection or amenities for the boater. Picnic tables and fire pits are available on shore on property belonging to Deception Pass State Park. Scuba divers flock here to enjoy the underwater park just south of Rosario Head. A statue of The Maiden of Deception Pass, carved from red cedar, watches over the park. Folklore says the maiden gave herself to the gods of the sea, and Ko-kwal-alwoot now safely guides mariners through the passage.

Notable -- Part of the bay is a marine preserve and, as such, no marine life should be taken or disturbed. The rest of the bay is privately owned; private homes and the Walla Walla College Marine Station facilities occupy the shore. A couple of restaurants and a market are along Route 20 across the bridge. Additional shops and services are about 8 miles north in the town of Anacortes or 10 miles south in Oak Bay.

Bowman Bay State Park

5175 N.S.H. 20; Oak Harbor, WA 98277

Tel: (360) 675-2417 **VHF: Monitor** n/a **Talk** n/a
Fax: n/a **Alternate Tel:** n/a
Email: n/a **Web:** www.parks.wa.gov/moorage/parks
Nearest Town: Anacortes *(10 mi.)* **Tourist Info:** (360) 675-9438

Navigational Information
Lat: 48°24.900' **Long:** 122°39.250' **Tide:** 14 ft. **Current:** n/a **Chart:** 18427
Rep. Depths (*MLW*): **Entry** 9 ft. **Fuel Dock** n/a **Max Slip/Moor** -/8 ft.
Access: South end of Fidalgo Island on Bowman Bay

Marina Facilities *(In Season/Off Season)*
Fuel: No
Slips: 0 Total, 0 Transient **Max LOA:** 45 ft. **Max Beam:** n/a
  Rate *(per ft.)*: **Day** n/a **Week** n/a **Month** n/a
  Power: 30 amp n/a, **50 amp** n/a, **100 amp** n/a, **200 amp** n/a
  Cable TV: No **Dockside Phone:** No
  Dock Type: Fixed, Alongside, Wood
Moorings: 5 Total, 5 Transient **Launch:** n/a, Dinghy Dock
  Rate: Day $16 **Week** n/a **Month** n/a
Heads: 2 Toilet(s)
Laundry: None **Pay Phones:** No
Pump-Out: No **Fee:** n/a **Closed Heads:** Yes

Marina Operations
Owner/Manager: Washington State Parks Department **Dockmaster:** n/a
In-Season: May-Oct, 6:30am-Dusk **Off-Season:** Nov-Apr, 8am-Dusk
After-Hours Arrival: Pay station
Reservations: No **Credit Cards:** No; Cash only
Discounts: None
Pets: Welcome, Dog Walk Area **Handicap Access:** No

Marina Services and Boat Supplies
Services - Trash Pick-Up **Supplies - 3+ mi:** Ice *(Cube)*, Ships' Store *(Marine Supply & Hardware 293-4014, 10 mi.)*, West Marine *(675-1976, 10 mi.)*

Boatyard Services
OnSite: Launching Ramp **OnCall:** Bottom Cleaning **3+ mi:** Travelift *(45T, 11 mi.)*, Engine mechanic *(gas, diesel)*, Hull Repairs *(11 mi.)*, Canvas Work *(The Canvas Riggers 672-4417, 11 mi.)*, Air Conditioning *(11 mi.)*, Refrigeration *(11 mi.)*, Divers *(11 mi.)*, Compound, Wash & Wax *(11 mi.)*, Propeller Repairs *(11 mi.)*, Woodworking *(11 mi.)*, Metal Fabrication *(11 mi.)*, Painting *(11 mi.)*, Awlgrip *(11 mi.)*, Yacht Broker *(11 mi.)*.
Nearest Yard: Oak Harbor Boat Works (360) 675-2659

Restaurants and Accommodations
1-3 mi: Restaurant *(Island Grill 679-3914, B $5-10, L $6-15)*, *(Deception Café & Grill 293-9250)*, *(Seabolt's Smokehouse 675-6485)* **3+ mi:** Motel *(Lake Campbell Lodge 293-5314, 4 mi.)*, Inn/B&B *(Autumn Leaves B&B 293-4920, 10 mi.)*, *(Anacortes Inn 293-3153, 10 mi.)*

Recreation and Entertainment
OnSite: Picnic Area, Grills, Park *(Conservation Corps Interpretive Center)*
1-3 mi: Golf Course *(Lam's Links 675-3412)*

Provisioning and General Services
1-3 mi: Market *(Whidbey Market 679-1384/Soundview Shopper 675-6132)*, Wine/Beer *(Nick's Deception Pass Saloon 675-6916)* **3+ mi:** Pharmacy *(Anacortes Pharmacy 299-8560, 10 mi.)*, Hardware Store *(Ace 293-3535, 10 mi.)*, Retail Shops *(Anacortes, 10 mi.)*

Transportation
OnCall: Rental Car *(Enterprise 293-4325)*, Taxi *(Anacortes City Cab 299-8294/ Oak Harbor 675-1244)* **3+ mi:** Ferry Service *(to San Juan Island & Sidney BC 293-6060, 10 mi.)* **Airport:** Anacortes *(10 mi.)*

Medical Services
911 Service **OnCall:** Ambulance **3+ mi:** Doctor *(Family Medicine 299-4211, 10 mi.)*, Dentist *(Huntsinger 299-9347, 8 mi.)*, Veterinarian *(Anacortes Animal Hospital 293-3431, 10 mi.)* **Hospital:** Island 299-1300 *(10 mi.)*

Setting -- Bowman Bay is located on the extreme southwest corner of Fidalgo Island. Enter between Rosario Head and Lighthouse Point, taking care to avoid Gull Rocks and Coffin Rocks in the middle of the entrance. Just around Lighthouse Point (also called Reservation Head) is Deception Pass. Set in a small bay surrounded by unspoiled, forested Bowman Bay State Park, the mooring field is frequented by herons and eagles that add a touch of wilderness.

Marina Notes -- Five moorings are available in this remote area. In addition to the mooring buoys, there is a fixed wooden dock that provides some side-tie access to the floating platform attached to it. Payment is on the honor system and should be dropped at the pay station. The heads are provided by the Park Service for the public and are a short distance from the boat launch.

Notable -- Bowman Bay State Park - a saltwater marine park - is popular with outdoor enthusiasts who enjoy diving and kayaking from its beaches, hiking or biking on mountain trails, and camping in its secluded areas. There is a fishing pier and a Conservation Corps Interpretive Center (Thursday-Sunday). Pass Lake, considered one of the best spots for fly-fishing in the region, is about 1.5 miles away. This area gets only half the rainfall that pelts Seattle annually, and tides here range from 4 to 14 feet. The closest restaurant is Deception Café & Grill, about 2 miles east of the landing. Following Route 20 farther north, some provisions can be found at Whidbey Market, and Lake Campbell Lodge offers 10 rooms on the NE side of the lake with the same name. Another market, two restaurants, a pub, and the 9-hole Lam's Links Golf Course are across the bridge. Anacortes and Oak Harbor are both about 10 miles away.

Navigational Information
Lat: 48°23.845' **Long:** 122°37.612' **Tide:** 12 ft. **Current:** n/a **Chart:** 18427
Rep. Depths (*MLW*): **Entry** 7 ft. **Fuel Dock** 7 ft. **Max Slip/Moor** 7 ft./-
Access: Enter from Deception Pass

Marina Facilities *(In Season/Off Season)*
Fuel: *Shell* - Gasoline, Diesel
Slips: 70 Total, 4 Transient **Max LOA:** 50 ft. **Max Beam:** 15 ft.
  Rate *(per ft.)*: **Day** $0.70/0.50 **Week** n/a **Month** n/a
  Power: 30 amp Incl., **50 amp** n/a, **100 amp** n/a, **200 amp** n/a
  Cable TV: No **Dockside Phone:** No
Dock Type: Floating, Long Fingers, Wood
Moorings: 0 Total, 0 Transient **Launch:** n/a, Dinghy Dock
  Rate: Day n/a **Week** n/a **Month** n/a
Heads: 2 Toilet(s)
Laundry: None, Book Exchange **Pay Phones:** Yes, 1
Pump-Out: No **Fee:** n/a **Closed Heads:** Yes

Marina Operations
Owner/Manager: Dundee Woods **Dockmaster:** Same
In-Season: Summer, 8am-7pm M-F **Off-Season:** Winter, 8am-4pm Tue-Fri
After-Hours Arrival: Call
Reservations: No **Credit Cards:** Visa/MC, Amex, Shell
Discounts: None
Pets: Welcome, Dog Walk Area **Handicap Access:** No

Deception Pass Marina

200 W. Cornet Bay Road; Oak Harbor, WA 98277

Tel: (360) 675-5411 **VHF: Monitor** Ch. 16 **Talk** Ch. 68
Fax: (360) 675-5411 **Alternate Tel:** n/a
Email: n/a **Web:** n/a
Nearest Town: Oak Harbor *(10 mi.)* **Tourist Info:** (360) 675-9438

Marina Services and Boat Supplies
Services - Docking Assistance, Trash Pick-Up, Dock Carts **Supplies -**
OnSite: Ice *(Block, Cube)*, Ships' Store, Bait/Tackle, Propane
3+ mi: West Marine *(675-1976, 10 mi.)*

Boatyard Services
Nearest Yard: Marine Services (360) 675-7900

Restaurants and Accommodations
1-3 mi: Restaurant *(Island Grill 679-3194, L $6-13, D $7-17)*, *(Seabolt's Smokehouse 675-6485)*, *(Nick's Deception Pass Saloon 675-6916)*
3+ mi: Restaurant *(Deception Café & Grill 293-9250, 5 mi.)*, Fast Food *(Taco Bell, McDonald's, 10 mi.)*, Motel *(Coachman Inn 675-0727, $60-165, 9 mi.)*, Hotel *(Best Western Harbor Plaza 679-4567, $72-119, 9 mi.)*, Inn/B&B *(Auld Holland Inn 675-2288, 9 mi.)*

Recreation and Entertainment
OnSite: Picnic Area **Near:** Park *(Deception Pass State Park)* **1-3 mi:** Golf Course *(Lam's Links 675-3412)* **3+ mi:** Dive Shop *(Whidbey Island Dive Center 675-1112, 9 mi.)*, Fitness Center *(Excaliber Gym 679-3347, 9 mi.)*, Horseback Riding *(Morning Star Riding & Learning 679-1960, 4 mi.)*, Movie Theater *(Oak Harbor Cinemas 279-2226, 10 mi.)*, Video Rental *(Blockbuster 675-1089, 9 mi.)*, Cultural Attract *(The Whidbey Playhouse 679-2237, 10 mi.)*, Special Events *(Whidbey Island Race Week - Jul, 10 mi.)*

Provisioning and General Services
OnSite: Convenience Store *("C" General Store)*, Wine/Beer **1-3 mi:** Market *(Soundview Shopper 675-6132)*, Bank/ATM **3+ mi:** Supermarket *(Safeway 679-3011, 10 mi.)*, Health Food *(Pilgrim's Natureway 675-0333, 10 mi.)*, Liquor Store *(WA Liquor 675-4343, 10 mi.)*, Farmers' Market *(seasonal, Thu 4-7pm at Oak Harbor Visitor Center 675-0472, 10 mi.)*, Post Office *(800-275-8777, 9 mi.)*, Catholic Church *(9 mi.)*, Protestant Church *(9 mi.)*, Library *(Oak Harbor 675-5115, 10 mi.)*, Beauty Salon *(Diana's 675-8081, 4 mi.)*, Barber Shop *(Top O' The Hill Barber Shop 675-9012, 9 mi.)*, Dry Cleaners *(Island Cleaners & Laundry 675-2731, 10 mi.)*, Bookstore *(Wind and Tide Bookshop 675-1342, 10 mi.)*, Pharmacy *(Saar's Market Place Food and Drug 675-3000, 9 mi.)*, Hardware Store *(Ace 679-3533, 10 mi.)*, Department Store *(Kmart 679-5545, Sears 675-0660, 10 mi.)*

Transportation
OnSite: Local Bus *(free shuttle throughout the island)* **OnCall:** Rental Car *(Enterprise 675-6052)*, Taxi *(Whidbey Island Taxi 279-9330)* **Airport:** Oak Harbor *(15 mi.)*

Medical Services
911 Service **3+ mi:** Doctor *(Compass Health Whidbey Island 675-6456, 7 mi.)*, Dentist *(Vasquez 675-3444, 10 mi.)*, Chiropractor *(Oak Harbor Chiropractic 240-0718, 10 mi.)*, Veterinarian *(North Whidbey Veterinary Hospital 679-3772, 10 mi.)* **Hospital:** Whidbey General 678-5151 *(20 mi.)*

Setting -- Nestled into Cornet Bay off Deception Pass, Deception Pass Marina lies on the port side, past the Deception Pass State Park docks. An inner breakwater made of pilings protects the open and covered docks. Overlooking the docks and the bay -- in front of the General Store - are a group of picnic tables shaded by umbrellas. Across the water and surrounding the marina, forested hills and sand dunes provide a serene, relaxing setting.

Marina Notes -- Founded in early '60s. Owned by the same family since 1990. Well-kept wooden docks have long finger piers and alongside space. Covered slips and boat houses may also be available. Provisions, including ice, beer and wine, food, books, bait (herring, chicken parts, night crawlers), and supplies are sold in the marina store. There are many long-term residents who graciously help guests find that good fishing spot. Fish-cleaning table on-site. Marine Services (675-7900) across the street provides some repair and boat services. There are no showers or laundry facilities on-site.

Notable -- Whidbey Island was named for Captain's Vancouver first mate, John Whidbey, when the two realized this was an island, and not a peninsula on the mainland. The discovery also prompted Vancouver to name the channel Deception Pass, since it had initially been mistaken for a bay. The Deception Pass Bridge, which looms 182 ft. above the water, was completed in less than a year by the Civilian Conservation Corps; it was inaugurated in 1935 and declared a National Historic Monument in 1982. A 3-room interpretive center tells the story of the bridge - an original worker may even be on duty. Explore this peaceful, natural wonderland, or catch a free bus to the towns of Oak Harbor, Coupeville or Langley, which offer shopping, entertainment, and special events.

Deception Pass State Park

Navigational Information
Lat: 48°24.018' **Long:** 122°37.411' **Tide:** 12 ft. **Current:** n/a **Chart:** 18427
Rep. Depths (*MLW*): **Entry** 7 ft. **Fuel Dock** n/a **Max Slip/Moor** 7 ft./7 ft.
Access: Whidbey Island's Cornet Bay off Deception Pass

Marina Facilities (*In Season/Off Season*)
Fuel: No
Slips: 15 Total, 15 Transient **Max LOA:** 50 ft. **Max Beam:** n/a
 Rate (*per ft.*): **Day** $0.50* **Week** n/a **Month** n/a
 Power: 30 amp n/a, 50 amp n/a, 100 amp n/a, 200 amp n/a
 Cable TV: No **Dockside Phone:** No
 Dock Type: Floating, Long Fingers, Wood
Moorings: 11 Total, 11 Transient **Launch:** n/a, Dinghy Dock
 Rate: Day $7 **Week** n/a **Month** n/a
Heads: 2 Toilet(s)
Laundry: None **Pay Phones:** No
Pump-Out: No **Fee:** n/a **Closed Heads:** Yes

Marina Operations
Owner/Manager: Washington State Parks Department **Dockmaster:** n/a
In-Season: Year-Round **Off-Season:** n/a
After-Hours Arrival: Tie up and pay at the pay station
Reservations: No **Credit Cards:** No
Discounts: None
Pets: Welcome **Handicap Access:** No

Deception Pass State Park

41229 N. State Highway 20; Oak Harbor, WA 98277

Tel: (360) 675-2417 **VHF: Monitor** n/a **Talk** n/a
Fax: (888) 226-7688 **Alternate Tel:** n/a
Email: n/a **Web:** www.parks.wa.gov/moorage/parks
Nearest Town: Oak Harbor (*10 mi.*) **Tourist Info:** (360) 675-9438

Marina Services and Boat Supplies
Supplies - Near: Ice (*Block, Cube*), Ships' Store (*Deception Pass Marina's General "C" Store*), Bait/Tackle, Propane **3+ mi:** West Marine (*675-1976, 11 mi.*)

Boatyard Services
Nearest Yard: Marine Services (360) 675-7900

Restaurants and Accommodations
1-3 mi: Restaurant (*Island Grill 679-3194, L $6-13, D $7-17*), (*Seabolt's Smokehouse 675-6485*), (*Nick's Deception Pass Saloon 675-6916*)
3+ mi: Restaurant (*Deception Café & Grill 293-9250, 5 mi.*), Fast Food , (*McDonald's, Taco Bell 10 mi.*), Motel (*Coachman Inn 675-0727, $60-165, 9 mi.*), Hotel (*Best Western Harbor Plaza 679-4567, $72-119, 9 mi.*), Inn/B&B (*Auld Holland Inn 675-2288, 9 mi.*)

Recreation and Entertainment
OnSite: Beach, Picnic Area, Grills, Jogging Paths, Park, Sightseeing
1-3 mi: Golf Course (*Lam's Links 675-3412*) **3+ mi:** Dive Shop (*Whidbey Is. Dive Center 675-1112, 10 mi.*), Fitness Center (*Excaliber Gym 679-3347, 10 mi.*), Horseback Riding (*Morning Star Riding & Learning - classes & trail rides; people with disabilities welcome 679-1960, 5 mi.*), Movie Theater (*Oak Harbor Cinemas 279-2226, 11 mi.*), Video Rental (*Blockbuster 675-1089, 10mi.*), Cultural Attract (*The Whidbey Playhouse 679-2237, 11 mi.*)

Provisioning and General Services
Near: Convenience Store (*General "C" Store at Deception Pass Marina 675-5411*), Wine/Beer (*Deception Pass Marina*) **1-3 mi:** Market (*Soundview Shopper 675-6132*), Bank/ATM **3+ mi:** Supermarket (*Safeway 679-3011, 11 mi.*), Health Food (*Pilgrim's Natureway 675-0333, 11 mi.*), Liquor Store (*WA State Liquor 675-4343, 11 mi.*), Farmers' Market (*seasonal, Thu 4-7pm at Oak Harbor Visitor Center 675-0472, 11 mi.*), Post Office (*800-275-8777, 10 mi.*), Catholic Church (*10 mi.*), Protestant Church (*10 mi.*), Library (*Oak Harbor Library 675-5115, 11 mi.*), Beauty Salon (*Diana's 675-8081, 5 mi.*), Barber Shop (*Top O' The Hill Barber Shop 675-9012, 6 mi.*), Dry Cleaners (*Island Cleaners & Laundry 675-2731, 11 mi.*), Bookstore (*Wide and Tide Book Store 675-1342, 11 mi.*), Pharmacy (*Rite Aid 679-3522, 10 mi.*), Hardware Store (*Ace 679-3533, 11 mi.*), Retail Shops (*Scotties Sporting Goods & Outfitters 679-8201, 11 mi.*), Department Store (*Wal-Mart 279-0665, Sears 675-0660, 11 mi.*)

Transportation
OnSite: Local Bus (*free shuttle around the island*) **OnCall:** Rental Car (*Enterprise 675-6052*), Taxi (*Whidbey Island Taxi 279-9330*)
Airport: OakHarbor (*16 mi.*)

Medical Services
911 Service 3+ mi: Doctor (*Compass Health Whidbey Island 675-6456, 10 mi.*), Dentist (*Pinson 675-3590, 11 mi.*), Chiropractor (*Oak Harbor Chiropractic 240-0718, 11 mi.*), Veterinarian (*Best Friends 336-3836, 10 mi.*)
Hospital: Whidbey General 678-5151 (*20 mi.*)

Setting -- Deception Pass State Park straddles the channel between Fidalgo and Whidbey Islands, covering over 4,000 acres on both islands. While the Pass is infamous for its strong tidal currents, Cornet Bay, tucked among mountains and hills, offers good protection. The Park's shoreline, with beaches and dunes backed by rugged cliffs and the quarter-mile long bridge presiding over the Pass, make the passage spectacular.

Marina Notes -- *$10 minimum. The Park's Cornet Bay facility consists of two L-shaped wooden piers and mooring buoys, but no services or amenities (other than heads) are available - making this a great destination for nature lovers who don't mind roughing it. There is no power at the docks. A marina host lives in an on-site RV and is available for questions. Pay at the pay station. Deception Pass Marina is a stone's throw away, and its General "C" Store offers ice, some groceries, beer & wine, and marine supplies. Heads are in the building at the end of the dock.

Notable -- The land that now comprises the park was acquired in parcels, the first in 1925 when an Act of Congress set it aside for recreation. Once the trails, buildings, and the bridge were built in the 1930s, this quickly became one of the state's most beloved parks, and its expansion continued until 1999. It is home to dozens of plant species from wild roses to old pines, land mammals, birds, waterfowl, and marine life. Hiking and fishing are extremely popular at the park, and over 250 campgrounds are provided for visitors. Annually, some 2 to 3 million people cross the steel cantilever truss bridge over Deception & Canoe Pass, enjoying the spectacular panoramas it offers. Island Grill is the closest restaurant, and a free shuttle ride makes the town of Oak Harbor accessible.

Navigational Information
Lat: 48°23.478' **Long:** 122°29.848' **Tide:** 10 ft. **Current:** 5 kts. **Chart:** 18427
Rep. Depths *(MLW)*: **Entry** 10 ft. **Fuel Dock** n/a **Max Slip/Moor** 12 ft./-
Access: On Swinomish Channel between Skagit Bay and Padilla Bay

Marina Facilities *(In Season/Off Season)*
Fuel: No
Slips: 6 Total, 6 Transient **Max LOA:** 110 ft. **Max Beam:** n/a
 Rate *(per ft.)*: **Day** $0.50 **Week** n/a **Month** n/a
 Power: 30 amp n/a, **50 amp** n/a, **100 amp** n/a, **200 amp** n/a
 Cable TV: No **Dockside Phone:** No
 Dock Type: Floating, Alongside, Wood
Moorings: 0 Total, 0 Transient **Launch:** n/a
 Rate: Day n/a **Week** n/a **Month** n/a
Heads: None
Laundry: None **Pay Phones:** No
Pump-Out: No **Fee:** n/a **Closed Heads:** Yes

Marina Operations
Owner/Manager: City of La Conner **Dockmaster:** Gordy Bell
In-Season: Year-Round, 8am-5pm **Off-Season:** n/a
After-Hours Arrival: Use the pay envelope at the dock
Reservations: No, Except for vessels over 100 ft. **Credit Cards:** No
Discounts: None
Pets: Welcome **Handicap Access:** No

La Conner Town Docks

PO Box 400; La Conner, WA 98257

Tel: (360) 466-3125 **VHF: Monitor** n/a **Talk** n/a
Fax: (360) 466-3901 **Alternate Tel:** n/a
Email: publicworks@townoflaconner.org **Web:** n/a
Nearest Town: La Conner **Tourist Info:** (888) 642-9284

Marina Services and Boat Supplies
Supplies - Near: Ice *(Block, Cube)* **Under 1 mi:** Marine Discount Store *(Boater's Discount Center 800-488-0245)*, Bait/Tackle, Live Bait, Propane

Boatyard Services
Near: Engine mechanic *(gas, diesel)*. **Under 1 mi:** Travelift *(Port of Skagit Co.)*, Forklift, Launching Ramp, Electrical Repairs, Electronics Repairs, Hull Repairs, Canvas Work *(Sail Locker 466-3909)*, Bottom Cleaning, Brightwork, Woodworking, Upholstery *(L & T Canvas and Upholstery 466-3295)*, Yacht Broker *(La Conner Yacht Sales 466-3300)*. **Nearest Yard:** La Conner Maritime Services (360) 466-3629

Restaurants and Accommodations
Near: Restaurant *(La Conner Tavern and Eatery L $5-8, D $9-15)*, *(Palmer's at the Lighthouse 466-3147, L $7-15, D $18-23, early bird $8-15)*, *(La Conner Seafood and Prime Rib 466-4014, L $7-16, D $14-19, early bird $10)*, *(Seeds 466-3280)*, *(Whiskers Waterfront Café 466-1008)*, *(Nell Thorn 466-4261, D $14-25)*, Pizzeria *(Village Pizza 466-4000)*, Hotel *(Hotel Planter 466-4710, $79-129)*, Inn/B&B *(Channel Lodge 466-1500)*
Under 1 mi: Inn/B&B *(Wild Iris Inn 466-1400, $115-180)*, *(Heron in La Conner 466-4626, $79-159)*

Recreation and Entertainment
Near: Picnic Area *(Pioneer Park)*, Grills, Playground, Video Rental *(Zimmerman's 466-3720)*, Park *(John Hammer Memorial Playground)*, Museum *(La Conner Quilt Museum 466-4288, Museum of the Northwest 466-4446, Skagit County Historical Museum 466-3365)*, Cultural Attract *(La Conner Institute of Performing Arts 466-2665)*, Sightseeing *(street sculptures*

walking tour), Special Events *(Skagit Valley Tulip Festival - Apr 428-5959, Opening Day Boat Parade - May 466-3125, Fireworks over the Channel - Jul 4th 466-4778, La Conner Classic Yacht & Car Show - Sep, Christmas Parade)*, Galleries *(Pacific Blue Gallery 466-0525, Class Act 466-2000)*
Under 1 mi: Boat Rentals *(Viking Cruises 466-2639)* **3+ mi:** Golf Course *(Similk Beach Golf Course 293-3444, 8 mi.)*

Provisioning and General Services
Near: Market *(Pioneer Market 466-0188, delivers)*, Gourmet Shop *(Ginger Grater 466-4161)*, Delicatessen *(Landing Deli 466-2950)*, Liquor Store *(WA Liquor 466-1828)*, Bakery *(Calico Cupboard 466-4451)*, Bank/ATM, Post Office *(466-5162)*, Catholic Church *(Sacred Heart)*, Protestant Church *(United Methodist)*, Library *(466-3352)*, Beauty Salon *(La Conner Hair Design 466-4321)*, Barber Shop *(La Conner Barber Shop 466-3060)*, Laundry, Bookstore *(The Next Chapter 466-2665)*, Hardware Store *(La Conner Hardware 466-5351)*, Clothing Store *(Cottons 466-5825)*
Under 1 mi: Farmers' Market *(466-2728 Sat & Sun 10am-Noon, 2nd St.)*, Fishmonger, Pharmacy *(La Conner Drug Store 466-3124)*

Transportation
OnCall: Rental Car *(Enterprise 757-7343)*, Taxi *(Yellow Cab 336-5500)*
Near: InterCity Bus *(SKAT)* **3+ mi:** Rail *(Mt. Vernon Amtrak, 11 mi.)*
Airport: Skagit Reional/Bellingham Int'l. *(10 mi./43 mi.)*

Medical Services
911 Service **OnCall:** Ambulance **Near:** Doctor *(La Conner Medical Clinic 466-3136)*, Dentist *(Town Square Dental 466-3196)*, Chiropractor *(Kress 466-4050)*, Veterinarian **Hospital:** Skagit Valley 424-4111 *(10 mi.)*

Setting -- The La Conner Town Docks are on the east bank of the Swinomish Channel, a short distance from the brightly colored Rainbow Bridge which connects the Swinomish Indian Reservation to town. The three simple floats are connected to the main bulkhead by a "well secured" ramp. Shoreside, picnic tables sit above the floats and waterfront eateries and shops flank the docks. The village's main street, and all its attractions, are only a few steps away.

Marina Notes -- *Reservations required for vessels over 100 feet. The three wooden docks are intended primarily for short visits. Overnight stays are allowed on a first-come-first-serve basis, but no facilities are onsite. Payment is on the honor system with pay envelopes located on the dock (display the receipt on the dockside window). Well-maintained public restrooms are in town. Note: Occasional strong tidal flows can boil along in excess of 10 knots.

Notable -- Little La Conner has a long, fascinating history, greatly influenced by the diking and dredging of the 11-mile Swinomish Channel. Learn about its past at the Skagit County Historical Museum; its hilltop location promises great views of the area. Today, the village is home to artists and galleries, unique shops, several B&Bs and eateries highly rated by regional critics. Everything is a short distance from the docks, and beautiful statues placed throughout town invite visitors to stroll. Famous artists - Morris Graves, Guy Anderson, Kenneth Callahan, and Mark Tobey - who are credited with establishing the Northwest School of Painting - once resided in La Conner. The town celebrates the work of local artists at the Museum of Northwest Art (daily 10am-5pm, seniors $5/4, under 12 free). Another unique destination is the La Conner Quilt Museum, one of only 12 in the U.S. (Wed-Sat 11am-4pm, Sun Noon-4pm, $5/under 12 free).

La Conner Marina

PO Box 1120; 613 North Second; La Conner, WA 98257

Tel: (360) 466-3118 **VHF: Monitor** Ch. 66A **Talk** Ch. 66A
Fax: (360) 466-3119 **Alternate Tel:** n/a
Email: n/a **Web:** www.portofskagit.com/marina
Nearest Town: La Conner (0.5 mi.) **Tourist Info:** (888) 642-9284

Navigational Information

Lat: 48°23.805' **Long:** 122°29.827' **Tide:** 10 ft. **Current:** 5 kt. **Chart:** 18427
Rep. Depths (*MLW*): **Entry** 10 ft. **Fuel Dock** 12 ft. **Max Slip/Moor** 12 ft./-
Access: In Swinomish Channel between Skagit Bay and Padilla Bay

Marina Facilities (In Season/Off Season)

Fuel: Gasoline, Diesel, High-Speed Pumps
Slips: 500 Total, 60 Transient **Max LOA:** 150 ft. **Max Beam:** n/a
Rate (per ft.): **Day** $0.80 **Week** $4.00 **Month** $6.90*
Power: 30 amp $0.07, **50 amp** $0.07, **100 amp** n/a, **200 amp** n/a
Cable TV: Yes **Dockside Phone:** No
Dock Type: Floating, Concrete, Wood
Moorings: 0 Total, 0 Transient **Launch:** n/a, Dinghy Dock
Rate: Day n/a **Week** n/a **Month** n/a
Heads: 25 Toilet(s), 14 Shower(s)
Laundry: 8 Washer(s), 8 Dryer(s) **Pay Phones:** Yes
Pump-Out: OnSite, Self Service, 1 Central **Fee:** Free **Closed Heads:** Yes

Marina Operations

Owner/Manager: Port of Skagit County **Dockmaster:** Russ Johnson
In-Season: Year-Round, 8am-5pm **Off-Season:** n/a
After-Hours Arrival: Pay with payment envelope on dock
Reservations: No **Credit Cards:** Visa/MC
Discounts: None
Pets: Welcome **Handicap Access:** Yes, Heads, Docks

Marina Services and Boat Supplies

Services - Security (24 hrs.), Dock Carts **Communication -** Fax in/out ($3/pg), Data Ports (Wi-Fi - Broadband Xpress), FedEx, DHL, UPS, Express Mail **Supplies - OnSite:** Ice (Block, Cube), Ships' Store, Marine Discount Store, Bait/Tackle, Propane, CNG

Boatyard Services

OnSite: Forklift, Crane, Engine mechanic (gas, diesel), Electrical Repairs, Electronic Sales, Electronics Repairs, Hull Repairs, Canvas Work, Bottom Cleaning, Brightwork, Compound, Wash & Wax, Propeller Repairs, Woodworking, Upholstery, Yacht Interiors, Metal Fabrication, Painting, Total Refits **OnCall:** Divers **Near:** Yacht Broker (La Conner Yacht Sales 466-3300). **Nearest Yard:** La Conner Maritime Services (360) 466-3629

Restaurants and Accommodations

Near: Restaurant (La Conner Tavern and Eatery L $5-8, D $9-15), (Palmer's at the Lighthouse 466-3147, L $6-15, D $18-23, early bird $8-15), (La Conner Seafood & Prime Rib 466-4014, L $8-16, D $14-19, early bird $10), (Kerstin's 466-9111, L $7-12, D $18-23), (La Conner Brewing Co. 466-1415), Pizzeria (Village Pizza 466-4000), Hotel (Hotel Planter 466-4710, $79-129), Inn/B&B (Channel Lodge 466-1500, $99-170) **Under 1 mi:** Inn/B&B (Wild Iris 466-1400, $115-180), (Heron in La Conner 466-4626, $79-159)

Recreation and Entertainment

OnSite: Pool, Picnic Area, Grills **Near:** Jogging Paths, Video Rental (Zimmerman's 466-3720), Park, Museum (Skagit County Historical Museum 466-3365, Museum of Northwest Art 466-4446, La Conner Quilt Museum 466-4288), Cultural Attract (La Conner Institute of Performing Arts

466-2665), Sightseeing, Special Events (Tulip Festival - Apr, Pioneer Picnic - Aug & July 4th Fireworks over the Channel), Galleries (Pacific Blue Gallery 466-0525, Nasty Jacks Antiques 466-3209) **3+ mi:** Golf Course (Similk Beach Golf 293-3444, 9 mi.)

Provisioning and General Services

Near: Convenience Store, Market (Pioneer Market 466-0188 - deliver sto the boat), Gourmet Shop (Olive Shoppe 466-4101), Delicatessen (La Conner Landing Deli & Espresso 466-2950), Wine/Beer (at Boater's Discount), Liquor Store (WA Liquor 466-1828), Bakery (Calico Cupboard 466-4451), Bank/ATM, Post Office, Catholic Church, Protestant Church, Library (466-3352), Beauty Salon (Hair Design 466-4321), Barber Shop (La Conner 466-3060), Bookstore (Next Chapter 466-2665), Hardware Store (La Conner 466-5351), Florist, Clothing Store, Copies Etc. **Under 1 mi:** Farmers' Market (Sat & Sun 10am-Noon, 2nd St. 466-2728) **1-3 mi:** Pharmacy (La Conner 466-3124) **3+ mi:** Department Store (Sears 757-5528, 10 mi.), Buying Club (Costco 757-5700, 10 mi.)

Transportation

OnCall: Rental Car (Enterprise 757-7343), Taxi (Yellow Cab 336-5500) **3+ mi:** Rail (Mt. Vernon Amtrak, 12 mi.) **Airport:** Skagit Reional/Bellingham Int'l. (11 mi./43 mi.)

Medical Services

911 Service **Near:** Doctor (La Conner Medical Clinic 466-3136), Dentist (Town Square Dental 466-3196), Chiropractor (Kress Chiropractic 466-4050) **Under 1 mi:** Holistic Services (Watergrass Day Spa 466-4626) **Hospital:** Skagit Valley 424-4111 (9 mi.)

Setting -- La Conner Marina is on the east side of the Swinomish Channel that separates Fidalgo Island from the mainland. The docks occupy two neighboring basins, both edged by long piers that provide alongside space for guests. Docks "A" through "F" are at the south basin, "G" through "K" at the north one. "F" and "G" docks, each 600 ft. long, front the main channel and are dedicated to side-tie guest moorage. Onshore a gray two-story blue-trimmed contemporary houses the office, facilities and the now-closed restaurant. A large grassy area includes a huge BBQ and picnic tables

Marina Notes -- *Covered slips $10.56/ft./mo. The alongside floats provide 2,400 linear feet of dockage and accomodate 50-60 visiting boats. A fuel dock and Boater's Discount chandlery separate the two large basins - the south one shared by the Swinomish Y.C.. Haul-out and most boatyard services are available. The small brick building nearthe BBQ is the laundry. The fully-tiled bathhouse at the south basin is open 6am to midnight; the north basin facilities require access cards. Note: The narrow channel typically has strong, unpredictable currents. The Marina Bistro & Bar has closed.

Notable -- The charming town of La Conner is a short walk from the marina. It was founded in the 1800s by John Conner, who named it L. A. Conner after his wife Louisa Ann. Initially a trading post, then a fishing village, it has become a popular destination for boaters and artists. It caters to tourists with three museums, galleries, boutiques, antique shops, and excellent restaurants. Beautiful sculptures are everywhere in town, and a walking tour is available to enjoy them. The Institute of Performing Arts has shows at Maple Hall in the winter, and holds a Water Music Festival in the summer.

Navigational Information
Lat: 48°23.860' **Long:** 122°29.786' **Tide:** 10 ft. **Current:** 5 kt. **Chart:** 18427
Rep. Depths (*MLW*)**: Entry** 10 ft. **Fuel Dock** n/a **Max Slip/Moor** 12 ft./-
Access: Swinomish Channel between Skagit Bay and Padilla Bay

Marina Facilities *(In Season/Off Season)*
Fuel: No
Slips: 4 Total, 4 Transient **Max LOA:** 150 ft. **Max Beam:** n/a
  Rate *(per ft.)*: **Day** $1.00 **Week** n/a **Month** n/a
  Power: 30 amp $3, **50 amp** n/a, **100 amp** n/a, **200 amp** n/a
  Cable TV: No **Dockside Phone:** No
  Dock Type: Floating, Concrete, Alongside
Moorings: 0 Total, 0 Transient **Launch:** n/a, Dinghy Dock
  Rate: Day n/a **Week** n/a **Month** n/a
Heads: 10 Toilet(s), 10 Shower(s)
Laundry: 5 Washer(s), 5 Dryer(s) **Pay Phones:** No
Pump-Out: No **Fee:** n/a **Closed Heads:** Yes

Marina Operations
Owner/Manager: Rod Crawford **Dockmaster:** Same
In-Season: May-Sep, 8am-6pm* **Off-Season:** Oct-Apr, 9am-4:30pm
After-Hours Arrival: Call ahead
Reservations: No **Credit Cards:** Visa/MC, Dscvr, Amex
Discounts: None
Pets: Welcome **Handicap Access:** No

Boater's Discount Center

PO Box 1590; 601 Dunlap Street; La Conner, WA 98257

Tel: (360) 466-3540; (800) 488-0245 **VHF: Monitor** Ch. 16 **Talk** n/a
Fax: (360) 466-5350 **Alternate Tel:** n/a
Email: boatersd@cnw.com **Web:** www.boathardtop.com
Nearest Town: La Conner *(0.5 mi.)* **Tourist Info:** (888) 642-9284

Marina Services and Boat Supplies
Services - Security *(Port security)*, Megayacht Facilities **Supplies -**
OnSite: Ice *(Block, Cube)*, Ships' Store, Bait/Tackle *(Herring, crab bait)*, Live
Bait, Propane, CNG

Boatyard Services
OnSite: Canvas Work, Compound, Wash & Wax, Propeller Repairs,
Inflatable Repairs, Yacht Broker *(Boater's Discount and Yacht Sales)*
OnCall: Engine mechanic *(gas, diesel)*, Brightwork, Divers, Woodworking,
Upholstery **Near:** Travelift, Railway, Forklift, Crane, Hull Repairs, Bottom
Cleaning, Metal Fabrication, Painting, Awlgrip. **Under 1 mi:** Launching
Ramp. **Nearest Yard:** La Conner MaritimeServices (360) 466-3632

Restaurants and Accommodations
Near: Restaurant *(La Conner Tavern and Eatery L $5-8, D $9-15)*, Inn/B&B
(Channel Lodge 466-1500, waterfront) **Under 1 mi:** Restaurant *(Palmer's
at the Lighthouse 466-3147, L $6-15, D $18-23)*, *(La Conner Seafood and
Prime Rib 466-4014, L $7-15, D $14-19)*, *(Kerstin's 466-9111)*, Pizzeria
(Village Pizza 466-4000), Hotel *(Hotel Planter 466-4710, $79-129)*
1-3 mi: Inn/B&B *(Wild Iris Inn 466-1400, $115-180)*, *(Heron In La Conner
466-4626, $79-159)*

Recreation and Entertainment
Near: Picnic Area, Grills, Playground, Jogging Paths, Cultural Attract *(La
Conner Institute of Performing Arts 466-2665)* **Under 1 mi:** Video Rental
(Zimmerman's 466-3720), Park, Museum *(La Conner Quilt Museum 466-
4288, Skagit County Historical Museum 466-3365)*, Sightseeing *(historic

downtown)*, Special Events *(Tulip Festival - Apr, Opening Day Boat Parade -
May, Jul 4th Fireworks, Classic Yacht & Car Show - Sep, Quilt Fest - Sep)*,
Galleries *(Earthen Works 466-4422, Courtyard Gallery 877-912-4632)*
3+ mi: Golf Course *(Similk Beach 293-3444, 9 mi.)*

Provisioning and General Services
OnSite: Convenience Store, Wine/Beer, Bookstore **Near:** Gourmet Shop,
Newsstand, Copies Etc. **Under 1 mi:** Market *(Pioneer Market 466-0188,
delivers)*, Delicatessen *(La Conner Landing Deli 466-2950)*, Liquor Store
(WA Liquor 466-1828), Bakery, Farmers' Market *(2nd St. Sat & Sun 10am-
Noon 466-2728)*, Fishmonger, Bank/ATM, Post Office, Catholic Church,
Protestant Church, Library *(466-3352)*, Beauty Salon *(La Conner Hair Design
466-4321)*, Barber Shop *(La Conner Barber Shop 466-3060)*, Hardware
Store *(La Conner Hardware 466-5351)*, Florist, Clothing Store *(Chez la Zoom
466-4546, Cottons 466-5825)* **1-3 mi:** Pharmacy *(La Conner Drug Store
466-3124)* **3+ mi:** Buying Club *(Costco 757-5700, 10 mi.)*

Transportation
OnCall: Rental Car *(Enterprise 757-7343)* **3+ mi:** Rail *(Mt. Vernon Amtrak,
12 mi.)* **Airport:** Skagit Regional/Bellingham Int'l. *(11 mi./44 mi.)*

Medical Services
911 Service **OnCall:** Ambulance **Under 1 mi:** Doctor *(La Conner Medical
Clinic 466-3136)*, Dentist *(Town Square Dental 466-3196)*, Chiropractor
(Kress Chiropractic 466-4050), Veterinarian *(La Conner Veterinary Hospital
466-3717)* **1-3 mi:** Holistic Services *(Watergrass Day Spa 466-4626)*
Hospital: Skagit Valley 424-4111 *(10 mi.)*

Setting -- Boater's Discount is located on the Swinomish Channel just north of the town of La Conner. It sits between the two basins of La Conner Marina. Its well-maintained 150-foot floating, concrete dock is adjacent to the store and across the channel from the Swinomish Indian Reservation.

Marina Notes -- *Closes at 3pm SundayBoater's Discount is a large, family-owned marine store with a dock. Space is primarily intended for shorter visits, but they welcome overnight guests when space is available. You can also stop by or dinghy in from neighboring marinas to stock up on supplies: marine hardware, books and charts, ice, bait, CNG and propane, limited groceries and gifts are all available. If you need it, it's probably here, but may not be in large , quantities. The service-oriented staff will try to meet your needs. Some boatyard services are also on-site, including hardtop extensions, VacuFlush heads, and davit installs. 10% discount for Vessel Assist. The large bathhouse and laundy at La Conner Marina are available to Boater's Discount guests.

Notable -- The historic town of La Conner has resisted change since the turn of the century. Victorian houses and buildings which now house restaurants, galleries, antique stores and boutiques are an easy walk from the dock. The Gaches Mansion on Second Street is home to the La Conner Quilt Museum, which sponsors an annual Quilt Festival in September. In the same area, high on a hilltop, the Skagit Historical Museum has exhibits on local history, a research library, videos available for rental, and beautiful views of the village. If you visit in April during the regional Tulip Festival, consider a ride with Sky Fly'n Helicopter for panoramic views of the gorgeous flower fields (888-377-4115, $40/pp, $60/two adults).

La Conner Maritime Service

920 West Pearl-Jensen Way; La Conner, WA 98257

Tel: (360) 466-3629 **VHF: Monitor** Ch. 10 **Talk** n/a
Fax: (360) 466-3632 **Alternate Tel:** (360) 466-5213
Email: service@laconnermaritime.com **Web:** www.laconnermaritime.com
Nearest Town: La Conner *(1 mi.)* **Tourist Info:** (360) 466-4778

Navigational Information
Lat: 48°24.104' **Long:** 122°29.770' **Tide:** 10 ft. **Current:** 5 kt. **Chart:** 18427
Rep. Depths *(MLW)*: **Entry** 10 ft. **Fuel Dock** n/a **Max Slip/Moor** 12 ft./-
Access: Swinomish Channel between Skagit Bay and Padilla Bay

Marina Facilities *(In Season/Off Season)*
Fuel: No
Slips: 4 Total, 2 Transient **Max LOA:** 100 ft. **Max Beam:** n/a
  Rate *(per ft.)*: **Day** $0.60 **Week** n/a **Month** n/a
  Power: 30 amp Incl., **50 amp** Incl., **100 amp** n/a, **200 amp** n/a
  Cable TV: No **Dockside Phone:** No
  Dock Type: Floating, Alongside, Wood
Moorings: 0 Total, 0 Transient **Launch:** n/a
  Rate: Day n/a **Week** n/a **Month** n/a
Heads: 1 Toilet(s)
Laundry: None **Pay Phones:** No
Pump-Out: No **Fee:** n/a **Closed Heads:** Yes

Marina Operations
Owner/Manager: Ed Oczkzewicz **Dockmaster:** Same
In-Season: Mon-Sat, 8am-4:30pm **Off-Season:** Oct-Mar, 8am-4:30pm
After-Hours Arrival: Find a spot and tie up
Reservations: No **Credit Cards:** Visa/MC
Discounts: None
Pets: Welcome **Handicap Access:** No

Marina Services and Boat Supplies
Communication - FedEx, DHL, UPS **Supplies - Near:** Ice *(Block)*, Ships' Store, Marine Discount Store *(Boater's Discount Center 466-3540)*, Bait/Tackle, Propane, CNG

Boatyard Services
OnSite: Travelift *(55T to 18 ft. beam & 85T to 25 ft. beam)*, Forklift, Crane *(12.5T)*, Hydraulic Trailer, Engine mechanic *(gas, diesel)*, Electrical Repairs, Electronic Sales, Hull Repairs, Rigger, Bottom Cleaning, Brightwork, Compound, Wash & Wax, Interior Cleaning, Woodworking, Yacht Interiors, Metal Fabrication, Painting, Awlgrip, Total Refits **OnCall:** Refrigeration *(Wyman's)*, Inflatable Repairs **Near:** Launching Ramp, Air Conditioning, Divers, Yacht Broker *(La Conner Yacht Sales 466-3300)*. **Under 1 mi:** Canvas Work. **3+ mi:** Propeller Repairs *(Prop Shop 425-745-1700, 37 mi.)*. **Dealer for:** Northern Lights, Volvo, OMC, Twin Disc, Naiad, Wesmar, Side Power. **Member:** ABYC - 2 Certified Tech(s) **Yard Rates:** $75-90/hr., Haul & Launch $5.50-14/ft. *(blocking $1.50-3/ft.)*, Power Wash $2.50-3/ft, Bottom Paint $25-70/ft. *(paint incl.)* **Storage:** On-Land $4.50-7.75/ft.

Restaurants and Accommodations
Under 1 mi: Restaurant *(La Conner Seafood and Prime Rib 466-4014, L $7-16, D $14-19, early bird $10)*, *(La Conner Tavern and Eatery L $5-8, D $9-15)*, *(Palmer's at the Lighthouse 466-3147, L $6-15, D $18-23, early bird $8-15; has a dock)*, *(Station House 466-4488)*, *(Nell Thorn 466-4261, D $14-25)*, Lite Fare *(La Conner Brewing Co. 466-1415)*, Pizzeria *(Village Pizza 466-4000)*, Hotel *(Hotel Planter 466-4710, $75-120)*, Inn/B&B *(Channel Lodge 466-1500)*, Condo/Cottage *(Estep Residences 466-2116, $135)*

Recreation and Entertainment
OnSite: Jogging Paths **OnCall:** Dive Shop **Near:** Picnic Area, Grills, Galleries *(Artful Living 466-4933)* **Under 1 mi:** Video Rental *(Zimmerman's 466-3720)*, Park, Museum *(Museums of the Northwest Art 466-4446, La Conner Quilt Museum 466-4288, Skagit County Historical Museum 466-3365)*, Cultural Attract *(Institute of Performing Arts 466-2665)*, Special Events *(Skagit Valley Tulip Fest - Apr 428-5959, Boat Show - Jun 466-3300, Arts Fest - Jun)* **3+ mi:** Golf Course *(Similk Beach 293-3444, 9 mi.)*

Provisioning and General Services
Under 1 mi: Market *(Pioneer Market 466-0188, delivers)*, Gourmet Shop *(Olive Shoppe 466-4101)*, Delicatessen *(Landing Deli & Espresso 466-2950)*, Liquor Store *(WA Liquor 466-1828)*, Farmers' Market *(2nd Street Sat & Sun 10am-Noon, 466-2728)*, Bank/ATM, Library *(466-3352)*, Beauty Salon *(Hair Design 466-4321)*, Barber Shop *(La Conner Barber 466-3060)*, Bookstore *(Next Chapter 466-2665)*, Hardware Store *(La Conner Hardware 466-5351)*, Copies Etc. *(Capt's Business Support 466-3381)* **1-3 mi:** Post Office, Catholic Church, Protestant Church, Pharmacy *(La Conner Drug Store 466-3124)* **3+ mi:** Buying Club *(Costco 757-5700, 10 mi.)*

Transportation
OnCall: Rental Car *(Enterprise 757-7343)*, Taxi *(Yellow Cab 336-5500)* **3+ mi:** Rail *(Amtrak, 12 mi.)* **Airport:** Skagit/Bellingham *(11 mi./44 mi.)*

Medical Services
911 Service **OnCall:** Ambulance **Under 1 mi:** Doctor *(La Conner Medical Center 466-3136)*, Dentist *(Town Sq. 466-3196)* **1-3 mi:** Holistic Services *(Watergrass Day Spa 466-4626)* **Hospital:** Skagit Valley 424-4111 *(10 mi.)*

Setting -- Follow the Swinomish Channel from Skagit Bay from the south or Padilla Bay from the north to reach the picturesque town of La Conner. This narrow waterway seems more like a river than a saltwater channel. If coming from the south, you will pass under the very orange, but majestic, Rainbow Bridge that signals you have almost arrived. The boatyard is on the starboard side, past the town docks, the yacht club, and the La Conner Marina docks.

Marina Notes -- This very large, full-service boatyard pledges to be the most helpful on the West Coast. If you need something they don't have, they will get it for you. A waiting dock can accommodate 1-2 transients, but space is limited so call ahead to inquire about availability. Two travelifts provide haulout for up to 85 tons, with certified techs on staff for repairs. Large work sheds and storage areas are on the shore, and stacked-storage is also available - customers call when they leave home and the boat is in the water when they arrive. The head lacks a shower; however, guests may use those at La Conner Marina next door for free. Note: Watch the changing currents and make allowances before landing.

Notable -- The picturesque village of La Conner lies among the flower fields and forests of Skagit Valley, at the foot of the Cascade Mountains. It attracts visiting boaters and artists, many of whom plan their trips around annual events in the area; the Skagit Valley Tulip Festival and the Boat Show & Swap Meet are both worth the trip. The marina is about a mile from the village, which offers wonderful restaurants, inns, and shops. To explore farther, Viking Cruises offers a variety of tours and nature trips on the Swinomish on an enclosed 58' boat (466-2639; 1-hour $12.50/pp, 3-hour trip with crab feast lunch $60/55).

Navigational Information
Lat: 48°17.235' **Long:** 122°37.970' **Tide:** 17 ft. **Current:** 3 kt. **Chart:** 18428
Rep. Depths (*MLW*): **Entry** 11 ft. **Fuel Dock** 4 ft. **Max Slip/Moor** 14 ft./-
Access: From Saratoga Passage, enter at red 2 channel marker

Marina Facilities (*In Season/Off Season*)
Fuel: Gasoline, Diesel
Slips: 424 Total, 70 Transient **Max LOA:** 60 ft. **Max Beam:** 15 ft.
 Rate (*per ft.*): **Day** $0.68* **Week** $175 **Month** $750
 Power: 30 amp Incl., 50 amp n/a, 100 amp n/a, 200 amp n/a
 Cable TV: No **Dockside Phone:** No
 Dock Type: Floating, Long Fingers, Alongside, Concrete
Moorings: 0 Total, 0 Transient **Launch:** n/a, Dinghy Dock (Free)
 Rate: Day n/a **Week** n/a **Month** n/a
Heads: 5 Toilet(s), 8 Shower(s)
Laundry: 3 Washer(s), 3 Dryer(s) **Pay Phones:** Yes, 2
Pump-Out: OnSite, Self Service, 3 Central **Fee:** Free **Closed Heads:** Yes

Marina Operations
Owner/Manager: City of Oak Harbor **Dockmaster:** Dave Williams
In-Season: Mem-LabDay, 8am-6pm **Off-Season:** Lab-MemDay, 8am-5pm**
After-Hours Arrival: Guest envelopes with gate cards at guest dock/office
Reservations: Not Required **Credit Cards:** Visa/MC
Discounts: Fuel 100+ gals. **Dockage:** n/a **Fuel:** $0.05/gal. **Repair:** n/a
Pets: Welcome, Dog Walk Area **Handicap Access:** Yes, Heads

Oak Harbor Marina

1401 SE Catalina Drive; Oak Harbor, WA 98277

Tel: (360) 279-4575 **VHF: Monitor** Ch. 16 **Talk** Ch. 68
Fax: (360) 240-0603 **Alternate Tel:** n/a
Email: ohmarina@whidbey.net **Web:** www.whidbey.com/ohmarina
Nearest Town: Oak Harbor (*0.5 mi.*) **Tourist Info:** (360) 675-3755

Marina Services and Boat Supplies
Services - Trash Pick-Up, Dock Carts **Communication -** Mail & Package Hold, Phone Messages, Fax in/out (*Free*), Data Ports (*Wi-Fi - Broadband Xpress*), FedEx, UPS (*Sat Del*) **Supplies - OnSite:** Ice (*Block, Cube*), Propane **Near:** Ships' Store **1-3 mi:** West Marine (*675-1976*)

Boatyard Services
OnSite: Crane (*$15/30 two-way*), Launching Ramp (*free*), Canvas Work (*Canvas Riggers 672-4417*) **OnCall:** Bottom Cleaning, Air Conditioning, Refrigeration, Divers **Near:** Travelift (*45T*), Engine mechanic (*gas, diesel*), Hull Repairs, Compound, Wash & Wax, Propeller Repairs, Woodworking, Metal Fabrication, Painting, Awlgrip, Yacht Broker. **Nearest Yard:** Mariner's Haven (360) 675-8828

Restaurants and Accommodations
Under 1 mi: Restaurant (*Zorba's 279-8322, L $6-10, D $10-16*), (*China City 279-8899, L $5-10, D $8-16*), (*Mi Pueblo 240-0193, L $5-12, D $8-12*), (*Erawan Thai & Sushi 679-8268*), (*PW Murphy's Restaurant & Pub 279-2528*), (*Dave's Pioneer Café 679-4860*), Coffee Shop (*Whidbey Coffee 679-1162*), Inn/B&B (*North Whidbey Inn 675-5911*) **1-3 mi:** Fast Food (*KFC, Subway, Wendy's*), Pizzeria (*Godfather's*), Motel (*Acorn Motor Inn 675-6646, $52-79*), Hotel (*Best Western Harbor Plaza 679-4567, $99-149*), Inn/B&B (*Auld Holland Inn 675-2288, $45-145*)

Recreation and Entertainment
OnSite: Picnic Area, Grills, Playground, Volleyball, Jogging Paths, Park, Special Events (*Whidbey Island Race Week - Jul, Old Fashioned Fourth of July 675-3535*) **Under 1 mi:** Beach, Tennis Courts, Bowling (*Oak Bowl*

679-2533*), Cultural Attract (*The Whidbey Playhouse 679-2237*)
1-3 mi: Heated Pool, Dive Shop (*Whidbey Island Dive 675-1112*), Movie Theater (*Plaza Cinema 279-2226*), Video Rental (*Blockbuster 675-1089*)
3+ mi: Golf Course (*Gallery Golf 257-2178, 5 mi.*)

Provisioning and General Services
Near: Library (*675-5115*), Bookstore (*Cardinal Bookstore 679-5313*) **Under 1 mi:** Bank/ATM, Post Office (*800-275-8777*), Catholic Church, Protestant Church, Beauty Salon (*Gallery Salon & Day Spa 679-9227*), Barber Shop (*Bay Watch 240-8820*), Dry Cleaners (*Island Cleaners & Laundry 675-2731*), Clothing Store (*J.C. Penney's 679-4171*) **1-3 mi:** Convenience Store (*7-Eleven 675-0109*), Market (*Saar's Marketplace 675-4511*), Supermarket (*Safeway 679-3011/Albertson's 279-8828*), Delicatessen (*Pot Belly Deli 675-5204*), Health Food (*Pilgrim's Natureway 675-0333*), Liquor Store (*WA Liquor 675-4343*), Farmers' Market (*Case Rd., Thu 4-7pm*), Fishmonger (*Seasonal Seafoods 679-4731*), Pharmacy (*Rite Aid 679-3522*), Hardware Store (*Ace 679-3533*), Retail Shops (*Scotties Sporting Goods & Outfitters 679-8201*), Department Store (*Kmart 679-5545, Sears 675-0660*)

Transportation
OnSite: Local Bus (*Island Transit 678-777, free bus throughout Whidbey*)
OnCall: Rental Car (*Enterprise 675-6052*), Taxi **3+ mi:** Ferry Service (*Clinton, 40 mi.*) **Airport:** Oak Harbor (*5 mi.*)

Medical Services
911 Service **OnCall:** Holistic Services **Under 1 mi:** Doctor (*Whidbey Medical Clinic 679-3161*), Chiropractor (*Island Chiropractic 675-4954*)
1-3 mi: Dentist **Hospital:** Whidbey General 678-5151 (*11 mi.*)

Setting -- Oak Harbor Marina lies between Blower's Bluff and Forbe's Point, tight against the Navy base at the east end of Oak Harbor. Bright blue boat sheds, a big, easy access fuel dock, and a two-story blue harbormaster's office make it easy to spot this extensive and well-maintained city-owned facility; it serves the largest community on Whidbey, and is surrounded by several marine services.

Marina Notes -- *Flat rate: under 20' $12/day, 20-24' $15, 25-28' $19, 29-37' $23, 38-45' $27, 46-55' $35, 56-65' $46, 66' and above $57. Sample weekly/monthly rates are for a 40 ft. boat. **Closed Sun Nov-Feb. Pay stations on the breakwater dock at F-27 and F-52, or at the Harbormaster's office. Run by a service-oriented staff. Constructed in 1974, expanded in 1987 and again in 1998. The guest docks are the farthest out, with easy access and a wide fairway - there are picnic alcoves with tables, and, at the end, a barge hosts convenient heads, pump-out and a porta-potty dump. Marine services include vessel, electronic, and engine repair, a chandlery, and a privately owned haulout crane. On-site Oak Harbor Y.C. may have reciprocal dockage available.

Notable -- The marina sponsors an old-fashioned Fourth of July Extravaganza. The annual Whidbey Race Week - a 5-day event rated by Sailing World Magazine as one of the Top 20 regattas in the world - attracts around 150 boats in July. Restaurants and shops are just under a mile away, around the bay. As is Oak Harbor Beach Park with asand beach, large oceanside playground, tennis courts, and playing fields. Free buses make is easy to go anywhere on the island, including Deception Pass, Fort Ebey and Fort Casey State Parks, the Victorian towns of Coupeville and Langley, and the Clinton Ferry Terminal.

Coupeville Wharf

PO Box 577; 24 Front Street; Coupeville, WA 98239

Tel: (360) 678-5020 **VHF: Monitor** n/a **Talk** n/a
Fax: (360) 678-7424 **Alternate Tel:** n/a
Email: execjim@verizon.net **Web:** n/a
Nearest Town: Coupeville **Tourist Info:** (360) 678-5434

Navigational Information
Lat: 48°13.369' **Long:** 122°41.065' **Tide:** 16 ft. **Current:** n/a **Chart:** 18441
Rep. Depths *(MLW)*: Entry 16 ft. **Fuel Dock** 8 ft. **Max Slip/Moor** 8 ft./20 ft.
Access: Enter Penn Cove off the Saratoga Passage Channel

Marina Facilities *(In Season/Off Season)*
Fuel: *Chevron* - Gasoline, Diesel, On Call Delivery
Slips: 10 Total, 10 Transient **Max LOA:** 180 ft. **Max Beam:** n/a
Rate *(per ft.)*: **Day** $0.40/0.20* **Week** n/a **Month** $7/3.50*
Power: 30 amp n/a, 50 amp n/a, 100 amp n/a, 200 amp n/a
Cable TV: No **Dockside Phone:** No
Dock Type: Floating, Alongside, Wood
Moorings: 4 Total, 4 Transient **Launch:** n/a, Dinghy Dock
Rate: Day $15 **Week** n/a **Month** n/a
Heads: 2 Toilet(s), 1 Shower(s)
Laundry: None **Pay Phones:** Yes, 1
Pump-Out: No **Fee:** n/a **Closed Heads:** Yes

Marina Operations
Owner/Manager: Port of Coupeville/James Patton **Dockmaster:** L. Bechard
In-Season: Jun-Sep, 9am-6pm **Off-Season:** Oct-May**
After-Hours Arrival: Moor at available space
Reservations: None Required **Credit Cards:** Visa/MC
Discounts: None
Pets: Welcome **Handicap Access:** No

Marina Services and Boat Supplies
Services - Docking Assistance **Supplies - OnSite:** Ships' Store **Near:** Ice *(Cube)* **3+ mi:** West Marine *(675-1976, 11 mi.)*

Boatyard Services
OnCall: Bottom Cleaning, Air Conditioning, Refrigeration, Divers
Near: Launching Ramp *(Capt. Thomas Coupe Park)*. **3+ mi:** Travelift *(45T, 5 mi.)*, Engine mechanic *(gas, diesel)*, Hull Repairs *(5 mi.)*, Canvas Work *(Canvas Riggers 672-4417, 5 mi.)*, Compound, Wash & Wax *(5 mi.)*, Propeller Repairs *(5 mi.)*, Metal Fabrication *(5 mi.)*, Painting *(5 mi.)*, Awlgrip *(5 mi.)* **Nearest Yard:** Oak Harbor Boat Works *(360) 675-2659*

Restaurants and Accommodations
OnSite: Restaurant *(Habor Store Café 678-6905, cooking classes, too)*
Near: Restaurant *(The Mad Crab 678-0241, L $6-16, D $6-16)*, *(Toby's Tavern 678-4222, L $5-7, D $11-16)*, *(Christopher's Front St. Café 678-5480, L, D)*, *(Oystercatcher 678-0683)*, *(Blue Goose Inn 678-4284, L & D Tue-Sat)*, Lite Fare *(Anna's Tea Room 678-5797)*, *(Bayleaf 678-6603, L, wine tasting Wed-Sun 11am-5pm)*, *(Kneed & Feed 678-5431, B & L)*, Pizzeria *(Great Times 678-5358)*, Hotel *(Tyee Hotel 678-6616)*, Inn/B&B *(Coupeville Inn 678-6668)*, *(Inn at Penn Cove 678-8000, $55-125)*, *(Anchorage Inn 678-5581, $85-140)*, *(The Blue Goose Inn 678-4284, $115-135)*

Recreation and Entertainment
OnSite: Boat Rentals *(Kayaks)*, Galleries *(Harbor Art Gallery on-site; nearby: Blue Heron 678-9052, Kwahu Kreations, Penn Cove 678-1176)* **Near:** Picnic Area, Grills, Playground, Tennis Courts, Jogging Paths, Video Rental *(Coupeville 678-4937)*, Park *(Coupeville Town, Thomas Coupe)*, Museum *(Island County Historical Society 678-3310)*, Sightseeing *(Historic Coupeville)*, Special Events *(Arts & Crafts Fest - 2nd wknd Aug, Music Festival, Concerts on the Cove in summer)* **Under 1 mi:** Fitness Center **3+ mi:** Golf Course *(Gallery Golf 257-2178, 15 mi.)*, Movie Theater *(Plaza Cinema 675-5667, 12 mi.)*

Provisioning and General Services
OnSite: Delicatessen **Near:** Gourmet Shop *(Bayleaf 678-6603)*, Bank/ATM *(Whidbey Island 648-4555)*, Post Office *(678-5353)*, Library *(Coupeville 678-4911)*, Beauty Salon *(Genesis 678-3117)*, Bookstore *(Kingfisher 678-8463)*, Florist *(Fresh Flower 678-8010)*, Clothing Store *(Collections 678-2100)* **Under 1 mi:** Market *(Coupeville Country Store 678-2249, The Prairie Center Red Apple Market 678-5611)*, Liquor Store *(WA Liquor 678-5511)*, Bakery *(Coupe's Village Bakery 678-4229)*, Catholic Church, Protestant Church, Pharmacy *(Coupeville 678-8882)* **3+ mi:** Supermarket *(Safeway 679-3011, 13 mi.)*, Hardware Store *(Ace 679-3533, 11 mi.)*, Department Store *(Sears 675-0660, Wal-Mart 279-0665, 13 mi.)*

Transportation
OnSite: Bikes, Local Bus *(Island Transit 678-7771, free service to all of Whidbey Island)* **OnCall:** Taxi *(Whidbey Cab Co. 678-1515)* **3+ mi:** Ferry Service *(Clinton, 28 mi.)* **Airport:** Oak Harbor *(8 mi.)*

Medical Services
911 Service **Under 1 mi:** Doctor *(Coupeville Clinic 678-6576)*, Dentist *(Coupeville Dental Clinic 678-8304)*, Chiropractor *(Peak Performance 678-5400)*, Holistic Services *(Island Medical Spa 678-6561)*, Veterinarian *(Leaman 678-6046)* **Hospital:** Whidbey General 678-5151 *(0.5 mi.)*

Setting -- The hamlet of Coupeville is on the south side of Penn Cove off Saratoga Passage, which runs between Whidbey and Camano Islands. The picturesque wharf's red two-story building can't be missed. Well-maintained floats on its east side provide alongside space for several boats and four mooring buoys are in deeper water. The Wharf leads ashore to shops and an eatery and then to the quaint village; it's worth changing course to visit.

Marina Notes -- *Flat rates & half price Oct-May: $7/ft. for boats under 15', $10 to 35', $15 to 40', $20 to 50', $35 over 50'. **Oct-May, moorage and fuel only. Depth at the dock is 5 to 7 ft. at zero tide. The wharf has been partially renovated and it now houses several businesses. There is no power or water, but everything is well-maintained. Picnic tables on the deck outside the store are available to the public. There is a public phone at the end of the wharf next to the Port of Coupeville building. A key is provided for the shower and for the heads after hours. Note: Watch for Penn Cove Mussels' pens st the entrance.

Notable -- In the wharf building, "Rosie," a three-year-old a gray whale's 30-foot skeleton, complete with baleen, hangs from the ceiling - the result of 200 volunteers contributing 2,000 hours. There's also Harbor Store Café, bicycle and boat rentals, a store and a gallery with photos, paintings, and fine gifts. The pier nestles up to Main Street, with an information booth adjacent - take a walking tour of Coupeville's Victorian buildings. The Island County Historical Museum (open May-October from 12-5pm) is nearby. Charming restaurants, cafés, shops and galleries are within walking distance. A nearby beachfront park has tennis courts and a playground. Coupeville hosts a Music Festival in the spring and a large Arts & Crafts Fest in Aug. Free buses go everywhere on Whidbey Island.

11. WA – North Puget Sound

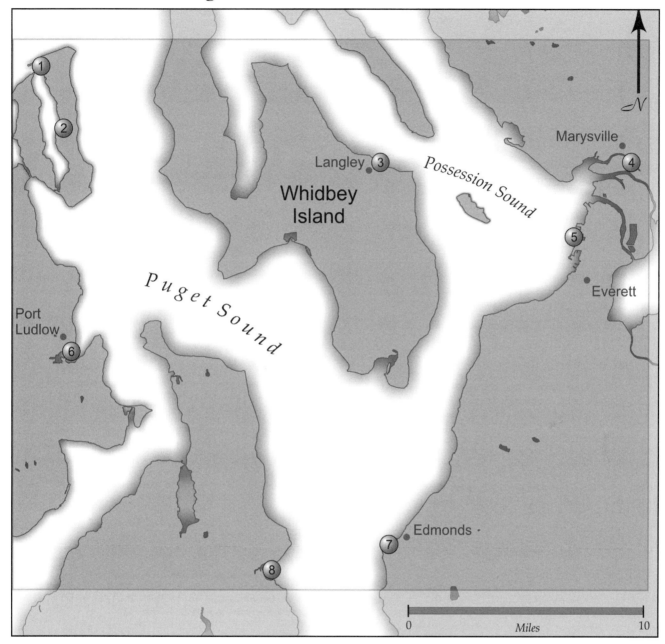

MAP	MARINA	HARBOR	PAGE	MAP	MARINA	HARBOR	PAGE
1	**Fort Flagler State Park**	*Kilisut Harbor*	166	5	**Port of Everett Marina and Boatyard**	*Port Gardner Bay*	170
2	**Mystery Bay State Park**	*Kilisut Harbor/Mystery Bay*	167	6	**Port Ludlow Marina**	*Ludlow Bay*	171
3	**Langley Boat Harbor**	*Langley Harbor*	168	7	**Port of Edmonds**	*Puget Sound/Edmonds*	172
4	**Geddes Marina**	*Ebay Slough*	169	8	**Port of Kingston**	*Apple Tree Cove*	173

▸ **Currency** — In Canadian Marina Reports, all prices are in Canadian dollars. In U.S. Marina Reports, all prices are in U.S. dollars.

▸ **"CCM"** — Denotes a Certified Clean Marina, a state/provincial award for environmental excellence. See page 298 for an explanation and page 299 for a list of Pump-Out facilities.

▸ **Ratings & Reviews** — An overview of the Atlantic Cruising Club's rating system is on page 6 and details on the content of each Marina Report are on pages 7 – 11.

▸ **Marina Report Updates** — Comments from boaters and new information from ACC reviewers and marinas are posted regularly on *www.AtlanticCruisingClub.com*.

Fort Flagler State Park

Fort Flagler State Park

10541 Flagler Road; Nordland, WA 98358

Tel: (360) 385-1259 **VHF: Monitor** n/a **Talk** n/a
Fax: (360) 379-1746 **Alternate Tel:** (360) 902-8844
Email: n/a **Web:** www.parks.wa.gov
Nearest Town: Nordland *(5 mi.)* **Tourist Info:** (360) 379-5380

Navigational Information
Lat: 48°05.467' **Long:** 122°43.183' **Tide:** 8.5 ft. **Current:** n/a **Chart:** 18464
Rep. Depths (*MLW*): **Entry** 6 ft. **Fuel Dock** n/a **Max Slip/Moor** 6 ft./6 ft.
Access: Located on the northern tip of Marrowstone Island

Marina Facilities *(In Season/Off Season)*
Fuel: No
Slips: 6 Total, 6 Transient **Max LOA:** 50 ft. **Max Beam:** n/a
  Rate *(per ft.)*: **Day** $0.50* **Week** n/a **Month** n/a
  Power: 30 amp n/a, **50 amp** n/a, **100 amp** n/a, **200 amp** n/a
  Cable TV: No **Dockside Phone:** No
  Dock Type: Floating, Alongside, Wood
Moorings: 7 Total, 7 Transient **Launch:** n/a, Dinghy Dock
  Rate: Day $10 **Week** $70 **Month** n/a
Heads: 2 Toilet(s), 2 Shower(s)
Laundry: None **Pay Phones:** No
Pump-Out: No **Fee:** n/a **Closed Heads:** Yes

Marina Operations
Owner/Manager: WA State Dept. of Parks **Dockmaster:** Mike Zimmerman
In-Season: MidMar - MidSep, 6:30am-10pm **Off-Season:** Closed
After-Hours Arrival: n/a
Reservations: No **Credit Cards:** No
Discounts: None
Pets: Welcome **Handicap Access:** No

Marina Services and Boat Supplies
Supplies - OnSite: Ice *(Cube)*, Bait/Tackle, Propane **Near:** Ice *(Block)*
3+ mi: Boater's World *(25 mi.)*

Boatyard Services
OnSite: Launching Ramp **OnCall:** Divers **1-3 mi:** Travelift
(60T/70T/300T), Forklift, Crane, Engine mechanic *(gas, diesel)*, Electrical
Repairs, Electronic Sales, Electronics Repairs, Hull Repairs, Rigger, Sail
Loft, Canvas Work, Bottom Cleaning, Brightwork, Air Conditioning,
Refrigeration, Propeller Repairs, Woodworking, Upholstery, Yacht Interiors,
Metal Fabrication, Painting, Awlgrip, Yacht Design, Yacht Building, Total
Refits. **Nearest Yard:** Port Townsend Boat Haven 3 mi. (360) 385-2355

Restaurants and Accommodations
OnSite: Snack Bar *(snacks & coffee)*, Condo/Cottage *(Fort Flagler Vacation
Houses 800-360-4240, $96-468, or 385-3701)* **3+ mi:** Restaurant *(Hadlock
House 385-3331, 11 mi.)*, *(Fiesta Jalisco 385-5285, 12 mi.)*, Lite Fare
(Village Baker 379-5310, 12 mi.), Pizzeria *(Ferino's 385-0840, 11 mi.)*,
Inn/B&B *(Inn at Port Hadlock 385-7030, 11 mi.)*, Condo/Cottage *(Oak Bay
Cottages 437-0380, 9 mi.)*, *(Honey Moon Cabin 385-4644, 5 mi.)*

Recreation and Entertainment
OnSite: Beach, Picnic Area, Grills *(fire pits)*, Jogging Paths, Museum *(Fort
Flagler Museum 11am-4pm - displays cover the Glory Days 1899-1919,
WWI, WWII, and the Korean War; the gun batteries that guarded the
entrance to Admiralty Inlet were never fired at a hostile target)*, Sightseeing
(hiking trails throughout the park), Special Events *(Strawberry Fest - Jun/
Norland Polar Dip - Jan 1st, Tractor Days - Mem Day, 5 mi.)* **1-3 mi:** Video
Rental **3+ mi:** Boat Rentals *(kayaks at Norland General Store 385-0777, 5
mi.)*, Galleries *(Little Island Gallery, 5 mi.)*

Provisioning and General Services
OnSite: Convenience Store *(snacks & espresso)* **1-3 mi:** Post Office *(800-
275-8777)* **3+ mi:** Market *(Nordland General Store 385-0777/Quality Food
Center 385-1070, 5 mi./11 mi.)*, Liquor Store *(WA Liquor 385-2316, 11 mi.)*,
Farmers' Market *(19 mi.)*, Fishmonger *(Carl Johnson's Clam & Oyster Co.
385-1508 or Snow Bay Clams & Oysters, 5 mi.)*, Bank/ATM *(Mariner Bank
344-3424, 11 mi.)*, Protestant Church *(Lutheran Church - The Redeemer
385-6977, 10 mi.)*, Library *(Jefferson County Library 385-6544, 11 mi.)*,
Beauty Salon *(Tresses 385-2044, 11 mi.)*, Pharmacy *(Port Hadlock
Pharmacy 385-1900, 11 mi.)*, Hardware Store *(True Value 385-1771,
11 mi.)*

Transportation
Airport: Hadlock/Jefferson County Int'l. *(12 mi./15 mi.)*

Medical Services
911 Service **3+ mi:** Dentist *(Dentistry Northwest 385-1000, 11 mi.)*,
Chiropractor *(Peninsula Chiropractic 385-4900, 11 mi.)*, Veterinarian
(Chimacum Valley Veterinary Hospital 385-4488, 11 mi.)
Hospital: Jefferson General 385-2200 *(19 mi.)*

Setting -- Fort Flagler State Park's pier and mooring field are on the northeast side of the entrance channel to Kilisut Harbor, which runs between
Marrowstone Island on the east and Indian Island on the west. At the end of the wharf, a seasonal float provides 256 feet of alongside space supplemented by
seven mooring buoys. Onshore, the 700-acre park sits on Marrowstone Point surrounded by water on three sides, providing spectacular panoramic views; the
view to the east is Kilisut Habor and Port Townsend with the Olympic Mountains in the background.

Marina Notes -- *$10 min. No power. Note the 6 ft. depth at low tide. Stays are limited to 3 consecutive days. The docks are removed from mid-September-
mid-March. Two boat ramps, picnic tables, and a long, wide beach popular with clam diggers are also on-site. A kitchen shelter without electricity is available
across the road. Annual moorage passes to WA State Parks are $3.50/ft. with a minimum of $50. Cinderblock heads and showers are quite usable.

Notable -- An on-site convenience store has ice, snacks, and basic supplies. Camping sites with water views and former military houses are available for
overnight stays at the park. Cottages house from 4 to 52 people, and can be reserved up to a year ahead (888-226-7688). Each has a fully equipped kitchen,
microwave, coffee pot and toaster. Guests bring their own towels, bedding, and food items. Once a fort guarding the entrance to Puget Sound, the Park
maintains both turn of the century and World War II base facilities. Its military history is told at the on-site museum. About 5 miles south, the community of
Nordland has a grocery store with kayak rentals, and fishmongers for fresh shellfish. Additional services are in Port Hadlock, on the other side of Indian Island.

Navigational Information
Lat: 48°03.270' **Long:** 122°41.480' **Tide:** 8.5 ft. **Current:** n/a **Chart:** 18464
Rep. Depths (*MLW*): **Entry** 9 ft. **Fuel Dock** 6 ft. **Max Slip/Moor** 6 ft./9 ft.
Access: On the east side of Kilisut Harbor on Morrowstone Island

Marina Facilities *(In Season/Off Season)*
Fuel: No
Slips: 18 Total, 18 Transient **Max LOA:** 50 ft. **Max Beam:** n/a
  Rate *(per ft.)*: **Day** $0.50* **Week** n/a **Month** n/a
  Power: 30 amp n/a, 50 amp n/a, 100 amp n/a, 200 amp n/a
  Cable TV: No **Dockside Phone:** No
  Dock Type: Floating, Alongside, Wood
Moorings: 7 Total, 7 Transient **Launch:** n/a, Dinghy Dock
  Rate: Day $10 **Week** $70 **Month** n/a
Heads: 2 Toilet(s)
Laundry: None **Pay Phones:** No
Pump-Out: OnSite, Self Service **Fee:** n/a **Closed Heads:** Yes

Marina Operations
Owner/Manager: WA State Dept. of Parks **Dockmaster:** Mike Zimmerman
In-Season: May-Sep, 6:30am-10pm **Off-Season:** Oct-Apr, 8am-5pm
After-Hours Arrival: Pay at the pay station
Reservations: No **Credit Cards:** No
Discounts: None
Pets: Welcome **Handicap Access:** No

Marina Services and Boat Supplies
Supplies - Under 1 mi: Ice *(Cube)*

Boatyard Services
OnSite: Launching Ramp *($5)* **OnCall:** Divers **3+ mi:** Travelift
(60T/70T/300T, 7 mi.). **Nearest Yard:** Port Townsend Boat Haven (360)
385-2355

Restaurants and Accommodations
Under 1 mi: Condo/Cottage *(Honey Moon Cabin 385-4644)*
1-3 mi: Condo/Cottage *(Beach Cottages on Marrowstone 385-3077,
$60-115, studios - 2 person, $60-75; cottages - 1/2 bedrooms, $90-115)*
3+ mi: Restaurant *(Hadlock House Restaurant 385-3331, 6 mi., B, L, D)*
, *(FlagshipGrill 344-3883, D $8-22, 6 mi.)*, Pizzeria *(Ferino's Pizzeria
385-0840, 6 mi.)*, Inn/B&B *(Inn at Port Hadlock 385-7030, 6 mi.)*

Recreation and Entertainment
OnSite: Beach *(rocky and marginal)*, Picnic Area, Jogging Paths, Park
Under 1 mi: Boat Rentals *(Kayaks at Nordland Central Store 385-0777)*,
Galleries *(Little Island Gallery)* **3+ mi:** Golf Course *(Port Townsend Golf
Club 385-4547, 11 mi.)*, Movie Theater *(Rose Theatre 385-1089, 11 mi.)*,
Video Rental *(Peninsula Video 385-5670, 6 mi.)*, Museum *(Fort Flagler State
Park Museum 385-1259, 5 mi.)*

Mystery Bay State Park

10541 Flagler Road; Nordland, WA 98358

Tel: (360) 385-1259 **VHF: Monitor** n/a **Talk** n/a
Fax: n/a **Alternate Tel:** n/a
Email: n/a **Web:** www.parks.wa.gov
Nearest Town: Nordland *(1 mi.)* **Tourist Info:** (360) 379-5380

Provisioning and General Services
Under 1 mi: Market *(Nordland General Store 385-0777)*, Wine/Beer,
Fishmonger *(Carl Johnson's Clam & Oyster Co. 385-1508 or Snow Bay
Clams & Oysters)* **1-3 mi:** Post Office *(385-2825)* **3+ mi:** Liquor Store *(WA
Liquor 385-2316, 6 mi.)*, Bakery *(Village Baker 379-5310, 7 mi.)*, Bank/ATM
(Frontier Bank 379-8255, 6 mi.), Protestant Church *(Lutheran, Evangelical,
7 mi.)*, Library *(Jefferson County Library 385-6544, 7 mi.)*, Beauty Salon
(Shear Attractions 385-3953, 6 mi.), Barber Shop *(Village Barber 385-6865,
6 mi.)*, Dry Cleaners *(Port Hadlock Wash and Dry Clean 385-6660, 6 mi.)*,
Laundry *(6 mi.)*, Pharmacy *(Port Hadlock Pharmacy 385-1900, 6 mi.)*,
Hardware Store *(True Value 385-1771, 6 mi.)*

Transportation
Airport: Jefferson County 385-2323 *(6 mi.)*

Medical Services
911 Service **OnCall:** Ambulance **3+ mi:** Dentist *(Dentistry Northwest 385-
1000, 6 mi.)*, Chiropractor *(Peninsula Chiropractic 385-4900, 6 mi.)*,
Veterinarian *(Oak Bay Animal Hospital 385-7297, 6 mi.)* **Hospital:** Jefferson
General 385-2200 *(10 mi.)*

Setting -- Mystery Bay is a small, shallow cove midway along the east side of Kilisut Harbor. This quiet little 10-acre park has 685 feet shoreline on Marrowstone Island near Norland. A single angled 683 foot float, populated with picnic tables, offers alongside tie-ups and dinghy dockage. The ramp leads to a pier; ashore are more picnic tables and a pavilion. Westward views of the Olympic Mountains over Indian Island are fantastic, especially at sunset.

Marina Notes -- *$10 minimum. No dockside power. Water has a high sodium level. Annual permits may be purchased at the State Parks headquarters in Olympia, regional offices, online and in person at the parks for $3.50/ft. with a $50 minimum. Daily permits are only available at the parks. Call 753-5771 for fee update. Stays are limited to 3 consecutive nights. The park also offers drinking water, a covered kitchen shelter, and a single-lane boat launch. Picnic tables, and a fire pit make for a leisurely barbequed meal. This is a day-use only park. Two small tan and brown structures house the vault toilets. Note: Use caution to avoid two concrete blocks located 20 to 30 feet off the east end of the State Park pier.

Notable -- Mystery Bay earned its name during Prohibition when boats smuggling liquor from Canada would "mysteriously" disappear along the area's wooded shores. The park attracts boaters as well as "car" visitors who forage for crabs, clams, and oysters at low tide. Two local businesses also retail Kilisut shellfish. Rent a kayak at the CentralGeneral Store and explore Mystery Bay. Fort Flagler State Park is about 5 miles of relatively flat land to the north. Large gun batteries and a self-guided military museum (385-1259) tell the story of the Fort, one of several built to guard the entrance to Puget Sound.

Navigational Information
Lat: 48°02.333' **Long:** 122°24.276' **Tide:** 10 ft. **Current:** n/a **Chart:** 18441
Rep. Depths *(MLW):* **Entry** 11 ft. **Fuel Dock** n/a **Max Slip/Moor** 30 ft./-
Access: Located off Saratoga Passage

Marina Facilities *(In Season/Off Season)*
Fuel: No
Slips: 32 Total, 20 Transient **Max LOA:** 75 ft. **Max Beam:** n/a
 Rate *(per ft.):* **Day** $0.60* **Week** n/a **Month** n/a
 Power: 30 amp $3.50** **50 amp** n/a, **100 amp** n/a, **200 amp** n/a
 Cable TV: No **Dockside Phone:** No
 Dock Type: Floating, Short Fingers, Alongside, Concrete
Moorings: 0 Total, 0 Transient **Launch:** n/a
 Rate: Day n/a **Week** n/a **Month** n/a
Heads: 4 Toilet(s), 2 Shower(s), Hair Dryers, Heated floors
Laundry: None **Pay Phones:** Yes
Pump-Out: OnSite, Self Service, 1 Central **Fee:** n/a **Closed Heads:** Yes

Marina Operations
Owner/Manager: City of Langley **Dockmaster:** n/a
In-Season: Year-Round, 24/7 **Off-Season:** n/a
After-Hours Arrival: n/a
Reservations: No **Credit Cards:** No
Discounts: None
Pets: Welcome, Dog Walk Area **Handicap Access:** No

Langley Boat Harbor

PO Box 366; 228 Wharf Street; Langley, WA 98260

Tel: (360) 914-1739 **VHF: Monitor** n/a **Talk** n/a
Fax: (360) 221-4265 **Alternate Tel:** n/a
Email: n/a **Web:** www.langleywa.org
Nearest Town: Langley *(0.2 mi.)* **Tourist Info:** (360) 221-6765

Marina Services and Boat Supplies
Services - Docking Assistance, Dock Carts **Supplies - Near:** Ice *(Block, Cube)*

Boatyard Services
Nearest Yard: Everett Shipyard (425) 259-0137

Restaurants and Accommodations
OnSite: Inn/B&B *(Boatyard Inn 221-5120, $125-210)*, *(Drake's Landing 221-3999)* **Near:** Restaurant *(Mike's Place Family Restaurant 221-6575, B $6-8, L $6-9, D $9-16)*, *(Langley Café 221-3090, L $7-9, D $11-18)*, *(The Dog House Tavern 221-9825, L $6-12, D $8-16)*, *(Edge Cliff 221-8899)*, *(Langley Tea Room & Sushi 221-6292)*, Pizzeria *(Village Pizzeria 221-3363)*, Inn/B&B *(Inn at Langley 221-3033, $225)*, *(The Whidbey Inn 221-7115, $120-175)*

Recreation and Entertainment
OnSite: Beach, Picnic Area, Park *(Phil Simon Memorial Park/ Lookout Park nearby)* **Near:** Fishing Charter, Movie Theater *(The Clyde Theatre 221-5525)*, Video Rental, Museum *(So. Whidbey Historical Society 221-6747)*, Cultural Attract *(Whidbey Island Center For Arts 221-8268)*, Sightseeing, Special Events *(Antique Show - May, Choochokam Arts Festival - Jul, Island County Fair - Aug)*, Galleries *(Artists Co-Op 221-7675, Isle of Art 221-8499)* **1-3 mi:** Tennis Courts *(Maxwelton Road near HS)*

Provisioning and General Services
Near: Market *(Star Store Grocery 221-5222)*, Gourmet Shop *(Island Gourmet and Gift Shoppe 221-3626)*, Health Food *(Living Green 221-8242)*, Liquor Store *(Langley Liquor 221-4520)*, Bakery *(Village Bakery 221-3525)*, Bank/ATM, Post Office, Catholic Church, Protestant Church, Library *(Langley Library 221-4383)*, Beauty Salon *(Paul's and Company 221-4140)*, Barber Shop *(Annalee's 221-2818)*, Bookstore *(Moonraker Books 221-6962)*, Pharmacy *(Linds Langley Drug 21-4369)*, Newsstand, Florist, Clothing Store *(Roberta's 221-1070)*, Retail Shops, Copies Etc. **Under 1 mi:** Laundry *(All Washed Up Laundromat 221-5828)* **1-3 mi:** Wine/Beer *(Whidbey Island Winery 221-2040)*, Hardware Store *(Locks and Hardware 221-6909)* **3+ mi:** Farmers' Market *(Bayview Hall Sat 10am-2pm, 5 mi.)*

Transportation
OnCall: Taxi *(Bayview Cab 321-4949)* **Near:** Local Bus *(Free buses on all of Whidbey Island)* **1-3 mi:** Airport Limo *(Sea-Tac Shuttle 877-679-4003)* **3+ mi:** Rail *(Amtrak 425-258-2458, 10 mi.)* **Airport:** Sea-Tac Int'l. *(40 mi.)*

Medical Services
911 Service **OnCall:** Ambulance **Near:** Doctor *(Langley Clinic 221-5272)*, Dentist *(Langley Professional Center 221-5060)*, Chiropractor *(Village Chiropractic 221-5141)*, Holistic Services *(Healing Garden Naturopathic Clinic 221-5596, Spa Essencia 221-0991)* **3+ mi:** Veterinarian *(So. Whidbey Animal Clinic 341-1200, 3 mi.)* **Hospital:** Whidbey General 678-5151 *(25 mi.)*

Setting -- Langley Boat Harbor is located off Saratoga Passage on the western side of Whidbey Island. Protection is provided by the southern end of Camano Island and a piling breakwater that surrounds the floating concrete docks on three sides. A long zig-zag wharf leads to the shore. A newer, 100 foot dock provides alongside visitor tie-ups. Shoreside views are of the delightful village; Mt. Baker and the Cascades can be seen from the bluff above the marina.

Marina Notes -- *Flat rates: 6-10' $5/nt., 11-20' $12, 21-30' $18, 31-40' $24, 41-50' $30, 51-60' $36, over 61' $42. $5 day use fee. Pay envelopes and drop box are at the top of the ramp. **Both 20 & 30 amp power, a boat ramp, and pump-out station on a floating barge. Most transients are on the 100 ft. dock but some slips may be available. Expect to be rafted during the busy summer season. This is also a working boat marina, so expect lots of crab pots. Picnic tables at the small Phil Simon Memorial Park provide a snack area off the boat. Note: A major expansion to provide moorage for 200 boats was approved in 2004.

Notable -- Langley is a lively arts community with galleries, fine antique shops, bookstores and shops featuring jewelry, glass and other hand crafted items. Enjoy the town's charming streets, sandy beaches and wide vistas. Bakeries, coffee shops, gourmet groceries and restaurants are a short walk from the marina. The Whidbey Island Center For Arts features year-round performances, and The Clyde Theater, a 1930s movies venue, is popular throughout the island (221-5525, admission $5/3). Langley boasts the highest density of B&Bs in the state, so this may be a good place for a night ashore. Daily, except Sunday, a free shuttle bus allows the boating visitor to see Oak Harbor, Coupeville, and other island attractions.

Navigational Information
Lat: 48°02.520' **Long:** 122°10.450' **Tide:** 13 ft. **Current:** 2 kt. **Chart:** 18443
Rep. Depths *(MLW)*: **Entry** 2 ft. **Fuel Dock** n/a **Max Slip/Moor** 15 ft./-
Access: Priest Pt. to north shore Ebey Slough to Marysville

Marina Facilities *(In Season/Off Season)*
Fuel: No
Slips: 98 Total, 2 Transient **Max LOA:** 50 ft. **Max Beam:** n/a
  Rate *(per ft.)*: **Day** $0.70 **Week** n/a **Month** $4.50
  Power: 30 amp Incl., **50 amp** n/a, **100 amp** n/a, **200 amp** n/a
  Cable TV: No **Dockside Phone:** No
  Dock Type: Floating, Wood
Moorings: 0 Total, 0 Transient **Launch:** n/a, Dinghy Dock
  Rate: Day n/a **Week** n/a **Month** n/a
Heads: None
Laundry: None **Pay Phones:** Yes
Pump-Out: No **Fee:** n/a **Closed Heads:** Yes

Marina Operations
Owner/Manager: Edward W. Geddes **Dockmaster:** Same
In-Season: May-Oct, 7:30am-6pm **Off-Season:** Oct-Apr, 9am-5pm
After-Hours Arrival: Raft
Reservations: Yes, Required **Credit Cards:** No
Discounts: None
Pets: Welcome, Dog Walk Area **Handicap Access:** No

Geddes Marina
1326 1st Street; Marysville, WA 98270

Tel: (360) 659-2575 **VHF: Monitor** n/a **Talk** n/a
Fax: (360) 653-3786 **Alternate Tel:** (360) 653-3782
Email: n/a **Web:** n/a
Nearest Town: Marysville **Tourist Info:** (888) 338-0976

Marina Services and Boat Supplies
Services - Security *(owner lives onsite)*, Trash Pick-Up **Supplies - Near:** Ice *(Block, Cube)* **Under 1 mi:** Bait/Tackle, Live Bait

Boatyard Services
OnSite: Travelift *(15T)*, Railway, Bottom Cleaning, Compound, Wash & Wax, Propeller Repairs *(Les's Propeller Service 659-7155)*, Woodworking
1-3 mi: Engine mechanic *(gas, diesel)*, Electrical Repairs, Electronic Sales, Electronics Repairs, Hull Repairs, Rigger, Divers, Interior Cleaning.

Restaurants and Accommodations
Near: Restaurant *(Village 659-2305, B $3-6, L $4-9, D $8-12)*, *(Maxwell's B $3-6, L $5-9, D $10-18)*, *(Don's 659-9555, B $3-6, L $4-8, D $7-12)*, *(Las Margaritas 653-4362)*, *(Pearl Garden 659-3992)*, Fast Food *(Jack in the Box 653-8344)*, Motel *(Village Motor Inn 659-0005, $49-130)* **Under 1 mi:** Fast Food *(Subway, McDonald's, Burger King)*, Motel *(BW Tulalip Inn 659-4488, $69)*, Hotel *(Holiday Inn Express 658-1339, $84)*

Recreation and Entertainment
Near: Park *(Comeford Park)* **Under 1 mi:** Tennis Courts, Fitness Center *(Gold's Gym 658-4653)*, Jogging Paths, Video Rental *(Hollywood Video 651-2585)*, Video Arcade, Special Events *(Marysville Strawberry Fest - Jun)*
1-3 mi: Picnic Area, Playground, Dive Shop *(Deep See 435-6696)*, Golf Course *(Bettle Creek 659-7931)*, Roller Blade/Bike Paths, Bowling *(Strawberry Lanes 659-7641)*, Group Fishing Boat *(Fishing Xtreme 651-0374)*, Movie Theater *(Regal Cinemas 659-1009)*

Provisioning and General Services
Near: Market *(4th Street Market & Deli 653-4444)*, Supermarket *(Albertson's 659-5841)*, Delicatessen *(First Stop Deli 659-1044)*, Wine/Beer, Bakery, Farmers' Market *(Sat 10am-3pm at Comeford Park)*, Bank/ATM, Catholic Church, Protestant Church, Beauty Salon *(Boston Hair Studio 658-7351)*, Barber Shop *(Joe's 653-8630)*, Dry Cleaners *(Elite Cleaners 659-3700)*, Bookstore *(Arlington Book Store 435-2742)*, Pharmacy *(Rite Aid 659-0492)*, Hardware Store *(Carr Hardware 659-2292)*, Clothing Store, Retail Shops *(Marysville Mall)*, Copies Etc. **Under 1 mi:** Convenience Store *(7-Eleven 659-2215)*, Health Food *(Sno-Isle Natural Foods 258-3798)*, Liquor Store *(Tulalip Liquor Store 651-3250)*, Post Office *(653-6379)*, Synagogue, Library *(Sno-Isle Regional 659-8447)*, Laundry *(PDQ Laundry Room 658-5188)*
1-3 mi: Gourmet Shop, Department Store *(Wal-Mart 657-1192)*

Transportation
OnCall: Rental Car *(Enterprise 653-2319)*, Taxi *(North County 658-7222)*
Near: Local Bus, InterCity Bus **3+ mi:** Rail *(Everett Amtrak 258-2458, 6 mi.)* **Airport:** Arlington Municipal/Sea-Tac Int'l. *(10 mi./46 mi.)*

Medical Services
911 Service **OnCall:** Ambulance **Near:** Doctor *(4th St. Medical 659-4141)*, Dentist *(McGary 659-3232)*, Chiropractor *(Keehn 659-8411)*
Hospital: Providence 425-261-3636 *(7 mi.)*

Setting -- Geddes Marina lies along a spur off Possession Sound on Ebey Slough right in the downtown area of Marysville. Getting to the marina requires some very tight maneuvering and careful timing. At low tide, there is only a two-foot entry depth. At high tide boats must clear under a highway bridge with 38 feet of vertical clearance. Very few large visiting sailboats attempt this approach. The boater will find a casual combination of open side-tie dockage and assorted boat houses

Marina Notes -- The marina dates back to the 1800s under various names. Ed Geddes and his wife have owned the marina since 2001 and are in the process of renovating it. The old wooden docks are gradually being replaced with newer, safer ones. The 98 slips are primarily rented as permanent moorage, so a phone call ahead is a must. Power and water are available, however, there are no other facilities for the boater. The nearby train track and freeway overpass make for a noisy environment. Les's Propeller Service (659-7155) is also located at the marina.

Notable -- This is a great shopping stop: Town Center Mall is directly across the street and department stores such as Penney's, Sears, and Gottschalks call it home, as does a well-stocked Albertson's supermarket. Many eateries are also convenient to the marina, including two delis, coffee shops, Mexican and Chinese restaurants, and fast food. Nearby Village Restaurant is known for its homemade pies. The Marysville Strawberry Festival is celebrated the third week of June.

Port of Everett Marina

PO Box 538; 1720 West Marine View Drive; Everett, WA 98201

Tel: (425) 259-6001; (800) 729-7678 **VHF: Monitor** Ch. 16 **Talk** Ch. 69
Fax: (425) 259-0860 **Alternate Tel:** n/a
Email: marina@portofeverett.com **Web:** www.portofeverett.com
Nearest Town: Everett **Tourist Info:** (425) 438-1487

Navigational Information
Lat: 47°59.795' **Long:** 122°13.520' **Tide:** 11 ft. **Current:** n/a **Chart:** 18443
Rep. Depths *(MLW)*: **Entry** 12 ft. **Fuel Dock** 12 ft. **Max Slip/Moor** 12 ft./-
Access: Off Port Gardner Channel at mouth of Snohomish River

Marina Facilities *(In Season/Off Season)*
Fuel: Gasoline, Diesel
Slips: 2050 Total, 50* Transient **Max LOA:** 100 ft. **Max Beam:** n/a
  Rate *(per ft.)*: **Day** $0.75/0.55 **Week** n/a **Month** $7.12/6.39
  Power: 30 amp Incl.**, 50 amp n/a, 100 amp n/a, 200 amp n/a
  Cable TV: No **Dockside Phone:** No
  Dock Type: Floating, Long Fingers, Alongside, Concrete
Moorings: 0 Total, 0 Transient **Launch:** n/a
  Rate: Day n/a **Week** n/a **Month** n/a
Heads: 6 Toilet(s), 2 Shower(s) *(with dressing rooms)*
Laundry: 6 Washer(s), 6 Dryer(s) **Pay Phones:** Yes
Pump-Out: OnSite, 3 Central **Fee:** Free **Closed Heads:** Yes

Marina Operations
Owner/Manager: Kim S. Buike **Dockmaster:** Jeff Lozeau
In-Season: Year-Round, 8am-5pm **Off-Season:** n/a
After-Hours Arrival: Call (425) 259-6001
Reservations: Yes **Credit Cards:** Visa/MC, Checks preferred
Discounts: Fuel **Dockage:** n/a **Fuel:** $.05 w/ 200gal **Repair:** n/a
Pets: Welcome, Dog Walk Area **Handicap Access:** Yes, Heads, Docks

Marina Services and Boat Supplies
Services - Docking Assistance, Security *(24 hrs.)*, Dock Carts
Communication - Data Ports *(Wi-Fi - Broadband Xpress)*, FedEx, UPS
Supplies - OnSite: Ice *(Block, Cube)*, Ships' Store *(Harbor Marine 259-3285)*, Bait/Tackle, Propane, CNG **Near:** West Marine *(303-1880)*
3+ mi: Boater's World *(646-9350, 6 mi.)*

Boatyard Services
OnSite: Travelift *(30/35T)*, Railway, Forklift, Engine mechanic *(gas, diesel)*, Hull Repairs, Rigger, Canvas Work, Bottom Cleaning, Woodworking, Inflatable Repairs, Life Raft Service, Upholstery **Near:** Launching Ramp *(10th St. Marine Park - 13 lanes)*, Yacht Broker *(Everett Yacht Sales 258-4655, Fairwinds 258-5318)*. **1-3 mi:** Propeller Repairs.

Restaurants and Accommodations
OnSite: Restaurant *(Anthony's Homeport 252-3333, D $15-25, Sun brunch $8-15)*, *(Anthony's Woodfire Grill 258-4000, L $8-14, D $8-15)*, *(Daruma Japanese 3-392-2307, L $5-11, D $9-14)*, *(Lomardi's Cucina 252-1886)*, Lite Fare *(Scuttlebutt Brewing 257-9414)*, Inn/B&B *(Inn at Port Gardner 252-6779, $79-249)* **Near:** Restaurant *(PK's Boathouse B $4-5, L $4-7, D $4-7)*
Under 1 mi: Restaurant *(Thai On Broadway 259-6406)*, Lite Fare *(Slim N Healthy 259-3544)* **1-3 mi:** Motel *(Travelodge 259-6141)*

Recreation and Entertainment
OnSite: Jogging Paths *(mile-long Marina Esplanade)*, Fishing Charter *(All Star 252-4188)*, Group Fishing Boat *(Mosquito Fleet 800-325-6722)*
Near: Park *(Grand Ave.)* **Under 1 mi:** Fitness Center *(Gold's 258-3862)*, Bowling *(Tyee Lanes 259-6161)*, Video Rental *(Blockbuster 259-3850)*

1-3 mi: Picnic Area, Grills, Playground, Tennis Courts, Golf Course *(Legion Mem.l 259-4653)*, Museum *(Snohomish County 259-2022, Children's 258-1006)*, Galleries *(Snohomish County 257-8380)* **3+ mi:** Dive Shop *(Underwater Sport 355-3338, 5 mi.)*, Movie Theater *(Everett 9 258-6766, 6 mi.)*, Sightseeing *(Museum of Flight Restoration 745-5150, Boeing Tour Center 800-464-146 - $10 taxi, 9 mi.)*

Provisioning and General Services
OnSite: Convenience Store, Delicatessen, Wine/Beer, Farmers' Market *(Jun-Sep Sun 11am-4pm, 921-3392)*, Beauty Salon *(Cutter's Cove 252-3737)*, Newsstand **Near:** Health Food *(Sno-Isle 259-3798)*
Under 1 mi: Market *(Hewitt 259-6092)*, Supermarket *(Safeway 252-9898)*, Liquor Store *(Everett 339-1997)*, Bakery *(La Gloria 304-0848)*, Fishmonger, Bank/ATM, Catholic Church, Protestant Church, Library *(388-9501)*, Barber Shop *(J&W 258-2812)*, Dry Cleaners *(Wetmore 252-7084)*, Laundry *(Bubbles 303-9712)*, Pharmacy *(Bartell 303-2583, Walgreens 252-5213)*, Hardware Store *(Tool Town 259-2590, True Value 259-3134, 2 mi.)* **3+ mi:** Retail Shops *(Everett Mall 355-1771, 6 mi.)*, Buying Club *(Costco 379-7487, 6 mi.)*

Transportation
OnCall: Taxi *(A-1 Taxi 435-5500)* **Near:** Ferry Service *(pristine Jetty Island)* **Under 1 mi:** Local Bus, InterCity Bus **1-3 mi:** Rail *(Amtrak 258-2458)* **Airport:** Sea-Tac Int'l. *(38 mi.)*

Medical Services
911 Service **OnSite:** Doctor *(339-5453)* **OnCall:** Ambulance
Under 1 mi: Dentist *(North Everett Dental 258-1764)*, Chiropractor, Holistic Services **1-3 mi:** Optician **Hospital:** Providence 261-3636 *(0.5 mi.)*

Setting -- Located in Port Gardner off Possession Sound - just north of the US Naval Station - this city-owned marina lies behind Jetty Island. Over 2000 slips and 1,800 feet of transient moorage make it the largest in the Pacific Northwest. A breakwater homes the fuel dock, pump-out and guest docks. Umbrella-topped picnic tables dot the floats and on shore the gray 3-story marina office sports a control tower. Private homes perch on a hill above the marina.

Marina Notes -- *1,800 ft. of linear moorage. **20 or 30 amp. 10-year $300 million North Marina redevelopment project - $30 million restructuring of NM dockage completed in 2006 - including 700 ft. of guest moorage on "A" dock & 40-50 ft. slips on "K". Guest moorage, including a dedicated ADA guest dock, are the first on either side of the entrance. The south dock is closer to restaurants and hotels and the north dock to marine services - but it's an easy dinghy ride to food and lodging establishments. Pay stations at the top of North dock and before the runway on South dock or in the office. Port of Everett Boat yard delivers the on-site yard services. Voted "Best Big Marina" by Sea Magazine. New 12th Street Yacht Basin opening in 2007. Both North & South Marinas have pleasant tiled bathhouses and laundries. Note: Watch for occasional debris from a nearby lumber mill.

Notable -- The Marina Village at the south dock is home to several eateries, an inn, and a seasonal farmers' market. More services are to the east in the mall called Landing at Port Gardner. Special events include Waterfront Evening Concerts on Thu from Jul-Aug, Jetty Island Days Wed-Sun Jun-Sep, and Fresh Paint Artists Fest in mid Aug (257-8300). Museums are downtown - about 1.5 miles south, and the Boeing Tour Center is about 9 miles away.

Navigational Information
Lat: 47°55.300' **Long:** 122°41.100' **Tide:** 11 ft. **Current:** n/a **Chart:** 18441
Rep. Depths *(MLW):* **Entry** 20 ft. **Fuel Dock** 19 ft. **Max Slip/Moor** 30 ft./-
Access: Admiralty Inlet to Ludlow Bay

Marina Facilities *(In Season/Off Season)*
Fuel: Gasoline, Diesel
Slips: 300 Total, 50 Transient **Max LOA:** 250 ft. **Max Beam:** n/a
  Rate *(per ft.):* **Day** $0.90/0.85 **Week** $4 **Month** $10
  Power: 30 amp $3, **50 amp** $6, **100 amp** n/a, **200 amp** n/a
  Cable TV: No **Dockside Phone:** No
  Dock Type: Floating, Long Fingers, Concrete
Moorings: 0 Total, 0 Transient **Launch:** n/a, Dinghy Dock
  Rate: Day n/a **Week** n/a **Month** n/a
Heads: 8 Toilet(s), 4 Shower(s)
Laundry: 2 Washer(s), 1 Dryer(s), Book Exchange **Pay Phones:** Yes, 1
Pump-Out: OnSite, 1 Central, 1 Port **Fee:** n/a **Closed Heads:** Yes

Marina Operations
Owner/Manager: Port Ludlow Associates **Dockmaster:** Kori Ward
In-Season: May-Oct, 8am-7pm* **Off-Season:** Oct-Apr, 8am-5pm
After-Hours Arrival: Choose an open slip on 'A' dock
Reservations: Yes, Preferred **Credit Cards:** Visa/MC, Amex
Discounts: None
Pets: Welcome, Dog Walk Area **Handicap Access:** No

Port Ludlow Marina

1 Gull Drive; Port Ludlow, WA 98365

Tel: (360) 437-0513; (800) 308-7991 **VHF: Monitor** Ch. 16 **Talk** Ch. 68
Fax: (360) 437-2428 **Alternate Tel:** n/a
Email: kward@ludlowbaymarina.com **Web:** www.ludlowbaymarina.com
Nearest Town: Port Ludlow *(0.5 mi.)* **Tourist Info:** (360) 437-9798

Marina Services and Boat Supplies
Services - Docking Assistance, Trash Pick-Up, Dock Carts, Megayacht Facilities **Communication -** Mail & Package Hold, Phone Messages, Fax in/out *(Free)*, FedEx, DHL, UPS **Supplies - OnSite:** Ice *(Block, Cube)*, Ships' Store, Propane, CNG **1-3 mi:** Bait/Tackle *(Jerry's Bait 437-2188)* **3+ mi:** West Marine *(379-1612, 13 mi.)*

Boatyard Services
OnCall: Divers **1-3 mi:** Launching Ramp. **3+ mi:** Travelift *(60T/70T/300T, 17 mi.).* **Nearest Yard:** Port Townsend Boat Haven (360) 385-2355

Restaurants and Accommodations
OnSite: Restaurant *(Harbormaster Restaurant 437-2222, L $8-14, D $8-24, 11am-9pm; kids eat free; prime rib special on wknds, Sat nights live jazz, Mon fish & chips)*, *(Fireside Lounge 437-2222, B $7-14, D $11-36, 7am-10/11am & 5pm on - dress is casual)*, Inn/B&B *(Inn at Port Ludlow 437-7000, $150-250)*, Condo/Cottage *(Pintail House & Guestroom 437-7000, $150-575, cottage or studio)* **Under 1 mi:** Restaurant *(Cucina Pazza 437-8200)*, *(Snug Harbor Café 437-8072)*, Pizzeria *(Cucina Pizza 437-8200)*

Recreation and Entertainment
OnSite: Beach *(half mile rocky)*, Picnic Area *(Popham shelter near store or at Burner Point marked by a totelm)*, Grills, Boat Rentals *(skiffs $20/hr., kayaks $10-15/hr.)*, Fishing Charter *(Captain Bry's 821-9056)*, Video Rental **Near:** Playground, Tennis Courts, Jogging Paths *(hiking trails to Ludlow Falls)*, Roller Blade/Bike Paths

Under 1 mi: Special Events *(Port Ludlow Fun Fest - Sep)* **1-3 mi:** Golf Course *(Port Ludlow Golf Club 437-0272 - 27 holes - 18/9 - $26/15 to $52/31; certified Audubon Cooperative Sanctuary)*, Cultural Attract *(Port Ludlow Arts Council 437-2208)*, Galleries *(Eagle Tree 437-2359)* **3+ mi:** Sightseeing *(Port Gamble - mill town, 10 mi.)*

Provisioning and General Services
OnSite: Convenience Store *(Port Ludlow Marina Store)*, Wine/Beer
Under 1 mi: Market *(Port Ludlow Village Market 437-9110)*, Farmers' Market *(Fridays Jul-Aug at Oak Bay & Paradise Bay Rd.)*, Bank/ATM, Beauty Salon *(Village Salon 437-9228)*, Barber Shop, Dry Cleaners, Pharmacy
1-3 mi: Delicatessen, Post Office, Catholic Church, Protestant Church
3+ mi: Supermarket *(QFC 385-1070, 11 mi.)*, Liquor Store *(WA Liquor Store 385-2316, 12 mi.)*, Hardware Store *(Hadock Building Supply 385-1771, 8 mi.)*, Retail Shops *(8 mi.)*

Transportation
OnSite: Bikes *($3-6/hr.)*, Local Bus, InterCity Bus *(Jefferson County Transit -Poulsbo to Port Townsend on Oak Bay Rd.)* **Near:** Ferry Service **3+ mi:** Rental Car *(Enterprise 779-6112, 20 mi.)* **Airport:** Jefferson County/SeaTac *(14 mi./86 mi.)*

Medical Services
911 Service **OnSite:** Holistic Services *(Resort At Port Ludlow Spa)*
OnCall: Ambulance **Under 1 mi:** Doctor, Dentist *(Port Ludlow Dentistry 437-9392)*, Chiropractor *(Brady 437-8008)* **3+ mi:** Veterinarian *(Oak Bay Animal 385-7297, 12 mi.)* **Hospital:** Jefferson General 385-2200 *(17 mi.)*

Setting -- Nestled below the Olympic Mountains, this beautiful resort and marina sit on peaceful, well protected Ludlow Bay off Admiralty Inlet at the mouth of Hood Canal. Upscale homes are tucked into the terraced hillside above a low-slung gray dock house that berths the office, store and amenities. On shore, overlooking the expanse of well-done moorage, is a pretty, covered picnic area, protected by glass windbreaks and edged with flowers. Adjacent to the docks, the Resort at Port Ludlow's sprawling Inn - a three-story, steel-roofed, west-coast contemporary - houses the eateries and spa.

Marina Notes -- *Fri & Sat May-Sep 8am-9pm. This is a popular destination marina so make reservations early. The on-site store offers beer and wine, groceries, souvenirs, gifts, marine supplies, and sundries. The fuel dock provides unleaded gasoline, #2 diesel, and CNG. If there is no attendant at the fuel dock, dial 322 on the phone at the fuel dock building. The pump-out station, porta-potty dump, and fuel and oil spill containment equipment are also located on the fuel dock. The inviting bathhouse and laundry are next to the marina store. A major remodel is scheduled for the entire waterfront area in the near future.

Notable -- Power skiffs, kayaks, and bicycles, well-suited to exploring the area, are rented on-site. The village is less than a mile away. An easy, interpretive hiking trail system wanders around the bay and includes a waterfall loop, Two restaurants, a full-service spa, 37 rooms and a half-mile of rocky beach are at the Resort. Packages that include room, spa services, meals, and golf are available - the championship Country Club golf course, a Certified Audubon Cooperative Sanctuary, is comprised of three nine-hole courses. The Port Ludlow Arts Council sponsors several shows and music events each year (437-2208).

Port of Edmonds

336 Admiral Way; Edmonds, WA 98020

Tel: (425) 774-0549 **VHF: Monitor** Ch. 69 **Talk** n/a
Fax: (425) 774-7837 **Alternate Tel:** (425) 508-7490
Email: info@portofedmonds.org **Web:** www.portofedmonds.org
Nearest Town: Edmonds *(0.5 mi.)* **Tourist Info:** (425) 670-3973

Navigational Information
Lat: 47°48.590' **Long:** 122°23.310' **Tide:** 17 ft. **Current:** 1 kt. **Chart:** 18473
Rep. Depths (*MLW***):** Entry 17 ft. **Fuel Dock** 14 ft. **Max Slip/Moor** 13 ft./-
Access: Adjacent to Puget Sound; waterfront at City of Edmonds

Marina Facilities *(In Season/Off Season)*
Fuel: Gasoline, Diesel
Slips: 720 Total, 50 Transient **Max LOA:** 74 ft. **Max Beam:** n/a
  Rate *(per ft.):* **Day** $0.70 **Week** n/a **Month** $14
  Power: 30 amp $3, **50 amp** n/a, **100 amp** n/a, **200 amp** n/a
  Cable TV: No **Dockside Phone:** No
  Dock Type: Floating, Long Fingers, Concrete
Moorings: 0 Total, 0 Transient **Launch:** n/a, Dinghy Dock
  Rate: Day n/a **Week** n/a **Month** n/a
Heads: 4 Toilet(s), 2 Shower(s) *(with dressing rooms)*
Laundry: 2 Washer(s), 2 Dryer(s) **Pay Phones:** Yes
Pump-Out: OnSite, 2 Central **Fee:** Free **Closed Heads:** Yes

Marina Operations
Owner/Manager: Chris Keuss **Dockmaster:** Marla Kempf
In-Season: Jun-Oct, 6am-7pm **Off-Season:** Nov-May, 7am-5pm
After-Hours Arrival: Guest moorage box on fuel dock
Reservations: No **Credit Cards:** No; Cash only
Discounts: None
Pets: Welcome **Handicap Access:** Yes, Heads

Marina Services and Boat Supplies
Services - Security *(24 hrs.)*, Dock Carts **Communication -** Data Ports *(Wi-Fi - Broadband Xpress)* **Supplies - OnSite:** Ice *(Block, Cube)*, Bait/Tackle *(Bud's: herring, crab)* **Under 1 mi:** West Marine *(670-2009)*, Boat/US *(670-2009)*

Boatyard Services
OnSite: Travelift *(50T Public)*, Launching Ramp *($22/pp)*, Yacht Broker *(Edmonds Yachts)* **OnCall:** Engine mechanic *(gas, diesel)*, Electrical Repairs, Electronic Sales, Electronics Repairs, Hull Repairs, Rigger, Sail Loft, Canvas Work, Brightwork, Divers, Compound, Wash & Wax *(Interior & Bottom Cleaning)*, Propeller Repairs, Woodworking, Inflatable Repairs, Upholstery, Metal Fabrication, Painting **Yard Rates:** $20-40/day, Haul & Launch $170+ **Storage:** On-Land $170-340/mo.

Restaurants and Accommodations
OnSite: Restaurant *(Anthony's Home Port 771-4400, D $15-25, Sun brunch $8-15)*, *(Anthony' Beach Café 771-4400, L $8-14, D $8-15)*, *(Arnie's Seafood 771-5688, L $8-15, D $14-30, kid's menu, Sun brunch $12-16)*, Seafood Shack *(Waterfront Café B $5-7, L $5-8)* **Near:** Restaurant *(Skippers 670-3393, L $5-9, D $5-9)*, *(Thai Park 771-3902)*, *(Rory's 776-9853)*, *(Las Brisas 672-5050)*, Inn/B&B *(Harbor Inn 771-5021)* **Under 1 mi:** Inn/B&B *(Edmonds B&B 778-1134, $80-120)* **1-3 mi:** Motel *(Travelodge 771-8008, $46-54)*

Recreation and Entertainment
OnSite: Fishing Charter, Special Events *(Waterfront Fest, Taste of Edmonds)* **Near:** Picnic Area, Grills, Playground, Dive Shop *(Underwater Sports 771-6322)*, Fitness Center *(Harbor Sq. 778-3546)*, Park *(Marina Beach Park, Olympic Beach)* **Under 1 mi:** Bowling *(Robin Hood 776-2101)*, Movie Theater *(Edmonds 778-4554)*, Museum *(Edmonds 774-0900)*, Galleries *(Gallery North 774-0946)* **1-3 mi:** Golf Course *(672-4653)*, Video Rental *(Blockbuster 776-8200)* **3+ mi:** Tennis Courts *(Central Park 822-2206, 13 mi.)*, Other *(Nile Country Club 766-5154, 3.2 mi.)*

Provisioning and General Services
OnSite: Newsstand **Near:** Laundry *(Maytag 774-1715)*
Under 1 mi: Market *(Petosa's 774-5244)*, Gourmet Shop *(Olives 771-5757)*, Delicatessen *(Engel's 275-0581)*, Wine/Beer *(Arista 771-7009)*, Bakery *(Edmonds 778-6811)*, Farmers' Market *(Sat 9am-3pm at Edmonds Museum)*, Bank/ATM, Post Office *(774-4077)*, Library *(Edmonds 771-1933)*, Beauty Salon *(Fayes Shear Design 771-2140)*, Bookstore *(Odyssey 672-9064)*, Pharmacy *(Cross Border 778-7852)* **1-3 mi:** Supermarket *(Albertson's 297-1800)*, Liquor Store *(WA 778-4001)*, Protestant Church, Retail Shops *(Bartells)* **3+ mi:** Hardware Store *(Home Depot 206-361-9600, 3.7 mi.)*, Department Store *(Alderwood Mall, 5 mi.)*, Buying Club *(Costco 206-546-0859, 3 mi.)*

Transportation
OnSite: Courtesy Car/Van *(Van 9am-5pm)* **OnCall:** Taxi **Near:** Local Bus *(353-RIDE)*, Rail *(Amtrak 778-3213)* **Under 1 mi:** Ferry Service *(Edmonds/Kingston)* **Airport:** Sea-Tac Int'l. *(33 mi.)*

Medical Services
911 Service **OnCall:** Ambulance **Near:** Dentist *(Harbor Square 778-7477)*, Chiropractor *(Downtown 712-9277)* **1-3 mi:** Doctor *(Walk-In Clinic 921-9474)*, Holistic Services *(Evergreen Homeopathic 206-542-5595)*, Veterinarian *(Westgate 774-8801)* **Hospital:** Stevens 640-4040 *(3 mi.)*

Setting -- Eight miles north of Lake Washington ship channel and seven miles south of Whidbey Island, Port of Edmonds faces Puget Sound, with the town of Edmonds to its back. A 2,400 foot steel and rock pile breakwater protects the concrete floats that host more than 700 moorage slips. Across from the entrance, next to the fuel dock, 50 guest slips lie on docks "I", "K", and "L." Onshore, the striking two-story gray, brown and glass contemporary houses the port offices and Anthony's restaurants. Natural and white railings edge the basin and planters add a colorful touch to the tables and benches that line the promenade.

Marina Notes -- Max. 7-day stay during the summer. This professional, full-service, shipshape public marina is managed by a helpful staff. Guests receive a "Destination Port of Edmonds" package. A large travelift makes repairs possible with ample on-the-hard area; a roster of technicians are on-call and do-it-yourselfers welcome. Covered slips and dry-stack storage are also options. Some groceries, bait & tackle, ice, and a soda machine are available on-site. Bud's Bait House, right on the dock, sells live crabs. A convenient recycling center is close by. Bathhouses are tile and well-cared for (showers $0.25/3 minutes).

Notable -- On-site are several eateries, a long fishing pier, and a public Weather Center. Marina Beach and Olympic Beach Parks border the marina. Great transportation makes this a convenient stop. The marina's courtesy van takes guests downtown. Amtrak's station is just steps away, and a bus stops nearby - connecting to a shopping mall, movie theaters, and downtown Seattle. Special events include an Arts Festival at the Frances Anderson Center on Father's Day weekend, Edmonds Waterfront Festival the weekend after Memorial Day (771-1744), and Taste of Edmonds in August (672-9112) .

Navigational Information
Lat: 47°47.711' **Long:** 122°29.810' **Tide:** 10 ft. **Current:** n/a **Chart:** 18446
Rep. Depths (*MLW*): **Entry** 20 ft. **Fuel Dock** 18 ft. **Max Slip/Moor** 20 ft./-
Access: Located about 1.5 miles south of Apple Cove Point

Marina Facilities *(In Season/Off Season)*
Fuel: *Chevron* - Gasoline, Diesel
Slips: 275 Total, 48 Transient **Max LOA:** 86 ft. **Max Beam:** n/a
  Rate (*per ft.*): **Day** $0.60 **Week** n/a **Month** n/a
  Power: 30 amp $3, **50 amp** n/a, **100 amp** n/a, **200 amp** n/a
  Cable TV: No **Dockside Phone:** No
  Dock Type: Floating, Long Fingers, Pilings, Alongside, Concrete
Moorings: 0 Total, 0 Transient **Launch:** n/a
  Rate: Day n/a **Week** n/a **Month** n/a
Heads: 10 Toilet(s), 4 Shower(s)
Laundry: 2 Washer(s), 2 Dryer(s) **Pay Phones:** Yes, 1
Pump-Out: OnSite, Self Service, 1 Central **Fee:** n/a **Closed Heads:** Yes

Marina Operations
Owner/Manager: Port of Kingston **Dockmaster:** Tom Berry
In-Season: Year-Round, 8am-5pm **Off-Season:** n/a
After-Hours Arrival: Use drop box by port office
Reservations: Yes, limited* **Credit Cards:** Visa/MC, Chevron
Discounts: None
Pets: Welcome **Handicap Access:** No

Port of Kingston

PO Box 559; 25864 Washington Blvd. ; Kingston, WA 98346

Tel: (360) 297-3545 **VHF: Monitor** Ch. 65 **Talk** n/a
Fax: (360) 297-2945 **Alternate Tel:** n/a
Email: ptkingston@aol.com **Web:** www.portofkingston.org
Nearest Town: Kingston *(0.25 mi.)* **Tourist Info:** (360) 297-3813

Marina Services and Boat Supplies
Services - Dock Carts **Communication -** Data Ports *(Wi-Fi - Broadband Xpress)* **Supplies - OnSite:** Ice *(Cube)*, Propane **3+ mi:** Boater's World *(25 mi.)*, Boat/US *(23 mi.)*

Boatyard Services
OnSite: Launching Ramp *(2-lane ramp, $2 for parking)* **Nearest Yard:** Port of Edmonds (425) 774-0549

Restaurants and Accommodations
Near: Restaurant *(Drifter's Sports Bar & Grill 297-7773, L $6-10, D $8-15)*, *(Luna Bella Ristorante 297-2220, L, D)*, Seafood Shack *(Skipper's Seafood n' Chowder 778-3290)*, Lite Fare *(Main St. Ale House 297-0440)*, Motel *(Smiley's Motel 297-3622)*, Inn/B&B *(Crabtree B&B 297-5015)* **1-3 mi:** Pizzeria *(Pizza Factory 297-7777)*

Recreation and Entertainment
OnSite: Picnic Area, Grills, Park *(Mike Wallace Memorial Park)* **Near:** Spa *(Blue Wind Massage 297-4774)*, Beach, Museum *(Kitsap Classic Cycling Museum - famous vintage bicycles from the late 19thC to mid 20thC, 297-7144)* **Under 1 mi:** Fitness Center *(Kingston Fitness 297-3336)*, Jogging Paths **1-3 mi:** Video Rental *(Peninsula Video 297-4707)*, Special Events *(Old Fashioned Fourth of July)* **3+ mi:** Golf Course *(Meadows Meer Golf Course, 20 mi.)*, Cultural Attract *(The Rovin Players 297-4751, 6 mi.)*, Sightseeing *(Point No Point Casino 297-0070, 4 mi.)*

Provisioning and General Services
OnSite: Farmers' Market *(9am-2pm on Sat, May-Oct)* **Near:** Gourmet Shop *(Natural Grounds Coffee Company 297-3364)*, Post Office *(297-3346)*, Protestant Church *(Unity Church 297-5100)*, Beauty Salon *(Bayside Salon 297-3247)*, Barber Shop *(Snippers Barber Shop 297-7566)*, Hardware Store *(Do It Best Henry Hardware 297-3366)*, Florist *(Kingston Thriftway Floral Shop 297-2260)*, Clothing Store, Retail Shops **Under 1 mi:** Convenience Store *(Kingston Arco 297-1717)*, Wine/Beer, Liquor Store *(Kingston Liquor Store 297-3370)*, Bank/ATM *(Bank of America 297-3318)* **1-3 mi:** Market *(Kountry Korner Grocery & Deli 297-2919)*, Supermarket *(Albertson's 297-1800)*, Delicatessen *(Streibel's Deli Mart 297-8066)*, Library *(Kitsap Regional Library 297-3330)*, Dry Cleaners *(Fabricare Cleaners 297-3382)*, Pharmacy, Newsstand *(Albretson's 297-1811)*, Copies Etc. *(Kingston Mail and Printing 297-2173)*

Transportation
OnSite: Ferry Service *(Kingston/Edmonds)* **3+ mi:** Rail *(Edmonds Amtrak via ferry 778-3213, 7 mi.)* **Airport:** Paine Field/SeaTac Int'l. *(13 mi./40 mi.)*

Medical Services
911 Service **Near:** Doctor *(Hanson Medical 297-1997)*, Dentist *(Apple Tree Cove Dental Center 297-3392)*, Veterinarian *(Apple Tree Cove Animal Hospital 297-2898)* **Under 1 mi:** Chiropractor *(Anchor Chiropractic 297-8111)* **1-3 mi:** Holistic Services *(Four Winds Natural Medicine 297-2975)* **Hospital:** Stevens Memorial 640-4000 *(7.5 mi.)*

Setting -- On the west side of Puget Sound, Kingston lies in Apple Tree Cove, about 1.5 miles south of Apple Cove Point. The marina is just around the corner from the busy ferry terminal. A stone breakwater, its end marked by a light, extends about 240 yards southwest between the two - providing shelter for the marina's docks. A long pier parallels the breakwater with slips on the inside for transients up to 50 feet; at the end of the pier, a float offers alongside dockage for larger boats. A modest two story brick and clapboard building houses the facilties and cement picnic tables and grills overlook the harbor.

Marina Notes -- *Reservations at this very busy marina can be made 3 days in advance - pay the 1st night with a credit card. Side-tie moorage for larger craft - 86 ft. on the outside, 75 ft. on the inside. The fuel dock with gas, diesel, and pump-out station is at the shore end of the main pier, with the office and facilities building nearby. Wooden picnic shelters edge the docks. Pay in the port office or at the pay station. Snacks and ice can be purchased on-site; Kingston Cove Y.C. makes its home here - reciprocal slips may be available. Bright bathhouse is cement and tile; the laundry area has a convenient folding table.

Notable -- At the foot of the main pier, grassy Mike Wallace Memorial Park hosts special events and the Saturday Farmers' Market. In the 1840s, having mistaken the dogwoods on shore for apple trees, explorers originally named Kingston "Apple Tree." The town, population 1600, is built around the cove. The ferry, which runs form Kingston to Edmonds, adds high traffic in the summer, as the town is en route to Olympic National Park. This is also a good base for island hopping. Enjoy the waterfront ambiance, cozy cafés, and the nearby Classic Cycling Museum.

Boaters' Notes

_____
_____
_____
_____
_____
_____
_____
_____
_____
_____
_____
_____
_____
_____
_____
_____
_____
_____
_____
_____
_____
_____
_____
_____
_____
_____
_____
_____

Add Your Ratings and Reviews at www.AtlanticCruisingClub.com

A complimentary six-month Silver Membership is included with the purchase of this _Guide_. Select "Join Now," then "Silver," then follow the instructions.

The AtlanticCruisingClub website provides updated Marina Reports, Destination and Harbor Articles and much more — including an option within each online Marina Report for boaters to add their ratings and comments regarding that facility. Please log on frequently to share your experiences — and to read other boaters' comments.

On the website, boaters may rate marinas in one or more of the following categories — on a scale of 1 (basic) to 5 (world class) — and also enter additional commentary.

▸ **Facilities & Services** (Fuel, Reservations, Concierge Services and General Helpfulness)

▸ **Amenities** (Pool, Beach, Internet Access, including Wi-Fi, Picnic/Grill Area, Boaters' Lounge)

▸ **Setting** (Views, Design, Landscaping, Maintenance, Ship-Shapeness, Overall Ambiance)

▸ **Convenience** (Access — including delivery services — to Supermarkets, other Provisioning Sources, Shops, Services, Attractions, Entertainment, Sightseeing, Recreation, including Golf and Sport Fishing & Medical Services)

▸ **Restaurants/Eateries/Lodgings** (Availability of Fine Dining, Snack Bars, Lite Fare, OnCall food service, and Lodgings ashore)

▸ **Transportation** (Courtesy Car/Vans, Buses, Water Taxis, Bikes, Taxis, Rental Cars, Airports, Amtrak, Commuter Trains)

▸ **Please Add Any Additional Comments**

12. WA – Hood Canal

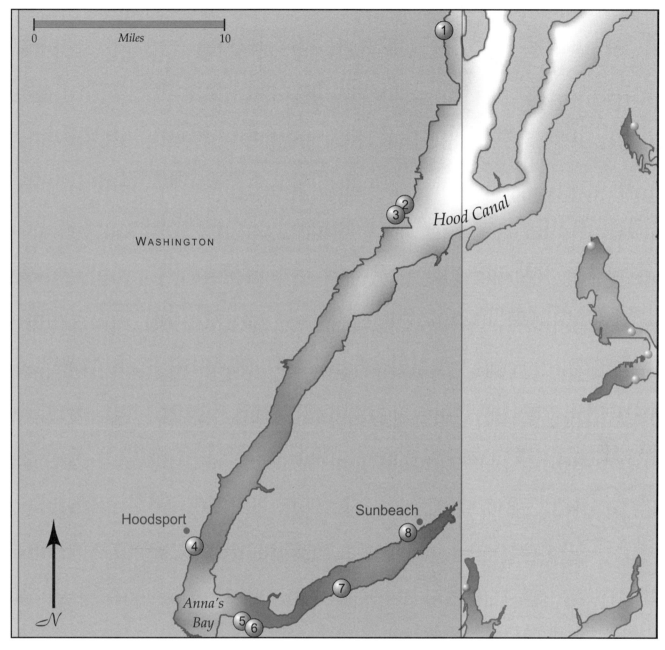

MAP	MARINA	HARBOR	PAGE	MAP	MARINA	HARBOR	PAGE
1	Quilcene Marina	Hood Canal/Quilcene Bay	176	5	Hood Canal Marina	Hood Canal	180
2	Pleasant Harbor State Park	Hood Canal/Pleasant Harbor	177	6	Alderbrook Resort and Spa	Hood Canal	181
3	Pleasant Harbor Marina	Hood Canal/Pleasant Harbor	178	7	Twanoh State Park	Hood Canal	182
4	Port of Hoodsport	Hood Canal	179	8	Port of Allyn North Shore Dock	Hood Canal	183

▶ **Currency** — In Canadian Marina Reports, all prices are in Canadian dollars. In U.S. Marina Reports, all prices are in U.S. dollars.

▶ **"CCM"** — Denotes a Certified Clean Marina, a state/provincial award for environmental excellence. See page 298 for an explanation and page 299 for a list of Pump-Out facilities.

▶ **Ratings & Reviews** — An overview of the Atlantic Cruising Club's rating system is on page 6 and details on the content of each Marina Report are on pages 7 – 11.

▶ **Marina Report Updates** — Comments from boaters and new information from ACC reviewers and marinas are posted regularly on *www.AtlanticCruisingClub.com*.

Quilcene Marina

PO Box 98; 1731 Linger Longer Road; Quilcene, WA 98376

Tel: (360) 765-3131 **VHF: Monitor** Ch. 16 **Talk** Ch. 9
Fax: n/a **Alternate Tel:** n/a
Email: n/a **Web:** www.portofpt.com
Nearest Town: Quilcene *(1.5 mi.)* **Tourist Info:** (360) 765-4999

Navigational Information
Lat: 47°48.060' **Long:** 122°51.920' **Tide:** 11 ft. **Current:** n/a **Chart:** 18476
Rep. Depths (*MLW*): Entry 4 ft. **Fuel Dock** 8 ft. **Max Slip/Moor** 12 ft./-
Access: West side of Quilcene Bay south of Port Townsend

Marina Facilities *(In Season/Off Season)*
Fuel: Gasoline, Diesel
Slips: 50 Total, 5 Transient **Max LOA:** 40 ft. **Max Beam:** n/a
  Rate *(per ft.)*: **Day** $0.60 **Week** n/a **Month** n/a
  Power: 30 amp $3, **50 amp** n/a, **100 amp** n/a, **200 amp** n/a
  Cable TV: No **Dockside Phone:** No
  Dock Type: Floating, Long Fingers, Alongside, Concrete
Moorings: 0 Total, 0 Transient **Launch:** n/a
  Rate: Day n/a **Week** n/a **Month** n/a
Heads: None
Laundry: None **Pay Phones:** No
Pump-Out: OnSite **Fee:** Free **Closed Heads:** Yes

Marina Operations
Owner/Manager: Port of Port Townsend **Dockmaster:** Jim Speer
In-Season: Year-Round, 8am-4:30pm **Off-Season:** n/a
After-Hours Arrival: Tie up and register in the morning
Reservations: Yes, Preferred **Credit Cards:** Visa/MC
Discounts: None
Pets: No **Handicap Access:** No

Marina Services and Boat Supplies
Services - Docking Assistance, Dock Carts **Supplies - OnSite:** Ice *(Cube)*

Boatyard Services
OnSite: Launching Ramp *($5/day)* **Nearest Yard:** Port of Townsend (360) 385-6211

Restaurants and Accommodations
1-3 mi: Restaurant *(Loggers Landing 765-3161)*, *(Twana Road House 765-6485)*, *(Whistling Oyster Café 765-9508)*, Coffee Shop *(Cruising 101 Café 765-3819, B, L, D)*, Motel *(Mount Walker Inn 765-3410, $50-70)*
3+ mi: Condo/Cottage *(Dabob Bay Cottage 765-3947, $100-125, 5 mi.)*

Recreation and Entertainment
OnSite: Beach, Picnic Area **Near:** Sightseeing *(Coast Seafoods Company's Shellfish Hatchery - 30-60 minute free tours to see the baby clams & oysters - require a minimum of two days advance notice 765-3345)*
Under 1 mi: Jogging Paths **1-3 mi:** Tennis Courts, Museum *(Quilcene Historical Museum 765-4848, Fri-Mon 1-5pm donations accepted)*, Special Events *(Quilcene Heritage Festival - 4th wknd Apr 765-4999, Brinnon Shrimpfest - 4th wknd May 796-4809, Olympic Music Festival - wknds Jul-Sep)*, Galleries *(Quilcene Art and Antiques 765-4447)*

3+ mi: Golf Course *(Port Ludlow Golf Club 437-0272, 18 mi.)*, Video Rental *(Linda's Gifts and Videos 796-4558, 13 mi.)*

Provisioning and General Services
Near: Oysterman *(Coast Seafoods 765-3474)* **1-3 mi:** Market *(Village Store 765-3433, Quilcene Grocery 765-0730)*, Supermarket *(Peninsula Food Store & deli 765-3389)*, Wine/Beer, Liquor Store *(WA Liquor Store 765-4550)*, Fishmonger *(Hood Canal Seafood 765-3288)*, Bank/ATM *(US Bank 765-3361)*, Post Office *(765-3760)*, Protestant Church *(First Presbyterian 765-3930)*, Beauty Salon *(Blue Bonnet Beauty Shop 765-3377)*, Pharmacy *(Coyle Drug 765-4508)*, Newsstand, Hardware Store *(Henery's Hardware 765-3113)*, Florist *(Olympic Rose Florist 765-3100)* **3+ mi:** Department Store *(13 mi.)*

Transportation
Airport: Jefferson County *(20 mi.)*

Medical Services
911 Service **OnCall:** Ambulance **1-3 mi:** Doctor *(South County Medical Clinic 765-3111)* **3+ mi:** Dentist *(Port Ludlow Dentistry 437-9392, 20 mi.)*
Hospital: Jefferson General 385-2200 *(23 mi.)*

Setting -- The western arm of Hood Canal, called Dabob Bay, splits at its northern end, with Quilcene Bay on the west side. Just inside the entrance to the Bay, on the east side, past the Coast Seafood's shellfish hatchery, a stone breakwater offers some protection to a small horseshoe of cement finger slips. Alongside space provides room for larger vessels. A swimming beach, with a float, is just on the other side of the breakwater. Landside, the marina sits at the end of aptly named Linger Longer Road; this is a quiet, peaceful place to tie up and unwind.

Marina Notes -- The marina offers fuel, pump-out and porta-potty dump, plus a launching ramp. The harbormaster is directly across the street. There are no boater amenities or heads although current plans for a new septic are anticipated for 2007. Note: The marina is open to southerly winds and parts of Dabob Bay may, at times, be restricted by the Navy - watch for red flashing lights.

Notable -- The Olympic Music Festival is held over 12 weekends from July to September. Inspiring chamber music performed by world class musicians in a century old barn outside Quilcene makes the festival a special event. Heritage Days are held in Quilcene the 4th weekend of April, and nearby Brinnon hosts its Shrimpfest the 4th weekend in May. Between the two festivals, the rhododendrons reach their peak bloom. The Hood Canal is famous for its shellfish - oysters, several kinds of clams (including geoducks), and shrimp. Take a tour of adjacent Coast Seafoods Company - the world's largest shellfish hatchery - that produces both clam and oyster larvae and seed. Alternatively, the Quilcene Ranger Station can provide information on harvesting this bounty yourself.

Navigational Information
Lat: 47°39.833' **Long:** 124°54.733' **Tide:** 12 ft. **Current:** n/a **Chart:** 18476
Rep. Depths (*MLW*): **Entry** 4 ft. **Fuel Dock** n/a **Max Slip/Moor** 8 ft./-
Access: 18 nm south of Hood Canal Floating Bridge on west shore

Marina Facilities (*In Season/Off Season*)
Fuel: No
Slips: 5 Total, 5 Transient **Max LOA:** 100 ft. **Max Beam:** n/a
 Rate (*per ft.*): **Day** $0.50* **Week** n/a **Month** n/a
 Power: 30 amp n/a, **50 amp** n/a, **100 amp** n/a, **200 amp** n/a
 Cable TV: No **Dockside Phone:** No
 Dock Type: Floating, Alongside, Wood
Moorings: 0 Total, 0 Transient **Launch:** n/a
 Rate: Day n/a **Week** n/a **Month** n/a
Heads: 1 Toilet(s)
Laundry: None **Pay Phones:** No
Pump-Out: No **Fee:** n/a **Closed Heads:** Yes

Marina Operations
Owner/Manager: Washington State Parks and Recreation **Dockmaster:** n/a
In-Season: Year-Round, 24/7 **Off-Season:** n/a
After-Hours Arrival: n/a
Reservations: No **Credit Cards:** No
Discounts: None
Pets: Welcome **Handicap Access:** No

Marina Services and Boat Supplies
Supplies - Near: Ice *(Cube)*, Ships' Store, Bait/Tackle

Boatyard Services
Nearest Yard: Quilcene Marine (360) 765-3131

Restaurants and Accommodations
Near: Snack Bar *(The Little Dinghy Deli 796-4611)*, Inn/B&B *(Houseboats for Two 796-4064, $195-240, couples only)*, Condo/Cottage *(The Harbor House 796-4064, $150 1st nt., $100 2nd nt., free 3rd nt., $500 a week)*
1-3 mi: Restaurant *(Half Way House Restaurant 796-4715, B $5-8, L $5-8, D $6-15)*, *(The Geoduck Restaurant and Lounge 796-4430, L, D)*, *(Halfway House Restaurant 796-4715)*, Motel *(Bayshore Motel 798-4220, $40-75)*, Inn/B&B *(Elk Meadows B&B 796-4886, $85)*

Recreation and Entertainment
OnSite: Picnic Area **Near:** Dive Shop *(Pacific Adventure 714-1482 - sightseeing and custom charters, up to 6 passengers)*, Boat Rentals, Park *(Dosewallips State Park 796-4415)* **1-3 mi:** Video Rental *(Linda's Gifts and Videos 796-4558)*, Special Events *(Brinnon ShrimpFest 796-4809)*

Pleasant Harbor State Park

Drawer K; Brinnon, WA 98320

Tel: (360) 796-4415 **VHF: Monitor** n/a **Talk** n/a
Fax: n/a **Alternate Tel:** n/a
Email: n/a **Web:** www.parks.wa.gov/moorage/parks
Nearest Town: Brinnon *(2.4 mi.)* **Tourist Info:** (360) 765-4999

3+ mi: Golf Course *(Port Ludlow Golf Club 437-0272, 21 mi.)*, Museum *(Quilcene Historical Museum 795-4848, 13 mi.)*

Provisioning and General Services
Near: Convenience Store *(Pleasant Harbor Marina 796-4611)*, Wine/Beer
1-3 mi: Market *(Brinnon General Store 796-4400)*, Liquor Store *(Cottage by the Bay 796-4613)*, Post Office *(796-4465)*, Protestant Church *(Brinnon Community Church 796-4462)*, Retail Shops *(Boone Trading Company 796-4330)* **3+ mi:** Supermarket *(Peninsula Food Store 765-3389, 13 mi.)*, Beauty Salon *(Blue Bonnet Beauty Shop 765-3377, 13 mi.)*, Pharmacy *(Coyle Drug 765-4508, 13 mi.)*, Hardware Store *(Henery's Hardware 765-3113, 13 mi.)*

Transportation
Airport: Jefferson County *(33 mi.)*

Medical Services
911 Service **OnCall:** Ambulance *(796-4711)* **3+ mi:** Doctor *(South Coast Medical Clinic 765-3111, 13 mi.)* **Hospital:** Jefferson General 385-2200 *(36 mi.)*

Setting -- Situated at the widest point on the Hood Canal, about 18 miles south of the Floating Bridge, Pleasant Harbor is a popular stop on the Olympic Peninsula side. The entrance to this small, protected harbor is shallow at low tide and about 100 feet wide (200 feet at high tide). Just inside the mouth, a 218-foot side-tie dock belongs to Pleasant Harbor State Park. From land, the view is pleasant and serene, with wooded hillsides.

Marina Notes -- *Dockge is $0.50/ft. with a $10 minimum: 26' and under $10/day, 27-34' $13/day, 35' and greater $16/day. Annual permits may be purchased at Washington State Parks' headquarters in Olympia, at regional offices, online and at the parks during hours. The cost is $3.50/ft. with a $50 minimum. The well-maintained dock has side-tie space on both sides, but no power. Pump-out is available, and a single picnic table is on the shore under the trees. A standard park toilet is also on-site. Note: Be careful not to stray into the adjacent private marina moorage area. Also Seabeck Harbor Marina, on the Kitsap Peninsula across the Canal, was closed at press time.

Notable -- This tiny state park covers less than a square mile, with about 200 feet of rocky beach on the quiet harbor. Pleasant Harbor Marina is nearby and has a general store, deli, and pizzeria. Given the modest cost, this is a good place to tie up, relax and still have access to provisions at the marina. Fishing and crabbing are touted as being some of the best and the beach calls on a warm day. If a night shore appeals, then make reservations at the nearby floating houseboats or at The Harbor House. The town of Brinnon, about 2 miles north, hosts a Shrimp Fest over Memorial Day weekend to kick off the boating season.

Pleasant Harbor Marina

308913 Highway 101; Brinnon, WA 98320

Tel: (360) 796-4611 **VHF: Monitor** Ch. 16 **Talk** Ch. 9 & 16
Fax: (866) 848-4612 **Alternate Tel:** n/a
Email: info@pleasantharbormarina.com **Web:** pleasantharbormarina.com
Nearest Town: Brinnon *(2.4 mi.)* **Tourist Info:** (360) 765-4999

Navigational Information

Lat: 47°39.745' **Long:** 124°54.974' **Tide:** 12 ft. **Current:** n/a **Chart:** 18476
Rep. Depths *(MLW)*: **Entry** 4 ft. **Fuel Dock** 20 ft. **Max Slip/Moor** 30 ft./-
Access: 18 nm south of the Hood Canal Floating Bridge on west shore

Marina Facilities *(In Season/Off Season)*

Fuel: Gasoline, Diesel
Slips: 312 Total, 43 Transient **Max LOA:** 220 ft. **Max Beam:** n/a
Rate *(per ft.)*: **Day** $0.75/0.25* **Week** n/a **Month** $10/7.5
Power: 30 amp Incl., 50 amp Incl., 100 amp n/a, 200 amp n/a
Cable TV: No **Dockside Phone:** No
Dock Type: Floating, Long Fingers, Alongside, Concrete, Wood
Moorings: 0 Total, 0 Transient **Launch:** n/a
Rate: Day n/a **Week** n/a **Month** n/a
Heads: 12 Toilet(s), 10 Shower(s)
Laundry: 3 Washer(s), 2 Dryer(s) **Pay Phones:** Yes
Pump-Out: OnSite, 1 Central, 1 Port **Fee:** Free **Closed Heads:** Yes

Marina Operations

Owner/Manager: Bill & Ryan Kaufman **Dockmaster:** Same
In-Season: May-Sep, 8am-7/8pm** **Off-Season:** Oct-Apr, 9am-5/6pm**
After-Hours Arrival: Register with night watchman
Reservations: Yes, Required **Credit Cards:** Visa/MC, Dscvr, Amex
Discounts: None
Pets: Welcome **Handicap Access:** Yes, Docks

Marina Services and Boat Supplies

Services - Security *(liveaboard security)*, Dock Carts **Communication -** Mail & Package Hold, Phone Messages, Fax in/out, Data Ports *(Wi-Fi - Broadband Xpress)*, FedEx, UPS, Express Mail **Supplies - OnSite:** Ice *(Cube)*, Ships' Store, Bait/Tackle

Boatyard Services

Nearest Yard: Port Townsend (360) 385-6211

Restaurants and Accommodations

OnSite: Snack Bar *(The Little Dinghy Deli 796-4611)*, Pizzeria *(Seabeck)*, Inn/B&B *(Houseboats for Two 796-3440, $195-240, couples only - no kids, pets, or guests)* **Near:** Restaurant *(Half Way House Restaurant 796-4715, B $5-8, L $5-8, D $6-15)*, Condo/Cottage *(The Harbor House 796-4064, $150 1st nt., $100 2nd nt., free 3rd nt., $500/wk.; children are welcome, no pets)* **1-3 mi:** Restaurant *(The Geoduck Restaurant and Lounge 796-4430, L, D)*, *(Halfway House Restaurant 796-4715)*, Motel *(Bayshore Motel 796-4220, $40-75)*, Inn/B&B *(Elk Meadows B&B 796-4886, $85)*, *(Brinnon Flats B&B 796-4935, $44-75)*

Recreation and Entertainment

OnSite: Heated Pool *(heated Apr-Oct)*, Spa *(hot tub open year-round)*, Picnic Area, Grills, Dive Shop *(Pacific Adventure 714-1482 - sightseeing and custom charters, up to 6 people)*, Boat Rentals **Near:** Park *(Dosewallips State Park 796-4415)* **1-3 mi:** Video Rental *(Linda's Gifts and Videos 796-4558)*, Special Events *(Brinnon's ShrimpFest - MemDay wknd)* **3+ mi:** Golf Course *(Port Ludlow Golf Club 437-0272, 21 mi.)*, Museum *(Quilcene Historical Museum 795-4848, 13 mi.)*

Provisioning and General Services

OnSite: Convenience Store *(groceries, supplies, gifts)*, Wine/Beer
1-3 mi: Market *(Brinnon General Store 796-4400)*, Liquor Store *(Cottage by the Bay 796-4613)*, Post Office *(796-4465)*, Protestant Church *(Brinnon Com-munityChurch 796-4462)*, Retail Shops *(Boone Trading Company 796-4330)*
3+ mi: Supermarket *(Peninsula Food Store 765-3389, 13 mi.)*, Beauty Salon *(Blue Bonnet Beauty Shop 765-3377, 13 mi.)*, Pharmacy *(Coyle Drug 765-4508, 13 mi.)*, Hardware Store *(Henery's Hardware 765-3113, 13 mi.)*

Transportation

Airport: Jefferson County *(33 mi.)*

Medical Services

911 Service **OnCall:** Ambulance *(796-4711)* **3+ mi:** Doctor *(South Coast Medical 765-3111, 10 mi.)* **Hospital:** Jefferson General 385-2200 *(36 mi.)*

Setting -- Pleasant Harbor is on the western shore of the Hood Canal, accessed through a relatively shallow and narrow entrance. Tree-covered hills rise steeply on both shores. Past the State Park dock, the marina's ten main piers lie perpendicular to the shore. At the top of the gangway, a boardwalk leads to the charming aqua and turquoise marina buildings amid wild rhodies and park-like grounds; they have commanding views of the harbor, docks and hillsides.

Marina Notes -- *Flat rate in-season/off-season: 35' or under $25/20, 36-44' $30/25, 45-51' $35/30, 52' and over $0.80/0.70 per ft. **May-Sep 8am-8pm wknds, Oct-Apr 9am-6pm wknds. Northern-most dock is "K", southern-most "A"; fuel is on "D", south of the covered slips. Reservations required summer & holiday wknds. Stationary & portable pump-outs. The amenities make this is a popular family destination. The Marina Store & Gift Shop provides a little of everything needed to restock the boat - produce, groceries, wine & beer, ice, clothing, gifts, and magazines. Book exchange and fish cleaning station on-site. Grills are near the heated pool. There's a roomy laundry room with folding table and three nicely detailed cinderblock bathhouses with fiberglass shower stalls.

Notable -- Seabeck Pizza in the marina store offers fresh pizza and ice cream. Also on-site, the Little Dinghy Deli features great sandwiches. There is excellent fishing on the canal - salmon plus, crabs, clams, oysters, and shrimp. One of the largest annual events in the area, the Brinnon ShrimpFest, is held over Memorial Day weekend. Besides craft and food booths, you can cheer for your favorite in the belt sander races. For a night ashore, House Boats for Two are floating houses for couples only -and nearby Harbor House offers two suites, both with great views of the harbor - .

Navigational Information

Lat: 47°24.512' **Long:** 123°08.165' **Tide:** 14 ft. **Current:** n/a **Chart:** 18448
Rep. Depths *(MLW)*: **Entry** 12 ft. **Fuel Dock** n/a **Max Slip/Moor** 12 ft./-
Access: South end of the Hood Canal on the eastern side

Marina Facilities *(In Season/Off Season)*

Fuel: No
Slips: 5 Total, 5 Transient **Max LOA:** 40 ft. **Max Beam:** n/a
  Rate *(per ft.)*: **Day** Free* **Week** n/a **Month** n/a
  Power: 30 amp n/a, **50 amp** n/a, **100 amp** n/a, **200 amp** n/a
  Cable TV: No **Dockside Phone:** No
  Dock Type: Floating, Short Fingers, Alongside, Wood
Moorings: 0 Total, 0 Transient **Launch:** n/a
  Rate: Day n/a **Week** n/a **Month** n/a
Heads: 2 Toilet(s)
Laundry: None **Pay Phones:** No
Pump-Out: No **Fee:** n/a **Closed Heads:** Yes

Marina Operations

Owner/Manager: Port of Hoodsport **Dockmaster:** n/a
In-Season: Year-Round, 9am-5pm **Off-Season:** n/a
After-Hours Arrival: Tie up
Reservations: No **Credit Cards:** No
Discounts: None
Pets: Welcome **Handicap Access:** No

Port of Hoodsport

PO Box 429; Highway 101; Hoodsport, WA 98548

Tel: (360) 877-9350 **VHF: Monitor** n/a **Talk** n/a
Fax: (360) 877-9350 **Alternate Tel:** n/a
Email: portmail@tctc.com **Web:** portofhoodsport.us
Nearest Town: Hoodsport **Tourist Info:** (360) 426-2021

Marina Services and Boat Supplies

Supplies - Near: Ice *(Cube)* **Under 1 mi:** Bait/Tackle

Boatyard Services

Nearest Yard: Belfair Marine (360) 275-5420

Restaurants and Accommodations

Near: Restaurant *(Hoodsport Marina Pub & Eatery 877-9059, B, L, D
Outdoor dining available)*, Fast Food *(Burger Stand 877-6122)*, Motel
(Creekside Inn 877-9686, $50-90) **Under 1 mi:** Restaurant *(Skipper's
John's Seafood 'n' Chowder House 877-5661)*, *(Model T 877-9883)*
1-3 mi: Fast Food *(Honey Bee Drive In 877-5612)*, Motel *(Glen Ayr
Motel and RV Park 877-9522, $45-75)*

Recreation and Entertainment

OnSite: Beach *(tidal)*, Picnic Area **Near:** Dive Shop *(Hood Sport & Dive
877-6818)*, Special Events *(Jul 4th - annual Street Fair & Fireworks)*
Under 1 mi: Sightseeing *(Fish Hatchery 877-6408, Hoodsport Winery
877-9894)* **1-3 mi:** Golf Course *(Lake Cushman Golf Course 877-5505)*
3+ mi: Boat Rentals *(Kayak Hood Canal 898-5925, 12 mi.)*, Movie
Theater *(Shelton Cinemas 426-1000, 20 mi.)*

Provisioning and General Services

Near: Market *(Hood Canal Grocery 877-9444)*, Wine/Beer, Liquor Store
(Hoodsport Liquor Store 877-5252), Post Office *(877-5552)*, Library
(Hoodsport Timberland Library 877-9339), Beauty Salon *(Connie's Hairport
877-6869)*, Barber Shop *(Hoodsport Mane Cut and Curl 877-9409)*,
Hardware Store *(G & M Hardware 877-9834)*, Retail Shops
Under 1 mi: Protestant Church *(Hood Canal Community Church 877-5751,
Saint Germain's Episcopal 877-9879)* **1-3 mi:** Bank/ATM *(West
Coast Bank 877-5272)* **3+ mi:** Delicatessen *(Alderbrook Deli Mart 898-7400,
12 mi.)*, Department Store *(Wal-Mart 427-6226, 20 mi.)*

Transportation

3+ mi: Rental Car *(Practical Rent-a-Car 426-5553, 20 mi.)*, Taxi *(Mason
County Taxi 426-5322, 20 mi.)* **Airport:** Sanderson Field/ Brementon Nat'l.
(13 mi./33 mi.)

Medical Services

911 Service **OnCall:** Ambulance *(Mason County Medic One 426-3403)*
Near: Doctor *(Hoodsport Family Clinic 877-0372)* **Under 1 mi:** Dentist
(Oleson 877-5151), Chiropractor *(Door Chiropractic Clinic 877-5979)*
3+ mi: Veterinarian *(Shelton Veterinary Clinic 426-2616, 20 mi.)*
Hospital: Mason General 426-1611 *(20 mi.)*

Setting -- Hoodsport lies on the west side of the Hood Canal, in the shadow of the Olympic Mountains. near the Great Bend in the fjord. It's about 4.5 miles south of the shallow bay at Lilliwaup. A gangway connects a small set of floating wood docks with a long stationery pier that leads to Hoodsport's main street. The town is only about four blocks long on the waterfront, but spreads inland, making this one of the largest communities on the Canal.

Marina Notes -- *Free moorage for 24 hours. The only amenities for boaters are two porta-potties. The atmosphere is friendly and very small town. Note: The floats tend to roll a great deal during high winds, so approach with caution. And the Hoodsport Inn restaurant is now closed.

Notable -- Hoodsport is a township; unincorporated and small, its downtown runs along the Canal. The people are welcoming, and the look and setting are typical of a small western seaside village. Visit the Hoodsport Winery Gift Shop and Wine Tasting. Open daily 10 am-6pm and located just south of town on Highway 101. July 3rd starts the annual street fair and fireworks display. Lake Cushman, in the foothills of the Olympic Mountains about five miles west of town, is a popular local spot for fishing and swimming. A world famous dive location is just a few miles north of Hoodsport - an underwater wall provides refuge for all kinds of sea life, the most famous being the legendary Hood Canal Octopi. The most popular activities are crabbing, digging for oysters, fishing for cod, hiking, biking, and boating.

Hood Canal Marina

Navigational Information
Lat: 47°21.425' **Long:** 123°05.860' **Tide:** 12 ft. **Current:** n/a **Chart:** 18476
Rep. Depths (*MLW*): **Entry** 35 ft. **Fuel Dock** n/a **Max Slip/Moor** 35 ft./-
Access: The hook of the Hood Canal, just east of Annas Bay on south shore

Marina Facilities *(In Season/Off Season)*
Fuel: No
Slips: 30 Total, 15 Transient **Max LOA:** 40 ft. **Max Beam:** n/a
  Rate *(per ft.)*: **Day** $0.75 **Week** n/a **Month** $5.50*
  Power: 30 amp $2, **50 amp** n/a, **100 amp** n/a, **200 amp** n/a
  Cable TV: No **Dockside Phone:** No
  Dock Type: Floating, Long Fingers, Alongside, Wood
Moorings: 0 Total, 0 Transient **Launch:** n/a
  Rate: Day n/a **Week** n/a **Month** n/a
Heads: 2 Toilet(s)
Laundry: None **Pay Phones:** No
Pump-Out: No **Fee:** n/a **Closed Heads:** Yes

Marina Operations
Owner/Manager: Martin Sun **Dockmaster:** Same
In-Season: Wed-Mon, 8am-5pm **Off-Season:** n/a
After-Hours Arrival: Call ahead
Reservations: Yes, Required **Credit Cards:** Visa/MC
Discounts: None
Pets: No **Handicap Access:** No

Hood Canal Marina

PO Box 86; 5101 E. State Route 106; Union, WA 98592

Tel: (360) 898-2252 **VHF: Monitor** n/a **Talk** n/a
Fax: (360) 898-8888 **Alternate Tel:** n/a
Email: n/a **Web:** n/a
Nearest Town: Union **Tourist Info:** (360) 426-2021

Marina Services and Boat Supplies
Supplies - OnSite: Ships' Store *(nautical gifts, too)* **Near:** Ice *(Cube)*, Bait/Tackle

Boatyard Services
Nearest Yard: Belfair Marine (360) 275-5420

Restaurants and Accommodations
Near: Restaurant *(Union Bay Café 898-2462, B $6-12, L $6-12, D $9-20)*, Condo/Cottage *(Kingfisher Cottage 898-8878, $150-195, 2 db, 2 bath cottage, up to 8 people)* **Under 1 mi:** Restaurant *(Robin Hood Restaurant & Pub 898-4400, D $10-25)*, Coffee Shop *(Villagers Press Espresso 898-2163, at Robin Hood)* **1-3 mi:** Restaurant *(Alderbrook Inn 898-5500, B $6-11, L $10-14, D $24-38, Sat-Sun brunch 7am-2pm)*, *(Par-Tee Café 898-1022, B, L, D)*, Lite Fare *(Union Square Deli 898-3354)*, Hotel *(Alderbrook Inn 898-2200, $153-240)*, Condo/Cottage *(Alderbrook Inn 898-2200, $261-290)*, *(Robin Hood Village 898-2163, $85-165)*

Recreation and Entertainment
OnSite: Boat Rentals *(Kayak Hood Canal 898-5925, $15/hr.)*
OnCall: Sightseeing *(Hood Canal Charters 877-9823, tours and custom trips - Union to Pleasant Harbor $80pp RT/ Hoodsport Winery 360-877-9894, 10 mi.)* **Under 1 mi:** Video Rental *(Friar Tuck's Grog Shoppe 898-2163)*
1-3 mi: Tennis Courts, Golf Course *(Alderbrook G.C. 898-2560)*, Galleries *(Hood Canal Gallery 898-8100)* **3+ mi:** Fitness Center *(Shelton Athletic Club 426-1388, 14 mi.)*, Movie Theater *(Shelton Cinemas 426-1000, 14 mi.)*

Provisioning and General Services
Near: Convenience Store, Market *(Union Country Store 898-2641)*
Under 1 mi: Wine/Beer, Post Office *(898-2653)* **1-3 mi:** Delicatessen *(Alderbrook Deli 898-7400)*, Liquor Store *(Robin Hood Village 898-2163)*, Protestant Church *(New Community Church of Union 898-7855)* **3+ mi:** Library *(Hoodsport Timberland Library 877-9339, 11 mi.)*, Beauty Salon *(Hoodsport Mane Cut & Curl 877-9409, 11 mi.)*, Pharmacy *(Medicine Shoppe 426-4272, 14 mi.)*, Hardware Store *(G & M Hardware 877-9834, 10 mi.)*, Department Store *(Wal-Mart 427-6226, 14 mi.)*

Transportation
3+ mi: Rental Car *(Practical Rent-a-Car 426-5553, 14 mi.)*, Taxi *(Mason County Taxi 426-5322, 14 mi.)* **Airport:** Sanderson Field/ Brementon Nat'l. *(14 mi./20 mi.)*

Medical Services
911 Service **OnCall:** Ambulance **1-3 mi:** Holistic Services *(Spa at Alderbrook 898-2200)* **3+ mi:** Doctor *(Hoodsport Family Clinic 877-0372, 11 mi.)*, Chiropractor *(Door Chiropractic Clinic 877-5979, 11 mi.)*, Veterinarian *(Belfair Animal Hospital 275-6008, 9 mi.)*
Hospital: Mason General 426-1611 *(14 mi.)*

Setting -- Just around the Great Bend of the Hood Canal, on the south shore, Hood Canal Marina is the first facility on the lower end of the fjord. The view from the docks is of the other side of the Hood Canal and, looming above, the snow-capped Olympic Mountains. A gated ramp leads to simple, wooden piers with long finger slips and some alongside tie-ups. On the hard is a large gray metal building that houses the marina office and a kayak rental operation.

Marina Notes -- *$110 min. monthly rate. The marina is used primarily by seasonal fishermen. Electricity is provided for a small fee. Depths are reported to be sufficient for small craft at all tides, but large craft should call ahead for conditions. The marina store offers nautical gifts and supplies for on-site convenience. Owner, Martin Sun, lives next to the marina and is available for unexpected emergencies. There is no shower or laundry, but the heads are well kept and quite usable.

Notable -- The community of Union on the east side of Hood Canal had grand hopes of becoming the western terminus of a major railroad. The railroad never arrived, but there is a legacy of some interesting old buildings here. Union has the distinction of having the oldest continuously operating post office in Mason County. Just under two miles is the Alderbrook Resort with its extraordinary waterfront dining, spa, and 18-hole championship golf course (Fri-Sun $40/18 holes, $20/9 holes; Mon-Thu $35/18 holes, $20/9 holes; twilight $20). The resort facilities are available to non-guests for $10 per eight hours. Alternatively, grab your fishing pole or a bucket for some great fishing and oyster and clam digging.

Navigational Information
Lat: 47°20.983' **Long:** 123°04.087' **Tide:** 12 ft. **Current:** n/a **Chart:** 18476
Rep. Depths (*MLW*): **Entry** 18 ft. **Fuel Dock** n/a **Max Slip/Moor** 18 ft./-
Access: Aprox. 2.5 mi. east of Annas Bay on south shore of the Hood Canal

Marina Facilities *(In Season/Off Season)*
Fuel: No
Slips: 38 Total, 38 Transient **Max LOA:** 100 ft. **Max Beam:** n/a
  Rate *(per ft.)*: **Day** $1.00/0.50* **Week** n/a **Month** n/a
  Power: 30 amp Incl., **50 amp** Incl., **100 amp** n/a, **200 amp** n/a
  Cable TV: No **Dockside Phone:** Yes, Incl.
  Dock Type: Floating, Alongside, Wood, Composition
Moorings: 0 Total, 0 Transient **Launch:** n/a
  Rate: Day n/a **Week** n/a **Month** n/a
Heads: 2 Toilet(s), 2 Shower(s) *(with dressing rooms)*
Laundry: None **Pay Phones:** Yes
Pump-Out: OnSite **Fee:** Incl. **Closed Heads:** Yes

Marina Operations
Owner/Manager: Janet Christopher, Dir. of Mktg. **Dockmaster:** Scott Glenn
In-Season: Year-Round, 24/7 **Off-Season:** n/a
After-Hours Arrival: Check in at the front desk
Reservations: Yes, Preferred **Credit Cards:** Visa/MC, Dscvr, Din, Amex
Discounts: None
Pets: Welcome, Dog Walk Area **Handicap Access:** Yes, Heads, Docks

Alderbrook Resort and Spa

10 East Alderbrook Drive; Union, WA 98592

Tel: ; (800) 622-9370 **VHF: Monitor** n/a **Talk** n/a
Fax: n/a **Alternate Tel:** n/a
Email: n/a **Web:** www.alderbrookresort.com
Nearest Town: Union **Tourist Info:** (360) 426-2021

Marina Services and Boat Supplies
Services - Docking Assistance, Trash Pick-Up, Dock Carts
Communication - Mail & Package Hold, Phone Messages, Fax in/out, Data Ports *(Incl.)*, FedEx, DHL, UPS, Express Mail *(Sat Del)* **Supplies -**
OnSite: Ice *(Cube)* **1-3 mi:** Bait/Tackle

Boatyard Services
Nearest Yard: Belfair Marine (360) 275-5420

Restaurants and Accommodations
OnSite: Restaurant *(Alderbrook Inn 898-2200, B $6-11, L $10-14, D $24-38, Sat-Sun brunch 7am-2pm - featuring fresh, local seafood plus a bar menu)*, Hotel *(Alderbrook Inn Resort 898-2200, $110-360)*, Condo/Cottage *(Alderbrook Inn Resort 898-2200, $261-290, fully equipped waterfront cottages)* **Under 1 mi:** Restaurant *(Robin Hood Restaurant, Pub and Oyster Bar 898-4400, D $10-25, Thu-Sun 5-9pm; reservations required)*, *(Par-Tee Café 898-1022, B, L)*, Condo/Cottage *(Robin Hood Village 898-2163, $115-130, waterfront cottages sleep 2-4)* **1-3 mi:** Restaurant *(Oyster Bay Café 898-2462, B $6-12, L $6-12, D $9-20)*, Lite Fare *(Union Square Deli 898-3354)*

Recreation and Entertainment
OnSite: Heated Pool, Spa, Beach, Picnic Area, Grills *(fire pit)*, Playground, Volleyball, Tennis Courts, Fitness Center, Jogging Paths, Boat Rentals *(kayaks - singles $25/hr., $20 each additional hr., doubles $40/hr., $35 each additional hr.; paddleboats 30/hr. or $20/half-hr., $25 each additional hr. 898-5925)* **OnCall:** Sightseeing *(Pavco Flight Adventures 800-645-3563, Hood Canal Charters 877-9823, tours and custom trips - Union to Pleasant*

Harbor $80pp RT) **Under 1 mi:** Golf Course *(Alderbrook Golf 898-2560 - one of the NW's top 25)* **1-3 mi:** Video Rental *(Friar Tuck's Grog Shoppe 898-2163)*, Special Events *(Bathtubs and Ballyhoo - Jul-Aug)*, Galleries *(Hood Canal Gallery 898-8100)* **3+ mi:** Bowling *(Timber Bowl 426-8452, 13 mi.)*, Movie Theater *(Shelton Cinemas 426-1000, 13 mi.)*

Provisioning and General Services
Near: Post Office *(898-2653)* **Under 1 mi:** Wine/Beer, Liquor Store *(Robin Hood Village 898-2163)*, Protestant Church *(New Community Church of Union 898-7855)* **1-3 mi:** Market *(Union Country Store 898-2641)*, Delicatessen *(Alderbrook Deli Mart 898-7400)* **3+ mi:** Library *(Hoodsport Timberland Library 877-9339, 13 mi.)*, Beauty Salon *(Hoodsport Mane Cut and Curl 877-9409, 13 mi.)*, Dry Cleaners *(Kneeland Plaza Cleaners 426-2607, 13 mi.)*, Bookstore *(Sage Book Store 426-6011, 13 mi.)*, Pharmacy *(Neil's Pharmacy 426-3327, 13 mi.)*, Hardware Store *(G & M Hardware 877-9834, 13 mi.)*, Department Store *(Wal-Mart 427-6226, 13 mi.)*, Copies Etc. *(Greg's Graphics and Printing 426-8628, 13 mi.)*

Transportation
OnCall: Taxi *(Shelton Cabs 426-9222)* **3+ mi:** Rental Car *(Practical Rent-a-Car 426-5553, 13 mi.)* **Airport:** Sanderson Field/ Brementon Nat'l. *(15 mi./18 mi.)*

Medical Services
911 Service **OnCall:** Ambulance **3+ mi:** Doctor *(Hoodsport Family Clinic 877-0372, 13 mi.)*, Chiropractor *(Door Chiropractic Clinic 877-5979, 13 mi.)*, Optician *(Shelton Veterinary Hospital 426-2616, 13 mi.)* **Hospital:** Mason General 426-1611 *(13 mi.)*

Setting -- Just past the Big Bend, Alderbrook's dramatic new wood and glass, three-story, craftsman-style main lodge sprawls along the Hood Canal's south shore. On the western horizon rise the snow-capped peaks of the Olympic Mountains. An integral part of the resort are fifteen hundred feet of wide, clean linear guest dockage that shares those spectacular views, a host of recreational amenities, and a service-focused staff standing by ready to assist.

Marina Notes -- *Flat rates vary by size. Under 24 ft. $25/night, 25-29' $30, 30-34' $35, 35-44' $40, over 45' $0.90/ft/night. 30 & 50 amp power included. Founded in 1913, multiple generations create a loyal clientele. All resort amentities are available to marina guests; moorage is free to guests of the hotel. Kayak and paddleboat rentals are on-site and a designated seaplane dock welcomes fly-ins. A $10 day fee provides 8 hrs. of facilities use to others - until 9 pm.

Notable -- Cruising the 65-mile Hood fjord requires careful planning, and many wonder if it is worth the effort. The "new" Alderbrook Resort & Spa answers with a resounding Yes! The beneficiary of a recent $12 million renovation, the resort features sharply peaked roofs, stone walls, massive pillars, and expansive green lawns spiked with rock outcroppings. Guests enjoy a glass-enclosed indoor swimming pool, jacuzzi, volleyball and tennis courts, a well-furnished exercise room, plus a top PGA golf course, world-class spa, 77 guest rooms and 17 two-bedroom cottages. Uninterrupted views of the Olympic Mountains and the Hood Canal fill large windows in the lobby and in the elegant, tiered restaurant, which is decorated in signature chocolate and sage. Pavco Flight Adventures picks up at the dock for scenic flights, airport service or charters. Fishing and shrimping are reportedly excellent - and canal oysters are considered among the best.

Twanoh State Park

12190 East Highway 106; Union, WA 98592

Tel: (360) 275-2222 **VHF: Monitor** n/a **Talk** n/a
Fax: n/a **Alternate Tel:** n/a
Email: n/a **Web:** www.parks.wa.gov/moorage/parks
Nearest Town: Union *(8 mi.)* **Tourist Info:** (360) 426-2021

Navigational Information
Lat: 47°22.850' **Long:** 122°58.180' **Tide:** 14 ft. **Current:** n/a **Chart:** 18476
Rep. Depths *(MLW)*: **Entry** 4 ft. **Fuel Dock** n/a **Max Slip/Moor** 4 ft./4 ft.
Access: Aprox. 8 mi. east of the Bend on the south shore of Hood Canal

Marina Facilities *(In Season/Off Season)*
Fuel: No
Slips: 9 Total, 9 Transient **Max LOA:** 75 ft. **Max Beam:** n/a
 Rate *(per ft.)*: **Day** $0.50* **Week** n/a **Month** n/a
 Power: 30 amp n/a, **50 amp** n/a, **100 amp** n/a, **200 amp** n/a
 Cable TV: No **Dockside Phone:** No
 Dock Type: Floating, Alongside, Wood
 Moorings: 7 Total, 7 Transient **Launch:** n/a, Dinghy Dock
 Rate: Day $10 **Week** n/a **Month** n/a
Heads: 4 Toilet(s), 4 Shower(s) *(with dressing rooms)*
Laundry: None **Pay Phones:** Yes, 1
Pump-Out: OnSite, Self Service **Fee:** n/a **Closed Heads:** Yes

Marina Operations
Owner/Manager: Washington State Parks and Recreation **Dockmaster:** n/a
In-Season: Year-Round **Off-Season:** n/a
After-Hours Arrival: n/a
Reservations: No **Credit Cards:** No
Discounts: None
Pets: Welcome **Handicap Access:** No

Marina Services and Boat Supplies
Supplies - Near: Ice *(Cube)* **3+ mi:** Bait/Tackle *(Union Country Store 898-2641, 7 mi.)*

Boatyard Services
OnSite: Launching Ramp **Nearest Yard:** Belfair Marine (360) 275-5420

Restaurants and Accommodations
OnSite: Snack Bar **3+ mi:** Restaurant *(China Capital 275-4444, 9 mi.)*, *(Alderbrook Inn 898-2200, B $6-11, L $10-14, D $24-38, 5 mi., Sat-Sun brunch 7am-2pm)*, *(Robin Hood Restaurant Pub 898-4400, D $10-25, 6 mi.)*, Coffee Shop *(Union Bay Café 898-2462, 7 mi.)*, Fast Food *(Subway, McDonald's 9 mi.)*, Lite Fare *(Sunset Deli 275-2072, 5 mi.)*, Pizzeria *(Pizza Factory 275-2000, 9 mi.)*, Hotel *(Alderbrook Spa and Resort 898-2200, $153-240, 5 mi.)*, Condo/Cottage *(Alderbrook Inn and Resort 898-2200, $261-290, 5 mi.)*, *(Robin Hood Village 898-2163, $85-165, 6 mi.)*

Recreation and Entertainment
OnSite: Beach, Picnic Area, Grills, Playground, Volleyball, Tennis Courts, Jogging Paths, Park **3+ mi:** Golf Course *(Alderbrook Golf Course 898-2560, 6 mi.)*, Boat Rentals *(Kayak Hood Canal 898-5925, 7 mi.)*, Video

Rental *(Friar Tuck's Grog Shoppe at Robin Hood Village 898-2163, 6 mi.)*, Sightseeing *(Hood Canal Charters 877-9823, tours and custom trips - Union to Pleasant Harbor $80pp RT, 7 mi.)*, Galleries *(6 mi.)*

Provisioning and General Services
3+ mi: Market *(Union Country Store 898-2641, 7 mi.)*, Supermarket *(QFC 275-9671, 9 mi.)*, Delicatessen *(Hood Canal Market & Deli 277-9081, 9 mi.)*, Liquor Store *(Robin Hood Village 898-2163, 6 mi.)*, Farmers' Market *(Belfair Elementary School, Sat 9am-3pm May-Oct, 275-0616, 9 mi.)*, Bank/ATM *(6 mi.)*, Post Office *(898-2653, 6 mi.)*, Protestant Church *(New Community Church of Union 898-7855, 6 mi.)*, Pharmacy *(Safeway 275-0953, 9 mi.)*, Hardware Store *(True-Value 275-0113, 9 mi.)*

Transportation
Airport: Brementon Nat'l. *(13 mi.)*

Medical Services
911 Service **OnCall:** Ambulance **3+ mi:** Doctor *(North Mason Medical Clinic 275-4084, 9 mi.)*, Dentist *(Belfair Dental 275-2855, 9 mi.)*, Chiropractor *(Ehresman Family Chiropractic 275-4401, 9 mi.)*, Veterinarian *(Belfair Animal Hospital 275-6008, 9 mi.)* **Hospital:** Mason General 426-1611 *(20 mi.)*

Setting -- Twanoh State Park lies about midway along the lower arm of the Hood Canal - at the Great Bend or "Fishhook Barb." It's eight miles east of Union, and seven miles before the end of the canal. This 182-acre marine and camping park features 3,167 feet of Hood Canal saltwater shoreline. Inland views are of the forested park; picnic tables are scattered along the shore, affording views of the Kitsap Peninsula on the opposite side. A ramp leads from a single floating wooden dock to a long pier. The park is a cut above the average when it comes to amenities and is just plain fun.

Marina Notes -- *Dock moorage is $0.50/ft. with a $10 minimum. Annual permits are available for purchase at the State Parks' headquarters in Olympia, regional offices, online and in person at the parks when staff is available. The cost is $3.50/ft.with a $50 minimum. Annual permits are valid for one year from date of purchase. Daily permits are only available at the parks. Heads are brick, cinderblock and basic. Note: Watch the tides; the fjord gets shallower here.

Notable -- Prior to becoming a park, the area was operated as a private resort which explains the abundance of facilities. Many of the buildings now used as restrooms, beach changing rooms, picnic shelters, outdoor kitchens, and a snack bar were built by the Civilian Conservation Corps during the '30s. Since the buildings are made of stone, brick and logs, they are still going strong 75 years later. A large unguarded swimming area is marked with floating rope buoys so visitors can test the claim that this is one of the warmest saltwater beaches in WA. Water-skiing, boating, shellfish gathering, fishing, bird watching and beachcombing are popular activities at Twanoh. Smelt dipping draws a crowd during the season. Playground areas with swings are available for children.

Navigational Information
Lat: 47°25.361' **Long:** 122°53.713' **Tide:** 12 ft. **Current:** n/a **Chart:** 18476
Rep. Depths (*MLW*): **Entry** 2 ft. **Fuel Dock** n/a **Max Slip/Moor** 2 ft./-
Access: Near the end of the Hood Canal on the north shore

Marina Facilities *(In Season/Off Season)*
Fuel: No
Slips: 10 Total, 10 Transient **Max LOA:** 45 ft. **Max Beam:** n/a
  Rate *(per ft.)*: **Day** $0.30* **Week** n/a **Month** n/a
  Power: 30 amp $2, **50 amp** n/a, **100 amp** n/a, **200 amp** n/a
  Cable TV: No **Dockside Phone:** No
Dock Type: Floating, Alongside, Wood
Moorings: 0 Total, 0 Transient **Launch:** n/a
  Rate: Day n/a **Week** n/a **Month** n/a
Heads: 1 Toilet(s)
Laundry: None **Pay Phones:** No
Pump-Out: OnSite **Fee:** $2 **Closed Heads:** Yes

Marina Operations
Owner/Manager: Bonnie Knight **Dockmaster:** John Williams
In-Season: Year-Round **Off-Season:** n/a
After-Hours Arrival: Pay box provided
Reservations: No **Credit Cards:** No
Discounts: None
Pets: Welcome, Dog Walk Area **Handicap Access:** No

Port of Allyn North Shore Dock

PO Box 1; Allyn, WA 98524

Tel: (360) 275-2430 **VHF: Monitor** n/a **Talk** n/a
Fax: (360) 275-2455 **Alternate Tel:** n/a
Email: portofallyn@aol.com **Web:** www.portofallyn.com
Nearest Town: Belfair *(5 mi.)* **Tourist Info:** (360) 275-1001

Marina Services and Boat Supplies
Supplies - 3+ mi: Ice *(Cube)*

Boatyard Services
OnSite: Launching Ramp **Nearest Yard:** Belfair Marine (360) 275-5420

Restaurants and Accommodations
Near: Inn/B&B *(Sheleh Inn 275-0916, $160-175)* **Under 1 mi:** Restaurant *(Belfair North Shore Inn 275-9337, B, L, D; Senior Dinner Show 2nd and 4th Sun at 5pm)* **1-3 mi:** Restaurant *(Pat's Little Red Barn 275-4441)*
3+ mi: Restaurant *(Teriyaki Wok 275-1111, 4 mi.)*, Fast Food *(McDonald's, Quiznos 4 mi.)*, Pizzeria *(Seabeck Pizza 275-2657, 4 mi.)*, Motel *(Belfair Motel 275-5443, 4 mi.)*

Recreation and Entertainment
Under 1 mi: Beach, Picnic Area *(Belfair State Park 275-0668)*, Grills, Playground, Boat Rentals *(Kayaks)*, Park *(Belfair)* **1-3 mi:** Video Rental *(Blockbuster Video 275-1249)*, Special Events *(Taste of Hood Canal - 2nd Sat in Aug)*, Galleries *(Candy's Wall of Fame 275-2773)* **3+ mi:** Golf Course *(275-6100, 8 mi.)*, Sightseeing *(The Theler Wetlands Center 275-4898, 4 mi.)*

Provisioning and General Services
1-3 mi: Convenience Store *(Short Stop Deli and Grocery 275-3024)*

3+ mi: Supermarket *(Safeway 426-9978, QFC 275-9671, 4 mi.)*, Delicatessen *(Sandy's Deli Mart 275-4787, 4 mi.)*, Health Food *(Dragon's Leir 277-9434, 4 mi.)*, Wine/Beer *(4 mi.)*, Liquor Store *(WA Liquor 275-4946, 4 mi.)*, Bakery *(Village Bakery and Cafe 275-0579, 4 mi.)*, Farmers' Market *(Belfair Elementary School Sat 9am-3pm May-Oct 275-0616, 4 mi.)*, Bank/ATM *(Kitsap Bank 275-0671, 4 mi.)*, Post Office *(4 mi.)*, Catholic Church *(Prince of Peace 275-8760, 5 mi.)*, Protestant Church *(North Mason United Methodist 275-3714, 4 mi.)*, Library *(North Mason Timberland Library 275-3232, 4 mi.)*, Beauty Salon *(Hair Shapers 275-3851, 4 mi.)*, Barber Shop *(Clipper Barber Shop 275-8188, 4 mi.)*, Dry Cleaners *(Belfair Maytag Laundry 277-0500, 4 mi.)*, Laundry *(Belfair Cleaners 277-0884, 4 mi.)*, Pharmacy *(Safeway 275-0953, 4 mi.)*, Hardware Store *(True Value 275-0113, 4 mi.)*, Florist *(Paul's Flowers 377-4422, 4 mi.)*, Copies Etc. *(L&M Stationers and Printers 275-4424, 4 mi.)*

Transportation
Near: Local Bus *(Mason County Dial-A-Ride 427-5033, Free)*
Airport: Bremerton Nat'l. *(8 mi.)*

Medical Services
911 Service **3+ mi:** Doctor *(North Mason Medical Clinic 275-4084, 4 mi.)*, Dentist *(North Mason Dental Center 275-4455, 4 mi.)*, Chiropractor *(Belfair Chiropractic Center 275-4411, 4 mi.)*, Veterinarian *(Belfair Animal Hospital 275-6008, 5 mi.)* **Hospital:** Mason General 426-1611 *(33 mi.)*

Setting -- The Port of Allyn North Shore Dock is located near the end of the Hood Canal, on the northern edge of the lower hook - approximately 13 miles from the Canal Bend. The single "T" shaped, wide, wooden dock is in a modest residential area that overlooks the Hood Canal Theler Wetlands - a mile west of Belfair State Park.

Marina Notes -- *Flat rate: Under 25' $8, 26-34' $10, 35' and over $12. A new dock was dedicated on Oct.14, 2003. Upgrades include a new gangway, which allows the float structure to reach 40 feet further out into the Canal for better access at low tides. At the same time, a pump-out unit was added. Limited 30 amp power is available, call ahead. A porta-potty is nearby.

Notable -- Belfair State Park is an easy one mile dinghy ride, It sports over 3,700 feet of shoreline on the canal, and an additional 3,000 feet of freshwater shoreline on Big Mission and Little Mission Creeks. Take care when visiting by dinghy, as this area can be shallow. The park has campsites, picnic tables, a horseshoe pit, and a swimming beach that is in an enclosed man-made saltwater basin controlled by a tide gate. Theler Wetlands, just seaward of Belfair, is an ideal place to learn about the unique ecology of the estuary - created by the Union River meeting the fjord waters. Opened in 1993, the park has viewing platforms along a 4-mile trail with opportunities for bird-watching; ospreys and eagles are the headliners here. For provisions and services, the town of Belfair is about 4 miles away. Few services are available to boaters without transportation.

Boaters' Notes

_____
_____
_____
_____
_____
_____
_____
_____
_____
_____
_____
_____
_____
_____
_____
_____
_____
_____
_____
_____
_____

Add Your Ratings and Reviews at www.AtlanticCruisingClub.com

A complimentary six-month Silver Membership is included with the purchase of this _Guide_. Select "Join Now," then "Silver," then follow the instructions.

The AtlanticCruisingClub website provides updated Marina Reports, Destination and Harbor Articles and much more — including an option within each online Marina Report for boaters to add their ratings and comments regarding that facility. Please log on frequently to share your experiences — and to read other boaters' comments.

On the website, boaters may rate marinas in one or more of the following categories — on a scale of 1 (basic) to 5 (world class) — and also enter additional commentary.

▸ **Facilities & Services** (Fuel, Reservations, Concierge Services and General Helpfulness)

▸ **Amenities** (Pool, Beach, Internet Access, including Wi-Fi, Picnic/Grill Area, Boaters' Lounge)

▸ **Setting** (Views, Design, Landscaping, Maintenance, Ship-Shapeness, Overall Ambiance)

▸ **Convenience** (Access — including delivery services — to Supermarkets, other Provisioning Sources, Shops, Services, Attractions, Entertainment, Sightseeing, Recreation, including Golf and Sport Fishing & Medical Services)

▸ **Restaurants/Eateries/Lodgings** (Availability of Fine Dining, Snack Bars, Lite Fare, OnCall food service, and Lodgings ashore)

▸ **Transportation** (Courtesy Car/Vans, Buses, Water Taxis, Bikes, Taxis, Rental Cars, Airports, Amtrak, Commuter Trains)

▸ **Please Add Any Additional Comments**

13. WA — West Central Puget Sound

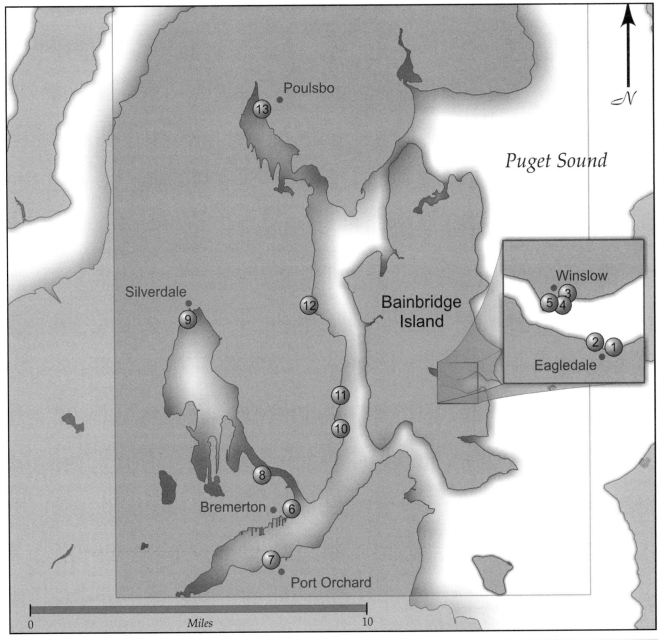

MAP	MARINA	HARBOR	PAGE	MAP	MARINA	HARBOR	PAGE
1	Bainbridge Island Marina	Eagle Harbor	186	8	Port Washington Marina	Washington Narrows	193
2	Eagle Harbor Marina	Eagle Harbor	187	9	Port of Silverdale	Dyes Inlet	194
3	Eagle Harbor Waterfront Park	Eagle Harbor	188	10	Illahee State Park	Port Orchard Bay	195
4	Winslow Wharf Marina	Eagle Harbor	189	11	Port of Illahee	Port Orchard Bay	196
5	Harbour Marina	Eagle Harbor	190	12	Port of Brownsville	Burke Bay	197
6	Bremerton Marina	Sinclair Inlet	191	13	Port of Poulsbo	Liberty Bay	198
7	Port Orchard Marina	Sinclair Inlet	192				

▸ **Currency** — In Canadian Marina Reports, all prices are in Canadian dollars. In U.S. Marina Reports, all prices are in U.S. dollars.

▸ **"CCM"** — Denotes a Certified Clean Marina, a state/provincial award for environmental excellence. See page 298 for an explanation and page 299 for a list of Pump-Out facilities.

▸ **Ratings & Reviews** — An overview of the Atlantic Cruising Club's rating system is on page 6 and details on the content of each Marina Report are on pages 7 – 11.

▸ **Marina Report Updates** — Comments from boaters and new information from ACC reviewers and marinas are posted regularly on *www.AtlanticCruisingClub.com*.

Bainbridge Island Marina

PO Box 1032; Eagle Harbor Drive; Bainbridge Island, WA 98110

Tel: (206) 842-9292 **VHF: Monitor** n/a **Talk** n/a
Fax: (206) 780-0955 **Alternate Tel:** n/a
Email: tallen@ci.bainbrifge-isl.wa.us **Web:** n/a
Nearest Town: Bainbridge Island *(3.5 mi.)* **Tourist Info:** (206) 842-3700

Navigational Information

Lat: 47°36.976' **Long:** 122°30.628' **Tide:** 11 ft. **Current:** n/a **Chart:** 18449
Rep. Depths *(MLW)*: **Entry** 30 ft. **Fuel Dock** n/a **Max Slip/Moor** 30 ft./-
Access: First marina on the south side as you enter Eagle Harbor

Marina Facilities *(In Season/Off Season)*

Fuel: No
Slips: 50 Total, 40 Transient **Max LOA:** 200 ft. **Max Beam:** n/a
  Rate *(per ft.)*: **Day** $1.00 **Week** n/a **Month** n/a
  Power: 30 amp Incl., **50 amp** n/a, **100 amp** n/a, **200 amp** n/a
  Cable TV: No **Dockside Phone:** No
  Dock Type: Floating, Long Fingers, Alongside, Wood
Moorings: 0 Total, 0 Transient **Launch:** n/a
  Rate: Day n/a **Week** n/a **Month** n/a
Heads: 6 Toilet(s), 6 Shower(s)
Laundry: 1 Washer(s), 1 Dryer(s) **Pay Phones:** No
Pump-Out: OnSite **Fee:** n/a **Closed Heads:** Yes

Marina Operations

Owner/Manager: Darrell McNabb **Dockmaster:** n/a
In-Season: Year-Round, 8am-5pm **Off-Season:** n/a
After-Hours Arrival: Call
Reservations: No **Credit Cards:** Cash only
Discounts: None
Pets: Welcome **Handicap Access:** No

Marina Services and Boat Supplies

Services - Security, Dock Carts **Supplies - OnSite:** Ice *(Cube)*
1-3 mi: Ships' Store *(The Chandlery 842-7245)*

Boatyard Services

Nearest Yard: Seaview Boatyard (206) 783-6550

Restaurants and Accommodations

Near: Inn/B&B *(Cedar Meadow B&B 842-6530)* **1-3 mi:** Restaurant *(Bistro 842-4347)*, *(Mandarin 855-1555)*, *(Streamliner Diner 842-8595)*, *(Doc's Marina Grill 842-8339, B $5-9, L $8-12, D $9-17)*, *(Bainbridge Thai Cuisine 780-2403, L $7-14, D $8-20, kids' B $2-3, L $2-3)*, *(Harbour Public House 842-0969, L $10-14, D $10-18, dock'n'dine)*, Pizzeria *(That's A Some Pizza 842-3848)*, Inn/B&B *(High Bank B&B 855-9763, $275)*, *(Seattle View B&B 855-0979)*, *(Buchanan Inn 780-9258, $115-159)*

Recreation and Entertainment

OnSite: Picnic Area **1-3 mi:** Grills, Playground, Dive Shop, Tennis Courts *(Waterfront Park)*, Fitness Center *(Island Health and Fitness 842-5720)*, Boat Rentals *(Bainbridge Island Boat Rentals 842-9229)*, Movie Theater *(Bainbridge Cinema 855-8169)*, Park *(Eagle Dale Park)*, Museum *(Bainbridge Island Museum 842-2773)*, Cultural Attract *(Playhouse 842-8569)*, Special Events, Galleries *(Rain Shadow Gallery 842-5450)* **3+ mi:** Heated Pool *(Bainbridge Aquatics Center - 6-lane, kids' pool, Jacuzzi, waterslide 842-2302, 4 mi.)*, Golf Course *(Wing Point Golf & Country Club 842-2688, 5 mi.)*, Video Rental *(Silver Screen 842-0261, 4 mi.)*, Sightseeing *(Bainbridge Island Winery 842-9463, 7 mi.)*

Provisioning and General Services

1-3 mi: Convenience Store, Market *(Town and Country Market 842-3849)*, Delicatessen *(Annie's Place 780-1206)*, Health Food *(Willow's Naturally 842-2759)*, Wine/Beer *(Winslow Wine Shop 780-8371)*, Liquor Store *(WA Liquor 842-5221)*, Bakery *(Blackbird Bakery 780-1322)*, Farmers' Market *(Sat 9am-1pm, Wed 4-7pm by City Hall 842-1500)*, Fishmonger, Bank/ATM, Post Office, Protestant Church, Synagogue, Beauty Salon *(Bainbridge Styling 842-6677)*, Barber Shop *(Cuts Above 842-7171)*, Dry Cleaners *(Clean Center 842-3612)*, Bookstore *(Fox Paw 842-7788)*, Pharmacy *(Vern's Winslow Drug 842-2652)*, Hardware Store *(Winslow Hardware and Mercantile 842-3101)*, Florist *(Flowering Around 842-0620)*, Clothing Store *(Blinx 842-3452)*, Copies Etc. *(Kitsap Xerographix 842-8283)*
3+ mi: Supermarket *(Safeway 842-7197, 4 mi.)*, Catholic Church *(St. Cecilia, 4 mi.)*, Library *(Bainbridge 842-4162, 4 mi.)*

Transportation

OnCall: Water Taxi *(Shiveley's Tugboat and Water Taxi 842-7595)*, Taxi *(Bainbridge Island Taxi 842-1021)* **1-3 mi:** Ferry Service *(Bainbridge-Seattle 842-2345)* **Airport:** Sea-Tac Int'l. *(15 mi.)*

Medical Services

911 Service **OnCall:** Ambulance **1-3 mi:** Doctor *(Winslow Medical Clinic 842-5632)*, Dentist *(Eagle Harbor Dental 842-2646)*, Chiropractor *(Eagle Harbor Chiropractic 842-2702)*, Holistic Services *(Bainbridge Natural Medicine 780-2628)*, Veterinarian *(Winslow Animal Clinic 842-6621)*, Optician *(Island Family Eyecare 842-2016)* **Hospital:** Harrison 360-337-8800 *(26 mi.)*

Setting -- Eagle Harbor indents the eastern shore of Bainbridge Island opposite Elliott Bay. B.I.M. is the first facility on the port side - the quiet side of the harbor. A two-story, cream-colored, teal-trimmed contemporary and a barn-red "farmhouse" sit on a wharf and house the office and amenities. Round, white picnic tables on a cement deck overlook the small grid of floating docks. The view across the harbor are dominated by the town of Bainbridge Island.

Marina Notes -- This large, casually maintained marina offers dockage primarily for long-term residents and seasonal fishermen who want to quickly get in and out of the harbor. However, slips may be available for transients. Owner and manager Darrell McNabb lives in Seattle (649-8525). An old ferry boat berths at the marina while waiting to be refurbished. Heads and showers with combination locks are located behind the office. Note: Eagle Harbor is two miles long and narrows at the head to 300 yards. The entrance is deep and the channel well marked, but the natural channel narrows between the reef south of Wing Point and the split on the west side of the channel entrance.

Notable -- The city of Bainbridge Island (formerly called Winslow) is the social hub of the island and lies directly across the bay. Dinghy to the town dock and explore downtown - critically acclaimed restaurants and cafés mingle with boutiques offering gifts and home furnishings. A good bookshop features many local authors. Dock 'n' dine is available at Harbour Marina's famous pub. Bainbridge has long boasted a thriving artistic community; an arts and crafts gallery is on Winslow Way. Stop by the City Hall and see the historic photos on permanent display in the building.

Navigational Information
Lat: 47°36.997' **Long:** 122°30.778' **Tide:** 11 ft. **Current:** n/a **Chart:** 18449
Rep. Depths *(MLW)*: **Entry** 9 ft. **Fuel Dock** n/a **Max Slip/Moor** 9 ft./-
Access: South side of Eagle Harbor - opposite Elliot Bay

Marina Facilities *(In Season/Off Season)*
Fuel: No
Slips: 107 Total, 5 Transient **Max LOA:** 66 ft. **Max Beam:** n/a
  Rate *(per ft.)*: **Day** $0.75 **Week** n/a **Month** $8
  Power: 30 amp $1.45, **50 amp** $1.45, **100 amp** n/a, **200 amp** n/a
  Cable TV: Yes **Dockside Phone:** Yes
  Dock Type: Floating, Long Fingers, Concrete
Moorings: 0 Total, 0 Transient **Launch:** n/a
  Rate: **Day** n/a **Week** n/a **Month** n/a
Heads: 4 Toilet(s), 3 Shower(s)
Laundry: 2 Washer(s), 2 Dryer(s), Book Exchange **Pay Phones:** Yes
Pump-Out: OnSite, 1 Port **Fee:** Free **Closed Heads:** Yes

Marina Operations
Owner/Manager: Tod Hornick **Dockmaster:** Same
In-Season: Year-Round, 9am-5pm* **Off-Season:** n/a
After-Hours Arrival: Call in Advance
Reservations: Yes, Preferred **Credit Cards:** No
Discounts: None
Pets: Welcome **Handicap Access:** No

Eagle Harbor Marina

5834 Ward Avenue NE; Bainbridge Island, WA 98110

Tel: (206) 842-4003 **VHF: Monitor** n/a **Talk** n/a
Fax: n/a **Alternate Tel:** n/a
Email: n/a **Web:** http://eagleharbormarina.com/
Nearest Town: Bainbridge Island *(3 mi.)* **Tourist Info:** (206) 842-3700

Marina Services and Boat Supplies
Services - Boaters' Lounge, Dock Carts **Communication -** Data Ports *(Wi-Fi - Broadband Xpress)*, FedEx, DHL, UPS **Supplies - OnSite:** Ice *(Cube)*

Boatyard Services
Nearest Yard: Seaview Boatyard (206) 783-6550

Restaurants and Accommodations
Near: Condo/Cottage *(Beach Cottage 842-6081)* **Under 1 mi:** Inn/B&B *(Cedar Meadow B&B 842-6530)* **1-3 mi:** Restaurant *(Doc's Marina Grill 842-8339, B $5-9, L $8-12, D $9-17, kids B $3-4, L $3-5)*, *(Bainbridge Thai Cuisine 780-2403, L $7-14, D $8-20)*, *(Harbour Public House 842-0969, L $10-14, D $10-18, dock'n'dine)*, *(Café Nola 842-3822)*, *(Isla Bonita 780-9644)*, Coffee Shop *(Pegasus Coffee House 780-9644)*, Pizzeria *(That's A Some Pizza 842-3848)*, Inn/B&B *(Buchanan Inn 780-9258, $115-159)*, *(High Bank B&B 855-9763, $275)*, *(Seattle View B&B 855-0979, $125)*

Recreation and Entertainment
OnSite: Fitness Center *(with Sauna)* **1-3 mi:** Picnic Area, Grills, Playground, Tennis Courts *(Waterfront Park)*, Boat Rentals *(Bainbridge Island Boat Rentals 842-9229)*, Movie Theater *(Bainbridge Cinema 855-8169)*, Museum *(Bainbridge Island Museum 842-2773)*, Cultural Attract *(Playhouse 842-8569)*, Special Events, Galleries *(Rain Shadow 842-5450)* **3+ mi:** Heated Pool *(Bainbridge Aquatics Center - 6-lane, kids' pool, Jacuzzi, waterslide 842-2302, 4 mi.)*, Golf Course *(Wing Point Golf & Country Club 842-2688, 4 mi.)*, Video Rental *(Silver Screen 842-0261, 4 mi.)*, Park *(Fort Ward 842-4041, 4 mi.)*, Sightseeing *(B.I.Winery 842-9463, 7 mi.)*

Provisioning and General Services
1-3 mi: Convenience Store *(Village Chevron Service Mart 842-8009)*, Market *(Town and Country Market 842-3849)*, Delicatessen *(Annie's Place 780-1206)*, Health Food *(Willow's Naturally 842-2759)*, Wine/Beer *(Winslow Wine Shop 780-8371)*, Liquor Store *(Washington Liquor Store 842-5221)*, Bakery *(Blackbird Bakery 780-1322)*, Farmers' Market *(Sat 9am-1pm, Wed 4-7pm by City Hall 842-1500)*, Bank/ATM, Post Office, Protestant Church, Synagogue, Beauty Salon *(Bainbridge Styling 842-6677)*, Barber Shop *(Cuts Above 842-7171)*, Dry Cleaners *(Clean Center 842-3612)*, Bookstore *(Fox Paw 842-7788)*, Pharmacy *(Vern's Winslow Drug 842-2652)*, Hardware Store *(Winslow Hardware 842-3101)*, Florist *(Flowering Around 842-0620)*, Clothing Store *(Blinx 842-3452)*, Retail Shops, Copies Etc. *(Kitsap Xerographix 842-8283)* **3+ mi:** Supermarket *(Safeway 842-7197, 4 mi.)*, Catholic Church *(4 mi.)*, Library *(Bainbridge 842-4162, 4 mi.)*

Transportation
OnCall: Water Taxi *(Shiveley Tugboat and Water Taxi 842-7595)*, Taxi *(Bainbridge Island Taxi 842-1021)* **1-3 mi:** Ferry Service *(Bainbridge-Seattle 842-2345)* **Airport:** Sea-Tac Int'l. *(15 mi)*

Medical Services
911 Service **OnCall:** Ambulance **1-3 mi:** Doctor *(Winslow Medical Clinic 842-5632)*, Dentist *(Eagle Harbor Dental 842-2646)*, Chiropractor *(Eagle Harbor Chiropractic 842-2702)*, Holistic Services *(Bainbridge Natural Medicine 780-2628)*, Veterinarian *(Winslow Animal Clinic 842-6621)*, Optician *(Island Family Eyecare 842-2015)* **Hospital:** Harrison 360-337-8800 *(26 mi.)*

Setting -- Eagle Harbor Marina's network of docks, backed by a two-story, weathered contemporary that houses its superb amenities, is on the south side of protected Eagle Harbor. This is the quiet side where gently rolling hills and a pristine shore create a tranquil, relaxing atmosphere. The views are spectacular - across the harbor to the town of Bainbridge Island (formerly Winslow) and the ferry dock and nine miles across the sound to the Seattle skyline.

Marina Notes -- CCM. *Closed Mondays during the season and Sunday and Monday during winter. This pleasant facility offers concrete floating docks with long finger slips, pump-out, 30 and 50 amp power on pedestals, plus cable, phone, and Wi-Fi. Shoreside, the comfortable clubhouse has a 2nd floor lounge with stunning views of Eagle Harbor and the Washington ferry terminal. The sitting area is furnished with white leather sofas, a fireplace, kitchen area, a dining or conference table, heads and a deck. Downstairs is a well-equipped exercise room with a sauna, plus more heads and showers and another deck. Note: The amp at low tide is quite steep - not a good time to haul laundry or groceries.

Notable -- In 1991, residents voted to expand the town of Winslow to an all-island city, and change its name to Bainbridge Island. Once the location of a famous shipyard, today the island is better known as home to more than 200 published authors and to a rich arts community that includes drama productions at the Playhouse, dancing at Island Center Hall, and a historic theater devoted to art films. Most of the activity is on the other side of the harbor, about 3 miles by road or under a mile by dinghy. Pick up a copy of the local newspaper, the "Bainbridge Review," for a detailed calendar of events and community activities.

PHOTOS ON DVD: 13

Eagle Harbor Waterfront Park

280 Madison Avenue N; Bainbridge Island, WA 98110

Tel: (206) 780-3733 **VHF: Monitor** n/a **Talk** n/a
Fax: (206) 780-0955 **Alternate Tel:** n/a
Email: tallen@ci.bainbridge-isl.wa.us **Web:** www.ci.bainbridge-isl.wa.us
Nearest Town: Bainbridge Island *(0.2 mi.)* **Tourist Info:** (206) 842-3700

Navigational Information
Lat: 47°37.298' **Long:** 122°31.208' **Tide:** 11 ft. **Current:** 2 kt. **Chart:** 18441
Rep. Depths *(MLW)*: **Entry** 18 ft. **Fuel Dock** n/a **Max Slip/Moor** 30 ft./40 ft.
Access: Eagle Harbor west of Seattle, near the Ferry Terminal

Marina Facilities *(In Season/Off Season)*
Fuel: No
Slips: 5 Total, 5* Transient **Max LOA:** 130 ft. **Max Beam:** n/a
  Rate *(per ft.)*: **Day** $0.25 **Week** n/a **Month** n/a
  Power: 30 amp n/a, 50 amp n/a, 100 amp n/a, 200 amp n/a
  Cable TV: No **Dockside Phone:** No
  Dock Type: Floating, Alongside, Concrete
Moorings: 6 Total, 6 Transient **Launch:** n/a, Dinghy Dock (Free)
  Rate: Day $0.25/ft. **Week** n/a **Month** n/a
Heads: 5 Toilet(s)
Laundry: None **Pay Phones:** Yes
Pump-Out: OnSite, Self Service, 1 Central **Fee:** Free **Closed Heads:** Yes

Marina Operations
Owner/Manager: City of Bainbridge Island **Dockmaster:** Tami Allen
In-Season: Year-Round, 24/7 **Off-Season:** n/a
After-Hours Arrival: Use Pay Envelopes
Reservations: No **Credit Cards:** No
Discounts: None
Pets: Welcome, Dog Walk Area **Handicap Access:** Yes, Heads

Marina Services and Boat Supplies
Services - Trash Pick-Up, Megayacht Facilities **Supplies - Near:** Ice *(Block, Cube, Shaved)*, Ships' Store *(The Chandlery at Winslow 842-7245)*, CNG *(The Chandlery)* **Under 1 mi:** Propane *(Chevron 842-8009)*

Boatyard Services
OnSite: Launching Ramp **Nearest Yard:** Seaview Boatyard (206) 783-6550

Restaurants and Accommodations
Near: Restaurant *(Doc's Marina Grill 842-8339)*, *(Bainbridge Thai 780-2403)*, *(Café Nola 842-3822)*, Snack Bar *(Island Ice Cream 842-2557)*, Pizzeria *(Westside Pizza 780-0755, delivers)*, Inn/B&B *(Captain's House 842-3557)* **Under 1 mi:** Restaurant *(Four Swallows 842-3397)*, *(San Carlos 842-1999)*, Hotel *(Island Country Inn 842-6861, $69-90)*, Inn/B&B *(Winslow Inn 842-3505)* **1-3 mi:** Fast Food *(McDonald's)*

Recreation and Entertainment
OnSite: Picnic Area, Grills, Playground, Tennis Courts, Jogging Paths, Boat Rentals *(B.I. Boat Rentals 842-9229 kayaks, swan boats)*, Park, Special Events *(Waterfront Park Concerts 842-2306; Jul 4th Fireworks; Arts Walk 842-7901- 1st Sunday in Feb., May, Aug. & Nov.)* **Near:** Dive Shop *(Exotic Aquatics 842-1980)*, Fitness Center *(Island Health & Fitness 842-5720)*, Movie Theater *(Lynwood 842-3080)*, Cultural Attract *(Playhouse 842-8569)*, Galleries *(Bainbridge Arts & Crafts 842-3132)* **Under 1 mi:** Video Rental *(Silver Screen 842-0261)* **1-3 mi:** Heated Pool *(Bainbridge Aquatics Center- 6-lane, kids' pool, Jacuzzi, sauna, lazy river, waterslide, trees, art, more 842-2302)*, Golf Course *(Wing Point 842-2688)*, Fishing Charter

(Seattle Charter Service 842-1923), Museum *(Bainbridge Island 842-2773)* **3+ mi:** Beach *(Fay Bainbridge State Pk. 842-3931, 6 mi.)*

Provisioning and General Services
OnSite: Fishmonger *(F/V Oceana 842-6036)* **Near:** Convenience Store, Market *(Town 7 Country Market 842-3848)*, Gourmet Shop, Delicatessen, Health Food, Wine/Beer *(Winslow Wine Shop 780-8371)*, Bakery *(Blackbird 780-1322)*, Farmers' Market *(Sat 9am-1pm, Wed 4-7pm by City Hall 842-1500)*, Bank/ATM, Post Office, Protestant Church, Beauty Salon *(Ericksen Ave. Salon 842-7216)*, Barber Shop *(Chuck's 842-2189)*, Dry Cleaners *(Clean Center 842-3612)*, Laundry, Bookstore *(Eagle Harbor Book Co. 842-5332)*, Pharmacy *(Vern's Winslow Drug 842-2652)*, Newsstand, Florist, Clothing Store *(Lindsey's 780-5808)*, Retail Shops, Copies Etc. *(Custom Printing 842-1606)* **Under 1 mi:** Liquor Store *(WA Liqour 842-5221)*, Catholic Church *(St. Cecilia)*, Library *(Bainbridge 842-4162)* **1-3 mi:** Supermarket *(Safeway 842-0127)*, Synagogue

Transportation
OnCall: Taxi *(Bainbridge Island Taxi 842-1021)* **Near:** Bikes *(Bainbridge Island Cycle 842-6413)*, Water Taxi *(shively Tug and Taxi 842-7595)*, Rental Car, Local Bus *(Kitsap Transit 842-8355)*, Ferry Service *(Bainbridge-Seattle 842-2345 - walk-on)* **Airport:** Sea-Tac Int'l. *(15 mi.)*

Medical Services
911 Service **OnCall:** Ambulance **Near:** Doctor *(Winslow Medical Clinic 842-5632)*, Dentist, Chiropractor *(Eagle Harbor Chiropractic 842-2702)*, Holistic Services **Under 1 mi:** Veterinarian *(Winslow Animal Clinic 842-6621)* **Hospital:** Harrison 360-337-8800 *(23 mi.)*

Setting -- Across Puget Sound from Seattle, Eagle Harbor is on the east side of rural Bainbridge Island. Past quiet harbors, rocky shoreline and densely forested hills, Eagle Harbor Waterfront Park is next to the ferry terminal on the starboard side. A narrow floating concrete dock provides 200 linear feet of moorage and a perpendicular dock berths Bainbridge Island Boat Rentals. Six mooring buoys float just off shore.

Marina Notes -- *100 foot side-tie dock - use both sides of the float. Alternatively, anchor in the harbor and dinghy in. The pay envelopes are at the bulletin board. No power, water, or boater services. Bainbridge Island Boat Rentals is on-site and offers a variety of small watercraft for rent. Tables and grills are scattered over a large lawn. Three porta-potties are available when the public restrooms are locked. Note: Caution is required as you enter the harbor - rocks and a shoal lie south of Wing Point and ferries have the right of way in the channel.

Notable -- Waterfront Park provides sheltered picnicking, a play area with swings and a jungle gym, tennis courts, a bandstand and, of course, waterfront on Eagle Harbor. A half-mile walking trail begins in the park. Rent a kayak or swan boat and explore the bay. Fee Wednesday evening concerts are in the park between 7-8:30pm. The Farmers' Market on Wednesdays and Saturdays is outside City Hall. A short distance from the dock, restaurants and shops line Winslow Way. For rainy days, there's the Pavilion with a 5-screen movie theater, fitness center, dining, and shopping - or visit the Bainbridge Island Historical Museum in City Hall with a photographic chronology. The historic 1936 Lynwood Theatre now shows foreign, classic and documentary films.

Navigational Information
Lat: 47°37.250' **Long:** 122°31.250' **Tide:** 11 ft. **Current:** 2 kt. **Chart:** 18449
Rep. Depths *(MLW)*: **Entry** 40 ft. **Fuel Dock** n/a **Max Slip/Moor** 40 ft./-
Access: North side of Eagle Harbor

Marina Facilities *(In Season/Off Season)*
Fuel: No
Slips: 238 Total, 5 Transient **Max LOA:** 50 ft. **Max Beam:** n/a
  Rate *(per ft.)*: **Day** $0.80* **Week** n/a **Month** n/a
  Power: 30 amp Incl., **50 amp** n/a, **100 amp** n/a, **200 amp** n/a
  Cable TV: No **Dockside Phone:** No
  Dock Type: Floating, Long Fingers, Alongside, Concrete
Moorings: 0 Total, 0 Transient **Launch:** n/a
  Rate: Day n/a **Week** n/a **Month** n/a
Heads: 2 Toilet(s), 2 Shower(s)
Laundry: 2 Washer(s), 4 Dryer(s) **Pay Phones:** Yes, 1
Pump-Out: OnSite **Fee:** Free **Closed Heads:** Yes

Marina Operations
Owner/Manager: Dave LeFave **Dockmaster:** Same
In-Season: Year-Round, 9am-5pm **Off-Season:** n/a
After-Hours Arrival: Call ahead; tie up in front of "B" dock
Reservations: Yes, Preferred **Credit Cards:** Visa/MC
Discounts: None
Pets: Welcome **Handicap Access:** No

Winslow Wharf Marina

PO Box 1029; 141 Parfitt Way SW; Bainbridge Island, WA 98110

Tel: (206) 842-4202 **VHF: Monitor** Ch. 9 **Talk** n/a
Fax: (206) 842-7785 **Alternate Tel:** n/a
Email: wwmcoa@seanet.com **Web:** n/a
Nearest Town: Winslow *(0.2 mi.)* **Tourist Info:** (206) 842-3700

Marina Services and Boat Supplies
Services - Security, Dock Carts **Communication -** Data Ports *(Wi-Fi - Broadband Xpress)* **Supplies - OnSite:** Ice *(Cube)*, Ships' Store *(The Chandlery)*, Bait/Tackle **Under 1 mi:** Propane *(Chevron 842-8009)*

Boatyard Services
Nearest Yard: Seaview Boatyard (206) 783-6550

Restaurants and Accommodations
OnSite: Restaurant *(Doc's Marina Grill 842-8339, B $5-9, L $8-12, D $9-17, kids B $2-3, L $2-3)* **Near:** Restaurant *(Pegasus Coffee House and Gallery 842-6725, adjacent)*, *(Bainbridge Thai Cusine 780-4303, L $7-14, D $8-20)*, *(Blue Water Diner 842-1151)*, *(Emmy's Vege House 855-2996)*, *(Harbor Public House 842-0969, L $10-14, D $10-18)*, Pizzeria *(Westside Pizza 780-0755)*, Inn/B&B *(Captain's House 842-3557)* **Under 1 mi:** Restaurant *(Island Grill 842-9037)*, Inn/B&B *(Island Country Inn 842-8429, $109-159)* **1-3 mi:** Inn/B&B *(High Bank B&B 855-9763, $275)*

Recreation and Entertainment
OnSite: Picnic Area *(Eagle Harbor Waterfront Park)* **Near:** Grills, Playground, Tennis Courts, Fitness Center *(Island Health & Fitness 842-5720)*, Boat Rentals *(kayaks at Harbour Marina 842-1980)*, Park *(Waterfront Park)*, Cultural Attract *(Playhouse 842-8569)*, Special Events *(Grand Ol' Fourth, 842-3700)* **Under 1 mi:** Movie Theater *(Bainbridge Cinema 855-8169)*, Video Rental *(Silver Screen 842-0261)*, Galleries *(Rain Shadow 842-5450)* **1-3 mi:** Heated Pool *(Bainbridge Aquatics Ctr. 842-2302)*, Golf Course *(Wing Point 842-2688)*, Museum *(Bainbridge Island Museum 842-2773)* **3+ mi:** Sightseeing *(Bainbridge Gardens 842-5888, 4 mi.)*

Provisioning and General Services
Near: Market *(Town and Country Market 842-3848)*, Delicatessen *(Annie's Place 780-1206)*, Health Food *(Willow's Naturally 842-2759)*, Wine/Beer *(Winslow Wine Shop 780-8371)*, Farmers' Market *(Sat 9am-1pm, Wed 4-7pm by City Hall 842-1500)*, Bank/ATM, Post Office, Protestant Church *(Eagle Harbor Congregational 842-4657)*, Beauty Salon *(Phoenix Rising Hair Salon 780-5099)*, Barber Shop *(Sandy's Barber Shop 855-7828)*, Bookstore *(Fox Paw 842-7788/Eagle Harbor Book Company 842-5332)*, Pharmacy *(Medicine Shoppe 842-1633)*, Hardware Store *(Winslow Hardware and Mercantile 842-3101)*, Florist *(Flowering Around 842-0620)*, Clothing Store *(Lindsley's Classic Clothing 780-5808)*, Retail Shops, Copies Etc. *(Custom Printing Co. 842-1606)* **Under 1 mi:** Convenience Store *(Village Chevron Service Mart 842-8009)*, Supermarket *(Safeway 842-7197)*, Liquor Store *(WA Liqour 842-5221)*, Catholic Church *(St. Cecelia 842-3594)*, Synagogue *(Congregaton Kol Shalom 855-0885)*, Library *(Bainbridge 842-4162)*, Dry Cleaners *(Clean Center 842-3612)*

Transportation
OnCall: Water Taxi *(Shiveley Tugboat 842-7595)*, Taxi *(Bainbridge Island 842-1021)* **Near:** Local Bus *(242-8355)* **Under 1 mi:** Ferry Service *(Bainbridge-Seattle 842-2345)* **Airport:** Sea-Tac Int'l. *(15 mi.)*

Medical Services
911 Service **OnCall:** Ambulance **Near:** Doctor *(Winslow Medical Clinic 842-5632)*, Chiropractor *(Eagle Harbor Chiropractic 842-2702)*, Holistic Services *(Bainbridge Natural Medicine 780-2680)* **Under 1 mi:** Dentist *(Eagle Harbor Dental 842-2646)*, Veterinarian *(Winslow Animal Clinic 842-6621)* **Hospital:** Harrison 360-337-8800 *(23 mi.)*

Setting -- Winslow Wharf is on the north shore of protected Eagle Harbor - past the ferry dock and the Eagle Harbor Waterfront Park. Flowers bloom enthusiastically and chairs are scattered about the well-landscaped grounds. A large teal shed-style building with a gray-shingled extension hosts the Chandlery and the amenities; picnic tables sit on the deck in front. Adjacent Doc's Marina Grill has a pretty patio overlooking the expansive network of floating docks.

Marina Notes -- *Flat rates - varies by size: 30' $22, 40' $32, 50' $40. A popular destination, so call ahead. If you haven't called for a slip assignment, tie up in front of "B" dock and locate the night watchman in the aluminum boat for assignment. Identifying slip letters are at the top of the pilings. The marina offers tidy concrete docks, dock carts, pump-out and a very well supplied chandlery. If you have a bike aboard, a bike rack in front of the office will keep it handy. The bathhouse and laundry facilities are located to the side of the office at the top of the pier (washers $0.75/dryers $1).

Notable -- The devoted clientele at Doc's Marina Grill love its Isle Farms Angus beef from nearby Vashon Island. And many more award-winning northwest eateries are a stroll away. Bainbridge Island is a compelling destination - art galleries and gift shops, theatre and summer concerts, and quiet spots to enjoy nature. Take the waterfront trail that begins in Waterfront Park and follow it for about half a mile - tennis courts and picnic tables dot the shore. Buy island-grown produce at the Farmer's Market held twice weekly in season, explore the Bainbridge Island Historical Museum (842-2773) or enjoy the fantastic town pool complex with a lap pool, kids' pool, waterslide, and Jacuzzi. Just over a mile east, the popular 18-hole Wing Point Golf Course offers great harbor views.

Harbour Marina

PO Box 1143; 233 Parfitt Way SW; Bainbridge Island, WA 98110

Tel: (206) 842-6502 **VHF: Monitor** n/a **Talk** n/a
Fax: (206) 842-5047 **Alternate Tel:** (206) 550-5340
Email: info@harbourpub.com **Web:** www.harbourpub.com
Nearest Town: Bainbridge Island *(0.2 mi.)* **Tourist Info:** (206) 842-3700

Navigational Information
Lat: 47°37.241' **Long:** 122°31.365' **Tide:** 11 ft. **Current:** 2 kt. **Chart:** 18449
Rep. Depths *(MLW):* **Entry** 25 ft. **Fuel Dock** n/a **Max Slip/Moor** 25 ft./-
Access: On the north side of Eagle Harbor

Marina Facilities *(In Season/Off Season)*
Fuel: No
Slips: 50 Total, 5 Transient **Max LOA:** 60 ft. **Max Beam:** n/a
 Rate *(per ft.):* **Day** $0.75* **Week** n/a **Month** n/a
 Power: 30 amp $3, **50 amp** n/a, **100 amp** n/a, **200 amp** n/a
 Cable TV: Yes ,available through Comcast **Dockside Phone:** No
 Dock Type: Floating, Long Fingers, Alongside, Wood
Moorings: 0 Total, 0 Transient **Launch:** n/a, Dinghy Dock
 Rate: Day n/a **Week** n/a **Month** n/a
Heads: 3 Toilet(s), 2 Shower(s)
Laundry: 2 Washer(s), 2 Dryer(s) **Pay Phones:** Yes
Pump-Out: OnSite **Fee:** Free **Closed Heads:** Yes

Marina Operations
Owner/Manager: Jeff Waite **Dockmaster:** Same
In-Season: Year-Round, 24/7 **Off-Season:** n/a
After-Hours Arrival: Call the dockmaster (206) 550-5340
Reservations: Yes, Preferred **Credit Cards:** No
Discounts: None
Pets: No **Handicap Access:** Yes

Marina Services and Boat Supplies
Services - Boaters' Lounge, Dock Carts **Communication -** Data Ports *(Wi-Fi through NohBel or BBX)*, FedEx, DHL, UPS **Supplies - Near:** Ice *(Cube)*, Ships' Store *(The Chandlery 842-7245)*, Bait/Tackle **Under 1 mi:** Propane *(Chevron 842-8009)*

Boatyard Services
Nearest Yard: Seaview Boatyard (206) 783-6550

Restaurants and Accommodations
OnSite: Restaurant *(Harbour Public House 842-0969, L $8-13, D $10-18)*
Near: Restaurant *(Winslow Way Café 842-0517)*, *(Bistro 842-4347)*, *(Bainbridge Sushi 780-9424)*, *(Streamliner Diner 842-8595)*, *(Pegasus Coffee House 842-6725)*, *(Doc's Marina Grill 842-8339)*, *(Bainbridge Thai 780-4303)*, Pizzeria *(Westside 780-0755)*, Inn/B&B *(Captain's House 842-3557)* **Under 1 mi:** Restaurant *(Four Swallows 842-3397)*, Inn/B&B *(Island Country Inn 842-6861, $69-89)*, *(Winslow Inn 842-3505)*

Recreation and Entertainment
OnSite: Picnic Area, Boat Rentals *(kayaks 842-1980)* **Near:** Grills, Playground, Tennis Courts *(Waterfront Park)*, Fitness Center *(Island Health & Fitness 842-5720)*, Park *(Waterfront Park)*, Cultural Attract *(Playhouse 842-8569)*, Special Events *(Grand Ol' Fourth - Jul 4 842-3700, summer concerts in the park)* **Under 1 mi:** Movie Theater *(Bainbridge 855-8169)*, Video Rental *(Silver Screen 842-0261)*, Galleries *(Rain Shadow 842-5450)* **1-3 mi:** Heated Pool *(Aquatics Center 842-2302)*, Golf Course *(Wing Point 842-2688)*, Museum *(Bainbridge Island Museum 842-2773)*, Sightseeing *(Bloedel Nature Reserve 842-7631, Fort Ward S.P. 842-4041)*

Provisioning and General Services
Near: Market *(Town & Country 842-3848)*, Delicatessen *(Colagreco's Deli 780-5354)*, Health Food *(Willow's Naturally 842-2759)*, Wine/Beer *(Winslow Wine Shop 780-8371)*, Bakery *(Blackbird 780-1322)*, Farmers' Market *(Sat 9am-1pm, Wed 4-7pm at City Hall Plaza 842-1500)*, Bank/ATM, Post Office, Protestant Church *(Eagle Harbor Congregational 842-4657)*, Beauty Salon *(Phoenix Rising 780-5099)*, Barber Shop *(Sandy 855-7828)*, Bookstore *(Fox Paw 842-7788)*, Pharmacy *(Medicine Shoppe 842-1633)*, Hardware Store *(Winslow Hardware and Mercantile 842-3101)*, Florist *(Flowering Around 842-0620)*, Clothing Store *(Lindsley's Classic Clothing 780-5808)*
Under 1 mi: Convenience Store *(Village Chrevron 842-8009)*, Supermarket *(Safeway 842-7197)*, Liquor Store *(WA Liqour 842-5221)*, Catholic Church *(St. Cecelia 842-3594)*, Synagogue *(Congregation Kol Shalom 855-0885)*, Library *(Bainbridge 842-4162)*, Dry Cleaners *(Clean Center 842-3612)*

Transportation
OnCall: Water Taxi *(Shiveley Tugboat and Water Taxi 842-7595)*, Taxi *(Bainbridge Island Taxi 842-1021)* **Under 1 mi:** Ferry Service *(Bainbridge-Seattle 842-2345)* **Airport:** Sea-Tac Int'l. *(15 mi.)*

Medical Services
911 Service **OnCall:** Ambulance **Near:** Doctor *(Winslow Medical Clinic 842-5632)*, Chiropractor *(Eagle Harbor Chiropractic 842-2702)*, Holistic Services *(Bainbridge Natural Medicine 780-2628)*, Optician *(Island Family Eyecare 842-2015)* **Under 1 mi:** Dentist *(Eagle Harbor Dental 842-2646)*, Veterinarian *(Winslow Animal Clinic 842-6621)* **Hospital:** Harrison 360-337-8800 *(23 mi.)*

Setting -- Harbour Marina is located on the north shore of Eagle Harbor, just beyond Winslow Wharf Marina. Downtown Bainbridge Island lies just to the north. On clear days, visitors can see Seattle and Mount Rainier to the east and the Olympic Mountains to the west. A long ramp leads from the well-cared-for docks to a raised wide boardwalk that skirts the shore. Picnic tables are scattered along the pier and a small russet, cream, and teal cottage houses the office.

Marina Notes -- *Flat rates based on boat size: under 32' $20, 32-39' $25, 40-48' $30, over 48' $35. Long wooden piers and pretty gangways lead to a set of docks with long finger slips; some alongside space is also available. Designated moorage for dinghies and kayaks. Space is reserved for dining guests at the end of "A" dock, on the outside. Kayak rentals are also on-site, perfect for exploring the waters near the head of the bay. The marina features an inviting bathhouse with tiled heads, glass-door showers and a laundry with a comfortable wicker-furnished boaters' lounge.

Notable -- The Harbour Public House, a popular pub, is on-site (free tieup for patrons) and has a dining deck with expansive views of the harbor. It serves "pub-fusion" fare - soups & salads, fish & chips, burgers, oyster and steak sandwiches with some organic and free-range choices plus 12 Northwest microbrews on tap, wine, too. First Tuesday each month is Open Mike Night starting at 8pm. Kayak around the harbor, explore the shops on Winslow Way, or enjoy free concerts at Waterfront Park on Wed nights, 7-8:30pm. On the other side of the island is Bloedel Reserve, a 150-acre private garden overlooking Agate Passage. Plant enthusiasts flock to Bainbridge just to visit the Reserve (Wed-Sun $6/4 842-7631).

Navigational Information
Lat: 47°33.760' **Long:** 122°37.390' **Tide:** 12 ft. **Current:** n/a **Chart:** 18452
Rep. Depths (*MLW*): **Entry** 35 ft. **Fuel Dock** n/a **Max Slip/Moor** 10 ft./-
Access: Across from Port Orchard at the mouth of Sinclair Inlet

Marina Facilities *(In Season/Off Season)*
Fuel: No
Slips: 45 Total, 12 Transient **Max LOA:** 60 ft. **Max Beam:** n/a
  Rate *(per ft.)*: **Day** $0.40* **Week** n/a **Month** n/a
  Power: 30 amp $4, **50 amp** n/a, **100 amp** n/a, **200 amp** n/a
  Cable TV: No **Dockside Phone:** No
  Dock Type: Floating, Long Fingers, Short Fingers, Alongside, Concrete
Moorings: 0 Total, 0 Transient **Launch:** n/a
  Rate: Day n/a **Week** n/a **Month** n/a
Heads: 6 Toilet(s), 6 Shower(s)
Laundry: 1 Washer(s), 1 Dryer(s) **Pay Phones:** No
Pump-Out: OnSite **Fee:** Free **Closed Heads:** Yes

Marina Operations
Owner/Manager: Port of Bremerton **Dockmaster:** Steve Sauer
In-Season: Year-Round, 8am-5pm **Off-Season:** n/a
After-Hours Arrival: Call in advance
Reservations: Yes, Preferred **Credit Cards:** Visa/MC, Dscvr, Amex
Discounts: None
Pets: Welcome **Handicap Access:** Yes, Heads

Bremerton Marina

102 Washington Avenue*; Bremerton, WA 98337

Tel: (360) 373-1035; (800) 462-3793 **VHF: Monitor** Ch. 66A **Talk** n/a
Fax: (360) 479-2928 **Alternate Tel:** n/a
Email: guest@portofbremerton.org **Web:** www.portofbremerton.org
Nearest Town: Bremerton **Tourist Info:** (360) 479-3579

Marina Services and Boat Supplies
Services - Security *(locked gates)*, Trash Pick-Up, Dock Carts
Communication - FedEx, DHL, UPS **Supplies - Near:** Ice *(Block, Cube)*
1-3 mi: Bait/Tackle, Propane **3+ mi:** West Marine *(479-2200, 4 mi.)*,
Boater's World *(478-4089, 4 mi.)*

Boatyard Services
Nearest Yard: Dockside Sales and Service (360) 876-9016

Restaurants and Accommodations
Near: Restaurant *(Park Avenue Diner 377-2457)*, *(Westside Burrito Connection 792-5288, L $4-9, D $5-14)*, *(Dodges Eatery 479-5599)*, *(City Limits 782-9700)*, *(Anthony's 377-5004, L $7-14, D $11-26)*, Coffee Shop *(Coffee Oasis 373-0461)*, Fast Food *(Burger King)*, Pizzeria *(Flippi's 373-9817)*, *(One World Bagel 377-3595)*, Hotel *(Hampton Inn 405-0200)*, Inn/B&B *(Highland 373-2235)* **Under 1 mi:** Restaurant *(La Fermata 373-5927, D $16-30)* **1-3 mi:** Motel *(Super 8 377-8881, $47-70)*, Hotel *(Best Western 405-1111, $71-96)*, Inn/B&B *(Oyster Bay 377-5510, $61-130)*

Recreation and Entertainment
OnSite: Picnic Area, Special Events *(Armed Forces Parade - May, Concerts on the Boardwalk - Fri 6-8pm Jul, Blackberry Fest in Sep)* **Near:** Museum *(USS Turner Joy 792-2457, Bremerton Naval 479-7447, Kitsap County 479-6226)*, Galleries *(Fish 479-0077)* **Under 1 mi:** Park *(Evergreen)* **1-3 mi:** Fitness Center *(YMCA 377-3741)*, Bowling *(Bremerton Lanes 782-2695)*, Movie Theater *(Bremerton 478-0577)*, Video Rental *(Blockbuster 415-9321)* **3+ mi:** Tennis Courts *(Bremerton Tennis & Athletic 692-8075, 4 mi.)*, Golf Course *(Rolling Hill 479-1212, 4 mi.)*

Provisioning and General Services
Near: Delicatessen *(Ship's Galley 479-0882)*, Bank/ATM, Post Office, Protestant Church, Library *(377-3955)*, Beauty Salon *(That Hair Place 479-3844)*, Barber Shop *(Tracey's 981-2388)*, Florist *(Flowers D'Amour 377-8888)* **Under 1 mi:** Convenience Store *(7-Eleven 479-4211)*, Market *(Midtown Market 373-8938)*, Wine/Beer, Bakery *(Larry and Kristi's Butter Bake 377-3296)*, Farmers' Market *(crafts & food market, Wed eve Jun-Oct in Café Destino's courtyard 782-0711)*, Catholic Church *(Our Lady of the Sea 479-3777)*, Synagogue, Copies Etc. *(PIP Printing 373-4523)*, Mosque *(Islamic Center Kitsap County 377-0902)* **1-3 mi:** Supermarket *(Safeway 792-9149)*, Health Food *(Helen's Healing Foods 377-1664)*, Liquor Store *(WA Liquor 478-4500)*, Fishmonger *(Jensen's Smoked Meats and Fish 377-4566)*, Dry Cleaners *(Eldon's Cleaners 373-0700)*, Laundry *(Maytag Laundry 373-9888)*, Pharmacy *(Safeway 792-9262)*, Hardware Store *(True Value 479-4414)* **3+ mi:** Department Store *(Wal-Mart 698-2889, 4 mi.)*

Transportation
OnCall: Rental Car *(Enterprise 377-1900/Budget 479-4500, 1 mi.)*, Taxi *(Troll's 478-8600)* **Near:** Local Bus *(Kitsap Transit 373-2877)*, Ferry Service *(to Seattle, plus foot ferry to Port Orchard)* **Airport:** Bremerton Nat'l. *(9 mi.)*

Medical Services
911 Service **Near:** Dentist *(Westside Dental 479-1600)*, Chiropractor *(Westside Chiropractic 405-9510)*, Holistic Services *(Aloha Spa Therapy 405-0676)* **Under 1 mi:** Doctor *(Peninsula Community Health Service 377-3776)*, Veterinarian *(Bremerton Animal Hospital 373-7333)* **1-3 mi:** Optician *(Kitsap Optical 479-6301)* **Hospital:** Harrison Memorial 377-3911 *(2 mi.)*

Setting -- The USS Turner Joy, one of Bremerton's main attractions, is directly north of the marina on the Bremerton Boardwalk - making it easy to spot. A breakwater protects the docks from backwash from the adjacent ferries. The on-going waterfront renovation moved the marina office to the new Harborside Development that sits at the top of the wide, concrete docks - next to the Kitsap Conference Center. Multi-level fountains spill down the plaza to the marina.

Marina Notes -- The Port of Bremerton maintains this excellent facility. Currently, moorage consists of 45 slips plus 500 feet of alongside space. The marina is scheduled to be expanded and rebuilt by 2008; it will triple its dock space and also offer permanent moorage. Portable pump-out on-site; fuel available at Port Orchard Marina. Gated access to the docks and a well maintained facilities building. *Mailing address: 8850 S.W. State Hwy.3, Port Orchard, WA 98367.

Notable -- Bremerton adjoins the Puget Sound Naval Shipyard and most of the city's business and affairs are keyed to the needs of the Navy establishment. Stroll the waterfront for great views of the harbor, and stop by the Visitors Bureau for maps. A new Hampton Inn and Convention Center opened on the waterfront, and an Anthony's Restaurant is now in Harborside. The Bremerton Naval Museum and Kitsap County Historical Museum are also nearby. Within blocks of the marina are several restaurants, coffee shops, a port office, boutiques, gift shops, and a JC Penney. Notice the beautiful mural at the downtown library. Special events are held all summer. Don't miss the seasonal Open Air Market on Wednesdays. A convenient bus stop makes the rest of Bremerton accessible, and a foot ferry connects to Port Orchard - every 1/2 hr. $1.25/0.50. Frequent car ferry service connects with Seattle.

Port Orchard Marina

8850 SW State Highway 3; Port Orchard, WA 98367

Tel: (360) 876-5535; (800) 462-3793 **VHF: Monitor** Ch. 66A **Talk** Ch. 66A
Fax: (360) 895-0291 **Alternate Tel:** n/a
Email: stevet@portofbremerton.org **Web:** www.portofbremerton.org
Nearest Town: Port Orchard **Tourist Info:** (360) 876-3505

Navigational Information
Lat: 47°32.588' **Long:** 122°38.275' **Tide:** 12 ft. **Current:** n/a **Chart:** 18449
Rep. Depths (MLW): Entry 30 ft. **Fuel Dock** 15 ft. **Max Slip/Moor** 40 ft./-
Access: Located on the south side of Sinclair Inlet

Marina Facilities *(In Season/Off Season)*
Fuel: 76 - Gasoline, Diesel
Slips: 340 Total, 50 Transient **Max LOA:** 52 ft. **Max Beam:** n/a
 Rate *(per ft.):* **Day** $0.58 **Week** n/a **Month** n/a
 Power: 30 amp $4, **50 amp** $4.50, **100 amp** n/a, **200 amp** n/a
 Cable TV: No **Dockside Phone:** No
 Dock Type: Floating, Alongside, Concrete
Moorings: 0 Total, 0 Transient **Launch:** Nearby
 Rate: Day n/a **Week** n/a **Month** n/a
Heads: 10 Toilet(s), 6 Shower(s)
Laundry: 4 Washer(s), 4 Dryer(s) **Pay Phones:** Yes
Pump-Out: OnSite, Self Service, 1 Central **Fee:** Free **Closed Heads:** Yes

Marina Operations
Owner/Manager: Port of Bremerton **Dockmaster:** Steve Slaton
In-Season: Jun-Sep, 8am-7pm **Off-Season:** Oct-May, 8am-6pm
After-Hours Arrival: Call in Advance
Reservations: Yes **Credit Cards:** Visa/MC, Dscvr, Din, Amex
Discounts: None
Pets: Welcome, Dog Walk Area **Handicap Access:** Yes, Heads

Marina Services and Boat Supplies
Services - Security *(locked gates)*, Trash Pick-Up, Dock Carts
Communication - Fax in/out *($1/pg)*, Data Ports *(Wi-Fi - Broadband Xpress)*, FedEx, DHL, UPS **Supplies - OnSite:** Ice *(Block, Cube)* **Under 1 mi:** Ships' Store *(Dockside 876-9016)*, Bait/Tackle *(Bay Street Outfitters 874-7880)* **1-3 mi:** Propane

Boatyard Services
Near: Launching Ramp. **Nearest Yard:** Dockside (360) 876-9016

Restaurants and Accommodations
Near: Restaurant *(Hiro Japanese 895-8591, L $6-14, D $8-20)*, *(Mako's 876-8124, B, L, D)*, *(Loscabos Grill 895-7878)*, Crab House *(Amy's On the Bay 876-1445, L $9-17, D $13-25)*, Lite Fare *(Bay Street Ale House 876-8030)*, Motel *(Comfort Inn 895-2666, $61-80)* **Under 1 mi:** Restaurant *(Bridge-water Seafood 997-9405)*, *(Tweten's Lighthouse 876-8464)*, Fast Food *(KFC)*, Pizzeria *(Godfather's Pizza 876-9296)* **1-3 mi:** Motel *(Days Inn 895-7818, $61-90)*, Hotel *(Best Western 405-1111, $70-96)*

Recreation and Entertainment
OnSite: Picnic Area, Playground, Park *(Waterfront Park)*, Sightseeing *(Kitsap Harbor Tours 876-1260)* **Near:** Museum *(Log Cabin Museum 876-3693, Sidney Museum 876-3693)*, Cultural Attract *(Performing Arts 769-7469)*, Special Events *(Seagull Calling Fest - 1st Sat May 876-3505, Concerts by the Bay - Thu eve midJul-Aug, Ball Racing - Aug)*
Under 1 mi: Dive Shop *(Taggerts 895-7860)*, Bowling *(Hi-Joy 876-8111)*
1-3 mi: Golf Course *(Village Greens 871-1222/ 6 courses w/in 11 mi.)*, Fitness Center *(Westcoast 874-2818)*, Video Rental *(Hollywood 874-0577)*

Provisioning and General Services
Near: Bakery *(Morningside Bread Company 876-1149)*, Farmers' Market *(May-Oct, Sat 9am-3pm at Waterfront Park)*, Bank/ATM *(Kitsap Bank 876-7800)*, Post Office, Protestant Church *(United Methodist 876-3975)*, Library *(Port Orchard 876-2224)*, Beauty Salon *(Artistry 876-2511)*, Bookstore *(Jomar Books 895-8462)*, Pharmacy *(Geiger Rexall Pharmacy 876-4021)*, Newsstand **Under 1 mi:** Market *(South Kitsap Grocery 876-2026)*, Health Food *(Natural Health 876-1134)*, Wine/Beer, Barber Shop *(Plaza Barber Shop 876-5467)*, Florist, Clothing Store, Retail Shops *(South Kitsap Mall 895-2112)* **1-3 mi:** Convenience Store, Supermarket *(Albertson's 895-2292)*, Delicatessen, Liquor Store *(Washington State Liquor Store 895-3835)*, Catholic Church *(St. Gabriel 876-2762)*, Synagogue *(Congregation Beth El Church 876-5581)*, Dry Cleaners *(1 Stop Cleaning 876-4656)*, Hardware Store *(Scott Mc Lendon's Hardware 876-8018)*, Department Store *(Wal-Mart 874-9060)*, Copies Etc. *(Office Depot 874-1996)*, Mosque

Transportation
OnSite: Ferry Service *(to Bremerton)* **OnCall:** Rental Car *(Enterprise 876-3799)*, Taxi *(Redtop 876-4949)* **Under 1 mi:** Local Bus *(Kitsap Transit 373-2877)* **1-3 mi:** Airport Limo *(Bremerton-Kitsap Airporter 876-1737)*
Airport: Bremerton Nat'l. *(8 mi.)*

Medical Services
911 Service **Under 1 mi:** Dentist *(Bay St. Dental 895-1401)*, Chiropractor *(Miller Chiropractic 876-1500)*, Optician *(Eye Designs 895-2020)*
1-3 mi: Doctor *(Community Health Clinic 779-1963)*, Holistic Services *(Integrative Medicine 871-4556)*, Veterinarian *(Kitsap Veterinary Hosp. 876-2021)* **Hospital:** Harrison Memorial 377-3911 *(10 mi.)*

Setting -- The marina is situated on the south shore of Sinclair Inlet at the southern end of the Port Orchard Waterway in the Old Town. A long dock wraps around the east and north sides, providing side-tie space for transients. Flowers and plants line the sidewalk at the top of the main ramp that connects to the pretty Waterfront Park - with tables and a new playground. Harbor views are of hillside homes, dock level businesses and the Puget Sound Naval Shipyard.

Marina Notes -- CCM. This excellent city facility is operated by Port of Bremerton. 3,000 feet of side-tie space and 50 slips are dedicated to transients. The fuel dock is near the shore on the west end of the marina, past the covered docks. The harbormaster's office is open daily with extended hours during the boating season. Pick up a welcome packet in the office. The security gates are locked from 10pm-7am in summer. Very spacious, upscale tiled bathhouse is about midpoint in the marina. Farther up the Inlet, the Port Orchard Yacht Club offers 380 ft. of dock space to members of reciprocal clubs (876-9010).

Notable -- The Old Town section of Port Orchard was built around the waterfront, making services easily accessible to boaters. It's a popular destination within a few hours' cruise of major cities, so it's often full in the summer. Many restaurants, shops, banks, the public library, and museums are within a short walk from the marina. The beautifully done waterfront invites a stroll, and the town has many nearby parks with tennis courts, playgrounds, and picnic tables. The waterfront is also home to fun annual events, plus Thursday evening concerts (6:30-8pm, free) and the Saturday Farmers' Market. Kitsap Harbor Tours operates from the marina in season, and the foot ferry to Bremerton is also on-site.

Navigational Information
Lat: 47°34.748' **Long:** 122°38.634' **Tide:** 12 ft. **Current:** 2 kt. **Chart:** 18449
Rep. Depths (*MLW*): **Entry** 6 ft. **Fuel Dock** n/a **Max Slip/Moor** 5 ft./-
Access: Located on the western side of the Washington Narrows

Marina Facilities *(In Season/Off Season)*
Fuel: No
Slips: 80 Total, 5 Transient **Max LOA:** 75 ft. **Max Beam:** n/a
  Rate *(per ft.)*: **Day** $1.25 **Week** n/a **Month** $5.75
  Power: 30 amp Incl., **50 amp** n/a, **100 amp** n/a, **200 amp** n/a
  Cable TV: No **Dockside Phone:** No
  Dock Type: Floating, Long Fingers, Wood
Moorings: 0 Total, 0 Transient **Launch:** n/a
  Rate: Day n/a **Week** n/a **Month** n/a
Heads: 4 Toilet(s), 4 Shower(s)
Laundry: 3 Washer(s), 3 Dryer(s) **Pay Phones:** No
Pump-Out: OnSite **Fee:** Free **Closed Heads:** Yes

Marina Operations
Owner/Manager: Steve Johnson **Dockmaster:** Same
In-Season: Tue-Sat, 10am-3pm **Off-Season:** n/a
After-Hours Arrival: Call ahead
Reservations: Yes, Required **Credit Cards:** Check or Money Order
Discounts: None
Pets: Welcome **Handicap Access:** No

Port Washington Marina

1805 Thompson Drive; Bremerton, WA 98337

Tel: (360) 479-3037 **VHF: Monitor** n/a **Talk** n/a
Fax: (425) 776-1447 **Alternate Tel:** n/a
Email: n/a **Web:** n/a
Nearest Town: Bremerton **Tourist Info:** (360) 479-3977

Marina Services and Boat Supplies
Services - Security, Dock Carts **Communication -** Mail & Package Hold, Phone Messages **Supplies - Near:** Ice *(Cube)* **Under 1 mi:** Bait/Tackle **1-3 mi:** West Marine *(479-2200)*

Boatyard Services
OnSite: Launching Ramp **Nearest Yard:** Dockside Sales and Service (360) 876-9016

Restaurants and Accommodations
Under 1 mi: Restaurant *(King's Pavilion 377-1558)*, Pizzeria *(Pizza Hut 373-0643)* **1-3 mi:** Restaurant *(Fiesta Mexican 415-9328)*, *(Noah's Ark 377-8100)*, *(Tony's Teriyake 792-0407)*, *(Keg Restaurant 373-8088)*, Fast Food *(Island Kitchen and Fast Food 377-8066)*, *(Burger King, KFC)*, Motel *(Dunes Motel 377-0093, $45-50)*, *(Flagship Inn 479-6566)*, Hotel *(Best Western 405-1111, $70-96)*, *(Hampton Inn 405-0200, downtown waterfront)*, Inn/B&B *(Oyster Bay Inn and Restaurant 377-5510, $61-130)*

Recreation and Entertainment
1-3 mi: Golf Course *(Kitsap Golf and Country Club 377-0166)*, Fitness Center *(Westcoast Fitness 377-5250)*, Bowling *(Bremerton Lanes and Casino 782-2695)*, Movie Theater *(Bremerton Cinema 478-0577)*, Video Rental *(Blockbuster Video 415-9321)*, Museum *(USS Turner Joy 792-2457)*, Cultural Attract *(Western Washington Center Arts 769-7469,)*, Special Events *(Boat Show - May, Armed Forces Parade - May, Concerts on the Boardwalk - Fri 6-8pm in Jul)*, Galleries *(Fish Gallery 479-0077, Arts Walk downtown 1st Fri of the month 5-8pm 373-4709)* **3+ mi:** Sightseeing *(Elandan Gardens 373-8260, 5 mi.)*

Provisioning and General Services
Under 1 mi: Convenience Store *(Am-Pm Mini Market 360-377-0816)*, Market *(Milltown Market 373-8938 Benjamin's Red Apple Market 479-4820377-3911)*, Wine/Beer, Bank/ATM *(Kitsap Community Federal 662-2000)*, Catholic Church, Protestant Church, Synagogue, Barber Shop *(Avenue Barber Shop 479-9164)*, Dry Cleaners *(Eldon's Cleaners 373-0700)*, Laundry *(Maytag Laundry and Cleaners 373-9888)*, Bookstore *(Elmo's Books 373-0551)*, Pharmacy *(Paul's Pharmacy and Gifts 373-0622)*, Florist *(Plum Creek Station 792-1100)*, Retail Shops, Mosque **1-3 mi:** Supermarket *(Safeway 792-9149)*, Delicatessen *(Handy Mart 377-1127)*, Health Food *(Helen's Healing Foods 377-1664)*, Liquor Store *(WA Liquor 792-6346)*, Bakery *(McGavin's Bakery 373-2414)*, Fishmonger *(Jensen's Smoked Meats and Fish 377-4566)*, Post Office, Library *(Kitsap Regional Library 377-3955)*, Beauty Salon *(Waves Ltd. 479-9800)*, Hardware Store *(Kitsap Lumber-True Value 479-4414)*, Copies Etc. *(Pip Printing 373-4523)*

Transportation
OnCall: Rental Car *(Enterprise 377-1900/Budget 479-4500, 1 mi.)*, Taxi *(Troll's Taxi 478-8600)* **Under 1 mi:** Local Bus *(Kitsap Transit 697-2877)* **1-3 mi:** Ferry Service *(Bremerton-Seattle; Port Orchard foot ferry)* **Airport:** Bremerton Nat'l. *(9 mi.)*

Medical Services
911 Service **OnCall:** Ambulance **1-3 mi:** Doctor *(Doctors Clinic 782-3700)*, Dentist *(Westside Dental 479-1600)*, Chiropractor *(Goodman Chiropractic Service 377-1626)*, Veterinarian *(Bremerton Animal Hospital 373-7333)*, Optician *(Olympic Peaks Optical 377-0097)* **Hospital:** Harrison 377-3911 *(2 mi.)*

Setting -- Three-mile-long Port Washington Narrows joins Dyes Inlet at its northern end, and Sinclair Inlet on the south. Port Washington Marina is located on the western side of the Port Washington Narrows, north of the Warren Avenue Bridge. A long floating pier parallels the shore anchoring the docks - long finger slips radiate from each. Perched on pilings above the docks is a two-story gray clapboard building with blue steel roof that houses the office and the bathhouse. White railings trim the surrounding walkways and decks. Private residences, businesses, and several petroleum distribution facilities can be seen along the shore.

Marina Notes -- CCM. The marina is geared for permanent moorages; however, if slips are available, they will take an occasional overnight guest. Office hours are limited, so call ahead. Quality power pedestals serve each slip. Because of the steep terrain, recreational areas are limited to the decks and walkways. The bathhouse promises sparkling heads, showers, and laundry. Note: Tidal currents in the narrows can reach 4 knots at times.

Notable -- The nearby area is residential; it's one mile to 6th Street and a Pizza Hut, Budget Rent-a-Car, and Am-Pm Mini-Mart. More restaurants and shops are along the same road, or downtown. Bremerton is in transition; projects are underway to revitalize the downtown and the waterfront. Galleries abound, and the first Friday of each month the art district organizes an Art Walk. Tour the USS Turner Joy at the Bremerton Boardwalk. This historic Navy destroyer serves as a floating museum, complete with a replica prisoner-of-war cell. Across Sinclair Inlet, Elandan Gardens features stunning Bonsai trees.

Port of Silverdale

PO Box 310; 3550 NW Byron Street; Silverdale, WA 98383

Tel: (360) 698-4918 **VHF: Monitor** n/a **Talk** n/a
Fax: (360) 698-2402 **Alternate Tel:** n/a
Email: portsil@tscnet.com **Web:** www.portofsilverdale.com
Nearest Town: Silverdale (0.5 mi.) **Tourist Info:** (360) 692-6800

Navigational Information
Lat: 47°38.564' **Long:** 122°41.526' **Tide:** 12 ft. **Current:** 4 kt. **Chart:** 18449
Rep. Depths (MLW): Entry 10 ft. **Fuel Dock** n/a **Max Slip/Moor** 10 ft./-
Access: Located off Sinclair Inlet through the Port Washington Narrows

Marina Facilities (In Season/Off Season)
Fuel: No
Slips: 32 Total, 32 Transient **Max LOA:** 75 ft. **Max Beam:** n/a
　Rate (per ft.): **Day** $0.25* **Week** n/a **Month** n/a
　Power: 30 amp Incl., **50 amp** n/a, **100 amp** n/a, **200 amp** n/a
　Cable TV: No **Dockside Phone:** No
　Dock Type: Floating, Long Fingers, Alongside, Wood
Moorings: 0 Total, 0 Transient **Launch:** Yes
　Rate: Day n/a **Week** n/a **Month** n/a
Heads: 4 Toilet(s), 1 Shower(s)
Laundry: None **Pay Phones:** Yes
Pump-Out: OnSite, Self Service **Fee:** Free **Closed Heads:** Yes

Marina Operations
Owner/Manager: Port of Silverdale **Dockmaster:** Theresa Haaland
In-Season: Year-Round, 24 hrs. **Off-Season:** n/a
After-Hours Arrival: Call in advance
Reservations: Clubs only **Credit Cards:** No
Discounts: None
Pets: Welcome, Dog Walk Area **Handicap Access:** Yes

Marina Services and Boat Supplies
Services - Security (cameras and foot patrol) **Supplies - Under 1 mi:** Ice (Block, Cube) **1-3 mi:** Bait/Tackle (Kitsap Sports 698-4808), Live Bait (Kitsap Sports)

Boatyard Services
OnSite: Launching Ramp (Free) **Nearest Yard:** Suldan's Boatworks (360) 876-4435

Restaurants and Accommodations
Near: Restaurant (Wok N Roll to Go 692-3085), (Waterfront Park Bakery and Café 698-2991), (Yacht Club Broiler 698-1601), (Lighthouse Café and Wine Bar 698-9462, L, D), (Old Town Pub 692-9132, L $5-10, D $7-15), Fast Food (Blimpie, McDonald's) **Under 1 mi:** Restaurant (Joy Teriyaki & Hamburgers 307-0902), (Blue Brick Bagel 692-1676), (Mariner Grill 692-0748, at Red Lion), Pizzeria (Pizza Time 698-1700), Motel (Cimarron Motel 692-7777, $59-79), (Poplars Motel 692-6126), Hotel (Red Lion 698-1000, $88-120)

Recreation and Entertainment
OnSite: Beach, Picnic Area, Grills, Playground, Park, Special Events (Whaling Days - last wknd Jul) **Near:** Fitness Center (Body Reform 662-0615), Video Rental (Total Video 698-4117), Museum (Kitsap County Historical Museum 692-1949) **Under 1 mi:** Movie Theater (Silverdale Cinemas 698-1510), Video Arcade (Kitsap Mall 698-2555), Cultural Attract (Center Stage Theater 692-9940), Galleries (Artists Edge 698-3113) **1-3 mi:** Jogging Paths, Bowling (All Star Lanes 692-5760) **3+ mi:** Tennis Courts (Bremerton Tennis & Athletic Club 692-8075, 5 mi.), Golf Course (Kitsap Golf and Country Club 377-0166, 4 mi.)

Provisioning and General Services
Near: Delicatessen (Stuart's Deli and Expresso 337-2206), Wine/Beer (Grape Expectations 698-0522), Bakery (Waterfront Park Bakery and Café 698-2991), Bank/ATM, Library (Silverdale 692-2779), Beauty Salon (Waterfront Hair Gallery 307-0514), Barber Shop (Old Town Barber Shop 698-8690), Dry Cleaners (Old Town Cleaners 307-0797), Pharmacy (Bogard's 692-2363) **Under 1 mi:** Market (Red Apple Market 613-2636), Supermarket (Safeway 692-4488, Albertsons 692-8088), Health Food (Helen's Health Food 698-1550), Liquor Store (WA Liquor 613-1642), Farmers' Market (Bucklin Hill Tue 11am-4pm MidApr-MidSep 830-9502), Protestant Church, Laundry (Suds-Eez Cleaners and Laundromat 307-8768), Bookstore (Pages Books 692-3352/Barnes and Noble 613-5352, 2 mi.) **1-3 mi:** Fishmonger (Down East Fish Mkt. 698-2306), Post Office (308-0301), Hardware Store (Home Depot 307-9200), Retail Shops (Silverdale Town Center 692-3536), Department Store (Sears, Macy's), Buying Club (Costco 692-1140), Copies Etc. (Office Depot 613-1920)

Transportation
OnCall: Rental Car (Enterprise 377-1900/Budget 479-4500, 1 mi.), Taxi (Super Kab 271-4012) **Near:** Local Bus (Kitsap Transit 373-2877)
Airport: Bremerton Nat'l. (14 mi.)

Medical Services
911 Service **OnCall:** Ambulance **Near:** Dentist (Dental Center 692-9560), Chiropractor (Back Health 692-0515), Holistic Services (Natural Options 698-4141), Veterinarian (Olympic Veterinary Hospital 692-0919), Optician (Silverdale Eyecare 692-6115) **Under 1 mi:** Doctor (Family Medical 692-2299) **Hospital:** Harrison 337-8800 (8 mi.)

Setting -- Access to Silverdale is through three-mile long Port Washington Narrows to the head of Dyes Inlet. The village of Silverdale sits on the west side at the head of the Inlet. The large "M" shaped, side-tie dock and the adjacent Silverdale Waterfront Park are in Silverdale's Old Town area, which offers a quiet, anti-mall shopping experience. Small unique shops, galleries and restaurants line the streets behind the waterfront. Heavily wooded hills rise above the inlet.

Marina Notes -- *Flat rate: up to 28' $5, 28' and over $10; 3-day limit. A wide pier leads out to the dock, accessed through the south end. It offers mostly alongside space, and larger boats can be accomodated. A payment box is at the shore end of the pier. Fresh water, power (included with dockage), and pump-out are on-site. There is no fee for the deep launch ramp. Security cameras at the pier are linked to the Port's website. Heads are at the park. Note: Tidal currents in the narrows attain speeds in excess of 4 kt. at times.

Notable -- The adjacent park has an inventive playground, waterfront benches, and a short walking path that will eventually connect with the popular Clear Creek Trail. Plans are for the trail to run all the way around Dyes Inlet and up to the Bangor base. A large picnic area sheltered under a pavilion provides lots of tables, a barbeque, and a sink area that has 4 electrical outlets. Bring your electric skillet and enjoy the outdoors. Silverdale, a destination shopping city, grew rapidly partly due to the construction of the submarine base at Bangor. Kitsap Mall, about 1 mile from the pier, is one of the largest in the area. The highlight of the summer is the Whaling Days celebration, with a parade, live entertainment, food, children's activities and fireworks.

Navigational Information
Lat: 47°35.979' **Long:** 122°35.584' **Tide:** 16 ft. **Current:** 3 kt. **Chart:** 18446
Rep. Depths (*MLW*): **Entry** 30 ft. **Fuel Dock** n/a **Max Slip/Moor** 30 ft./30 ft.
Access: West shore of Port Orchard Bay between Bremerton & Brownsville

Marina Facilities (*In Season/Off Season*)
Fuel: No
Slips: 9 Total, 9** Transient **Max LOA:** 120 ft. **Max Beam:** 25 ft.
  Rate (*per ft.*): **Day** $0.50* **Week** n/a **Month** n/a
  Power: 30 amp n/a, 50 amp n/a, 100 amp n/a, 200 amp n/a
  Cable TV: No **Dockside Phone:** No
  Dock Type: Floating, Alongside, Wood
Moorings: 5 Total, 5 Transient **Launch:** n/a, Dinghy Dock
  Rate: Day $10 **Week** n/a **Month** n/a
Heads: 4 Toilet(s), 2 Shower(s)
Laundry: None **Pay Phones:** Yes, 2
Pump-Out: No **Fee:** n/a **Closed Heads:** Yes

Marina Operations
Owner/Manager: Washington State Parks **Dockmaster:** Steve Kendall
In-Season: Year-Round, 8am-Dusk **Off-Season:** n/a
After-Hours Arrival: Self registration
Reservations: No **Credit Cards:** No
Discounts: None
Pets: Welcome **Handicap Access:** Yes, Heads, Docks

Illahee State Park

3540 Bahia Vista; Bremerton, WA 98310

Tel: (360) 478-6460 **VHF: Monitor** n/a **Talk** n/a
Fax: (360) 792-6067 **Alternate Tel:** n/a
Email: n/a **Web:** www.parks.wa.gov/moorage/parks
Nearest Town: Bremerton (*1 mi.*) **Tourist Info:** (360) 479-3979

Marina Services and Boat Supplies
Services - Security (*law enforcement on duty*) **Supplies - Under 1 mi:** Ice (*Block, Cube*) **1-3 mi:** Ships' Store, West Marine (*479-2200*), Boater's World (*478-4089*), Bait/Tackle, Propane (*Union 76 479-5855*), CNG (*Ferrellgas 373-2515*)

Boatyard Services
1-3 mi: Travelift (*Kitsap Marine 895-2193*), Forklift, Engine mechanic (*gas, diesel*), Launching Ramp, Electrical Repairs, Electronic Sales, Electronics Repairs, Bottom Cleaning, Propeller Repairs, Upholstery.
Nearest Yard: Kitsap Marine (360) 895-2193

Restaurants and Accommodations
Under 1 mi: Restaurant (*China House 377-8913*) **1-3 mi:** Restaurant (*Pat's Restaurant and Bakery 479-1717*), (*Shari's 373-1768, B $5-9, L $6-10, D $8-15*), (*Boat Shed 377-2600, L $8-13, D $12-22*), (*Sizzler 479-5748*), (*Barrio Fiesta 377-2611*), (*Sunny Teriyaki 377-3193*), Fast Food (*McDonald's, Burger King*), Pizzeria (*Domino's 373-2020*), Motel (*Midway Inn 479-2909*), Inn/B&B (*Illahee Manor B&B 698-7555, $115-195, waterfront*) **3+ mi:** Hotel (*Best Western Bremerton Inn 405-1111, $61-85, 4 mi.*), (*Hampton Inn 405-0200, 4 mi., waterfront*)

Recreation and Entertainment
OnSite: Beach, Picnic Area, Grills, Playground, Volleyball, Park
1-3 mi: Heated Pool, Dive Shop, Tennis Courts (*Bremerton Tennis 692-8075*), Golf Course (*Rolling Hills Golf Course 479-1212*), Fitness Center (*YMCA 377-3741*), Video Rental (*Hollywood Video 373-0899*), Museum (*USS Turner Joy 792-2457, Bremerton Naval Museum 479-7447*), Cultural

Attract (*Community Theatre 373-5152*), Galleries (*Fish Gallery 479-0077*)
3+ mi: Movie Theater (*Bremerton Cinema 478-0577, 4 mi.*)

Provisioning and General Services
Under 1 mi: Market (*Red Apple Market 377-3908*), Wine/Beer, Post Office, Catholic Church, Beauty Salon (*A&C Styling Salon & Barber Shop 479-9446*), Barber Shop, Newsstand, Retail Shops **1-3 mi:** Supermarket (*Albertson's 377-0220*), Delicatessen (*Village Bakery and Deli 377-1000*), Health Food (*Helen's Healing Foods 377-1664*), Liquor Store (*WA Liquor 373-6816*), Bakery, Bank/ATM, Protestant Church, Library (*Bremerton 405-9100*), Dry Cleaners, Laundry (*TCL Cleaners and Laundry 377-2399*), Pharmacy (*Walgreens 782-0901*), Hardware Store (*Lowe's 405-6270*), Department Store (*Wal-Mart 692-0923*), Copies Etc. (*Pak Mail 377-1372*)
3+ mi: Meat Market (*Sweeney's Country Style Meats & Seafood 692-8802/ Jensen's Smoked Meats and Fish 692-8802, 4 mi.*), Bookstore (*Barnes and Noble 613-5352, 9 mi.*), Buying Club (*Costco 692-1140, 9 mi.*)

Transportation
OnCall: Rental Car (*Enterprise 377-1900/Budget 479-4500, 4 mi.*), Taxi (*Arrow 479-5676*) **Near:** Local Bus (*Kitsap Transit 373-2877*) **3+ mi:** Ferry Service (*Bremerton-Seattle, 4 mi.*) **Airport:** Bremerton Nat'l. (*10.8 mi.*)

Medical Services
911 Service OnCall: Ambulance **Under 1 mi:** Dentist (*Dr. Pitcher 479-4380*), Veterinarian (*Relling Small Animal Hospital 373-1467*)
1-3 mi: Doctor (*Doctors Clinic 782-3300*), Chiropractor (*Alpine Chiropractic Clinic 479-2144*), Holistic Services, Optician (*Kitsap Optical 479-6301*)
Hospital: Harrison Memorial 377-3911 (*2 mi.*)

Setting -- Ilahee State Park is about 3 miles north of Bremerton and one mile south of Illahee, along Port Orchard Bay. The 75-acre park includes over 1700 feet of ocean frontage. The floating wood dock offers 360 feet of guest moorage - a gangway leads up to a sturdy pier. The floating breakwater provides some protection on the northern side. Remarkably, despite the presence of a realtively nearby town, there is a marked sense of wilderness.

Marina Notes -- *$10 minimum. **Alongside dockage. 5 mooring buoys are available with designated dinghy space on the inside of the dock. Annual dockage permits for Washington State Parks are available - the fee for pass holders is $3.50/ft.with a $50 minimum. Daily permits are only available at the parks. A tiled, well maintained bathhouse is nearby and includes a seasonal outdoor shower to rinse off the sand.

Notable -- Illahee means "earth" or "country" in the local Native American language. The park has a playground, two horseshoe pits, a softball field, and two volleyball fields. There are three kitchen shelters with electricity which can be reserved. Picnic tables and grills are near the pier overlooking the bay, and more are sprinkled throughout the park. Diving, shellfish harvesting and beachcombing are prime activities in this park. Oysters and clams are planted in shallow water in beds marked with yellow plastic tubes - they are abundant. Foragers are encouraged to shuck them on-site so the shells will provide an anchoring point for the next generation. Enticing, wooded hiking trails start near the water. Downtown Bremerton is about three miles away - catch a bus or a cab. Stroll the waterfront boardwalk, and visit the USS Turner Joy, a retired Vietnam-era destroyer (792-2457).

Navigational Information
Lat: 47°36.750' **Long:** 122°35.700' **Tide:** 12 ft. **Current:** n/a **Chart:** 18449
Rep. Depths *(MLW):* **Entry** 10 ft. **Fuel Dock** n/a **Max Slip/Moor** 10 ft./-
Access: On the west shore of Port Orchard about 3 mi. south of Battle Point

Marina Facilities *(In Season/Off Season)*
Fuel: No
Slips: 5 Total, 5 Transient **Max LOA:** 40 ft. **Max Beam:** n/a
 Rate *(per ft.):* **Day** Free* **Week** n/a **Month** n/a
 Power: 30 amp n/a, **50 amp** n/a, **100 amp** n/a, **200 amp** n/a
 Cable TV: No **Dockside Phone:** No
 Dock Type: Floating, Alongside, Wood
 Moorings: 0 Total, 0 Transient **Launch:** n/a
 Rate: Day n/a **Week** n/a **Month** n/a
Heads: 1 Toilet(s)
Laundry: None **Pay Phones:** Yes, 1
Pump-Out: No **Fee:** n/a **Closed Heads:** Yes

Marina Operations
Owner/Manager: Port of Illahee **Dockmaster:** Don Deitch
In-Season: Year-Round, 24/7 **Off-Season:** n/a
After-Hours Arrival: Tie up
Reservations: No **Credit Cards:** No
Discounts: None
Pets: Welcome **Handicap Access:** No

Port of Illahee

PO Box 2357; Allview Blvd NE; Bremerton, WA 98310

Tel: (360) 479-0216 **VHF: Monitor** n/a **Talk** n/a
Fax: n/a **Alternate Tel:** n/a
Email: n/a **Web:** n/a
Nearest Town: Illahee **Tourist Info:** (360) 479-3979

Marina Services and Boat Supplies
Supplies - 1-3 mi: West Marine *(479-2200)*, Boater's World *(478-4089)*, Bait/Tackle, Propane, CNG *(Ferrellgas 373-2515)*

Boatyard Services
1-3 mi: Travelift *(Kitsap Marine 895-2193)*, Railway, Engine mechanic *(gas, diesel)*, Launching Ramp, Electrical Repairs, Electronic Sales, Electronics Repairs, Bottom Cleaning, Propeller Repairs, Upholstery. **Nearest Yard:** Kitsap Marine (360) 895-2193

Restaurants and Accommodations
Under 1 mi: Inn/B&B *(Illahee Manor B&B 698-7555, $115-195, waterfront estate)* **1-3 mi:** Restaurant *(China House 377-8913, L $6-12, D $7-19)*, *(Shari's 373-1768, B $5-9, L $6-10, D $8-15)*, *(Putters Restaurant & Lounge 377-7077, B, L, D)*, *(Outback Steakhouse 479-4676)*, *(Spiro's Pizza & Pasta 377-6644)*, *(Sizzler 479-5748)*, *(Loscabos Grill 373-1320)*, Fast Food *(Burger King, Quiznos)*, Motel *(Midway Inn 479-2909)* **3+ mi:** Hotel *(Best Western Bremerton Inn 405-1111, $61-85, 5 mi.)*, *(Edgewood Villas 874-8300, 4 mi.)*

Recreation and Entertainment
1-3 mi: Dive Shop *(Sound Dive 373-6141)*, Tennis Courts *(Bremerton Tennis and Athletic 692-8075)*, Golf Course *(Rolling Hills Golf Course 479-1212)*, Fitness Center *(YMCA 377-3741)*, Video Rental *(Hollywood Video 373-0899)*, Park *(Illahee State Park)* **3+ mi:** Heated Pool *(Bayshore Pool 373-9962, 5 mi.)*, Movie Theater *(Charleston Cinema 373-6093, 6 mi.)*, Museum *(Bremerton Historic Ships 792-2457, Kitsap Co. Historical Society 479-6226, Bremerton Naval Museum 479-7447, 5 mi.)*, Cultural Attract *(Community Theatre 373-5152, 4 mi.)*, Galleries *(Fish Gallery 479-0077, 4 mi.)*

Provisioning and General Services
1-3 mi: Market *(Red Apple Market 377-3908)*, Supermarket *(Safeway 373-8911)*, Delicatessen, Health Food *(Helen's Healing Foods 377-1664)*, Wine/Beer, Liquor Store *(WA Liquor 373-6816)*, Bakery *(Village Bakery and Deli 377-1000)*, Meat Market *(Sweeney's Country Style Meats and Seafood 692-8802)*, Bank/ATM, Post Office, Catholic Church, Protestant Church, Synagogue *(Congregation Kol Shalom 855-0885)*, Library *(Bremerton 405-9100)*, Beauty Salon *(Trends 373-6325)*, Barber Shop *(C Barber Shop and Styling 792-1441)*, Dry Cleaners, Laundry *(TCL Cleaners and Laundry 377-2399)*, Bookstore *(Black Ink Books 377-9822)*, Pharmacy *(Medicine Shoppe 479-4830)*, Hardware Store *(Lowe's 405-6270)*, Department Store *(Wal-Mart 698-2889)*, Copies Etc. *(Pak Mail 377-1372)* **3+ mi:** Buying Club *(Costco 692-1140, 7 mi.)*

Transportation
OnCall: Rental Car *(Enterprise 377-1900/Hertz 373-0405, Budget 479-4500, 5 mi.)*, Taxi *(479-5676)* **Under 1 mi:** Local Bus *(Kitsap Transit 373-2877)* **3+ mi:** Ferry Service *(Bremerton to Seattle, 5 mi.)* **Airport:** Bremerton Nat'l. *(14 mi.)*

Medical Services
911 Service **OnCall:** Ambulance **1-3 mi:** Doctor *(Doctors Clinic 782-3400)*, Dentist *(Gentlecare Dental 377-9800)*, Chiropractor *(First Choice Chiropractic 377-6335)*, Veterinarian *(Alder Trail Animal Hosp. 377-3971)*, Optician *(Kitsap Optical 479-6301)* **Hospital:** Harrison Memorial 377-3911 *(3 mi.)*

Setting -- Illahee is a small remote community on the western shore of the Port Orchard Waterway, about a mile north of Illahee State Park. A gangway leads from two side-tie floats to a wide wooden pier. A fishing hole has been cut into the top of the deck and protected by a grate - it allows fishermen to sit on the bench and fish right on the dock. Private homes are perched up on the hillside surrounding the pier. Looking across the water, the visitor has a clear view of Bainbridge Island. Landside is the residential area of Illahee.

Marina Notes -- *Free. 3-day limit. There is no power, water, pump-out or boater amenities. The pay phone and porta-potty are across the street near the grocery store - which was closed at the time of our visit. Note: A fish haven, marked by buoys, extends about 140 feet from the end of the wharf.

Notable -- The neighboring area is residential, with no services. The nearest restaurant is Putters at Rolling Hills Golf Course (they serve three meals daily, starting at 8:30am during the week, 7:30 on weekends). Catch a cab to Route 303 in Bremerton for additional eateries and good shopping; Safeway and West Marine are about 2 miles away. In downtown Bremerton, the Admiral Theatre, a pre-WW II cinema, has been recently remodeled into an art deco performing arts center that books regional and national productions. Within walking distance of the theater are numerous art galleries, the Kitsap Historical Museum on Fourth Street, and the Bremerton Naval Museum on Pacific Avenue. The Vietnam-era destroyer "USS Turner Joy" is a floating museum on the Bremerton boardwalk- operated by the Bremerton Historic Ships Assoc. For a peaceful respite, take the boat or dinghy south to Illahee State Park.

Navigational Information
Lat: 47°39.035' **Long:** 122°36.811' **Tide:** 12 ft. **Current:** n/a **Chart:** 18446
Rep. Depths *(MLW)*: **Entry** 15 ft. **Fuel Dock** 5 ft. **Max Slip/Moor** 15 ft./-
Access: In Buke Bay on the western shore of Port Orchard Waterway

Marina Facilities *(In Season/Off Season)*
Fuel: *Shell* - Gasoline, Diesel
Slips: 310 Total, 30 Transient **Max LOA:** 100 ft. **Max Beam:** n/a
  Rate *(per ft.)*: **Day** $0.33* **Week** n/a **Month** n/a
  Power: 30 amp $3, 50 amp n/a, 100 amp n/a, 200 amp n/a
  Cable TV: No **Dockside Phone:** Yes
  Dock Type: Floating, Long Fingers, Alongside, Concrete, Wood
Moorings: 0 Total, 0 Transient **Launch:** n/a
  Rate: Day n/a **Week** n/a **Month** n/a
Heads: 6 Toilet(s), 4 Shower(s)
Laundry: 4 Washer(s), 4 Dryer(s) **Pay Phones:** Yes, 1
Pump-Out: OnSite, Self Service **Fee:** Free **Closed Heads:** Yes

Marina Operations
Owner/Manager: Port of Brownsville **Dockmaster:** Jerry Rowland
In-Season: Year-Round, 8am-5pm **Off-Season:** n/a
After-Hours Arrival: Self registration
Reservations: Groups only **Credit Cards:** Visa/MC, Dscvr, Amex
Discounts: None
Pets: Welcome, Dog Walk Area **Handicap Access:** Yes, Heads, Docks

Port of Brownsville

9790 Ogle Road NE; Bremerton, WA 98311

Tel: (360) 692-5498 **VHF: Monitor** Ch. 16 **Talk** Ch. 68
Fax: (360) 698-8023 **Alternate Tel:** n/a
Email: pob@portofbronsville.org **Web:** www.portofbrownsville.org
Nearest Town: Silverdale *(3 mi.)* **Tourist Info:** (360) 692-6800

Marina Services and Boat Supplies
Services - Security, Trash Pick-Up, Dock Carts **Communication -** Phone Messages, Fax in/out, Data Ports *(Wi-Fi - Broadband Xpress)* **Supplies -** **OnSite:** Ice *(Cube)*, Ships' Store *(Brownsville)*, Bait/Tackle, Propane **3+ mi:** West Marine *(479-2200, 4 mi.)*, Boater's World *(478-4089, 4 mi.)*

Boatyard Services
OnSite: Launching Ramp **Nearest Yard:** Port Orchard (360) 876-5535

Restaurants and Accommodations
OnSite: Snack Bar *(Brownsville Marine & Deli 692-4127, B $2-6, L $3-6, D $3-6)* **1-3 mi:** Restaurant *(Lynn's Kitchen 698-0127)*, *(Lin's Teriyaki Fish & Chips 698-1094)*, *(Golden Mum 692-3377, L $6-10, D $8-20)*, Inn/B&B *(Bird's Eye View B&B 698-2448)* **3+ mi:** Restaurant *(Mariner Grill 692-0748, 4 mi., waterfront, at Red Lion)*, *(Spiro's Pizza & Pasta 698-4800, 4 mi.)*, *(Bahn Thai 698-3663, 4 mi.)*, *(Osaka 698-7266, 4 mi.)*, Fast Food *(Subway, McDonald's 4 mi.)* Pizzeria *(Fairgrounds Pizzeria 698-0073, 4 mi.)*, Hotel *(Oakwood Silverdale Ridge 687-3322, 4 mi.)*, *(Red Lion 698-1000, $87-110, 4 mi.)*, *(Best Western 405-1111, $70-100, 9 mi.)*

Recreation and Entertainment
OnSite: Picnic Area *(Burke Bay Overlook Park)*, Grills **Near:** Park **1-3 mi:** Tennis Courts, Fitness Center *(Bremerton Tennis 692-8075)* **3+ mi:** Golf Course *(Rolling Hills 479-1212, 4 mi.)*, Bowling *(Bay Bowl 377-2103, 6 mi.)*, Movie Theater *(Kitsap Mall Cinema 692-4421, 4 mi.)*, Video Rental *(Hollywood Video 373-0899, 4 mi.)*, Museum *(Keyport Naval Undersea Museum 396-4148/Brementon Naval Museum 479-7447, 4 mi./6 mi.)*, Cultural Attract *(Center Stage Theater 692-9940, 4 mi.)*

Provisioning and General Services
OnSite: Convenience Store *(Strommes One-Stop)*, Delicatessen *(Brownsville Marine 692-4127)*, Wine/Beer, Copies Etc. **Under 1 mi:** Meat Market *(Sweeney's Country Style Meats 692-8802)* **1-3 mi:** Market *(Handy Andy's 692-7667)*, Liquor Store *(WA Liquor 613-16642)*, Bank/ATM, Post Office, Protestant Church, Beauty Salon *(Manila Hair Salon 308-8796)*, Barber Shop *(Michael's Head Quarters 698-1116)* **3+ mi:** Supermarket *(Safeway 692-4488, 4 mi.)*, Health Food *(Helen's Health Foods 698-1550, 4 mi.)*, Farmers' Market *(Silverdale Bucklin Hill Tue 11am-4pm 830-9502, 4 mi.)*, Catholic Church *(4 mi.)*, Library *(Silverdale 692-2779, 4 mi.)*, Dry Cleaners *(Silverdale Cleaners 698-2027, 4 mi.)*, Bookstore *(Barnes & Noble 613-5352, 4 mi.)*, Pharmacy *(Save-On Drugs 692-8088, 4 mi.)*, Hardware Store *(Lowe's 405-6270, 4 mi.)*, Department Store *(Kitsap Mall - Macy's, Sears, 4 mi.)*, Buying Club *(Costco 692-1140, 4 mi.)*

Transportation
OnCall: Rental Car *(Enterprise 779-6112/Budget 479-4500, 4 mi.)*, Taxi *(Taxis and Tours 842-7660)* **1-3 mi:** Local Bus *(Kitsap Transit 373-2877)* **Airport:** Bremerton Nat'l. *(15 mi.)*

Medical Services
911 Service **OnCall:** Ambulance **3+ mi:** Doctor *(Doctors Clinic 782-3100, 4 mi.)*, Dentist *(Heinemann 692-3030, 4 mi.)*, Chiropractor *(Mariner Chiro Center 692-5350, 4 mi.)*, Holistic Services *(Natural Options Health Clinic 698-4141, 4 mi.)*, Veterinarian *(Animal Hospital Center Kitsap 692-6162, 4 mi.)*, Optician *(Vista Optical Center 692-1337, 4 mi.)* **Hospital:** Harrison 337-8800 *(6 mi.)*

Setting -- Port of Brownsville sits in a rural area at the mouth of Burke Bay on the west side of the Port Orchard waterway - about 4.5 miles south of the Highway 305 bridge and 4.5 miles north of Point White at the tip of Bainbridge Island. A long dock borders the north and east sides of the marina, acting as a breakwater with side-tie moorage. Spiffy gazebos protect access to the floating concrete docks, many covered by boathouses. Just west of the docks a two-story blue-and-gray contemporary sports two new eating decks and houses the chandler-snack-bar-deli, harbormaster's office, and the new main bathhouse.

Marina Notes -- CCM. *Flat rates: 24' $9, 25-32' $11, 33-42' $14, 43-50' $18, over 50' $20. Dedicated transient space is along the north and east breakwaters, plus a number of 24' slips. Space is first-come, first-served; however, boat clubs can reserve moorage on the breakwater. Fish cleaning station. An electric lift on the gangway makes the north dock handicap accessible. The "Waterloo" is a state-of-the-art floating head conveniently located on the dock.

Notable -- The marina is well-suited for quiet getaways or rendezvous. Perched above the docks is Burke Bay Overlook Park - scenic views of the marina and bay coupled with covered picnic tables, barbeques, a fire pit, and multihued potted plants make this a delightful place to gather. The lower pavilion leads out to one of the guest docks and offers picnic tables and historical interpretive signs drawn by Larry Eifert. Brownsville Marine & Deli (Tuesday-Sunday 8am-8pm) carries groceries, boat supplies, wine & beer, and sandwiches. It also has a book exchange, video rentals, and espresso. Stromme's Store, in one of Brownsville's original buildings, has an ATM and groceries. The nearest bus stop is a 1.5 mile trek; a bus runs from East Bremerton to Kitsap Mall hourly.

Navigational Information
Lat: 47°43.984' **Long:** 122°38.780' **Tide:** 13 ft. **Current:** n/a **Chart:** 18446
Rep. Depths *(MLW)*: **Entry** 12 ft. **Fuel Dock** 5 ft. **Max Slip/Moor** 12 ft./-
Access: Located on the north shore of Liberty Bay off Manzanita Bay

Marina Facilities *(In Season/Off Season)*
Fuel: Gasoline, Diesel
Slips: 266 Total, 130 Transient **Max LOA:** 75 ft. **Max Beam:** n/a
  Rate *(per ft.)*: **Day** $0.50* **Week** n/a **Month** n/a
  Power: 30 amp Incl., **50 amp** n/a, **100 amp** n/a, **200 amp** n/a
  Cable TV: No **Dockside Phone:** No
  Dock Type: Floating, Long Fingers, Concrete
Moorings: 0 Total, 0 Transient **Launch:** n/a
  Rate: Day n/a **Week** n/a **Month** n/a
Heads: 5 Toilet(s) 6 Shower(s)
Laundry: 2 Washer(s), 2 Dryer(s) **Pay Phones:** No
Pump-Out: OnSite, 1 Central **Fee:** Free **Closed Heads:** Yes

Marina Operations
Owner/Manager: Port of Poulsbo **Dockmaster:** Kirk Stickels
In-Season: Year-Round, 8am-8pm **Off-Season:** n/a
After-Hours Arrival: Use pay envelope or pay in the morning
Reservations: Groups, Limited Individual **Credit Cards:** Visa/MC, Amex
Discounts: None
Pets: Welcome, Dog Walk Area **Handicap Access:** No

Port of Poulsbo

PO Box 732; 18721 Front Street; Poulsbo, WA 98370

Tel: (360) 779-3505 **VHF: Monitor** Ch. 66A **Talk** n/a
Fax: (360) 779-8090 **Alternate Tel:** n/a
Email: portofpoulsbo@yahoo.com **Web:** www.poulsbo.net/portofpoulsbo
Nearest Town: Poulsbo **Tourist Info:** (360) 779-4848

Marina Services and Boat Supplies
Services - Security *(gated docks)*, Trash Pick-Up, Dock Carts
Communication - Mail & Package Hold, Phone Messages, Data Ports *(Wi-Fi - Broadband Xpress)*, FedEx, DHL, UPS, Express Mail **Supplies -**
OnCall: Ice *(Cube)* **Near:** Bait/Tackle

Boatyard Services
OnSite: Launching Ramp **Nearest Yard:** Dockside (360) 786-9016

Restaurants and Accommodations
Near: Restaurant *(Poulsbo Pasta Co 697-7278, L $6-10, D $9-10)*, *(Bayside Broiler 779-9076, L $8-10, D $16-21)*, *(Golden Dragon 779-7673, L $5-10, D $5-10)*, *(Mor Mor Bistro 697-3449, L $6-10, D $12-22, 11-9pm - great reviews)*, Seafood Shack *(JJ's Fishouse 779-6609)*, Coffee Shop *(Gazebo Café 697-1447, L $5-8)*, Lite Fare *(Sheila's Bay Café 779-3921, B $5-8, L $6-8, adjacent to the docks - kids' menu $2-3; evenings - Portside Pub)*, Pizzeria *(Pizza Prima 697-3400)*, Inn/B&B *(Murphy House 779-1600)*
Under 1 mi: Restaurant *(Chung's Teriyaki 697-7703)*, *(Vege 697-2538)*, Fast Food *(Burger King, FKC)*, Motel *(Holiday Inn 697-2119, $89-149)*, Inn/B&B *(Poulsbo Inn 779-3921, $81-120)*

Recreation and Entertainment
OnSite: Picnic Area **Near:** Boat Rentals *(Olympic Outdoor Center 697-6095 - kayaks and canoes)*, Park *(waterfront with 600-foot boardwalk)*, Cultural Attract *(Marine Science Center 779-5949 $4/3/2; Jewel Box Theatre 779-9688)*, Special Events *(Viking Fest - May, Passagemaker's TrawlerFest - Jun, Arts by the Bay - Aug 10am-5pm 779-2098, Classic Yacht Assoc. Rendevous - Sep 297-3281)* **Under 1 mi:** Fitness Center *(Poulsbo Athletic Club 779-3285)*, Movie Theater *(Regal 697-3080)*, Video Rental *(Poulsbo Market Place 779-9778)*, Galleries *(Olympic Inn 697-5677; Things Northwest 779-7180)* **3+ mi:** Golf Course *(Meadowmeer 842-2218, 10 mi.)*

Provisioning and General Services
OnSite: Market *(Marina Market 779-8430)* **Near:** Delicatessen *(Poulsbo Country 779-2763)*, Bakery *(Sluy's 779-2798 - pastries, bread, & passionate devotees)*, Bank/ATM, Post Office *(779-1353)*, Protestant Church, Library *(Poulsbo 779-2915)*, Beauty Salon *(Head Hunter 394-1777)*, Bookstore *(Liberty Bay 779-5909, Book Stop 779-9773)*, Pharmacy *(Poulsbo 779-2737)*, Retail Shops *(Liberty Bay Mall)*, Copies Etc. *(Kitsap Printing 697-2286)* **Under 1 mi:** Convenience Store *(Poulsbo Foodmart 779-8022)*, Supermarket *(Albertson's 697-2302)*, Wine/Beer, Liquor Store *(WA Liquor 779-2816)*, Farmers' Market *(Sat. 9am-1pm Iverson & 7th Ave./Wed 12-4pm at Hostmark St. & Hwy. 305)*, Catholic Church, Dry Cleaners, Laundry *(Suds-Eez Cleaners & Laundromat 394-7768)*, Hardware Store *(Coast Do It Best 779-2000)*, Florist *(Jacqui's 779-2110)*

Transportation
OnCall: Rental Car *(Enterprise 779-6112)*, Taxi *(Bainbridge 697-4442)*
Near: Local Bus *(697-2877)* **Airport:** Brementon Nat'l. *(25 mi.)*

Medical Services
911 Service **Near:** Holistic Services *(Sound Naturopathic Clinic 598-6999)*
Under 1 mi: Doctor *(Peninsula Community Health 779-1963)*, Dentist *(Village Dental 697-4610)*, Chiropractor *(Wunderful Health 779-0555)*, Veterinarian *(North Kitsap Vet Clinic 779-3414)* **1-3 mi:** Optician *(Pacific Optical 779-9300)* **Hospital:** Harrison 337-8800 *(11 mi.)*

Setting -- Port of Poulsbo is near the head of Liberty Bay fjord - a narrow inlet extending about four miles north of Manzanita Bay. The harbor is protected on the south and west sides by an angled, timbered breakwater marked by private lights. The Port's two main guest piers, hosting 230 floating docks, are on the east shore, past Liberty Bay Marina and Poulsbo Yacht Club. Services are housed in a handful of gray clapboard structures highlighted by bright blue trim. An adjacent park has a picnic area with views of the harbor.

Marina Notes -- *$15/day min. 130 visitor slips, 60 reservable for groups of 15 or more boats; some individual reservations. The dedicated staff has an office on the dock for easy payment and slip assignment. A multi-purpose room behind the pay station can be rented for special events. Well-maintained bathhouse with tiled heads, showers (fee), and laundry. Note: Liberty Bay Marina does not have transient slips; Poulsbo Y.C. offers some reciprocal moorage.

Notable -- Poulsbo is commonly referred to as "Little Norway on the Fjord;" its early settlers were from Scandinavia and were attracted to the area by its resemblance to their homeland. Today special events still include Viking Fest in May and Scandia Midsommarfest. Downtown Poulsbo's restaurants, outdoor murals and shops - their facades decorated with rosemaling - are a pleasant five minute walk from the docks; Poulsbo Village Shopping Center is a half mile. The waterfront Poulsbo Marine Science Center features exhibits on Puget Sound flora and fauna, touch tanks, a whale skeleton, and gift shop (daily 11am-5pm). And yet, sea lions still relax on the docks; eagles scour the bay for snacks, and king salmon make the fall run north to Dogfish Creek.

14. WA – Seattle Area

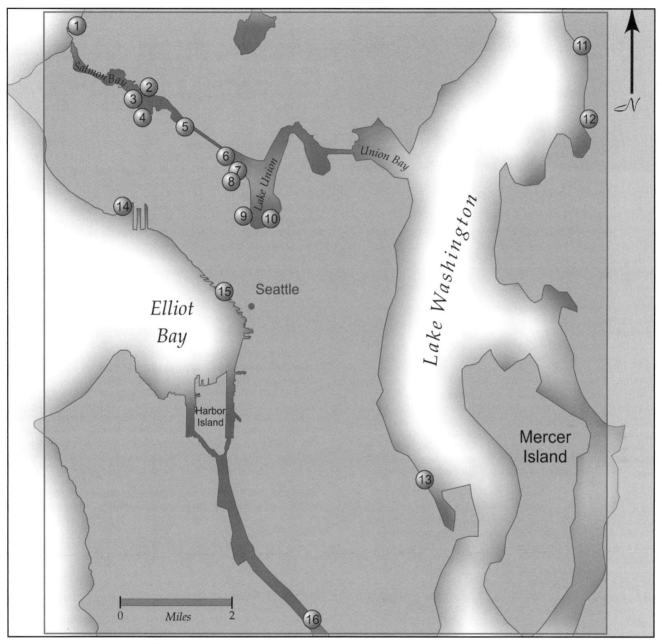

MAP	MARINA	HARBOR	PAGE	MAP	MARINA	HARBOR	PAGE
1	**Shilshole Bay Marina**	*Shilshole Bay*	*200*	9	**AGC Marina**	*Lake Union*	*208*
2	**Ballard Mill Marina**	*Lake Washington Ship Canal*	*201*	10	**Fairview Marinas**	*Lake Union*	*209*
3	**Salmon Bay Marina**	*Lake Wash. Ship Canal/Salmon Bay*	*202*	11	**Kirkland Marina Park**	*Lake Washington*	*210*
4	**Fishermen's Terminal**	*Lake Wash. Ship Canal/Salmon Bay*	*203*	12	**Carillon Point Marina**	*Lake Washington*	*211*
5	**Ewing Street Moorings**	*Lake Washington Ship Canal*	*204*	13	**Lakewood Moorage**	*Lake Washington/Andrews Bay*	*212*
6	**Lee's Landing**	*Lake Union*	*205*	14	**Elliott Bay Marina**	*Elliott Bay*	*213*
7	**Commercial Marine**	*Lake Union*	*206*	15	**Bell Harbor Marina**	*Elliott Bay*	*214*
8	**Nautical Landing**	*Lake Union*	*207*	16	**South Park Marina**	*Duwamish River*	*215*

▸ **Currency** — In Canadian Marina Reports, all prices are in Canadian dollars. In U.S. Marina Reports, all prices are in U.S. dollars.

▸ **"CCM"** — Denotes a Certified Clean Marina, a state/provincial award for environmental excellence. See page 298 for an explanation and page 299 for a list of Pump-Out facilities.

▸ **Ratings & Reviews** — An overview of the Atlantic Cruising Club's rating system is on page 6 and details on the content of each Marina Report are on pages 7 – 11.

▸ **Marina Report Updates** — Comments from boaters and new information from ACC reviewers and marinas are posted regularly on *www.AtlanticCruisingClub.com*.

Shilshole Bay Marina

7001 Seaview Avenue NW; Seattle, WA 98117

Tel: (206) 728-3006; (800) 426-7817 **VHF: Monitor** Ch. 17 **Talk** Ch. 17
Fax: (206) 728-3391 **Alternate Tel:** (206) 601-4089
Email: sbm@portseattle.org **Web:** www.portseattle.org
Nearest Town: Seattle *(5 mi.)* **Tourist Info:** (206) 784-9705

Navigational Information
Lat: 47°40.717' **Long:** 122°24.475' **Tide:** 17 ft. **Current:** n/a **Chart:** 18446
Rep. Depths *(MLW)*: **Entry** 15 ft. **Fuel Dock** 20 ft. **Max Slip/Moor** 30 ft./-
Access: Puget Sound just north of Lake Washington Ship Canal

Marina Facilities *(In Season/Off Season)*
Fuel: *Bio-Diesel available* - Gasoline, Diesel
Slips: 1476 Total, 90 Transient **Max LOA:** 130 ft. **Max Beam:** 30 ft.
 Rate *(per ft.)*: **Day** $1.00/.75* **Week** n/a **Month** n/a
 Power: 30 amp $3, **50 amp** $5, **100 amp** Inq., **200 amp** Inq.
 Cable TV: Yes every slip **Dockside Phone:** No
 Dock Type: Floating, Long Fingers, Alongside
Moorings: 0 Total, 0 Transient **Launch:** n/a
 Rate: Day n/a **Week** n/a **Month** n/a
Heads: 20 Toilet(s), 20 Shower(s)
Laundry: 6 Washer(s), 6 Dryer(s), Book Exchange **Pay Phones:** Yes, 6
Pump-Out: OnSite, 3 Central **Fee:** Free **Closed Heads:** Yes

Marina Operations
Owner/Manager: Port of Seattle **Dockmaster:** Sharon Briggs
In-Season: Year-Round, 24/7** **Off-Season:** n/a
After-Hours Arrival: Ch. 17 or 206-601-4089. Pay station at office
Reservations: No, until renovation complete **Credit Cards:** Visa/MC, Amex
Discounts: None
Pets: Welcome, Dog Walk Area **Handicap Access:** Yes, Heads, Docks

Marina Services and Boat Supplies
Services - Docking Assistance, Security *(24 hrs., staff onsite)*, Dock Carts
Communication - Mail & Package Hold, Fax in/out *($0.50/pg)*, Data Ports
(Wi-Fi - Broadband Xpress), FedEx, DHL, UPS, Express Mail **Supplies -**
OnSite: Ice *(Cube)*, Bait/Tackle, Propane **Near:** West Marine *(789-4640)*

Boatyard Services
OnSite: Travelift *(35T)*, Crane *(4T)*, Engine mechanic *(gas, diesel)*,
Launching Ramp *(2-lane)*, Electrical Repairs, Hull Repairs, Rigger, Bottom
Cleaning, Brightwork, Compound, Wash & Wax, Interior Cleaning, Propeller
Repairs, Woodworking, Painting, Awlgrip, Yacht Broker *(Sailboats at
Shilshole 789-8044)* **Near:** Electronic Sales, Electronics Repairs, Sail Loft
(Schattauer Sails 783-2400), Canvas Work. **Under 1 mi:** Air Conditioning,
Refrigeration, Divers, Inflatable Repairs, Life Raft Service, Upholstery.

Restaurants and Accommodations
OnSite: Lite Fare *(Little Coney 782-6598, L & D $4-10)* **Near:** Restaurant
(Anthony's Homeport 783-0780), *(Ray's Boathouse 782-0094, L & D at the
Café $10-18; D in restaurant, $20-38)* **Under 1 mi:** Restaurant *(Totem
House Seafood 784-2300)*, *(Lockspot Fish & Burger 789-4865)*
1-3 mi: Pizzeria *(Pizza Hut 783-3700)*, Motel *(Star Light Motel 784-8304)*

Recreation and Entertainment
OnSite: Playground *(with a fountain)*, Boat Rentals *(Wind Works Sailing 784-
9386)* **Near:** Beach, Picnic Area, Grills, Park *(Golden Gardens)*
Under 1 mi: Jogging Paths, Video Rental *(Rain City Video 789-0132)*, Museum *(Nordic Heritage Museum 789-5707)* **1-3 mi:** Heated Pool *(Ballard Pool
684-4094)*, Fitness Center *(Curves 789-1965)*, Fishing Charter *(Adventure

Charters 789-8245)*, Movie Theater *(Majestic Bay 781-2229)*, Sightseeing
(Ballard Locks & Botanical Gardens - free tours Mar-Nov 783-7059)
3+ mi: Golf Course *(Green Lake 632-2280, 5 mi.)*

Provisioning and General Services
OnSite: Convenience Store, Wine/Beer, Bank/ATM **Near:** Bookstore
(Abraxus Books 789-8262) **Under 1 mi:** Market *(Sunset Hill Greenmarket
784-7594)*, Liquor Store *(WA Liquor 706-5892)*, Lobster Pound,
Crabs/Waterman, Beauty Salon *(Sweet Lilly 297-0222)*, Barber Shop
(Sunset Hill Barber 782-6820), Dry Cleaners **1-3 mi:** Supermarket *(Safeway
782-7464)*, Gourmet Shop, Delicatessen *(Ballard Locks Grocery & Deli 784-
5751)*, Health Food *(Pilgrim's Natureway 782-6377)*, Bakery *(Blue Ribbon
783-4055)*, Farmers' Market, Fishmonger *(Fresh Fish Co. 782-1632)*, Post
Office *(784-5410)*, Library *(Ballard 684-4089)*, Laundry, Pharmacy *(Bartell
Drugs 783-3050)*, Hardware Store *(Ballard Hardware 783-6626)*, Clothing
Store, Department Store *(Fred Meyer One Stop Shopping 784-3140)*, Copies
Etc. *(Kinko's 784-0061)* **3+ mi:** Buying Club *(Sam's 362-6700, 6 mi.)*

Transportation
OnCall: Rental Car *(Enterprise 783-5565)*, Taxi *(Redtop Taxi 789-4949)*
Near: Local Bus *(Metro Trans. 800-542-7876)* **3+ mi:** Rail *(Amtrak 382-
4125, 8 mi.)* **Airport:** Sea-Tac Int'l. *(17 mi.)*

Medical Services
911 Service **OnCall:** Ambulance **Near:** Dentist *(Gentle Dental 782-9183)*
1-3 mi: Doctor *(Olympic Medical Center 782-1133)*, Chiropractor *(Vitality
Chiropractic 297-2792)*, Veterinarian, Optician **3+ mi:** Holistic Services
(Seattle Holistic 525-9035, 5 mi.) **Hospital:** Northwest 364-0500 *(6 mi.)*

Setting -- Shilshole Bay Marina lies at the foot of Seattle's Sunset Hill area - north of the Lake Washington ship canal, a few blocks from the Hiram M. Chittenden Locks. A 4400-foot rock breakwater parallels the shore; lights at each end indicate the access points. A mile-long promenade, ending at 88-acre Golden Garden State Park, skirts 24 concrete floating piers that host nearly 1500 slips. All have spectacular views of the Olympic Mountains. Shoreside, the yellow & white observation tower, labeled "Shilshole," rises above Port of Seattle's long, green, glass two-story office building.

Marina Notes -- CCM. *$1.00/ft. under 49', over 50' $1.50/ft. **Staffed 24 hrs. Office 8:30am-4:30pm Mon-Fri, 8am-3pm Sat, Closed Sun. Fuel dock 8am-8pm summer, 9am-5pm winter. A store on the dock provides groceries, ice, beer, wine, and bait. A massive $80 million project is replacing and widening all docks, adding 4,000 linear feet of moorage and expanding dedicated transient space - expected completion May '08. Guest moorage limited in the interim - call ahead. At the south end is Seaview West B.Y. (783-6550). At the north end, the Corinthian Yacht Club (789-1919) and Seattle Sailing Club (782-5100) may have reciprocal dockage. Brand new bathhouses are at several locations. Additional heads are in the new one-story, 12,000 square foot marina building.

Notable -- At the north end of the marina, Little Coney has served burgers, chowder, fish & chips since 1957 (10am-10pm). A new dockside Anthony's Restaurant opens summer '07. Additional eateries are south of the marina, including Ray's Boathouse, a Seattle institution, as is a West Marine on Seaview Ave. At the north end of the marina, popular Golden Gardens Park features a sandy beach, fishing pier, forested trails, picnic areas, and off-leash dog area.

Navigational Information

Lat: 47°39.770' **Long:** 122°22.793' **Tide:** n/a **Current:** 2 kt. **Chart:** 18447
Rep. Depths (MLW): Entry 30 ft. **Fuel Dock** n/a **Max Slip/Moor** 30 ft./-
Access: Lake Washington's Ship Canal, inside the locks on north side

Marina Facilities (In Season/Off Season)

Fuel: No
Slips: 115 Total, 5 Transient **Max LOA:** 55 ft. **Max Beam:** n/a
  Rate (per ft.): **Day** $1.00 **Week** n/a **Month** $8.13-9.72*
  Power: 30 amp Incl., **50 amp** n/a, **100 amp** n/a, **200 amp** n/a
  Cable TV: No **Dockside Phone:** No
  Dock Type: Floating, Long Fingers, Alongside
Moorings: 0 Total, 0 Transient **Launch:** n/a
  Rate: Day n/a **Week** n/a **Month** n/a
Heads: 4 Toilet(s), 4 Shower(s)
Laundry: 6 Washer(s), 6 Dryer(s), Iron, Iron Board **Pay Phones:** Yes, 1
Pump-Out: OnSite, 1 Central **Fee:** Free **Closed Heads:** Yes

Marina Operations

Owner/Manager: Willy Jenkins **Dockmaster:** Same
In-Season: Year-Round, 9am-5pm **Off-Season:** n/a
After-Hours Arrival: Call ahead
Reservations: No **Credit Cards:** Visa/MC, Amex
Discounts: None
Pets: Welcome, Dog Walk Area **Handicap Access:** Yes, Heads, Docks

Ballard Mill Marina

4733 Shilshole Avenue NW; Seattle, WA 98107

Tel: (206) 789-4777 **VHF: Monitor** n/a **Talk** n/a
Fax: (206) 706-0405 **Alternate Tel:** n/a
Email: bmm@surfbest.net **Web:** www.portseattle.org
Nearest Town: Seattle **Tourist Info:** (206) 389-7200

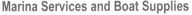

Marina Services and Boat Supplies

Services - Dock Carts **Supplies - Near:** Ships' Store (Open Boat Shop 909-3625), Bait/Tackle (LFS Marine 789-8110/ Neptune Marine 789-3790) **Under 1 mi:** Propane (U-Haul 784-2100) **1-3 mi:** West Marine (789-4640)

Boatyard Services

Nearest Yard: Seaview East (206) 789-3030

Restaurants and Accommodations

Near: Restaurant (Bad Albert's Tap and Grill 782-9623, B $6-9, L $6-15, D $6-15, B Sat-Sun only), (Louie's Cuisine of China 782-8855, L $6-10, D $8-17), (Salmon Bay Café 782-5539), (Hattie's Hat 784-0175), Motel (Star Light 784-8304) **Under 1 mi:** Restaurant (Thai Ku 706-7807), (India Bistro 783-5080), (Market St. Urban Grill 789-6766), (Lombardi's Cucina 783-0055), Coffee Shop (Starbucks), Fast Food (Subway), Lite Fare (Vera's 782-9966, voted best breakfast) **1-3 mi:** Pizzeria (Pizza Hut 783-3700), Motel (Howard Johnson Express Plaza 284-1900), (Lake Union 632-2550), (Bridge Motel 632-7835), Inn/B&B (Inn of Twin Gables 284-3979, $100-160)

Recreation and Entertainment

Near: Fitness Center (Olympic Athletic Club 789-5010) **Under 1 mi:** Boat Rentals (Seattle Boat Charters 782-3003), Movie Theater (Majestic Bay Theatres 781-2229), Video Rental (Rain City Video 783-8367) **1-3 mi:** Golf Course (Greenlake 632-2280), Fishing Charter (Anchor Bay Charters 781-0709), Group Fishing Boat (Sea Joy Charters and Marine 781-4627), Museum (Nordic Heritage Museum 789-5707, History House 675-8875), Sightseeing (Woodland Park Zoo 684-4800), Galleries (The Glass Eye 441-3221) **3+ mi:** Cultural Attract (Seattle Aquarium 386-4320, 4 mi.)

Provisioning and General Services

Near: Convenience Store (7-Eleven 782-5270), Farmers' Market (Ballard Ave. Sun 10am-4pm 282-5706), Bank/ATM, Newsstand, Florist
Under 1 mi: Market (Dish Urban Market 297-1852), Supermarket (Safeway 782-1831), Gourmet Shop (Gourmet Food Worlds 782-3206), Delicatessen (Mr. J's Deli Market 789-3900), Health Food, Wine/Beer, Liquor Store (WA Liquor 706-5892), Bakery (Tall Glass Bakery 706-0991), Green Market, Fishmonger (Wild Salmon Seafood Market 283-3366), Meat Market (Butcher Shoppe 783-0454), Shrimper (Fresh Fish Co. 782-1632), Post Office (784-5410), Catholic Church, Protestant Church, Library (Ballard 684-4089), Beauty Salon (Cloud 9 782-4950), Barber Shop (Charlie's 284-6477), Dry Cleaners (Sparkle 782-7826), Laundry (Fremont Ave. 632-8924), Bookstore (Ballard Books 782-0820), Pharmacy (Walgreens 781-0054), Hardware Store (Ballard Hardware 783-6626), Retail Shops (Scandinavian Gifts 784-9370, Radio Shack 784-8986) **3+ mi:** Department Store (Sears 364-9000, 6 mi.), Buying Club (Sam's 362-6700, 6 mi.)

Transportation

OnCall: Airport Limo (Airport Express 568-1972/Seattle Airporter 547-4700, 5 mi.) **Under 1 mi:** Local Bus (800-542-7876) **1-3 mi:** Taxi (Redtop Taxi 789-4949) **3+ mi:** InterCity Bus (Greyhound 628-5526, 5 mi.), Ferry Service (Bainbridge & Bremerton, 6 mi.) **Airport:** Sea-Tac Int'l. (18 mi.)

Medical Services

911 Service **OnCall:** Ambulance **Under 1 mi:** Doctor, Dentist (Family Dentisty 781-9204), Chiropractor (Ballard 784-3895), Veterinarian (Ballard Animal Hosp. 782-4222) **1-3 mi:** Holistic Services (Emerald City Naturopath 781-2206) **Hospital:** Northwest 364-0500 (6 mi.)

Setting -- Ballard Mill Marina is located a quarter mile above the Hiram M. Chittenden Locks and before the Ballard Bridge, on the north shore of fresh water Salmon Bay. It is tucked between two commercial marinas, in what is primarily a working waterfront area - surrounded by boatyards. A barbeque, table and chairs offer the potential of an outdoor meal overlooking the marina. Within an easy walk are many businesses and restaurants.

Marina Notes -- *Monthly rate varies from $8.13-$9.72/ft/mo. by slip size and location. Very limited transient moorage. The docks are clean and uncluttered. While the marina offers the boater few amenities, downtown Ballard and the Seattle area offer exciting sightseeing adventures. Bathhouse and laundy. Seaview East Boatyard, immediately next door, provides complete marine repair services and a good ship's chandlery. Tide is not an issue above the locks.

Notable -- The town of Ballard has a distinctive Scandinavian feel, reflecting the heritage of those who originally settled the area. Today it hosts an eclectic mix of artists, mill workers and fishermen. Nearby Ballard Avenue offers several shops, pubs, restaurants and a motel. It's also home to a Sunday Farmers' Market that is within walking distance of the marina. To the north, Market Street is lined with services, restaurants, galleries, banks, plus a movie theater. Just over a mile east is Seattle's Woodland Park Zoo, open daily year-round ($10.50/7.50). Downtown Seattle is within easy reach via Metro Bus or cabs. Don't miss Pike Place Market, with a huge assortment of stores and the famous fish market "where guys throw the fish" (tours by res. Wed-Sun $8/6, 774-5249).

Salmon Bay Marina

Salmon Bay Marina

2100 W Commodore Way; Seattle, WA 98199

Tel: (206) 282-5555 **VHF: Monitor** n/a **Talk** n/a
Fax: (206) 282-8482 **Alternate Tel:** n/a
Email: sales@salmonbaymarina.com **Web:** salmonbaymarina.com
Nearest Town: Seattle **Tourist Info:** (206) 284-5836

Navigational Information
Lat: 47°39.574' **Long:** 122°22.953' **Tide:** n/a **Current:** 2 kt. **Chart:** 18447
Rep. Depths (MLW): Entry 20 ft. **Fuel Dock** n/a **Max Slip/Moor** 20 ft./-
Access: South side of Ship Canal above the locks

Marina Facilities (In Season/Off Season)
Fuel: No
Slips: 170 Total, 2 Transient **Max LOA:** 65 ft. **Max Beam:** n/a
  Rate (per ft.): **Day** $1.00 **Week** n/a **Month** $8-13.25*
  Power: 30 amp Incl.**, **50 amp** n/a, **100 amp** n/a, **200 amp** n/a
  Cable TV: No **Dockside Phone:** No
  Dock Type: Fixed, Floating, Long Fingers, Short Fingers, Concrete, Wood
Moorings: 0 Total, 0 Transient **Launch:** n/a
  Rate: Day n/a **Week** n/a **Month** n/a
Heads: 2 Toilet(s), 1 Shower(s), Phone for local calls
Laundry: 2 Washer(s), 2 Dryer(s) **Pay Phones:** No
Pump-Out: No **Fee:** n/a **Closed Heads:** Yes

Marina Operations
Owner/Manager: Leslie Campbell **Dockmaster:** Same
In-Season: Year-Round, 8am-5pm **Off-Season:** n/a
After-Hours Arrival: Make arrangements with office
Reservations: Yes, Required **Credit Cards:** Visa/MC
Discounts: None
Pets: Welcome **Handicap Access:** No

Marina Services and Boat Supplies
Services - Trash Pick-Up, Dock Carts **Communication -** Mail & Package Hold, Fax in/out, FedEx, DHL, UPS, Express Mail (Sat Del) **Supplies - Near:** Ice (Block, Cube), Ships' Store (Seattle Marine & Fishing Supply 285-5010) **1-3 mi:** Bait/Tackle, CNG **3+ mi:** West Marine (789-4640, 4 mi.)

Boatyard Services
OnCall: Electrical Repairs **Near:** Travelift, Crane, Engine mechanic (gas, diesel), Hull Repairs, Bottom Cleaning, Brightwork, Air Conditioning, Refrigeration, Compound, Wash & Wax, Interior Cleaning, Woodworking, Metal Fabrication, Painting, Awlgrip, Yacht Design, Yacht Building, Total Refits. **1-3 mi:** Launching Ramp, Rigger, Sail Loft, Canvas Work, Divers, Propeller Repairs, Inflatable Repairs, Life Raft Service, Upholstery, Yacht Interiors. **Nearest Yard:** Salmon Bay Boat Works (206) 633-4400

Restaurants and Accommodations
Near: Restaurant (Sunny Teriyake 281-9339), Pizzeria (Pudge Brothers Pizza 213-0333) **Under 1 mi:** Restaurant (Katina's Kitchen 284-0370), (La Palma 284-1001, Mexican), (Crab King 283-2722), (Romio's Pizza & Pasta 284-5420), Seafood Shack (Little Chinook's Fish & Chips 283-4665, L & D $7-10), Coffee Shop (Discovery Espresso & Juice 286-1481), Fast Food (Subway, McDonald's, Wendy's), (Jack in the Box), Lite Fare (Bay Café 282-3435), Inn/B&B (Inn of Twin Gables 284-3979, $100-160)
1-3 mi: Restaurant (Paragon 283-4548), Pizzeria (Pizza Hut 783-3700), Motel (Howard Johnson 284-1900), (BW International 283-4140)

Recreation and Entertainment
Under 1 mi: Park (Commodore Park or Discovery Park, 1.5 mi.)

1-3 mi: Golf Course (Interbay Family Golf Center 285-4195), Fitness Center (Bodyfit Balance 378-0091), Movie Theater (Majestic Bay Theaters 781-2229), Video Rental (Blockbuster 782-0645), Sightseeing (Woodland Park Zoo 684-4800) **3+ mi:** Video Arcade (Quarters 625-9196, 4 mi.)

Provisioning and General Services
Near: Wine/Beer, Bank/ATM, Dry Cleaners (Crowne Cleaners 213-0826)
Under 1 mi: Convenience Store (7-Eleven 782-5270, Fishermen's Grocery 281-7818), Delicatessen, Liquor Store (WA Liquor 545-6648), Fishmonger (Wild Salmon 283-3366), Post Office, Library (Magnolia 386-4225), Beauty Salon (Fashion Beauty 782-5060), Barber Shop (Charlie's 284-6477), Hardware Store (Ace 283-6060) **1-3 mi:** Supermarket (Safeway 784-6480, QFC 283-3600), Farmers' Market (Sat 10am-2pm 34th Ave. W), Meat Market (Butcher Shoppe 783-0454), Catholic Church (St. Margaret's 282-1804), Protestant Church, Laundry (Fremont Ave. 632-9924), Bookstore (Ballard Books 782-0820), Pharmacy (QFC 283-0366), Department Store (Fred Meyer 784-3140) **3+ mi:** Buying Club (Costco 621-9997, 7 mi.)

Transportation
OnCall: Rental Car (Enterprise 783-5565), Taxi (Yellow Cab 455-4999)
Near: Local Bus (Metro Trans. 800-542-7876) **3+ mi:** Rail (Amtrak 382-4125, 5 mi.), Ferry Service (Bainbridge Ils./Bremerton 464-6400, 5 mi.)
Airport: Sea-Tac Int'l. (17 mi.)

Medical Services
911 Service **OnCall:** Ambulance **Near:** Dentist (Yoder 285-1338) **Under 1 mi:** Doctor (Swedish Medical Center 781-6341), Chiropractor (Chiropractic Centre 285-1068), Veterinarian **Hospital:** Virginia Mason 624-1144 (6 mi)

Setting -- Fresh water Salmon Bay is located about 0.8 mile inside the Hiram M. Chittenden Locks on the south shore of the Canal, before Fishermen's Terminal and Ballard Bridge. There are numerous piers & floats with extensive small-craft facilities on the bay. Salmon Bay Marina, a basic facility in a fairly industrial area, is one of the very few that offers overnight transient moorage. A small white office building anchors the docks.

Marina Notes -- *Monthly rates range from $8-13.25/ft. **Monthly electric is a minimum $17/mo. Built in '61, this family owned and operated facility offers open and covered slips in a protected area. As available, the marina will use the service dock or vacant slips for guests. Call ahead to reserve. They also sell boats and yachts through their marina brokerage. Limited liveaboard space and do-it-yourself yard. Seattle Marine & Fishing Supply is adjacent. The heads and the laundry are bright and inviting. There are few dockside amenities, but Seattle's attractions are a short bus ride away.

Notable -- East of the marina is Seattle's Magnolia neighborhood, home to Discovery Park. 1.5 miles away, it is accessible by bus and features 530 acres of gardens, forest trails, beaches, and dunes overlooking the sound. A short walk west is historic Fishermen's Terminal, which berths Seattle's commercial fishing fleet, as well as a fish market and some eateries. Across the Canal is Ballard, settled by Scandinavian fishermen and known as a "town within a city." Free tours of the locks are offered in season (783-7059). Also on the north shore, about 3 miles away, is Woodland Park Zoo (daily from 9:30am, $10/$7.50). A Zoo Evening Adventure or Zoo Overnight Adventure lets guests enjoy the park after hours, as the animals prepare for sleep.

PHOTOS ON DVD: 9

Navigational Information

Lat: 47°39.373' **Long:** 122°22.785' **Tide:** n/a **Current:** n/a **Chart:** 18447
Rep. Depths (*MLW*): **Entry** 25 ft. **Fuel Dock** 25 ft. **Max Slip/Moor** 28 ft./-
Access: South side of Lake Washington's Ship Canal

Marina Facilities *(In Season/Off Season)*

Fuel: Gasoline, Diesel
Slips: 528 Total, 20 Transient **Max LOA:** 250 ft. **Max Beam:** n/a
 Rate *(per ft.)*: **Day** $0.75* **Week** n/a **Month** n/a
 Power: 30 amp Incl.***, **50 amp** n/a, **100 amp** n/a, **200 amp** n/a
 Cable TV: No **Dockside Phone:** No
 Dock Type: Floating, Pilings, Alongside, Wood
Moorings: 0 Total, 0 Transient **Launch:** n/a
 Rate: Day n/a **Week** n/a **Month** n/a
Heads: 6 Toilet(s), 5 Shower(s) *(with dressing rooms)*
Laundry: 2 Washer(s), 2 Dryer(s), Book Exchange **Pay Phones:** Yes, 2
Pump-Out: OnSite, Full Service **Fee:** Free **Closed Heads:** Yes

Marina Operations

Owner/Manager: Port of Seattle **Dockmaster:** Ken Lyles
In-Season: Year-Round, 7am-4:30pm** **Off-Season:** Sat, 7am-3:30pm
After-Hours Arrival: Pay envelopes outside of the office
Reservations: Yes, Preferred **Credit Cards:** Visa/MC, Amex
Discounts: None
Pets: Welcome **Handicap Access:** No

Fishermen's Terminal

3919 18th Avenue West; Seattle, WA 98119

Tel: (206) 728-3395; (800) 426-7817 **VHF: Monitor** Ch. 17 **Talk** n/a
Fax: (206) 728-3393 **Alternate Tel:** n/a
Email: ft@portseattle.org **Web:** www.portseattle.org
Nearest Town: Seattle **Tourist Info:** (206) 784-9705

Marina Services and Boat Supplies

Services - Security *(24 hrs.)*, Trash Pick-Up **Communication -** FedEx, DHL, UPS, Express Mail **Supplies - Near:** Ice *(Cube)*, Ships' Store *(Seattle Marine & Fishing Supply 285-5010)*, Bait/Tackle **3+ mi:** West Marine *(789-4640, 3 mi.)*

Boatyard Services

OnSite: Travelift, Railway *(300T)*

Restaurants and Accommodations

OnSite: Restaurant *(Chinook's at Salmon Bay 283-4665, L $8-19, D $12-25, B 7:30am wknds only - an "Anthony's Homeport")*, Seafood Shack *(Chinook's Fish and Chips 283-4665, L $7-10, D $7-10)*, Lite Fare *(Bay Café 282-3435, B, L)* **Near:** Restaurant *(Sunny Teriyaki 281-9339)*, Pizzeria *(Pudge Brothers Pizza 213-0333)* **Under 1 mi:** Restaurant *(Katina's Kitchen 284-0370)*, *(La Palma 284-1001, Mexican)*, *(Romio's Pizza & Pasta 284-5420)*, Coffee Shop *(Java Jazz 282-2321)*, Lite Fare *(Red Mill Burgers 284-6363)*, Inn/B&B *(Inn of Twin Gables 284-3979)* **1-3 mi:** Restaurant *(Taco Del Mar 545-8001, L $5-12, D $8-19)*, Motel *(Hillside Motel 285-7860)*, *(BW International 283-4140, from $99)*, *(Howard Johnson 284-1900)*

Recreation and Entertainment

OnSite: Cultural Attract *(Fishermen's Memorial 782-3395)*, Sightseeing *(Show Me Seattle Tour Buses 633-CITY)*, Galleries *(Afishionado Gallery 283-5078)* **Near:** Special Events *(Maritime Festival - May)* **1-3 mi:** Golf Course *(Interbay Golf Center 285-2200)*, Fitness Center *(Bodyfit Balance 378-0091)*, Movie Theater *(Majestic Bay Theaters 781-2229)*, Video Rental *(Blockbuster 782-0645)*, Park *(Interbay Athletic Field, Discovery Park)*

Provisioning and General Services

OnSite: Market *(Fishermen's Grocery 281-7818)*, Fishmonger *(Wild Salmon 283-3366)*, Copies Etc. *(The Station Business Center)* **Near:** Delicatessen, Bank/ATM *(Bank of America 800-288-4408)*, Barber Shop *(Charlie's 284-6477)*, Dry Cleaners *(Crowne Cleaners 213-0826)*, Newsstand, Retail Shops **Under 1 mi:** Wine/Beer, Post Office, Catholic Church *(St. Margaret's 282-1804)*, Protestant Church, Pharmacy *(QFC 283-0366)*, Hardware Store *(Ace 283-6060)* **1-3 mi:** Supermarket *(QFC 283-3600)*, Liquor Store *(WA Liquor 545-6648)*, Farmers' Market *(Magnolia Community Center Sat 10am-2pm Jun-Sep)*, Synagogue, Library *(Magnolia 386-4225)*, Beauty Salon *(Glynn's Full Service Salon 298-3354)*, Laundry, Bookstore *(Ballard Books 782-0820)*, Florist *(Ballard Blosom 782-4213)* **3+ mi:** Buying Club *(Costco 621-9997, 7 mi.)*

Transportation

OnCall: Rental Car *(Enterprise 783-5565/ Hertz 784-8405, 1 mi.)*, Taxi *(Redtop Taxi 789-4949)*, Airport Limo *(City Towncar 383-0836)* **Near:** Local Bus *(Metro Trans. 800-542-7876)* **3+ mi:** InterCity Bus *(Grayhound 628-5555, 5 mi.)*, Rail *(Amtrak 382-4125, 5 mi.)*, Ferry Service *(WA State 464-6400, 5 mi.)* **Airport:** Sea-Tac Int'l. *(17 mi.)*

Medical Services

911 Service **OnCall:** Ambulance **Under 1 mi:** Doctor *(Swedish Medical Center 781-6341)*, Dentist *(Yoder 285-1338)*, Chiropractor *(Chiropractic Centre 285-1068)*, Veterinarian *(Interbay Animal Hospital 282-1961)* **1-3 mi:** Holistic Services, Optician *(Optical Shoppe 789-8694)* **Hospital:** Virginia Mason 624-1144 *(6 mi.)*

Setting -- Fishermen's Terminal, operated by the Port of Seattle, is immediately west of the Ballard Bridge on the Lake Washington Ship Canal. Home to the North Pacific fishing fleet since 1913, it berths more than 700 commercial fishing and workboats plus a variety of pleasure craft. An enormous, shed-like gray building (trimmed in blue) stretches along the shore; it houses the office, some eateries, a fish market, chandlery, marine services and other businesses.

Marina Notes -- *80-125' $1.00/ft., over 125' $1.25/ft.. **Staffed 24 hrs. Office is open 7-4:30, Mon-Fri, closed Sun. Off-season, closed Sat too. ***15, 20 & 30A service. Upgrades are underway, which include the installation of new concrete docks. When completed in May '08, the Terminal will offer 3,000 ft. of linear dockage and 170 new slips 40-100 ft. Most moorage is Mediterranean-style; boats dock stern-to or nose-in, and tie off on pilings. Welcomes flotillas. Well-maintained heads and laundry facilities are near the port office. A washer and dryer are located within each of the women's and men's bathhouses.

Notable -- A nicely landscaped area, featuring the bronze and stone Fishermen's Memorial statue, dedicated in 1988, adds charm to the boardwalk. The annual Blessing of the Fleet and Memorial Service is held here in May, and a Fall Fest takes place in September. Popular Chinooks' at Salmon Bay features window walls and an outside deck overlooking the docks. Seattle Tours offers a variety of excursions: the waterfront, locks, downtown, plus more distant ones like Boeing and National Parks (800-305-9617). Their colorful Show Me Seattle minibus (633-CITY) stops at the marina. Seattle's Citywide Concierge (461-5888) has tickets to shows and attractions, will make dinner reservations, arrange transportation, and even locate flowers, spas, and short-term child care.

PHOTOS ON DVD: 10

Ewing Street Moorings

624 Ewing Street West; Seattle, WA 98119

Tel: (206) 283-1075 **VHF: Monitor** n/a **Talk** n/a
Fax: (206) 283-1075 **Alternate Tel:** n/a
Email: mikew@ewingstreet.com **Web:** www.ewingstreet.com
Nearest Town: Seattle **Tourist Info:** (206) 784-9705

Navigational Information
Lat: 47°39.193' **Long:** 122°21.840' **Tide:** n/a **Current:** n/a **Chart:** 18447
Rep. Depths (*MLW*): Entry 8 ft. **Fuel Dock** n/a **Max Slip/Moor** 8 ft./-
Access: Lake Wash. Ship Canal between Lake Union & the Ballard Locks

Marina Facilities *(In Season/Off Season)*
Fuel: No
Slips: 60 Total, 2 Transient **Max LOA:** 120 ft. **Max Beam:** n/a
 Rate *(per ft.):* **Day** $1.00 **Week** n/a **Month** $10
 Power: 30 amp $5, **50 amp** n/a, **100 amp** n/a, **200 amp** n/a
 Cable TV: No **Dockside Phone:** No
 Dock Type: Floating, Alongside, Wood
Moorings: 0 Total, 0 Transient **Launch:** n/a
 Rate: Day n/a **Week** n/a **Month** n/a
Heads: 1 Toilet(s)
Laundry: None **Pay Phones:** Yes
Pump-Out: No **Fee:** n/a **Closed Heads:** Yes

Marina Operations
Owner/Manager: Mike Wollaston **Dockmaster:** Same
In-Season: Year-Round, 7am-5pm **Off-Season:** n/a
After-Hours Arrival: Call ahead
Reservations: Yes, Preferred **Credit Cards:** No; Cash only
Discounts: None
Pets: Welcome **Handicap Access:** No

Marina Services and Boat Supplies
Services - Security *(owner lives onsite)* **Communication -** Mail & Package Hold, Phone Messages, FedEx, UPS **Supplies - 1-3 mi:** West Marine *(292-8663)*, Bait/Tackle *(Ballard Bait and Tackle 784-3016)*

Boatyard Services
Nearest Yard: Seaview East Boatyard (206) 789-3030

Restaurants and Accommodations
Under 1 mi: Restaurant *(Yasuko's Teriyaki 283-9152)*, *(La Palma 284-1001)*, *(Chinook's at Salmon Bay 283-4665, L $8-19, D $12-25, B 7:30am wknds only)*, *(Ponti Seafood Grill 284-3000, L $10-18, D $21-40, bar menu $6-15, happy hour daily 4-6:30pm & after 9pm: $5 specials)*, Coffee Shop *(Java Jazz 282-2321)*, Fast Food *(Sub Shop 286-0800)*, Lite Fare *(Red Mill Burgers 284-6363)*, Pizzeria *(Zeek's Pizza 285-6046)*, Motel *(Dravus Court Suites 285-9698)*, Inn/B&B *(Inn of Twin Gables 284-3979)* **1-3 mi:** Restaurant *(Hoki's Teriyaki Hut 634-1128, L $5-9, D $7-20)*, Motel *(Howard Johnson Express 284-1900)*, Hotel *(MarQueen Hotel $144-230)*

Recreation and Entertainment
Under 1 mi: Video Rental *(Rain City Video 545-3539)*, Sightseeing *(Fishermen's Terminal 782-3395, Show Me Seattle bus tours 633-CITY, Kenmore Air sightseeing flights 425-486-1257)*, Galleries *(History House 675-8875)* **1-3 mi:** Golf Course *(Interbay Golf Center 285-2200)*, Fitness Center *(Sound Mind & Body 547-3470)*, Boat Rentals *(The Electric Boat Co. 223-7476 - 21 ft. Duffys $79/hr., 2 hr. min.)*, Fishing Charter *(Adventure Charters 789-8245)*, Movie Theater *(Oak Tree Cinemas 527-1748)*, Special Events *(Maritime Fest - May, Fremont Fair - Jun 694-6706)*

Provisioning and General Services
Near: Bank/ATM *(US Bank 217-0019)*, Protestant Church *(First Methodist 281-2240)*, Dry Cleaners *(Clean-M-Rite 283-4042)* **Under 1 mi:** Convenience Store *(7-Eleven 285-0496)*, Supermarket *(QFC 283-3600)*, Delicatessen *(Sal's Out to Lunch 283-9104)*, Health Food *(PCC Natural Markets 632-6811)*, Wine/Beer, Fishmonger *(Wild Salmon Seafood Market 283-3366)*, Post Office, Catholic Church *(St. Margaret's 282-1804)*, Hardware Store *(Five Corners Hardware 282-5000)*, Florist *(Hansen's 632-9330)*, Copies Etc. *(Kinko's 352-9211)* **1-3 mi:** Liquor Store *(WA Liquor 545-6781)*, Farmers' Market *(Magnolia Community Center Sat 10am-2pm)*, Meat Market *(A & J Meats and Seafood 284-3885)*, Synagogue *(Temple International 297-2090)*, Library *(Queen Anne 386-4227)*, Beauty Salon *(Christy Carner Salon 548-8224)*, Barber Shop *(Rudy's 547-0818)*, Laundry *(Fremont Ave. 632-8924)*, Bookstore *(Fremont Place Book Co. 547-5970)*, Pharmacy *(QFC 283-0366)*, Department Store *(Fred Meyer 297-4333)*

Transportation
OnCall: Rental Car *(Enterprise 783-5565)*, Taxi *(Farwest 622-1717)*, Airport Limo *(Airport Taxi 362-2000)* **Near:** InterCity Bus *(Metro Trans. 800-542-7876)* **3+ mi:** Rail *(Amtrak 382-4125, 5 mi.)*, Ferry Service *(WA State 464-6400, 4 mi.)* **Airport:** Sea-Tac Int'l. *(16 mi.)*

Medical Services
911 Service OnCall: Ambulance **Under 1 mi:** Chiropractor *(Chiropractic Centre 285-1068)*, Holistic Services *(Water's Edge Natural Health 283-1383)* **1-3 mi:** Doctor *(Swedish Physicians 215-6300)*, Dentist *(Fremont Dental 675-0366)*, Veterinarian *(Interbay 282-1961)* **Hospital:** Virginia Mason 624-1144 *(5 mi.)*

Setting -- Lake Washington Ship Canal extends from Puget Sound through Shilshole Bay, Salmon Bay, Lake Union, Portage Bay, and Union Bay to deep water in Lake Washington. Ewing Street Moorings lies immediately east of the Ballard Bridge after passing through Salmon Bay. This rustic marina berths commercial vessels, houseboats and many classics in various stages of rejuvenation. A small gray single-story structure houses the office and amenities.

Marina Notes -- Guest moorage is on an "as available" basis. Docks can be uneven and electricity is at a premium and may not be available. Founded in 1953 by Capt. M.C. Reaber. The current owner, Mike Wollaston, has a passion for old boats. He offers restored classic yachts and project boats for sale and owns or has owned several of the boats. In addition to these classics, he also deals in hard-to-find boat parts and marine equipment, including antique collectibles; he also offers vessel charter, delivery, inspection and repair services. Houseboats with liveaboards are interspersed throughout the marina. Note: Depths above the locks reference low water of the lakes which is 20 ft. above the plane of mean low water of Puget Sound.

Notable -- Hop on the free bus to downtown for some premier sightseeing. Pike Place Market offers a wide variety of shops, restaurants, and galleries. The colorful, high-energy bazaar stimulates all senses. Kenmore Air (866) 435-9524 offers 20 minute sightseeing trips in a seaplane. Taking off and landing from the water is a quintessential Pacific Northwest experience. Flights are $75 but are sometimes discounted at the last minute. Group and child rates available.

PHOTOS ON DVD: 7

Navigational Information

Lat: 47°38.740' **Long:** 122°20.838' **Tide:** n/a **Current:** n/a **Chart:** 18447
Rep. Depths *(MLW)*: **Entry** 30 ft. **Fuel Dock** n/a **Max Slip/Moor** 30 ft./-
Access: Ship Canal through Fremont Bridge, S shore under Aurora Bridge

Marina Facilities *(In Season/Off Season)*

Fuel: No
Slips: 40 Total, 5 Transient **Max LOA:** 50 ft. **Max Beam:** n/a
  Rate *(per ft.)*: **Day** $0.60* **Week** n/a **Month** n/a
  Power: 30 amp Incl., **50 amp** Incl., **100 amp** n/a, **200 amp** n/a
  Cable TV: No **Dockside Phone:** No
  Dock Type: Fixed, Long Fingers, Short Fingers, Pilings, Alongside, Concrete
Moorings: 0 Total, 0 Transient **Launch:** n/a
  Rate: Day n/a **Week** n/a **Month** n/a
Heads: 2 Toilet(s), 2 Shower(s) *(with dressing rooms)*, Hair Dryers
Laundry: 1 Washer(s), 1 Dryer(s), Iron Board **Pay Phones:** No
Pump-Out: OnCall *(Sani-Tug)* **Fee:** $15-20 **Closed Heads:** Yes

Marina Operations

Owner/Manager: Lee Van Divort **Dockmaster:** Same
In-Season: Year-Round, 24/7 **Off-Season:** n/a
After-Hours Arrival: Call in Advance
Reservations: Yes, Required **Credit Cards:** No; Cash only
Discounts: None
Pets: Welcome **Handicap Access:** No

Lee's Landing

2900 Westlake Avenue North; Seattle, WA 98109

Tel: (206) 285-6040 **VHF: Monitor** n/a **Talk** n/a
Fax: (206) 285-8407 **Alternate Tel:** n/a
Email: n/a **Web:** n/a
Nearest Town: Seattle **Tourist Info:** (206) 283-6976

Marina Services and Boat Supplies

Services - Security *(owner lives onsite)* **Communication -** FedEx, UPS, Express Mail **Supplies - 1-3 mi:** Ships' Store *(Marine & Fishing Supply 285-5010)*, West Marine *(292-8663)*, Bait/Tackle

Boatyard Services

OnCall: Canvas Work, Divers, Propeller Repairs **1-3 mi:** Travelift *(88T)*, Forklift *(2)*, Crane, Electrical Repairs, Hull Repairs, Rigger, Bottom Cleaning, Brightwork, Compound, Wash & Wax, Interior Cleaning, Woodworking, Metal Fabrication, Painting, Awlgrip, Total Refits. **Nearest Yard:** Seaview East Boatyard *(206) 789-3030*

Restaurants and Accommodations

Near: Restaurant *(Ponti Seafood Grill 284-3000, L $10-18, D $21-40, bar menu $6-15, happy hour daily 4-6:30pm & after 9pm $5 specials)*, *(Canlis 283-3313, D $27-70, formal, since 1950; prix fixe $80)*, Hotel *(Continental Plaza 284-1900)* **Under 1 mi:** Restaurant *(Paragon 283-4548, L $7-10, D $15-22)*, *(Sergio's Mexican Cuisine 632-6685, L $5-10, D $6-17)*, *(Chinoise Café 284-6671, L $8-12, D $9-15, sushi bar)*, *(Banjara 282-7752, Indian)*, *(Thai Kitchen 285-8424)*, *(Ototo Sushi 691-3838)*, Pizzeria *(Eliott Bay Pizza 285-0500)*, Motel *(Hillside 285-7860)* **1-3 mi:** Restaurant *(Fire and Ice Lounge 443-1921, D $17-22)*, *(Scarlet Tree 523-7153, B, L, D)*, Motel *(Comfort Suites 282-2600)*, Hotel *(Residence Inn 624-6000, $110-150)*

Recreation and Entertainment

Under 1 mi: Fitness Center *(Meridian Fitness 283-9519)*, Video Rental *(Rain City Video 545-3539)*, Park, Sightseeing *(Kenmore Air sightseeing flights 425-486-1257)* **1-3 mi:** Golf Course *(Interbay Golf 285-2200)*, Movie

Theater *(Oak Tree 527-1748)*, Museum *(Puget Sound Maritime 624-3028, Center for Wooden Boats 382-2628)*, Cultural Attract *(Mercer Arts Arena 684-7200, Center on Contemporary Art 728-1980)*

Provisioning and General Services

Near: Delicatessen *(Esperanto 283-9104)* **Under 1 mi:** Convenience Store *(7-Eleven 284-8166)*, Market *(Lyons Grocery 284-1410)*, Supermarket *(Safeway 282-8090)*, Wine/Beer, Bakery *(Bite Me 286-7519)*, Bank/ATM, Protestant Church, Beauty Salon *(We Hair Design 285-0844)*, Barber Shop *(Bostonian 286-1477)*, Dry Cleaners *(Barg French 448-0808)*, Bookstore *(Armchair Sailor 283-0858)*, Pharmacy *(Safeway 282-8090)*, Hardware Store *(Ace 282-0055)*, Copies Etc. *(Perfect Copy 547-2679)* **1-3 mi:** Health Food *(PCC Natural 632-6811)*, Liquor Store *(WA Liquor 545-6781)*, Fishmonger *(A & J 284-3885)*, Post Office, Catholic Church, Synagogue, Library *(Queen Anne 386-4227)*, Laundry *(Queen Anne 282-6645)*, Florist *(Flower Lady 325-5751)*, Department Store *(Westlake Ctr 467-1600)*

Transportation

OnCall: Rental Car *(Enterprise 382-1051)*, Taxi *(Yellow Cabs 622-6500)*
Under 1 mi: Local Bus *(Metro Trans. 800-542-7876)* **1-3 mi:** Ferry Service *(WA State 464-6400)*, Airport Limo *(Seattle Airporter 547-4700)* **3+ mi:** Rail *(Amtrak 382-4125, 4 mi.)* **Airport:** Sea-Tac Int'l. *(15 mi.)*

Medical Services

911 Service **OnCall:** Ambulance **Under 1 mi:** Doctor *(Queen Anne Family Medicine 284-6132)*, Dentist *(Queen Anne Dental 284-7812)*, Chiropractor *(Whole Health Chiropractic 378-5755)*, Veterinarian *(Jones 284-2148)*
Hospital: Virginia Mason 624-1144 *(4 mi.)*

Setting -- From Salmon Bay the Lake Washington Ship Canal leads southeast to Lake Union, which is about a mile long and about a half mile wide. Lee's Landing is located east of the Fremont bascule bridge and almost directly under the George Washington (Aurora) Bridge on the south side. Across the bridge is the funky and fun community of Fremont. The buttoned-up wooden docks sit in front of an attractive three-story tan contemporary edged with railed decks.

Marina Notes -- *Rates vary, call ahead. This is primarily a long-term marina, with transient space available when tenants are away. Wooden docks with long finger piers and quality power pedestals, plus alongside space. All the basics are provided. The bathouse and laundry are located on the bottom floor of the owner's home next to the office. The tiled head/shower combination is well-maintained and includes a stocked magazine rack ($0.50 for showers). The laundry room has a folding table and ironing board. Note: Depths range generally from 32-49'. A 7-knot speed limit is enforced, except for seaplanes.

Notable -- Everything is within walking distance, with most services strung along Queen Anne Avenue, southeast of the marina. Numerous eateries (coffee shops, juice bars, bagels, all flavors of Asian cuisine, pizzerias, etc.), a 7-Eleven, Safeway supermarket, hardware store and banks are within easy reach. Two upscale restaurants are also nearby: Ponti Seafood is to the west, and Canlis, a Seattle institution, is south on Aurora Ave. At the south end of the lake is Seattle's Center for Wooden Boats. The Museum is open May-October 16, 10am-6pm, with shorter hours January-April (donations). You can also rent some of the boats - call for hours and to arrange for instruction. Free classic boat rides every Sunday 2-3pm, plus seminars, workshops, and an annual boat show.

Commercial Marine

2540 Westlake Ave North; Seattle, WA 98109

Tel: (206) 352-2241 **VHF: Monitor** n/a **Talk** n/a
Fax: (206) 352-2211 **Alternate Tel:** n/a
Email: cmccmarina@msn.com **Web:** n/a
Nearest Town: Seattle **Tourist Info:** (206) 283-6876

Navigational Information
Lat: 47°38.559' **Long:** 122°20.575' **Tide:** n/a **Current:** n/a **Chart:** 18447
Rep. Depths (MLW): Entry 30 ft. **Fuel Dock** n/a **Max Slip/Moor** 6 ft./-
Access: Ship canal through locks, past Aurora Bridge

Marina Facilities (In Season/Off Season)
Fuel: No
Slips: 84 Total, 2 Transient **Max LOA:** 60 ft. **Max Beam:** n/a
   Rate (per ft.): **Day** $0.50* **Week** $13 **Month** $9.00
   Power: 30 amp Incl., **50 amp** n/a, **100 amp** n/a, **200 amp** n/a
   Cable TV: No **Dockside Phone:** No
   Dock Type: Floating, Alongside, Concrete
Moorings: 0 Total, 0 Transient **Launch:** n/a
   Rate: Day n/a **Week** 3 **Month** 9.50
Heads: 1 Toilet(s)
Laundry: None **Pay Phones:** No
Pump-Out: No **Fee:** n/a **Closed Heads:** Yes

Marina Operations
Owner/Manager: Gene Mitchell **Dockmaster:** Same
In-Season: Year-Round **Off-Season:** n/a
After-Hours Arrival: Contact manager
Reservations: Yes **Credit Cards:** Visa/MC
Discounts: None
Pets: No **Handicap Access:** No

Marina Services and Boat Supplies
Services - Trash Pick-Up, Dock Carts **Communication -** Mail & Package Hold, Fax in/out *($1/pg)*, FedEx, UPS **Supplies - Near:** Ships' Store *(Fisheries Supply 632-4462)* **1-3 mi:** West Marine *(292-8663)*, Bait/Tackle *(Happy Hooker 284-0441)*

Boatyard Services
OnCall: Canvas Work **Near:** Divers, Propeller Repairs, Yacht Broker *(Compass Point Yachts 625-1580)*. **1-3 mi:** Travelift *(88T)*, Engine mechanic *(gas, diesel)*. **Nearest Yard:** Seaview East (206) 789-3030

Restaurants and Accommodations
Near: Hotel *(Continental Plaza 284-1900)* **Under 1 mi:** Restaurant *(Canlis 283-3313, D $27-70, formal)*, *(Ponti Seafood Grill 284-3000, L $10-18, D $21-40, waterfront; $5 happy hour specials daily)*, *(China Harbor 286-1688)*, *(Rock Salt Steak House 284-1047, L $6-12, D $13-40, waterfront, Sun brunch 11am-3pm $7-17)*, Hotel *(Courtyard Lake Union 213-0100)* **1-3 mi:** Restaurant *(I Love Sushi 625-9604, L $6, D $8-25)*, *(BluWater Bistro 447-0769, L $8-13, D $12-21)*, *(Paragon 283-4548, L $7-10, D $15-22)*, *(Thai Kitchen 285-8424)*, *(Perugia 282-5385)*, Pizzeria *(Vincenzo's Pizza & Pasta 217-4433, L $5-10, D $7-20, delivers)*, Hotel *(Residence Inn 624-6000, $104-209)*, *(Comfort Suites Downtown $79-119)*

Recreation and Entertainment
Near: Boat Rentals *(Northwest Outdoor Center 281-9694 - kayaks $12-17/hr., $24-52/4-hrs., $60-80 all day)* **Under 1 mi:** Fitness Center *(Meridian 283-9519)* **1-3 mi:** Golf Course *(Gold Mountain 464-1175)*, Movie Theater *(Oak Tree Cinemas 527-1748)*, Video Rental *(Video Isle 285-9511)*, Park

(South Lake Union), Museum *(Puget Sound Maritime 624-3028, Center for Wooden Boats 282-2628, Science Fiction 724-3428)*, Cultural Attract *(Repertory Theatre 443-2210, Intiman Theatre 269-1901)*, Sightseeing *(Seattle Center 684-7200)*

Provisioning and General Services
Near: Market *(Lyon's 284-1410)*, Delicatessen *(Dexter's 284-5508)*
Under 1 mi: Convenience Store *(7-Eleven 284-3245)*, Supermarket *(Safeway 282-8090)*, Gourmet Shop *(Oliver's 381-1418)*, Wine/Beer, Bakery *(Simply Desserts 633-2671)*, Bank/ATM, Beauty Salon *(We Hair Design 285-0844)*, Bookstore *(Armchair Sailor 283-0858)*, Hardware Store *(Ace 282-0055)*
1-3 mi: Health Food *(Mother Nature's 284-4422)*, Liquor Store *(WA 298-4616)*, Farmers' Market *(Pike Place daily 8-5)*, Fishmonger, Post Office, Catholic Church, Protestant Church, Library *(Queen Anne 386-4227)*, Barber Shop *(Bostonian 286-1477)*, Dry Cleaners *(Jewell 284-9116)*, Laundry *(Queen Anne Maytag 282-6645)*, Pharmacy *(Safeway 284-4226)*, Department Store *(Westlake Center 467-1600)*, Copies Etc. *(Documart 583-0304)*

Transportation
OnCall: Rental Car *(Enterprise 382-1051)*, Taxi *(Yellow Cabs 622-6500)*
Under 1 mi: Local Bus *(Metro Trans. 800-542-7876)*, InterCity Bus *(Greyhound 628-5526)* **1-3 mi:** Ferry Service *(WA State 464-6400)*
3+ mi: Rail *(Amtrak 382-4125, 3.5 mi.)* **Airport:** Sea-Tac Int'l. *(15 mi.)*

Medical Services
911 Service **OnCall:** Ambulance **Under 1 mi:** Dentist *(Westlake Landing Denistry 284-4505)*, Chiropractor *(Krosen 547-2992)* **1-3 mi:** Doctor *(Queen Anne 284-6132)* **Hospital:** Virginia Mason 624-1144 *(3.5 mi.)*

Setting -- Scattered among the commercial enterprises on Lake Union are a good many quiet little marinas, not obvious from the city streets. Most cater to long-term customers and offer no guest facilities. Commercial Marine is a bit of an exception. This tiny marina, sandwiched between two interesting "over-water" buildings offers good protection on three sides. It can be spotted by the orange windsock flying at the end of the dock and the owner's brown three story contemporary West Coast-style boathouse. Large trees border the shoreline keeping the marina totally private.

Marina Notes -- *In-season $15 min. The concrete docks are largely devoted to liveaboards. Manager Gene Mitchell doesn't have dedicated transient slips, but will always try to find a spot - call ahead. Space is alongside, no finger piers. The owner lives above the docks with his shop below, at the same level as the boats - several whimsical sculptures in the shop reveal his delightful sense of humor. There you will find one head, old but maintained, there are no shower or laundry facilities. A Sea Ray dealer is next door, as are Boat Electric (281-7570) and Hansen Yachts (298-9990).

Notable -- Westlake Avenue follows the lakefront for a mile down to South Lake Union Park. Queen Anne, on the other side of Aurora Ave., offers a multitude of services and casual restaurants. Seattle Center is about 2 miles south - home to the Space Needle, several museums, theaters, and the Seattle Opera House. A must for boating enthusiasts, the Center for Wooden Boats (382-2628) is at the tip of the lake, near Northwest Seaport (447-0800). The Go Seattle Card, which grants admission to many attractions, plus restaurant and shop discounts, can be purchased ahead for 1, 2, 3, 5 and 7 days (800-887-9103).

Navigational Information
Lat: 47°38.555' **Long:** 122°20.572' **Tide:** n/a **Current:** n/a **Chart:** 18447
Rep. Depths (*MLW*): Entry 40 ft. **Fuel Dock** n/a **Max Slip/Moor** 40 ft./-
Access: West shore of Lake Union

Marina Facilities (*In Season/Off Season*)
Fuel: *Northstar Fuel Dock* - Diesel
Slips: 7 Total, 7 Transient **Max LOA:** 300 ft. **Max Beam:** 40 ft.
 Rate (*per ft.*): **Day** $1.00* **Week** n/a **Month** $10
 Power: 30 amp kwh, **50 amp** kwh, **100 amp** kwh**, **200 amp** n/a
 Cable TV: No **Dockside Phone:** No
 Dock Type: Fixed, Floating, Pilings, Alongside, Concrete
Moorings: 0 Total, 0 Transient **Launch:** n/a
 Rate: Day n/a **Week** n/a **Month** n/a
Heads: None
Laundry: None **Pay Phones:** No
Pump-Out: OnCall **Fee:** n/a **Closed Heads:** Yes

Marina Operations
Owner/Manager: Bob & Molly Cadranell **Dockmaster:** Same
In-Season: Mon-Fri, 10am-4pm **Off-Season:** n/a
After-Hours Arrival: Call ahead
Reservations: Yes, Required $100 deposit **Credit Cards:** No
Discounts: None
Pets: Welcome **Handicap Access:** No

Nautical Landing

2500 Westlake Avenue North; Seattle, WA 98109

Tel: (206) 284-2308 **VHF: Monitor** n/a **Talk** n/a
Fax: (206) 285-6998 **Alternate Tel:** n/a
Email: info@nautical-landing.com **Web:** www.nautical-landing.com
Nearest Town: Seattle **Tourist Info:** (206) 283-6876

Marina Services and Boat Supplies
Services - Docking Assistance, Crew Lounge, Security (*locked gate + pair of Schipperkes*), Megayacht Facilities, 3 Phase **Communication -** Mail & Package Hold, Phone Messages, Fax in/out, FedEx, DHL, UPS, Express Mail (*Sat Del*) **Supplies - Near:** Ships' Store (*Fisheries Supply 632-4462*)
1-3 mi: West Marine (*292-8663*)

Boatyard Services
OnSite: Engine mechanic (*diesel*), Electronics Repairs, Canvas Work, Brightwork, Air Conditioning, Refrigeration **Nearest Yard:** Seaview East Boatyard (206) 789-3030

Restaurants and Accommodations
OnCall: Restaurant (*Terentino Carry-Out 283-1796*), (*Vincenzo's Pizza & Pasta 217-4433, L $5-10, D $7-20*) **Near:** Restaurant (*Canlis 283-3313, D $27-70*), Hotel (*Howard Johnson 284-1900, $68-81*) **Under 1 mi:** Restaurant (*Ponti Seafood 284-3000, L $10-18, D $21-40, happy hour 4-6:30 & 9pm+*), (*Perugia 282-5385*), Lite Fare (*Savi Subs 217-9350*), Hotel (*Courtyard Lake Union 213-0100, $93-119*), Inn/B&B (*Silver Cloud Inn 447-9500, $113-137*) **1-3 mi:** Restaurant (*Paragon 283-4548, L $7-10, D $15-22, Sun brunch $7-10*), (*The Bistro 547-6285, L $7-8, D $7-13*), (*Orrapin Thai 283-7118*), (*Queen Anne Café 285-2060, B, L & D $6-14 - pizza*)

Recreation and Entertainment
Near: Boat Rentals (*Northwest Outdoor Center 281-9694 - kayaks $12-17/hr., $24-52/4-hrs., $60-80 all day*) **Under 1 mi:** Fitness Center (*Meridian 283-9519*), Movie Theater (*Oak Tree 527-1748*), Video Rental (*Video Isle 285-9511*), Park (*Gas Works Park*)

1-3 mi: Golf Course (*Greenlake G.C. 632-2280*), Museum (*Experience Music 770-2702, Science Fiction Museum 724-3428, Center for Wooden Boats 282-2628*), Sightseeing (*Citywide Conciege 461-5888*)

Provisioning and General Services
Near: Convenience Store (*Lyon's 284-1410*), Delicatessen (*Deli and Espresso 286-6610*), Bakery (*Bite Me 286-7519*), Beauty Salon (*We Hair Design 285-0844*) **Under 1 mi:** Wine/Beer, Bank/ATM, Post Office, Bookstore (*Armchair Sailor 283-0858*), Florist (*Mari 217-0462*)
1-3 mi: Supermarket (*Safeway 282-8090*), Health Food (*PCC Natural 632-6811*), Liquor Store (*WA Liquor 545-6781*), Farmers' Market (*Pike Place daily 8-5*), Fishmonger (*Wild Salmon 283-3366*), Catholic Church, Protestant Church, Library (*386-4227*), Barber Shop (*Bostonian 286-1477*), Dry Cleaners (*Barg 448-0808*), Laundry (*Queen Anne 282-6645*), Pharmacy (*Bartell 284-1353*), Hardware Store (*Five Corners 282-5000*), Retail Shops (*Westlake 467-1600*), Department Store (*Nordstrom 628-21111*)

Transportation
OnCall: Rental Car (*Enterprise 382-1051/Avis 448-1700, 1 mi.*), Taxi (*Yellow Cabs 622-6500*) **1-3 mi:** Local Bus (*Metro Trans. 800-542-7876*), InterCity Bus, Ferry Service (*WA State 464-6400*) **3+ mi:** Rail (*Amtrak 382-4125, 3.5 mi.*) **Airport:** Sea-Tac Int'l. (*15 mi.*)

Medical Services
911 Service OnCall: Ambulance **Under 1 mi:** Dentist (*Westlake Landing Denistry 284-4505*), Chiropractor (*Krosen 547-2992*) **1-3 mi:** Doctor (*Queen Anne 284-6132*), Veterinarian (*Emerald City Emergency Clinic 634-9000*) **Hospital:** Virginia Mason 624-1144 (*3.5 mi.*)

Setting -- Nautical Landing is on the west shore of Lake Union, about three miles east of the Chittenden Locks through three bascule bridges and the fixed George Washington Aurora Bridge. This megayacht-only facility rests at the foot of a multistory building housing various businesses, including the marina office. Wide concrete piers offer alongside tie-ups and allow vehicles shipside access for deliveries and service. Views across Lake Union are of Gas Works Park.

Marina Notes -- *Note: Min LOA 80 feet - prefers a one week min. **Electricity is charged for actual usage. Power to 100 amp 470 3-phase. 300 ft. long pier. The marina is designed for and caters specifically to megayachts. Professionally managed by the owners, the marina offers services aimed at professional crews: a crew desk for phone service, directories, copier, and fax. LaCasse Maritime provides crew placement services (632-6858). The seaplane terminal is just south of the marina, allowing easy access for owners & crew. No heads, showers or laundry services are on-site; it is assumed that these needs are met aboard the yacht - but there are heads in the building at the top of the ramp. Note: A 7 kts. speed limit is enforced on the lake.

Notable -- Practically any service visiting owners, guests or crew could need can be arranged through Citywide Concierge Service, located in the Seattle Visitors & Convention Bureau. Within walking distance of the marina are repair facilities, Fisheries Supply - an excellent chandlery, Armchair Sailor for charts and reads, kayak rentals, restaurants and delis. Grocery deliveries can be arranged with Homegrocer (425-201-7500). Several museums and theaters are near the Space Needle at Seattle Center (684-7200). Seattle's Center on Contemporary Art (728-1980) is in the same neighborhood (Wed-Sun 12-5pm).

AGC Marina

1200 Westlake Avenue N, Suite 504; Seattle, WA 98109

Tel: (206) 284-4204 **VHF: Monitor** n/a **Talk** n/a
Fax: (206) 286-1111 **Alternate Tel:** n/a
Email: n/a **Web:** n/a
Nearest Town: Seattle **Tourist Info:** (206) 283-6976

Navigational Information
Lat: 47°37.789' **Long:** 122°20.369' **Tide:** n/a **Current:** n/a **Chart:** 18447
Rep. Depths (*MLW*): **Entry** 25 ft. **Fuel Dock** n/a **Max Slip/Moor** 12 ft./-
Access: West shore of Lake Union

Marina Facilities (In Season/Off Season)
Fuel: No
Slips: 94 Total, 2 Transient **Max LOA:** 70 ft. **Max Beam:** n/a
Rate (*per ft.*): **Day** $0.60* **Week** n/a **Month** n/a
Power: 30 amp kwh** **, 50 amp** kwh** **, 100 amp** n/a, **200 amp** n/a
Cable TV: No **Dockside Phone:** No
Dock Type: Fixed, Long Fingers, Pilings, Alongside, Wood
Moorings: 0 Total, 0 Transient **Launch:** n/a
Rate: Day n/a **Week** n/a **Month** n/a
Heads: 2 Toilet(s), 2 Shower(s) (*with dressing rooms*)
Laundry: 1 Washer(s), 1 Dryer(s) **Pay Phones:** No
Pump-Out: No **Fee:** n/a **Closed Heads:** Yes

Marina Operations
Owner/Manager: Paula Jackson **Dockmaster:** Same
In-Season: Year-Round, 8:30am-5pm **Off-Season:** n/a
After-Hours Arrival: Call ahead
Reservations: Yes, Required **Credit Cards:** No
Discounts: None
Pets: Welcome **Handicap Access:** No

Marina Services and Boat Supplies
Services - Security (*locked gates*), Dock Carts **Communication -** FedEx
Supplies - Under 1 mi: Ships' Store (*Fisheries Supply 632-4462*), West Marine (*292-8663*)

Boatyard Services
OnCall: Canvas Work, Divers, Propeller Repairs **1-3 mi:** Travelift (*88T*).
Nearest Yard: Seaview East Boatyard (206) 789-3030

Restaurants and Accommodations
OnSite: Restaurant (*McCormick & Schmick's Harborside 270-9052, L $8-15, D $12-30, fresh menu daily*) **Near:** Restaurant (*Rock Salt Steak House 284-1047, L $6-12, D $13-40, Sun brunch 11am-3pm $7-17*), (*Buca di Beppo 244-2288, Italian, L Sat-Sun, D daily; takeout*), (*Bonefish Grill 405-2663*), Lite Fare (*Savi Subs 217-9350*), (*Taco Del Mar 281-9968*), Pizzeria (*Pasta Freska 283-1515*), Hotel (*Courtyard Lake Union 213-0100, $101-269*)
Under 1 mi: Restaurant (*I Love Sushi 625-9604*), (*Bamboo Garden 282-6616, Vegetarian*), Motel (*BW Loyal Inn 682-0200*), (*Travelodge Space Needle 441-7878, $105-135*), Hotel (*Comfort Suites Downtown 282-2600*)

Recreation and Entertainment
OnSite: Sightseeing (*Argosy Cruises 623-1445*) **Near:** Jogging Paths
Under 1 mi: Fitness Center (*Meridian Fitness 283-9519*), Movie Theater (*Oak Tree 527-1748*), Video Rental (*Blockbuster 352-9525*), Park, Museum (*Northwest Seaport 447-0800, Pacific Science Center 443-2001, Contemporary Art 728-1980*), Cultural Attract (*Seattle Opera 389-7676*)
1-3 mi: Golf Course (*Gold Mountain 632-2280*), Boat Rentals (*Northwest Outdoor Center 281-9694 - kayaks; Center for Wooden Boats 382-2628*)

Provisioning and General Services
OnSite: Bank/ATM **Near:** Delicatessen (*Dexter Deli 284-5508*), Copies Etc. (*CEO Quick Print and Mail 285-3062*) **Under 1 mi:** Market (*Lyon's Grocery 284-1410, Plaid Pantry Market 284-9550*), Gourmet Shop (*Pasta and Co. 283-1182*), Wine/Beer, Liquor Store (*WA Liquor 298-4616*), Bakery (*Bite Me 286-7519*), Catholic Church, Protestant Church, Beauty Salon (*Egos 448-7778*), Dry Cleaners (*Queens Cleaners 284-1490*), Bookstore (*Armchair Sailor 283-0858*), Pharmacy (*Bartell Drug 284-1353*), Florist (*Flower Shop 622-1467*) **1-3 mi:** Supermarket (*Fred Meyer 323-6586*), Health Food (*Mother Nature's Natural Foods 284-4422*), Farmers' Market (*Pike Place daily 8am-5pm*), Fishmonger (*A & J Meats and Seafood 284-3885*), Meat Market, Post Office, Synagogue, Library (*Queen Anne 386-4227*), Barber Shop (*Scissors 284-8582*), Laundry (*Crystal Clean Laundry 323-4969*), Hardware Store (*Ace 282-0055*), Clothing Store (*REI 223-1944*), Department Store (*Nordstrom 628-21111*)

Transportation
OnSite: Local Bus (*Metro Trans. 800-542-7876*) **OnCall:** Rental Car (*Enterprise 382-1051*), Taxi (*Yellow Cabs 622-6500*) **1-3 mi:** InterCity Bus (*Greyhound 628-5526*), Rail (*Amtrak 382-4125*), Ferry Service (*WA State 464-6400*) **Airport:** Sea-Tac Int'l. (*14 mi.*)

Medical Services
911 Service **OnCall:** Ambulance **Under 1 mi:** Doctor (*Emerald City 281-1616*), Dentist (*Westlake Landing Dentristy 284-4505*), Chiropractor (*Queen Anne Chiropractic 282-8275*), Optician (*Coliazzo Opticians 285-7212*)
1-3 mi: Veterinarian (*Queen Anne Animal Clinic 284-2148*)
Hospital: Virginia Mason 624-1144 (*2 mi.*)

Setting -- AGC Marina is near the southern end of Lake Union on the western shore. A tall gray building topped with the gold initials "AGC" rises behind the network of wooden docks - making it easy to spot. On the south side of the building, McCormick & Schmick's Harborside restaurant occupies two stories and provides short-term side-tie space for dining guests - along with views of the lake. This is a convenient location, less than a mile from Seattle's Center.

Marina Notes -- *Flat rate $25 up to 40', $35 over 40'. Call ahead for rates and availability. **Electricity is metered. Wooden docks with long finger piers house a variety of power and sailboats. Locked gates. Several other businesses share office space here, including the Starbucks corporate offices and many other amenities are found nearby. The marina provides a pleasant bathhouse and laundry.

Notable -- McCormick and Schmick's Harborside restaurant offers inside and outdoor deck dining. Menus are printed daily, - the dinner menu is available at 3:30pm. The marina serves as departure point for a year-round, 2-hour narrated tour of Lake Union and Lake Washington offered by Argosy Cruises (800-642-7816). One of the sights is the houseboat community where Tom Hanks was "Sleepless in Seattle." Reserve ahead or stop by the ticket booth (adults $26.65 in-season, $21.14 off-season, kids $9.19/8.27). Additional Argosy cruises, including a locks tour and dinner cruise, depart from other locations in Seattle. Just around the corner is the Center for Wooden Boats, a hands-on museum that also rents wooden boats. For breathtaking views from 520 ft. above Seattle, the Space Needle (905-2111) is within walking distance - it has a gift shop, fine dining restaurant, and a pavilion.

Navigational Information
Lat: 47°37.800' **Long:** 122°19.840' **Tide:** n/a **Current:** 0 kt. **Chart:** 18447
Rep. Depths (MLW): Entry 25 ft. **Fuel Dock** n/a **Max Slip/Moor** 20 ft./-
Access: South end of Lake Union on the east side

Marina Facilities *(In Season/Off Season)*
Fuel: No
Slips: 50 Total, 2 Transient **Max LOA:** 120 ft. **Max Beam:** n/a
  Rate *(per ft.)*: **Day** $1.00 **Week** n/a **Month** $8.50-9.50
  Power: 30 amp Inq., **50 amp** Inq., **100 amp** n/a, **200 amp** n/a
  Cable TV: No **Dockside Phone:** No
Dock Type: Floating, Long Fingers, Concrete
Moorings: 0 Total, 0 Transient **Launch:** n/a
  Rate: Day n/a **Week** n/a **Month** n/a
Heads: 4 Toilet(s), 4 Shower(s)
Laundry: None **Pay Phones:** No
Pump-Out: OnSite **Fee:** Free **Closed Heads:** Yes

Marina Operations
Owner/Manager: Steven Agnew **Dockmaster:** Same
In-Season: Year-Round **Off-Season:** n/a
After-Hours Arrival: Call in advance
Reservations: Yes, Required **Credit Cards:** No
Discounts: None
Pets: Welcome **Handicap Access:** No

Fairview Marinas

PO Box 3955; 1109 Fairview Avenue N.; Seattle, WA 98124

Tel: (888) 673-1118 **VHF: Monitor** n/a **Talk** n/a
Fax: n/a **Alternate Tel:** n/a
Email: seattleboatslips@hotmail.com **Web:** n/a
Nearest Town: Seattle **Tourist Info:** (206) 389-7200

Marina Services and Boat Supplies
Communication - FedEx, DHL, UPS **Supplies - Near:** Ships' Store, West Marine *(293-8663)* **Under 1 mi:** Ice *(Block)*, Bait/Tackle *(Outdoor Emporium 624-6550)* **1-3 mi:** Propane

Boatyard Services
Near: Travelift. **1-3 mi:** Launching Ramp. **Nearest Yard:** Fish Marina (206) 623-3233

Restaurants and Accommodations
OnCall: Pizzeria *(Toscana 325-0877)* **Near:** Restaurant *(I Love Sushi 625-9604, L $6, D $8-25)*, *(BluWater Bistro 447-0769, L $8-13, D $12-21)*, *(Duke's Chowder House 382-9963, L & D $10-20)*, *(Daniel's Broiler 621-8262, L $13-22, D $25-80, L Mon-Fri, D daily, happy hour daily 4-6:30pm)*, *(Chandler's Crabhouse 223-2722, L $16-20, D $20-60)*, *(Cucina Cucina 447-2782)*, Hotel *(Residence Inn Downtown 624-6000)*, Inn/B&B *(Silver Cloud Inn 447-9500, $99-159)* **Under 1 mi:** Restaurant *(McCormick and Schmick's Harborside 270-9052, L $8-15, D $12-30)*, *(Buca di Beppo 244-2288)*, Lite Fare *(Taco Del Mar 624-2114)*, Hotel *(Comfort Suites 282-2600, $79-129)*, *(Courtyard Lake Union 213-0100, $101-269)* **1-3 mi:** Restaurant *(Space Needle Sky City 905-2100, L $23-30, D $35-52, wknd brunch $43)*

Recreation and Entertainment
Near: Boat Rentals *(Moss Bay Rowing and Kayaking 682-2031 - kayaks, sail boats, rowing shells)*, Museum *(Puget Sound Maritime 634-3028, Wooden Boat Center 382-2628/ Asian Art Museum 654-3100, 2 mi.)*, Special Events **Under 1 mi:** Playground *(Cascade Playground)*, Fitness Center *(Pro Sport Club 332-1873)*, Video Rental *(Video Quest 325-9301)*

1-3 mi: Golf Course *(Gold Mountain 464-1175)*, Movie Theater *(Cinerama 441-3653)*, Sightseeing *(Seattle Center, Pioneer Square, Pike Place)*

Provisioning and General Services
Near: Fishmonger *(Chandler's Crabhouse & Fresh Fish 223-2722)*, Bank/ATM, Beauty Salon *(Lake Union 343-9218)* **Under 1 mi:** Convenience Store *(Pete's)*, Market *(Broadway Grocery 324-0711)*, Wine/Beer, Liquor Store, Bakery, Post Office, Catholic Church, Protestant Church, Library *(Capitol Hill 684-4715)*, Laundry *(Crystal Clean 323-4969)*, Pharmacy *(Bartell 323-2830)*, Florist *(Blooms on Broadway 324-8845)*, Copies Etc.
1-3 mi: Supermarket *(Safeway 323-1190, QFC 322-8200)*, Gourmet Shop *(Oliver's 381-1418)*, Farmers' Market *(Broadway & 10th Sun 11am-3pm, 632-5234)*, Meat Market, Crabs/Waterman *(A & J Meats & Seafood 284-3885)*, Shrimper, Dry Cleaners *(Four Seasons 324-4341)*, Bookstore, Hardware Store *(Ace 282-0055)*, Retail Shops *(Westlake Center 467-1600)*, Mosque **3+ mi:** Buying Club *(Costco 622-3136, 5 mi.)*

Transportation
OnCall: Rental Car *(Enterprise 382-1051)*, Taxi *(Yellow Cabs 622-6500)* **Under 1 mi:** Local Bus *(Metro Trans. 800-542-7876)*, Airport Limo *(Advanced Limo 941-0323)* **1-3 mi:** Rail *(Amtrak 382-4125)*, Ferry Service *(WA State 464-6400)* **Airport:** Sea-Tac Int'l. *(14 mi.)*

Medical Services
911 Service **OnCall:** Ambulance **Near:** Dentist *(East Lake Dental Care 325-7456)* **Under 1 mi:** Veterinarian *(Easltake 328-2675)* **1-3 mi:** Doctor *(Health South Medical 682-7418)*, Chiropractor *(Seattle Chiropractic Health 623-6800)* **Hospital:** Virgina Mason 624-1144 *(1.5 mi.)*

Setting -- Nestled among several ship and boat moorages, Fairview Marinas is located at the south end of Lake Union on the eastern shore. Onshore is a dramatic two-story wood and glass contemporary backed by the spires of the downtown business district. At the top of the gangway, a deck, lined with plants and comfortable benches, encourages a stop to enjoy the scene. From the docks, the views across the lake are of hillside homes and marine businesses.

Marina Notes -- Guest moorages on Lake Union are rare, so call ahead for reservations. There is no dedicated marina staff, but manager Steven Agnew picks up cell phone calls immediately. If unavailable, leave a message. Floating concrete docks with narrow finger piers host both sail and power vessels. There is power on pedestals and a side-tie moorage for the pump-out station. The marina offers few amenities for the boater, but many attractions and restaurants are within easy walking distance. The heads are located inside the two-story building behind the deck and require a key for entrance.

Notable -- The immaculate 120-foot private yacht, "Thea Foss," built in 1930 for John Barrymore, calls the marina home. Right next to the marina, popular, well-reviewed BluWater Bistro offers good food, moderate prices, and optional outside dining. The Silver Cloud Inn is directly across the street with all the amenities of a three-star hotel. Further south is the Center for Wooden Boats - free classic boat rides every Sunday 2-3pm (admission to museum by donation). The Maritime Heritage Museum (624-3028) is an easy walk from the marina. Volunteer Park is a bit of a hike, but the views from Capitol Hill are worth the trip. The park is also home to the Asian Art Museum (654-310) and the Conservatory garden and greenhouses (684-4743).

Kirkland Marina Park

25 Lakeshore Plaza; Kirkland, WA 98033

Tel: (425) 587-3340 **VHF: Monitor** n/a **Talk** n/a
Fax: n/a **Alternate Tel:** (425) 587-3300
Email: n/a **Web:** www.ci.kirkland.wa.us
Nearest Town: Kirkland **Tourist Info:** (425) 822-7066

Navigational Information
Lat: 47°40.500' **Long:** 122°12.580' **Tide:** n/a **Current:** n/a **Chart:** 18447
Rep. Depths (*MLW*)**: Entry** 15 ft. **Fuel Dock** n/a **Max Slip/Moor** 15 ft./-
Access: Via Lake Washington Ship Canal from Seattle

Marina Facilities (*In Season/Off Season*)
Fuel: No
Slips: 77 Total, 77 Transient **Max LOA:** 100 ft. **Max Beam:** n/a
  Rate (*per ft.*)**: Day** $0.40* **Week** n/a **Month** n/a
  Power: 30 amp n/a, 50 amp n/a, 100 amp n/a, 200 amp n/a
  Cable TV: No **Dockside Phone:** No
  Dock Type: Fixed, Long Fingers, Alongside, Wood
Moorings: 0 Total, 0 Transient **Launch:** n/a
  Rate: Day n/a **Week** n/a **Month** n/a
Heads: 9 Toilet(s)
Laundry: None **Pay Phones:** Yes, 1
Pump-Out: No **Fee:** n/a **Closed Heads:** Yes

Marina Operations
Owner/Manager: City of Kirkland **Dockmaster:** Sudie Elkayssi
In-Season: Year-Round, 8am-5pm **Off-Season:** n/a
After-Hours Arrival: Pay daily in advance at pay station
Reservations: No **Credit Cards:** No; Cash only
Discounts: None
Pets: Welcome **Handicap Access:** Yes, Docks

Marina Services and Boat Supplies
Communication - FedEx, DHL, UPS, Express Mail **Supplies - OnSite:** Ice
(*Block, Cube*) **1-3 mi:** Propane (*U-Haul 822-4134*) **3+ mi:** West Marine
(*641-4065, 6 mi.*), Boater's World (*646-9350, 6 mi.*)

Boatyard Services
OnSite: Launching Ramp **OnCall:** Canvas Work, Divers, Propeller
Repairs **Nearest Yard:** Seaview East Boatyard (206) 789-3030

Restaurants and Accommodations
Near: Restaurant (*Marina Park Grill 889-9000, D $18-28*), (*Jalisco 822-3355,
L $6-8, D $11-14*), (*Ristorante Paradiso 889-8601, L $8-10, D $12-19*),
(*Icatus 893-7999, L $8-10, D $10-15*), (*Anthony's 822-0225, L&D*), (*Tokyo
Grill 822-3473*), (*Paradiso 899-8601, L $8-15, D $12-22*), (*21 Central 822-
1515, D $15-40, steaks & seafood*), (*Pasta Ya Gotcha 899-1511*), (*Lai Thai
739-9747, L & D $10-17, delivery & takeout*), (*Lynn's 889-2808, L $9-14,
D $15-25, closed Mon*), Lite Fare (*World Wraps 827-9727*), Pizzeria (*Coyote
Creek 822-2226*) **1-3 mi:** Hotel (*The Woodmark 822-3700, $140-365*),
(*LaQuinta Inn 828-6585, $89-114*), (*BW Kirkland Inn 822-2300, $68-135*)

Recreation and Entertainment
OnSite: Beach, Picnic Area, Jogging Paths, Cultural Attract (*"Music in the
Park"* - Tue & Thu pm Wed am; near Kirkland Performance Ctr. 893-9900),
Sightseeing (*M.V. Kirkland, Argosy Cruises 206-623-4252*) **Near:** Pool
(*Peter Kirk*), Playground, Tennis Courts (*Peter Kirk*), Fitness Center (*24 Hour
Fitness 889-2582*), Video Rental (*Blockbuster 889-9300*), Park (*Peter Kirk*),
Museum (*Rovzar Art Gallery 889-4627*), Special Events (*TASTE! waterfront
fest - 3rd week Sept; hydroplane races - 1st wknd. Aug*), Galleries

(*Berozkina 803-5032*) **Under 1 mi:** Fishing Charter (*Anglers Rendevous
833-0288*) **1-3 mi:** Dive Shop (*Underwater Sports 821-7200*), Golf Course
(*Heron Links 883-1200*), Movie Theater (*Parkplace Cinema 6 822-5100*)

Provisioning and General Services
Near: Convenience Store (*Market on Central 8890711*), Delicatessen
(*GST605-8300*), Bakery (*Sweet Cakes 821-6565*), Farmers' Market (*Park
Lane May-Oct Wed 10am-4pm*), Bank/ATM, Beauty Salon (*Salon Remeek
822-2123*), Barber Shop (*The Shop 739-9881*), Dry Cleaners (*Kirkland Park
Lane 822-2550*), Laundry, Florist (*Moss Bay 828-3660*), Retail Shops,
Copies Etc. (*UPS Store 889-8900*) **Under 1 mi:** Supermarket (*QFC 827-
2205*), Wine/Beer (*The Grape Choice 827-7551*), Fishmonger (*Ohana 576-
1887*), Shrimper, Post Office, Protestant Church, Library (*Kirkland 822-
2459*), Bookstore (*Park Place 828-6546*), Pharmacy (*Bartell 827-3934*)
1-3 mi: Health Food (*PCC Natural 828-4622*), Liquor Store (*WA Liquor 822-
2880*), Catholic Church, Hardware Store (*Ace 376-0327*), Department Store
(*Fred Meyer 820-3233*), Buying Club (*Costco 822-0414*)

Transportation
OnCall: Rental Car (*Enterprise 820-8555*), Taxi (*Golden Ride 576-9100*),
Airport Limo (*First Chioce 825-9030*) **Near:** Bikes (*Kirkland Bicycle 329-
7333*) **3+ mi:** Rail (*Amtrak 382-4125, 13 mi.*), Ferry Service (*WA State
464-6400, 12 mi.*) **Airport:** Sea-Tac Int'l. (22 mi.)

Medical Services
911 Service **OnCall:** Ambulance **Near:** Doctor (*Lakeshore 827-9711*),
Dentist (*Kirkland 827-3097*), Chiropractor, Holistic Services (*Park Lane
Wellness 827-3102*) **Hospital:** Evergreen 899-1000 (4 mi.)

PHOTOS ON DVD: 11

Setting -- Kirkland sits on the eastern shore of Lake Washington in Moss Bay. Just north of the Kirkland Yacht Club, the Marina Park's docks are surrounded by a waterfront park with whimsical statues, a large gazebo, grassy amphitheater, picnic tables and beach area. Vibrant downtown Kirkland - a pretty, arts-centric, upscale community - is just a few steps from the boat. Clear days bring views of the Seattle skyline with the Olympic Mountains in the distance.

Marina Notes -- *Flat Fee: Under 20' $7/night, 21-30' $8, 31-34' $12, 35-40' $15, 41-50' $18, over 51' $25/night. No reservations. Payment is on the honor system and there is a three-day limit on stays. The city-owned marina has wide wooden piers with slips and side-ties. The docks are well-lit with large round lamps. No power or water. A deep launch ramp is on-site. Ice, heads and a public phone are located at the top of the dock. (In '05, the porta-potties were replaced with real heads.) Note: There is no breakwater, so it can get rolly at times.

Notable -- The park is surrounded by low-rise condo buildings and the charming nearby streets provide most services and provisions. Dozens of excellent restaurants, sidewalk cafés, galleries, boutiques, and several salons are within an easy stroll. Summer concerts are held on-site at the Pavilion. Two blocks inland is Peter Kirk Park with tennis, basketball, skateboarding, the library, 400-seat live performance center, swimming pool, and daily baseball games open to all. Kirkland's ArtWalk (889-8212) is every second Thursday 6-9 pm and showcases artists in more than a dozen galleries (889-8212). Take a 1.2 mile walk in either direction - south past mansions, parks and art installations to Carillon Point or north past historic homes to 144-acre Juanita Bay wildlife preserve.

Navigational Information
Lat: 47°39.450' **Long:** 122°12.393' **Tide:** n/a **Current:** n/a **Chart:** 18447
Rep. Depths (MLW): Entry 30 ft. **Fuel Dock** n/a **Max Slip/Moor** 20 ft./-
Access: Via Lake Washington Ship Canal from Seattle

Marina Facilities (In Season/Off Season)
Fuel: No
Slips: 200 Total, 2 Transient **Max LOA:** 60 ft. **Max Beam:** n/a
  Rate (per ft.): **Day** $1.50/1.00 **Week** n/a **Month** $595*
  Power: 30 amp $3-5**, **50 amp** $3-5**, **100 amp** n/a, **200 amp** n/a
  Cable TV: Yes, Incl **Dockside Phone:** Yes, Incl
Dock Type: Floating, Long Fingers, Concrete
Moorings: 0 Total, 0 Transient **Launch:** n/a
  Rate: Day n/a **Week** n/a **Month** n/a
Heads: 6 Toilet(s), 6 Shower(s) (with dressing rooms), Hair Dryers
Laundry: None **Pay Phones:** Yes, 2
Pump-Out: OnSite, Self Service **Fee:** Free **Closed Heads:** Yes

Marina Operations
Owner/Manager: Shelley Taylor **Dockmaster:** Ramon Pin Pin
In-Season: Jul-Sep, 8am-5pm **Off-Season:** Oct-Jun, 8am-5pm
After-Hours Arrival: Call ahead
Reservations: Yes, Required **Credit Cards:** Visa/MC
Discounts: None
Pets: Welcome **Handicap Access:** No

Carillon Point Marina

3240 Carillon Point; Kirkland, WA 98033

Tel: (425) 822-1700 **VHF: Monitor** n/a **Talk** n/a
Fax: (425) 828-3094 **Alternate Tel:** n/a
Email: shelley@carillonprop.com **Web:** www.carillon-point.com
Nearest Town: Kirkland (1.2 mi.) **Tourist Info:** (425) 822-7066

Marina Services and Boat Supplies
Services - Docking Assistance, Security (24 hr., 4pm-7am), Trash Pick-Up, Dock Carts **Communication -** Data Ports (Wi-Fi - Broadband Xpress)
Supplies - 1-3 mi: West Marine (641-4065)

Boatyard Services
OnCall: Canvas Work, Brightwork, Divers, Propeller Repairs
3+ mi: Travelift (8 mi.) **Nearest Yard:** Seaview East (206) 789-3030

Restaurants and Accommodations
OnSite: Restaurant (Yarrow Bay Grill and Beach Café 889-0303, L $10-18, D $12-19), (Waters Lakeside Bistro 803-5595, B $7-15, L $8-15, D $20-25, wknd brunch $9-16), (Cucina Cucina Italian Café 822-4000, L $8-14, D $8-19), Snack Bar (Starbucks 827-2130, B, L), (Poppinjoy 828-3049, L $5-10), Hotel (The Woodmark 822-3700, $140-365) **Under 1 mi:** Restaurant (Foghorn 827-0654, L $8-13, D $13-22, seafood), (Keg Steakhouse & Bar 822-5131, L Mon-Fri, D daily, kids' menu), Hotel (LaQuinta Inn 828-6585, $89-114) **1-3 mi:** Restaurant (Marina Park Grill 889-9000, D $18-28), (21 Central 822-1515, D $15-40, steakhouse)

Recreation and Entertainment
OnSite: Fitness Center (Hotel Fitness Center available to marina guests), Sightseeing (cruise aboard The Woodmark II, 1956, 28' Chris-Craft)
Near: Picnic Area, Playground, Boat Rentals (Yarrow Bay Marina 822-6066 - 19-22 ft. runabouts) **Under 1 mi:** Beach **1-3 mi:** Dive Shop (Underwater Sports 821-7200), Horseback Riding (Bridle Trails State Park - beautiful, wooded riding trails & equestrian shows on summer weekends 822-7647), Fishing Charter (Anglers Rendevous 833-0288), Movie Theater (Parkplace

822-5100), Video Rental (Blockbuster 889-9300), Park (Bridle Trails St. Park), Cultural Attract (Performance Center 893-9900, Arts Center 822-7161), Galleries (Lakeshore 827-0606) **3+ mi:** Golf Course (Bellevue Municipal 452-7250, 4 mi.)

Provisioning and General Services
OnSite: Bank/ATM, Post Office, Beauty Salon (David Michael 822-2990), Dry Cleaners (hotel valet), Clothing Store (Mainsail 284-2484), Retail Shops **Under 1 mi:** Market (Houghton 822-9211), Wine/Beer, Liquor Store (WA Liquor 822-2880), Farmers' Market (Park Lane Wed 10am-4pm), Protestant Church, Barber Shop (Earl's 822-4274) **1-3 mi:** Supermarket (QFC 822-4154), Delicatessen (Meze Mediterranean 828-3923), Health Food (PCC Natural Markets 828-4622), Bakery (Noah's NY Bagels 827-7382), Fishmonger (Ohana Seafood 576-1887), Catholic Church, Library (Kirkland 822-2459), Laundry (Waves of Suds 828-9737), Bookstore (Park Place 828-6546), Pharmacy (Bartell 827-3934), Hardware Store, Buying Club (Costco 822-0414), Copies Etc. (Kinko's 889-2290)

Transportation
OnCall: Rental Car (Enterprise 820-8555), Taxi (Golden Ride 576-9100), Airport Limo (First Chioce 825-9030) **3+ mi:** Rail (Amtrak 382-4125, 11 mi.), Ferry Service (WA State 464-6400, 10 mi.) **Airport:** Sea-Tac Int'l. (20 mi.)

Medical Services
911 Service **OnSite:** Holistic Services (Spa at the Woodmark 803-9000)
OnCall: Ambulance **Near:** Dentist (Sakuma 827-0426), Chiropractor (Denton 576-4242) **1-3 mi:** Doctor (Lakeshore Clinic 827-9711)
Hospital: Overlake 688-5000 (3 mi.)

Setting -- Carillon Point is a mini-destination. The modern brick and granite Woodmark Hotel, rising behind the docks, presents a striking vista from the water. Blue awnings top the waterfront dining decks, and colorful potted plants add to the upscale ambience. Five main piers and a long T-head host 200 vessels - protected by a double breakwater. Above the docks, Carillon Heights' townhouses overlook the lake. And to the west, the view is Seattle's skyline.

Marina Notes -- CCM. *Monthly flat rate - annual month + 20% (i.e. 40' $595) incl. power. Weekly - prorated monthly + 20% incl. power. 2 hrs. free dockage for shoppers or diners. **Trans power $3 under 40', $5 over 40'. Complex is built on the Seattle Seahawks training site - which was the Lake Washington Shipyard. Today a wide pier, acting as a breakwater, wraps the perfectly maintained cement docks which sport long finger piers and high-end power pedestals. Transients dock along the pier on the north side - to port as you enter. Designated areas for recycling. Fuel and other marine services are available within 100 yards. Sparkling, fully tiled heads, locker and shower facilities, with fresh white curtains, are on the 1st floor of the project's 5000 Building.

Notable -- Carillion Point is an elegantly executed mixed use property - offices, a hotel, restaurants, parks, shops, and a marina surrounded by manicured gardens and art installations - on the shores of Lake Washington. The four-star Woodmark Hotel dominates the waterfront, and offers several good restaurants and a topflight spa. During the summer, lunchtime concerts feature a variety of musical acts. From Carillon Point, it is about 1.2 miles to downtown Kirkland, a charming, arts-centered, vibrant town and one of several communities on the lake that comprise the area known as the Eastside.

Lakewood Moorage

4500 Lake Washington Blvd. S; Seattle, WA 98118

Tel: (206) 722-3887 **VHF: Monitor** n/a **Talk** n/a
Fax: (206) 760-5301 **Alternate Tel:** n/a
Email: n/a **Web:** www.seattle.gov/parks/boats/lakewoodmoorage.htm
Nearest Town: Seattle *(5 mi.)* **Tourist Info:** (206) 632-1500

Navigational Information
Lat: 47°33.813' **Long:** 122°16.005' **Tide:** n/a **Current:** n/a **Chart:** 18447
Rep. Depths *(MLW)*: **Entry** 15 ft. **Fuel Dock** n/a **Max Slip/Moor** 15 ft./-
Access: West side of Lake Washington across from Mercer Island

Marina Facilities *(In Season/Off Season)*
Fuel: No
Slips: 3 Total, 3 Transient **Max LOA:** 60 ft. **Max Beam:** n/a
 Rate *(per ft.)*: **Day** $0.75* **Week** n/a **Month** $6.75
 Power: 30 amp $3, **50 amp** $3, **100 amp** n/a, **200 amp** n/a
 Cable TV: No **Dockside Phone:** No
 Dock Type: Fixed, Floating, Alongside, Wood
Moorings: 0 Total, 0 Transient **Launch:** n/a
 Rate: Day n/a **Week** n/a **Month** n/a
Heads: 3 Toilet(s)
Laundry: 1 Washer(s), 1 Dryer(s) **Pay Phones:** Yes, 1
Pump-Out: No **Fee:** n/a **Closed Heads:** Yes

Marina Operations
Owner/Manager: Seattle Parks & Rec. **Dockmaster:** Ell & Kathie Schober
In-Season: Year-Round, 10am-5pm **Off-Season:** n/a
After-Hours Arrival: Call ahead
Reservations: Yes, Required **Credit Cards:** No
Discounts: None
Pets: Welcome **Handicap Access:** No

Marina Services and Boat Supplies
Services - Dock Carts **Supplies - OnSite:** Ice *(Block, Cube)*, Ships' Store

Boatyard Services
OnCall: Propeller Repairs **3+ mi:** Yacht Broker *(Executive Yacht Sales 286-1441, 5 mi.)*. **Nearest Yard:** Mercer Marine (425) 641-2090

Restaurants and Accommodations
Near: Restaurant *(Lee's 725-6309, L $5-10, D $7-18)*, *(Both Ways Café 722-5799)*, Coffee Shop *(Starbucks)* **Under 1 mi:** Restaurant *(Susan's 5100 Bistro 721-6308, B $4-8, L $6-9, D $14-23, weekend. Brunch 8am-3pm)*, *(Pizzuto's Italian Café 722-6395, closed Sun)*, Pizzeria *(Pizza Rio 723-0445, closed Mon; delivery w/$10 order; 12 in. $10-14, 16 in. $12-17)* **1-3 mi:** Restaurant *(Leschi Lakeside GBB Restaurant & Bar 524-0061, L $9-14, D $15-23)*, *(Daniel's Broiler 329-4191, D $23-40)*, *(Fasica 723-1971,Ethiopian; closed Mon, L Fri-Sat, D Tue-Sun $6-12)*, Lite Fare *(Columbia City Ale House 723-5123)* **3+ mi:** Motel *(Travelodge Mercer Island 232-8000, $51-109, 5 mi.)*, *(Ontario 343-7958, 4 mi.)*

Recreation and Entertainment
OnSite: Picnic Area, Grills, Jogging Paths, Roller Blade/Bike Paths, Special Events *(Seafarer Weekend - 1st wknd of Aug)* **Near:** Park *(Genesee Park & Playfield/ Seward Park)* **1-3 mi:** Golf Course *(Jefferson Park Golf Club 762-4513)*, Fitness Center *(Club Emerald 232-7080)*, Boat Rentals *(Speed on the Beach 721-7433)*, Bowling *(AMF Imperial Lanes 325-2525)*, Video Rental *(Hollywood Video 723-0197)*, Museum *(Georgetown Power Plant Museum 763-2542)*, Cultural Attract *(Rainier Valley Cultural Center 725-7517)*, Galleries *(Sasak Gallery and Imports 763-7374)*

3+ mi: Movie Theater *(Seattle Imax Dome Theater 622-1868, 5 mi.)*

Provisioning and General Services
OnSite: Convenience Store **Near:** Provisioning Service, Dry Cleaners *(QC Cleaners 721-5000)* **Under 1 mi:** Bank/ATM, Post Office, Protestant Church, Beauty Salon *(Creative Styles 725-4355)*, Barber Shop *(Big John's Barber Shop 722-9902)*, Pharmacy *(Columbia 723-5233/ Walgreens 760-6134)*, Florist *(D & F Flowers 723-5202)* **1-3 mi:** Market *(Rainier Food Market 722-2828)*, Supermarket *(Safeway 725-9575)*, Health Food *(PCC Natural Market 723-2720)*, Wine/Beer, Bakery, Fishmonger *(Mutual Fish Company 322-4368)*, Meat Market *(Bob's Quality Meats 725-1221)*, Crabs/Waterman, Shrimper, Catholic Church, Library *(New Holly Branch 386-1908)*, Laundry *(Super Suds Coin Laundry 722-8606)*, Bookstore *(Pathfinder Book Store 323-1755)*, Hardware Store *(Lowes 760-0832)*, Clothing Store **3+ mi:** Buying Club *(Costco 622-3136, 4 mi.)*

Transportation
OnCall: Rental Car *(Enterprise 723-1959)*, Taxi *(Seattle Taxi Cab 709-2003)* **Under 1 mi:** Local Bus *(Sound Transit Ranier Valley 723-7900)* **3+ mi:** Rail *(Amtrak 382-4125, 5 mi.)*, Ferry Service *(WA State 464-6400, 6 mi.)* **Airport:** Sea-Tac Int'l. *(11 mi.)*

Medical Services
911 Service **OnCall:** Ambulance **Under 1 mi:** Doctor *(Ranier Park Medical Clinic 461-6957)* **1-3 mi:** Dentist *(Seward Park Family Dental 721-7880)*, Chiropractor *(Seward Park Chiropractic Center 324-3791)*, Holistic Services *(Natural Family Medicine Clinic 723-4891)*, Veterinarian *(Four Paws 760-5200)*, Optician **Hospital:** Virginia Mason 624-1144 *(6 mi.)*

Setting -- Lake Washington is 16 miles long and lies on the eastern edge of Seattle. The bridges have vertical clearances ranging from 33 ft. under I-90 to 57 ft. under SR-520. The marina is at the north end of Andrews Bay on the western shore. From the docks, visitors are treated to magnificent views of Mt. Baker to the west, Mt. Rainier to the east and Mercer Island to the northeast. Crowded with luxury estates, it is home to Seattle's rich and famous, including Bill Gates.

Marina Notes -- *$10 min. The 60 ft. guest dock is on the east pier with side-ties. Step off the boat into a parklike setting with lawns, benches and a gazebo. (Beavers live near the locked gate and the deck is lined with flower boxes.) The store offers food, ice, some ships' supplies and sundries. There are no showers, pump-out or fuel on-site, but a washer & dryer are available. The only fuel dock on Lake Washington is located on the east side. Seattle Parks & Recreation also manages Leschi Moorage, about 3 miles north - for boats 38 feet or less, reserve ahead (325-3730, $0.75/ft.). It is exposed to (sometimes substantial) northerly and easterly winds.

Notable -- Lakewood Moorage is located in the Genesee residential neighborhood of Seattle with two great parks nearby. Just up the hill are Lee's Chinese Restaurant, Both Ways Café, and QC Cleaners. Most shops and restaurants are along Rainier Avenue, just over a mile away; a few restaurants are south of the marina. If you dinghy to Leschi to provision, you'll find Leschi Food Mart with fresh meat & seafood (322-0700), Pert's Deli (625-0277), and Lakeside GBB (Good, Better, Best) Restaurant and Bar - the bar is cozy with a fireplace and the warmth of wood-paneled walls.

Navigational Information
Lat: 47°37.866' **Long:** 122°23.238' **Tide:** 14 ft. **Current:** 0 kt. **Chart:** 18449
Rep. Depths (MLW): Entry 30 ft. **Fuel Dock** 30 ft. **Max Slip/Moor** 30 ft./-
Access: Elliott Bay in the central part of Puget Sound

Marina Facilities (In Season/Off Season)
Fuel: Gasoline, Diesel
Slips: 1200 Total, 50 Transient **Max LOA:** 400 ft. **Max Beam:** n/a
 Rate (per ft.): **Day** $1.00/0.50* **Week** n/a **Month** $330*
 Power: 30 amp $3, **50 amp** $10, **100 amp** $75**, **200 amp** n/a
 Cable TV: Yes, Incl. **Dockside Phone:** Yes, Incl. jack
Dock Type: Floating, Concrete
Moorings: 0 Total, 0 Transient **Launch:** n/a
 Rate: Day n/a **Week** n/a **Month** n/a
Heads: 6 Toilet(s), 4 Shower(s)
Laundry: 4 Washer(s), 4 Dryer(s), Iron, Iron Board **Pay Phones:** Yes, 4
Pump-Out: Onsite (Slip-Side), Full Service **Fee:** $17.50 **Closed Heads:** Yes

Marina Operations
Owner/Manager: Dwight Jones **Dockmaster:** Dan Park
In-Season: Year-Round, 24/7 **Off-Season:** n/a
After-Hours Arrival: Call on cell phone or VHF
Reservations: Yes, Preferred **Credit Cards:** Visa/MC, Amex
Discounts: None
Pets: Welcome **Handicap Access:** Yes, Heads, Docks

Elliott Bay Marina
2601 West Marina Place; Seattle, WA 98199

Tel: (206) 285-4817 **VHF: Monitor** Ch. 78A **Talk** Ch. 78
Fax: (206) 282-0626 **Alternate Tel:** n/a
Email: dan@elliottbaymarina.net **Web:** www.elliottbaymarina.net
Nearest Town: Seattle **Tourist Info:** (206) 284-5836

Marina Services and Boat Supplies
Services - Docking Assistance, Concierge, Security (24/7), Dock Carts, Megayacht Facilities, 3 Phase **Communication -** Fax in/out, Data Ports (Wi-Fi - Broadband Xpress), FedEx, DHL, UPS **Supplies - OnSite:** Ice (Cube), Bait/Tackle **Under 1 mi:** Propane **1-3 mi:** West Marine (293-8663), Marine Discount Store (Seattle Marine Supply 285-5010), CNG

Boatyard Services
OnSite: Electrical Repairs (Yacht Care 285-2600), Electronic Sales (Emerald 285-3632), Electronics Repairs, Rigger, Air Conditioning, Refrigeration, Divers, Woodworking **Nearest Yard:** Seaview West (206) 783-6550

Restaurants and Accommodations
OnSite: Restaurant (Palisade 285-1000, L $11-17, D $19-45, L Mon-Fri, D daily, Sun brunch 9:30am- 2pm, happy hour daily 4-6pm; before 6pm $30 3-course meal), (Maggie Bluff's Grill 283-8322, B $5-10, L $8-10, D $8-15)
OnCall: Restaurant (Pronto Pizza & Pasta 283-6005, delivers 10am-11pm; pizzas $8-20, pasta $7-11, sandwiches & burgers $7-8, fish & chips $9), (Chen's Village 281-8838) **Under 1 mi:** Restaurant (Kinnaree Thai 285-4460), Fast Food (Burger King, Subway), Lite Fare (Niko's Gyros 285-4778)
1-3 mi: Restaurant (Paragon 283-4548, L $7-10, D $15-22), Motel (BW International 283-4140), Hotel (MarQueen 282-7407, $120-200), Inn/B&B (Mediterranean 428-4700, $89-139), (Queen Anne 217-9719)

Recreation and Entertainment
OnSite: Jogging Paths **Near:** Beach, Park (5-acre Smith Cove & 12-acre Magnolia with gorgeous madrona trees) **Under 1 mi:** Fitness Center (Mikeo's Fitness 286-9130), Video Rental (Hollywood Video 285-2611)

1-3 mi: Playground, Dive Shop, Golf Course (Interbay 285-2200), Boat Rentals (Northwest Outdoor Center 281-9694 - kayaks), Movie Theater (Uptown Cinemas 285-1022), Museum (Puget Sound Maritime Museum 624-3028, Contemporary Art 728-1980, Experience Music 770-2702, Pacific Science Center 443-2001, Odyssey Maritime Discovery 374-4000), Sightseeing (Argosy Cruises 623-1445, Emerald City Charters 624-3931)

Provisioning and General Services
OnSite: Convenience Store, Wine/Beer, Copies Etc. **Under 1 mi:** Supermarket (Albertson's 283-9434), Liquor Store (WA 298-4615), Bakery (Upper Crust 283-1003), Bank/ATM, Post Office, Catholic Church, Protestant Church, Beauty Salon (Hair Masters 282-9598), Dry Cleaners (Victoria 283-0229), Pharmacy (Bartell 282-2880), Hardware Store (Ace 282-1916)
1-3 mi: Market (Ken's 282-1100), Delicatessen (Manhatten Express 285-5727), Health Food (Mother Nature's 284-4422), Farmers' Market (Magnolia Comm. Cntr. Sat 10am-2pm Jun-Sep), Fishmonger (Wild Salmon Seafood 283-3366), Library (Magnolia 386-4225), Laundry (A-1 441-1570), Bookstore (Magnolia 283-1062), Retail Shops (Highland Hills 654-1630)

Transportation
OnCall: Rental Car (Enterprise 389-8650), Taxi (Yellow Cabs 622-6500)
3+ mi: Rail (Amtrak 382-4125, 5 mi.) **Airport:** Sea-Tac Int'l. (16 mi.)

Medical Services
911 Service **OnSite:** Doctor (OnSite Docs 624-6050) **OnCall:** Ambulance **Under 1 mi:** Veterinarian (Urbanvet 352-6900) **1-3 mi:** Dentist (Magnolia Family Dentistry 284-5660), Chiropractor (Magnolia Village Chiropractic 283-9860) **Hospital:** Virginia Mason 223-6877 (4 mi.)

Setting -- Elliott Bay indents the east shore of Puget Sound just north of Duwamish Head. The marina commands spectacular views of the Seattle skyline, Olympic Mountains and Mt. Rainier. 14 main piers hosting 1200 slips edge the shoreline and are protected by a 2700-foot stone breakwater. Landside, the view is dominated by the brown shingled, barrel-fronted home of the Palisade Restaurant. The surrounding grounds have swaths of lawn and a plethora of flowers flanked by two parks. The location - outside the locks and on Magnolia's south bluff - makes it a convenient stop for a Seattle visit.

Marina Notes -- CCM. *Under 65 ft. $1.00/ft., over 65 ft. $1.50/ft.; monthly 32-62' flat rate $331-810; 63'+ $14.06/ft. **100 amp 3-phase 480V $75/day. 150 amp 3-phase 480V $125/day. Opened in 1991 as a state-of-the-art ecologically-atuned facility, it's also certified "clean" of stray voltage - zinc testing is offered free to incoming boats. Strict adherence to environmental rules is required - hazardous materials are collected & disposed of free; recycle bins provided. Marina Biz Center offers conference rooms, fax, Internet, printers, notary, etc. Pink and blue all-tile bathhouses and spacious laundries are at each end of the marina.

Notable -- Much of the marina's design was intended specifically to protect marine plants and animals - especially salmon; exploring the infrastructure is fascinating and instructive. The highly-touted, dramatic Palisade Restaurant features a saltwater pond, cobblestone bridge, breathtaking views and some of the best food in Seattle. More casual Maggie Bluff's Grill offers outdoor dining when the weather allows. Enjoy Seattle Center's huge variety of activities - pro basketball game, a Shakespearean play, the Children's Museum & the Pacific Science Center, two IMAX theaters, and a planetarium (443-2001).

Navigational Information
Lat: 47°36.600' **Long:** 122°20.900' **Tide:** 14 ft. **Current:** n/a **Chart:** 18447
Rep. Depths (*MLW*): **Entry** 45 ft. **Fuel Dock** n/a **Max Slip/Moor** 45 ft./-
Access: Located in the northeast corner of Elliott Bay off Puget Sound

Marina Facilities (*In Season/Off Season*)
Fuel: No
Slips: 36 Total, 5 Transient **Max LOA:** 120 ft. **Max Beam:** 28 ft.
  Rate (*per ft.*): **Day** $1.25/1.00* **Week** n/a **Month** n/a
  Power: 30 amp $3, **50 amp** $5, **100 amp** $10, **200 amp** n/a
  Cable TV: No **Dockside Phone:** No
  Dock Type: Floating, Long Fingers, Alongside, Concrete
Moorings: 0 Total, 0 Transient **Launch:** n/a
  Rate: Day n/a **Week** n/a **Month** n/a
Heads: 2 Toilet(s), 2 Shower(s)
Laundry: None **Pay Phones:** Yes, 1
Pump-Out: OnSite, 1 Central **Fee:** Free **Closed Heads:** Yes

Marina Operations
Owner/Manager: Port of Seattle **Dockmaster:** Ramel W. Fuentes
In-Season: May-Sep, 7am-5:30/7pm **Off-Season:** Sep-Apr, 7am-5:30pm
After-Hours Arrival: Call ahead or VHF Ch. 66A after hours
Reservations: Yes, Preferred** **Credit Cards:** Visa/MC
Discounts: None
Pets: Welcome **Handicap Access:** Yes, Heads, Docks

Bell Harbor Marina

PO Box 1209; 2203 Alaskan Way; Seattle, WA 98121

Tel: (206) 615-3952; (800) 426-7817 **VHF: Monitor** Ch. 66A **Talk** Ch. 66A
Fax: (206) 615-3965 **Alternate Tel:** (206) 615-3951
Email: harbor.B@portseattle.org **Web:** www.portseattle.org
Nearest Town: Seattle **Tourist Info:** (206) 956-9243

Marina Services and Boat Supplies
Services - Docking Assistance, Security (*24 hrs., locked gates & cameras*),
Dock Carts **Communication** - Data Ports (*Broadband Wi-Fi, Free*), FedEx,
DHL, UPS, Express Mail (*Sat Del*) **Supplies - Near:** Ice (*Block, Cube*),
Bait/Tackle **1-3 mi:** Ships' Store, West Marine (*292-8663*)

Boatyard Services
Nearest Yard: Seaview West Boatyard (206) 789-6550

Restaurants and Accommodations
OnSite: Restaurant (*Anthony's Pier 66 448-6688, L $8-12, D $12-21, Also
Anthony's Fish Bar*), (*Anthony's Bell Street Diner 448-6688, L $6-10, D $8-
16*) **Near:** Restaurant (*Fish Club 256-1040, B $8-13, L $11-18, D $20-28,
L Mon-Fri, D daily, B buffet $17, Sat-Sun brunch 6:30am-2:30pm $11-15*),
(*Six Seven at the Edgewater 269-4575, B & L Mon-Sat, D daily, Sun brunch
10am-2:30pm*), (*La Fontana Siciliana 441-1045, L Tue & Thu, D daily*),
(*Waterfront Seafood Grill 956-9171, D $23-70*), Pizzeria (*Belltown 441-
2653*), Hotel (*The Edgewater 728-7000, $179-325, waterfront*), (*Marriott
Waterfront 443-5000, $169-221*) **Under 1 mi:** Restaurant (*Wasabi 441-
6044*), (*Marrakesh Moroccan 234-6700*), (*Koji Oskayo 583-0980*), Motel
(*Days Inn 448-6366*), Hotel (*St. Regis 448-6366*)

Recreation and Entertainment
OnSite: Museum (*Odyssey Maritime Discovery 374-400 - within 1 mile:
Seattle Art Museum 654-3100, Coast Guard Museum 217-6993*)
Near: Fitness Center (*X Gym 728-9496*), Jogging Paths, Roller Blade/Bike
Paths, Video Rental (*Belltown Video 4335435*), Park **Under 1 mi:** Picnic
Area, Grills, Golf Course (*Greater Seattle Golf Complex 464-1175*), Movie

Theater (*Cinerama 411-3653*), Video Arcade, Cultural Attract (*Seattle Symphony
215-4747, Oliver McCaw Hall 389-7600, Paramount Theatre 682-1414*),
Sightseeing (*Pike Place Market 956-9391*), Galleries (*Fireworks Fine Crafts
682-6462, Glayy Eye 441-3221, Gallery Mikhel 292-2000*)

Provisioning and General Services
Near: Convenience Store (*Site 17 441-5484*), Delicatessen (*Arbor 728-
1960*), Bakery (*Macrina 448-4032*), Farmers' Market (*Pike Place daily*),
Fishmonger (*Pike Place 628-7181*), Bank/ATM, Beauty Salon (*Gary Manuel
728-9933*), Barber Shop (*Belltown 441-4738*), Dry Cleaners (*A-1 441-1570*),
Hardware Store (*Rainier 448-9415*) **Under 1 mi:** Market, Health Food (*Pike
Place Natural 623-2231*), Wine/Beer, Post Office, Bookstore (*Lamplight 652-
5554*), Pharmacy (*Rite Aid 41-8790*), Department Store (*Westlake Center*),
Copies Etc. **1-3 mi:** Supermarket (*QFC 285-5491*), Liquor Store (*WA Liquor
464-7910*), Catholic Church, Protestant Church, Synagogue, Library (*Central
264-1120*), Buying Club (*Costco 621-9997*)

Transportation
OnCall: Rental Car (*Enterprise 389-8650*), Taxi (*Orange 522-8800*)
Near: Local Bus (*Metro Trans. 800-542-7876/ Waterfront Streetcar Line
624-PASS*) **Under 1 mi:** Ferry Service (*WA State 464-6400*) **1-3 mi:** Rail
(*Amtrak 382-4125*) **Airport:** Sea-Tac Int'l. (*13 mi.*)

Medical Services
911 Service **OnCall:** Ambulance **Near:** Doctor (*Pike Market Medical Clinic
728-4143*) **Under 1 mi:** Dentist (*Pike place Dental 625-1267*), Chiropractor
(*Marketplace Chiropractic 441-0109*), Holistic Services **1-3 mi:** Veterinarian
(*Urbanvet 352-6900*) **Hospital:** Virginia Mason 223-6877 (*1 mi.*)

Setting -- Elliott Bay is on the eastern shore of Puget Sound, with West Point on the north and Alki Point 5 miles south. The bay proper, lying east of a line between Magnolia Bluff and Duwanish Head, is about two miles wide. Bell Harbor marina lies at Pier 66, less than 1 mile north of the State Ferry Dock - right in downtown Seattle, with all it has to offer. Shoreside, Pier 66 consists of eleven acres of large, modern glass buildings that house the Odyssey Maritime Discovery Center, a waterfront Conference Center, restaurants, waterfront plazas, and office space. The Waterfront Marriott is across the street.

Marina Notes -- CCM. *In-season $1.25/ft. Sun-Thu for all vessels; Fri-Sat under 50' $1.25/ft., over 50' $1.50. Off-season $1.00/ft. Sun-Thu, Fri-Sat boats over 50' $1.50. All boats $1.75/ft. during holidays, year-round. Short stays: up to 4 hrs., $10 for all boats, additional hours $5 each. **Non-refundable deposit required with reservations. The concrete docks and their steel pilings are almost totally enclosed by a concrete breakwater. Slips can accommodate larger boats, and side-tie is also available.This Port of Seattle marina is well-maintained and run by a helpful staff. Tight security; water and pump-out included.

Notable -- Explore the on-site Puget Sound's Odyssey Maritime Discovery Center's 44 hands-on exhibits. Climb aboard a real kayak. Guide a freighter into Elliott Bay ($7/5/2 - Wed-Thu 10-3, Thu 10-4, Sat & Sun 11-5). Most of downtown Seattle is within walking distance - the shopping district, fine hotels, restaurants and attractions - including the Seattle Aquarium. Seattle Center, home of the Space Needle and a wealth of cultural activities, lies just to the north. For provisioning, eateries and just plain fun, Pike Place Market is a few blocks south. The George Benson Waterfront Streetcar Line stops in front of the marina.

Navigational Information
Lat: 47°31.630' **Long:** 122°18.686' **Tide:** 11 ft. **Current:** 1 kt. **Chart:** 18450
Rep. Depths *(MLW):* **Entry** 8 ft. **Fuel Dock** 8 ft. **Max Slip/Moor** 8 ft./-
Access: 5 mi. south of Elliott Bay up Duwamish River

Marina Facilities *(In Season/Off Season)*
Fuel: No
Slips: 165 Total, 2 Transient **Max LOA:** 60 ft. **Max Beam:** 16 ft.
  Rate *(per ft.):* **Day** $1.11 **Week** n/a **Month** $7
  Power: 30 amp n/a, 50 amp n/a, 100 amp n/a, 200 amp n/a
  Cable TV: No **Dockside Phone:** No
  Dock Type: Floating, Long Fingers, Alongside, Wood
Moorings: 0 Total, 0 Transient **Launch:** n/a
  Rate: Day n/a **Week** n/a **Month** n/a
Heads: 2 Toilet(s), 2 Shower(s)
Laundry: 2 Washer(s), 2 Dryer(s) **Pay Phones:** Yes, 1
Pump-Out: No **Fee:** n/a **Closed Heads:** Yes

Marina Operations
Owner/Manager: Guy Crow **Dockmaster:** same
In-Season: Year-Round, 8:30am-5pm **Off-Season:** n/a
After-Hours Arrival: Call ahead
Reservations: Yes **Credit Cards:** Visa/MC, Dscvr
Discounts: None
Pets: Welcome **Handicap Access:** Yes, Heads, Docks

South Park Marina

8604 Dallas Ave. S; Seattle, WA 98108

Tel: (206) 762-3880 **VHF: Monitor** n/a **Talk** n/a
Fax: (206) 767-3066 **Alternate Tel:** n/a
Email: crow45@aol.com **Web:** www.southparkmarina.com
Nearest Town: Seattle *(3 mi.)* **Tourist Info:** (206) 632-1500

Marina Services and Boat Supplies
Services - Trash Pick-Up, Dock Carts, Megayacht Facilities
Communication - Mail & Package Hold, Phone Messages, Fax in/out
($1/pg), FedEx, DHL, UPS, Express Mail *(Sat Del)* **Supplies - OnSite:** Ice
(Shaved) **Near:** Ice *(Cube),* Propane **1-3 mi:** Bait/Tackle *(Xstream Fly
Fishing & Sport Gear 762-6170)*

Boatyard Services
OnSite: Travelift *(15T),* Crane, Engine mechanic *(gas, diesel),* Launching
Ramp, Electrical Repairs, Electronic Sales, Electronics Repairs, Hull
Repairs, Rigger, Sail Loft, Bottom Cleaning, Brightwork, Compound, Wash &
Wax, Interior Cleaning, Propeller Repairs, Woodworking, Painting, Awlgrip
Near: Canvas Work, Air Conditioning, Refrigeration, Divers, Metal
Fabrication. **1-3 mi:** Upholstery, Yacht Interiors, Total Refits.
3+ mi: Inflatable Repairs *(5 mi.),* Life Raft Service *(5 mi.),* Yacht Broker
(Executive Yacht Sales 286-1441, 6 mi.), Yacht Building *(5 mi.).*
Yard Rates: Haul & Launch $6/ft., Power Wash $3.25/ft.
Storage: In-Water $7 /ft./mo., On-Land $6 /ft./mo.

Restaurants and Accommodations
Near: Restaurant *(County Line 762-7370, B $5-6, L $5-8, D $5-8),* *(Jalisco
Mexican Restaurant 767-1943, L $4-9, D $6-12),* *(Scent of Asia 767-5880, L
$5-9, D $6-12, specialty is teriyaki),* *(Pattaya 763-7781, Asian),* *(Napoli
Pizzeria & Ristorante 768-9615)* **Under 1 mi:** Fast Food *(Subway,
McDonald's)* **1-3 mi:** Motel *(La Hacienda Motel 762-2460, $57-80),* Hotel
(Red Lion Hotel 762-0300, $64-89), Inn/B&B *(Georgetown Inn 762-2233)*
3+ mi: Restaurant *(Juan O'Riley's 622-JUAN, B $4-8, L $6-10, D $6-10,
5 mi.)*

Recreation and Entertainment
Under 1 mi: Playground, Tennis Courts, Jogging Paths, Horseback Riding,
Park, Museum *(Museum of Flight 764-5720, $14/$6.50/free)* **1-3 mi:** Picnic
Area, Grills, Fitness Center *(United Gym 244-1333),* Boat Rentals *(Pacific
Water Sports 246-9385 - kayaks & canoes),* Roller Blade/Bike Paths, Video
Rental *(Video Galore 723-0199),* Cultural Attract *(Prop Gallery West 762-
6808)* **3+ mi:** Golf Course *(West Seattle Municipal 285-2200, 5 mi.)*

Provisioning and General Services
Near: Convenience Store *(South Park Arco 763-4145),* Market *(South Park
Food Center 763-7896),* Beauty Salon *(Primavera 767-2737),* Dry Cleaners
(Regency Cleaners 768-9200) **Under 1 mi:** Delicatessen *(Breaktime Deli
and Cafe 764-5068),* Wine/Beer, Bank/ATM, Catholic Church, Protestant
Church, Copies Etc. **1-3 mi:** Liquor Store *(WA Liquor 764-6421),* Bakery,
Post Office, Library *(South Park Branch),* Barber Shop, Laundry, Pharmacy
(Sea-Mar Pharmacy 762-3730), Newsstand, Hardware Store *(McLendon
Hardware 762-4090),* Retail Shops *(Ranier Mall 723-5966),* Buying Club
(Costco 674-1220) **3+ mi:** Supermarket *(QFC 935-0585, 4 mi.)*

Transportation
OnCall: Rental Car *(Enterprise 933-1750),* Taxi *(Seattle Shuttle 510-6244)*
Near: Local Bus, InterCity Bus **3+ mi:** Rail *(Amtrak 382-4125, 6 mi.),* Ferry
Service *(WA State 464-6400, 6 mi.)* **Airport:** Sea-Tac Int'l. *(7 mi.)*

Medical Services
911 Service **OnCall:** Ambulance **Near:** Doctor *(Sea Mar Community
Health Center 762-3730),* Dentist *(Sea Mar Dental 762-3263)*
Hospital: Highline Community 248-4530 *(3 mi.)*

Setting -- South Park Marina is about 5 miles south of Elliott Bay in the Duwamish River, on the western shore next to the 16th Avenue South Bridge. Factories, warehouses, and commercial buildings surround the facility. Three long wooden piers parallel the shore providing alongside moorage and a small number of finger piers. An abandoned Boeing plant lies across the river from the marina. This is not a tourist neighborhood.

Marina Notes -- The marina, founded in 1970, has a boatyard, an ample boat storage and repair area, and a launch ramp. Most moorages are occupied by long-term tenants. Docks are labeled "A" through "F-line." Convenient guest dockage is along "A-line" and finger piers are available for smaller boats on the inside of the pier closest to shore ("F-line"). A locked gate is at the top of the ramp. On-site, Evergreen Boat Transport Co. provides transport for power and sail boats 20-45' (762-5144). Note: Spokane St. Bridge is 55 ft. at center (684-7443). 1st Ave. South Bridge 48 ft. at center (764-4160), and 16th Ave. South Bridge 34 ft. at center (762-2530).

Notable -- Most recreational activities require a 2-3 mile walk or a cab. Three miles north at Terminal 107 is Duwamish Public Access, a 7.2-acre park with a walking path for wildlife observation, bike path, picnic tables, benches and interpretive information. It is worth the cab fare to explore the magnificent Seattle Aquarium & Waterfront Park 386-4320. The Museum of Flight at Boeing Field shows the only Concorde on the west coast. The first jet-powered Air Force One which carried Kennedy, Johnson, and Nixon is open to walk through. A simulator lets you pilot a spy plane. Open daily, 10am-5pm, except holidays.

Boaters' Notes

Add Your Ratings and Reviews at www.AtlanticCruisingClub.com

A complimentary six-month Silver Membership is included with the purchase of this *Guide*. Select "Join Now," then "Silver," then follow the instructions.

The AtlanticCruisingClub website provides updated Marina Reports, Destination and Harbor Articles and much more — including an option within each online Marina Report for boaters to add their ratings and comments regarding that facility. Please log on frequently to share your experiences — and to read other boaters' comments.

On the website, boaters may rate marinas in one or more of the following categories — on a scale of 1 (basic) to 5 (world class) — and also enter additional commentary.

▸ **Facilities & Services** (Fuel, Reservations, Concierge Services and General Helpfulness)

▸ **Amenities** (Pool, Beach, Internet Access, including Wi-Fi, Picnic/Grill Area, Boaters' Lounge)

▸ **Setting** (Views, Design, Landscaping, Maintenance, Ship-Shapeness, Overall Ambiance)

▸ **Convenience** (Access — including delivery services — to Supermarkets, other Provisioning Sources, Shops, Services, Attractions, Entertainment, Sightseeing, Recreation, including Golf and Sport Fishing & Medical Services)

▸ **Restaurants/Eateries/Lodgings** (Availability of Fine Dining, Snack Bars, Lite Fare, OnCall food service, and Lodgings ashore)

▸ **Transportation** (Courtesy Car/Vans, Buses, Water Taxis, Bikes, Taxis, Rental Cars, Airports, Amtrak, Commuter Trains)

▸ **Please Add Any Additional Comments**

15. WA – Southeast Puget Sound

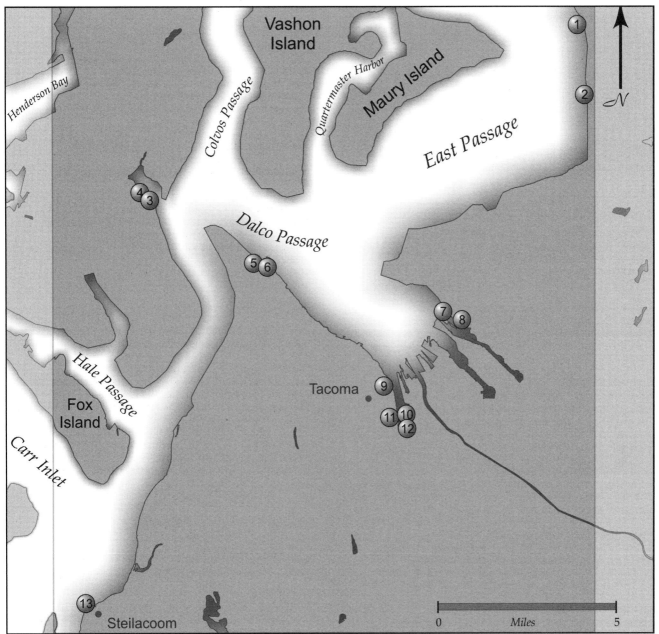

MAP	MARINA	HARBOR	PAGE	MAP	MARINA	HARBOR	PAGE
1	Des Moines Marina	East Passage	218	8	Chinook Landing Marina	Hylebos Waterway	225
2	Saltwater State Park	Poverty Bay	219	9	Foss Waterway Marina	Thea Foss Waterway	226
3	Jerisich Dock	Gig Harbor	220	10	Johnny's Dock	Thea Foss Waterway	227
4	Arabella's Landing	Gig Harbor	221	11	Dock Street Marina	Thea Foss Waterway	228
5	Point Defiance Boathouse Marina	Dalco P./Commencement Bay	222	12	Foss Landing Marina & Storage	Thea Foss Waterway	229
6	Breakwater Marina	Dalco P./Commencement Bay	223	13	Steilacoom Marina	Puget Sound/Gordon Point	230
7	Ole and Charlie's Marina	Hylebos Waterway	224				

▸ **Currency** — In Canadian Marina Reports, all prices are in Canadian dollars. In U.S. Marina Reports, all prices are in U.S. dollars.

▸ **"CCM"** — Denotes a Certified Clean Marina, a state/provincial award for environmental excellence. See page 298 for an explanation and page 299 for a list of Pump-Out facilities.

▸ **Ratings & Reviews** — An overview of the Atlantic Cruising Club's rating system is on page 6 and details on the content of each Marina Report are on pages 7 – 11.

▸ **Marina Report Updates** — Comments from boaters and new information from ACC reviewers and marinas are posted regularly on www.AtlanticCruisingClub.com.

Des Moines Marina

Navigational Information
Lat: 47°24.096' **Long:** 122°19.903' **Tide:** 14 ft. **Current:** n/a **Chart:** 18474
Rep. Depths (MLW): Entry 10 ft. **Fuel Dock** 10 ft. **Max Slip/Moor** 10 ft./-
Access: East Passage across northern end of Maury Island

Marina Facilities *(In Season/Off Season)*
Fuel: Gasoline, Diesel
Slips: 840 Total, 65 Transient **Max LOA:** 55 ft. **Max Beam:** n/a
  Rate *(per ft.)*: **Day** $0.60* **Week** n/a **Month** n/a
  Power: 30 amp $3, **50 amp** n/a, **100 amp** n/a, **200 amp** n/a
  Cable TV: Yes, $2/night **Dockside Phone:** No
  Dock Type: Floating, Long Fingers, Concrete, Wood
Moorings: 0 Total, 0 Transient **Launch:** n/a
  Rate: Day n/a **Week** n/a **Month** n/a
Heads: 6 Toilet(s), 2 Shower(s)
Laundry: None **Pay Phones:** Yes, 3
Pump-Out: OnSite, Full Service **Fee:** Free **Closed Heads:** Yes

Des Moines Marina

22307 Dock Avenue South; Des Moines, WA 98198-4

Tel: (206) 824-5700 **VHF: Monitor** n/a **Talk** Ch. 16
Fax: (206) 878-5940 **Alternate Tel:** n/a
Email: n/a **Web:** www.desmoineswa.gov
Nearest Town: Des Moines *(0.2 mi.)* **Tourist Info:** (206) 878-7000

Marina Operations
Owner/Manager: City of Des Moines **Dockmaster:** Joe Dusenbury
In-Season: Year-Round, 6am-8pm **Off-Season:** Nov-Feb, 8am-5pm
After-Hours Arrival: Self check-in at fuel dock
Reservations: Yes **Credit Cards:** Visa/MC
Discounts: None
Pets: Welcome **Handicap Access:** Yes, Heads

Marina Services and Boat Supplies
Services - Docking Assistance, Security *(roving at night)*, Dock Carts
Communication - FedEx, DHL, UPS, Express Mail **Supplies - OnSite:** Ice *(Block, Cube)*, Bait/Tackle, Propane **Under 1 mi:** Ships' Store **3+ mi:** Boater's World *(575-1920, 7 mi.)*

Boatyard Services
OnSite: Travelift *(37T)*, Engine mechanic *(gas, diesel)*, Launching Ramp, Electrical Repairs, Hull Repairs, Rigger, Bottom Cleaning, Brightwork, Compound, Wash & Wax, Woodworking, Inflatable Repairs, Upholstery, Yacht Interiors, Metal Fabrication, Painting, Awlgrip, Yacht Broker *(Classic Yachts 824-1200)*, Total Refits

Restaurants and Accommodations
OnSite: Restaurant *(Milluzzo's Sul Aqua 878-7719, B $4-9, L $5-15, D $5-15, Italian)*, *(Anthony's Home Port 824-1947, L $7-14, D $11-26, Sun brunch 10am-2pm)* **Near:** Restaurant *(China Sea 878-2593)*, *(New Toyko Teriyaki 824-0327)*, Seafood Shack *(Wally's Chowder & Broiler House 878-8140)*, Fast Food *(Jack in the Box)*, *(Red Robin 824-2214)*, Pizzeria *(Boston Pizza and Pasta 824-7333)*, Hotel *(Ramada Limited 824-9920, $82-90)* **Under 1 mi:** Restaurant *(Café Debra at Marine View 824-6672)*, *(Sushi Saki 824-1776)* **1-3 mi:** Motel *(Sleep Inn 878-3600, $58-71)*, *(Garden Suites 878-3020)*, Hotel *(Best Western 878-3300, $59-109)*

Recreation and Entertainment
OnSite: Beach, Picnic Area, Grills, Park *(Des Moines Beach Park)*, Special Events *(Waterland - last week of Jul)* **Near:** Jogging Paths, Movie Theater *(Des Moines Cinema 878-1540)*, Galleries *(Art Collectors 878-8879)*

Under 1 mi: Fitness Center *(Marine View Health and Fitness 870-3074)* **1-3 mi:** Golf Course *(Tyee Golf Course 878-3540)*, Video Rental *(Blockbuster 870-7800)* **3+ mi:** Bowling *(Hi-Line Lanes 244-2272, 4 mi.)*, Museum *(Hydriokabe & Race Boat Museum 764-9453, 5 mi.)*

Provisioning and General Services
OnSite: Farmers' Market *(Sat 10am-2pm Jun15-Oct)*, Newsstand **Near:** Convenience Store *(ABC Convenience Grocery 878-4144, 7-Eleven 878-8308)*, Fishmonger *(B&E Meat and Seafood 878-3700)*, Bank/ATM, Beauty Salon, Dry Cleaners *(The Cleaners 824-6002)*, Pharmacy *(Medicine Shoppe 824-4127)* **Under 1 mi:** Supermarket *(QFC 824-6610)*, Delicatessen *(Hoagy's Corner Deli 878-8308)*, Wine/Beer, Bakery, Catholic Church, Protestant Church, Library *(Des Moines 824-6066)*, Barber Shop *(Bill's Barber Shop 878-2205)*, Laundry *(Des Moines Highlander 824-7850)*, Clothing Store *(Benson's Apparel 878-3120)* **1-3 mi:** Liquor Store *(WA Liquor 872-4068)*, Post Office, Bookstore *(Book World 824-9422)*, Hardware Store *(Ace 941-8870)*, Copies Etc. *(Kinko's 878-5043)* **3+ mi:** Department Store *(Target 575-0682, 7 mi.)*, Buying Club *(Costco 575-4575, 5 mi.)*

Transportation
OnCall: Rental Car *(Enterprise 878-9103)*, Taxi *(Yellow 941-8178)*, Airport Limo *(R&M 244-8141)* **Near:** Local Bus **Airport:** Sea-Tac Int'l. *(5 mi.)*

Medical Services
911 Service **OnCall:** Ambulance **Near:** Doctor *(Emergency Medical 824-9997)*, Dentist *(Des Moines Dental 824-2804)*, Chiropractor *(Today's Chiropractic 878-2225)*, Veterinarian *(Marine View Vet 878-7616)*
Hospital: Highline Community 244-9970 *(4 mi.)*

Setting -- Des Moines Marina is located on the east side of East Passage north of Poverty Bay. It lies across from Vashon Island and the northern end of Maury Island. A long rock breakwater shelters the wide, uncluttered, well-scrubbed docks; enter at the north end betwen the breakwater and the long fishing pier. From the sea, one sees a line-up of apartment buildings dotting the hillside - each unit straining to get the best view. An attractive two-story white clapboard contemporary office building rises above the docks just behind the waterfront promenade. A spiffy varnished railing adds a nautical touch.

Marina Notes -- CCM. *Guest moorage flat rates per night: under 20' $13, 21-25' $15, 26-30' $18, 31-35' $21, 36-40' $24, 41-45' $27, 46-50' $30, 51-55' $33, 56' and over $40. This large, full-service marina was built in 1970 and has been run by the City since 1992. The easily accessed 2-station fuel dock, near the breakwater, also sells ice, propane, and bait. Turn the knob next to the window on the side of the fuel shack and hear the current weather report. A fish cleaning sink moves the messy cleaning outside. The marina is also home to Des Moines Yacht Club (878-7220). Boatyard services are provided by CSR Marine (878-4414). The sparkling heads are finished with blue and white tile. Showers are free to guest moorage customers.

Notable -- On-site are two popular restaurants, Anthony's Home Port and Milluzzo's Sul Aqua (which replaced Breakers). Just north of the fishing pier is Des Moines Beach Park with a beach, trails, and a great playground. Everything in the village is within walking distance. Special events include the Waterland Festival in July and a haunted house in October. A seasonal Saturday Farmers' Market is held right on the waterfront.

Des Moines Marina

PHOTOS ON DVD: 15

Navigational Information
Lat: 47°22.475' **Long:** 122°19.540' **Tide:** 12 ft. **Current:** n/a **Chart:** 18474
Rep. Depths (*MLW*): **Entry** n/a **Fuel Dock** n/a **Max Slip/Moor** -/70 ft.
Access: East side of East Passage at Poverty Bay

Marina Facilities *(In Season/Off Season)*
Fuel: No
Slips: 0 Total, 0 Transient **Max LOA:** n/a **Max Beam:** n/a
  Rate *(per ft.):* **Day** n/a **Week** n/a **Month** n/a
  Power: 30 amp n/a, 50 amp n/a, 100 amp n/a, 200 amp n/a
  Cable TV: No **Dockside Phone:** No
  Dock Type: n/a
Moorings: 5 Total, 5 Transient **Launch:** None
  Rate: Day $10 **Week** $35 **Month** $150
Heads: 4 Toilet(s), 1 Shower(s)
Laundry: None **Pay Phones:** Yes, 1
Pump-Out: No **Fee:** n/a **Closed Heads:** Yes

Marina Operations
Owner/Manager: Washington State Parks and Recreation **Dockmaster:** n/a
In-Season: May-Sep, 8am-Dusk **Off-Season:** Oct-Apr, 8am-Dusk
After-Hours Arrival: Tie up and pay in the morning
Reservations: No **Credit Cards:** No
Discounts: None
Pets: Welcome, Dog Walk Area **Handicap Access:** Yes, Heads

Saltwater State Park

25205 6th Place South; Des Moines, WA 98198

Tel: (253) 661-4956; (800) 233-0321 **VHF: Monitor** n/a **Talk** n/a
Fax: n/a **Alternate Tel:** n/a
Email: n/a **Web:** www.parks.wa.gov
Nearest Town: Des Moines *(2 mi.)* **Tourist Info:** (206) 878-4595

Marina Services and Boat Supplies
Supplies - 1-3 mi: Ice *(Block, Cube)*, Ships' Store, Bait/Tackle, Live Bait,
Propane

Boatyard Services
OnSite: Launching Ramp **1-3 mi:** Travelift, Yacht Broker, Yacht Building.
Nearest Yard: CSR Marine (206) 878-4414

Restaurants and Accommodations
1-3 mi: Restaurant *(Mandarin Kitchen 878-8511)*, *(Milluzzo's Sul Aqua 878-7719)*, *(Anthony's Homeport 824-1947, L $7-14, D $11-26)*, *(Silk Thai Café 941-0109)*, *(Viva Mexico 839-1903)*, *(Redondo Grill 839-5040)*, Fast Food *(Taco Bell, Subway)*, Motel *(Travelodge 878-3020)*, *(Ramada 824-9920, $67-100)*, Hotel *(Crossland Seattle Kent 946-1744, $39-50)*

Recreation and Entertainment
OnSite: Beach, Picnic Area, Grills, Playground *(an intriguing wooden structure)*, Jogging Paths, Park **1-3 mi:** Fitness Center *(Marine View Health and Fitness 206-870-3074)*, Movie Theater *(Des Moines Cinema 206- 878-1540)*, Video Rental *(Blockbuster 206-870-7800)*, Galleries *(Art Collectors Gallery 206-878-8879)* **3+ mi:** Golf Course *(Riverbend Golf 859-4000, 5 mi.)*, Boat Rentals *(Action Rentals 284-9284, 5 mi.)*, Cultural Attract *(Burien Little Theater 206-242-5180, 8 mi.)*

Provisioning and General Services
1-3 mi: Convenience Store *(Am-Pm 839-7411)*, Supermarket *(Safeway 946-8087, QFC 206-824-6610)*, Delicatessen *(Hogy's Corner Deli 206-878-8308)*, Fishmonger *(B&E Meats and Seafoods 206-878-3700)*, Bank/ATM, Post Office, Catholic Church, Protestant Church, Library *(Des Moines Library 839-0121)*, Beauty Salon *(Great Clips 529-8733)*, Barber Shop *(Fantastic Sam's 839-4071)*, Dry Cleaners *(Cleaner's Number 1 839-7700)*, Laundry *(DC Laundry 529-0840)*, Pharmacy *(Bartell 839-2583)*, Hardware Store *(Ace 941-8870)*, Florist *(Des Moines Florist 206-824-5920)*, Copies Etc. *(UPS Store 941-4450)* **3+ mi:** Bookstore *(Barnes & Noble 839-2535, 4 mi.)*, Department Store *(Sea Tac Mall 529-7782, 5 mi.)*, Buying Club *(Costco 874-3652, 5 mi.)*

Transportation
OnCall: Rental Car *(Enterprise 661-9866)*, Taxi *(Farwest Taxi 852-2500, Cuddy's 569-5729)* **Near:** Local Bus *(King County Transit 206-553-3000)*, InterCity Bus **Airport:** Sea-Tac Int'l. *(7 mi.)*

Medical Services
911 Service 1-3 mi: Dentist *(Holmes 575-4840)*, Chiropractor *(Mac Dermott 206-824-0107)*, Holistic Services *(Aruba J Mind Body Wellness Center 206-592-1423)*, Veterinarian *(Marine View Veterinary Hospital 206-878-7616)* **3+ mi:** Doctor *(Marina Medical 206-878-8600, 5 mi.)*, Optician *(Sears Optical 529-8314, 5 mi.)* **Hospital:** Highline Community 206-244-9970 *(6 mi.)*

Setting -- Saltwater State Park, located two miles south of Des Moines on Poverty Bay, is one of 120 state parks scattered throughout Washington. Approaching from the north, enter via East Passage. From the south, travel through Dalco Passage, past Commencement Bay and the city of Tacoma. The Park faces west and looks out to nearby Maury Island, which gives the views from the mooring field a lake-like feeling. Plan on a spectacular sunset when the weather is clear.

Marina Notes -- For boaters, 5 mooring buoys are available on a first-come first-serve basis for a $10 flat fee. Annual permits are available for purchase at State Parks headquarters in Olympia, regional offices, online and in person at the parks when staff is available. The cost of an annual pass is $3.50/ft.with a $50 minimum. Daily permits are only available at the parks. Heads are basic campsite cinderblock and an outdoor shower doubles for scuba gear.

Notable -- The park is situated halfway between Seattle and Tacoma, and the two cities buried a real hatchet here in a symbolic gesture. Dinghy ashore and enjoy the nearly 1500 feet of saltwater shoreline. The rock-bound beach is the main attraction - but there're also tidal pools, fishing, picnicking, jogging and hiking trails, bird-watching, volleyball, and scuba diving (this is one of the rare parks with an underwater, artificial reef and a scuba rinse station). There are interpretive activities during the summer months. Picnic tables are on the waterfront and scattered throughout the park - two covered cooking shelters without electricity are available (to reserve, call 253-661-4956). The park also offers 52 campsites and two large fire circles that invite an evening of fireside stories.

Jerisich Dock

3211 Harborview Drive; Gig Harbor, WA 98332

Tel: (253) 851-8145 **VHF: Monitor** n/a **Talk** n/a
Fax: n/a **Alternate Tel:** n/a
Email: n/a **Web:** www.gigharborguide.com
Nearest Town: Gig Harbor **Tourist Info:** (253) 857-4842

Navigational Information
Lat: 47°19.900' **Long:** 122°34.830' **Tide:** 12 ft. **Current:** n/a **Chart:** 18474
Rep. Depths (*MLW***): Entry** 6 ft. **Fuel Dock** n/a **Max Slip/Moor** 6 ft./-
Access: West side of Gig Harbor southeast of Arabella's Landing

Marina Facilities *(In Season/Off Season)*
Fuel: No
Slips: 10 Total, 10 Transient **Max LOA:** 50 ft. **Max Beam:** n/a
 Rate *(per ft.)*: **Day** Free* **Week** n/a **Month** n/a
 Power: 30 amp n/a, 50 amp n/a, 100 amp n/a, 200 amp n/a
 Cable TV: No **Dockside Phone:** No
 Dock Type: Floating, Alongside, Concrete, Wood
Moorings: 0 Total, 0 Transient **Launch:** n/a, Dinghy Dock
 Rate: Day n/a **Week** n/a **Month** n/a
Heads: 2 Toilet(s)
Laundry: None **Pay Phones:** No
Pump-Out: OnSite **Fee:** Free **Closed Heads:** Yes

Marina Operations
Owner/Manager: City of Gig Harbor **Dockmaster:** Dave Brereton
In-Season: Year-Round **Off-Season:** n/a
After-Hours Arrival: Call ahead
Reservations: No **Credit Cards:** No
Discounts: None
Pets: Welcome **Handicap Access:** No

Marina Services and Boat Supplies
Supplies - Near: West Marine *(West Marine Express 858-6250)*
Under 1 mi: Ships' Store *(Ship to Shore 858-6090)* **3+ mi:** Boater's
World *(472-3393, 10 mi.)*

Boatyard Services
Nearest Yard: Gig Harbor Marina and Boatyard (253) 851-7157

Restaurants and Accommodations
Near: Restaurant *(Tides Tavern 858-3982, L $6-15, D $13-30, breakfast Sat
& Sun 8-11am; moorage available)*, *(Harbor Inn 851-5454, B $7-10, L $8-12,
D $15-24)*, *(Anthony's Restaurant 853-6353)*, *(Brix 25° 858-6626, D $19-27)*,
(Tokyo Teriyaki 853-3232), Lite Fare *(IsaMira Gourmet Cheese & Café 857-
7511, 7am-7pm soups, sandwiches, cheese platters $5-20; D to go only,
serves 6-8 $25-35; outdoor deck)*, Pizzeria *(Spiro's Pizza & Pasta 851-9200)*,
Inn/B&B *(Maritime Inn 858-1818, $90-160)*, *(Rose of Gig Harbor 853-7990)*
Under 1 mi: Fast Food *(Burger King)*, Hotel *(BW Wesley Inn 858-9690,
$107-169)*, Inn/B&B *(Waterfront Inn 857-0770, $150-190)* **1-3 mi:** Inn/B&B
(The Inn at Gig Harbor 858-1111, $99-195)

Recreation and Entertainment
OnSite: Picnic Area, Park *(Peninsula Park)* **Near:** Fitness Center *(Harbor
Health & Wellness 857-2687)*, Video Rental *(Harbor Video 851-5866)*,
Special Events *(Maritime Gig Festival - 1st wknd Jun, Garden Tour - Jun, Art
Fest - Jul, Heritage Row - Aug, 1st Sat Art Walk year-round 1-5pm)*

Under 1 mi: Boat Rentals *(Gig Harbor Rent-A-Boat 858-7341 - kayaks &
power boats)*, Museum *(Gig Harbor Historical Society 858-6722)*, Galleries
(Harbor Gallery 851-8626, Ebb Tide Gallery 851-5293) **3+ mi:** Golf Course
(Madrona Links Golf Course 851-5193, 3 mi.)

Provisioning and General Services
Near: Supermarket *(QFC 858-2400)*, Delicatessen *(Susanne's Bakery and
Deli 853-6220)*, Health Food *(Whole Foods Market 851-8120)*, Bakery,
Bank/ATM, Post Office, Catholic Church *(St. Nicholas 851-8850)*, Beauty
Salon *(Athena's Salon & Spa 851-2883)*, Bookstore *(Mostly Books 851-
3219)*, Pharmacy *(Rexall 858-9908)*, Retail Shops, Copies Etc. *(Copy It Mail
It 858-9915)* **Under 1 mi:** Farmers' Market *(Sat 8:30am-2:30pm Apr-Sep
at Stroh's Harbor Field)*, Protestant Church *(United Methodist 851-2625)*,
Barber Shop *(Harbor Barber 858-3613)*, Hardware Store *(Ace 851-6169)*,
Florist *(Posie Patch Florist 851-7673)* **1-3 mi:** Gourmet Shop *(Gourmet
Essentials 858-7711)*, Liquor Store *(WA Liquor 857-5966)*, Library *(Peninsula
Branch 851-3793)*, Dry Cleaners *(Peninsula 858-3141)*

Transportation
OnCall: Rental Car *(Enterprise 853-3882)* **Airport:** Tacoma Narrows *(6 mi.)*

Medical Services
911 Service **Under 1 mi:** Dentist *(Gig Harbor Dental 858-3457)*,
Chiropractor *(Gig Harbor Spine & Posture 858-2474)* **1-3 mi:** Doctor *(Gig
Harbor Urgent Care 459-7570)*, Holistic Services *(Gig Harbor Naturopathic
851-7550)*, Veterinarian *(Evergreen Animal Hosp 851-9195)*, Optician
(Harbor Optical 851-7895) **Hospital:** Tacoma General 403-1000 *(12 mi.)*

Setting -- Gig Harbor is a mile-long inlet on the west side of the south entrance to Colvos Passage abreast Point Defiance. After entering the harbor, the dock
is off the port side backed by.Jerisich Park. An expansive bright green lawn and a large, well-deisgned wooden deck - populated by picnic tables and the
Fishermen's Monument - overlook the 400 linear feet of quality floating concrete moorage. Majestic snow-capped Mt. Ranier looms above the park, which also
hosts outdoor concerts, movies, and festivals. Beyond this quaint tourist town - and around the bay are pretty homes peeking out from the forested shore.

Marina Notes -- *This well-done public facility offers complimentary, nicely maintained, no-frills side-tie guest moorage in a beautiful setting. No power,
water, or showers, but pump-out is provided; in '07, the harbor will have a new fuel dock. Quarry tile and stianless steel public heads at the top of the ramp.

Notable -- Gig Harbor was discovered in 1841 by Capt. Charles Wilkes and named "Gig" after the gig (small boat) he used to explore the harbor. It was
settled in 1867 and fisherman Sam Jerisich was one of the first European colonists. Wahington's Maritime Village - as Gig Harbor bills itself - still supports an
historic fishing fleet, dating from 19th C. Croatian immigrants. Learn a bit of northwest history by following the historic markers along the History Walk - a self-
guided tour is available at the new Visitors Center (3125 Judson). For history buffs, another option is the Gig Harbor Peninsula Historical Society & Museum
(858-6722) which features exhibits about the pioneers who built this community along with videos of the old Tacoma Narrows Bridge collapse. Time your visit
for the Maritime Gig Festival (851-6865) always held the first full weekend in June. Or join the award-winning Saturday Art Walk held the first of each month.

Navigational Information
Lat: 49°19.934' **Long:** 122°34.870' **Tide:** 12 ft. **Current:** n/a **Chart:** 18474
Rep. Depths (MLW): Entry 6 ft. **Fuel Dock** n/a **Max Slip/Moor** 6 ft./-
Access: West side of Gig Harbor just beyond the Bayview Dock sign

Marina Facilities *(In Season/Off Season)*
Fuel: No
Slips: 57 Total, 12 Transient **Max LOA:** 160 ft. **Max Beam:** n/a
  Rate *(per ft.):* **Day** $1.00/0.75* **Week** n/a **Month** n/a
  Power: 30 amp Incl., **50 amp** Incl., **100 amp** n/a, **200 amp** n/a
  Cable TV: No **Dockside Phone:** No
  Dock Type: Floating, Long Fingers, Alongside, Concrete, Wood
Moorings: 0 Total, 0 Transient **Launch:** n/a
  Rate: Day n/a **Week** n/a **Month** n/a
Heads: 4 Toilet(s), 2 Shower(s)
Laundry: 2 Washer(s), 2 Dryer(s) **Pay Phones:** No
Pump-Out: OnSite, Self Service, 1 Central **Fee:** $1 **Closed Heads:** Yes

Marina Operations
Owner/Manager: Judy Stearns **Dockmaster:** Same
In-Season: Year-Round, M-Th 8am-4pm, F&S 9am-6pm** **Off-Season:** n/a
After-Hours Arrival: See instructions on the pay envelope
Reservations: Yes, Preferred **Credit Cards:** Visa/MC
Discounts: None
Pets: Welcome **Handicap Access:** Yes

Arabella's Landing

3323 Harborview Drive; Gig Harbor, WA 98332

Tel: (253) 851-1793 **VHF: Monitor** n/a **Talk** n/a
Fax: (253) 851-1793 **Alternate Tel:** n/a
Email: n/a **Web:** www.arabellaslanding.com
Nearest Town: Gig Harbor **Tourist Info:** (253) 857-4842

Marina Services and Boat Supplies
Services - Docking Assistance, Concierge, Boaters' Lounge, Security, Dock Carts **Communication -** Data Ports *(Wi-Fi - Broadband Xpress)*, FedEx, DHL, UPS **Supplies - Near:** West Marine *(West Marine Express 858-6250)* **Under 1 mi:** Ships' Store *(Ship to Shore 858-6090)* **3+ mi:** Boater's World *(Boater's World 472-3393, 10 mi.)*

Boatyard Services
Nearest Yard: Gig Harbor Marina and Boatyard (253) 851-7157

Restaurants and Accommodations
Near: Restaurant *(Harbor Inn 851-5454, B $7-10, L $8-12, D $15-24)*, *(Anthony's 853-6353)*, *(El Pueblito 858-9077)*, Pizzeria *(Spiro's Pizza & Pasta 851-9200)*, Inn/B&B *(Maritime Inn 858-1818, $90-160)* **Under 1 mi:** Restaurant *(Tides Tavern 858-3982, L, D; moorage available)*, *(Harbor Kitchen 853-6040)*, *(Thai Hut 858-8523)*, Fast Food *(Quiznos, KFC)*, Hotel *(Best Western 858-9690, $107-169)*, Inn/B&B *(Waterfront Inn 857-0770, $150-190)* **1-3 mi:** Restaurant *(Bistro Satsuma 858-5151, sushii)*, Inn/B&B *(The Inn at Gig Harbor 858-1111, $99-195)* **3+ mi:** Restaurant *(The Beachouse at Purdy 858-9900, D $13-30, 4 mi., early bird special $17)*

Recreation and Entertainment
OnSite: Picnic Area, Grills **Near:** Fitness Center *(Harbor Health & Wellness 857-2687)*, Video Rental *(Harbor Video 851-5866)*, Park, Special Events *(Maritime Gig Festival - 1st wknd Jun, Garden Tour - Jun, Art Fest - Jul, Heritage Row - Aug)*, Galleries *(Fire & Light Gallery 279-2580, Gallery Row 851-6020)* **Under 1 mi:** Dive Shop *(Tagerts Dive Locker 857-3660)*, Boat Rentals *(Gig Harbor Rent-A-Boat 858-7341)*, Museum *(Gig Harbor*

Historical Society 858-6722) **1-3 mi:** Movie Theater *(Paradise Theater 851-7529)*, Cultural Attract *(Encore Theater Co. 858-2282)* **3+ mi:** Golf Course *(Madrona Links Golf Course 851-5193, 3 mi.)*

Provisioning and General Services
Near: Supermarket *(QFC 858-2400)*, Delicatessen *(Susanne's Bakery and Deli 853-6220)*, Health Food *(Whole Foods Market 851-8120)*, Wine/Beer, Bakery, Bank/ATM, Post Office *(858-7262)*, Catholic Church *(St. Nicholas 851-8850)*, Beauty Salon *(Style-Maker 851-2887)*, Bookstore *(Mostly Books 851-3219)*, Pharmacy *(Belfair Drugs 851-6688)*, Retail Shops, Copies Etc. *(Copy It Mail It 858-9915)* **Under 1 mi:** Farmers' Market *(Stroh's Harbor Field Sat 8:30am-2:30pm May-Sep)*, Protestant Church *(United Methodist 851-5798)*, Barber Shop *(Harbor Barber 858-3613)*, Hardware Store *(Ace 851-6169)* **1-3 mi:** Gourmet Shop *(Gourmet Essentials 858-7711)*, Liquor Store *(WA Liquor 857-5966)*, Library *(Peninsula Branch 851-3793)*, Dry Cleaners *(Peninsula Laundry & Dry Cleaners 858-3141)*

Transportation
OnCall: Rental Car *(Enterprise 853-3882)* **Near:** Local Bus *(Town Around, Sat 10-4 857-4842)* **Airport:** Tacoma Narrows *(6 mi.)*

Medical Services
911 Service **Under 1 mi:** Dentist *(Gig Harbor Dental 858-3457)*, Chiropractor *(Chiropractics Northwest 853-3353)* **1-3 mi:** Doctor *(Gig Harbor Urgent Care 459-7570)*, Holistic Services *(Gig Harbor Naturopathic Medicine 851-7550)*, Veterinarian *(Evergreen Animal Hospital 851-9195)*, Optician *(Harbor Optical 851-7895)* **Hospital:** Tacoma General 403-1000 *(12 mi.)*

Setting -- Just beyond Bayview Dock, Arabella's Landing's lush, park-like grounds slope gently toward the impeccably-mantained floating docks. Potted geraniums trim the rails of a deck furnished with umbrellaed tables. And colorful pansies surround the well-used flag pole - all connected by brick walkways. Lovely houses hug the opposite shore with Mount Rainier looming above. Quaint Old Town is just behind the marina, so everything is close at hand.

Marina Notes -- *$22.50 minimum. $45 flat rate for 60 ft. slips. **Summer hours 8am-4pm Mon-Thu., 9am-7pm Fri-Sat, 9am-6pm Sun. Off-season Mon-Thu only. Most guest moorage is alongside, but slips (to 60 ft.) are also available. Arabella's Landing welcomes guests with helpful staff and dock hands, and a well-equipped boater's lounge with comfortable microsuede seating areas, a kitchenette with fresh Starbucks coffee, and cozy fireplace. Menus from nearby restaurants are available and the staff will provide concierge services. Access to the docks is gated. Bright, inviting tiled bathhouse and laundry facility.

Notable -- Provisions, boutiques and galleries are a short walk from the marina. Numerous eateries are nearby, offering everything from ice cream to elegant meals. Stroll along the waterfront and absorb the sights and smells of the fishing industry that underpins the local economy. Enjoy a show at the Encore Theater ($12/8/5, 858-2282) or at the Paradise Dinner Theater ($30-38, 851-7528), or an outdoor movie at Skansie Brothers Park in the summer (853-3554). Don't miss the Art Walk held the first Sat of each month from 1 to 5pm. The Gig Harbor Historical Society & Museum invites visitors to explore local history (Tue-Sat 10am-4pm, 858-6722). Summer '06 was the inauguration of the trial "Town Around" shuttle that runs every half-hour Sat 10am-4pm, making 8 convenient stops.

Point Defiance Boathouse Marina

Pt. Defiance Boathouse Marina

5912 N. Waterfront Drive; Tacoma, WA 98407

Tel: (253) 591-5325 **VHF: Monitor** n/a **Talk** n/a
Fax: n/a **Alternate Tel:** n/a
Email: boathouse@tacomaparks.com **Web:** www.metroparkstacoma.org
Nearest Town: Tacoma *(3 mi.)* **Tourist Info:** (253) 627-2175

Navigational Information

Lat: 47°18.380' **Long:** 122°30.918' **Tide:** 16 ft. **Current:** 2 kt. **Chart:** 18453
Rep. Depths *(MLW)*: **Entry** 24 ft. **Fuel Dock** n/a **Max Slip/Moor** 22 ft./-
Access: Dalco Passage to Point Defiance

Marina Facilities *(In Season/Off Season)*

Fuel: *Temp Suspended*** - Gasoline, Diesel
Slips: 5 Total, 5 Transient **Max LOA:** 45 ft. **Max Beam:** n/a
  Rate *(per ft.)*: **Day** $0.50* **Week** n/a **Month** n/a
  Power: 30 amp n/a, 50 amp n/a, 100 amp n/a, 200 amp n/a
  Cable TV: No **Dockside Phone:** No
  Dock Type: Floating, Alongside, Concrete, Wood
Moorings: 10 Total, 10 Transient **Launch:** n/a, Dinghy Dock
  Rate: Day Free*** **Week** n/a **Month** n/a
Heads: 2 Toilet(s)
Laundry: None **Pay Phones:** No
Pump-Out: OnSite, Self Service, 1 Central **Fee:** Free **Closed Heads:** Yes

Marina Operations

Owner/Manager: Tacoma Metro Parks **Dockmaster:** Tim Hartman
In-Season: Year-Round, 6am-6pm **Off-Season:** n/a
After-Hours Arrival: Call ahead
Reservations: No **Credit Cards:** Visa/MC, Dscvr
Discounts: None
Pets: Welcome **Handicap Access:** No

Marina Services and Boat Supplies

Communication - FedEx, UPS, Express Mail **Supplies - OnSite:** Ice *(Cube)*, Ships' Store, Bait/Tackle *(Boathouse Tackle & Gift Shop)* **3+ mi:** West Marine *(926-2533, 11 mi.)*, Boater's World *(472-3393, 8 mi.)*

Boatyard Services

OnSite: Launching Ramp *(new 4-lane)* **Near:** Engine mechanic *(gas, diesel)*, Electrical Repairs, Electronic Sales, Electronics Repairs, Hull Repairs, Bottom Cleaning, Brightwork, Air Conditioning, Refrigeration, Divers, Woodworking, Upholstery, Yacht Interiors, Yacht Broker *(Breakwater)*. **Nearest Yard:** Breakwater Marina (253) 752-6663

Restaurants and Accommodations

OnSite: Restaurant *(Anthony's at Pt. Defiance 752-9700, L $7-13, D $10-30)* **Under 1 mi:** Restaurant *(Ruston Inn 752-3288, B $4-8, L $6-9, D $10-14)*, *(Unicorn Bar & Grill 752-5939, L $5-12, D $7-14)*, *(Point Defiance Café 756-5101)*, *(Sar's Oriental Cuisine 761-2727)*, Lite Fare *(Antique Sandwich 752-4069)* **1-3 mi:** Restaurant *(Gateway to India 761-1266)*, *(Lobster Shop 759-2165, L $8-15, D $17-33, waterfront; twilight menu Sun-Thu 4:30-5:30pm $17, Brunch buffet $20)* **3+ mi:** Hotel *(Silver Cloud 272-1300, $109-299, 4 mi.)*, Inn/B&B *(Green Cape Cod 752-1977, $125-155, 4 mi.)*

Recreation and Entertainment

OnSite: Picnic Area, Boat Rentals, Park, Sightseeing *(Pt. Defiance Zoo and Aquarium 591-5337)*, Special Events *(at the park: Pt. Defiance Flower Show - Jun, Taste of Tacoma - end of Jun)* **Near:** Beach, Playground, Jogging Paths, Roller Blade/Bike Paths **Under 1 mi:** Grills, Tennis Courts, Museum *(Fort Nisqually; Camp 6 Logging Museum 591-5339)*

1-3 mi: Golf Course *(Highlands 759-3622)*, Fitness Center *(YMCA 564-9622)*, Video Rental *(Hollywood Video 759-9923)* **3+ mi:** Bowling *(Tower Lanes 564-8853, 4 mi.)*, Movie Theater *(Blue Mouse Theater 752-9500, 4 mi.)*

Provisioning and General Services

OnSite: Convenience Store **Near:** Hardware Store *(Defiance Hardware 759-4642)* **Under 1 mi:** Market *(Don's Ruston Market & Delli 759-8151)*, Delicatessen, Liquor Store *(Ruston Liquor 752-3007)*, Catholic Church, Protestant Church, Beauty Salon *(Sue's Hair Solution 761-9456)*, Barber Shop *(Mane Place Barber 759-7955)* **1-3 mi:** Bank/ATM, Post Office *(778-3447)*, Bookstore *(Park Bench Books 752-4848)*, Pharmacy *(Rite Aid 879-0140)* **3+ mi:** Supermarket *(Safeway 752-5570, 4 mi.)*, Health Food *(Whole Foods 565-0188, 5 mi.)*, Library *(Wheelock Library 591-5640, 4 mi.)*, Laundry *(New Era 759-3501, 4 mi.)*, Buying Club *(Costco 475-2093, 7 mi.)*

Transportation

OnCall: Rental Car *(Enterprise 573-1055)*, Taxi *(Orange Cab 779-4444)*, Airport Limo *(Bayview Limousine 272-7880)* **Near:** Bikes, Local Bus *(Pierce Transit 581-8000)*, Ferry Service *(WA State Ferry to Vashon Island 464-6400)* **3+ mi:** Rail *(Amtrak 627-8141, 8 mi.)* **Airport:** Tacoma Narrows/Sea-Tac Int'l. *(7 mi./26 mi.)*

Medical Services

911 Service **OnCall:** Ambulance **1-3 mi:** Doctor *(Multi Care Urgent Care Center 459-7130)*, Dentist *(Winters 756-1600)*, Veterinarian *(Westgate Animal Hospital 752-6161)* **3+ mi:** Chiropractor *(Proctor Chiropractor 756-7500, 4 mi.)* **Hospital:** Tacoma General 403-1000 *(6 mi.)*

Setting -- Between the south end of Colvos Passage and the north end of Tacoma Narrows, is 700-acre Point Defiance Park. Along its northeast side is the Boathouse Marina with two docking areas separated by an eigth mile. One is for day use. Next to the boat launch, a floating breakwater protects a pair of piers that berth transient slips. An attractive gray shingled, pyramid roofed building houses the office and store; Anthony's restaurant is nearby.

Marina Notes -- *Flat rate based on size: under 19' $10, 19-29' $15, over 30' $20. Honor system. **Fuel service temporaily suspended while fuel lines are replaced. ***10 mooring buoys are free. 72-hour limit. Marina caters mostly to park guests, including many fishing boats drawn to Puget Sound's salmon fishing captial - it's primarily a dry storage facility for boats 17 ft. and under. Larger boat transient moorage is alongside old, but well-kept concrete docks. No power or water. The well-supplied marina store sells fishing equipment, live bait, fuel, snacks, gifts, souvenirs, and some grocery items. 2

Notable -- Pt. Defiance is one of the 20 largest urban parks in the U.S., and hosts about two million people each year. Attractions include an old growth forest, formal gardens, hiking trails, picnic areas, beaches with great views, a restaurant, go-carts, and special events. The park is also home to Point Defiance Zoo & Aquarium (591-5333); summer hours are 9:30am-6pm, closing at 4/5pm from fall-spring ($10/$9 seniors/$8 kids, under 3 free). Another attracton is Fort Nisqually, the first European settlement on Puget Sound; two of the original buildings remain, one housing a museum with antique logging tools and equipment, plus an old train used to haul the logs. A lodge is available for events (305-1010, up to 100 people). Bus service to the rest of the city.

PHOTOS ON DVD: 17

Navigational Information
Lat: 47°18.300' **Long:** 122°30.690' **Tide:** 16 ft. **Current:** 2 kt. **Chart:** n/a
Rep. Depths (*MLW*): **Entry** 24 ft. **Fuel Dock** 20 ft. **Max Slip/Moor** 22 ft./-
Access: Dalso Passage to Point Defiance

Marina Facilities *(In Season/Off Season)*
Fuel: *Chevron* - Slip-Side Fueling, Gasoline, Diesel
Slips: 180 Total, 10 Transient **Max LOA:** 110 ft. **Max Beam:** 24 ft.
  Rate *(per ft.)*: **Day** $0.75 **Week** n/a **Month** n/a
  Power: 30 amp Incl., **50 amp** n/a, **100 amp** n/a, **200 amp** n/a
  Cable TV: No **Dockside Phone:** No
Dock Type: Floating, Concrete, Wood
Moorings: 0 Total, 0 Transient **Launch:** n/a, Dinghy Dock
  Rate: Day n/a **Week** n/a **Month** n/a
Heads: 2 Toilet(s), 1 Shower(s)
Laundry: 2 Washer(s), 2 Dryer(s) **Pay Phones:** No
Pump-Out: OnSite **Fee:** $5 **Closed Heads:** Yes

Marina Operations
Owner/Manager: Michael Marchetti **Dockmaster:** Dave Goughnour
In-Season: May-Sep, 8am-8pm **Off-Season:** Sep-May, 9am-5pm
After-Hours Arrival: Call ahead
Reservations: Yes **Credit Cards:** Visa/MC, Dscvr, Chevron
Discounts: Quantity Fuel* **Dockage:** n/a **Fuel:** $0.03 **Repair:** n/a
Pets: Welcome **Handicap Access:** No

Breakwater Marina

5603 N Waterfront Drive; Tacoma, WA 98407

Tel: (253) 752-6663 **VHF: Monitor** n/a **Talk** n/a
Fax: (253) 752-8291 **Alternate Tel:** n/a
Email: n/a **Web:** www.breakwatermarina.com
Nearest Town: Tacoma *(3 mi.)* **Tourist Info:** (253) 627-2175

Marina Services and Boat Supplies
Services - Docking Assistance, Security *(24 hrs., locked gate)*, Trash Pick-Up, Dock Carts **Communication -** Data Ports *(Wi-Fi - Broadband Xpress)*
Supplies - OnSite: Ice *(Block, Cube)*, Propane **Near:** Bait/Tackle *(herring)*
3+ mi: West Marine *(926-2533, 11 mi.)*, Boater's World *(472-3393, 8 mi.)*

Boatyard Services
OnSite: Engine mechanic *(gas, diesel)*, Electrical Repairs, Electronic Sales, Electronics Repairs, Hull Repairs, Bottom Cleaning, Brightwork, Air Conditioning, Refrigeration, Divers, Compound, Wash & Wax, Woodworking, Yacht Broker *(Breakwater Marine Yacht Sales)* **Near:** Launching Ramp, Upholstery, Yacht Interiors. **Member:** ABYC - 3 Certified Tech(s)

Restaurants and Accommodations
Near: Restaurant *(Anthony's at Pt. Defiance 752-9700, L $7-13, D $10-23)*
Under 1 mi: Restaurant *(Ruston Inn 752-3288, B $4-8, L $6-9, D $10-14)*, *(Unicorn Bar and Grill 752-5939, L $5-12, D $7-14)*, *(Sar's Oriental Cuisine 761-2727)*, Lite Fare *(Antique Sandwich Co. 752-4069)* **1-3 mi:** Restaurant *(Gateway to India 761-1266)*, *(Lobster Shop 759-2165, L $8-15, D $17-33, twilight menu Sun-Thu 4:30-5:30pm $17, Brunch buffet $20)* **3+ mi:** Pizzeria *(Abella Pizzeria 779-0769, 6 mi.)*, Hotel *(Silver Cloud 272-1300, $109-299, 4 mi.)*, *(Courtyard 591-9100, 6 mi.)*, *(Sheraton Tacoma 572-3200, 6 mi.)*, Inn/B&B *(Commencement Bay B&B 752-8175, $100-130, 4 mi.)*

Recreation and Entertainment
Near: Playground, Jogging Paths, Boat Rentals, Park, Museum *(Camp 6 Logging Museum at Pt. Defiance Park 591-5339)*, Sightseeing *(Pt. Defiance Zoo and Acquarium 591-5337)* **Under 1 mi:** Beach, Picnic Area, Grills,

Tennis Courts **1-3 mi:** Golf Course *(Highlands 759-3622)*, Fitness Center *(YMCA 564-9622)*, Video Rental *(Hollywood Video 759-9923)* **3+ mi:** Movie Theater *(Blue Mouse Theater 752-9500, 4 mi.)*

Provisioning and General Services
Near: Convenience Store **Under 1 mi:** Market *(Don's Ruston Market & Delli 759-8151)*, Liquor Store *(Ruston Liquor Store 752-3007)*, Catholic Church, Beauty Salon *(Sue's Hair Solution 761-9456)*, Barber Shop *(Mane Place 759-7955)*, Hardware Store *(Defiance Hardware 759-4642)* **1-3 mi:** Supermarket *(Safeway 752-5500)*, Bank/ATM, Post Office *(778-3447)*, Protestant Church, Synagogue, Bookstore *(Park Bench Books 752-4848)*, Pharmacy *(Rite Aid 879-0140)* **3+ mi:** Health Food *(Whole Foods Market 565-0188, 5 mi.)*, Library *(Wheelock Library 591-5640, 4 mi.)*, Dry Cleaners *(New Era Cleaners & Laundry 759-3101, 4 mi.)*, Department Store *(Tacoma Central 627-0988, 6 mi.)*, Buying Club *(Costco 475-2093, 7 mi.)*

Transportation
OnCall: Rental Car *(Enterprise 573-1055)*, Taxi *(Yellow Cab 472-3303)*, Airport Limo *(Bayview Limousine 272-7280)* **Near:** Bikes, Local Bus *(Pierce Transit 581-8000)*, Ferry Service *(WA State Ferry to Vashon Island 464-6400)* **3+ mi:** Rail *(Amtrak 627-8141, 8 mi.)* **Airport:** Tacoma Narrows/Sea-Tac Int'l. *(7 mi./26 mi.)*

Medical Services
911 Service **OnCall:** Ambulance **1-3 mi:** Doctor *(Multi Care Urgent Care Center 403-7177)*, Dentist *(Winters 756-1600)*, Veterinarian *(Westgate Animal Hospital 752-6161)* **3+ mi:** Chiropractor *(Proctor Chiropractor 756-7500, 4 mi.)* **Hospital:** Tacoma General 403-1000 *(6 mi.)*

Setting -- Breakwater Marina is located off Dalco Passage, past the Pt. Defiance piers - and just beyond the Washington State Ferry dock that delivers passengers to Vashon Island. From the sea, the mariner sees a hillside thick with trees and ferns. Tacoma Yacht Club is on the small spit just across the harbor The view from land is of the floating concrete docks and a thicket of boathouses. But take a few steps to port and Vashon Island comes into view. The small blue-trimmed gray Cape Cod at the top of the main gangway houses the marina offices.

Marina Notes -- CCM. *Fuel discount of $0.03 for 100+ gallons, if paid with cash, check, or Chevron card. Offers repairs and most boatyard services, although there is no haulout mechanism on-site - only a tidal grid. Boats needing haul-out are moved to a yard in the area. Open slips and covered boat houses with an even mix of power and sailboats. Recycling receptacles. Marina staff is available 8am-8pm during the peak season summer months. Gas dock and the electrical system were upgraded - including removal of the overhead wires. The free shower and two rustic, unisex heads are right on the dock.

Notable -- The lower boundaries of Pt. Defiance Park are across the street from the marina. Endless outdoor attractions, a 29-acre zoo and aquarium, a logging museum, beaches, and gorgeous views of Puget Sound, Mount Ranier, and the Olympic Mountains await visitors. The Boathouse at Pt. Defiance provides limited groceries, bait, tackle, and small boat rentals. Also within the park is Anthony's at Pt. Defiance, part of the restaurant chain. Two miles away, along the waterfront, the Lobster Shop boasts the best seafood in the Pacific Northwest. A bus stop at the ferry dock brings downtown Tacoma into close range.

PHOTOS ON DVD: 11

Ole and Charlie's Marina

4224 Marine View Drive; Tacoma, WA 98422

Tel: (253) 272-1173 **VHF: Monitor** n/a **Talk** n/a
Fax: (253) 572-1622 **Alternate Tel:** n/a
Email: ocmarina@comcast.net **Web:** www.ocmarinas.com
Nearest Town: Federal Way *(5 mi.)* **Tourist Info:** (253) 627-2175

Navigational Information
Lat: 47°16.932' **Long:** 122°24.136' **Tide:** 13 ft. **Current:** 0 kt. **Chart:** 18453
Rep. Depths *(MLW):* **Entry** 30 ft. **Fuel Dock** n/a **Max Slip/Moor** 14 ft./-
Access: Enter Hylebos Waterway from Commencement Bay

Marina Facilities *(In Season/Off Season)*
Fuel: No
Slips: 200 Total, 10 Transient **Max LOA:** 75 ft. **Max Beam:** n/a
  Rate *(per ft.):* **Day** $0.50 **Week** n/a **Month** $5.75
  Power: 30 amp Incl., **50 amp** n/a, **100 amp** n/a, **200 amp** n/a
  Cable TV: No **Dockside Phone:** No
  Dock Type: Floating, Pilings, Wood
Moorings: 0 Total, 0 Transient **Launch:** n/a
  Rate: Day n/a **Week** n/a **Month** n/a
Heads: 2 Toilet(s)
Laundry: None **Pay Phones:** No
Pump-Out: OnSite **Fee:** n/a **Closed Heads:** Yes

Marina Operations
Owner/Manager: Steve Olson **Dockmaster:** Same
In-Season: Year-Round, Mon-Fri 8am-6pm, Sat & Sun 8-8 **Off-Season:** n/a
After-Hours Arrival: n/a
Reservations: No **Credit Cards:** No; Cash only
Discounts: None
Pets: Welcome **Handicap Access:** No

Marina Services and Boat Supplies
Services - Security *(8 hrs., property walks)*, Dock Carts **Supplies - OnSite:** Ice *(Cube)* **Near:** Ships' Store *(Chinook Landing)* **Under 1 mi:** Bait/Tackle, Live Bait **3+ mi:** West Marine *(926-2533, 6 mi.)*, Boater's World *(472-3393, 10 mi.)*.

Boatyard Services
OnSite: Travelift *(10T sling lift)*, Railway, Canvas Work *(Trophy Boat Tops 572-1725)*, Upholstery

Restaurants and Accommodations
OnCall: Restaurant *(The Private Pantry 988-4418, will ship, cook, and clean up on your boat)*, Pizzeria *(Garlic Jim's Gourmet Pizza 838-7744)* **1-3 mi:** Restaurant *(On the Greens Restaurant 927-7439)*, *(Cliff House 927-0400, L $11-15, D $17-36, waterfront)*, *(Tropical Tides 383-1144, at Hylebos Marina)*, *(Howard's Corner 927-1162)*, Pizzeria *(Brown's Point Pizza 942-4337)* **3+ mi:** Restaurant *(Hot Mama's Thai 835-3430, 4 mi.)*, Motel *(Econo Lodge 922-9520, 6 mi.)*, *(BW Dome 272-7737, 6 mi.)*, Inn/B&B *(Dashaway B&B 927-7275, 4 mi.)*

Recreation and Entertainment
OnSite: Galleries *(SA Olson - displays the artwork of Manager Steve Olson)* **Near:** Beach **1-3 mi:** Golf Course *(North Shore Golf Course 838-3660)*, Park **3+ mi:** Dive Shop *(Lighthouse Diving 627-7617, 7 mi.)*, Boat Rentals *(Boat and Watersports Rental 272-7979, 8 mi.)*, Movie Theater *(Grand Tacoma Cinema 593-4474, 8 mi.)*, Museum *(WA State History Museum 272-9747, Tacoma Art Museum 272-4258, Museum of Glass 284-4719, 8 mi.)*, Cultural Attract *(Broadway Performing Arts 591-5894, Tacoma Little Theatre 272-2481, 8 mi.)*, Special Events *(Gardens of Tacoma Tour - Jun 474-0400, Art Walk - Jun-Aug 591-5191, 5 mi.)*

Provisioning and General Services
Near: Convenience Store *(Chinook Landing)* **1-3 mi:** Market *(Crescent Heights 952-9225)*, Supermarket *(QFC 925-5040)*, Liquor Store *(Brown's Point Liquor 927-7348)*, Bank/ATM, Protestant Church, Library *(Kobetich Branch 591-5630)*, Beauty Salon *(Hair Masters 925-5763)*, Barber Shop *(EZ Barbershop 661-5560)*, Hardware Store *(Ace 927-9633)*, Florist *(Dana's Daisies 568-6936)* **3+ mi:** Farmers' Market *(Tacoma Farmer's Market 272-7077, 8 mi.)*, Fishmonger *(Johnny's 627-2158, 8 mi.)*, Meat Market *(Novak's Meats 272-3110, 6 mi.)*, Laundry *(Fox Cleaners 838-1961, 4 mi.)*, Bookstore *(Tacoma Book Center 572-8248, 8 mi.)*, Pharmacy *(Walgreens 838-1290, 4 mi.)*, Retail Shops *(4 mi.)*, Department Store *(Seatac Mall 475-4565, 7 mi.)*, Buying Club *(Cotsco 475-2093, 10 mi.)*

Transportation
OnCall: Rental Car *(Enterprise 926-3950)*, Taxi *(Cascade Taxi 942-8773)*, Airport Limo *(A1 Max Out Limousine 922-5297)* **1-3 mi:** Local Bus *(Greyhound Bus Lines 383-4621)* **3+ mi:** Rail *(Amtrak 627-8141, 7 mi.)* **Airport:** Sea-Tac Int'l. *(17 mi.)*

Medical Services
911 Service **OnCall:** Ambulance **1-3 mi:** Doctor *(Multi Care Medical 925-1744)*, Dentist *(Zimmer 927-8040)*, Chiropractor *(Brown's Point Chiropractic 927-9325)*, Veterinarian *(Brown's Point Veterinary Clinic 927-1323)* **3+ mi:** Optician *(Lenscrafters 473-2511, 8 mi.)* **Hospital:** Tacoma General 403-1000 *(8 mi.)*

Setting -- Ole and Charlie's Marina sits near the entrance to Hylebos Waterway at the northeast corner of Commencement Bay, where a number of marinas, boat builders and boatyards crowd the industrial shoreline. Many of the floating wood docks are covered by boat sheds and boathouses. Looming above them - on clear days - cargo loaders and other icons of commercial operations compete with spectacular views of Mt. Ranier. Onshore, a steep hill covered in evergreens rises above the handful of rust-colored buildings that house the marina services.

Marina Notes -- One of the few area facilities to offer guest moorage, this family-owned operation has very limited transient slips during the summer months - the best prospects occur in the off-season. Primarily a dry storage facility, it offers year-round enclosed sheds as well as wet slips. A small boat launch sling lift and a railway manage haul-out. Dedicated staff accommodates boaters. One of the few "amenities," the nearby Port Administraiton's observation tower provides a view over the entire harbor. Reasonable heads on the premises, but no showers or laundry. Note: Be aware of lumber mill and log boom activity.

Notable -- Most services are over two miles from the marina, with transportation limited to taxi cabs. Downtown Tacoma is eight miles by land, or a three-mile dinghy ride south. Two waterfront restaurants are about two miles from the marina along Marine Drive: Cliff House, with fantastic views of Tacoma, is to the east. To the west is Tropical Tides at Hylebos Marina. Get a cab or rent a car to visit the Working Waterfront Museum (272-2750); it displays historic boats and artifacts, and you can view restoration projects underway in the museum's workshop. The Emerald Casino (594-7777) will pick up a group.

Navigational Information
Lat: 47°16.846' **Long:** 122°23.582' **Tide:** 13 ft. **Current:** 0 kt. **Chart:** 18453
Rep. Depths (MLW): Entry 30 ft. **Fuel Dock** n/a **Max Slip/Moor** 15 ft./-
Access: Follow Hylebos Waterway off Commencement Bay

Marina Facilities (In Season/Off Season)
Fuel: No
Slips: 213 Total, 12 Transient **Max LOA:** 100 ft. **Max Beam:** n/a
  Rate (per ft.): **Day** $1.00/0.75 **Week** n/a **Month** n/a
  Power: 30 amp $5 min., **50 amp** $5 min., **100 amp** n/a, **200 amp** n/a
  Cable TV: Yes **Dockside Phone:** No
  Dock Type: Floating, Long Fingers, Concrete, Wood
Moorings: 0 Total, 0 Transient **Launch:** n/a
  Rate: Day n/a **Week** n/a **Month** n/a
Heads: 1 Toilet(s), 1 Shower(s)
Laundry: 1 Washer(s), 1 Dryer(s) **Pay Phones:** No
Pump-Out: OnSite **Fee:** $5 **Closed Heads:** Yes

Marina Operations
Owner/Manager: Dennis LaPointe **Dockmaster:** William Dillon
In-Season: Year-Round, 8:30am-5pm **Off-Season:** n/a
After-Hours Arrival: Call ahead
Reservations: No **Credit Cards:** Visa/MC, Dscvr, Amex
Discounts: None
Pets: Welcome **Handicap Access:** Yes

Chinook Landing Marina

3702 Marine View Drive; Tacoma, WA 98422

Tel: (253) 627-7676 **VHF: Monitor** Ch. 79 **Talk** n/a
Fax: (253) 779-0576 **Alternate Tel:** n/a
Email: clm@puyallupinternational.com **Web:** n/a
Nearest Town: Tacoma (7 mi.) **Tourist Info:** (253) 627-2175

Marina Services and Boat Supplies
Services - Security (24 hrs.), Trash Pick-Up, Dock Carts **Supplies -**
OnSite: Ice (Cube), Ships' Store **Under 1 mi:** Bait/Tackle, Live Bait
3+ mi: West Marine (926-2533, 5 mi.), Boater's World (472-3393, 10 mi.)

Boatyard Services
Near: Travelift (85T), Railway (100T), Woodworking, Metal Fabrication,
Yacht Building. **Nearest Yard:** Modutech Marine (253) 272-9319

Restaurants and Accommodations
OnSite: Lite Fare (Lunch & Expresso Bar) **OnCall:** Restaurant (The Private
Pantry 988-4418, will ship, cook, and clean up on your boat) **1-3 mi:**
Restaurant (On the Greens 927-7439), (Howard's Corner 927-1162), (Think
Teriyaki 874-5160), (Tropical Tides 383-1144), (Cliff House 927-0400,
L $11-15, D $17-36), Fast Food (Quiznos), Pizzeria (Pizza Hut 927-7400)
3+ mi: Motel (La Quinta Inn 383-0146, $71-129, 6 mi., 5 mi.), (Travelodge
922-0550, 5 mi.), (Days Inn 922-3500, 5 mi.)

Recreation and Entertainment
OnSite: Picnic Area **Near:** Beach **1-3 mi:** Golf Course (North Shore Golf
Course 927-1375) **3+ mi:** Dive Shop (Lighthouse Diving 627-7617, 7 mi.),
Boat Rentals (Boat and Watersports Rental 272-7979, 8 mi.), Movie Theater
(Grand Tacoma Cinema 593-4474, 8 mi.), Museum (WA State History
Museum 272-9747, Tacoma Art Museum 272-4258, Museum of Glass
284-4719, Working Waterfront Museum 272-2750, 8 mi.), Cultural Attract
(Broadway Center for the Performing Arts 591-5894, 8 mi.), Special
Events (Tacoma Jazz Fest - Apr, Freedom Fair - Jul 4, Art Walk
Jun-Aug 591-5191, 8 mi.)

Provisioning and General Services
OnSite: Convenience Store **1-3 mi:** Market (Crescent Heights Grocery 952-
9225), Supermarket (QFC 925-5040, Albertson's 838-8988), Liquor Store
(Brown's Point Liquor 927-7348), Bank/ATM, Catholic Church, Protestant
Church, Library (Kobetich Memorial 591-5630), Beauty Salon (Gloria's
Cutting 941-1663), Barber Shop (Valerie's Barber Shop 952-4495),
Hardware Store (Ace 927-9633), Florist (Dana's Daisies 568-6936)
3+ mi: Farmers' Market (Tacoma's Market 272-7077, 8 mi.), Fishmonger
(Johnny's Sea Food Co. 627-2158, 8 mi.), Meat Market (Novak's Meats
272-3110, 6 mi.), Laundry (Norge Village Laundromat 874-9603, 4 mi.),
Bookstore (Tacoma Book Center 572-8248, 8 mi.), Pharmacy (Walgreens
838-1290, 4 mi.), Retail Shops (Seatac Mall 529-7782, 8 mi.), Department
Store (Target 627-2112, 10 mi.), Buying Club (Costco 475-2093, 10 mi.)

Transportation
OnCall: Rental Car (Enterprise 926-3950), Taxi (Cascade Taxi 942-8773),
Airport Limo (A1 Max Out Limousine 922-5297) **1-3 mi:** Local Bus
(Greyhound 383-4621) **3+ mi:** Rail (Amtrak 627-8141, 6 mi.)
Airport: Sea-Tac Int'l. (17 mi.)

Medical Services
911 Service **OnCall:** Ambulance **1-3 mi:** Doctor (Multi Care Medical 925-
1744), Dentist (Zimmer 927-8040), Chiropractor (Brown's Point Chiropractic
Clinic 927-9325), Veterinarian (Brown's Point Veterinary Clinic 927-1323)
Hospital: Tacoma General 403-1000 (8 mi.)

Setting -- On the north shore of the Hylebos Waterway, just beyond Ole & Charlie's Marina, this modern new marina is not altogether expected in such a heavily industrialized area. The network of high-end floating concrete docks sprawls along the shore, and a ramp leads to a pretty two-story, natural log and shingled longhouse-style building trimmed with hanging pots of flowers. More flowers decorate the deck which hosts picnic tables and dock carts.

Marina Notes -- CCM. Everything at $3.6 million-dollar Chinook Landing is very well cared for. 430 feet of transient moorage. Excellent docks with power, water, and lots of dedicated transient moorage. The marina's general store has a limited assortment of groceries, plus boat supplies, charts, clothing, and an espresso and snack bar. Comfortable chairs and picnic tables on the deck outside have views of the marina and, on clear days, of Mt. Olympia. 24-hr. security means you can catch a ride into Tacoma without worrying about your boat. Marina staff will assist with transportation if necessary.

Notable -- A steep hill rises behind the marina, separating Marine View Drive from the residential and shopping areas to the north. A car is required to get to any sevices. Downtown Tacoma is across Commencement Bay, or 8 miles by land. It offers excellent museums, galleries, theaters and year-round special events. The Museum of Glass rotates exhibits on a regular basis (284-4750). Leading to it is the Chihuly Bridge of Glass, and an esplanade open 24-hours. For a mixture of art on paper, textiles and sculpture, try the American Art Company at 1126 Broadway Plaza (800-753-2278). It represents a number of artists in the prime of their careers, and the prices reflect that. The gallery puts an emphasis on pieces meant for the home rather than the corporate world.

Foss Waterway Marina

821 Dock Street; Tacoma, WA 98402

Tel: (253) 272-4404 **VHF: Monitor** n/a **Talk** n/a
Fax: (253) 272-0367 **Alternate Tel:** n/a
Email: john@fosswaterwaymarina.com **Web:** fosswaterwaymarina.com
Nearest Town: Tacoma **Tourist Info:** (253) 627-2175

Navigational Information
Lat: 47°15.411' **Long:** 122°26.132' **Tide:** 12 ft. **Current:** 0 kt. **Chart:** 18445
Rep. Depths (*MLW*): **Entry** 10 ft. **Fuel Dock** n/a **Max Slip/Moor** 10 ft./-
Access: Commencement Bay to Thea Foss Waterway

Marina Facilities (*In Season/Off Season*)
Fuel: No
Slips: 465 Total, 2 Transient **Max LOA:** 96 ft. **Max Beam:** n/a
   Rate (*per ft.*): **Day** $0.75 **Week** n/a **Month** n/a
   Power: 30 amp $.07 kwh, **50 amp** $.07 kwh, **100 amp** n/a, **200 amp** n/a
   Cable TV: Yes **Dockside Phone:** Yes
   Dock Type: Floating, Long Fingers, Concrete, Composition
Moorings: 0 Total, 0 Transient **Launch:** n/a
   Rate: Day n/a **Week** n/a **Month** n/a
Heads: 6 Toilet(s), 6 Shower(s)
Laundry: 3 Washer(s), 3 Dryer(s) **Pay Phones:** Yes
Pump-Out: OnSite **Fee:** $5 **Closed Heads:** Yes

Marina Operations
Owner/Manager: Tracy McKendry **Dockmaster:** John Campbell
In-Season: Year-Round **Off-Season:** n/a
After-Hours Arrival: Call ahead
Reservations: Yes, Preferred **Credit Cards:** Visa/MC
Discounts: None
Pets: Welcome **Handicap Access:** No

Marina Services and Boat Supplies
Services - Security (*locked gates and patrol*), Trash Pick-Up, Dock Carts
Communication - Mail & Package Hold, Phone Messages **Supplies -**
OnSite: Ice (*Block, Cube*), Ships' Store, Bait/Tackle, Propane **1-3 mi:** West
Marine (*926-2533*) **3+ mi:** Boater's World (*472-3393, 5 mi.*)

Boatyard Services
OnSite: Travelift (*2 hoists*), Yacht Broker (*Quinnsboats 272-5828*)

Restaurants and Accommodations
OnSite: Snack Bar **Near:** Restaurant (*Dock Street Landing 272-5004, B $7-8, L $2-9, D $3-10, daily specials*), Pizzeria (*Round Table Pizza 572-2008*)
Under 1 mi: Restaurant (*Altezzo Ristorante 591-4155, D $8-17, top of Sheraton*), (*Galanga Thai 272-3393*), (*The Blue Olive 383-7275, L $8-15, D $14-30*), (*Fujiya 627-5319*), (*Sea Grill 272-5656, L $9-17, D $20-35, L Mon-Fri, D daily*), Coffee Shop (*Penelopes Espresso*), Fast Food (*Subway, McDonald's*), Lite Fare (*Hall of a Sub 272-2933, B $3-5, L $4-7*), (*Varsity Grill 627-1229, dine and watch games; 17-seat theater with projection screen*), Hotel (*Sheraton Tacoma 572-3200, $79-209*), (*Courtyard 591-9100*)
1-3 mi: Hotel (*Silver Cloud Inn 272-1300, $109-299*), Inn/B&B (*Chinaberry Hill Victorian B&B 272-1292, $120-195*)

Recreation and Entertainment
OnSite: Picnic Area, Boat Rentals (*Boats and Watersports Rentals 272-7979*) **Near:** Museum (*Working Waterfront Museum 272-2750, WA State History Museum 888-238-4373, Museum of Glass 627-2555*)
Under 1 mi: Fitness Center (*YMCA 597-6444*), Movie Theater (*Grand Tacoma 593-4474*), Park (*Fireman's*), Cultural Attract (*Broadway Center*

for the Performing Arts 591-5894*) **1-3 mi:** Pool (*People's Pool 591-5323*), Video Rental (*Sunshine 627-1124*), Galleries (*American Art Co. 272-4327*)
3+ mi: Golf Course (*Allenmore Public Golf Course 627-1161, 3 mi.*), Fishing Charter (*Capt. Jerry's Charters 752-1100, 5 mi.*)

Provisioning and General Services
OnSite: Convenience Store **Near:** Wine/Beer (*Wine Merchants 779-8258*), Fishmonger (*Johnny's Seafood 627-2158*), Bank/ATM, Copies Etc. (*Minuteman Press 383-4377*) **Under 1 mi:** Market (*Mini Market & Deli 383-3213*), Delicatessen (*Dock Street Sandwhich Co. 627-5882*), Farmers' Market (*Broadway Thu 10am-2/3pm*), Post Office, Catholic Church, Protestant Church, Library (*Tacoma PL 591-5666*), Beauty Salon (*Christine's Day Spa 627-8887*), Dry Cleaners (*All City Cleaners 383-1541*), Hardware Store (*Contract Hardware 531-5855*) **1-3 mi:** Supermarket (*Safeway 627-8840*), Liquor Store (*WA Liquor 593-2080*), Synagogue, Laundry (*Simmons Coin-op 272-9756*), Pharmacy (*Rite Aid 779-0601*) **3+ mi:** Health Food (*Whole Foods 565-0188, 6 mi.*), Buying Club (*Costco 475-2093, 4 mi.*)

Transportation
OnCall: Rental Car (*Enterprise 573-1055*), Taxi (*Orange Cab 779-4444*), Airport Limo (*PCI 572-6900*) **Under 1 mi:** Local Bus (*Sound Transit 888-889-6368*) **1-3 mi:** Rail (*Amtrak 627-8141*) **Airport:** Tacoma Narrows/Sea-Tac Int'l. (*8 mi./22 mi.*)

Medical Services
911 Service **OnCall:** Ambulance **Under 1 mi:** Chiropractor (*Woon 272-9959*) **1-3 mi:** Doctor (*Downtown Clinic 597-3813*), Dentist (*Northwest Dental Service 383-3001*) **Hospital:** Tacoma General 403-1000 (*2 mi.*)

Setting -- Located in downtown Tacoma near the entrance to the historic Thea Foss Waterway, Foss Waterway Marina provides excellent moorage and easy access to the city's tourist attractions. Strung along the shore, about 20 excellent composite docks host 465 slips. Onshore umbreallaed picnic tables overlook the docks and front a long, low, light-blue building with a burgundy horizonal stripe; the city skyline looms dramatically overhead.

Marina Notes -- CCM. This shipshape facility has very limited space for guests; contact the friendly staff ahead of your visit. One of the biggest marinas on the Waterway, it can accomodate boats up to 96' on the end dock. Beautiful docks with full pedestals that offer metered electricity ($.07/kwt) plus cable TV and phone lines to each slip. Liveaboards are allowed. Recycling and hazardous waste collection sites. Pump-out is $5 for visitors. Two large stationary hoists are on-site. The marina store offers a variety of supplies, food, bait and tackle, ice, and sundries. All-tile head/showers and a laundry room.

Notable -- Commencement Bay has a bit of everything - wooded bluffs at Point Defiance, the "downtown" feel of the Thea Foss Waterway, and the intense activity of loading and off-loading ships in one of the busiest ports in the Northwest. Explore the city from this great location. The Working Waterfront Museum (272-2750) is very close, and well worth a visit. Once on the other side of the rail tracks, you'll find lots of shops and restaurants. Less than a mile away are the Childrens' Museum (627-6031), the old Rialto Theater (591-5894), now home to live performances, Tacoma Art Museum (272-4258), housed in a spectacular building, the Museum of Glass (284-4750) and the 500 foot long pedestrian glass bridge that links it to the Washington State History Museum (272-3500).

Navigational Information
Lat: 47°14.725' **Long:** 122°25.890' **Tide:** 13 ft. **Current:** 0 kt. **Chart:** 18445
Rep. Depths *(MLW)*: **Entry** 10 ft. **Fuel Dock** n/a **Max Slip/Moor** 15 ft./-
Access: Commencement Bay to the Thea Foss Waterway

Marina Facilities *(In Season/Off Season)*
Fuel: No
Slips: 60 Total, 3 Transient **Max LOA:** 50 ft. **Max Beam:** n/a
  Rate *(per ft.)*: **Day** $0.50* **Week** n/a **Month** n/a
  Power: 30 amp Incl., **50 amp** n/a, **100 amp** n/a, **200 amp** n/a
  Cable TV: No **Dockside Phone:** No
  Dock Type: Floating, Long Fingers, Wood
Moorings: 0 Total, 0 Transient **Launch:** n/a
  Rate: Day n/a **Week** n/a **Month** n/a
Heads: 6 Toilet(s)
Laundry: None **Pay Phones:** No
Pump-Out: No **Fee:** n/a **Closed Heads:** Yes

Marina Operations
Owner/Manager: David Bingham, John Crabill **Dockmaster:** n/a
In-Season: Year-Round, 10am-? **Off-Season:** n/a
After-Hours Arrival: Call ahead
Reservations: No **Credit Cards:** Visa/MC, Dscvr, Din, Amex
Discounts: None
Pets: No **Handicap Access:** Yes, Heads, Docks

Johnny's Dock

1900 East D Street; Tacoma, WA 98421

Tel: (253) 627-3186 **VHF: Monitor** n/a **Talk** n/a
Fax: n/a **Alternate Tel:** n/a
Email: winston@yahoo.com **Web:** www.johnnysdock.com
Nearest Town: Tacoma *(1 mi.)* **Tourist Info:** (253) 627-2175

Marina Services and Boat Supplies
Services - Security *(locked gates)* **Communication -** Data Ports *(Wi-Fi - Broadband Xpress)* **Supplies - Under 1 mi:** Ice *(Cube)*, Bait/Tackle
1-3 mi: Ice *(Block)*, West Marine *(926-2533)*, Boater's World *(472-3393)*

Boatyard Services
Nearest Yard: Modutech Marine (253) 272-9319

Restaurants and Accommodations
OnSite: Restaurant *(Johnny's Dock Restaurant and Marina 627-3186, L $8-15, D $15-30, 7 days from 10am)* **Under 1 mi:** Restaurant *(Dock Street Landing 272-5004, B $6-8, L $2-9, D $3-10, tie up at city dock)*, *(El Gaucho 882-0009)*, *(Sea Grill 272-5656, L $9-17, D $20-35, L Mon-Fri, D daily)*, *(Indochine 272-8200)*, *(Melting Pot 535-3939)*, Lite Fare *(Dock St. Sandwich Co. 627-5882)*, *(Taco Del Mar 572-8393)*, Hotel *(BW Dome 272-7737)*
1-3 mi: Hotel *(Sheraton Tacoma 572-3200, $99-229)*, *(La Quinta Inn 383-0146, $69-125)*, *(Days Inn 475-5900, $55-90)*, *(Courtyard 591-9100)*

Recreation and Entertainment
Under 1 mi: Boat Rentals *(Boats and Watersports Rental 272-7979)*, Fishing Charter *(Capt Jerry's Charters 752-1100)*, Museum *(Museum of Glass 396-1768, Tacoma Art Museum 272-4258, Working Waterfront Museum 272-2750, Washington State History Museum 888-238-4373)*, Cultural Attract *(Broadway Center for the Performing Arts 591-5894, Tacoma Dome 272-3663)*, Special Events *(Jazz Fest - Apr, Freedom Fair - Jul 4, Ethnic Fest - Jul, Maritime Fest - Sep)*, Galleries **1-3 mi:** Heated Pool, Spa, Beach, Picnic Area, Playground, Golf Course *(Allenmore 627-1161)*, Fitness Center *(YMCA 597-6444)*, Movie Theater *(Grand Tacoma Cinema 593-4474)*

Provisioning and General Services
Near: Bank/ATM **Under 1 mi:** Convenience Store, Delicatessen, Bakery *(Peggy's Cinnamon Rolls 627-5560)*, Farmers' Market *(Freighthouse Square Tue 3-7pm)*, Fishmonger *(Johnny's Seafood 627-2158)*, Post Office, Beauty Salon *(Ozone Hair Design 627-1212)*, Bookstore *(University Bookstore 272-8080)*, Retail Shops *(Freighthouse Square 305-0678)* **1-3 mi:** Supermarket *(Albertson's 474-0946, Safeway 627-8840)*, Liquor Store *(WA Liquor 593-2452)*, Catholic Church, Protestant Church, Library *(Tacoma PL 591-5666)*, Dry Cleaners *(All City Cleaners 383-1541)*, Laundry *(Simmons Coin-op 272-9756)*, Pharmacy *(Rite Aid 779-0601)*, Hardware Store *(Contract Hardware 531-5855)*, Buying Club *(Costco 475-2093)*, Copies Etc. *(FedEx Kinkos 474-4997)* **3+ mi:** Health Food *(Whole Foods 565-0188, 6 mi.)*

Transportation
OnCall: Rental Car *(Enterprise 573-1055/ Budget 383-4944, 2 mi.)*, Taxi *(Yellow Cab 472-3303)*, Airport Limo *(A 1 Max Out Limousine 922-5297)*
Near: Local Bus *(Sound Transit 888-889-6368, Light Rail "Sounder" to Seattle and Everett)*, InterCity Bus *(Greyhound 383-4621)* **1-3 mi:** Bikes, Rail *(Amtrak 627-8141)* **Airport:** Tacoma Narrows/Sea-Tac Int'l. *(10 mi./20 mi.)*

Medical Services
911 Service **OnCall:** Ambulance **Under 1 mi:** Doctor *(Community Health Care 597-4550)* **1-3 mi:** Dentist *(Family Dental Clinic 573-0070)*, Chiropractor *(Estabrok 627-9151)*, Veterinarian *(Button Veterinary Hospital 383-5531)*, Optician *(Harbor Optical 272-7685)* **Hospital:** St. Joseph Hospital 426-4101 *(1.5 mi.)*

Setting -- One of the most attractively sited marinas on the Foss Waterway, Johnny's all-glass, Japanese-inspired restaurant overlooks the floating docks and Tacoma skyline. For alfresco dining, there is a large inviting deck with umbrella-topped white tables and chairs. Thanks to a SuperFund cleanup, Tacoma's waterfront has become a desirable destination for visiting boaters, and Johnny's Dock, on the Waterway's eastern shore, is one of the few transient facilities.

Marina Notes -- *Flat rate $20/night, but fee usually waived for those who dine at the restaurant. Check-out time 10am, with a 2-night maximum stay. Mostly seasonal tennants, with limited guest space. John Meaker founded the restaurant in 1953, built this facility in 1977, and today it is run by his grandson. Well-maintained, uncluttered docks - a combination of wood and cement with power on pedestals. Heads are shared with the restaurant. Note: After 100 years of abuse and pollution, Thea Foss Waterway was designated a Superfund Clean Up Site in the '80s. A $90 million restoration was finally completed in 2006.

Notable -- The restaurant offers an eclectic menu, fireplace, and a west facing deck with fabulous views of the city skyline. The city is conveniently within reach - either walk less than a mile south and catch the Link Light Rail to downtown or take your boat or dinghy to the new Dock Street Marina, right across from Johnny's Dock. The Glass Museum, one of Tacoma's most popular, has a Hot Shop where you can watch artisans at work. The Art Museum and State History Museum are close by, as is the University of Washington at Tacoma. Along Dock Street is the Working Waterfront Museum, where boat restoration is always in progress. Concerts and events are often held at the Tacoma Dome. Lots of shops and casual eateries occupy the Freighthouse Square area.

PHOTOS ON DVD: 13

Dock Street Marina

1817 Dock Street; Tacoma, WA 98402

Tel: (253) 272-4352; (253) 572-2524 **VHF: Monitor** Ch. 78A **Talk** Ch. 78A
Fax: (253) 572-2768 **Alternate Tel:** (253) 250-1906
Email: info@dockstreetmarina.com **Web:** www.dockstreetmarina.com
Nearest Town: Tacoma **Tourist Info:** (253) 627-2175

Navigational Information
Lat: 47°14.711' **Long:** 122°25.990' **Tide:** 13 ft. **Current:** 2 kt. **Chart:** 18445
Rep. Depths (MLW): Entry 10 ft. **Fuel Dock** n/a **Max Slip/Moor** 15 ft./-
Access: 1 mi. south of Commencement Bay on west shore

Marina Facilities *(In Season/Off Season)*
Fuel: No
Slips: 89 Total, 35 Transient **Max LOA:** 130 ft. **Max Beam:** 30 ft.
 Rate *(per ft.):* **Day** $1.00/1.00 **Week** n/a **Month** $9-11
 Power: 30 amp $3, **50 amp** $5, **100 amp** $10, **200 amp** n/a
 Cable TV: Yes Incl. **Dockside Phone:** No
 Dock Type: Floating, Long Fingers, Concrete
 Moorings: 0 Total, 0 Transient **Launch:** n/a, Dinghy Dock
 Rate: Day n/a **Week** n/a **Month** n/a
Heads: 4 Toilet(s), 2 Shower(s) *(with dressing rooms)*
Laundry: 1 Washer(s), 1 Dryer(s), Book Exchange **Pay Phones:** No
Pump-Out: OnSite, Full Service **Fee:** n/a **Closed Heads:** Yes

Marina Operations
Owner/Manager: Foss Waterway Dev. Authority **Dockmaster:** Mike Norman
In-Season: Year-Round, 24 hrs. **Off-Season:** n/a
After-Hours Arrival: Call in advance
Reservations: Yes, Preferred **Credit Cards:** Visa/MC, Amex
Discounts: Groups* **Dockage:** n/a **Fuel:** n/a **Repair:** n/a
Pets: Welcome **Handicap Access:** Yes, Heads, Docks

Marina Services and Boat Supplies
Services - Docking Assistance, Security *(24 hrs., roving patrol)*, Trash Pick-Up, Dock Carts, Megayacht Facilities **Communication -** Data Ports *(Wi-Fi - Broadband Xpress)* **Supplies - Near:** Ice *(Shaved)* **Under 1 mi:** Ships' Store *(Maddens Marine 779-0900)*, Propane **1-3 mi:** West Marine *(926-2533)*, Boater's World *(472-3393)*

Boatyard Services
Near: Travelift. **Nearest Yard:** Tacoma Marine Repair (253) 272-9014

Restaurants and Accommodations
Near: Restaurant *(Melting Pot 535-3939)*, *(Blue Olive 383-7275, L $8-15, D $14-30)*, *(Harmon Brewing Co. 383-2739)*, *(Renaissance Café 572-1029)*, *(Grassi's Garden Café 627-1216, B $5-12, L $8-12, boxed L to go also)*, *(Old Spaghetti Factory 383-2214)*, Coffee Shop *(Starbucks)*, Fast Food *(Subway, Taco Del Mar)*, Lite Fare *(Hotrod Dog 593-6030)*, Pizzeria *(Bella Pizzeria 779-0769)* **Under 1 mi:** Restaurant *(Altezzo Ristorante 591-4155, D $8-17)*, *(Paya Thai Fish & Chips 627-8432)*, Hotel *(Best Western Dome 272-7737)*, *(Sheraton 572-3200, $79-209)* **1-3 mi:** Hotel *(Courtyard 591-9100)*, Inn/B&B *(Chinaberry Hill B&B 272-1282, $120-195)*

Recreation and Entertainment
OnSite: Museum *(Museum of Glass 396-1768; Nearby: Tacoma Art Museum 272-4258, Washington State History Museum 888-238-4373)*, Special Events *(boating events/ Jazz Fest - Apr, Freedon Fair - Jul 4, Ethnic Fest - Jul, Maritime Fest - Sep, nearby)* **Near:** Galleries *(Chihuly at Union Station 572-9310)* **Under 1 mi:** Fitness Center *(YMCA 597-6444)*, Boat Rentals *(Foss Waterway Marina 272-7979)*, Movie Theater *(Grand Tacoma Cinema 593-4474)*, Cultural Attract *(Broadway Center for the Performing Arts 591-5894, Tacoma Dome 272-3663)* **1-3 mi:** Pool *(People's Pool 591-5323)*, Golf Course *(Allenmore 627-1161)*

Provisioning and General Services
OnSite: Wine/Beer **Near:** Convenience Store, Delicatessen *(Dock Street Sandwich Co. 627-5882)*, Fishmonger *(Johnny's Seafood 627-2158)*, Bank/ATM, Beauty Salon *(Synergy Salon 573-1308)*, Dry Cleaners *(Fifteenth St. Drive-In Laundry & Cleaners 272-3313)*, Laundry, Bookstore *(University Bookstore 272-8080)* **Under 1 mi:** Market *(Downtown Market 383-0671)*, Bakery *(Peggy's Cinnamon Rolls 627-5560)*, Farmers' Market *(Broadway Thu 9am-2pm)*, Post Office *(275-8777)*, Catholic Church, Protestant Church, Library *(Tacoma PL 591-5666)*, Barber Shop *(Noggins 383-5788)*, Pharmacy *(Century Plaza 591-6920)*, Hardware Store *(Western Builders Supply 383-4423)* **1-3 mi:** Supermarket *(Safeway 627-8840)*, Liquor Store *(WA Liquor 593-2080)*, Buying Club *(Costco 671-6000)* **3+ mi:** Department Store *(Tacoma Mall 475-4565, 4 mi.)*

Transportation
OnSite: Local Bus *(Link Light Rail/ Sound Transit 888-889-6368)*
OnCall: Rental Car *(Enterprise 573-1055)*, Taxi *(Orange Cab 779-4444)*
Under 1 mi: InterCity Bus *(Greyhound 383-4621)*, Rail *(Amtrak 627-8141)*
Airport: Tacoma Narrows/Sea-Tac Int'l. *(9 mi./21 mi.)*

Medical Services
911 Service **1-3 mi:** Doctor *(Downtown Clinic 597-3813)*, Dentist *(Family Dental 573-0070)*, Chiropractor *(Estabrok 627-9151)*, Veterinarian *(6th Ave. Vet 752-6448)* **Hospital:** Tacoma General 403-1000 *(2 mi.)*

Setting -- Tacoma's Renaissance involved a fabulous waterfront redevelopment with the new Dock Street Marina built as its "centerpiece." Conveniently situated right in front of the striking Glass Museum building, the marina offers state of the art docks for transients as well as long-term tenants. A new 1.5 mile esplanade leads visitors along the waterfront, from the Dome District just near the suspension bridge to the mouth of the waterway at Commencement Bay.

Marina Notes -- CCM. *Discounts for groups of 6 or more boats from Mon-Wed. The Thea Foss Development Authority was responsible for the Waterway cleanup and completed construction of the marina in 2004. Excellent concrete piers have long finger piers on both sides, the largest being 60 ft. Yachts up to 130 ft. long and 30 ft. wide can tie up to the pier ends; there is no extra charge for catamarans. The nonslip docks offer 30 and 50 amp power, cable TV, and in-slip pump-out. 100 amp is available on T-heads. Fuel is available nearby. A trolley takes guest from the marina to the museum, theater and gallery districts.

Notable -- The revitalized Tacoma waterfront has already hosted a very popular Tall Ships event in 2005 and a Classic Wooden Yacht Rendezvous in 2006. Check with the marina for future events, including a return of the tall ships in 2008. The popular Museum of Glass is on-site, open Mon-Sat 10am-5pm, Sun 12-5pm ($10/$8 Seniors/$4 Kids). Free admission and hours extended to 8pm every 3rd Thu. Walk across the Glass Bridge to more museums, shops, galleries, and dozens of restaurants. About a mile north, the Working Waterfront Museum (272-2750) tells the story of Thea Foss, the Norwegian woman who inspired the "Tugboat Annie" series. The museum is open, though its historic building is currently being renovated and new exhibits are being added (to open in 2008).

Navigational Information

Lat: 47°14.708' **Long:** 122°25.888' **Tide:** 13 ft. **Current:** 0 kt. **Chart:** 18445
Rep. Depths *(MLW):* **Entry** 10 ft. **Fuel Dock** n/a **Max Slip/Moor** 15 ft./-
Access: Commencement Bay to the Thea Foss Waterway, at the south end

Marina Facilities *(In Season/Off Season)*

Fuel: No
Slips: 25 Total, 5 Transient **Max LOA:** 75 ft. **Max Beam:** n/a
  Rate *(per ft.):* **Day** $1.00 **Week** n/a **Month** n/a
  Power: 30 amp $5, **50 amp** $5, **100 amp** n/a, **200 amp** n/a
  Cable TV: Yes **Dockside Phone:** Yes
  Dock Type: Floating, Long Fingers, Wood
Moorings: 0 Total, 0 Transient **Launch:** n/a, Dinghy Dock
  Rate: Day n/a **Week** n/a **Month** n/a
Heads: 1 Toilet(s), 1 Shower(s)
Laundry: None **Pay Phones:** No
Pump-Out: Full Service, 1 Port, 10 InSlip **Fee:** n/a **Closed Heads:** Yes

Marina Operations

Owner/Manager: Tim Curry **Dockmaster:** Audrey Hamel
In-Season: Year-Round, 8am-5pm **Off-Season:** n/a
After-Hours Arrival: Call ahead to make arrangements
Reservations: Yes, Preferred **Credit Cards:** Visa/MC
Discounts: None
Pets: Welcome **Handicap Access:** Yes, Heads, Docks

Foss Landing Marina & Storage

1940 East D Street; Tacoma, WA 98421

Tel: (253) 627-4344 **VHF: Monitor** n/a **Talk** n/a
Fax: (253) 627-4878 **Alternate Tel:** n/a
Email: info@fosslanding.com **Web:** www.fosslanding.com
Nearest Town: Tacoma **Tourist Info:** (253) 627-2175

Marina Services and Boat Supplies

Services - Docking Assistance, Concierge, Boaters' Lounge, Security (8 hrs., onsite patrols 10pm-6am), Trash Pick-Up, Dock Carts, Megayacht Facilities **Communication -** Mail & Package Hold, Phone Messages, Fax in/out *(Free)*, Data Ports *(Wi-Fi - Broadband Xpress)*, FedEx, DHL, UPS **Supplies - Under 1 mi:** Ice *(Cube)*, Bait/Tackle, Live Bait **1-3 mi:** Ice *(Block)*, West Marine *(926-2533)*, Boater's World *(472-3393)*, Propane

Boatyard Services

Nearest Yard: Modutech Marine (253) 272-9319

Restaurants and Accommodations

Near: Restaurant *(Johnny's Dock Restaurant and Marina 627-3186, L $8-15, D $15-30)* **Under 1 mi:** Restaurant *(Whistle Stop Diner 627-1277)*, *(Blue Olive 383-7275, L $8-15, D $14-30)*, *(Harmon Brewing Co. 383-2739, B, L, D)*, Coffee Shop *(Commencement Bay Coffee Co. 274-1173)*, Fast Food *(Jack-in-the-Box 627-2907)*, Pizzeria *(Rock Pasta & Brick Oven Pizza 627-7625)*, Hotel *(BW Dome 272-7737)* **1-3 mi:** Restaurant *(Altezzo Ristorante 591-4155, D $8-17, top of Sheraton)*, Fast Food *(Fantastic Burger 572-2510)*, Motel *(Days Inn 475-5900, $55-90)*, Hotel *(Courtyard 591-9100)*, *(Sheraton Tacoma 572-3200, $99-229)*

Recreation and Entertainment

Under 1 mi: Park *(McKinley)*, Museum *(Shanaman Sports 627-5857, Museum of Glass 396-1768, WA State History 272-3500, Tacoma Art 272-4258, Working Waterfront Museum 272-2750)*, Cultural Attract *(Tacoma Dome 272-3663, Broadway Center for the Performing Arts 591-5894)*, Special Events **1-3 mi:** Heated Pool, Spa, Beach, Picnic Area, Playground,

Golf Course *(Allenmore 627-1161)*, Fitness Center *(YMCA 597-6444)*, Boat Rentals *(Boats and Watersports Rental 272-7979)*, Bowling *(Lincoln Lanes 474-1800)*, Movie Theater *(Grand Tacoma Cinema 593-4474)*

Provisioning and General Services

Near: Bank/ATM **Under 1 mi:** Convenience Store, Supermarket *(Safeway 627-8840, Albertson's 474-0946)*, Delicatessen *(Dock Street Sandwich Co. 627-5882)*, Bakery *(Peggy's Cinnamon Rolls 627-5560)*, Farmers' Market *(Freighthouse Square Tue 3-7pm)*, Fishmonger *(Johnny's Seafood Co. 627-2158)*, Post Office, Protestant Church, Beauty Salon *(Synergy Salon 573-1308)*, Bookstore *(University Bookstore 272-8080)*, Retail Shops *(Freighthouse Shopping Center 305-0678)* **1-3 mi:** Liquor Store *(WA Liquor 593-2452)*, Catholic Church, Library *(Tacoma PL 591-5666)*, Barber Shop *(Noggins 383-5788)*, Dry Cleaners *(All City Cleaners 383-1541)*, Laundry *(Simmons Coin-op 272-9756)*, Pharmacy *(Rite Aid 779-0601)*, Hardware Store *(Contract Hardware 531-5855)*, Buying Club *(Costco 671-6000)*, Copies Etc. *(Ex-Press 272-6785)*

Transportation

OnCall: Rental Car *(Enterprise 573-1055)*, Taxi *(Orange Cab 779-4444)*, Airport Limo *(PCI 572-6900)* **Near:** Local Bus *(Sound Transit 888-889-6368)*, InterCity Bus *(Greyhound 383-4621)* **1-3 mi:** Rail *(Amtrak 627-8141)* **Airport:** Tacoma Narrows/Sea-Tac Int'l. *(10 mi./20 mi.)*

Medical Services

911 Service **OnCall:** Ambulance **Under 1 mi:** Doctor *(Community Health Care 597-4550)* **1-3 mi:** Dentist *(Family Dental 573-0070)*, Chiropractor, Veterinarian **Hospital:** St. Joseph Hospital 426-4101 *(1.5 mi.)*

Setting -- Located at the south end of Thea Foss Waterway, on the east shore, Foss Landing's docks are backed by a large brown and green storage shed. To the south, the dramatic cable-stay SR-509 Bridge looms above the slips. A stone's throw away are the Tacoma Dome, home to sporting events & concerts, and the Freighthouse Square Shopping Center. Across the water, the new Dock Street Marina lies in front of the extraordinary Tacoma Glass Museum.

Marina Notes -- CCM. *$1.50/ft. for boats 75' or over. Formerly Picks Cove Marine Center. The Foss Landing Project is part of a citywide renaissance - including an effort to offer boaters better facilities. In every slip: two 50A/250v outlets, slip-side pump-out, phone, cable TV & internet access. 10 new 70 & 75 ft. slips with were added in '05. The service-oreinted staff will assist with dockage and help plan your stay. Indoor heated dry stack storage for 200 boats is served by a heavy duty travelift - but there are no boatyard services. An adjacent warehouse, part of the FLP, is being converted into high-end office space.

Notable -- The Tacoma Dome (0.6 miles) hosts Shanaman Sports Museum and, during events, McKinley's Grill. Even closer is the Freighthouse Square Farmers' Market. Transportation, including the Amtrak station and Greyhound Bus terminal is easily accessible. The Link Light Rail is a short walk and offers free rides to downtown. Visit Tacoma's top notch museums, University of Washington and Theater District. Every 3rd Thursday is an Art Walk (5-8pm) - meet the artists and watch demonstrations (272-4327). The Glass, Art, and WA State Museums are free during this time (walk the Chihuly Bridge of Glass!). The landmark Union Station, former terminus of the Northern Pacific Railroad, houses an amazing Chihuly exhibit as well (free tours at 1pm Tuesday & Friday).

Steilacoom Marina

402 First Street; Steilacoom, WA 98388

Tel: (253) 582-2600 **VHF: Monitor** n/a **Talk** n/a
Fax: (253) 581-0159 **Alternate Tel:** n/a
Email: n/a **Web:** n/a
Nearest Town: Steilacoom *(0.25 mi.)* **Tourist Info:** (253) 582-4204

Navigational Information
Lat: 47°10.095' **Long:** 122°36.836' **Tide:** 20 ft. **Current:** 2 kt. **Chart:** 18448
Rep. Depths (*MLW*): Entry n/a **Fuel Dock** n/a **Max Slip/Moor** 30 ft./-
Access: Through The Narrows to Steilacoom at Gordon Pt.

Marina Facilities *(In Season/Off Season)*
Fuel: No
Slips: 30 Total, 10 Transient **Max LOA:** 40 ft. **Max Beam:** n/a
  Rate *(per ft.)*: **Day** $0.60 **Week** n/a **Month** n/a
  Power: 30 amp $3*, **50 amp** n/a, **100 amp** n/a, **200 amp** n/a
  Cable TV: No **Dockside Phone:** No
  Dock Type: Floating, Wood
Moorings: 0 Total, 0 Transient **Launch:** n/a
  Rate: Day n/a **Week** n/a **Month** n/a
Heads: 1 Toilet(s)
Laundry: None **Pay Phones:** Yes
Pump-Out: No **Fee:** n/a **Closed Heads:** Yes

Marina Operations
Owner/Manager: Shirley Wang **Dockmaster:** Same
In-Season: Year-Round, 8am-7:30pm **Off-Season:** n/a
After-Hours Arrival: Call; management onsite
Reservations: No **Credit Cards:** No
Discounts: None
Pets: Welcome **Handicap Access:** No

Marina Services and Boat Supplies
Services - Dock Carts **Communication -** FedEx, DHL, UPS **Supplies - OnSite:** Ice *(Cube)*, Bait/Tackle **3+ mi:** Boater's World *(472-3393, 10 mi.)*

Boatyard Services
Nearest Yard: Chambers Bay Marine & Storage (253) 582-6325

Restaurants and Accommodations
Under 1 mi: Restaurant *(Bair Restaurant 588-9778, B daily, L Mon-Sat, D Wed-Sat)*, *(ER Rogers Restaurant 582-0280, D $20-31, D, Sun brunch)*, Inn/B&B *(Above the Sound B&B 589-1441, $90-75)* **1-3 mi:** Restaurant *(Hunan Garden 589-2100, L $5-9, D $7-15)*, Coffee Shop *(Lacoste Coffee 589-8358)*, Fast Food *(McDonald's)*, Pizzeria *(Pacific Pizza 589-1111)*
3+ mi: Motel *(Best Value Inn 589-8800, 6 mi.)*, Hotel *(Best Western 584-2212, $65-76, 5 mi.)*

Recreation and Entertainment
OnSite: Beach, Picnic Area, Grills, Boat Rentals **Near:** Park *(Salters Point Park)* **Under 1 mi:** Museum *(Steilacoom Historical Museum 584-4133)*, Cultural Attract *(Tribal Cultural Center and Museum 584-6308 $3/1)*, Special Events *(Salmon Bake - last Sun Juy Noon-4pm, Apple Squeeze - 1st Sun Oct)*, Galleries *(Haskett Art Gallery 582-0581)* **1-3 mi:** Pool *(Swimming Pool & Fitness Center 964-6678)*, Spa, Golf Course *(Fort Steilacoom Golf Course 588-0613)*, Fitness Center *(Swimming Pool & Fitness Center 964-6678)*, Sightseeing *(Nathaniel Orr Pioneer Home Site $2, Sat-Sun 1-4pm May-Oct)* **3+ mi:** Bowling *(Bowlero Lanes 584-0212, 6 mi.)*, Movie Theater *(Lakewood Cinemas 581-5055, 6 mi.)*

Provisioning and General Services
Under 1 mi: Market *(Steilacoom Grocery 588-7855)*, Delicatessen *(Steilacoom Deli & Pub 584-7693)*, Bank/ATM *(Key Bank 582-0900)*, Post Office, Protestant Church *(Oberlin Congregational 584-4623)*, Beauty Salon *(Steilacoom Styling Salon 581-0439)*, Barber Shop *(Steilacoom Barber Shop 582-5564)*, Clothing Store, Retail Shops **1-3 mi:** Convenience Store *(7-Eleven 531-7537)*, Supermarket *(Albertson's 582-5844)*, Catholic Church *(St. John Bosco 582-1028)*, Library *(588-1452)*, Pharmacy *(Walgreens 581-0196)*, Hardware Store *(Ace 588-8586)*, Florist *(Crane's Creations 584-1400)*, Copies Etc. *(Justrite Printing 584-8088)* **3+ mi:** Health Food *(Whole Foods Market 565-0188, 8 mi.)*, Liquor Store *(WA Liquor 983-2139, 6 mi.)*, Meat Market *(A&G Meats 472-8150, 6 mi.)*, Dry Cleaners *(Carriage Cleaners 584-6291, 4 mi.)*, Laundry *(Spin City Laundry & Cleaners 983-1987, 4 mi.)*, Bookstore *(Barnes & Nobles 983-0852, 6 mi.)*, Department Store *(Target 581-7171, 6 mi.)*, Buying Club *(Costco 475-2093, 10 mi.)*

Transportation
OnCall: Rental Car *(Enterprise 582-8240)* **Near:** Ferry Service *(Anderson & Ketron Islands 798-2766)* **3+ mi:** Taxi *(Yellow Cab 472-3303, 10 mi.)*
Airport: Tacoma Narrows *(7 mi.)*

Medical Services
911 Service **OnCall:** Ambulance **Under 1 mi:** Dentist *(McAvoy 582-3106)* **1-3 mi:** Optician *(Rainier Econoline Optical 582-9089)* **3+ mi:** Doctor *(Johnson Family Practice 222-5568, 3 mi.)*, Chiropractor *(Tillotson Chiropractic Clinic 584-2414, 4 mi.)*, Veterinarian *(Lakewood Veterinary Hospital 584-2114, 4 mi.)* **Hospital:** Western State 582-8900 *(3 mi.)*

Setting -- Steilacoom Marina lies north of Cormorant Passage which runs between the mainland and Ketron Island. North of the marina is a ferry dock with connections to the three nearby islands - Anderson & Ketron. The third, McNeil Island to the west, is home to a state penitentiary. A hillside rises behind the marina and a small park with a cooking shelter and a gravel beach is nearby. Train tracks run between the marina and the nearest street.

Marina Notes -- *$3 for 20 amp electricity. Plans are underway to remodel this marina, pending approval of the permits. As of publication, permits had not been acquired - so call and ask. The current wooden docks are rough and uneven and accommodate smaller boats. The marina is exposed to wind and weather as there is no protecting breakwater. The only head is inside the office and not available after hours; there is no shower or laundry.

Notable -- This marina is accessible by land only after crossing a foot bridge over the railroad tracks and walking along the beach to the marina. Although there is a market close by in town, all supplies would have to be carried approximately 300 yards from the parking area to the boat. It's about a mile to downtown Steilacoom, the oldest incorporated town in Washington. Many of its buildings, dating back to the 1850s, are now on the National Registry of Historic Places. Walking tours are available. Town Hall is home to The Steilacoom Historical Museum, which focuses on the European experience in the area. The Steilacoom Tribal Cultural Center presents the Native American story. Before Tacoma took over as the county seat, Steilacoom was expected to become the major population center in this part of Washington.

16. WA – Southwest Puget Sound

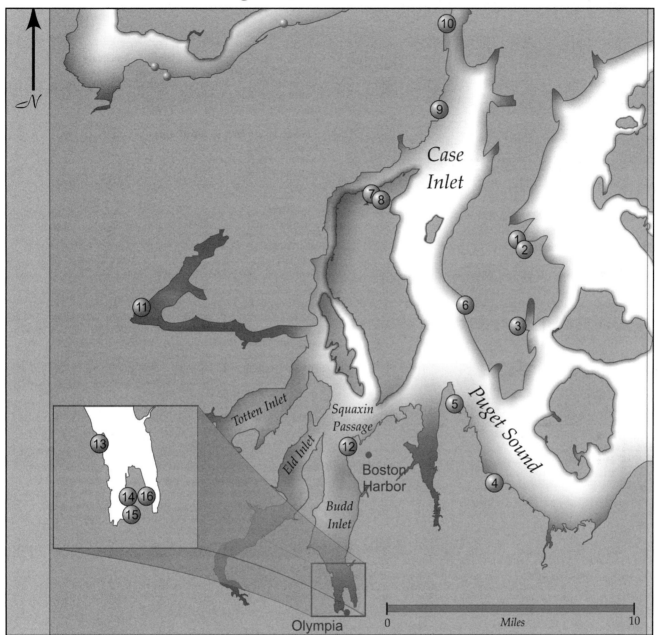

MAP	MARINA	HARBOR	PAGE	MAP	MARINA	HARBOR	PAGE
1	Penrose Point State Park	Carr Inlet/Mayo Cove	232	9	Fair Harbor Marina	Case Inlet/Fair Harbor	240
2	Lakebay Marina	Carr Inlet/Mayo Cove	233	10	Port of Allyn Waterfront Park	Case Inlet/North Bay	241
3	Longbranch Marina	Filucy Bay	234	11	Port of Shelton	Oakland Bay	242
4	Tolmie State Park	Nisqually Reach	235	12	Boston Harbor Marina	Boston Harbor	243
5	Zittel's Marina	Johnson Point	236	13	West Bay Marina	Budd Inlet	244
6	Joemma Beach State Park	Case Inlet/Whiteman Cove	237	14	Port Plaza Guest Dock	Budd Inlet	245
7	Jarrell's Cove Marina	Pickering Passage/Jarrell Cove	238	15	Percival Landing Moorage	Budd Inlet	246
8	Jarrell Cove Marine State Park	Pickering Passage/Jarrell Cove	239	16	Swantown Marina & Boatworks	Budd Inlet	247

▸ **Currency** — In Canadian Marina Reports, all prices are in Canadian dollars. In U.S. Marina Reports, all prices are in U.S. dollars.

▸ **"CCM"** — Denotes a Certified Clean Marina, a state/provincial award for environmental excellence. See page 298 for an explanation and page 299 for a list of Pump-Out facilities.

▸ **Ratings & Reviews** — An overview of the Atlantic Cruising Club's rating system is on page 6 and details on the content of each Marina Report are on pages 7 – 11.

▸ **Marina Report Updates** — Comments from boaters and new information from ACC reviewers and marinas are posted regularly on *www.AtlanticCruisingClub.com*.

Penrose Point State Park

321 158th Avenue KPS; Lakebay, WA 98349

Tel: (253) 884-2514 **VHF: Monitor** n/a **Talk** n/a
Fax: (253) 884-2526 **Alternate Tel:** n/a
Email: n/a **Web:** www.parks.wa.gov
Nearest Town: Home *(2 mi.)* **Tourist Info:** (360) 902-8844

Navigational Information
Lat: 47°15.510' **Long:** 122°45.526' **Tide:** 13 ft. **Current:** n/a **Chart:** 18448
Rep. Depths *(MLW):* **Entry** 30 ft. **Fuel Dock** n/a **Max Slip/Moor** 6 ft./30 ft.
Access: Western side of Mayo Cove off Carr Inlet

Marina Facilities *(In Season/Off Season)*
Fuel: No
Slips: 7 Total, 7** Transient **Max LOA:** 40 ft. **Max Beam:** n/a
 Rate *(per ft.):* **Day** $0.50* **Week** n/a **Month** n/a
 Power: 30 amp n/a, 50 amp n/a, 100 amp n/a, 200 amp n/a
 Cable TV: No **Dockside Phone:** No
 Dock Type: Floating, Alongside, Wood
Moorings: 8 Total, 8 Transient **Launch:** None, Dinghy Dock
 Rate: Day $10 **Week** n/a **Month** n/a
Heads: 5 Toilet(s), 2 Shower(s)
Laundry: None **Pay Phones:** Yes
Pump-Out: OnSite, Self Service, 1 Central **Fee:** Free **Closed Heads:** Yes

Marina Operations
Owner/Manager: Washington State Parks and Recreation **Dockmaster:** n/a
In-Season: Year-Round **Off-Season:** n/a
After-Hours Arrival: Pay station at the top of the pier
Reservations: No **Credit Cards:** Visa/MC, ***Limited Hours, In Season
Discounts: Annual permits **Dockage:** n/a **Fuel:** n/a **Repair:** n/a
Pets: Welcome **Handicap Access:** No

Marina Services and Boat Supplies
Supplies - Under 1 mi: Ice *(Cube)*, Ships' Store *(Lakebay Marina)*,
Propane

Boatyard Services
Nearest Yard: Gig Harbor or Olympia

Restaurants and Accommodations
1-3 mi: Restaurant *(Home Port Restaurant and Lounge 844-3743)*
3+ mi: Restaurant *(Blondie's on the Bluff 884-1300, 8 mi.)*, *(Huckleberry Inn Restaurant 884-3707, 8 mi.)*, Pizzeria *(Pizza Plus 884-9599, 8 mi.)*, Motel *(Westwind Motel 857-4047, 15 mi.)*, Inn/B&B *(BW Wesley Inn 858-9690, 20 mi., Gig Harbor)*

Recreation and Entertainment
OnSite: Beach *(and horseshoe pit)*, Picnic Area, Grills, Jogging Paths, Park

Provisioning and General Services
Under 1 mi: Convenience Store *(Lakebay Marina 884-3350 or Texaco convenience in Home 884-3828, 2 mi.)*, Wine/Beer *(Lakebay Marina)*

1-3 mi: Market *(Home Country Store 884-2106)*, Post Office *(884-2389)*, Protestant Church *(Lakebay Community Church 884-3899)*, Laundry *(Home Laundromat)* **3+ mi:** Supermarket *(Peninsula Market 884-3325, 8 mi.)*, Delicatessen *(Lisa's Fresh Express Deli & Coffee House 884-3354, 8 mi.)*, Liquor Store *(Key Center Liquor 884-2101, 8 mi.)*, Bank/ATM *(8 mi.)*, Library *(Key Center Library 864-2242, 8 mi.)*, Beauty Salon *(Pretty People Styling Salon 884-5250, 3 mi.)*, Pharmacy *(Albertson's 853-4750, 18 mi.)*, Hardware Store *(Capitol Lumber 884-2311, 8 mi.)*

Transportation
1-3 mi: Local Bus *(Bus Plus 984-8155)* **3+ mi:** Rental Car *(Enterprise 853-3882, 21 mi.)* **Airport:** Tacoma Narrows *(26 mi.)*

Medical Services
911 Service **OnCall:** Ambulance *(Key Peninsula Ambulance 884-2222)*
3+mi: Doctor *(Key Medical Center 884-9221, 8 mi.)*, Dentist *(Olsson 884-9455, 8 mi.)*, Chiropractor *(Key Center Chiropractic 884-3040, 8 mi.)*
Hospital: Tacoma General 403-1000 *(22 mi.)*

Setting -- 162-acre Penrose Point State Park sits on the east side of Key Peninsula. The park occupies almost the entire south shore of Mayo Cove off Carr Inlet with nearly two miles of saltwater frontage and engaging, wooded trails. A single float - at the end of a stationary pier - and 8 mooring buoys are available for boaters. Picnic tables sit near the dock overlooking the water. Lakebay Marina is a short hop across the bay; the small village of Lakebay is to the south.

Marina Notes -- *Dock moorage $10 minimum. Annual permits are $3.50/ft. with a $50 minimum. Annual permits are valid for the calendar year. Daily permits available at the park by self-registration on arrival. **No slips, dockage is alongside tie-ups with 138 ft. on one side of the float and 132 ft. on the other. There are also 8 mooring buoys, 3 in Mayo Cove and 5 in Delano Bay on the outside of the point. ***Visa & Mastercard accepted only during limited hours in the summer season. Pay phones are at the park entrance. Pump-out and porta-potty station available on-site. One vault toilet is located near the dock. Four additional heads are located throughout the park, with flush toilets and running water; two of these have hot showers (summer only; fee). Note: If you're on the float, beware of shallow water at low tide - a tide table is conveniently displayed on the bulletin board at the top of the pier.

Notable -- The park's beaches, hiking trails and a nature trail with interpretive signs attract lots of visitors. Clams and oysters are popular sportfishing targets during the open season; a shellfish license is required. Camping at the Cascadia Marine Trail site, $10/night, is permited only for those arriving in human-powered beachable watercraft. The trailhead for the interpretive trail is within a quarter mile.

Navigational Information

Lat: 47°15.457' **Long:** 122°45.381' **Tide:** 13 ft. **Current:** n/a **Chart:** 18448
Rep. Depths (*MLW*): **Entry** n/a **Fuel Dock** n/a **Max Slip/Moor** 6 ft./-
Access: Western side of Carr Inlet north of Penrose Point State Park

Marina Facilities *(In Season/Off Season)*

Fuel: Gasoline
Slips: 50 Total, 10 Transient **Max LOA:** 50 ft. **Max Beam:** n/a
  Rate *(per ft.)*: **Day** $0.50 **Week** n/a **Month** n/a
  Power: 30 amp Incl.*, **50 amp** n/a, **100 amp** n/a, **200 amp** n/a
  Cable TV: No **Dockside Phone:** No
  Dock Type: Floating, Short Fingers, Alongside, Concrete, Wood
Moorings: 0 Total, 0 Transient **Launch:** n/a
  Rate: Day n/a **Week** n/a **Month** n/a
Heads: 2 Toilet(s)
Laundry: None **Pay Phones:** No
Pump-Out: No **Fee:** n/a **Closed Heads:** Yes

Marina Operations

Owner/Manager: Lee Hostetler **Dockmaster:** Same
In-Season: Year-Round, 9am-6pm **Off-Season:** n/a
After-Hours Arrival: Call ahead
Reservations: Yes, Required **Credit Cards:** No; Cash or Check only
Discounts: None
Pets: Welcome **Handicap Access:** No

Lakebay Marina

15 Lorenz Road N.; Lakebay, WA 98349

Tel: (253) 884-3350 **VHF: Monitor** n/a **Talk** n/a
Fax: n/a **Alternate Tel:** n/a
Email: n/a **Web:** n/a
Nearest Town: Home *(1.5 mi.)* **Tourist Info:** (360) 876-1801

Marina Services and Boat Supplies

Services - Dock Carts **Supplies - OnSite:** Ice *(Cube)*, Ships' Store, Propane

Boatyard Services

OnSite: Launching Ramp **Nearest Yard:** Gig Harbor or Olympia

Restaurants and Accommodations

1-3 mi: Restaurant *(Home Port Restaurant and Lounge 844-3743)*
3+ mi: Restaurant *(Huckleberry Inn Restaurant 884-3707, 7 mi.)*, *(Blondie's on the Bluff 884-1300, 7 mi.)*, Pizzeria *(Pizza Plus 884-9599, 7 mi.)*, Motel *(Westwind Motel 857-4047, 15 mi.)*, Inn/B&B *(Maritime Inn 858-1818, $90-160, 20 mi.)*

Recreation and Entertainment

OnSite: Picnic Area **Under 1 mi:** Beach, Jogging Paths, Park *(Penrose Point State Park 884-2514)* **3+ mi:** Playground *(Key Peninsula Park 884-9240, 5 mi.)*

Provisioning and General Services

OnSite: Convenience Store, Wine/Beer **1-3 mi:** Market *(Home Country Store 884-2106)*, Post Office *(884-2389)*, Protestant Church *(Lakebay Community Church 884-3899)*, Laundry **3+ mi:** Supermarket *(Peninsula Market 884-3325; chain supermarket in Gig Harbor, 7 mi.)*, Delicatessen *(Lisa's Fresh Express Deli & Coffee House 884-3354, 7 mi.)*, Liquor Store *(Key Center Liquor 884-2101, 7 mi.)*, Bank/ATM *(7 mi.)*, Library *(Key Center Library 884-2242, 7 mi.)*, Beauty Salon *(Pretty People Styling Salon 884-5250, 3 mi.)*, Pharmacy *(Albertson's 853-4750, 18 mi.)*, Hardware Store *(Capitol Lumber 884-2311, 7 mi.)*

Transportation

1-3 mi: Local Bus *(Bus Plus 984-8155)* **3+ mi:** Rental Car *(Enterprise 853-3882, 21 mi.)* **Airport:** Tacoma Narrows *(24 mi.)*

Medical Services

911 Service **OnCall:** Ambulance *(Key Peninsula Ambulance 884-2222)*
3+ mi: Doctor *(Key Medical Center 884-9221, 7 mi.)*, Dentist *(Olsson 884-9455, 7 mi.)*, Chiropractor *(Key Center Chiropractic 884-3040, 7 mi.)*
Hospital: Tacoma General 403-1000 *(21 mi.)*

Setting -- Lakebay, at the head of Mayo Cove on the southwest shore of Carr Inlet, is a tiny village with a store and several small private piers. Views of neighboring homes greet boaters as they enter the cove. Rustic Lakebay Marina's office and amenities are in a white, two-story building suspended above the water on pilings. Gangways lead from the docks to a wide, long pier - with no guardrails - that connects to the building. The pier's planks are spaced apart so the water below shows between them, but it's sturdy enough for cars to pass.

Marina Notes -- *Power is 15 amp. The fuel dock provides gasoline, but the channel to the pier can be difficult to navigate, especially at low tide. The docks are a combination of wood and concrete with both slips and side-tie space. The casual marina store provides a variety of items including fishing supplies, limited groceries and ice. A picnic table on the dock is a convenient place to have lunch. A check must accompany a request for a reservation. Note: Low tide can make for very steep gangways.

Notable -- Other than a stroll down a country road, there is not much to do immediately around the marina. Hop in the dinghy and explore the picturesque bay or cross to neighboring Penrose Point State Park - with beaches, hiking trails and a nature trail with interpretive signs. Crabs and oysters are popular in this area; get a license and try your hand. About 1.5 miles from the marina, the town of Home offers a restaurant and convenience store. Additional services are about 7-8 miles north. Accommodations and large chain supermarkets are more than 15 miles away in Gig Harbor.

Longbranch Marina

PO Box 111; Lakebay, WA 98349

Tel: (253) 884-5137 **VHF: Monitor** n/a **Talk** n/a
Fax: n/a **Alternate Tel:** n/a
Email: see below **Web:** longbranchimprovementclub.org
Nearest Town: Home *(5 mi.)* **Tourist Info:** (253) 851-6865

Navigational Information

Lat: 47°12.600' **Long:** 122°45.490' **Tide:** 12 ft. **Current:** 0 kt. **Chart:** 18448
Rep. Depths (MLW): Entry 25 ft. **Fuel Dock** n/a **Max Slip/Moor** 120 ft./-
Access: The Narrows to Filucy Bay behind McNeil & Anderson Islands

Marina Facilities *(In Season/Off Season)*

Fuel: No
Slips: 65 Total, 20* Transient **Max LOA:** 65 ft. **Max Beam:** n/a
  Rate *(per ft.)*: **Day** $0.50 **Week** n/a **Month** n/a
  Power: 30 amp $4.50, **50 amp** n/a, **100 amp** n/a, **200 amp** n/a
  Cable TV: No **Dockside Phone:** No
  Dock Type: Floating, Long Fingers, Alongside, Wood
Moorings: 0 Total, 0 Transient **Launch:** n/a, Dinghy Dock (See dockmaster)
  Rate: Day n/a **Week** n/a **Month** n/a
Heads: 4 Toilet(s)
Laundry: None, Book Exchange **Pay Phones:** No
Pump-Out: No **Fee:** n/a **Closed Heads:** Yes

Marina Operations

Owner/Manager: Longbranch Improvement Club **Dockmaster:** M. Runions
In-Season: Year-Round, 24/7 **Off-Season:** n/a
After-Hours Arrival: Honor system
Reservations: No **Credit Cards:** No
Discounts: None
Pets: Welcome, Dog Walk Area **Handicap Access:** No

Marina Services and Boat Supplies

Services - Docking Assistance, Security *(locked gates)*, Trash Pick-Up, Dock Carts **Supplies - OnSite:** Ice *(Block, Cube)* **3+ mi:** Bait/Tackle *(Lakebay Marina 884-3350, 4 mi.)*, Propane *(4 mi.)*

Boatyard Services

Nearest Yard: Olympia or Gig Harbor

Restaurants and Accommodations

3+ mi: Restaurant *(Home Port Restaurant 887-3743, 5 mi.)*, *(Huckleberry Inn Restaurant 884-3707, 12 mi.)*, Lite Fare *(Lisa's Fresh Express Deli & Coffee House 884-3354, 12 mi.)*, Pizzeria *(Pizza Plus 884-9599, 12 mi.)*, Motel *(Westwind Motel 857-4047, 20 mi.)*

Recreation and Entertainment

OnSite: Picnic Area, Grills **Near:** Special Events *(Opening Day Breakfast - 1st wknd May; MemDay and LabDay Dance. Shuttle service provided from dock to dance hall)* **3+ mi:** Park *(Joemma Beach State Park 884-1944/ Penrose Point State Park 884-2514, 3 mi./4 mi.)*

Provisioning and General Services

Near: Protestant Church **3+ mi:** Convenience Store *(Home Texaco 884-3828, 5 mi.)*, Market *(Home Country Store 884-2106, 5 mi.)*, Wine/Beer *(Lakebay Marina 884-3350, 4 mi.)*, Liquor Store *(Key Center Liquor 884-2101, 12 mi.)*, Post Office *(884-2389, 5 mi.)*, Library *(Key Center Library 884-2242, 12 mi.)*, Laundry *(Home Laundromat, 5 mi.)*, Hardware Store *(Capitol Lumber 884-2311, 12 mi.)*

Transportation

3+ mi: Rental Car *(Enterprise 853-3882, 25 mi.)*, Local Bus *(Bus Plus 984-8155, 5 mi.)* **Airport:** Tacoma Narrows *(30 mi.)*

Medical Services

911 Service **OnCall:** Ambulance *(Key Peninsula Ambulance 884-2222)* **3+ mi:** Doctor *(Key Medical Center 884-9221, 12 mi.)*, Dentist *(Olsson 884-9455, 12 mi.)*, Chiropractor *(Key Center Chiropractic 884-3040, 12 mi.)* **Hospital:** Tacoma General 403-1000 *(27 mi.)*

Setting -- Pristine Filucy Bay sits on the eastern shore of Key Peninsula, near the southern tip of the mainland. It is tucked behind McNeil and Anderson Islands, opposite Balch Passage. This beautiful bay is about 1.5 miles long with an entrance 0.4 mile wide. Numerous homes line the shoreline and perch in the hills. Longbranch, a village in the small cove opposite the entrance, has a guest pier and slips for local residents - hanging baskets of flowers liven the marina with color. The area is serene, filled with the sounds of chirping birds and blessed with absolutely stunning views of Mt. Ranier "when the mountains are out."

Marina Notes -- *760 ft. of side-tie transient moorage. Founded in 1921, the non-profit Longbranch Improvement Club maintains the the docks and clubhouse. First come, first serve - tie up anywhere on the yellow rails. 36 slips and 7 boathouses are reserved for residents. Wide, wooden, well-cared for docks connect to a large concrete pier - an abandoned ferry landing. Some private buoys sprinkle the bay. A book exchange and ice are available at the pavilion. 2-4 portable heads are on-site, depending on the season. Email: lic@longbranchimprovementclub.org. Note: McNeil Island is a federal prison.

Notable -- On the dock, a covered pavilion with long tables invites boaters to gather and picnic - and enjoy the view. Plastic walls can be rolled up or down depending on the weather. Holiday and special events weekends are especially popular; boats are often rafted to make room. A half-mile from the docks, the large Longbranch Clubhouse is a 1939 steeply pitched A-frame; a former gymnasium, it's main room is now an oak-floored ballroom and it's available for rent. There are no services in Longbranch. The town of Home has a restaurant and convenience store, plus bus service to Gig Harbor across the bridge.

Navigational Information
Lat: 47°07.267' **Long:** 122°46.524' **Tide:** 14 ft. **Current:** n/a **Chart:** 18448
Rep. Depths (*MLW*): **Entry** 30 ft. **Fuel Dock** n/a **Max Slip/Moor** -/30 ft.
Access: Across Nisqually Reach from Anderson Island

Marina Facilities *(In Season/Off Season)*
Fuel: No
Slips: 0 Total, 0 Transient **Max LOA:** n/a **Max Beam:** n/a
  Rate *(per ft.)*: **Day** n/a **Week** n/a **Month** n/a
  Power: 30 amp n/a, 50 amp n/a, 100 amp n/a, 200 amp n/a
  Cable TV: No **Dockside Phone:** No
  Dock Type: n/a
Moorings: 5 Total, 5 Transient **Launch:** None
  Rate: Day $10 **Week** n/a **Month** n/a
Heads: 6 Toilet(s)
Laundry: None **Pay Phones:** No
Pump-Out: No **Fee:** n/a **Closed Heads:** Yes

Marina Operations
Owner/Manager: WA State Parks and Recreation **Dockmaster:** n/a
In-Season: May-Sep, Sun-Sat **Off-Season:** Oct-Apr, Wed-Sun only
After-Hours Arrival: All boaters must self-register
Reservations: No **Credit Cards:** No
Discounts: None
Pets: Welcome **Handicap Access:** No

Tolmie State Park

PO Box 4265; 7150 Cleanwater Lane; Olympia, WA 98504

Tel: (360) 456-6464 **VHF: Monitor** n/a **Talk** n/a
Fax: n/a **Alternate Tel:** n/a
Email: n/a **Web:** www.parks.wa.gov
Nearest Town: Olympia *(11 mi.)* **Tourist Info:** (360) 357-3362

Marina Services and Boat Supplies
Supplies - 3+ mi: Ice *(Cube)*, Ships' Store *(dinghy, 4 mi.)*, West Marine *(352-1244, 12 mi.)*, Boater's World *(754-1834, 12 mi.)*, Bait/Tackle *(Zittel's Marina 459-1950, dinghy, 4 mi.)*

Boatyard Services
Nearest Yard: Swantown Boatworks (360) 528-8059

Restaurants and Accommodations
Under 1 mi: Inn/B&B *(Puget View Guesthouse 413-9474, $99-129, featured in "Best Places to Stay in the Northwest")* **3+ mi:** Restaurant *(Hawks Prairie Inn 459-0901, 6 mi.)*, *(Emperor's Palace 923-2323, 7 mi.)*, *(Season's Teriyaki 456-8919, 6 mi.)*, *(Seoul Restaurant 459-3364, 7 mi.)*, *(Ruby Tuesday 438-6655, 6 mi., By boat)*, Snack Bar *(Boston Harbor Marina 357-5670, 5 mi., By boat)*, Coffee Shop *(Mocha Magic 455-1648, 6 mi.)*, Fast Food *(Burger King, McDonald's 6 mi.)*, Lite Fare *(Norma's Burger Hut 456-6547, 7 mi.)*, Pizzeria *(Godfather's Pizza 456-1400, 6 mi.)*, Motel *(King Oscar 438-3333, 6 mi.)*, *(Days Inn 493-1991, 8 mi.)*

Recreation and Entertainment
OnSite: Beach *(and underwater park for divers)*, Picnic Area, Grills, Playground, Jogging Paths, Park, Sightseeing **3+ mi:** Pool *(Tanglewilde Pool 491-3907, 8 mi.)*, Golf Course *(Hawks Prairie - 2 18-hole courses 355-8383, 7 mi.)*, Fitness Center *(Powerhouse Gym 459-5400, 8 mi.)*, Movie Theater *(Capitol Theater 754-5378, 11 mi.)*, Video Rental *(Hollywood Video 412-1591, 6 mi.)*, Museum *(Olympia museums and Capitol tours, 11 mi.)*

Provisioning and General Services
1-3 mi: Convenience Store *(Zittel's Marina 459-1950, by boat)* **3+ mi:** Market *(Buddie's Grocery & Deli 456-4955, 7 mi.)*, Supermarket *(Safeway 438-3914, Mega Foods 456-5353, 6 mi.)*, Delicatessen *(Lylan Sandwich 491-4068, 8 mi.)*, Wine/Beer *(Boston Harbor Marina by boat 357-5670, 5 mi.)*, Bakery *(Bavarian Corner 456-5066, 7 mi.)*, Fishmonger *(at Boston Harbor Marina - by boat)*; Bank/ATM, Post Office *(8 mi.)*, Catholic Church *(Sacred Heart 491-0890, 8 mi.)*, Protestant Church *(St. Benedict's Episcopal 456-2240, 8 mi.)*, Library *(Lacey 491-3860)*, Beauty Salon *(Hairmasters 438-1793, 6 mi.)*, Barber Shop *(Hut Barbers 459-1945, 7 mi.)*, Dry Cleaners*(College 459-5339, 6mi.)*, Laundry *(Spin City 456-3362, 8 mi.)*, Pharmacy *(Rite Aid 456-0444, 6 mi.)*, Hardware Store *(Home Depot 459-4256, 6 mi.)*, Department Store *(Office Depot 923-1099, 8 mi.)*

Transportation
OnCall: Rental Car *(Enterprise 438-1278)*, Taxi *(Red Top 357-3700)*, Airport Limo *(Capital West 923-5818)* **3+ mi:** Local Bus *(786-1881, 7 mi.)*, Rail *(Amtrak 923-4602, 13 mi.)* **Airport:** Olympia Regional *(12 mi.)*

Medical Services
911 Service **OnCall:** Ambulance **3+ mi:** Doctor *(Sea Mar Community Health Care 491-1399, 6 mi.)*, Dentist *(Hawk's Prairie Dental 456-7070, 6 mi.)*, Chiropractor *(Back On Track Chiropractic 456-4954, 6 mi.)*, Veterinarian *(Tanglewilde Vet Hospital 491-4691, 7 mi.)* **Hospital:** Providence St. Peter 491-9180 *(10 mi.)*

Setting -- This 105-acre State Park is located on Nisqually Reach 2.4 miles north and west of Nisqually Head. Look for it one mile west of green buoy #3. The buoys are located quite far out due to shallow waters near the shore. There is no dock, but dinghies can be beached and dragged ashore. An arched wooden bridge crosses a saltwater lagoon and leads from the beach to the picnic area. The park features three miles of hiking trails as well as a short nature trail.

Marina Notes -- Five mooring buoys are located off the 1,800-foot beach. The moorings are available year-round, though the park closes Mondays and Tuesdays from October through April. Fees are charged for mooring from 3pm-8am. All moorage buoys are $10 a night. Annual permits are also available: boats under 26' are $60 per year, 26-34' $90, and boats 35' and over $110 a year. They can be purchased at some marine parks, on the State Parks website (www.parks.wa.gov) and at the State Parks Olympia headquarters (902-8608). No trash cans are provided at the park. Portable toilets are available at the beach, and well-maintained flush toilets are in the upper park; no showers.

Notable -- A beautiful, quiet beach - part sand and part pebble - stretches along the waterfront. Three barges have been sunk offshore to provide a fish haven to enhance the diving and fishing at the park. Year-round clamming and oyster gathering is available with a Washington State shellfish license. Good views of Anderson and McNeill Islands as well as Mt. Ranier and the Olympic Mountains are a feature of the bay. Forested hiking trails provide opportunities to enjoy the birds. The picnic areas offer fresh water and charcoal grills, and two shelters with electricity are available.

Navigational Information

Lat: 47°09.887' **Long:** 122°48.552' **Tide:** 15 ft. **Current:** 1 kt. **Chart:** 18448
Rep. Depths (*MLW*): **Entry** n/a **Fuel Dock** 10 ft. **Max Slip/Moor** 8 ft./-
Access: Located on the east side of Johnson Point

Marina Facilities *(In Season/Off Season)*

Fuel: *ValvTect* - Gasoline, Diesel
Slips: 200 Total, 8** Transient **Max LOA:** 48 ft. **Max Beam:** n/a
  Rate *(per ft.)*: **Day** $0.75* **Week** n/a **Month** $7.40-10.65
  Power: 30 amp Incl., **50 amp** n/a, **100 amp** n/a, **200 amp** n/a
  Cable TV: No **Dockside Phone:** No
  Dock Type: Floating, Long Fingers, Wood
Moorings: 0 Total, 0 Transient **Launch:** n/a
  Rate: Day n/a **Week** n/a **Month** n/a
Heads: 2 Toilet(s)
Laundry: None **Pay Phones:** No
Pump-Out: OnSite **Fee:** $7.50 + 0.50/gal. **Closed Heads:** Yes

Marina Operations

Owner/Manager: Mike Zittel **Dockmaster:** Same
In-Season: Year-Round, 8am-5pm **Off-Season:** n/a
After-Hours Arrival: n/a
Reservations: Yes **Credit Cards:** Visa/MC
Discounts: None
Pets: Welcome, Dog Walk Area **Handicap Access:** No

Zittel's Marina

9144 Gallea Street NE; Olympia, WA 98516

Tel: (360) 459-1950 **VHF: Monitor** n/a **Talk** n/a
Fax: (360) 459-8984 **Alternate Tel:** n/a
Email: n/a **Web:** www.zittelsmarina.com
Nearest Town: Olympia *(8 mi.)* **Tourist Info:** (360) 357-3362

Marina Services and Boat Supplies

Services - Security *(owners onsite)*, Trash Pick-Up, Dock Carts **Supplies -**
OnSite: Ice *(Cube)*, Ships' Store, Bait/Tackle **3+ mi:** West Marine *(352-1244, 13 mi.)*, Boater's World *(754-1834, 13 mi.)*

Boatyard Services

OnSite: Travelift *(Sling launch $20-65)*, Hydraulic Trailer *(25T)*, Engine mechanic *(gas, diesel)*, Launching Ramp *($10 - wide, concrete)*, Bottom Cleaning **Yard Rates:** $70/hr., Power Wash $30 + $1.75/ft., Bottom Paint $27-33 *(paint incl.)* **Storage:** On-Land $11/ft.

Restaurants and Accommodations

OnSite: Snack Bar *(Zittel's Marina 459-1950)* **3+ mi:** Restaurant *(Skipper's Seafood N Chowder House 456-4133, 9 mi.)*, *(Casa Mia 459-0440, 9 mi.)*, *(Cebu 455-9128, 9 mi., authentic Filipino food)*, *(Mei Wei Seafood 456-2333, 9 mi., Chinese)*, Fast Food *(Wendy's, Taco Bell 9 mi.)*, Lite Fare *(Paul's Burger & Teriyaki 455-3328, 9 mi.)*, Pizzeria *(Brewery City Pizza 491-6630, 9 mi.)*, Motel *(Comfort Inn 456-6300, $72-86, 9 mi.)*, Hotel *(Holiday Inn 412-1200, $58-85, 9 mi.)*, *(Ameritel 459-8866, $90-190, 9 mi.)*

Recreation and Entertainment

OnSite: Picnic Area, Boat Rentals *(power boats 4 hrs. $135, 8 hrs. $215)*
Near: Jogging Paths **1-3 mi:** Park *(Tolmie State Park, Joemma State Park - by boat)*, Sightseeing *(Nisqually Fishery)* **3+ mi:** Golf Course *(Hawks Prairie - 2 18-hole courses 355-8383, 10 mi.)*, Fitness Center *(Bally Total Fitness 438-2800, 9 mi.)*, Bowling *(Aztec Lanes 357-8808, 11 mi.)*, Movie Theater *(Capitol Theater 754-5378, 12 mi.)*, Video Rental *(Gull Harbor Mercantile 352-4014, 5 mi.)*, Galleries *(Art by Beth Rowley 534-9159, 9 mi.)*

Provisioning and General Services

OnSite: Convenience Store **3+ mi:** Market *(Gull Harbor Mercantile 352-4014, 5 mi.)*, Supermarket *(Safeway 438-3914, 9 mi.)*, Delicatessen *(South Bay Market and Deli 491-1114, 7 mi.)*, Health Food *(J-Vee Health Foods 491-1930, 9 mi.)*, Liquor Store *(Tanglewilde Liquor Store 491-5522, 10 mi.)*, Bakery *(Gai's/Franz Bakery 491-1900, 10 mi.)*, Fishmonger *(fresh seafood at Boston Harbor Marina 357-5670, by boat, 7 mi.)*, Bank/ATM *(Bank of America 753-9816, 9 mi.)*, Catholic Church *(Sacred Heart 491-0890, 9 mi.)*, Protestant Church *(8 mi.)*, Library *(Lacey 491-3860, 9 mi.)*, Beauty Salon *(South Bay Family Salon 491-0272, 7 mi.)*, Barber Shop *(Hair Perfect 438-3391, 9 mi.)*, Dry Cleaners *(College Cleaners 459-5339, 10 mi.)*, Laundry *(Spin City Laundry 456-3362, 10 mi.)*, Bookstore *(Fireside Book Store 352-4006, 12 mi.)*, Pharmacy *(Rite Aid 491-4111, 9 mi.)*, Hardware Store *(Lowe's 486-0856, 8 mi.)*, Florist *(Ivy Trellis 943-4726, 7 mi.)*, Department Store *(Target 486-8920, 9 mi.)*, Copies Etc. *(Office Depot 923-1099, 9 mi.)*

Transportation

OnCall: Rental Car *(Enterprise 438-1278)*, Taxi *(Capitol City Taxi 357-4949)* **3+ mi:** Local Bus *(in Lacey 786-1881, 9 mi.)*, Rail *(Amtrak 923-4602, 14 mi.)* **Airport:** Olympia Regional *(18 mi.)*

Medical Services

911 Service **OnCall:** Ambulance **3+ mi:** Doctor *(Express Urgent Care 923-1111, 10 mi.)*, Dentist *(Willows Pond Dental Center 459-8348, 9 mi.)*, Chiropractor *(Hawkes Prairie Chiropractic 459-5990, 9 mi.)*, Holistic Services *(Healing Touch 705-3373, 8 mi.)*, Veterinarian *(South Bay Pet Clinic 456-7387, 7 mi.)*, Optician *(Lilly Road Optical 456-1930, 9 mi.)*
Hospital: Providence St. Peter 491-9480 *(9 mi.)*

Setting -- Johnson Point shoulders its way between Nisqually Reach and Henderson Inlet. The Point looks up Case Inlet toward the Olympic Mountains. Zittel's is located in a quiet cove on the eastern shore of the point and is protected by a log boom. An attractive three-story gray-sided contemporary dominates the shore; on its first level, an inviting deck is populated with umbrella-topped tables and chairs and barrels of blooms - with views of the floating wood docks.

Marina Notes -- *$.65/ft. up to 35', over 35' $0.75/ft., $12.50 min. Monthly open slips: 20-28' $13.35 ft./mo., over 28' + 5%, over 39' + 10%. Monthly covered slips: 24' $7.50 ft., 30' $8.75-9.20, 34' $9.90-10.20, 40' $10.65. **300 ft. of side-tie moorage - Guest moorage is on an as-available basis, so call ahead. Family owned for more than 45 years, the marina is managed by the Zittels and their 5 children - who live upstairs. Covered and uncovered slips with end and side-ties for larger boats plus dry storage. A hydraulic trailer and sling launch is supported by basic boatyard services. The store sells bait, some marine supplies, charts, and limited groceries and rents 17- and 18-foot open bow runabouts with 90 horsepower outboards for area day trips. Basic heads; no showers or laundry.

Notable -- Nisqually Reach is very popular for fishing and crabbing, and the marina staff is happy to share the latest reports. Grab a seat at the small coffee bar inside the store and get updates. Explore Tolmie State Park, about 3 miles south, or the Nisqually River Delta, approximately 5 miles away. A 3,000 acre National Wildlife Refuge was established there in 1974 to protect migratory birds. Though too shallow for larger boats, the refuge is a great trip for the dinghy or one of the rental boats from the marina. The surrounding area is residential and offers no services. For provisions, the town of Lacey is about 9 miles south.

Navigational Information
Lat: 47°13.290' **Long:** 122°48.280' **Tide:** 14 ft. **Current:** n/a **Chart:** 18448
Rep. Depths *(MLW)*: **Entry** 15 ft. **Fuel Dock** n/a **Max Slip/Moor** 15 ft./15 ft.
Access: Case Inlet across from Harstine Island

Marina Facilities *(In Season/Off Season)*
Fuel: No
Slips: 12 Total, 12** Transient **Max LOA:** 60 ft. **Max Beam:** n/a
  Rate *(per ft.)*: **Day** $0.50* **Week** n/a **Month** n/a
  Power: 30 amp n/a, **50 amp** n/a, **100 amp** n/a, **200 amp** n/a
  Cable TV: No **Dockside Phone:** No
Dock Type: Floating, Alongside, Wood
Moorings: 3 Total, 3 Transient **Launch:** No, Dinghy Dock
  Rate: Day $10 **Week** n/a **Month** n/a
Heads: 5 Toilet(s)
Laundry: None **Pay Phones:** No
Pump-Out: No **Fee:** n/a **Closed Heads:** Yes

Marina Operations
Owner/Manager: WA State Parks & Recreation **Dockmaster:** n/a
In-Season: MemDay-LabDay, 8am-Dusk **Off-Season:** Winter, 8am-6:30pm
After-Hours Arrival: None
Reservations: No **Credit Cards:** No
Discounts: None
Pets: Welcome **Handicap Access:** No

Joemma Beach State Park

PO Box 898; Lakebay, WA 98349

Tel: (253) 884-1944 **VHF: Monitor** n/a **Talk** n/a
Fax: (253) 884-2526 **Alternate Tel:** n/a
Email: n/a **Web:** www.parks.wa.gov/moorage/parks
Nearest Town: Lakebay *(5 mi.)* **Tourist Info:** (360) 357-3362

Marina Services and Boat Supplies
Supplies - 3+ mi: Ice *(Block, Cube)*, Bait/Tackle *(Lakebay Marina 884-3350, 4 mi.)*, Propane *(4 mi.)*

Boatyard Services
OnSite: Launching Ramp *($5)* **Nearest Yard:** Gig Harbor or Olympia

Restaurants and Accommodations
3+ mi: Restaurant *(Home Port Restaurant and Lounge 887-3743, 5 mi.)*, *(Huckleberry Inn Restaurant 884-3707, 12 mi.)*, Lite Fare *(Lisa's Fresh Express Deli & Coffee 884-3663, 12 mi.)*, Pizzeria *(Pizza Plus 884-9599, 12 mi.)*, Motel *(Westwind Motel 857-4047, 20 mi.)*

Recreation and Entertainment
OnSite: Beach *(3,000 ft. of saltwater frontage)*, Picnic Area *(7 open picnic tables and 1 picnic shelter reservable for up to 50 people 888-226-7688)*, Grills, Park *(122-acre Joemma Beach)*

Provisioning and General Services
3+ mi: Convenience Store *(Home Texaco 884-3828, 5 mi.)*, Market *(Home Country Store 884-2106, 5 mi.)*, Supermarket *(Peninsula Market 884-3325, 12 mi.)*, Wine/Beer *(Lakebay Marina 884-3350, 4 mi.)*, Liquor Store *(Key Center Liquor 884-2101, 12 mi.)*, Bank/ATM *(12 mi.)*, Post Office *(884-2389, 5 mi.)*, Protestant Church *(Lakebay Community Church 884-3899, 5 mi.)*, Beauty Salon *(Pretty People Styling Salon 884-5250, 6 mi.)*, Laundry *(Home Laundromat, 5 mi.)*, Pharmacy *(Albertson's 853-4750, 22 mi.)*, Hardware Store *(Capitol Lumber 884-2311, 12 mi.)*

Transportation
3+ mi: Rental Car *(Enterprise 853-3882, 25 mi.)*, Local Bus *(Bus Plus 984-8155, 5 mi.)* **Airport:** Tacoma Narrows *(30 mi.)*

Medical Services
911 Service **OnCall:** Ambulance *(Key Peninsula Ambulance 884-2222)*
3+ mi: Doctor *(Key Medical Center 884-9221, 12 mi.)*, Dentist *(Olsson 884-9455, 12 mi.)*, Chiropractor *(Key Center Chiropractic 884-3040, 12 mi.)*
Hospital: Tacoma General 403-1000 *(26 mi.)*

PHOTOS ON DVD: 9

Setting -- Case Inlet, a popular spot for sportfishing, extends some 16 miles north from Johnson Point. On the eastern side of the inlet, fewer than three miles from Johnson Point, is Whiteman Cove. The lovely Joemma Beach State Park on Key Peninsula is just to the north of the cove. A tidal rocky beach extends on both sides of the sturdy, wide, wooden "E-shaped" float. From the floats, an aluminum gangway, steep at low tide, leads to a long stationery pier that ends on the wooded shore. Just beyond, picnic tables and a covered picnic shelter perch on a small hill overlooking the water; a quarter-mile hiking trail leads inland.

Marina Notes -- *$10 minimum. Limited stays - maximum 3-day limit. **500 ft. of alongside tie-ups. Annual permits available at the state parks' headquarters in Olympia, online and from park staff - $3.50/ft.with a $50 minimum. A boat ramp is near the dock. Two portapotties and five vault toilets serve as heads.

Notable -- Originally called R. F. Kennedy Park, it was renamed Joemma when it became a state park in 1995. Joe and Emma Smith (JoEmma) lived here from 1917 to 1932 and helped settle the Key Peninsula. Joe was a well-known newspaper publisher/editor; the park honors their contribution. With 3,00 feet of saltwater frontage, Joemma Beach is a premier site for ocean beachcombing. Rock crabbing is reportedly excellent. The park covers 122 acres - mostly forested, with fir, madrona and maple trees - creating wonderful hiking and nature trails for appreciating the rhododendrons and wildlife. During the summer 19 overnight camping tent sites ($15) are sprinkled throughout the park. Two water trail sites ($10) are accessible only by human-powered vessels. The covered picnic shelter is great for group barbeques and can be reserved ahead. Harstine Island is clearly in view as you look across the inlet.

Jarrell's Cove Marina

Jarrell's Cove Marina

220 E. Wilson Road; Shelton, WA 98584

Tel: (360) 426-8823; (800) 362-8823 **VHF: Monitor** n/a **Talk** n/a
Fax: (360) 432-8494 **Alternate Tel:** n/a
Email: n/a **Web:** n/a
Nearest Town: Shelton *(15 mi.)* **Tourist Info:** (360) 426-2021

Navigational Information
Lat: 47°17.145' **Long:** 122°53.262' **Tide:** 14 ft. **Current:** 1 kt. **Chart:** 18448
Rep. Depths *(MLW)*: **Entry** 30 ft. **Fuel Dock** n/a **Max Slip/Moor** 25 ft./-
Access: Located on the northern part of Harstine Island

Marina Facilities *(In Season/Off Season)*
Fuel: *Chevron* - Gasoline, Diesel
Slips: 70 Total, 5 Transient **Max LOA:** 100 ft. **Max Beam:** n/a
  Rate *(per ft.)*: **Day** $0.65 **Week** n/a **Month** n/a
  Power: 30 amp $3.60, **50 amp** n/a, **100 amp** n/a, **200 amp** n/a
  Cable TV: No **Dockside Phone:** No
  Dock Type: Floating, Long Fingers, Short Fingers, Alongside, Wood
Moorings: 0 Total, 0 Transient **Launch:** n/a, Dinghy Dock
  Rate: Day n/a **Week** n/a **Month** n/a
Heads: 2 Toilet(s), 2 Shower(s) *(with dressing rooms)*
Laundry: 2 Washer(s), 2 Dryer(s), Iron Board **Pay Phones:** No
Pump-Out: OnSite, Self Service **Fee:** Free **Closed Heads:** Yes

Marina Operations
Owner/Manager: Gary & Lorna Hink **Dockmaster:** Same
In-Season: MemDay-LabDay, 8am-6pm **Off-Season:** Sep-May, by appt.
After-Hours Arrival: Call ahead
Reservations: No **Credit Cards:** Visa/MC, Amex, Chevron
Discounts: 200+ gal.* **Dockage:** n/a **Fuel:** $0.05 fuel **Repair:** n/a
Pets: Welcome **Handicap Access:** No

Marina Services and Boat Supplies
Services - Security, Dock Carts **Communication** - Phone Messages, Fax in/out **Supplies - OnSite:** Ice *(Block, Cube, Shaved)*, Ships' Store, Bait/Tackle, Propane

Boatyard Services
OnCall: Bottom Cleaning, Divers, Compound, Wash & Wax, Upholstery
Nearest Yard: Pickering Marine (360) 427-7876

Restaurants and Accommodations
3+ mi: Restaurant *(Spencer Lake Resort Restaurant 426-2505, B $4-8, L $7-27, D $7-27, 7 mi.)*, *(Sierra's Restaurant 432-2777, 15 mi.)*, Fast Food *(Taco Bell, Burger King 17 mi.)*, Pizzeria *(Pizza Hut 426-2700, 15 mi.)*, Motel *(Shelton Super 8 526-1654, $48-65, 15 mi.)*, Inn/B&B *(Shelton Inn 426-4468, 15 mi.)*

Recreation and Entertainment
OnSite: Beach, Picnic Area, Grills **Near:** Jogging Paths, Park *(Jarrell Cove State Park 426-9226, dinghy)* **3+ mi:** Golf Course *(Bayshore Golf Course 426-1271, 12 mi.)*, Fitness Center *(Shelton Athletic Club 426-1388, 15 mi.)*, Bowling *(Timber Bowl 426-8452, 15 mi.)*, Video Rental *(Deer Creek Store 426-3671, 10 mi.)*, Galleries *(Harstine Island Gallery 426-8840, 6 mi.)*

Provisioning and General Services
OnSite: Convenience Store, Wine/Beer **1-3 mi:** Farmers' Market *(Harstine Community Hall Sat 10am-Noon 427-1017)* **3+ mi:** Market *(Spencer Lake Grocery 426-1901, 6 mi.)*, Supermarket *(Safeway 426-9978, 15 mi.)*, Delicatessen *(Mickey's Grocery & Deli 426-7662, 15 mi.)*, Health Food *(Nature's Best Natural Foods 426-7474, 15 mi.)*, Liquor Store *(Shelton Liquor 427-2168, 15 mi.)*, Bank/ATM *(Bank of America 426-8295, 15 mi.)*, Library *(Timberland 426-1362, 15 mi.)*, Dry Cleaners *(15 mi.)*, Laundry *(Shelton Laundry and Cleaners 426-4812, 15 mi.)*, Bookstore *(Sage Book Store 426-6011, 15 mi.)*, Pharmacy *(Money Savers 426-2666, 15 mi.)*, Newsstand *(Ace 426-2411, 15 mi.)*, Department Store *(Wal-Mart 427-6226, 17 mi.)*

Transportation
Under 1 mi: Local Bus *(Mason County Transit Dial A Ride 427-5033)*
3+ mi: Rental Car *(Practical Rent-A-Car 426-5553, 15 mi.)*, Taxi *(Shelton Cab 426-9222, 15 mi.)* **Airport:** Bremerton Nat'l./Olympia Regional *(42 mi.)*

Medical Services
911 Service **OnCall:** Ambulance *(Mason County Fire Department 426-3348)* **3+ mi:** Doctor *(Olympic Physicians 426-2500, 17 mi.)*, Dentist *(Shelton Family Dentistry 426-2631, 16 mi.)* **Hospital:** Mason General 426-1611 *(17 mi.)*

PHOTOS ON DVD: 7

Setting -- At the northwest end of Harstine Island, Jarrell Cove is carved into small, quiet Case Inlet and surrounded by densely wooded hills. The setting is idyllic. Well-manicured landscape greets visitors as they arrive at the marina. The network of well-kept, uncluttered docks is connected to the shore by a pier lined with small potted trees and attractive railings. Picnic tables with grills sit under old pines, backed by a couple of modest houses and the store.

Marina Notes -- *Fuel discount of $0.05 for purchases over 200 gal. with cash, check, or Chevron card. Guests are asked to stop by the covered fuel dock for slip assignment. Family-owned and operated for over 20 years, Jarrell's Cove Marina is a beautiful place to relax and enjoy the parklike setting. The on-site store has ice, marine hardware, bait & tackle, propane, books, fishing licenses and limited groceries, including beer. A pump-out station is at the easy-access fuel dock. The service-oriented owners and staff are available to help with whatever you need. An RV parking area at the top of the hill is hidden in the woods. The tiled, inviting bathhouse and laundry include a book exchange.

Notable -- This entire area is an appealing, peaceful place to kick back and watch the parade go by. The rock/gravel beach is not comfortable for sunbathing, but don an old pair of tennis shoes and enjoy the water. Get a fishing license and try your luck. Enjoy the dedicated picnic area under the trees for a family barbeque. Or dinghy directly across the inlet to one of Jarrell Cove State Park's two docks. (note: the small float on the park's northwest side goes aground at low tide). Most of the island is completely rural - undeveloped and covered by timber farms. Services are 15 miles away in Shelton.

Navigational Information
Lat: 47°16.920' **Long:** 122°53.220' **Tide:** n/a **Current:** n/a **Chart:** 18448
Rep. Depths *(MLW)*: **Entry** 30 ft. **Fuel Dock** n/a **Max Slip/Moor** 10 ft./-
Access: Located on the northwestern side of Harstine Island

Marina Facilities *(In Season/Off Season)*
Fuel: No
Slips: 16 Total, 16* Transient **Max LOA:** 60 ft. **Max Beam:** n/a
  Rate *(per ft.)*: **Day** $0.50* **Week** n/a **Month** n/a
  Power: 30 amp n/a, **50 amp** n/a, **100 amp** n/a, **200 amp** n/a
Cable TV: No **Dockside Phone:** No
Dock Type: Floating, Alongside, Wood
Moorings: 14 Total, 14 Transient **Launch:** None, Dinghy Dock
  Rate: Day $10 **Week** n/a **Month** n/a
Heads: Toilet(s), Shower(s)
Laundry: None **Pay Phones:** No
Pump-Out: OnSite, Self Service **Fee:** Free **Closed Heads:** Yes

Marina Operations
Owner/Manager: WA State Parks & Recreation **Dockmaster:** n/a
In-Season: May-Sep, 6:30am-10pm **Off-Season:** Oct-Apr, 8am-5pm
After-Hours Arrival: n/a
Reservations: No **Credit Cards:** No
Discounts: None
Pets: Welcome **Handicap Access:** No

Jarrell Cove Marine State Park

391 East Wingert Road; Shelton, WA 98584

Tel: (360) 426-9226 **VHF: Monitor** n/a **Talk** n/a
Fax: n/a **Alternate Tel:** n/a
Email: n/a **Web:** www.parks.wa.gov
Nearest Town: Shelton *(17 mi.)* **Tourist Info:** (360) 426-2021

Marina Services and Boat Supplies
Supplies - Under 1 mi: Ice *(Block, Cube, Shaved)*, Ships' Store *(Jarrel's Cove Marina 426-8826, dinghy)*, Bait/Tackle, Propane

Boatyard Services
OnSite: Launching Ramp *($5)* **Nearest Yard:** Pickering Marine (360) 427-7876

Restaurants and Accommodations
3+ mi: Restaurant *(Spencer Lake Resort Restaurant 426-2505, B $5-8, L $5-17, D $5-17, 7 mi.)*, *(Pine Tree 426-2604, 16 mi.)*, Pizzeria *(Pizza Hut 426-2700, 16 mi.)*, Motel *(Shelton Super 8 526-1654, $48-55, 16 mi.)*, Inn/B&B *(Shelton Inn 426-4468, 16 mi.)*

Recreation and Entertainment
OnSite: Beach, Picnic Area *(4 sheltered and 10 unsheltered tables plus 2 kitchen shelters - 888-CAMPOUT to reserve.)*, Grills, Volleyball, Park
Under 1 mi: Jogging Paths **3+ mi:** Golf Course *(Bayshore Golf Course 426-1271, 12 mi.)*, Fitness Center *(Shelton Athletic Club 426-1388, 16 mi.)*, Bowling *(Timber Bowl 426-8453, 16 mi.)*, Movie Theater *(Shelton Cinemas 426-1000, 16 mi.)*, Video Rental *(Deer Creek Store 426-3671, 10 mi.)*, Museum *(Mason County Historical Society 426-1020, 16 mi.)*, Galleries *(Harstine Island Gallery 426-8840, 6 mi.)*

Provisioning and General Services
Under 1 mi: Convenience Store *(Jarrell's Cove Marina)*, Wine/Beer **1-3 mi:** Farmers' Market *(Harstine Community Hall 10am-Noon Sat)* **3+ mi:** Market *(Spencer Lake Grocery 426-7662, 7 mi.)*, Supermarket *(Safeway 426-9978, 16 mi.)*, Health Food *(Nature's Best Natural Foods 426-7474, 16 mi.)*, Liquor Store *(Shelton Liquor 427-2168, 16 mi.)*, Bank/ATM *(16 mi.)*, Library *(Timberland Library 426-1362, 16 mi.)*, Dry Cleaners *(16 mi.)*, Laundry *(Shelton Laundry and Cleaners 426-4812, 16 mi.)*, Bookstore *(Sage Book Store 426-6011, 16 mi.)*, Pharmacy *(Money Savers 426-2666, 16 mi.)*, Hardware Store *(Ace Hardware 426-2411, 16 mi.)*, Department Store *(Wal-Mart 427-6226, 18 mi.)*

Transportation
Under 1 mi: Local Bus *(Mason County Transit Dial A Ride 427-5033)* **3+ mi:** Rental Car *(Practical Rent-A-Car 426-5553, 16 mi.)*, Taxi *(Shelton Cab 426-9222, 16 mi.)* **Airport:** Bremerton Nat'l./Olympia Regional *(42 mi.)*

Medical Services
911 Service **OnCall:** Ambulance *(Mason County Fire Department 426-3348)* **3+ mi:** Doctor *(Olympic Physicians 426-2500, 16 mi.)*, Dentist *(Shelton Family Dentistry 426-2631, 16 mi.)* **Hospital:** Mason General 426-1611 *(16 mi.)*

Setting -- Access to Jarell Cove is off Case Inlet, via Pickering Passage. Harstine Island is quiet and rustic, and most visitors come here for the solitude. 43-acre Jarrell Cove Marine State Park is located on the east side of the inlet, across the cove from Jarrell's Cove Marina and features 3500 feet of shoreline. The view toward land is of deep forests that hide 22 campsites. The state park has two very nice wooden docks with steel gangways: one at its northwest end, and a second, larger one farther south. Together they provide 682 feet of side-tie moorage. Fourteen mooring balls float in the quiet cove.

Marina Notes -- *$10 minimum. Alongside side-tie moorage and mooring buoys. Annual permits are available at the park - $3.50/ft. with a $50 minimum. Daily permits are also available. Pump-out and porta-potty dump sites are available at the park. Fuel, groceries and some boat supplies are at Jarrell's Cove Marina. Pets not permitted on beach and must be leashed. Note: the small float on the park's northwest side goes aground at low tides.

Notable -- There are campsites near the docks (1 ADA accessible) as well as in interior grassy clearings - $15-21 per night - plus four sheltered and ten uncovered picnic tables, two fire-circles and two kitchen shelters (reservable). Most campers arrive by boat. The park beach is not suitable for swimming but the waters are popular for scuba diving - there is a marine trail. Badminton and volleyball courts and horseshoe pits are available but do not include equipment). Mile-long hiking trails offer bird-watching opportunities for novices and experts alike. Common feathered residents iinclude Bald Eagles, Red-tailed Hawks, Pigeon Guillemot colonies, Glaucous-Winged Gulls, Canada Geese, and Great Blue Herons. Pick up a bird list at the park; it also offers hints for better birding.

Fair Harbor Marina

Fair Harbor Marina

PO Box 160; 5050 E Grapeview Loop Rd.; Grapeview, WA 98546

Tel: (360) 426-4028 **VHF: Monitor** n/a **Talk** n/a
Fax: n/a **Alternate Tel:** n/a
Email: info@fairharbormarina.us **Web:** www.fairharbormarina.us
Nearest Town: Grapeview *(1 mi.)* **Tourist Info:** (360) 275-1001

Navigational Information
Lat: 47°20.020' **Long:** 122°49.800' **Tide:** 14 ft. **Current:** n/a **Chart:** 18448
Rep. Depths (MLW): Entry 25 ft. **Fuel Dock** 25 ft. **Max Slip/Moor** 25 ft./-
Access: West side of Case Inlet north of Stretch Island

Marina Facilities *(In Season/Off Season)*
Fuel: *Chevron* - Gasoline
Slips: 78 Total, 9 Transient **Max LOA:** 120 ft. **Max Beam:** n/a
  Rate *(per ft.):* **Day** $0.60 **Week** n/a **Month** $6
  Power: 30 amp $4*, **50 amp** n/a, **100 amp** n/a, **200 amp** n/a
  Cable TV: No **Dockside Phone:** No
  Dock Type: Floating, Pilings, Wood
Moorings: 0 Total, 0 Transient **Launch:** n/a
  Rate: Day n/a **Week** n/a **Month** n/a
Heads: 2 Toilet(s), 2 Shower(s) *(with dressing rooms)*, Hair Dryers
Laundry: None, Book Exchange **Pay Phones:** No
Pump-Out: No **Fee:** n/a **Closed Heads:** Yes

Marina Operations
Owner/Manager: Susan & Vern Nelson **Dockmaster:** Same
In-Season: May-Sep, 8am-7pm **Off-Season:** Oct-Apr, by appt.
After-Hours Arrival: Tie up and pay in the morning
Reservations: Yes **Credit Cards:** Visa/MC, Dscvr, Din, Amex, Chevron
Discounts: None
Pets: Welcome, Dog Walk Area **Handicap Access:** Yes, Heads, Docks

Marina Services and Boat Supplies
Services - Docking Assistance, Trash Pick-Up, Dock Carts
Communication - Mail & Package Hold, Phone Messages **Supplies -**
OnSite: Ice *(Cube)*, Propane

Boatyard Services
OnSite: Launching Ramp **OnCall:** Divers **Nearest Yard:** Pickering Marine
(360) 427-7876

Restaurants and Accommodations
OnSite: Coffee Shop *(Expresso Bar - coffee concoctions & Italian sodas)*
3+ mi: Restaurant *(Leonard K's Restaurant and Lounge 275-6060, 4 mi.)*,
(Allyn Inn 275-5422, 5 mi.), *(Lakeside Bistro 275-5603, 5 mi.)*, Lite Fare
(Big Bubba's Burgers 275-6000, 5 mi.), Motel *(Belfair 275-4485, 10 mi.)*

Recreation and Entertainment
OnSite: Beach, Picnic Area, Grills, Special Events *(Grapeview Water
Festival and Wooden Boat Show - last Sat in Jul)* **1-3 mi:** Museum
*(Maritime Museum of Puget Sound 858-7971 - Housed in an old winery, the
museum has relics from the Mosquito Fleet)* **3+ mi:** Golf Course *(Lakeland
Village Golf Course 275-6100, free shuttle, 5 mi.)*

Provisioning and General Services
OnSite: Convenience Store *(Floating Country Store)*, Wine/Beer
Near: Post Office *(275-8553)* **1-3 mi:** Beauty Salon *(Gwen's Special Touch
275-3161)* **3+ mi:** Market *(Market Place 275-3826, 5 mi.)*, Supermarket *(QFC
275-9672, 10 mi.)*, Delicatessen *(Ben's Deli Mart 275-3798, 5 mi.)*, Health
Food *(Dragoun's Leir 277-9434, 10 mi.)*, Liquor Store *(WA Liquor 275-3512,
5 mi.)*, Farmers' Market *(Belfair Elementary School 275-0616 May-Oct Sat
9am-3pm, 10 mi.)*, Bank/ATM *(Kitsap Bank 874-7114, 5 mi.)*, Library *(North
Mason Timberland Library 275-3232, 10 mi.)*, Barber Shop *(Lois' Barber
Shoppe 275-5766, 5 mi.)*, Laundry *(Belfair Maytag Laundry 277-0500, 10
mi.)*, Pharmacy *(Safeway 275-0953, 10 mi.)*, Hardware Store *(True Value
275-0113, 10 mi.)*, Copies Etc. *(Copy It Mail It 277-0652, 10 mi.)*

Transportation
OnSite: Courtesy Car/Van *(to Lakeland Village Golf Course and Museum of
Puget Sound)* **Airport:** Bremerton Nat'l. *(14 mi.)*

Medical Services
911 Service **3+ mi:** Doctor *(North Mason Medical Clinic 275-4084, 10 mi.)*,
Dentist *(Belfair Dental Clinic 275-2855, 10 mi.)*, Chiropractor *(Nelson
Chiropractic 275-0670, 5 mi.)*, Veterinarian *(Belfair Animal Clinic 275-6008,
10 mi.)* **Hospital:** Mason General 426-1611 *(17 mi.)*

Setting -- Elegant Fair Harbor Marina sits on the west side of Case Inlet, just north of Stretch Island. Upscale docks sprawl along the shore anchored by the attractive floating Country Store and Espresso Bar - topped by a French blue hipped roof. A blue gangway leads to a stationary pier edged with white railings; colorful pots of flowers hang from the lampposts. Onshore, the landscaping and new pavilion are wedding-venue quality - which this is.

Marina Notes -- *350 feet of guest moorage. Monthly $6/ft. + $20 power. Covered $272 for 38' & $360 for 50'. **20 & 30 amp power. Family operated marina; attention to detail is visible everywhere. Can accomodate boats to 120 ft.. Well-kept, secure docks feature a live camera to permit tenants to check on their boats via the Internet. Fuel is on-site, and minor repairs can be arranged. The Country Store offers sundries, wine and beer, ice, marine supplies, bait, plus a wide selection of tasteful, nautically-themed gifts, and small furniture pieces. An espresso bar gives a perfect excuse to linger at the tables outside. At the top of the dock is an immaculate, energy efficient bathhouse ($0.25/min) with handicap-accessible showers.

Notable -- Tucked away, with pretty grounds and upscale facilities, this is a popular destination for special events. A plush new pavilion - complete with chandeliers, an outdoor kitchen, and serving buffet - accommodates 100 seated. Designed for weddings and other large affairs, complimentary use is extended to club cruises booking a minimum of 1,000 ft. of prepaid moorage (check the website for details). Boaters are offered courtesy transportation to the 27-hole Lakeland Village Golf Course in Allyn and, with prior reservation, to the Museum of Puget Sound on Stretch Island. Downtown Allyn is a short dinghy ride north.

Navigational Information
Lat: 47°23.037' **Long:** 122°49.607' **Tide:** 13 ft. **Current:** n/a **Chart:** 18448
Rep. Depths (MLW): Entry 6 ft. **Fuel Dock** n/a **Max Slip/Moor** 6 ft./-
Access: North on Case Inlet almost to the end, on west side

Marina Facilities *(In Season/Off Season)*
Fuel: No
Slips: 10 Total, 10 Transient **Max LOA:** 40 ft. **Max Beam:** n/a
  Rate *(per ft.):* **Day** $0.33* **Week** n/a **Month** n/a
  Power: 30 amp n/a, **50 amp** n/a, **100 amp** n/a, **200 amp** n/a
  Cable TV: No **Dockside Phone:** No
  Dock Type: Floating, Alongside, Wood
Moorings: 0 Total, 0 Transient **Launch:** n/a
  Rate: Day n/a **Week** n/a **Month** n/a
Heads: 2 Toilet(s)
Laundry: None **Pay Phones:** No
Pump-Out: OnSite **Fee:** $2 - in quarters **Closed Heads:** Yes

Marina Operations
Owner/Manager: Bonnie Knight **Dockmaster:** John Williams
In-Season: Year-Round, 11am-4:30pm** **Off-Season:** n/a
After-Hours Arrival: Dockage open anytime. Pay box provided
Reservations: No **Credit Cards:** No
Discounts: None
Pets: Welcome, Dog Walk Area **Handicap Access:** No

Port of Allyn Waterfront Park

PO Box 1; 18560 East State Route 3; Allyn, WA 98524

Tel: (360) 275-2430 **VHF: Monitor** n/a **Talk** n/a
Fax: (360) 275-2455 **Alternate Tel:** (360) 275-8436
Email: portofallyn@aol.com **Web:** www.portofallyn.com
Nearest Town: Allyn *(0 mi.)* **Tourist Info:** (360) 275-1001

Marina Services and Boat Supplies
Supplies - Near: Ice *(Cube)* **3+ mi:** Bait/Tackle *(McLendon's 275-0113, 5 mi.),* Propane *(4 mi.)*

Boatyard Services
OnSite: Launching Ramp *(double - $3)* **Nearest Yard:** Pickering Marine (360) 427-7876

Restaurants and Accommodations
Near: Restaurant *(Leonard K's Restaurant & Lounge 275-6060),* Lite Fare *(Big Bubba's Burgers 275-6000)* **Under 1 mi:** Restaurant *(Lakeside Bistro 275-5603, 4pm-9pm - near golf course)* **3+ mi:** Motel *(Belfair Motel 275-4485, 5 mi.),* Inn/B&B *(Shelah Inn 275-0916, $160-175, 9 mi., free transportation to and from the dock with prior arrangement), (Cady Lake Manor 372-2673, $125-185, 19 mi., free transportation to and from the dock with prior arrangement)*

Recreation and Entertainment
OnSite: Beach, Picnic Area, Playground, Park, Special Events *(Allyn Days - 3rd wknd Jul, Chainsaw Carving Contest, 3 annual art shows at Port building - call for info)* **Near:** Video Rental *(or Blockbuster 275-1249, 5 mi.)* **Under 1 mi:** Golf Course *(Lakeland Village Golf Course 275-6100 Wkdys $22/27 holes, wknds $27/18 holes)* **1-3 mi:** Galleries *(The Old Cedar Forge 275-6769 - blacksmithing demos; George Kenny School of Chainsaw Carving 888-775-7414 - classes)* **3+ mi:** Sightseeing *(The Theler Wetlands Center 275-4898, 4 mi.)*

Provisioning and General Services
Near: Convenience Store *(Ben's Deli Mart 275-3798),* Market *(Market Place 275-3826),* Delicatessen *(Ben's Deli-Mart 275-3798),* Wine/Beer, Liquor Store *(WA Liquor 275-3512),* Bank/ATM *(Kitsap Bank 874-7114),* Post Office *(275-6504),* Protestant Church *(St. Hugh's Episcopal 275-8450, Allyn Baptist Church 275-6992),* Barber Shop *(Lois' Barber Shoppe 275-5766),* Retail Shops **1-3 mi:** Beauty Salon *(Gwen's Special Touch 275-3161),* Florist *(Paul's Flowers 277-3355)* **3+ mi:** Supermarket *(QFC 275-9672 / Safeway 275-0953, 5 mi.),* Health Food *(Dragon's Leir 277-9434, 5 mi.),* Bakery *(Village Bakery & Cafe 275-0579, 5 mi.),* Farmers' Market *(Belfair Elementary School Sat May-Oct 9am-3pm 275-0616, 5 mi.),* Catholic Church *(Prince of Peace 275-8760, 7 mi.),* Library *(North Mason Timberland Library 275-3232, 5 mi.),* Dry Cleaners *(Belfair Cleaners 277-0500, 5 mi.),* Laundry *(Belfair Maytag Laundry 277-0500, 5 mi.),* Pharmacy *(Safeway 275-0953, 5 mi.),* Hardware Store *(True Value 275-0113, 5 mi.),* Copies Etc. *(Copy It Mail It 277-0652, 5 mi.)*

Transportation
Near: Local Bus *(Mason County Dial-A-Ride 427-5033, Free)* **3+ mi:** Ferry Service *(Bremerton - Seattle, 14 mi.)* **Airport:** Bremerton Nat'l. *(9 mi.)*

Medical Services
911 Service **OnCall:** Ambulance **Near:** Chiropractor *(Nelson Chiropractic 275-0670)* **3+ mi:** Doctor *(North Mason Medical Clinic 275-4084, 5 mi.),* Dentist *(Belfair Dental Clinic 275-2855, 5 mi.),* Veterinarian *(Belfair Animal Clinic 275-6008, 5 mi.)* **Hospital:** Mason General 275-8614 *(5 mi.)*

Setting -- The village of Allyn is near the head of Case Inlet, on the west side, about a half mile north of Sherwood Creek. A gangway leads from the float to a long stationary pier that ends at Allyn Waterfront Park. Arbors, draped with climbing roses and fragrant honeysuckle vines, lead to a large, fetching shingle-roofed pavilion edged with a dozen round, concrete tables and benches. Rhododendrons abound and oyster beds cover the flats.

Marina Notes -- *Same flat rates in-season/off-season: under 25' $8, 26-34' $10, over 35' $12. **Off-season open Mon, Wed, and Fri. (use pay box when closed) Two-week limit during the summer, longer stays possible off-season. A long wooden pier and an 80-foot gangway make the 10 slips easily accessible during all tides. At minus 2.5' tide, there is 4 feet of water on the finger pier, and at least 2 feet of water at all points around the floats. There is no power; a marine pump-out is on the dock, and two boat ramps are on-site. The current heads consist of two porta-potties.

Notable -- The Park stretches along the shore between the pier and the launching ramps. A long, narrow beach is bordered by a walkway that passes the pavilion (reservable for group functions), a playground with jungle gym, and acres of well-trimmed lawn sheltered by mature shade trees and studded with picnic tables. Several businesses, including a bank, restaurant, grocery store, liquor store and medical clinic are within a block. The shallow waters are great for kayaking-- Native American petroglyphs can be seen at low tide. A quarter mile outside of town is the popular, affordable golf course at Lakeland Village (27 holes cover 9,000 yards). Theler Center, at the end of the Hood Canal, has an excellent wetland trail system and interpretive center - reachable by bus.

Port of Shelton

PO Box 2270; 701 E Pine Street; Shelton, WA 98584

Tel: (360) 426-9476 **VHF: Monitor** n/a **Talk** n/a
Fax: (360) 427-0231 **Alternate Tel:** n/a
Email: n/a **Web:** www.sheltonyachtclub.com
Nearest Town: Shelton *(0.5 mi.)* **Tourist Info:** (360) 426-2021

Navigational Information
Lat: 47°12.850' **Long:** 123°05.100' **Tide:** 20 ft. **Current:** 5 kt. **Chart:** 18457
Rep. Depths *(MLW):* Entry 13 ft. **Fuel Dock** n/a **Max Slip/Moor** 13 ft./13 ft.
Access: Squaxin Passage to nnd of Hammersley Inlet

Marina Facilities *(In Season/Off Season)*
Fuel: No
Slips: 106 Total, 6 Transient **Max LOA:** 45 ft. **Max Beam:** n/a
  Rate *(per ft.):* **Day** $0.25 **Week** $1.75 **Month** $2.55*
  Power: 30 amp $3, **50 amp** n/a, **100 amp** n/a, **200 amp** n/a
  Cable TV: No **Dockside Phone:** No
  Dock Type: Floating, Long Fingers, Alongside, Concrete, Wood
Moorings: 0 Total, 0 Transient **Launch:** n/a, Dinghy Dock
  Rate: Day n/a **Week** n/a **Month** n/a
Heads: 1 Toilet(s)
Laundry: None **Pay Phones:** Yes, 1
Pump-Out: OnSite, Self Service, 1 Central **Fee:** Free **Closed Heads:** Yes

Marina Operations
Owner/Manager: Port of Shelton **Dockmaster:** Mike Byrne
In-Season: Year-Round **Off-Season:** n/a
After-Hours Arrival: Self check-in
Reservations: No **Credit Cards:** No; Cash only
Discounts: None
Pets: Welcome, Dog Walk Area **Handicap Access:** No

Marina Services and Boat Supplies
Services - Trash Pick-Up **Communication -** FedEx, DHL, UPS **Supplies - Under 1 mi:** Ice *(Block, Cube)* **1-3 mi:** Bait/Tackle, Propane *(Wal-Mart 427-0500)* **3+ mi:** Ships' Store *(Verle's Marine Center 426-0933, 4 mi.)*

Boatyard Services
OnCall: Divers, Propeller Repairs, Yacht Broker **1-3 mi:** Engine mechanic *(gas, diesel).* **Nearest Yard:** Swantown Boatworks (360) 528-8049

Restaurants and Accommodations
Under 1 mi: Restaurant *(El Sarape 426-4294), (Timbers 426-8757), (Kobe Teriyaki 432-0533), (Pine Tree 426-2604, American), (Sierra's Family Restaurant 432-2777), (Steven's Fine Dining 426-4407),* Seafood Shack *(Xinh's Clam and Oyster House 427-8709),* Fast Food *(Dairy Queen),* Lite Fare *(Travaglione's Italian Deli 427-3844),* Pizzeria *(Pizza Hut 426-2700), (Domino's 427-8700),* Motel *(Super 8 526-1654, $50-65), (City Center 426-3397),* Inn/B&B *(Shelton Inn 426-4468)*

Recreation and Entertainment
OnSite: Jogging Paths **Under 1 mi:** Movie Theater *(Shelton Cinemas 426-1000),* Video Rental *(Hollywood Video 427-6188),* Park *(Kneeland Park),* Museum *(Mason County Historical Museum 426-1020; Squaxin Island Museum - Native American history 432-3839, 8 mi.),* Special Events *(Mason Co. Forest Festival, Heritage Festival early Jun; Music in the Park at Post Office Park - Thu evenings Jul & Aug; Oysterfest early Oct)* **1-3 mi:** Fitness Center *(Shelton Athletic Club 426-1388),* Bowling *(Timber Bowl 426-8253)* **3+ mi:** Golf Course *(Bayshore Golf Course 426-1271, 4 mi.)*

Provisioning and General Services
Under 1 mi: Convenience Store *(Shop & Hop 427-1277),* Market, Supermarket *(Safeway 426-9978),* Delicatessen *(Mickey's Grocery & Deli 426-7662),* Health Food *(Nature's Best Natural Foods 426-7474),* Wine/Beer, Liquor Store *(Shelton Liquor 427-2168),* Bakery *(Olympic Bakery 462-2253),* Bank/ATM, Post Office *(426-1476),* Catholic Church, Protestant Church, Library *(William Reed 426-1362),* Beauty Salon *(A Hair Affair 426-2141),* Barber Shop *(Abstract Barber 426-6114),* Dry Cleaners *(Kneeland Plaza Cleaners 426-2607),* Laundry *(Shelton Laundry & Cleaners 426-4812),* Bookstore *(Sage Book Store 426-6011),* Pharmacy *(Money Saver 426-2666; Neil's Pharmacy 426-3327),* Newsstand, Hardware Store *(Tozier's True Value 426-2411),* Florist *(Ferguson Flowers 426-8502),* Clothing Store, Retail Shops, Copies Etc. *(Office Supply 426-6102)* **1-3 mi:** Fishmonger *(Oyster Dave's Seafood 427-1648),* Department Store *(Wal-Mart 427-0500)*

Transportation
OnCall: Taxi *(Shelton Cab 426-9222)* **1-3 mi:** Rental Car *(Practical Rent-a-Car 426-5553 /Enterprise Olympia 956-3714, 20 mi.),* Local Bus *(Mason County Transportation 427-5033)* **3+ mi:** Rail *(Amtrak Olympia 923-4602, 20 mi.)* **Airport:** Olympia Regional *(20 mi.)*

Medical Services
911 Service **OnCall:** Ambulance **1-3 mi:** Doctor *(Health Care Center 426-9717),* Dentist *(Shelton Family Dentristy 426-2631),* Chiropractor *(Sound View Chiropractic Center 427-9013),* Holistic Services, Veterinarian *(Shelton Veterinary Hospital 426-2616),* Optician **Hospital:** Mason General 426-1611 *(2 mi.)*

Setting -- The town of Shelton is on the western side of Oakland Bay at the end of seven mile long Hammersley Inlet. The marina's small network of concrete docks includes a mix of open slips, boathouses, and side-tie space. On the shore is Shelton Yacht Club's light blue home - with its sixties-style slightly pitched roof. Surrounding the Port is a fairly industrial upland, including an adjacent lumber operation.

Marina Notes -- *Monthly off-season rate; minimum: $78. This quiet little marina, owned by the Port of Shelton and run by the Shelton Yacht Club, has permanent and visitor moorage and emergency haulouts. Pump-out has alongside tie-up with a life ring near the pump. Use the pay station at the top of the ramp. The head is a porta-potty near the ramp. Note: The inlet is obstructed by shoals, particularly at its mouth.

Notable -- The Oakland Bay region and small town of Shelton lie in the shadow of the Olympia National Forest. The community is rooted in logging and supported by sawmills, Christmas tree farming, and shellfish harvesting - plywood and oysters. Some of that story is told at the Shelton/Mason County Historical Museum. Initially built in 1914 as the town hall, it is now on the National Register of Historic Places. It is downtown (a half mile) at Fifth Street and Railroad Avenue where one can also find marine supplies, restaurants, and hotels. Annual events include Shelton Old Time Fiddle Fest, held the 1st weekend in April since 1985. In June, The Mason County Forest Festival features a carnival, logging show, Fun Run/Walk, fireworks, a car show and the Paul Bunyan Parade. Music in the Park in July and August on Thursday evenings. Oysterfest in October is a major event in the community (427-6959).

Navigational Information
Lat: 47°08.401' **Long:** 122°54.308' **Tide:** 15 ft. **Current:** n/a **Chart:** 18448
Rep. Depths *(MLW)*: **Entry** 12 ft. **Fuel Dock** n/a **Max Slip/Moor** 12 ft./12 ft.
Access: Puget Sound south, off Dofflemyer Point at entrance to Budd Inlet

Marina Facilities *(In Season/Off Season)*
Fuel: *Chevron* - Gasoline, Diesel
Slips: 110 Total, 10 Transient **Max LOA:** 50 ft. **Max Beam:** n/a
  Rate *(per ft.)*: **Day** $0.35* **Week** $1.75 **Month** $6.25
  Power: 30 amp $2**, **50 amp** n/a, **100 amp** n/a, **200 amp** n/a
  Cable TV: No **Dockside Phone:** No
  Dock Type: Floating, Long Fingers, Wood
Moorings: 2 Total, 2 Transient **Launch:** None, Dinghy Dock (Free)
  Rate: Day $0.35 **Week** $1.75 **Month** $5.50
Heads: 2 Toilet(s)
Laundry: None, Book Exchange **Pay Phones:** Yes
Pump-Out: No **Fee:** n/a **Closed Heads:** Yes

Marina Operations
Owner/Manager: Pam McHugh **Dockmaster:** Don McHugh
In-Season: Jun-Sep, 8am-7pm **Off-Season:** Oct-May, 9am-6pm
After-Hours Arrival: Call ahead
Reservations: Yes **Credit Cards:** Visa/MC, Dscvr, Din, Amex, Chevron
Discounts: None
Pets: Welcome, Dog Walk Area **Handicap Access:** Yes, Heads

Boston Harbor Marina

312 73rd Avenue NE; Olympia, WA 98506

Tel: (360) 357-5670 **VHF: Monitor** Ch. 16 **Talk** Ch. 68
Fax: (360) 352-2816 **Alternate Tel:** n/a
Email: bhm@bostonharbormarina.com **Web:** bostonharbormarina.com
Nearest Town: Olympia (8 mi.) **Tourist Info:** (360) 357-3352

Marina Services and Boat Supplies
Services - Docking Assistance, Trash Pick-Up, Dock Carts
Communication - Phone Messages, Fax in/out *($2/pg)*, FedEx, DHL, UPS, Express Mail **Supplies - OnSite:** Ice *(Block, Cube)*, Ships' Store, Bait/Tackle, Live Bait, CNG **1-3 mi:** Propane **3+ mi:** West Marine *(352-1244, 9 mi.)*, Boater's World *(754-1834, 10 mi.)*

Boatyard Services
OnSite: Launching Ramp *(Free)* **OnCall:** Inflatable Repairs, Life Raft Service **3+ mi:** Travelift *(77T, 6 mi.)*, Forklift *(6 mi.)*, Crane *(6 mi.)*, Yacht Broker *(Capital City Yachts 352-2007, 6 mi.)*. **Nearest Yard:** Swantown Boatworks (360) 528-8059

Restaurants and Accommodations
OnSite: Lite Fare *(Boston Harbor Marina 357-5670, deli in the store - sandwiches, soups, ice cream, coffee; Sun brunch)* **3+ mi:** Restaurant *(Budd Bay Café 357-6963, 8 mi.)*, Pizzeria *(Pizza Time 956-9020, 8 mi.)*, Motel *(Econo Lodge 943-4710, 8 mi.)*, Inn/B&B *(Lighthouse Bungalow 754-0389, $125-475, 6 mi.)*, *(Ramada Inn 352-7700, 8 mi.)*

Recreation and Entertainment
OnSite: Beach, Picnic Area, Grills, Boat Rentals *(day-long rentals: kayaks $45-60, sailboats $80-90, small runabouts $75, rowboats, canoes, pedal boats, hytro bikes all $25 plus hourly rentals, tours and sailing & kayaking classes)* **Near:** Jogging Paths, Roller Blade/Bike Paths, Sightseeing *(Doflemyer Point Lighthouse)* **Under 1 mi:** Playground, Park *(Burfoot - saltwater beach, hiking trails, playground)* **1-3 mi:** Video Rental *(Gull Harbor Mercantile 352-4014)*, Galleries *(Art by Beth Rowley 534-9159)*

3+ mi: Pool *(YMCA 357-6609, 8 mi.)*, Golf Course *(Turnwater Municipal 943-9500, 11 mi.)*, Bowling *(Westside Lanes 943-2400, 9 mi.)*, Museum *(Bigelow House Museum 534-9159, State Capital Museum 753-2580, 8 mi.)*

Provisioning and General Services
OnSite: Convenience Store *(with a Deli)*, Wine/Beer, Fishmonger *(oysters, clams, mussels, crab, salmon)* **1-3 mi:** Bank/ATM *(Gull Harbor Mercantile 352-4014)*, Post Office, Protestant Church, Florist *(Ivy Trellis 943-4726)* **3+ mi:** Market *(Steamboat General Store 866-0836, 4 mi.)*, Supermarket *(Olympia Food Co-op 357-1106, Safeway 943-0144, 8 mi.)*, Gourmet Shop *(Butterfields Etc. 943-9094, 8 mi.)*, Liquor Store *(8 mi.)*, Farmers' Market *(Olympia Farmer's Market 352-9096, 8 mi.)*, Meat Market *(Western Meat Co. 357-6601, 9 mi.)*, Library *(Olympia 352-0595, 8 mi.)*, Beauty Salon *(Donna's Beauty Salon 357-4681, 4 mi.)*, Dry Cleaners *(Capitol Cleaners & Laundry 943-6170, 8 mi.)*, Bookstore *(Fireside Book Store 352-4006, 7 mi.)*, Pharmacy *(Safeway 943-0144, 8 mi.)*, Hardware Store *(Home Depot 459-5538, 7 mi.)*, Retail Shops *(Capitol Mall 754-8017, 9 mi.)*

Transportation
OnCall: Rental Car *(Enterprise 736-8222)*, Taxi *(Capital City 357-4949)* **3+ mi:** Rail *(Amtrak 923-4602, 14 mi.)* **Airport:** Olympia Regional (13 mi.)

Medical Services
911 Service **OnCall:** Ambulance **3+ mi:** Doctor *(Capital Medical Center 754-5858, 8 mi.)*, Dentist *(Capitol Dental Center 943-9260, 8 mi.)*, Chiropractor *(Family Chiropractic Olympia 352-8112, 8 mi.)*, Veterinarian *(Healthy Pets Animal Hospital 943-8900, 8 mi.)* **Hospital:** Providence 491-9180 *(9 mi.)*

Setting -- Likeable Boston Harbor sits at the mouth of Budd Inlet; Dofflemyer Point Light marks its western shore. The view from the sea is of a well-kept, rural residential area with the marina's docks flowing out into the bay - backed, on a clear day, by the Olympic Mountains. At the top of the gangway, a blue building houses the marine store and cafe; its deck, furnished with umbrella-topped tables and chairs, is a gathering place for locals and boaters. A shaded grouping of Adirondack chairs sits dockside. Well-tended gardens and pots overflowing with blooms step up the whole atmosphere.

Marina Notes -- *$7/day min. **20 amp power. Family owned and operated, the marina offers friendly service and an intimate, community feel. Wooden docks have long finger piers and a fuel dock floats at the entrance. 24 hr. card lock fuel system. Since downtown (Olympia, 7.5 mi. south) is not readily accessible, the marina has tried to meet most basic needs. The store sells "logo" clothing, locally crafted gifts, nautical items, food, boat supplies, sandwiches, and fresh seafood. It's also the only CNG exchange south of Tacoma. Sailboats, small powerboats, rowboats, pedal boats and kayaks available for rent are perfect for exploring the bay. The dock and deck are handicap accessible, the store is not. Basic cinderblock public heads across the street; no showers.

Notable -- Want peace and quiet in a laid-back atmosphere? This is the place. Beautiful views of the water, cascades of flowers, and a sunny deck create a tranquil, relaxed setting. The store specializes in freshly caught clams, mussels, oysters, salmon, crab, and geoduck. Be sure to try their smoked salmon sandwiches and geoduck chowder. During the summer, Sunday brunch is served from 9-11am. An adjacent beach area is great for swimming.

West Bay Marina

West Bay Marina

2100 West Bay Drive NW; Olympia, WA 98502

Tel: (360) 943-2022; (800) 884-2080 **VHF: Monitor** n/a **Talk** n/a
Fax: (360) 753-4773 **Alternate Tel:** n/a
Email: neilfalkenburg@hotmail.com **Web:** www.westbay-marina.com
Nearest Town: Olympia *(1.5 mi.)* **Tourist Info:** (360) 357-3362

Navigational Information
Lat: 47°03.880' **Long:** 122°54.943' **Tide:** 17 ft. **Current:** 2 kt. **Chart:** 18456
Rep. Depths *(MLW):* **Entry** 10 ft. **Fuel Dock** n/a **Max Slip/Moor** 10 ft./-
Access: Puget Sound south to west side of Budd Inlet

Marina Facilities *(In Season/Off Season)*
Fuel: No
Slips: 400 Total, 8 Transient **Max LOA:** 70 ft. **Max Beam:** n/a
  Rate *(per ft.):* **Day** $0.40* **Week** n/a **Month** $6.85-7.45
  Power: 30 amp $5, **50 amp** n/a, **100 amp** n/a, **200 amp** n/a
  Cable TV: No **Dockside Phone:** No
  Dock Type: Floating, Long Fingers, Alongside, Wood
Moorings: 0 Total, 0 Transient **Launch:** n/a
  Rate: Day n/a **Week** n/a **Month** n/a
Heads: 5 Toilet(s), 6 Shower(s)
Laundry: 3 Washer(s), 3 Dryer(s) **Pay Phones:** Yes, 1
Pump-Out: OnSite, Full Service **Fee:** Free **Closed Heads:** Yes

Marina Operations
Owner/Manager: Jerry Barrufi **Dockmaster:** Neil Falkenburg
In-Season: Year-Round, 9am-5pm **Off-Season:** n/a
After-Hours Arrival: Call ahead
Reservations: Yes **Credit Cards:** Dscvr
Discounts: None
Pets: Welcome **Handicap Access:** No

Marina Services and Boat Supplies
Services - Dock Carts **Supplies - OnSite:** Ice *(Cube)* **Under 1 mi:** Ice *(Block)* **1-3 mi:** West Marine *(352-1244)*, Boater's World *(754-1834)*

Boatyard Services
1-3 mi: Crane, Launching Ramp, Electrical Repairs, Electronic Sales, Hull Repairs, Rigger, Bottom Cleaning, Brightwork, Divers, Compound, Wash & Wax, Interior Cleaning, Propeller Repairs, Woodworking, Metal Fabrication, Painting, Yacht Design, Yacht Broker *(Capital City Yachts 352-2007)*.
Nearest Yard: Swantown Boatworks (360) 528-8059

Restaurants and Accommodations
OnSite: Restaurant *(Tugboat Annie's 943-1850, B $5-10, L $6-12, D $11-19, B Sat-Sun, L & D daily; Kids' Menu $4-5 - Entertain.)* **Under 1 mi:** Restaurant *(Seven Gables 352-2349, waterfront)*, *(Rosey's on Rogers 352-1103)* **1-3 mi:** Restaurant *(Budd Bay Café 357-6963, L $7-13, D $14-31)*, *(Anthony's Homeport 357-9700, L $9-12, D $16-24)*, *(Koibito 352-4751, Japanese)*, *(Little Saigon 943-8013)*, Pizzeria *(Vic's Pizzeria 943-8044)*, Motel *(BW Aladdin Motor Inn 352-7200, $63-80)*, Hotel *(Red Lion 943-4000, $77-110)*, *(Phoenix Inn Suites 570-0555, $99-179)*, *(Ramada Inn 352-7700)*

Recreation and Entertainment
OnSite: Boat Rentals *(Pedalcraft & Annie's Kayaks)* **1-3 mi:** Fitness Center *(YMCA 357-6609)*, Jogging Paths, Bowling *(Westside Lanes 943-2400)*, Movie Theater *(Capitol Theater 754-5378)*, Video Rental *(Hollywood Video 357-9519)*, Park, Museum *(Hands-On Children's Museum 956-0818, Bigelow House Museum 753-1215)*, Cultural Attract *(WA Center for Performing Arts*

753-8585), Sightseeing *(Tours of the Capitol 586-TOUR)*, Special Events *(Wooden Boat Fair - May)*, Galleries *(The Artist's Gallery 357-6920)*
3+ mi: Golf Course *(Turnwater Municipal 943-9500, 5 mi.)*

Provisioning and General Services
Under 1 mi: Market *(Olympia Food Co-op 764-7666)*, Protestant Church **1-3 mi:** Supermarket *(Safeway 956-3782)*, Gourmet Shop *(Butterfields Etc. 943-9094)*, Delicatessen *(Bayview Deli and Bakery 352-48997)*, Wine/Beer, Liquor Store *(Olympia Liquor 586-2397)*, Bakery, Farmers' Market *(Capitol Way Tue-Sun 10am-3pm 352-9096)*, Fishmonger, Bank/ATM, Post Office, Catholic Church, Library *(Timberland 352-0595)*, Beauty Salon *(Charles Austin Salon 705-1306)*, Barber Shop *(Hair Care and Associates 352-7113)*, Dry Cleaners *(Pacific Cleaners 352-7016)*, Laundry *(Westside Laundry 943-3857)*, Bookstore *(Fireside Book Store 352-4006)*, Pharmacy *(Rite Aid Pharmacy 754-8014)*, Hardware Store *(True Value 754-6659)*, Florist *(House of Roses 754-3949)*, Department Store *(Westfield Mall 754-8017)*

Transportation
OnCall: Rental Car *(Enterprise 956-3714)*, Taxi *(Capital City Taxi 357-4949)* **1-3 mi:** Local Bus *(Intercity Transit 786-1881)*, InterCity Bus *(Greyhound 357-5541)* **3+ mi:** Rail *(Amtrak 923-4602, 10 mi.)*
Airport: Olympia Regional *(8 mi.)*

Medical Services
911 Service **OnCall:** Ambulance **1-3 mi:** Doctor *(West Olympia 753-7212)*, Dentist *(Flemmings Dental 705-4636)*, Chiropractor *(Bay Center Chiropractic 786-6322)*, Holistic Services *(Northwest Center - Natural Med 754-7775)*, Veterinarian **Hospital:** Providence St. Peter 491-9480 *(5 mi.)*

Setting -- Budd Inlet is at the southern-most point of Puget Sound; it is entered between Dofflemyer Point and the narrow Copper Point. West Bay Marina sits about 5 miles up the Inlet, on the western shore. It's easily recognizable by the extensive network of covered docks, and the blue-roofed, tan building that houses Tugboat Annie's second floor restaurant. On clear days, the Capitol Dome is visible to the south. Log booms and commercial facilities share the area.

Marina Notes -- *Overnight is flat rate - up to 30' $10; over 30' $15. Monthly rate is per foot - up to 35' $6.85/ft., over 35' $7.45/ft. Only occasionally has an overnight slip, the marina serves primarily long-term tennants, with a good mix of power and sailboats, plus some houseboats - with opened and covered slips and alongside space which can accommodate larger vessels. Power is on pedestals and a pump-out station is available. The marina owners converted over 2,000 ft. of retail space into a large conference room - "Viewoint" - with 3 walls of glass, making this a useful facility for gatherings and reunions. Kayak rentals are on-site. The inviting bathhouse and large laundry are on the first floor in the same building.

Notable -- The marina is located in a quiet, residential area with no nearby services. Upstairs, casual Tugboat Annie's Restaurant offers views of the docks and inlet, a nice bar area with a pool table, and a large deck for outside dining. It's open daily for lunch and dinner, with breakfast on Sat-Sun only (Happy Hour 3-6, M-F). They will also cater on-site functions. If you are looking for action, travel a mile and a half to Olympia where there is a wealth of entertainment options, plus restaurants, specialty shops, bookstores, antique shops, and art galleries. Many historic buildings also await visitors in Washinton's capital.

Navigational Information
Lat: 47°03.037' **Long:** 122°54.301' **Tide:** 20 ft. **Current:** n/a **Chart:** 18456
Rep. Depths *(MLW)*: **Entry** 10 ft. **Fuel Dock** n/a **Max Slip/Moor** 10 ft./-
Access: In West Bay at southern end of Budd Inlet

Marina Facilities *(In Season/Off Season)*
Fuel: No
Slips: 12 Total, 12 Transient **Max LOA:** 95 ft. **Max Beam:** n/a
  Rate *(per ft.)*: **Day** $0.30* **Week** n/a **Month** n/a
  Power: 30 amp n/a, 50 amp n/a, 100 amp n/a, 200 amp n/a
  Cable TV: No **Dockside Phone:** No
  Dock Type: Floating, Long Fingers, Alongside, Concrete
Moorings: 0 Total, 0 Transient **Launch:** n/a
  Rate: Day n/a **Week** n/a **Month** n/a
Heads: 2 Toilet(s)
Laundry: None **Pay Phones:** Yes, 1
Pump-Out: No **Fee:** n/a **Closed Heads:** Yes

Marina Operations
Owner/Manager: Port of Olympia WA **Dockmaster:** Cheryl Maynard
In-Season: Year-Round, 24/7 **Off-Season:** n/a
After-Hours Arrival: Fee station
Reservations: Yes, $5 **Credit Cards:** No
Discounts: None
Pets: Welcome **Handicap Access:** Yes, Heads, Docks

Port Plaza Guest Docks

701 NW Columbia Street; Olympia, WA 98501

Tel: (360) 528-8049 **VHF: Monitor** n/a **Talk** n/a
Fax: n/a **Alternate Tel:** n/a
Email: See Below **Web:** www.portolympia.com/swantown_marina.html
Nearest Town: Olympia **Tourist Info:** (360) 357-3362

Marina Services and Boat Supplies
Supplies - Near: Ice *(Block, Cube)* **1-3 mi:** West Marine *(352-1244)*,
Boater's World *(754-1834)*, Bait/Tackle *(Fishy Business 352-0383)*

Boatyard Services
OnCall: Inflatable Repairs, Life Raft Service, Yacht Interiors
Under 1 mi: Travelift, Forklift, Crane, Engine mechanic *(gas, diesel)*,
Launch Ramp,Electrical Repairs, Electronics Repairs, Hull Repairs, Rigger,
Bottom Cleaning, Divers, Interior Cleaning, Propeller Repairs, Wood
working, Metal Fabrication, Painting, Yacht Broker *(Capital City Yachts
352-2007)* **Nearest Yard:** Swantown Boatworks (360) 528-8059

Restaurants and Accommodations
OnSite: Restaurant *(Anthony's Homeport 357-9700, L $9-12, D $16-24)*
Near: Restaurant *(Oyster House 753-7000, L $8-14, D $14-24)*, *(Budd Bay
Café 357-6963, B $5-11, L $7-13, D $14-31, Sun brunch 9:30am-1pm, voted
Best Brunch in Olympia)*, *(Mercato Ristorante 528-3663, L $6-10, D $12-21)*,
(Luna's 956-7189), *(Lemon Grass 705-1832)*, *(King Solomon's Reef 357-
5552)*, *(Gardner's Seafood & Pasta 786-8466)*, Coffee Shop *(Coffee Plant)*,
Lite Fare *(Heyday Café 709-0953)*, Pizzeria *(Old School Pizza 786-9640)*,
Hotel *(Phoenix Inn Suites 570-0555, $99-179)* **Under 1 mi:** Motel *(BW
Aladdin Moror Inn 352-7200, $63-80)*, Inn/B&B *(Ramada Inn 352-7700)*

Recreation and Entertainment
Near: Picnic Area, Playground, Movie Theater *(Capital Mall Cinemas 754-
8777)*, Park, Cultural Attract *(WA Center for Performing Arts 753-8585)*,
Special Events, Galleries *(Sidedoor Studio 705-1818)* **1-3 mi:** Fitness
Center *(Gold's Gym 352-2533)*, Jogging Paths, Video Rental *(Hollywood

Video 357-9519)*, Museum *(State Capital Museum 753-2580, Bigelow House
Museum 753-1215/ Olympic Flight Museum 705-3925, 6 mi.)* **3+ mi:** Golf
Course *(Turnwater 943-9500, 3 mi.)*

Provisioning and General Services
Near: Market *(Washington St. Market 754-9349)*, Delicatessen *(Whale's Tail
956-1928)*, Bakery *(Otto's 352-8640)*, Farmers' Market *(Tue-Sun 10am-3pm
352-9096)*, Fishmonger *(Olympia Seafood 570-8816)*, Bank/ATM, Beauty
Salon *(Hairaway By Phyllis 352-7113)*, Bookstore *(Barnes & Noble 754-6782)*,
Newsstand, Copies Etc. *(UPS Store 754-6800)* **Under 1 mi:** Supermarket
(Safeway 943-0144), Wine/Beer, Liquor Store *(WA Liquor 753-4937)*,
Protestant Church, Library *(352-0595)*, Dry Cleaners *(Capitol Cleaners
& Laundry 943-6170)*, Laundry, Pharmacy *(Safeway 943-0144)*, Hardware
Store *(True Value 357-6659)* **1-3 mi:** Health Food *(Good Life 786-1500)*,
Post Office, Catholic Church, Department Store *(Westfield Mall 754-8017)*
3+ mi: Buying Club *(Costco 670-6410, 4 mi.)*

Transportation
OnCall: Rental Car *(Enterprise 956-3714/ Hertz 786-5665, 2 mi.)*, Taxi *(DC
Cab 786-5226)* **Near:** Local Bus *(786-1881)* **Under 1 mi:** InterCity Bus
3+ mi: Rail *(Amtrak 923-4602, 8 mi.)* **Airport:** Olympia Regional *(6 mi.)*

Medical Services
911 Service **OnCall:** Ambulance **Under 1 mi:** Dentist *(Bailey 352-9391)*,
Chiropractor *(Community Chiropractic 705-1116)* **1-3 mi:** Doctor *(Westside
Medical 705-1980)*, Holistic Services *(Holistic Choices 352-1868)*,
Veterinarian **Hospital:** Providence St. Peter 491-9480 *(3 mi.)*

PHOTOS ON DVD: 8

Setting -- At the southern end of Budd Inlet, a small peninsula splits it into two bays. On the west side of the peninsula, Port Plaza Guest Moorage is just north of Olympia's Percival Landing - putting visitors within walking distance of all of Olympia's attractions. A clear view of the Capitol Dome and the Olympic Mountains dominates the skyline. Paralleling the shore, a concrete dock provides approximately 500 linear feet of moorage. A long gangway leads to a stationary boardwalk behind Anthony's Homeport Restaurant. This is the kick-off point for a mile of shops and eateries along the Percival Landing boardwalk.

Marina Notes -- *This Port Olympia facility offers free moorage for up to 4 hours during the day. If staying over 4 hours or overnight, boats must register. The rate is $7 for boats under 20'; larger vessels pay an additional $0.30/ft. The payment schedule is on the back of the fee envelopes available on-site. Stays are limited to 7-days. **500 ft. of side-tie moorage is on a first-come, first-serve basis - reservations possible for a $5 nonrefundable fee. Potable water but no shore power. Heads are public and located at Percival Landing a short distance from the dock. Email: marina@portolympia.com

Notable -- A pleasant stroll from the dock will net restaurants, shops, galleries, and the Farmers' Market, located at the end of the Percival Landing Boardwalk. Budd Bay Café and Whale's Tale (956-1928) offer lunches near the water; several eateries feature fresh seafood - the Olympia Oyster House is a perennial favorite. Less than a mile south, tours of the 12-acre Olmsted-designed Capitol Campus include the 1892 "Old Capitol" and soaring 1928 Romanesque-style Legislative Building. The popular Hands-On Children's Museum (956-0818) features a "working waterfront."

Percival Landing Moorage

222 Columbia St. N; Olympia, WA 98501

Tel: (360) 753-8380 **VHF: Monitor** n/a **Talk** n/a
Fax: n/a **Alternate Tel:** n/a
Email: n/a **Web:** www.ci.olympia.wa.us/par/percivallanding/
Nearest Town: Olympia **Tourist Info:** (360) 357-3362

Navigational Information

Lat: 47°02.833' **Long:** 122°54.236' **Tide:** 20 ft. **Current:** n/a **Chart:** 18456
Rep. Depths (MLW): Entry 10 ft. **Fuel Dock** n/a **Max Slip/Moor** 6 ft./-
Access: In West Bay at the southern tip of Budd Inlet

Marina Facilities *(In Season/Off Season)*

Fuel: No
Slips: 37 Total, 37** Transient **Max LOA:** 50 ft. **Max Beam:** n/a
 Rate *(per ft.)*: **Day** $0.33* **Week** n/a **Month** n/a
 Power: 30 amp *, 50 amp** n/a, **100 amp** n/a, **200 amp** n/a
 Cable TV: No **Dockside Phone:** No
 Dock Type: Floating, Long Fingers, Alongside, Concrete
Moorings: 0 Total, 0 Transient **Launch:** n/a, Dinghy Dock
 Rate: Day n/a **Week** n/a **Month** n/a
Heads: 4 Toilet(s), 4 Shower(s)
Laundry: None **Pay Phones:** Yes, 1
Pump-Out: OnSite, 1 Central **Fee:** Free **Closed Heads:** Yes

Marina Operations

Owner/Manager: City of Olympia **Dockmaster:** Scott Riber
In-Season: Year-Round, 24/7 **Off-Season:** n/a
After-Hours Arrival: Fee station next to the restroom
Reservations: Yes, Oct-Mar & groups of 5 or more **Credit Cards:** Visa/MC
Discounts: None
Pets: Welcome **Handicap Access:** No

Marina Services and Boat Supplies

Supplies - Near: Ice *(Block, Cube)* **Under 1 mi:** Bait/Tackle, Live Bait, Propane **1-3 mi:** West Marine *(352-1244)*, Boater's World *(754-1834)*

Boatyard Services

OnCall: Inflatable Repairs, Life Raft Service **Under 1 mi:** Travelift, Engine mechanic *(gas, diesel)*, Electrical Repairs, Electronics Repairs, Hull Repairs, Rigger, Bottom Cleaning, Brightwork, Divers, Compound, Wash & Wax, Interior Cleaning, Propeller Repairs, Woodworking, Metal Fabrication, Painting. **Nearest Yard:** Swantown Boatworks (360) 528-8059

Restaurants and Accommodations

Near: Restaurant *(Urban Onion Restaurant 943-9242, L, D)*, *(Budd Bay Café 357-6963, B $5-11, L $7-13, D $14-31)*, *(Anthony's Homeport 357-9700, L $9-12, D $16-24)*, *(Mercato Ristorante 528-3663, L $6-10, D $12-21)*, *(King Solomon's Reef 357-5552)*, *(Gardner's Seafood & Pasta 786-8466)*, *(Oyster House 753-7000, L $8-14, D $14-24, Oyster Feast $39)*, *(Thai Pavilion 943-9093)*, Coffee Shop *(Starbucks)*, Lite Fare *(Orange Julius 754-9114)*, *(Dockside Deli & Pizza 956-1928)*, Hotel *(Phoenix Inn Suites 570-0555, $99-179)* **Under 1 mi:** Inn/B&B *(Olympia Inn 352-8533)*

Recreation and Entertainment

OnSite: Picnic Area, Playground *(with several jungle gyms)*, Jogging Paths, Boat Rentals **Near:** Movie Theater *(Capital Mall Cinemas 754-8777)*, Park, Cultural Attract *(WA Center for Performing Arts 753-8585)*, Special Events *(Wooden Boat Fair - May)* **Under 1 mi:** Heated Pool, Fitness Center *(YMCA 357-6609)*, Museum *(Bigelow House 753-1215, Hands-On Children's Museum 956-0818, tours of the Capitol 586-TOUR)*, Galleries *(The Artist's Gallery 357-6920)* **1-3 mi:** Video Rental *(Hollywood Video 357-9519)* **3+ mi:** Golf Course *(Turnwater Municipal 943-9500, 3 mi.)*, Sightseeing *(Olympic Flight Museum 705-3925 - Gathering of Warbirds in June, 6 mi.)*

Provisioning and General Services

Near: Market *(Bay View Grocery; Capital Lake Grocery 753-1052)*, Delicatessen *(Dockside 956-1928)*, Bakery *(Boston Harbor Pies 705-3180)*, Farmers' Market *(Tue-Sun 10am-3pm, 352-9096 - northern end of Capitol Way)*, Fishmonger *(Olympia Seafood Co. 570-8816)*, Bank/ATM, Beauty Salon *(Hairaway By Phyllis 352-7113)*, Bookstore *(Barnes & Noble 754-6782)*, Copies Etc. *(UPS Store 754-6800)* **Under 1 mi:** Supermarket *(Safeway 943-0144)*, Wine/Beer, Liquor Store *(WA Liquor 753-4937)*, Post Office, Protestant Church, Library *(Olympia 352-0595)*, Dry Cleaners *(Capitol Cleaners & Laundry 943-6170)*, Laundry, Pharmacy *(Safeway 943-0144)*, Hardware Store *(True Value 357-6659)* **1-3 mi:** Health Food *(Good Life 786-1500)*, Catholic Church, Department Store *(Westfield Mall 754-8017)* **3+ mi:** Buying Club *(Costco 357-9743, 4 mi.)*

Transportation

OnCall: Rental Car *(Enterprise 956-3714/ Avis 943-7346, 1 mi.)*, Taxi *(Red Top 357-3700)* **Near:** Local Bus *(Intercity Transit 786-1881)* **Under 1 mi:** InterCity Bus *(Greyhound 357-5541)* **3+ mi:** Rail *(Amtrak 923-4602, 8 mi.)* **Airport:** Olympia Regional *(6 mi.)*

Medical Services

911 Service **OnCall:** Ambulance **Under 1 mi:** Dentist *(Bailey 352-9391)*, Chiropractor *(Rosser 754-6499)* **1-3 mi:** Doctor *(Westside Medical 705-1980)* **Hospital:** Providence St. Peter 491-9480 *(3 mi.)*

Setting -- Percival Landing is located at the head of Budd Inlet - past commercial Port of Olympia and Port Plaza Guest Docks. 1100 feet of moorage are an integral part of this delightful 3.4 acre waterfront park - which offers convenient access to downtown Olympia's tourist attractions. The public space includes a large playground, picnic area, sculptures, art installations and the famous mile-long boardwalk. On clear days, the Olympic Mountains decorate the horizon.

Marina Notes -- *Flat Rate. No fee for daytime stays of up to 4 hours. Longer and overnight stays require registration. Nightly transient flat rate: under 29' $11; 30-39' $12; 40-49' $13; 50' plus $14. **All transient dock space is alongside. ***Power hook-ups on 600 ft. of dockage, but electrical service was terminated in '06 pending replacement of 30-year-old wiring (solutions are at hand, call to inquire status). Maximum stay 7-days in 30-day period. Register at the Olympia Center one block east of "E" dock. Reservations are taken for groups of 5 or more boats. Very modern, tile and bright stainless public restrooms open from 7:30am-8pm. Note: Olympia Yacht Club's Main Station is adjacent as is their grid of 330 slips, including about 190 boathouses.

Notable -- Originally constructed in the mid-seventies, Percival Landing was named for the Percival family who built and operated a steamship wharf on this site in the early 1860s. Today, it continues to evolve with plans for a $9+ million dollar redesign and rebuild of the Percival Landing Boardwalk - a mile of shops, restaurants and parks. Pheonix Inn hotel is just across the street and many restaurants, delis, and coffee shops are close by. A daily Farmers' Market is near, as is the Washington Center for the Performing Arts. A half-mile south, past many historic buildings, are the 12-acre, Olmsted-designed Capitol grounds.

Navigational Information
Lat: 47°03.411' **Long:** 122°53.860' **Tide:** 20 ft. **Current:** 1 kt. **Chart:** 18456
Rep. Depths *(MLW):* **Entry** 12 ft. **Fuel Dock** n/a **Max Slip/Moor** 12 ft./-
Access: In East Bay at the southern tip of Budd Inlet

Marina Facilities *(In Season/Off Season)*
Fuel: No
Slips: 656 Total, 70 Transient **Max LOA:** 110 ft. **Max Beam:** n/a
  Rate *(per ft.):* **Day** $0.50* **Week** n/a **Month** $6.30-7.65
  Power: 30 amp Incl., **50 amp** Incl., **100 amp** n/a, **200 amp** n/a
  Cable TV: Yes **Dockside Phone:** Yes
  Dock Type: Floating, Long Fingers, Concrete
Moorings: 0 Total, 0 Transient **Launch:** n/a, Dinghy Dock (Free)
  Rate: Day n/a **Week** n/a **Month** n/a
Heads: 16 Toilet(s), 10 Shower(s) *(with dressing rooms)*
Laundry: 10 Washer(s), 10 Dryer(s) **Pay Phones:** Yes, 3
Pump-Out: OnSite, Self Service, 1 Central **Fee:** Free **Closed Heads:** Yes

Marina Operations
Owner/Manager: Bruce Marshall, Dir. **Dockmaster:** Cheryl Maynard
In-Season: Apr-Oct, 8am-6pm **Off-Season:** Oct-Apr, 8am-5:30pm
After-Hours Arrival: Tie to "A" dock and report in the morning
Reservations: Yes, $5 **Credit Cards:** Visa/MC
Discounts: Groups **Dockage:** 10% **Fuel:** n/a **Repair:** n/a
Pets: Welcome, Dog Walk Area **Handicap Access:** Yes, Heads, Docks

Swantown Marina & Boatworks

1022 Marine Drive NE; Olympia, WA 98501

Tel: (360) 528-8049 **VHF: Monitor** Ch. 65A **Talk** n/a
Fax: (360) 528-8094 **Alternate Tel:** n/a
Email: marina@portolympia.com **Web:** www.portolympia.com
Nearest Town: Olympia *(0.4 mi.)* **Tourist Info:** (360) 357-3362

Marina Services and Boat Supplies
Services - Security *(24 hrs., armed roving guard)*, Dock Carts, Megayacht
Facilities **Communication -** Mail & Package Hold, FedEx, DHL, UPS
Supplies - OnSite: Ice *(Cube)*, Ships' Store **Near:** Bait/Tackle, Live Bait
Under 1 mi: Ice *(Block)*, Propane **1-3 mi:** West Marine *(352-1244)*,
Boater's World *(754-1834)*

Boatyard Services
OnSite: Travelift *(77T)*, Forklift, Crane, Engine mechanic *(gas, diesel)*,
Launching Ramp *($5 - two lane)*, Electrical Repairs, Electronics Repairs,
Hull Repairs, Rigger, Bottom Cleaning, Brightwork, Air Conditioning,
Refrigeration, Divers, Compound, Wash & Wax, Interior Cleaning, Prop
Repairs, Woodworking, Upholstery, Painting, Awlgrip, Total Refits
OnCall: Inflatable Repairs, Life Raft Service, Yacht Design
Under 1 mi: Electronic Sales, Sail Loft, Canvas Work, Metal Fabrication

Restaurants and Accommodations
Near: Restaurant *(Anthony's 357-9700, L $9-12, D $16-24)*, *(Luna's 956-
7189)*, *(Gardner's Seafood & Pasta 786-8466)*, Lite Fare *(Heyday Café
709-0953)*, Inn/B&B *(Phoenix Inn570-0555, $99-179)* **Under 1 mi:** Rest-
aurant *(Budd Bay Café 357-6963, L $7-13, D $14-31)*, *(Santosh 943-
3442)*, *(Trinacria Ristorante 352-8892)*, *(New Moon Café 357-3452)*, Lite
Fare *(Fishbowl Brew Pub & Café 943-3650)*, Motel *(BW Aladdin Motor Inn
352-7200, $63-80)*, Hotel *(Ramada Inn 352-7700, $73-170)*

Recreation and Entertainment
OnSite: Picnic Area, Special Events *(Boat Swap & Chowder Challenge -
May)* **Near:** Playground, Park, Museum *(State Capital Museum in Lord*

Mansion 753-2580; Bigelow House Museum 753-1215)* **Under 1 mi:** Heat-
ed Pool, Fitness Center *(Gold's Gym 352-2533)*, Movie Theater *(Capital
Mall 754-8777)*, Cultural Attract *(WA Center for Performing Arts 753-8585)*,
Galleries *(State of Arts 705-0317)* **1-3 mi:** Golf Course *(Turnwater
Municipal 943-9500)*, Video Rental *(Hollywood 357-9519)*

Provisioning and General Services
OnSite: Bank/ATM **Near:** Convenience Store *(K&J Mini Mart 754-4737)*,
Delicatessen *(Dockside Deli & Pizza 956-1928)*, Farmers' Market *(Tue-Sun
10am-3pm 352-9096)*, Fishmonger *(Olympia Seafood Co. 570-8816)*, Beauty
Salon *(Premiere Salon & Spa 353-3299)* **Under 1 mi:** Supermarket
(Safeway 943-0144), Wine/Beer, Liquor Store *(WA Liquor 753-4937)*, Post
Office, Protestant Church, Library *(Olympia 352-0595)*, Dry Cleaners *(Capitol
Cleaners & Laundry 943-6170)*, Laundry, Bookstore *(Browsers' 357-7462)*,
Pharmacy *(Safeway 943-0144)*, Hardware Store *(True Value 357-6659)*,
Copies Etc. *(UPS Store 754-6800)* **1-3 mi:** Health Food *(Good Life 786-
1500)*, Catholic Church, Department Store *(Westfield Mall 754-8017)*

Transportation
OnCall: Rental Car *(Enterprise 956-3714)*, Taxi *(Red Top 357-3700)*, Airport
Limo *(Capital Aeroporter 754-7113)* **Under 1 mi:** Local Bus *(Intercity
Transit 786-1881)*, InterCity Bus **3+ mi:** Rail *(Amtrak 923-4602, 8 mi.)*
Airport: Olympia Regional *(6 mi.)*

Medical Services
911 Service **OnCall:** Ambulance **Under 1 mi:** Dentist *(Bailey 352-9391)*,
Chiropractor *(Rosser 754-6499)* **1-3 mi:** Doctor *(Westside Medical 705-
1980)* **Hospital:** Providence St. Peter 491-9480 *(3 mi.)*

Setting -- Swantown Marina is Port Olympia's full-service facility, located on East Bay at the end of Budd Inlet - on the east side of the narrow peninsula. A
large network of well-maintained concrete floats is protected by a breakwater that extends from the point. Grassy areas and lots of trees create a parklike
setting, with picninc tables near the docks. In the immediate area are walking trails, and marine and general business services. Practically at the foot of Mt.
Ranier and within sight of the Olympic mountains, the docks are in a perfect location for sightseeing and shopping.

Marina Notes -- *Transient rate $10 min., for boats to 20'; over 20' add $0.50/ft. Built in 1984, boatyard in 1999 and expanded in 2003. Reservations are
available with a $5 non-refundable fee. Dedicated transient space is on "A" dock, just past the breakwater and launch ramp. Guest moorage limited to 29 days
in 12 months. 157 slips added in 2003; boosting the number to around 650. Water & power at each slip - requires 30/50A twist lock plug. Pump-out and a
recycling center are available. On-site Swantown Boatworks (528-8059) has a large haul-out capacity and offers a full range of yard services. Plans for new
upland facilities and a 4-acre "marina village." A modern, concrete bathhouse hosts a plethora of impeccable heads with showers and laundry.

Notable -- Complete shoreside facilities make this an ideal place to begin or end a Puget Sound cruise. Provisions and restaurants are a short distance from
the marina, including the state's second largest Farmer's Market. Walk to nearby Percival Landing's boardwalk, and stroll Capitol Way, lined with shops and
galleries, to the historic 12 square block Capitol Campus historic district. The Washington State Capital Museum describes Olympia's political history.

PHOTOS ON DVD: 17

Boaters' Notes

Add Your Ratings and Reviews at www.AtlanticCruisingClub.com

A complimentary six-month Silver Membership is included with the purchase of this *Guide*. Select "Join Now," then "Silver," then follow the instructions.

The AtlanticCruisingClub website provides updated Marina Reports, Destination and Harbor Articles and much more — including an option within each online Marina Report for boaters to add their ratings and comments regarding that facility. Please log on frequently to share your experiences — and to read other boaters' comments.

On the website, boaters may rate marinas in one or more of the following categories — on a scale of 1 (basic) to 5 (world class) — and also enter additional commentary.

▸ **Facilities & Services** (Fuel, Reservations, Concierge Services and General Helpfulness)

▸ **Amenities** (Pool, Beach, Internet Access, including Wi-Fi, Picnic/Grill Area, Boaters' Lounge)

▸ **Setting** (Views, Design, Landscaping, Maintenance, Ship-Shapeness, Overall Ambiance)

▸ **Convenience** (Access — including delivery services — to Supermarkets, other Provisioning Sources, Shops, Services, Attractions, Entertainment, Sightseeing, Recreation, including Golf and Sport Fishing & Medical Services)

▸ **Restaurants/Eateries/Lodgings** (Availability of Fine Dining, Snack Bars, Lite Fare, OnCall food service, and Lodgings ashore)

▸ **Transportation** (Courtesy Car/Vans, Buses, Water Taxis, Bikes, Taxis, Rental Cars, Airports, Amtrak, Commuter Trains)

▸ **Please Add Any Additional Comments**

17. WA – Strait of Juan de Fuca

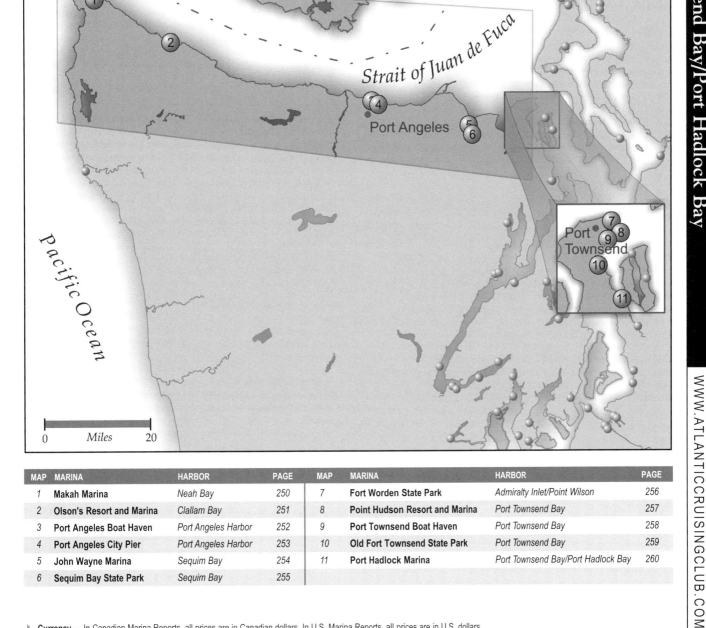

MAP	MARINA	HARBOR	PAGE	MAP	MARINA	HARBOR	PAGE
1	Makah Marina	Neah Bay	250	7	Fort Worden State Park	Admiralty Inlet/Point Wilson	256
2	Olson's Resort and Marina	Clallam Bay	251	8	Point Hudson Resort and Marina	Port Townsend Bay	257
3	Port Angeles Boat Haven	Port Angeles Harbor	252	9	Port Townsend Boat Haven	Port Townsend Bay	258
4	Port Angeles City Pier	Port Angeles Harbor	253	10	Old Fort Townsend State Park	Port Townsend Bay	259
5	John Wayne Marina	Sequim Bay	254	11	Port Hadlock Marina	Port Townsend Bay/Port Hadlock Bay	260
6	Sequim Bay State Park	Sequim Bay	255				

▹ **Currency** — In Canadian Marina Reports, all prices are in Canadian dollars. In U.S. Marina Reports, all prices are in U.S. dollars.

▹ **"CCM"** — Denotes a Certified Clean Marina, a state/provincial award for environmental excellence. See page 298 for an explanation and page 299 for a list of Pump-Out facilities.

▹ **Ratings & Reviews** — An overview of the Atlantic Cruising Club's rating system is on page 6 and details on the content of each Marina Report are on pages 7 – 11.

▹ **Marina Report Updates** — Comments from boaters and new information from ACC reviewers and marinas are posted regularly on *www.AtlanticCruisingClub.com*.

Makah Marina

PO Box 137; Neah Bay, WA 98357

Tel: (360) 645-3015 **VHF: Monitor** Ch. 16 **Talk** Ch. 66A
Fax: (360) 645-3016 **Alternate Tel:** n/a
Email: mtcport@centurytel.net **Web:** www.makah.com/marina.html
Nearest Town: Neah Bay **Tourist Info:** (360) 374-2531

Navigational Information

Lat: 48°22.147' **Long:** 124°37.732' **Tide:** 8 ft. **Current:** n/a **Chart:** 18484
Rep. Depths (*MLW*): **Entry** 15 ft. **Fuel Dock** 15 ft. **Max Slip/Moor** 15 ft./-
Access: 5 miles east of Cape Flattery on the southwest shore of Neah Bay

Marina Facilities (*In Season/Off Season*)

Fuel: Gasoline, Diesel
Slips: 200 Total, 100 Transient **Max LOA:** 130 ft. **Max Beam:** n/a
  Rate (*per ft.*): **Day** $0.60/0.55 **Week** n/a **Month** n/a
  Power: 30 amp $2.50, **50 amp** $3, **100 amp** n/a, **200 amp** n/a
  Cable TV: No **Dockside Phone:** No
  Dock Type: Floating, Long Fingers, Alongside, Concrete
Moorings: 0 Total, 0 Transient **Launch:** n/a
  Rate: Day n/a **Week** n/a **Month** n/a
Heads: 5 Toilet(s), 4 Shower(s) (*with dressing rooms*)
Laundry: None **Pay Phones:** Yes, 1
Pump-Out: OnSite **Fee:** Free **Closed Heads:** Yes

Marina Operations

Owner/Manager: Robert P. Buckingham **Dockmaster:** Same
In-Season: Year-Round, 8am-5pm **Off-Season:** n/a
After-Hours Arrival: Tie up and check in the morning
Reservations: No **Credit Cards:** Visa/MC
Discounts: None
Pets: Welcome **Handicap Access:** Yes, Heads, Docks

Marina Services and Boat Supplies

Services - Dock Carts **Communication -** Mail & Package Hold, Phone Messages, FedEx, DHL, UPS, Express Mail **Supplies - Near:** Ice (*Block, Cube*), Bait/Tackle (*Washburn's 645-2211*)

Boatyard Services

OnSite: Launching Ramp **3+ mi:** Travelift (*70T, 80 mi.*), Engine mechanic (*gas, diesel*), Hull Repairs (*80 mi.*). **Nearest Yard:** Port Angeles Boat Haven (360) 457-4505

Restaurants and Accommodations

OnSite: Restaurant (*Warm House 645-2924, overlooking the marina; seafood and Angus steaks*) **Near:** Restaurant (*Makah Maiden Pantry 645-2124, B $5-9, L $5-8, D $10-12*), (*Bebee's Café 645-2872, L $5-9, D $7-10*), Snack Bar (*Cedar Shack Espresso 645-2380*), Pizzeria (*Natalie's Pizza 645-2670*), Motel (*Cape Motel and RV Park 645-2250, $50-65*), (*Hobuck Beach Resort 645-2339, $45-$85*), (*Tyee Motel 645-2223, $45-69*), (*Bays Best Lodging 645-2019, $55-65, kitchen facilities avail. for $10 extra*)

Recreation and Entertainment

OnSite: Beach, Cultural Attract (*Olympic Coast National Marine Sanctuary Visitors' Center Tue-Sat 11am-4pm*) **Near:** Museum (*Makah Museum/Makah Cultural & Research Center 645-2711, MemDay-Sep 15, 7 days, Sep 16-MemDay, Wed-Sun, $5/$4 - 1pm-4pm*), Special Events (*Makah Days - 4th wknd Aug*), Galleries (*Raven's Corner 645-2426 - Native art and crafts*) **Under 1 mi:** Playground (*Dakwas Playground*), Fishing Charter (*Big Salmon Fishing Resort 645-2374*) **3+ mi:** Park (*Olympic National Park*), Sightseeing (*Cape Flattery - most northwest point of the lower 48 states; 3/4 mi. hiking trail, 4 observaton decks with fabulous views of Olympic Coast Marine Sanctuary & Tatoosh Island, 7 mi.*)

Provisioning and General Services

Near: Convenience Store (*Mini-mart 645-2802*), Market (*Washburn's General Store 645-2211*), Delicatessen (*The Cedar Shack 645-2380*), Farmers' Market, Protestant Church (*Presbyterian Church 645-2276/Makah Lutheran Church 645-2523/Assembly of God 645-2761/Apostlic Faith Church 645-2524*), Beauty Salon (*Longhouse Beauty 645-2270*), Laundry, Retail Shops (*Washburn's General Merchandise 645-2211*), Copies Etc. (*Mini-Mart 645-2802 - copy, fax, and Internet hotspot*) **3+ mi:** Liquor Store (*Washington State Liquor Store 963-2323, 20 mi.*)

Transportation

Near: Local Bus (*Makah Transit connects with Clallam Transit to Port Angeles 452-4511*) **Airport:** Sekiu/Fairchild Int'l. (*17 mi./67 mi.*)

Medical Services

911 Service **OnCall:** Ambulance **1-3 mi:** Doctor (*Indian Health Services Clinic*), Dentist **Hospital:** Olympic Memorial 417-7000 (*67 mi.*)

Setting -- About five miles east of Cape Flattery, the buoyed entrance to easily accessed and protected Neah Bay lies between Waadah Island and Baada Point. Just past the Makah Nation's T-head pier, a tall, angled breakwater protects the network of five quality floating piers housing over 200 slips. Inviting sandy beaches flank the marina and on shore, the gangway leads to a large modern pale-gray building with aqua steel roof that houses the administration offices and amenities. The striking Warm House restaurant - a neo-northwest Indian design - overlooks the docks and bay.

Marina Notes -- Built in 1997, the marina has 200 slips from 30 ft to 70 ft., and accommodates vessels to 130' on T-heads. It is home to a very sizeable fleet of commercial fishing boats, most of them quite large. Concrete docks have power on pedestals and fresh water near the finger piers. Fuel and diesel are available on-site. In the main building, clean, bright red, blue and yellow restrooms have cement floors and fiberglass showers. A laundry is across the street. Note: Neah Bay is protected from all but eastern weather and is a harbor of refuge for small boats in foul weather. And Neah Bay is "dry."

Notable -- Watch for seals, sea lions, and shore birds in the surrounding waters. Several businesses, including restaurants and lodges, are near the marina. Right across the street, the fantastic Makah Museum features exhibits on the pre-contact life of the Makah Nation. Archaeological excavation at the site of Ozette, a Native village buried under mud in the 1700s, uncovered thousands of well-preserved artifacts which are displayed here. At the end of August, Makah Days celebrate the tribe with traditional dances, songs, games, and fireworks (visitors $10). Try your luck at Makah Tribal Bingo at 7pm; $10 buy-in (645-2264).

Navigational Information
Lat: 48°15.870' **Long:** 124°17.978' **Tide:** 8 ft. **Current:** n/a **Chart:** 18460
Rep. Depths (MLW): Entry 24 ft. **Fuel Dock** 6 ft. **Max Slip/Moor** 6 ft./-
Access: On the western shore of Clallam Bay off the Strait of Juan De Fuca

Marina Facilities (In Season/Off Season)
Fuel: Gasoline, Diesel
Slips: 150 Total, 150 Transient **Max LOA:** 40 ft. **Max Beam:** n/a
  Rate (per ft.): **Day** $0.50* **Week** n/a **Month** n/a
  Power: 30 amp n/a, **50 amp** n/a, **100 amp** n/a, **200 amp** n/a
  Cable TV: No **Dockside Phone:** No
  Dock Type: Floating, Wood
Moorings: 0 Total, 0 Transient **Launch:** n/a
  Rate: Day n/a **Week** n/a **Month** n/a
Heads: 20 Toilet(s), 14 Shower(s) (with dressing rooms)
Laundry: 2 Washer(s), 3 Dryer(s) **Pay Phones:** Yes, 1
Pump-Out: No **Fee:** n/a **Closed Heads:** Yes

Marina Operations
Owner/Manager: Arlen and Donnalynn Olson **Dockmaster:** Same
In-Season: Apr-Sep, 5:30am-10pm **Off-Season:** Oct-Mar, closed
After-Hours Arrival: Tie up and register in the morning
Reservations: No **Credit Cards:** Visa/MC, Dscvr, Amex
Discounts: None
Pets: No **Handicap Access:** No

Olson's Resort and Marina

PO Box 216; 444 Front Street; Sekiu, WA 98381

Tel: (360) 963-2311 **VHF: Monitor** n/a **Talk** n/a
Fax: n/a **Alternate Tel:** n/a
Email: info@olsonsresort.com **Web:** www.olsonsresort.com
Nearest Town: Sekiu **Tourist Info:** (360) 963-2339

Marina Services and Boat Supplies
Services - Security, Dock Carts **Supplies - OnSite:** Ice (Cube),
Bait/Tackle

Boatyard Services
OnSite: Launching Ramp **Nearest Yard:** Port Angeles Boat Haven (360)
457-4505

Restaurants and Accommodations
OnSite: Motel (Olson's Resort 963-2311, $55-70, some rooms with
kitchenettes), Condo/Cottage (Olson's Resort $78-110) **Near:** Restaurant
(Cove Restaurant & Lounge 963-2881, B $5-9, L $5-10, D $7-18, B, L, D),
Motel (Curley's Resort Motel 963-2281, $48-80), (Van Riper's Resort 963-
2334, $60-160, 1-3 db, kitchens) **1-3 mi:** Restaurant (Spring Tavern 963-
2855), Lite Fare (Weel Road Deli 963-2777)

Recreation and Entertainment
OnSite: Picnic Area, Fishing Charter (Tommycod Charters 800-283-8900),
Group Fishing Boat (Olson's Resort and Marina), Special Events (Salmon
Derby, Halibut Derby - Jun, "No Fin, You Win" - Sep) **Near:** Beach, Dive
Shop (Curley's Resort & Dive Center 963-2281), Jogging Paths, Boat

Rentals (Curley's Resort and Dive Center 963-2281 Kayaks), Sightseeing
(Clallam Bay Park) **3+ mi:** Cultural Attract (Makah Museum & Cultural
Center, Neah Bay 645-2711, 19 mi.)

Provisioning and General Services
OnSite: Convenience Store (or Al's Mini Mart 963-2925, near by dinghy or
3 mi. by land), Wine/Beer **Near:** Market (Ray's Grocery 963-2261), Post
Office (963-2510) **1-3 mi:** Delicatessen (Weel Rd. Deli 963-2777), Liquor
Store (WA Liquor Store 963-2323), Bank/ATM (ATM at Weel Rd. Deli 963-
2777), Library (Clallam Bay 963-2414), Beauty Salon (Robyn's Nest Salon
963-3199) **3+ mi:** Protestant Church (Clallam Bay Church of Christ 963-
2603, 10 mi.)

Transportation
1-3 mi: Local Bus (Clallam Transit 452-4511; service to Port Angeles
available) **Airport:** Sekiu/Fairchild Int'l. (0.5 mi./50 mi.)

Medical Services
911 Service **OnCall:** Ambulance **3+ mi:** Doctor (Clallam Bay Medical Clinic
963-2202, 3 mi.), Veterinarian (Forks Animal Hospital 374-8882, 30 mi.)
Hospital: Forks Community 374-6271/Olympic Memorial 417-7000
(30 mi./50 mi.)

Setting -- About 23 miles east of Cape Flattery, Clallam Bay lies on the southern shore of the Strait. The resort and sport fishing community of Sekiu is on the
west side of the Bay, south of Sekiu Point. Watch for the Three Sisters, the landmark sea stacks to starboard, as you enter the harbor. A rock breakwater
protects the docks, and onshore a rustic flat-roofed, single-story building houses a well-supplied store. Further upland are a modern two-story motel, a row of
sweet little gray camping cabins, and an RV park.

Marina Notes -- *Boats 21' and less are $11; over 21' add $0.50/ft. Family owned since 1936. The structures and docks show their age, but are well-
maintained. A fuel dock is next to the launch ramp. The on-site store carries gifts, groceries, ice, bait and tackle, and a good assortment of fishing gear. Older
but cared-for heads are located in the store and in a separate bathhouse, with dressing rooms, up the hill. The airy laundry is locked at 9pm - last load is 7pm.
Note: A reef projects west of Sekiu Point. The docks are brought in during the winter, but the cove offers a protected anchorage year-round.

Notable -- Sekiu is predominantly a fishing village and tournaments abound: Sekiu's Halibut Derby is in June (entry fee is $15/person); the weigh-in is at Van
Riper's Resort. The "No Fin, You Win" salmon derby is held every September (963-2311). Besides halibut and salmon, the area is abundant with lingcod and
bottomfish. Kayak the straits or charter a boat for fishing, diving, whale and wildlife tours. You'll find a full service dive shop, gift shops, espresso and more, in
town. The nearby Cove pub and restaurant boast water views. Limited services are also available in Clallam Bay (3 miles around the cove or dinghy across).

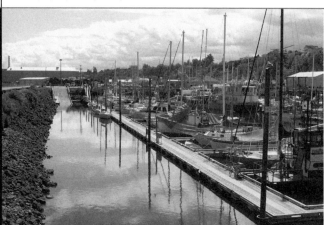

Navigational Information

Lat: 48°07.479' **Long:** 123°27.042' **Tide:** 7 ft. **Current:** n/a **Chart:** 18468
Rep. Depths *(MLW)*: Entry 19 ft. **Fuel Dock** 15 ft. **Max Slip/Moor** 14 ft./-
Access: The facility is located on the southwest shore of Port Angeles harbor

Marina Facilities *(In Season/Off Season)*

Fuel: Gasoline, Diesel
Slips: 520 Total, 16 Transient **Max LOA:** 150 ft. **Max Beam:** n/a
  Rate *(per ft.)*: **Day** $1.10* **Week** n/a **Month** n/a
  Power: 30 amp $4** , **50 amp** n/a, **100 amp** n/a, **200 amp** n/a
  Cable TV: No **Dockside Phone:** No
  Dock Type: Floating, Long Fingers, Wood
Moorings: 0 Total, 0 Transient **Launch:** n/a
  Rate: Day n/a **Week** n/a **Month** n/a
Heads: 10 Toilet(s), 2 Shower(s)
Laundry: None **Pay Phones:** Yes, 3
Pump-Out: OnSite, Self Service, 2 Central **Fee:** Free **Closed Heads:** Yes

Marina Operations

Owner/Manager: Port of Port Angeles **Dockmaster:** Chuck Faires, HM
In-Season: Year-Round, 8am-5pm **Off-Season:** Oct-Mar
After-Hours Arrival: Call ahead
Reservations: No **Credit Cards:** Visa/MC, Dscvr, Amex
Discounts: None
Pets: Welcome **Handicap Access:** No

Port Angeles Boat Haven

832 Boat Haven Drive; Port Angeles, WA 98363

Tel: (360) 457-4505 **VHF: Monitor** n/a **Talk** n/a
Fax: (360) 457-4921 **Alternate Tel:** n/a
Email: n/a **Web:** www.portofpa.com
Nearest Town: Port Angeles **Tourist Info:** (360) 452-2363

Marina Services and Boat Supplies

Services - Security *(Nightly)*, Dock Carts **Communication -** FedEx, DHL, UPS **Supplies - OnSite:** Ships' Store **Near:** Ice *(Cube)*, Bait/Tackle *(Waters West 417-0937)*

Boatyard Services

OnSite: Travelift *(70T, 300T)*, Railway *(200T)*, Crane, Engine mechanic *(gas, diesel)*, Launching Ramp *(2-lane summer, 1-lane year-round; $8)*, Hull Repairs **OnCall:** Electronics Repairs

Restaurants and Accommodations

Near: Restaurant *(High Tide Sea Foods 452-8488, L, D)*, *(Sabai Thai 452-4505)*, *(Hilltop Chinese 565-8476)*, *(Castaways 452-1177)*, Inn/B&B *(Clarks' Harbor 457-9891)* **Under 1 mi:** Restaurant *(Landing's 457-6768, D)*, *(Bella Italia 457-5442, D daily after 4pm; entrees $17-28, pizza & pasta from $11)*, Lite Fare *(Crazy Fish 457-1944, Mexican)*, *(Matay Lunch & Lattes 457-0970)*, Motel *(Quality Inn 457-9434)*, Hotel *(Downtown 565-1125, $40-70)*, *(Red Lion 452-9215, $130-160)* **1-3 mi:** Pizzeria *(Gordy's 457-5056)*

Recreation and Entertainment

OnSite: Picnic Area **Near:** Movie Theater *(Lincoln Theater 457-7997)*, Video Rental *(Prime Time 457-8535)*, Special Events *(Arts in Action Street Fair and Sand Sculpture - Jul, Dungeness Crab & Seafood Fest - Oct 457-6110)*, Galleries *(Wildfire 417-6800)* **Under 1 mi:** Pool *(Shore 417-4595)*, Dive Shop *(Scuba Supplies 457-3190)*, Fitness Center *(Fitness West 452-1118)*, Boat Rentals *(Sound Kayaks 457-1240 $40/day)*, Museum *(Clallam County 452-2662)*, Sightseeing *(Seventh Wave 808-0505 - harbor cruises)* **1-3 mi:** Bowling *(P.A. Lanes 457-7570)*, Park *(Olympic Nat'l. 457-4265)*, Cultural Attract *(P.A. Symphony 457-5579)* **3+ mi:** Golf Course *(Sky Ridge 683-3673, 15 mi.)*

Provisioning and General Services

OnSite: Fishmonger **Under 1 mi:** Convenience Store *(Jackpot 457-7800)*, Market *(Grandview Grocery 457-6575)*, Supermarket *(Safeway 457-0788)*, Delicatessen *(Round the Clock 452-1777)*, Health Food *(Country-Aire 452-7175)*, Wine/Beer, Liquor Store *(WA Liquor 417-1737)*, Bakery *(Star's Northwest 457-3279)*, Farmers' Market *(Sat 10am-2pm, 4th and Peabody St.)*, Crabs/Waterman, Bank/ATM, Protestant Church, Beauty Salon *(Laurel 457-9102)*, Barber Shop, Laundry *(Speed Klean 457-4044)*, Bookstore *(Odyssey 457-1045)*, Hardware Store *(Do It Best 457-3369)*, Department Store *(Gottschalks 452-4571, Sears 457-9481)*, Copies Etc. *(Copies Plus 452-4748)* **1-3 mi:** Post Office *(417-2940)*, Catholic Church, Synagogue *(Olympic B'Nai Shalom Havarah 452-2471)*, Library *(452-9253)*, Pharmacy *(Rite Aid 452-9784)*

Transportation

OnCall: Rental Car *(Enterprise 417-3083/Budget 452-4774, 1 mi.)*, Taxi *(Acme 452-5262)* **Near:** Local Bus *(Clallam Transit 582-3736)* **Under 1 mi:** Bikes *(Sound Bikes & Kayaks 457-1240 $30/day)*, Ferry Service *(Coho Ferry 457-4491/ Express 452-8088)* **Airport:** Fairchild Int'l. *(3 mi.)*

Medical Services

911 Service **OnCall:** Ambulance **Under 1 mi:** Doctor *(Family Medicine 452-7891)*, Dentist *(Laurel Dental 452-9744)*, Chiropractor *(Olympic Chiro Clinic 457-8292)* **1-3 mi:** Veterinarian *(Blue Mountain Animal Clinic 457-3842)* **Hospital:** Olympic Memorial 417-7000 *(2 mi.)*

Setting -- Accessed between the three-mile-long Ediz Hook and the main shore, Port Angeles plays frequent host to very large vessels which are refueling, awaiting orders or a tug, or weather-bound. Port Angeles Boat Haven is located toward the west end of the 2.5-mile-long harbor in a nearly enclosed basin protected by its own rock & piling breakwater. This 16-acre facility berths over 500 vessels - pleasure and commercial craft plus 75 houseboats - divided into two basins each with three long main piers. The marina is backed by a steep foliage-covered hill dotted with houses perched high above the docks.

Marina Notes -- *Open slips: up to 30' $0.55/ft./day, over 30' $1.10/ft./day; Covered: up to 15 days $1.10/ft./day, 16+ days $2.20/ft./day. **20 amp service available for $4/day. Large slips & covered boathouses available. The office is in a small gray building topped by a shingled, pyramidal roof. Fuel, pump-out, boat ramps, parking area and do-it-yourself boatyard are on-site. Arrow Marine (800-255-2948) provides services at Port Angeles Boat Yard. Conveniently adjacent are several marine businesses, including marine supplies, charter service, bait & tackle, restaurants. Platypus Marine, near the east end of the marina, hauls up to 300 tons (417-0709). Inviting white cinderblock bathhouses with fiberglass shower stalls ($0.50/5 min.) are located at both ends of the basin.

Notable -- Port Angeles means recreation. Monthly festivals, exciting tours, museums, golf, hiking - and the list goes on. Grab a tour brochure at the Visitors' Center and take the self-guided Art on the Town tour. Rent a car at nearby Budget and do the 17 mi. drive to the Visitor Center in Olympic National Park (May-October). The park offers some truly breathtaking views of the Strait of Juan de Fuca. July's Sand Sculpture competition is a major event.

Navigational Information
Lat: 48°07.206' **Long:** 123°25.674' **Tide:** 7 ft. **Current:** n/a **Chart:** 18468
Rep. Depths *(MLW):* **Entry** 19 ft. **Fuel Dock** n/a **Max Slip/Moor** 19 ft./-
Access: Located on the south shore of Port Angeles harbor

Marina Facilities *(In Season/Off Season)*
Fuel: No
Slips: 5 Total, 5 Transient **Max LOA:** 55 ft. **Max Beam:** n/a
Rate *(per ft.):* **Day** $0.25* **Week** $1.75* **Month** n/a
Power: 30 amp n/a, **50 amp** n/a, **100 amp** n/a, **200 amp** n/a
Cable TV: No **Dockside Phone:** No
Dock Type: Floating, Long Fingers, Alongside, Wood
Moorings: 0 Total, 0 Transient **Launch:** n/a
Rate: Day n/a **Week** n/a **Month** n/a
Heads: 4 Toilet(s)
Laundry: None **Pay Phones:** No
Pump-Out: No **Fee:** n/a **Closed Heads:** Yes

Marina Operations
Owner/Manager: City of Port Angeles **Dockmaster:** n/a
In-Season: MemDay-Oct **Off-Season:** Nov-May, closed
After-Hours Arrival: Pay station at the top of the ramp
Reservations: No **Credit Cards:** No
Discounts: None
Pets: Welcome **Handicap Access:** No

Port Angeles City Pier

321 East 5th Street; Port Angeles, WA 98363

Tel: (360) 417-4550 **VHF: Monitor** n/a **Talk** n/a
Fax: (360) 417-4559 **Alternate Tel:** n/a
Email: recreation@cityofpa.us **Web:** cityofpa.us/parkrec-citypier.htm
Nearest Town: Port Angeles **Tourist Info:** (360) 452-2363

Marina Services and Boat Supplies
Supplies - Near: Ice *(Cube)*, Bait/Tackle *(Waters West 417-0937)*
Under 1 mi: Ships' Store

Boatyard Services
OnCall: Electronics Repairs **Under 1 mi:** Travelift *(70T)*, Engine mechanic
(gas, diesel), Hull Repairs. **Nearest Yard:** Port Angeles Boat Yard (360)
452-4444

Restaurants and Accommodations
Near: Restaurant *(Downriggers 452-2700)*, *(Golden Gate 457-6944)*, *(Port
Angeles Crab House 457-0424, L, D, Sun brunch)*, *(Thai Pepper 452-4995,
L $7-15, D $7-20)*, *(India Oven 452-5170, L, D)*, *(Delaney's 452-8456)*,
(Bella talia 457-5442), *(Harbor Café 452-8683)*, *(On the Boardwalk 457-
9554)*, Fast Food *(Fire Mountain Express 452-0736)*, Motel *(Riviera Inn
417-3955)*, Hotel *(Downtown 565-1125, $40-70)* **Under 1 mi:** Pizzeria *(All
About Pizza 417-1234)*, Hotel *(Red Lion 452-9215, $130-160)*, Inn/B&B
(Clarks' Harbor View 457-9891)

Recreation and Entertainment
OnSite: Beach, Picnic Area, Playground, Jogging Paths *(The Waterfront
Trail 6.5 mi.)*, Museum *(Marine Life Center 417-6254)*, Cultural Attract *(City
Pier Pavilion Wed. Nite Concerts 6-8pm; or Playhouse 452-6651 1 mi.)*
Near: Pool *(Shore 417-4595)*, Fitness Center *(Fitness West 452-1118)*, Boat
Rentals *(Sound Bikes & Kayaks 457-1240 $40/day)*, Roller Blade/Bike Paths,
Movie Theater *(Lincoln Theater 457-7997)*, Video Rental *(Blockbuster 417-
0616)*, Galleries *(Art Space 457-8115)* **Under 1 mi:** Special Events *(Arts in
Action Street Fair & Sand Sculpture - Jul, Dungeness Crab & Seafood*

Fest - Oct) **1-3 mi:** Park *(Olympic Nat'l. 457-4265)* **3+ mi:** Golf Course
(Sky Ridge 683-3673, 15 mi.)

Provisioning and General Services
Near: Convenience Store *(Chevron 457-3746)*, Supermarket *(Safeway 457-
0788)*, Health Food *(Country Aire Natural Foods 452-7175)*, Wine/Beer,
Bakery *(Star's Northwest 457-3279)*, Bank/ATM *(Bank of America 457-
1121)*, Post Office *(417-2940)*, Beauty Salon *(Julie's 457-7993)*, Barber Shop
(Laurel 457-9212), Dry Cleaners *(Jiffy 452-9309)*, Bookstore *(Port Book &
News 452-6367)*, Hardware Store *(Do It Best 457-3369)*, Department Store
(Gottschalks 452-4571), Copies Etc. *(Pen-Print 457-3404)* **Under 1 mi:** Deli-
catessen *(Olympic Bagel 452-9100)*, Liquor Store *(WA Liquor 417-1737)*,
Farmers' Market *(Sat 10am-2pm year-round, corner of 4th & Peabody St.)*,
Fishmonger *(Port Angeles Boat Haven)*, Catholic Church *(Queen of Angels
452-2351)*, Library *(452-9253)*, Laundry *(Speed Klean 457-4044)*, Pharmacy
(Rite Aid 452-9784), Florist *(A Corner-Copia of Flowers 565-1256)*
1-3 mi: Synagogue *(Olympic B'Nai Shalom Havarah 452-2471)*

Transportation
OnCall: Rental Car *(Enterprise 417-3083/Budget 452-4774, nearby)*, Taxi
(Blue Top 452-2223) **Near:** Local Bus *(Clallam Transit 452-4511)*, Ferry
Service *(Coho 457-4491/ Express 452-8088)* **Airport:** Fairchild Int'l. *(4 mi.)*

Medical Services
911 Service **OnCall:** Ambulance **Near:** Doctor *(Family Medicine 452-
7891)*, Dentist *(Peninsula Dental 452-3808)* **Under 1 mi:** Chiropractor
(Olympic Chiro Clinic 457-8292), Veterinarian *(Port Angeles Veterinary Clinic
452-5541)* **Hospital:** Olympic Memorial 417-7000 *(1 mi.)*

Setting -- The Port Angeles City Pier is located east of the ferry dock, toward the eastern end of the harbor. A landmark 50-foot bright blue observation tower on the end of the pier is an unmistakable beacon. A main float parallels the pier, with long fingers on the other side. The pier itself, shaped like an upside-down "L," offers some protection. Onshore, a wooden picnic deck offers beautiful views of the harbor and the adjacent sandy beach. The city's popular waterfront trail, part of the Olympic Discovery Trail, extends in both directions from the pier; walk or bike the 6.5 miles around the waterfront.

Marina Notes -- *Flat rate - $10/day. Limit on the pier is 24 hours. Very sturdy, well-maintained docks have wide finger piers and are connected to the main pier by aluminum gangways. There is no power, or other amenities, available for boaters, but the location makes up for this. The floats are seasonal, available only until October. The restrooms are on the pier and open to the public. The nearby city pool has showers.

Notable -- Enjoy the picnic area that overlooks the harbor or the tables located right by the beach. Summer concerts are held on-site on Wednesday nights. The covered pavilion used as a stage can be rented for group functions. A small play area is nearby. Also at the pier is the Art Feiro Marine Life Center, a working lab open to the public. The Pier provides an excellent location for exploring downtown Port Angeles. Everything is close by, including restaurants, laundry, groceries, and the Chamber of Commerce. The town is also the gateway to Olympic National Park, and car rentals are nearby. Ferry service across the Strait: the Coho ferry carries cars and passengers year-round (457-4491); Victoria Express, for passengers only, operates May-September (800-633-1589).

17. WA - STRAIT OF JUAN DE FUCA

PHOTOS ON DVD: 15

John Wayne Marina

John Wayne Marina

2577 West Sequim Bay Road; Sequim, WA 98382

Tel: (360) 417-3440 **VHF: Monitor** n/a **Talk** n/a
Fax: (360) 417-3442 **Alternate Tel:** n/a
Email: rona@portofpa.com **Web:** portofpa.com/marinas
Nearest Town: Sequim, WA *(3 mi.)* **Tourist Info:** (360) 683-6198

Navigational Information
Lat: 48°03.661' **Long:** 123°02.446' **Tide:** 9 ft. **Current:** n/a **Chart:** 18471
Rep. Depths *(MLW):* **Entry** 12 ft. **Fuel Dock** 12 ft. **Max Slip/Moor** 12 ft./-
Access: Located on the western side of Sequim Bay at Pitship Point

Marina Facilities *(In Season/Off Season)*
Fuel: Gasoline, Diesel
Slips: 302 Total, 22 Transient **Max LOA:** 164 ft. **Max Beam:** n/a
 Rate *(per ft.):* **Day** $0.88* **Week** n/a **Month** $4.93-6.45**
 Power: 30 amp $5.85, **50 amp** n/a, **100 amp** n/a, **200 amp** n/a
 Cable TV: No **Dockside Phone:** Yes, $33.75
 Dock Type: Floating, Long Fingers, Alongside, Concrete, Wood
Moorings: 0 Total, 0 Transient **Launch:** yes (8.)
 Rate: Day n/a **Week** n/a **Month** n/a
Heads: 14 Toilet(s), 4 Shower(s)
Laundry: 2 Washer(s), 2 Dryer(s) **Pay Phones:** Yes, 1
Pump-Out: OnSite **Fee:** Free **Closed Heads:** Yes

Marina Operations
Owner/Manager: Port of Port Angeles **Dockmaster:** Ron Amundson
In-Season: Year-Round, 8am-5pm **Off-Season:** n/a
After-Hours Arrival: Register at the office
Reservations: No **Credit Cards:** No
Discounts: None
Pets: Welcome **Handicap Access:** Yes

Marina Services and Boat Supplies
Services - Security *(nights)*, Trash Pick-Up **Communication -** Mail & Package Hold, Fax in/out, Data Ports *(Wi-Fi - Broadband Xpress)*, FedEx, UPS, Express Mail **Supplies - OnSite:** Ships' Store *(Bosun's Locker 683-6521)* **1-3 mi:** Ice *(Cube)*, Propane *(True Value 683-4111)*

Boatyard Services
OnSite: Launching Ramp *($8)* **3+ mi:** Travelift *(60T/70T/300T, 28 mi.)*.
Nearest Yard: Port Townsend Boat Haven (360) 385-2355

Restaurants and Accommodations
Under 1 mi: Inn/B&B *(Greywolf B&B 683-5889, $85-135)* **1-3 mi:** Restaurant *(Dynasty Chinese 683-6511, D $8-14)*, *(Las Palomas 681-3842)*, Coffee Shop *(Hurricane Coffee Co. 681-6008)*, Motel *(Ramada 683-1775, $95-140)*, *(Econo Lodge 683-7113)* **3+ mi:** Restaurant *(Khu Larb Thai 681-8550, 3 mi.)*, *(Riptide 683-7244, 3 mi.)*, *(Oak Table Café 683-2179, B $7-12, L $9-12, 3 mi., B daily all day, L Mon-Sat after 11)*, *(Tarcisio 683-5809, 3 mi.)*, Fast Food *(Burger King, Subway 3 mi.)*, Pizzeria *(The Unknown Pizza Co. 683-9600, 3 mi.)*, Motel *(Sundowner 683-5532, $49-69, 4 mi.)*

Recreation and Entertainment
OnSite: Beach, Picnic Area, Group Fishing Boat *(Venture Charters 895-5424)* **1-3 mi:** Video Rental *(Hollywood Video 582-1533)*, Park *(Sequim Bay State Park 683-4235)*, Special Events *(Juan de Fuca Fest - MemDay 457-5411, Lavender Fest - Jul 683-3334, Rotary Salmon Bake - Aug 683-3840)* **3+ mi:** Golf Course *(Sky Ridge 683-3673, 8 mi.)*, Fitness Center *(Sequim Gym 681-2555, 3 mi.)*, Museum *(Museum and Arts Center 683-8110, 3 mi.)*, Galleries *(Blue Whole 681-6033, Native Expressions 681-4640, 3 mi.)*

Provisioning and General Services
OnSite: Convenience Store **1-3 mi:** Supermarket *(QFC 683-1151)*, Bank/ATM, Beauty Salon *(Kathy's Hair Styling 683-5092)*, Pharmacy *(QFC 683-1156)* **3+ mi:** Market *(D&L Grocery Deli 683-3065, 3 mi.)*, Delicatessen *(Jean's Deli 683-6727, 3 mi.)*, Health Food *(Sunny Farms 683-6056, 3 mi.)*, Liquor Store *(Liquor Store 683-3879, 4 mi.)*, Farmers' Market *(Cedar St. Sat 9am-3pm, 3 mi.)*, Post Office *(681-7833, 3 mi.)*, Catholic Church *(St. Joseph's 683-6076, 3 mi.)*, Protestant Church *(St. Luke's Episcopal 683-4862, 3 mi.)*, Library *(Sequim 683-1161, 4 mi.)*, Barber Shop *(Northwest Haircut 681-3583, 3 mi.)*, Dry Cleaners *(Sequim Dry Cleaning Laundry 683-2642, 3 mi.)*, Bookstore *(The Good Book 683-3600, 3 mi.)*, Hardware Store *(True Value 683-4111, 3 mi.)*, Florist *(Garden Florist 683-4709, 4 mi.)*, Clothing Store *(Mad Maggi Clothing Boutique 683-2239, 4 mi.)*, Buying Club *(Costco 681-2031, 7 mi.)*, Copies Etc. *(Staples 681-5493, 3 mi.)*

Transportation
OnSite: Bikes **OnCall:** Taxi *(Sun Taxi 683-1872)* **Under 1 mi:** Local Bus *(Jefferson Transit 385-4777)* **3+ mi:** Rental Car *(Enterprise 417-3083, 18 mi.)* **Airport:** Sequim Valley/ Fairchild Int'l. *(9 mi./19 mi.)*

Medical Services
911 Service **OnCall:** Ambulance *(683-3347)* **1-3 mi:** Chiropractor *(West Bay 683-8111)*, Veterinarian *(Greywolf Vet 683-2106)* **3+ mi:** Doctor *(North Olympic Family Practice 683-7246, 3 mi.)*, Dentist *(Family Dentistry 683-4850, 3 mi.)*, Holistic Services *(Naturopathic Wellness 683-1110, 3 mi.)* **Hospital:** Olympic Memorial 417-7000 *(19 mi.)*

Setting -- Serene Sequim ("Skwim") Bay lies approximately 6 miles southeast of Dungeness Bay. The bay is separated from the Strait by long Travis Spit which creates a narrow entry channel. John Wayne Marina is shaped like a horseshoe and very protected - a rock breakwater wraps around the east and south sides of the impeccable concrete docks. 22 gorgeous parklike acres are studded with picnic spots and a lovely gray and French-blue two-story contemporary trimmed with decks houses an office, store and banquet facility. On a clear day, Victoria, B.C. can be seen across the straits.

Marina Notes -- *Flat rate: under 20' $15/day, 20-29' $25, 30-39' $30, 40-49' $35, over 50' $40/day. **Monthly fees $4.93-6.45/ft./mo., depending on size. Built in 1985 on land donated by the John Wayne family. The actor often visited the bay on his "Wild Goose" and imagined a facility on its west side. Operated by the Port of Port Angeles. A large meeting room with dance floor is ideal for club cruises. Bosun's Locker sells marine hardware and supplies, charts, clothing and gifts and John Wayne memorabilia and rents powerboats, kayaks, and fishing gear. Liveaboards permitted. The bathhouses are cinderblock with sparkling all-tiled showers ($0.25 for 2.5 min.). On-site Sequim Bay Y.C. offers reciprocal moorage (683-1338). Note: Onsite restaurant is temporarily closed - Inquire.

Notable -- Manicured lawns with shade trees, picnic areas and benches invite relaxation. Rent a crab pot and drop it right off the dock for your own Dungeness crab. You can also dig clams on the beach, but call the shellfish safety hotline first (800-562-5632). Explore the bay in a kayak for some inspired bird-watching. Do the drive or walking tour in the Olympic Game Farm (683-4295). For shops and restaurants, Sequim is just over 3 miles away.

Navigational Information
Lat: 48°02.400' **Long:** 123°01.499' **Tide:** 9 ft. **Current:** n/a **Chart:** 18471
Rep. Depths (*MLW*): **Entry** 10 ft. **Fuel Dock** n/a **Max Slip/Moor** 6 ft./9 ft.
Access: Located at the southwest end of Sequim Bay

Marina Facilities *(In Season/Off Season)*
Fuel: No
Slips: 10 Total, 10** Transient **Max LOA:** 50 ft. **Max Beam:** n/a
Rate *(per ft.)*: **Day** $0.50* **Week** n/a **Month** n/a
Power: 30 amp n/a, 50 amp n/a, 100 amp n/a, 200 amp n/a
Cable TV: No **Dockside Phone:** No
Dock Type: Floating, Alongside, Wood
Moorings: 6 Total, 6 Transient **Launch:** n/a
Rate: Day $10 **Week** n/a **Month** n/a
Heads: 6 Toilet(s)
Laundry: None **Pay Phones:** Yes, 2
Pump-Out: No **Fee:** n/a **Closed Heads:** Yes

Marina Operations
Owner/Manager: WA State Parks and Recreation **Dockmaster:** n/a
In-Season: Year-Round, 8am-Dusk **Off-Season:** n/a
After-Hours Arrival: Pay at the pay station
Reservations: Yes, preferred **Credit Cards:** No
Discounts: None
Pets: Welcome **Handicap Access:** Yes, Heads, Docks

Sequim Bay State Park

269035 Highway 101; Sequim, WA 98382

Tel: (360) 683-4235 **VHF: Monitor** n/a **Talk** n/a
Fax: n/a **Alternate Tel:** n/a
Email: n/a **Web:** www.parks.wa.gov
Nearest Town: Sequim *(5 mi.)* **Tourist Info:** (360) 683-6198

Marina Services and Boat Supplies
Supplies - 1-3 mi: Ships' Store *(John Wayne Marina 417-3440)*,
Bait/Tackle **3+ mi:** Propane *(True Value 683-4111, 5 mi.)*

Boatyard Services
OnSite: Launching Ramp **3+ mi:** Travelift *(60T/70T/300T, 27 mi.)*.
Nearest Yard: Port Townsend Boat Haven (360) 385-2355

Restaurants and Accommodations
Under 1 mi: Restaurant *(Xanadu Grill 681-0928)*, Snack Bar *(Daisy Donut 683-6518)*, Motel *(Sequim Bay Lodge 683-0691)* **1-3 mi:** Inn/B&B *(Greywolf 683-5889, $85-135)* **3+ mi:** Restaurant *(Dynasty Chinese 683-6511, D $8-14, 4 mi.)*, *(Gwennie's 683-4157, 4 mi.)*, *(Oak Table Café 683-2179, B $7-12, L $9-12, 5 mi., B daily all day, L Mon-Sat after 11)*, *(El Cazador 683-4788, L $6-10, D $8-13, 5 mi., Mexican)*, Fast Food *(Subway, Burger King 5 mi.)*, Pizzeria *(Unknown Pizza Co. 683-9600, 5 mi.)*, *(Dominos 582-1600, 5 mi.)*, Motel *(Ramada 683-1775, $95-140, 4 mi.)*, *(Econo Lodge 683-7113, 4 mi.)*

Recreation and Entertainment
OnSite: Beach, Picnic Area, Grills, Playground, Tennis Courts, Park, Cultural Attract *(Interpretive Center Mon-Sun 9am-9pm)* **1-3 mi:** Boat Rentals *(Bosun's Locker 693-6521 $35-75/hr.)* **3+ mi:** Golf Course *(Sky Ridge 683-3673, 10 mi.)*, Fitness Center *(Sequim Gym 681-2555, 5 mi.)*, Bowling *(Olympic Lanes 683-3500, 5 mi.)*, Video Rental *(Hollywood Video 582-1533, 4 mi.)*, Museum *(Museum and Arts Center 683-8110, 5 mi.)*, Sightseeing *(Olympic Game Farm 683-4295, 9 mi.)*, Special Events *(Irrigation Fest - May, Lavender Fest - Jul, Salmon Bake - Aug, 5 mi.)*, Galleries *(Wildlight 582-9900, Blue Whole 681-6033, 5 mi.)*

Provisioning and General Services
3+ mi: Convenience Store *(B's Lil Store 683-7475, 5 mi.)*, Supermarket *(QFC 683-1151, 4 mi.)*, Gourmet Shop *(Alaska Catch 683-4030 - wild salmon, 5 mi.)*, Delicatessen *(Jean's Deli 683-6727, 5 mi.)*, Health Food *(Sunny Farms 683-6056, 5 mi.)*, Liquor Store *(Liquor Store 683-3879, 6 mi.)*, Farmers' Market *(Cedar St. Sat 9am-3pm, 5 mi.)*, Bank/ATM *(Olympic Bank 683-9855, 5 mi.)*, Post Office *(681-7833, 5 mi.)*, Catholic Church *(St. Joseph's 683-6076, 5 mi.)*, Protestant Church *(St. Luke's 683-4862, 5 mi.)*, Library *(Sequim 683-1161, 6 mi.)*, Beauty Salon *(Reflections 683-2233, 5 mi.)*, Barber Shop *(Northwest 681-3583, 5 mi.)*, Dry Cleaners *(Sequim 683-2642, 5 mi.)*, Bookstore *(Pacific Mist 683-1396, 5 mi.)*, Pharmacy *(QFC 683-1157, 4 mi.)*, Hardware Store *(Rookard Hardware 683-3939, 5 mi.)*, Clothing Store *(5 mi.)*, Buying Club *(Costco 681-2031, 10 mi.)*, Copies Etc. *(Staples 681-5493, 5 mi.)*

Transportation
Under 1 mi: Local Bus *(Jefferson Transit 385-4777)* **1-3 mi:** Bikes *(John Wayne Marina)* **3+ mi:** Rental Car *(Enterprise 417-3083, 20 mi.)*, Ferry Service *(Port Angeles - Victoria, 20 mi.)* **Airport:** Sequim Valley/ Fairchild Int'l. *(11 mi./21 mi.)*

Medical Services
911 Service **OnCall:** Ambulance *(683-3347)* **3+ mi:** Doctor *(North Olympic 683-7246, 5 mi.)*, Dentist *(Family Dentistry 683-4850, 5 mi.)*, Chiropractor *(West Bay 683-8111, 4 mi.)*, Holistic Services *(Naturopathic Wellness 683-1110, 5 mi.)*, Veterinarian *(Sequim Animal Hospital 683-7286, 5 mi.)*
Hospital: Olympic Memorial 417-7000 *(20 mi.)*

Setting -- Sequim Bay State Park is located toward the end of Sequim Bay, on the western shore. This year-round marine park offers 4,909 feet of saltwater coast and 92 acres of forest to explore. A stationary wooden pier reaches out into the bay, and a seasonal dock floats at the end. Mooring buoys are placed farther out. From the dock, the view up the bay is 1.5 miles north to John Wayne Marina or shoreside to the forested hill ablaze with wild rhododendrons.

Marina Notes -- *$10 minimum at the dock, and $10 flat rate for buoys. 36-hr. limit. Annual permits (good for the calendar year) are available at the park when staff is available, or purchase online or at WA State Parks' offices. Fee for an annual permit is $3.50/ft. with a $50 minimum. **The wide wooden dock provides 424 ft. of alongside space during the summer. Some picnic tables are conveniently located right on the float. 6 deep water buoys offer additional moorage. A wide launching ramp is also on-site. If you bring a car, there's a $5 fee for the day or $10 overnight. Two kitchen shelters without electricity plus 20 sheltered and 15 open picnic tables are reservable (888-226-7688). Basic stone & cinderblock heads are near the pier.

Notable -- Two overlapping sandbars protect the bay - Sequim means "quiet waters." This engaging park offers campgrounds, some RV sites, picnic areas, hiking trails, a beach, swings, horseshoe pits, plus a tunnel that connects to a baseball field and tennis courts. The Interpretive Center has information on local fishing and shellfish harvesting. The town of Sequim is 5 miles away; self-guided tours are available. The Museum and Arts Center has a mastodon exhibit and local history. (Tuesday-Saturday 8am-4pm, free). The Blue Whole Gallery features local art - meet the artists the 1st Friday of every month from 5:30-7:30pm.

PHOTOS ON DVD: 12

Fort Worden State Park

Fort Worden State Park

200 Battery Way; Port Townsend, WA 98368

Tel: (360) 344-4400; (800) 452-5687 **VHF: Monitor** n/a **Talk** n/a
Fax: (360) 385-7248 **Alternate Tel:** n/a
Email: fwinfo@fortworden.org **Web:** www.fortworden.org
Nearest Town: Port Townsend *(2 mi.)* **Tourist Info:** (360) 385-7869

Navigational Information
Lat: 48°08.150' **Long:** 122°45.670' **Tide:** 9 ft. **Current:** n/a **Chart:** 18441
Rep. Depths *(MLW)*: **Entry** 9 ft. **Fuel Dock** n/a **Max Slip/Moor** 3 ft./20 ft.
Access: Located on Admiralty Inlet just south of Point Wilson

Marina Facilities *(In Season/Off Season)*
Fuel: No
Slips: 5 Total, 5** Transient **Max LOA:** 55 ft. **Max Beam:** n/a
  Rate *(per ft.)*: **Day** $0.50* **Week** n/a **Month** n/a
  Power: 30 amp n/a, **50 amp** n/a, **100 amp** n/a, **200 amp** n/a
  Cable TV: No **Dockside Phone:** No
  Dock Type: Floating, Alongside, Concrete, Wood
Moorings: 8 Total, 8 Transient **Launch:** n/a
  Rate: Day $10 **Week** n/a **Month** n/a
Heads: 4 Toilet(s)
Laundry: None **Pay Phones:** Yes
Pump-Out: No **Fee:** n/a **Closed Heads:** Yes

Marina Operations
Owner/Manager: WA Sate Parks and Recreation **Dockmaster:** n/a
In-Season: Year-Round, 8am-Dusk **Off-Season:** n/a
After-Hours Arrival: Pay at the pay station
Reservations: Yes, Preferred **Credit Cards:** No
Discounts: None
Pets: Welcome **Handicap Access:** No

Marina Services and Boat Supplies
Supplies - Near: Ice *(Cube)*, Bait/Tackle **1-3 mi:** West Marine *(379-1612)*

Boatyard Services
OnSite: Launching Ramp *($5)* **OnCall:** Divers **1-3 mi:** Travelift *(60T/70T/300T)*, Engine mechanic *(gas, diesel)*, Electrical Repairs, Electronics Repairs, Hull Repairs, Rigger, Sail Loft, Canvas Work, Air Conditioning, Refrigeration, Propeller Repairs. **Nearest Yard:** Port Townsend Boat Haven (360) 385-2355

Restaurants and Accommodations
OnSite: Lite Fare *(Hot Java 344-4411, B $6-9)*, *(Cable House Canteen 385-3419, burgers)*, *(Commons Coffee & Snacks B $8.50, L $10, D $14.50, 7am-5pm, Kids' 4-8 1/2 price)*, Motel *(Fort Worden Dormitories 344-4400, $10-16, $34-40/nt. incl. meals)*, Condo/Cottage *(Fort Worden Victorian Houses 385-7248, $102-359, elegant officers' row; houses - 2 -11 bdrms)* **Near:** Motel *(Olympic 344-4400, $7-17, couples add $8)* **Under 1 mi:** Motel *(Big Red Barn 385-4837)* **1-3 mi:** Restaurant *(Public House Grill & Ales 385-9708, L $7-12, D $9-18, crab chowder)*, *(Landfall 385-5814, waterfront; Thu-Mon 7am-9pm, Tue-Wed 7am-3pm)*, *(Shanghai 385-4810)*, *(Galatea Café & Tapas Bar 385-5225, Wed-Sat from 5pm, Sun brunch 10:30am)*, Pizzeria *(Waterfront 385-6629)*

Recreation and Entertainment
OnSite: Beach, Picnic Area, Grills, Boat Rentals *(Paddle & Pedal 344-4520 $45-60/half day or Kayak Port Townsend 385-6240 2 mi.)*, Museum *(Coast Artillery Museum 385-0373, Commanding Officers Quarters 344-4400; Rothschild House - $11 for all Ft. Worden sites May-Sep 11-4/5)*, Cultural

Attract *(Marine Science Center 385-5582 - Apr-Oct)*, Special Events *(Concerts - Fri Noon; Chamber Music Fest - Jun; Jazz Fest - Jul)* **1-3 mi:** Golf Course *(Port Townsend 385-4547)*, Fitness Center *(Gym 385-3674)*, Movie Theater *(Uptown 385-3883)*, Video Rental *(Video Mart 385-4443)*

Provisioning and General Services
OnSite: Convenience Store *(Cable House Canteen 385-3419)*
Under 1 mi: Beauty Salon *(Gina Holeman 385-7147)*, Newsstand
1-3 mi: Market *(Penny Saver Mart 385-2695)*, Supermarket *(Safeway 385-2806)*, Delicatessen *(Lehani's 385-3961)*, Health Food *(Food Co-Op 385-2883)*, LiquorStore *(WA Liquor 379-5050)*, Bakery *(Pane D'Amore 385-1199)*, Farmers' Market *(Lawrence & Tyler St. Sat 9:30am-1:30pm)*, Fishmonger *(Key City 379-5516)*, Bank/ATM, Post Office, Catholic Church, Protestant Church, Library *(Port Townsend 385-3181)*, Dry Cleaners *(Dockside 385-4585)*, Laundry *(Carol's 385-5508)*, Bookstore *(Phoenix Rising 385-4464)*, Pharmacy *(Don's 385-2622)*, Hardware Store *(Swain's 385-1313)*

Transportation
OnSite: Bikes *(Mt. Bike Port Townsend 344-4520)* **OnCall:** Taxi *(Peninsula Taxi 385-1872)* **1-3 mi:** Rental Car *(Budget 385-7766)*, Ferry Service *(to Keystone 888-808-7977/ Puget Sound Express to Friday Harbor 385-5288)*
Airport: Jefferson County Int'l. *(8 mi.)*

Medical Services
911 Service **OnCall:** Ambulance **1-3 mi:** Doctor *(Monroe St. Medical Clinic 385-5658)*, Dentist *(Uptown Dental 385-4700)*, Chiropractor *(Port Townsend Chiropractic 379-0800)*, Holistic Services *(Olympic Naturopathic 385-2107)*
Hospital: Jefferson General 385-2200 *(3 mi.)*

Setting -- Fort Worden State Park occupies the northeast end of the Olympic Peninsula, where the Straight of Juan de Fuca meets Admiralty Inlet at the historic Point Wilson lighthouse. 434 acres of woods, trails and restored structures - plus two miles of waterfront and sandy beach that leads to a lighthouse - make this popular destination. Perched on a high bluff are a museum, old officers' quarters converted to accommodate visitors, and numerous historic sites and buildings. A long, L-shaped wood wharf, that hosts the Marine Science Center, leads to the moorage floats; 8 buoys are available as well.

Marina Notes -- *$10 minimum. Buoys are $10/night. Annual pass holder fee: $3.50/ft., $50 minimum. **235 feet of alongside tie-ups on the float. The sturdy pier helps cut down the wake from boat traffic and the crossing ferries. 3-night limit for boaters. A cement launching ramp and dinghy dock are on-site, but no power or pump-outs are available. A convenience store and casual eateries are in the park as are kayak and bike rentals. To book event space call 344-4435. Public heads are nearby. "An Officer and a Gentleman" was shot here. Port Townsend is two miles south.

Notable -- Housed in a small tan building on the wharf, the Marine Science Center's Lab and Aquarium offers exhibits and interpretive programs popular with kids (April-October). On the upland are the Natural History building and outdoor exhibits. The Coast Artillery Museum describes the fort's history; gun batteries guard the entrance to Puget Sound. The Centrum Institute organizes fantastic events, held in the amphitheater or in the open - from poetry readings to lawn concerts (385-3102). The Park's conference center maintains various-sized converted buildings - including the exquisite Commons that accommodates 600.

Navigational Information

Lat: 48°06.983' **Long:** 122°45.039' **Tide:** 13 ft. **Current:** 2 kt. **Chart:** 18464
Rep. Depths (*MLW*): Entry 9 ft. **Fuel Dock** n/a **Max Slip/Moor** 12 ft./-
Access: Located on the west shore of Port Townsend Bay

Marina Facilities (*In Season/Off Season*)

Fuel: No
Slips: 45 Total, 5 Transient **Max LOA:** 65 ft. **Max Beam:** n/a
  Rate (*per ft.*): **Day** $1.00 **Week** n/a **Month** $7.50
  Power: 30 amp $3, **50 amp** $5, **100 amp** n/a, **200 amp** n/a
  Cable TV: No **Dockside Phone:** No
  Dock Type: Floating, Long Fingers, Pilings, Alongside, Wood
Moorings: 0 Total, 0 Transient **Launch:** n/a
  Rate: Day n/a **Week** n/a **Month** n/a
Heads: 6 Toilet(s), 6 Shower(s)
Laundry: 4 Washer(s), 4 Dryer(s) **Pay Phones:** Yes
Pump-Out: OnSite **Fee:** n/a **Closed Heads:** Yes

Marina Operations

Owner/Manager: Ken Radon **Dockmaster:** Chris Wenger
In-Season: May-Sep, 8am-5pm **Off-Season:** Oct-Apr, 8am-4:30pm*
After-Hours Arrival: Self registration at office
Reservations: Yes, only for boats under 40' **Credit Cards:** Visa/MC
Discounts: None
Pets: Welcome, Dog Walk Area **Handicap Access:** No

Point Hudson Resort and Marina

PO Box 1180; 103 Hudson Street; Port Townsend, WA 98368

Tel: (360) 385-2828 **VHF: Monitor** Ch. 9 **Talk** n/a
Fax: (360) 385-7331 **Alternate Tel:** n/a
Email: n/a **Web:** www.portofpt.com/point_hudson.htm
Nearest Town: Port Townsend (*0.1 mi.*) **Tourist Info:** (360) 385-7869

Marina Services and Boat Supplies

Communication - Fax in/out, Data Ports (*Wi-Fi - Broadband Xpress*)
Supplies - OnSite: Ice (*Block, Cube*), Ships' Store, Propane (*Fleet Marine 385-4000*), CNG **Under 1 mi:** West Marine (*379-1612*), Bait/Tackle (*Port Townsend Angler 379-3763*)

Boatyard Services

OnSite: Travelift (*35T*), Launching Ramp (*$5/day*)

Restaurants and Accommodations

OnSite: Restaurant (*Otter Crossing Café 379-0592, B $4-7, L $6-8*), (*Shanghai 385-4810, L $6-7, D $7-16*) **Near:** Restaurant (*Public House Grill and Ales 385-9708, L $7-12, D $9-18*), (*Landfall 385-5814, Thu-Mon 7am-9pm, Tue-Wed 7am-3pm*), (*Water St. Brewing House 379-6438*), (*Silverwater Café 385-6448, L $7-13, D $11-20*), Lite Fare (*Jordini's Subs 385-2037*), Pizzeria (*Waterfront Pizza 385-6629*), Hotel (*Water Street Hotel 385-5467, $125-195*), (*Bishop & Swan 385-1718*), Inn/B&B (*Inn at Waterfront Place 385-6957, $110-125*) **Under 1 mi:** Restaurant (*The Wild Coho 379-1030, D $18-22, Tue-Sat; Thu small plates $5-6*), Hotel (*Port Townsend Inn 385-2211, $48-98*), Inn/B&B (*Blue Gull B&B 379-3241, $95-200*)

Recreation and Entertainment

OnSite: Beach (*north of the RV park*), Picnic Area, Special Events (*2-day Wooden Boat Fest - Sep 385-3628, boat rentals too*) **Near:** Playground, Fitness Center (*Port Townsend Athletic Center 385-6560*), Boat Rentals (*Kayak Port Townsend 385-6240*) **Under 1 mi:** Dive Shop (*Dive Shop 379-3635*), Golf Course (*Port Townsend 385-4547 18 holes $40/cart $20*), Movie Theater (*Rose Theater 385-1089*), Sightseeing (*PS Express 385-5288 whale

watching, scenic charters*), Galleries (*Port Townsed Gallery 379-8110; Gallery Walk 1st Sat each month*) **1-3 mi:** Video Rental (*Hollywood 379-2580*), Park, Museum (*Marine Science Center 385-5582, Coast Artillery 385-0373*), Cultural Attract (*Centrum Institute 344-4435/385-3102*)

Provisioning and General Services

Near: Convenience Store, Bank/ATM, Beauty Salon (*Hair Studio 385-4240*)
Under 1 mi: Market (*Aldrich's 385-0500*), Delicatessen (*Lehani's Deli & Coffee 385-3961*), Health Food (*The Food Co-Op 385-2883*), Liquor Store (*WA Liquor 379-5050*), Bakery (*Sweet Laurette Patisserie 385-4886*), Farmers' Market (*Lawrence & Tyler St. Sat 9:30am-1:30pm*), Post Office, Catholic Church, Protestant Church, Library (*385-3181*), Barber Shop (*Jim's Barber 385-2266*), Dry Cleaners (*Dockside Cleaners 385-4585*), Laundry (*Carol's 385-5508*), Bookstore (*The Imprint 385-3643*), Pharmacy (*Don's 385-2622*), Hardware Store (*Swain's 385-1313*) **1-3 mi:** Supermarket (*Safeway 385-2806*), Fishmonger (*Key City Fish Co. 379-5516*)

Transportation

OnCall: Taxi (*Peninsula Taxi 385-1872*) **Near:** Local Bus (*Jefferson County Transit 385-3020*) **Under 1 mi:** Ferry Service (*WA Ferry to Keystone 888-808-7977/ Puget Sound Express to Friday Harbor 385-5288*) **1-3 mi:** Rental Car (*Budget 385-7766*) **Airport:** Jefferson County Int'l. (*7 mi.*)

Medical Services

911 Service **OnCall:** Ambulance **Near:** Doctor (*Monroe St. Clinic 385-5658*) **Under 1 mi:** Dentist (*Uptown Dental 385-4700*), Chiropractor (*Carlson Chiropractic 385-0322*) **1-3 mi:** Veterinarian (*Sherwood 385-0512*) **Hospital:** Jefferson General 385-2200 (*2 mi.*)

Setting -- At the northwest end of Admiralty Inlet, Point Hudson Resort and Marina sits just around Point Hudson, facing Marrowstone and Indian Islands. A man-made basin shaped like a deep "U" hosts the docks; lighted jetties protect the entrance. On the east side, finger slips lie perpendicular to the shore. On the west shore, a long dock berths the pump-out station and provides additional side-tie space. The basin is edged by a series of white clapboard buildings housing the office and amenities, Wooden Boat Foundation with the cupola, and Otter Crossing with outdoor seating overlooking the docks.

Marina Notes -- *Closed Sun in the off-season. Formerly a Coast Guard station, this quiet, protected facility is managed by Port of Port Townsend. Reserveable slips accommodate boats to 40 feet; larger vessels tie up alongside 800 feet of linear dockage (first-come, first-serve), rafting up to 5 permitted. Fleet Marine, at the head of the basin, provides a travelift, store and some yard services (385-4000). On-site Wooden Boat Foundation (385-3628) offers seminars, sailing classes, and a small store; beautiful wooden craft can always be seen at the docks, and their annual festival is a do-not-miss for wooden boat enthusiasts. The bathhouse is shared by the residents of the adjacent RV park on the point. Note: Construction of a redesigned facility began Oct. '06.

Notable -- Point Hudson Resort lies on the northeast edge of the Victorian seaport and arts community of Port Townsend. Dozens of late 19th C. historic buildings have been restored and converted into unique galleries, a wealth of restaurants, and bed-and-breakfasts. With everything within walking distance, it's easy to enjoy the town. Or head to Fort Worden's Marine Science Center, Coast Artillery Museum, or any of the Centrum Institute's fabulous cultural events.

Navigational Information

Lat: 48°06.411' **Long:** 122°46.225' **Tide:** 9 ft. **Current:** n/a **Chart:** 18464
Rep. Depths (MLW): Entry 12 ft. **Fuel Dock** 12 ft. **Max Slip/Moor** 12 ft./-
Access: Located on Port Townsend Bay just west of town

Marina Facilities *(In Season/Off Season)*

Fuel: Gasoline, Diesel
Slips: 400 Total, 25 Transient **Max LOA:** 100 ft. **Max Beam:** n/a
  Rate *(per ft.):* **Day** $0.65/0.60 **Week** n/a **Month** $9
  Power: 30 amp $3, **50 amp** $3-5*, **100 amp** n/a, **200 amp** n/a
  Cable TV: No **Dockside Phone:** No
  Dock Type: Floating, Long Fingers, Alongside, Concrete, Wood
Moorings: 0 Total, 0 Transient **Launch:** n/a
  Rate: Day n/a **Week** n/a **Month** n/a
Heads: 6 Toilet(s), 4 Shower(s)
Laundry: 2 Washer(s), 3 Dryer(s) **Pay Phones:** Yes
Pump-Out: OnSite **Fee:** n/a **Closed Heads:** Yes

Marina Operations

Owner/Manager: Port of Port Townsend **Dockmaster:** Ken Radon
In-Season: Mon-Fri, 8am-4:30pm **Off-Season:** Sat-Sun, 8am-6:30pm
After-Hours Arrival: Check map outside marina office for available slips
Reservations: Yes, Preferred **Credit Cards:** Visa/MC
Discounts: None
Pets: Welcome **Handicap Access:** No

Port Townsend Boat Haven

PO Box 1180; 2601 Washington St.; Port Townsend, WA 98368

Tel: (360) 385-2355 **VHF: Monitor** Ch. 9 **Talk** Ch. 66A
Fax: (360) 379-8205 **Alternate Tel:** n/a
Email: info@portofpt.com **Web:** www.portofpt.com
Nearest Town: Port Townsend *(0.7 mi.)* **Tourist Info:** (360) 385-7869

Marina Services and Boat Supplies

Services - Trash Pick-Up, Dock Carts **Communication -** Data Ports *(Wi-Fi - Broadband Xpress)* **Supplies - OnSite:** Ice *(Cube)*, Ships' Store **Near:** West Marine *(379-1612)*, Bait/Tackle *(Port Townsend Angler 379-3763)* **1-3 mi:** Propane *(Fleet Marine 385-4000)*, CNG

Boatyard Services

OnSite: Travelift *(60T/70T/330T- to 150 ft. long x 30 ft. wide)*, Forklift, Crane, Engine mechanic *(gas, diesel)*, Launching Ramp *($5/day)*, Electrical Repairs, Electronic Sales, Electronics Repairs, Hull Repairs, Rigger, Sail Loft, Canvas Work, Bottom Cleaning, Brightwork, Air Conditioning, Refrigeration, Propeller Repairs, Woodworking, Upholstery, Yacht Interiors, Metal Fabrication, Painting, Awlgrip, Yacht Design, Yacht Building, Total Refits
OnCall: Divers **Yard Rates:** $25/hr., Haul & Launch $9.50/ft. *(blocking 75% of haul-out rate)* **Storage:** On-Land $0.50/ft/day

Restaurants and Accommodations

Near: Restaurant *(Silverwater Café 385-6448, L $7-13, D $11-20)*, *(T's 385-0700)*, *(Sea J's Café 385-6312)*, *(Castagno's 385-6778)*, Fast Food *(McD's)*, Motel *(Tides Inn 385-0595)*, Hotel *(Harborside Inn 385-7909, $75-150)*, Inn/B&B *(Old Consulate 385-6753, $91-195)* **Under 1 mi:** Restaurant *(Public House Grill & Ales 385-9708, L $7-12, D $9-18, crab chowder; free wireless)*, *(Ichikawa 379-4000)*, *(Bayview 385-1461)*, Pizzeria *(Pizza Factory 385-7223)*, Hotel *(Palace 385-0773, $59-109)*

Recreation and Entertainment

OnSite: Picnic Area, Jogging Paths *(Larry Scott Mem. Trail)* **Near:** Video Rental *(Hollywood Video 379-2580)*, Galleries *(Northwind 379-1086)*

Under 1 mi: Dive Shop *(Port Townsend Dive Shop 379-3635)*, Golf Course *(Port Townsend 385-4547)* **1-3 mi:** Fitness Center *(Gym 385-1465)*, Boat Rentals *(Kayak Port Townsend 385-6240)*, Movie Theater *(Rose Theatre 385-1039)*, Park *(Fort Worden)*, Museum *(Coast Artillery Museum 385-0373)*, Special Events *(Wooden Boat Fest - Sep 385-3628, Port Townsend Film Fest - Sep 379-1333)*, Sightseeing *(PS Express Whale Watching 385-5288)*

Provisioning and General Services

OnSite: Convenience Store **Near:** Supermarket *(Safeway 385-2806)*, Delicatessen *(Portside Deli & Espresso 379-6683)*, Health Food *(The Food Co-Op 385-2883)*, Fishmonger *(New Day Fish Market 379-3603)*, Bank/ATM, Post Office *(379-2996)*, Beauty Salon *(Marta's 385-9636)*, Laundry *(Carol's 385-5508)*, Hardware Store *(Marine Exchange 385-4237)* **Under 1 mi:** Liquor Store *(WA Liquor 379-5050)*, Bakery *(Port Townsend Pie 344-3306)*, Farmers' Market *(Lawrence & Tyler St. Sat 9:30am-1:30pm)*, Library *(385-3181)*, Dry Cleaners *(Dockside 385-4585 - delivery)*, Pharmacy *(Don's 385-2622)*, Copies Etc. *(Cyber Bean Cafe 385-9773)*

Transportation

OnCall: Taxi *(Peninsula Taxi 385-1872)* **Near:** Local Bus *(Jefferson County Transit 385-3020)* **Under 1 mi:** Rental Car *(Budget 385-7766)*, Ferry Service *(WA Ferry to Keystone 888-808-7977/ Puget Sound Express to Friday Harbor 385-5288)* **Airport:** Jefferson County Int'l. *(6 mi.)*

Medical Services

911 Service **OnCall:** Ambulance **Near:** Dentist *(Gray 385-2411)*, Optician *(Pacific Eye Care 385-5386)* **Under 1 mi:** Doctor *(Family Physicians 385-3500)*, Chiropractor **Hospital:** Jefferson 385-2200 *(1 mi.)*

Setting -- Enter Port Townsend Bay between Point Hudson and Marrowstone Point. Massive Port Townsend Boat Haven is three quarters of a mile west of the ferry dock and Port Townsend's downtown historic district. A long rock breakwater protects the recreational and shipyard moorage; the entrance is at the southeast corner. Pass the commercial docks, then the Coast Guard station and tie up on the west side of the fuel/registration dock.

Marina Notes -- *Electric: $5 for boats over 55 ft. Port-operated marina capable of berthing 400 boats, with slips from 20 to 50 ft. Ample alongside moorage for larger boats on a long dock parallel to the breakwater. Fuel dock offers a waste pump-out station and stove oil in addition to gas and diesel. On-site shipyard, completed in 1997, has haul-out capacity up to 300T and can handle major refits. An expansive area onshore is occupied by the yard buildings, with space for do-it-yourselfers, and 10 acres of dry storage for up to 200 boats. 60 on-site marine trades offer supplies, parts, and charts. Port Townsend Y. C. offers reciprocal moorage; contact them directly. Spacious, modern bathhouse has stainless enclosures. (Note: Occasional odors from a distant pulp mill.)

Notable -- A great provisioning stop, there's a deli, Safeway supermarket, and Food Co-op close by. West Marine, and a few restaurants and B&Bs are also an easy walk. Downtown galleries, shops, a farmers' market, and more than 20 restaurants are 0.7 mile west - walk or take the local bus. Enjoy the Victorian homes built in the early days when Port Townsend planned to become the largest harbor on the west coast. Today the "Victorian Village by the Sea" offers entertainment and events year-round - like arts and crafts shows, a gallery walk, music and quilting events, a film festival, and a popular wooden boats fest.

Navigational Information
Lat: 48°04.600' **Long:** 122°47.110' **Tide:** 8 ft. **Current:** n/a **Chart:** 18464
Rep. Depths (MLW): Entry 8 ft. **Fuel Dock** n/a **Max Slip/Moor** -/8 ft.
Access: Located between Glen Cove and Kala Point

Marina Facilities *(In Season/Off Season)*
Fuel: No
Slips: 0 Total, 0 Transient **Max LOA:** 60 ft. **Max Beam:** n/a
 Rate *(per ft.)*: **Day** n/a **Week** n/a **Month** n/a
 Power: 30 amp n/a, **50 amp** n/a, **100 amp** n/a, **200 amp** n/a
 Cable TV: No **Dockside Phone:** No
 Dock Type: n/a
Moorings: 4 Total, 4 Transient **Launch:** n/a
 Rate: Day $10 **Week** n/a **Month** n/a
Heads: Toilet(s), Shower(s)
Laundry: None **Pay Phones:** Yes, 1
Pump-Out: No **Fee:** n/a **Closed Heads:** Yes

Marina Operations
Owner/Manager: WA State Parks and Recreation **Dockmaster:** n/a
In-Season: Apr-Oct, 8am-Dusk **Off-Season:** Nov-Mar, closed
After-Hours Arrival: Use pay envelope at pay station
Reservations: No **Credit Cards:** Cash Only
Discounts: None
Pets: Welcome **Handicap Access:** No

Old Fort Townsend State Park

1370 Old Fort Townsend Road; Port Townsend, WA 98368
Tel: (360) 385-3595 **VHF: Monitor** n/a **Talk** n/a
Fax: n/a **Alternate Tel:** n/a
Email: n/a **Web:** www.parks.wa.gov
Nearest Town: Port Townsend *(5 mi.)* **Tourist Info:** (360) 385-7869

Marina Services and Boat Supplies
Supplies - 3+ mi: West Marine *(379-1612, 5 mi.)*, Bait/Tackle *(Port Townsend Angler 379-3763, 5 mi.)*, Propane *(Fleet Marine 385-4000, 7 mi.)*, CNG *(7 mi.)*

Boatyard Services
3+ mi: Yacht Broker *(Mahina Yachts 379-4980, 5 mi.)*. **Nearest Yard:** Port Townsend Boat Haven (360) 385-2355

Restaurants and Accommodations
3+ mi: Restaurant *(T's Restaurant 385-0700, 5 mi.)*, *(Public House Grill and Ales 385-9708, L $7-12, D $9-18, 6 mi.)*, *(Hadlock House Restaurant 385-3331, 4 mi.)*, *(Zhang's Garden 385-6175, 4 mi.)*, *(Tony's Café 379-1619, 4 mi.)*, Coffee Shop *(1012 Coffee Bar 379-1012, 5 mi.)*, Fast Food *(McDonald's 385-0244, 5 mi.)*, Pizzeria *(Pizza Factory 385-7223, 6 mi.)*, Hotel *(Harborside Inn 385-7909, $75-150, 6 mi.)*, *(Palace Hotel 385-0773, $59-109, 6 mi.)*, Inn/B&B *(Manresa Castle 385-5750, $75-175, 5 mi.)*, *(Blue Gull Inn Bed and Breakfast 379-3241, $95-200, 5 mi.)*, *(Tides Inn 385-0595, $58-115, 5 mi.)*

Recreation and Entertainment
OnSite: Beach, Picnic Area, Grills *(fire pits)*, Playground, Jogging Paths, Park **1-3 mi:** Fitness Center *(Fountain's Fitness 385-3036)* **3+ mi:** Dive Shop *(Port Townsend Dive Shop 379-3635, 5 mi.)*, Golf Course *(Port Townsend Golf Club 385-4547 18 hole $20/cart $20, 6 mi.)*, Boat Rentals *(Kayak Port Townsend 385-6240, 7 mi.)*, Movie Theater *(Rose Theatre 385-1039, 6 mi.)*, Video Rental *(Hollywood Video 379-2580, 5 mi.)*, Museum *(Jefferson County Historical Society 385-1003, 5 mi.)*, Special Events *(Wooden Boat Festival - Sep 385-3628, 6 mi.)*, Galleries *(Stepping*

Stones 379-6910, Port Townsend Gallery Co-Op 379-8110, 6 mi.)

Provisioning and General Services
3+ mi: Convenience Store *(Sea Breeze Gas & Grocery 385-0440, 4 mi.)*, Supermarket *(Safeway 385-2806, 5 mi.)*, Delicatessen *(Lehani's Deli and Coffee 385-3961, 6 mi.)*, Health Food *(The Food Co-Op 385-2883, 5 mi.)*, Liquor Store *(WA Liquor 379-5050, 5 mi.)*, Bakery *(Port Townsend Pie and Bakery Company 344-3306, 6 mi.)*, Farmers' Market *(Port Townsend - Lawrence & Tyler St. Sat 9:30am-1:30pm, 6 mi.)*, Fishmonger *(New Day Fish Market 379-3603, 5 mi.)*, Bank/ATM *(Frontier Bank 385-9911, 4 mi.)*, Post Office *(6 mi.)*, Catholic Church *(St. Mary's Star of the Sea 385-3700, 6 mi.)*, Protestant Church *(Church of Christ 385-7834, 5 mi.)*, Synagogue *(Bet Shira Congregation 379-3068, 4 mi.)*, Library *(Port Townsend 385-3181, 6 mi.)*, Beauty Salon *(Today's Hair 385-3946, 4 mi.)*, Dry Cleaners *(Dockside Cleaners 385-4585, 6 mi.)*, Laundry *(Carol's Laundromat 385-5508, 5 mi.)*, Bookstore *(Imprint Books 385-3643, 6 mi.)*, Pharmacy *(Don's Pharmacy 385-2622, 6 mi.)*, Hardware Store *(Henery Do It Best 385-5900, 5 mi.)*

Transportation
OnCall: Taxi *(Peninsula Taxi 385-1872)* **3+ mi:** Rental Car *(Budget 385-7766, 6 mi.)*, Ferry Service *(WA Ferry to Keystone 888-808-7977/ Puget Sound Express to Friday Harbor 385-5288, 6 mi.)* **Airport:** Jefferson County Int'l. *(4 mi.)*

Medical Services
911 Service **OnCall:** Ambulance **3+ mi:** Doctor *(Olympic Primary Care 379-8031, 5 mi.)*, Dentist *(Family Dental 385-7000, 4 mi.)*, Chiropractor *(Masci 385-0280, 4 mi.)* **Hospital:** Jefferson General 385-2200 *(5 mi.)*

Setting -- Located about four miles from the entrance to Port Townsend Bay, Old Fort Townsend State Park covers 376 acres, with miles of hiking trails, camping and picnic areas, an amphitheater, and over 3, 900 ft. of shoreline. Four mooring buoys provide boater access to this marine park. From the buoys, the visitor can see Indian Island across the Bay and Port Townsend off to the left.

Marina Notes -- *Flat rate, cash only. 4 buoys are available for visiting boaters. Annual and daily permits to WA state parks are available online and in person at the parks when staff is available. The steep climb to the park at low tide is enough reason to have an annual pass displayed rather than using the pay station. There is no dock - just the remnants of the park's old wharf. This is a satellite park to Fort Worden. Heads and showers are available in the summer only, when the park is open for camping. Note: the park is next to a pulp mill with its accompanying aromas.

Notable -- A ranger offers interpretive programs on summer Saturdays. Stop by the information center to pick up a self-guided tour map for the nature and historical trails and the other 6.5 miles of hiking trails that wind through natural forest. Three picnic shelters and 43 tables are available on a first-come, first-serve basis. Call a taxi for the 5-mile trip to downtown historic Port Townsend. Galleries, shops, restaurants, and activities abound in this historic seaside town. The overall flavor is late 19th century Victorian and the residents strive to maintain the step-back-in-time atmosphere. On the National Register of Historic Places, Port Townsend is one of only three Victorian seaports in the U.S.

Port Hadlock Marina

310 Alcohol Loop Road; Port Hadlock, WA 98339

Tel: (360) 385-6368; (800) 785-7030 **VHF: Monitor** Ch. 16 **Talk** Ch. 68
Fax: (360) 385-6955 **Alternate Tel:** n/a
Email: phharbormaster@hotmail.com **Web:** innatporthadlock.com
Nearest Town: Port Hadlock *(1 mi.)* **Tourist Info:** (360) 379-5380

Navigational Information
Lat: 48°01.877' **Long:** 122°44.725' **Tide:** 9 ft. **Current:** n/a **Chart:** 18464
Rep. Depths *(MLW)*: **Entry** 25 ft. **Fuel Dock** n/a **Max Slip/Moor** 20 ft./-
Access: Southern tip of Port Townsend Bay

Marina Facilities *(In Season/Off Season)*
Fuel: No
Slips: 164 Total, 10 Transient **Max LOA:** 70 ft. **Max Beam:** n/a
  Rate *(per ft.)*: **Day** $0.85 **Week** n/a **Month** n/a
  Power: 30 amp $3, **50 amp** $3, **100 amp** n/a, **200 amp** n/a
  Cable TV: No **Dockside Phone:** No
  Dock Type: Floating, Long Fingers, Concrete
Moorings: 0 Total, 0 Transient **Launch:** n/a
  Rate: Day n/a **Week** n/a **Month** n/a
Heads: 4 Toilet(s), 2 Shower(s) *(with dressing rooms)*
Laundry: 1 Washer(s), 1 Dryer(s) **Pay Phones:** No
Pump-Out: OnSite, Self Service **Fee:** Free **Closed Heads:** Yes

Marina Operations
Owner/Manager: Vista Hosp/Paul Christiansen **Dockmaster:** Jerry Spencer
In-Season: Year-Round, 8:30am-5pm **Off-Season:** n/a
After-Hours Arrival: Check the office wall for instructions
Reservations: Yes, Required **Credit Cards:** Visa/MC, Dscvr, Amex
Discounts: None
Pets: Welcome **Handicap Access:** No

Marina Services and Boat Supplies
Services - Security *(locked gates)*, Dock Carts **Supplies - 3+ mi:** Ships' Store *(Nordland Marine and Mechanical 385-7779, 5 mi.)*

Boatyard Services
OnCall: Divers **3+ mi:** Travelift *(60T/70T/300T, 9 mi.)*, Propeller Repairs *(9 mi.)*. **Nearest Yard:** Port Townsend Boat Haven (360) 385-2355

Restaurants and Accommodations
OnSite: Restaurant *(Nemo's 379-1009, B $6, L $10-16, D $13-26, B Mon-Fri 7-11am, Sat-Sun B Buffet 7-10:30am; L daily 11am-4pm; D 4-9pm Sun-Thu, 4-10pm Fri-Sat)*, Hotel *(Inn at Port Hadlock 385-7030, $89-359)*
Under 1 mi: Pizzeria *(Ferino's Pizzeria 385-0840)* **1-3 mi:** Restaurant *(Hadlock House 385-3331, B, L, D)*, *(Fiesta Jalisco 385-5285)*, *(Scampi & Halibuts 385-0161, D $8-22, L & D)*, *(Ajax Café 385-3450 - Hats galore! Entertainment, too)*, Motel *(Hadlock Motel 385-3111, $50-80)*, Inn/B&B *(Old Church Inn 732-7552)*, Condo/Cottage *(Oak Bay 437-0380, $60-115)*

Recreation and Entertainment
OnSite: Beach, Picnic Area, Grills, Special Events *(Hadlock Days Festival - second wknd Jul; Chimacum celebrates Wild Olympic Salmon - Oct)*, Galleries *(The Art Mine at the Inn 379-8555 Tue-Thu & Sun 10am-6pm, Fri & Sat 10am-8pm - hosts an artists' studio program)*, Sightseeing *(Tours on sloop "SingaWing;" PS Exress Whale Watch)* **1-3 mi:** Video Rental *(Peninsula 385-5670)* **3+ mi:** Golf Course *(Discovery Bay G.C. 385-0704 - formerly Chevy Chase, and Port Ludlow G.C. 437-0272, 8 mi.)*, Fitness Center *(Evergreen 385-3036, 6 mi.)*, Museum *(Jefferson County Historical 379-6673, 5 mi.)*

Provisioning and General Services
Under 1 mi: Supermarket *(QFC 385-1070)*, Wine/Beer, Liquor Store *(WA Liquor 385-2316)*, Protestant Church *(Lutheran Church 385-6977)*, Laundry *(Port Hadlock Laundromat 385-6660)*, Pharmacy *(Port Hadlock Pharmacy 385-1900)* **1-3 mi:** Delicatessen *(Easy Times Espresso 385-9505)*, Bakery *(Village Baker 379-5310)*, Bank/ATM *(Frontier Bank 379-8255)*, Post Office *(385-0605)*, Library *(385-6544)*, Beauty Salon *(Hello Gorgeous Salon 385-6007)*, Barber Shop *(Dupras Barber Shop 385-6865)*, Dry Cleaners, Hardware Store *(Hadlock Building Supply 385-1771)*, Florist *(Carousel Florist 385-5562)*, Clothing Store *(Good Sports 385-3996)*, Copies Etc. *(Chimacum Creek Printing 379-3807)* **3+ mi:** Market *(Nordland General Store 385-0777, 5 mi.)*

Transportation
1-3 mi: Local Bus *(Jefferson Transit 385-4777)* **3+ mi:** Rental Car *(Budget 385-7766, 10 mi.)*, Taxi *(Peninsula Taxi 385-1872, 10 mi.)*, Ferry Service *(WA Ferry to Keystone 888-808-7977/ Puget Sound Express to Friday Harbor 385-5288, 10 mi.)* **Airport:** Jefferson County Int'l. *(5 mi.)*

Medical Services
911 Service **OnCall:** Ambulance **Near:** Veterinarian *(Oak Bay Animal Hospital 385-7297)* **Under 1 mi:** Dentist *(Dentistry Northwest 385-1000)* **1-3 mi:** Doctor *(Port Hadlock Medical Care 379-6737)*, Chiropractor *(Peninsula Chiro Clinic 385-4900)* **Hospital:** Jefferson General 385-2200 *(8 mi.)*

Setting -- Port Hadlock Marina and Inn at Port Hadlock are located 0.4 mi. southwest of the north entrance to the Port Townsend-Oak Bay Canal. In addition to the marina, a mooring float is maintained during the summer by the Port of Port Townsend. A beautiful pier extends far out into the bay and leads to the docks. There are over 160 slips, protected by 700 feet of breakwater. Shoreside, the Inn's building sits on the bluff netting panoramic views. Picnic tables and grills are near the water. Port Hadlock is surrounded by jagged mountain peaks and lush evergreen forests.

Marina Notes -- This inn and marina are newly renovated. The goal is to offer boaters hotel facilities and amenities as they are developed. Quality concrete docks have finger piers for boats to 64 ft. Transient space is based on monthly resident departures. After hours, check the wall for instructions and where to find the pay envelopes. The marina office is conveniently located on the pier and features an outdoor whimsical "Boaters' Lounge" complete with a picnic table and potted geraniums. A recent float plane dock has also been added. The facilities can also be rented for private functions, and the restaurant will cater.

Notable -- Completed in 2004, nine years and four million dollars were invested in turning an old, abandoned alcohol plant into the current resort. The inn offers 47 waterfront rooms and suites, and elegant dining at the restaurant - with views of the beach and Skunk Island's wildlife habitat. 4,000 square foot Art Mine Gallery displays rotating exhibits of local artists and functions as an artists' colony within the inn. Port Hadlock is also equidistant to two championship golf courses. This land was once a coastal Salish tribal village and the resort is committed to preserving that history. The town of Port Hadlock is less than a mile.

18. WA – South Coast

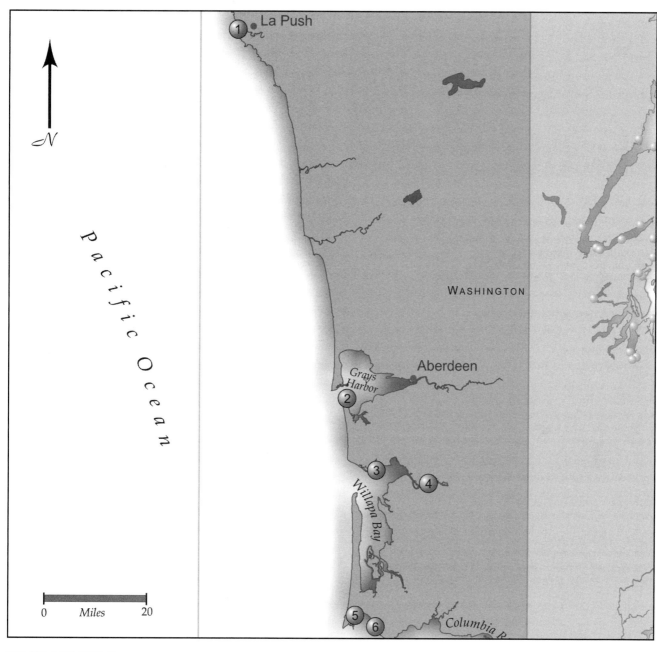

MAP	MARINA	HARBOR	PAGE		MAP	MARINA	HARBOR	PAGE
1	**Quileute Marina**	*Quillayute River*	262		4	**Port of Willapa Harbor**	*Willapa Harbor*	265
2	**Westport Marina**	*Grays Harbor*	263		5	**Port of Ilwaco**	*Columbia River/Illwaco Harbor*	266
3	**Tokeland Marina**	*Willapa Harbor*	264		6	**Port of Chinook**	*Columbia River/Baker Bay*	267

▸ **Currency** — In Canadian Marina Reports, all prices are in Canadian dollars. In U.S. Marina Reports, all prices are in U.S. dollars.

▸ **"CCM"** — Denotes a Certified Clean Marina, a state/provincial award for environmental excellence. See page 298 for an explanation and page 299 for a list of Pump-Out facilities.

▸ **Ratings & Reviews** — An overview of the Atlantic Cruising Club's rating system is on page 6 and details on the content of each Marina Report are on pages 7 – 11.

▸ **Marina Report Updates** — Comments from boaters and new information from ACC reviewers and marinas are posted regularly on *www.AtlanticCruisingClub.com*.

Navigational Information

Lat: 47°54.640' **Long:** 124°38.305' **Tide:** 9 ft. **Current:** n/a **Chart:** 18480
Rep. Depths (MLW): Entry 12 ft. **Fuel Dock** 12 ft. **Max Slip/Moor** 12 ft./-
Access: On the east bank and about 0.4 mi above mouth of Quillayute River

Marina Facilities *(In Season/Off Season)*

Fuel: Gasoline, Diesel
Slips: 96 Total, 80 Transient **Max LOA:** 50 ft. **Max Beam:** n/a
  Rate *(per ft.):* **Day** $0.80 **Week** n/a **Month** n/a
  Power: 30 amp Incl., **50 amp** n/a, **100 amp** n/a, **200 amp** n/a
  Cable TV: No **Dockside Phone:** No
  Dock Type: Floating, Long Fingers, Alongside, Wood
Moorings: 0 Total, 0 Transient **Launch:** n/a
  Rate: Day n/a **Week** n/a **Month** n/a
Heads: 1 Toilet(s)
Laundry: None **Pay Phones:** Yes
Pump-Out: OnSite, Self Service **Fee:** Free **Closed Heads:** Yes

Marina Operations

Owner/Manager: Arnold Black Jr., Harbormaster **Dockmaster:** Same
In-Season: Year-Round, 6am-5pm **Off-Season:** n/a
After-Hours Arrival: Call ahead (360) 375-7762
Reservations: No **Credit Cards:** Visa/MC
Discounts: None
Pets: Welcome **Handicap Access:** Yes, Docks

Quileute Marina

PO Box 279; 71 Main Street; La Push, WA 98350

Tel: (360) 374-5392 **VHF: Monitor** Ch. 80 **Talk** n/a
Fax: (360) 374-2542 **Alternate Tel:** n/a
Email: qmarina01@centurytel.net **Web:** n/a
Nearest Town: La Push **Tourist Info:** (360) 374-2460

Marina Services and Boat Supplies

Services - Dock Carts **Supplies - Near:** Ice *(Cube)*, Propane

Boatyard Services

OnSite: Launching Ramp **Nearest Yard:** Port Angeles Boat Haven (360)
457-4505

Restaurants and Accommodations

Near: Restaurant *(Quileute River's Edge 374-5777, B $4-18, L $5-10, D $10-20, ask about a traditional salmon bake, catering and takeout lunches)*, Motel *(Thunderbird & Whale 374-5267, $55-80, quite basic, two-story 1940s structure right on First Beach)*, Condo/Cottage *(Quileute Oceanside Resort 374-5267, $30-220, five types of 1-3 bedroom cabins, from rustic campers to new ocean view cabins)*, *(Shoreline Resort 374-5267)*

Recreation and Entertainment

OnSite: Fishing Charter *(All Ways Fishing 374-2052 - salmon & deep sea; Top Notch Ocean Charters 374-26602)*, Group Fishing Boat **Near:** Beach *(First, Second, Third and Rialto Beaches offer varying experiences)*, Playground *(At the Community Center)*, Sightseeing *(James Island or Akalat - dramatic Top of the Rock - spiritual home of the Quileute people)*, Special Events *(Quileute Days - midJul)* **Under 1 mi:** Jogging Paths

Provisioning and General Services

Near: Market *(Lonesome Creek Store 374-4338)*, Wine/Beer, Crabs/Waterman *(fresh crabs in season at Lonesome Creek Store)*, Post Office, Protestant Church *(Assembly of God 374-4080)* **3+ mi:** Supermarket *(Forks Outfitters 374-6161, 16 mi.)*, Liquor Store *(WA Liquor 374-6391, 16 mi.)*, Library *(Forks Memorial 374-6402, 15 mi.)*, Pharmacy *(Chinook Pharmacy & Variety 374-2294, 15 mi.)*, Hardware Store *(Ace Hardware 374-5564, 16 mi.)*

Transportation

1-3 mi: Local Bus *(Clallam Transit 452-4511; Rte. 15 to Forks, 3 times/day Mon-Fri, 2 times/day Sat-Sun)* **Airport:** Quillayute State/ Forks Municipal *(10 mi./17 mi.l)*

Medical Services

911 Service **OnCall:** Ambulance **Near:** Doctor *(Quileute Public Health Clinic 374-9035)* **3+ mi:** Dentist *(Robert 374-2288, 16 mi.)*, Veterinarian *(Forks Animal Hospital 374-8882, 16 mi.)* **Hospital:** Forks Community Hospital 374-6271 *(16 mi.)*

Setting -- The Quileute Reservation and its Quileute Oceanside Resort are located in the small village of La Push - about 33 miles south of Cape Flattery. A tall piling and rock breakwater extends from the marina's north end, protecting the network of wood docks supported by black plastic floats - entry is on the south side. The Quileute Tribe's colorful commercial fishing boats occupy about half of the marina and sportfishing and recreational vessels the other.

Marina Notes -- If arriving after hours, call Harold or Arnold (374-7762) for a slip assignment. The docks have long finger piers and older pedestals with power & water. Fish processing on-site, too. The Lonesome Creek Convenience Store carries basic groceries, beer, ice, propane, Native crafts & books, as well as deli sandwiches & fresh crab in season. A laundry, shower facilities, picnic tables and fire pits are at nearby Quileute RV Park. More Showers are a block away at the High Tide Seafood Company. Note: A Coast Guard station is on-site; if you see a blinking light, call for entrance conditions.

Notable -- It's a fisherman's paradise - charter and head boats leave right from the marina. Visitors can also kayak the Quilayute River estuary, whale watch, surf, and watch for wildlife and birds. Popular nearby beaches offer varying kinds of experiences: mystical First Beach is great for surfing and whale watching (and is home to Quileute RV Park). Photographers love Second Beach; a 0.7 mi. trek on a forested trail leads to wooden stairs down the cliff. Third Beach is a 1.5 mi. hike through level forest. And Rialto Beach is the gateway to Olympic National Park's coastal strip. A new Visitor Information & Conference Center is close to the highway - quite far from the marina. Bus service connects to Forks, 16 mi. away, for additional shops and services.

Navigational Information
Lat: 46°54.501' **Long:** 124°06.642' **Tide:** 10 ft. **Current:** n/a **Chart:** 18502
Rep. Depths (*MLW*): **Entry** 16 ft. **Fuel Dock** 12 ft. **Max Slip/Moor** 16 ft./-
Access: 40 miles north of Cape Disappointment off the Pacific Ocean

Marina Facilities *(In Season/Off Season)*
Fuel: *Exxon Mobil* - Gasoline, Diesel
Slips: 650 Total, 150 Transient **Max LOA:** 200 ft. **Max Beam:** n/a
 Rate *(per ft.):* **Day** $0.25* **Week** $1.50 **Month** $4-5
 Power: 30 amp $3, 50 amp $11**, 100 amp n/a, 200 amp n/a
 Cable TV: No **Dockside Phone:** No
Dock Type: Floating, Long Fingers, Alongside, Concrete, Wood
Moorings: 0 Total, 0 Transient **Launch:** n/a
 Rate: Day n/a **Week** n/a **Month** n/a
Heads: 2 Toilet(s), 2 Shower(s)
Laundry: None **Pay Phones:** Yes, 1
Pump-Out: OnSite **Fee:** Free **Closed Heads:** Yes

Marina Operations
Owner/Manager: Port of Grays Harbor **Dockmaster:** Robin Leraas
In-Season: Jun-Sep, 8am-5pm **Off-Season:** Oct-May, 9am-4pm
After-Hours Arrival: Pay envelopes
Reservations: Yes, Preferred **Credit Cards:** Visa/MC
Discounts: None
Pets: Welcome **Handicap Access:** Yes

Westport Marina
PO Box 1601; 326 E Lamb Street; Westport, WA 98595

Tel: (360) 268-9665 **VHF: Monitor** Ch. 71 **Talk** n/a
Fax: (360) 268-9413 **Alternate Tel:** n/a
Email: rleraas@portgrays.org **Web:** www.portofgraysharbor.com
Nearest Town: Westport **Tourist Info:** (360) 268-9422

Marina Services and Boat Supplies
Services - Security, 3 Phase **Communication -** Mail & Package Hold, Phone Messages, FedEx, UPS **Supplies - Near:** Ice *(Cube)*, Bait/Tackle *(Angler 268-1030)* **1-3 mi:** Ships' Store *(Englund Marine 268-9311)*

Boatyard Services
OnSite: Launching Ramp *($5 RT, Year pass $50)* **Nearest Yard:** Port of Ilwaco (360) 642-3143

Restaurants and Accommodations
OnSite: Lite Fare *(Islander Lounge 268-9166, B, L, D)*, Hotel *(Harbor Resort 268-0169, $69-180, 7 waterfront rooms & 7 cottages - with kitchens)*
Near: Restaurant *(Barbara's by the Sea 268-1329)*, *(Westwind 268-1315)*, *(Anthony's 268-1609)*, *(Totem Drive Inn 268-0909)*, Lite Fare *(Castaways Café & Expresso 268-1412)*, Motel *(Albatros 268-9233)* **Under 1 mi:** Pizzeria *(Original House of Pizza 268-0901)*, Motel *(Breakers 268-0848)*, Hotel *(Chateau Westport 268-9101, $74-307, rooms & suites)*, Inn/B&B *(Historic Glenacres 268-0958, $75-195)* **1-3 mi:** Restaurant *(King's 268-2556, B $5-9, L $6-12, D $8-20)*, *(The Diner 268-6097, B, L, D)*

Recreation and Entertainment
OnSite: Fishing Charter *(e.g: Chas'n A Dream 921-3334 - across from float #6, also Deep Sea Charters 268-9300, Angler 268-1030)*, Museum *(Westport Maritime 268-0078 $3/1)* **Near:** Picnic Area, Boat Rentals *(Winds of Westport 268-1760)*, Sightseeing *(Grays Harbor Lighthouse 268-0078)*, Special Events *(Rusty Scupper's Pirate Daze - Jun, Windriders Kite Fest - Jul, Art Fest - Aug, Seafood Fest - Sep)* **Under 1 mi:** Beach, Fitness Center *(Curves 268-0165)*, Park *(Westhaven State - walk the paved dune trail)*

1-3 mi: Bowling *(Viking Bowl & Rest. 268-2622)* **3+ mi:** Golf Course *(Ocean Shores 289-3357, via ferry, 5 mi.)*, Movie Theater *(Ocean Shores Cinemas 289-1234, via ferry, 5 mi.)*, Cultural Attract *(Ocean Shores Interpretive Center 289-4617, 5 mi.)*

Provisioning and General Services
OnSite: Fishmonger *(off the boats, or nearby Merino's Seafood Market 268-5009 - also prepares sandwiches & lunches to go)* **Near:** Convenience Store *(Harbor Gift and Grocery 268-0621)*, Wine/Beer, Bakery *(Little Richards - Donuts 268-9733)*, Bank/ATM, Clothing Store *(Mermaid's Closet 268-0124)*, Retail Shops **Under 1 mi:** Liquor Store *(Westport 268-9550)*, Post Office, Catholic Church *(St. Paul's 268-9625)*, Protestant Church *(Coastal Community 268-0177)* **1-3 mi:** Market *(Hungry Whale 268-0136, Ted's Red Apple 268-9650)*, Delicatessen *(Midtown268-1070)*, Library *(Westport 268-0521)*, Barber Shop *(Red's 589-1273)*, Laundry *(Westport 268-1555)*, Pharmacy *(Twin Harbor 268-0505)*, Hardware Store *(Ace 268-9111)*

Transportation
OnSite: Ferry Service *(Westport-Ocean Shores Passenger Ferry 268-0047, $10/RT, kids under 4 free; daily Jun 18-LabDay, wknds only May-Jun17 & Sep)* **1-3 mi:** Local Bus *(Grays Harbor 532-2770)* **3+ mi:** Rental Car *(Enterprise 533-1094, 22 mi.)* **Airport:** Westport Municipal *(1 mi.)*

Medical Services
911 Service **OnCall:** Ambulance **Near:** Doctor *(Clinic at South Beach 268-0195)*, Dentist *(South Beach Dental Clinic 268-6225)* **Hospital:** Grays Harbor Community 532-8330 *(24 mi.)*

Setting -- The town of Westport sits just inside Grays Harbor, within the southern spit that protects the entrance. The Fishing Boardwalk is on the point and a two-story lookout tower - topped by a metal pyramid roof - sits in the elbow of the spit. The 20 main piers and 650 slips home a sizeable commercial fleet as well as some pleasure craft. Divided by the Westport Shipyard, one side of the marina is heavily industrial and the other side is focused more on recreational boating and charter fishing - shops and several casual restaurants line the bulkhead. Guest moorage is to starboard on Float 21 adjacent to the boat ramp.

Marina Notes -- *Flat rate depending on LOA: up to 14' $6/day, 15-24' $8, 25-29' $9, 30-39' $10, 40-49' $12, 50-59' $14/day; weekly flat rates from $36-84, monthly from $174-270. 60' and over $0.25/ft., $1.50/ft. weekly, $4-5/ft. monthly. **50A 3-phase $13. Float 6, boat launch, and marina office (across from Float 5) have pay envelopes. Float 20 has handicap access. Gas, diesel, and lubricants available at the fuel dock (8-5 Mon-Sat, in-season; Mon-Fri, off-season). Berths the largest charter fishing fleet in the state and there is a shoreside processing plant for the catch. On-site Westport Shipyard (268-0117) builds megayachts - 98-164 ' and offers some services. Two basic, cinderblock public restrooms. Showers available at the Port office during office hours.

Notable -- Westport Maritime Museum (MemDay-LabDay, 10-4 daily, off-season Thu-Mon 12-4pm) features a glass-sided outbuilding that houses a huge whale skeleton with adjacent picnic tables. Nearby is the Fresnel lens building and Grays Harbor Lighthouse - the tallest working light on the west coast. The Fishing Boardwalk is a popular spot; rent crab rings and fishing gear nearby. Westport, Grayland, North Cove, and Tokeland comprise the South Beach region.

Tokeland Marina

PO Box 592; 3242 Front Street; Tokeland, WA 98590

Tel: (360) 267-2888 **VHF: Monitor** n/a **Talk** n/a
Fax: n/a **Alternate Tel:** n/a
Email: n/a **Web:** www.portofwillapaharbor.com
Nearest Town: Tokeland **Tourist Info:** (360) 267-2003

Navigational Information
Lat: 46°42.486' **Long:** 123°58.093' **Tide:** 9 ft. **Current:** n/a **Chart:** 18504
Rep. Depths (*MLW*): **Entry** 15 ft. **Fuel Dock** n/a **Max Slip/Moor** 10 ft./-
Access: Northern side of Willapa Harbor, 5 mi. from the Pacific Ocean

Marina Facilities *(In Season/Off Season)*
Fuel: No
Slips: 50 Total, 10 Transient **Max LOA:** 100 ft. **Max Beam:** n/a
  Rate *(per ft.)*: **Day** $0.30 **Week** n/a **Month** n/a
  Power: 30 amp n/a, 50 amp n/a, 100 amp n/a, 200 amp n/a
  Cable TV: No **Dockside Phone:** No
  Dock Type: Floating, Long Fingers, Pilings, Alongside, Wood
Moorings: 0 Total, 0 Transient **Launch:** n/a
  Rate: Day n/a **Week** n/a **Month** n/a
Heads: 2 Toilet(s)
Laundry: None **Pay Phones:** No
Pump-Out: No **Fee:** n/a **Closed Heads:** Yes

Marina Operations
Owner/Manager: Port of Willapa **Dockmaster:** Bob Cordova, HM
In-Season: Year-Round, 8am-5pm **Off-Season:** n/a
After-Hours Arrival: Tie up and register in the morning
Reservations: No **Credit Cards:** No
Discounts: None
Pets: Welcome, Dog Walk Area **Handicap Access:** No

Marina Services and Boat Supplies
Services - Trash Pick-Up **Communication** - FedEx, UPS, Express Mail
Supplies - **OnSite:** Ice *(Cube)*, Bait/Tackle

Boatyard Services
OnSite: Launching Ramp **Nearest Yard:** Port of Ilwaco (360) 642-3143

Restaurants and Accommodations
Under 1 mi: Restaurant *(Tokeland Hotel and Restaurant 267-7006, B $5-10, L $5-10, D $10-18)*, Hotel *(Tokeland Hotel 267-7006, $44-65)* **3+ mi:** Motel *(Tradewinds on the Bay 267-7500, $50-80, 4 mi.)*, *(Grayland Motel and Cottages 267-2395, $49-85, 7 mi.)*

Recreation and Entertainment
Near: Beach, Special Events *(Fourth of July Parade)*, Galleries *(Bay Side Gallery 267-0606)* **1-3 mi:** Cultural Attract *(Shoalwater Bay Tribal Center)* **3+ mi:** Movie Theater *(Raymond Theater 942-5536, 22 mi.)*, Video Rental *(Beachcomber Grocery and Deli 267-7716, 12 mi.)*

Provisioning and General Services
OnSite: Crabs/Waterman *(off Gail Force 267-4194, or at Nelson Crab 267-2911, nearby)* **Near:** Convenience Store *(in the RV Park)*, Fishmonger **1-3 mi:** Market *(Shoalwater Bay Grocery 267-3745)*, Wine/Beer, Post Office *(267-1498)* **3+ mi:** Liquor Store *(WA Liquor 268-9550, 12 mi.)*, Bank/ATM *(Harbor Community Bank 267-5481, 11 mi.)*, Protestant Church *(Lutheran Church Twin Harbors 267-3124, 7 mi.)*, Beauty Salon *(Hair FX 267-6063, 7 mi.)*, Barber Shop *(Red's Barbershop 589-1273, 13 mi.)*, Pharmacy *(Sagen's Pharmacy 942-2634, 22 mi.)*, Hardware Store *(Grayland True Value 267-2594, 10 mi.)*

Transportation
Airport: Westport Municipal *(17 mi.)*

Medical Services
911 Service **OnCall:** Ambulance **1-3 mi:** Doctor *(Shoalwater Bay Medical Clinic 267-0119)*, Dentist *(Shoalwater Bay Medical Clinic 267-0119)* **3+ mi:** Veterinarian *(Willapa Veterinary Service 267-4444, 8 mi.)* **Hospital:** Willapa Harbor 875-5526 *(27 mi.)*

Setting -- The Willapa Bay entrance is about 24 miles north of the Columbia River and is used primarily by oyster and fishing boats. At the north end of Willapa Bay, Tokeland on Toke Point is a summer community with seasonal cottages and RV parks. A two-story weathered shingle, hip roofed office overlooks fifty slips at the end of the single pier. Adjacent is a new concrete launch ramp. Timbered bluffs, rolling hills, and ridges are visible across the bay.

Marina Notes -- Established in 1965 and managed by Port of Willapa, this marina is primarily used by fishermen. The marina was established in 1965 and updated in 2002 with new six-foot-wide floats with wood decking. The Port keeps the water dredged to 10 ft. MLW providing easy access to the ocean. Inexpensive year-round moorage at only $10 per foot per year. The marina crab pots line the roadway during crab season. Just across the road, the Tokeland RV park store has ice, bait and tackle, crab trap rentals, and snacks. Two porta-potties serve as the only restrooms.

Notable -- The Port also maintains the nearby 30-site Tokeland RV Park and two industrial buildings that house three seafood processing plants. The area offers great fishing opportunities, and fresh crab is sold from the boat, Gail Force, daily at 4pm ($6 each; 267-4194 or 267-CRAB). Up the street is the Tokeland Hotel & Restaurant, built in 1886 - the only restaurant within walking distance of the marina (daily 8am-8pm). The hotel, once a private home, offers rooms year-round. Few services are available in this quiet area. Just inside the Willapa bar is reportedly the best spots for large Chinook salmon. The Shoalwater Bay Indian Reservation and Casino is less than three miles north of the marina and oceanfront Grayland Beach State Park (267-4301) is about 9 miles.

Navigational Information
Lat: 46°41.024' **Long:** 123°45.290' **Tide:** 9 ft. **Current:** n/a **Chart:** 18504
Rep. Depths (*MLW*): **Entry** 19 ft. **Fuel Dock** n/a **Max Slip/Moor** 19 ft./-
Access: Located on the Willapa River off Willapa Bay

Marina Facilities *(In Season/Off Season)*
Fuel: No
Slips: 40 Total, 10* Transient **Max LOA:** 100 ft. **Max Beam:** n/a
  Rate *(per ft.)*: **Day** $0.30/0.20 **Week** n/a **Month** $2
  Power: 30 amp Incl., **50 amp** Incl., **100 amp** n/a, **200 amp** n/a
  Cable TV: No **Dockside Phone:** No
  Dock Type: Floating, Alongside, Wood
Moorings: 0 Total, 0 Transient **Launch:** n/a
  Rate: Day n/a **Week** n/a **Month** n/a
Heads: 2 Toilet(s)
Laundry: None **Pay Phones:** No
Pump-Out: No **Fee:** n/a **Closed Heads:** Yes

Marina Operations
Owner/Manager: Rebecca Chaffe **Dockmaster:** Dawn King
In-Season: Year-Round, 7:30am-4:30pm **Off-Season:** n/a
After-Hours Arrival: Tie up and register in the morning
Reservations: No **Credit Cards:** No
Discounts: None
Pets: Welcome **Handicap Access:** No

Port of Willapa Harbor

1725 Ocean Avenue; Raymond, WA 98577

Tel: (360) 942-3422 **VHF: Monitor** n/a **Talk** n/a
Fax: (360) 942-5865 **Alternate Tel:** n/a
Email: portofwh@willapabay.org **Web:** www.portofwillapaharbor.com
Nearest Town: Raymond *(1.5 mi.)* **Tourist Info:** (360) 942-5419

Marina Services and Boat Supplies
Services - Trash Pick-Up **Communication -** FedEx, UPS, Express Mail
Supplies - OnSite: Propane *(Airgas 943-5374)* **Near:** Ice *(Block, Cube)*
Under 1 mi: Live Bait

Boatyard Services
OnSite: Forklift, Crane *(5000 lb.)*, Yacht Design *(Pedigree Cats 942-2870)*,
Yacht Broker, Yacht Building **Under 1 mi:** Launching Ramp. **1-3 mi:** Hull
Repairs *(South Bend Boat Yard)*, Air Conditioning *(Sunset Air 875-4070)*,
Refrigeration *(Sunset Air)*. **Nearest Yard:** South Bend Boat Shop (360)
875-5712

Restaurants and Accommodations
Near: Restaurant *(Café Omni 875-6555)* **Under 1 mi:** Inn/B&B *(Riverview
Inn B&B 942-5271)* **1-3 mi:** Restaurant *(Las Maracas Mexican Restaurant
942-6151, L $5-10, D $7-15)*, *(The Barge 942-5100)*, *(Corner Café 942-3607,
B, L, D)*, *(Slaters Diner 942-5109, B, L, D)*, *(Eastern Garden 942-2079, L,D)*,
Fast Food *(Dairy Queen 942-3103)*, Pizzeria *(Willapa Pizza 942-6022)*, Motel
(Mountcastle Motel 942-5571, $55-80), Condo/Cottage *(Summerhouse 942-
2843, $55-65)*

Recreation and Entertainment
Near: Jogging Paths *(6 mi. Willapa Hills Trail runs past marina)*, Sightseeing
(Riverfront Park), Galleries *(Willapa River Art Studio 942-4989)* **Under 1
mi:** Fitness Center *(Willapa Harbor Fitness Center 942-0073)* **1-3 mi:** Pool
(Nevitt Memorial Pool 942-4151), Heated Pool, Beach *(Tokeland/Grayland)*,
Playground *(8th Street Park)*, Tennis Courts *(8th Street Park)*, Movie Theater
(Hannan Playhouse 942-5477), Video Rental *(Everyone's Video 942-2213)*,

Park *(8th Street Park)*, Museum *(Willapa Seaport Museum 942-5666 Thu-
Sun 10am-4pm Donations, Carriage Museum 942-4150 Wed-Sat 10am-4pm
$3/1)*, Cultural Attract, Special Events *(Willapa Harbor Fest - Aug)*
3+ mi: Golf Course *(Willapa Harbor Golf Course 942-2392, 4 mi.)*, Bowling
(Riverfront Lanes 875-6773, 6 mi.)

Provisioning and General Services
1-3 mi: Market *(Corner Grocery 942-2993)*, Supermarket *(Everybody's
Market Place 942-2100)*, Wine/Beer, Liquor Store *(WA Liquor 942-2304)*,
Bakery *(Sweet Dreams 942-4962)*, Farmers' Market *(Willapa Public Market
942-4700, seasonal; Alder St. Wed-Sat 10am-5pm)*, Bank/ATM *(Harbor
Community Bank 942-2401)*, Post Office, Catholic Church *(St. Lawrence
942-3000)*, Protestant Church *(First Baptist 942-315)*, Library *(Raymond 942-
2408)*, Beauty Salon *(Linda's Salon 942-5732)*, Barber Shop *(Kay's Place
942-5054)*, Bookstore *(South Bend Book Store 875-6675)*, Pharmacy
(Raymond Drug 942-2153), Hardware Store *(Dennis Co Ace 942-2427)*,
Florist *(Flowers by Lynne 942-2110)*, Clothing Store

Transportation
Under 1 mi: Local Bus *(Pacific Transit System 875-9418)* **Airport:** Willapa
Harbor/Westport Municipal *(8 mi./35 mi.)*

Medical Services
911 Service **OnCall:** Ambulance *(Raymond Fire Dept. 911)* **1-3 mi:** Doctor
(Riverview Health Center 942-3040), Dentist *(Deep River Dental 942-3600)*,
Chiropractor *(Riverview Chiropractic Clinic 942-3232)*, Veterinarian *(Vetters
Animal Hospital 942-3440)* **Hospital:** Willapa Harbor 875-5526 *(2 mi.)*

Setting -- Raymond is located at the navigable head of the Willapa River, 3 miles above South Bend and 15 miles from the Willapa Bay bar. Port of Willapa Harbor manages Indian Park, a 15-acre industrial complex that includes the moorage. Paralleling the shoreline, a long floating dock connects to a substantial stationary pier that hosts gray and blue work sheds. At its head is a small blue single-story marina office. A forested, undeveloped hill rises in the background.

Marina Notes -- *600-foot pier and 750-foot float with linear moorage on both sides - able to accommodate large vessels due to its deep water location. The Port was originally founded in 1928 and served as landing for the large ships carrying logs. Although there are no boater amenities, the harbor is calm and protected. A 5,000-lb. crane is on-site for tenant use but does not do haul outs. Also in the complex are: Airgas Nor-Pac Inc., Jackpot Industries - a fishing vessel maintenance facility, Pedigree Cats - a catamaran builder, an aquaculture training center, and Vanson HaloSource, a manufacturer of water treatment products. Note: Contact the Coast Guard before attempting to cross the bar. The shoals are ever-changing and tricky, according to the Port office.

Notable -- Incorporated in 1907, Raymond is known for its wonderful Wildlife-Heritage Sculpture Corridor - sculptures and murals throughout the downtown depicting the heritage of the area. Be sure to stop by the Willapa Seaport Museum (942-5666) which displays marine artifacts and has information about the local shipbuilding and logging industry. The Carriage Museum (942-4150) interprets the general history of horse-drawn vehicles. The area is rich in oysters, clams, crabs and fish for the sport angler. Oysters are particularly abundant; Willapa Bay produces as much as 15 percent of the national crop.

Port of Ilwaco

PO Box 307; 165 Howerton Avenue; Ilwaco, WA 98624

Tel: (360) 642-3143 **VHF: Monitor** Ch. 16 **Talk** Ch. 69
Fax: (360) 642-3148 **Alternate Tel:** n/a
Email: n/a **Web:** www.portofilwaco.com
Nearest Town: Ilwaco **Tourist Info:** (360) 642-2400

Navigational Information
Lat: 46°18.282' **Long:** 124°02.347' **Tide:** 8 ft. **Current:** n/a **Chart:** 18521
Rep. Depths (MLW): Entry 18 ft. **Fuel Dock** 12 ft. **Max Slip/Moor** 12 ft./-
Access: On the Columbia River, 6 nautical miles upstream from the mouth

Marina Facilities *(In Season/Off Season)*
Fuel: Gasoline, Diesel
Slips: 800 Total, 100 Transient **Max LOA:** 125 ft. **Max Beam:** n/a
  Rate *(per ft.):* **Day** $0.38* **Week** n/a **Month** $6
  Power: 30 amp $2.75** **50 amp** $4-6, **100 amp** n/a, **200 amp** n/a
  Cable TV: No **Dockside Phone:** No
  Dock Type: Floating, Concrete, Wood
Moorings: 0 Total, 0 Transient **Launch:** n/a
  Rate: Day n/a **Week** n/a **Month** n/a
Heads: 24 Toilet(s), 3 Shower(s) *(with dressing rooms)*
Laundry: None **Pay Phones:** Yes
Pump-Out: OnSite, Self Service **Fee:** n/a **Closed Heads:** Yes

Marina Operations
Owner/Manager: Mack Funk **Dockmaster:** Mark Elliott
In-Season: Year-Round **Off-Season:** n/a
After-Hours Arrival: Pay envelopes outside the port office
Reservations: No **Credit Cards:** Visa/MC
Discounts: None
Pets: Welcome **Handicap Access:** No

Marina Services and Boat Supplies
Supplies - OnSite: Ice *(Cube)* **Near:** Marine Discount Store *(Englund Marine Supply 642-2308)* **Under 1 mi:** Live Bait **1-3 mi:** Bait/Tackle *(Ed's Bait and Tackle 642-2248)*

Boatyard Services
OnSite: Travelift *(50T)*, Launching Ramp *($5 round trip)*, Bottom Cleaning, Divers, Compound, Wash & Wax *(Wash only)* **Near:** Engine mechanic *(gas, diesel)*, Electrical Repairs, Hull Repairs, Brightwork, Propeller Repairs, Woodworking, Metal Fabrication, Awlgrip. Other Certifications: Self service boatyard **Yard Rates:** $30-75/hr., Haul & Launch $5/ft., 30' min., Power Wash $1.25/ft. **Storage:** On-Land $2.25/ft./mo.

Restaurants and Accommodations
OnSite: Restaurant *(Harbor Lights 642-3196, B $6-11, L $5-11, D $12-17)*, *(Port Bistro 642-8447, L $7-10, D $11-16)*, *(Imperial Schooner 642-8667)* **Near:** Restaurant *(Sea Hag Bar and Grill 642-5899, B, L, D)*, Coffee Shop *(Ole Bob's Coffee Net Café)*, Motel *(Haciendas 101 642-8459, $30-95)*, *(Coho Motel 642-3333)*, Inn/B&B *(Belltower Inn 642-8686, $100-175)* **Under 1 mi:** Restaurant *(Lightship Restaurant 642-3252, D $8-25)*, Pizzeria *(Don's Portside Café and Pizzeria 642-3477, B, L, D, delivery)* **1-3 mi:** Restaurant *(The Shoalwater 642-4142, L $7-20, D $14-30, at the Shelburne Inn, daily)*, Inn/B&B *(Shelburne Inn 642-2442, $135-195)*

Recreation and Entertainment
OnSite: Fishing Charter, Group Fishing Boat *(COHO 800-239-2646, Pacific Salmon 642-3466, Sea Breeze 642-2300)*, Special Events *(Blessing of the Fleet - Apr, fishing derbys, World's Longest Garage Sale - May,*

Fireworks 1st Sat in Jul, Art Walks - Jul-Sep) **Near:** Galleries *(Wade Gallery 642-2291, Shoalwater Cove Gallery)* **Under 1 mi:** Fitness Center *(Anybody's 642-3649)* **1-3 mi:** Park *(Cape Disappointment State Park 642-3078)*, Museum *(Ilwaco Heritage Museum 642-3446)*, Sightseeing *(Lighthouses/ Lewis and Clark Interpretive Center at Cape Disappointment)* **3+ mi:** Beach *(5 mi.)*, Golf Course *(Peninsula 642-2828, 8 mi.)*

Provisioning and General Services
OnSite: Farmers' Market *(Sat May-Sep)*, Fishmonger *(Ole Bob's Seafood 642-4332)* **Near:** Convenience Store *(Ilwaco Market & Deli 642-2413)*, Wine/Beer, Bank/ATM, Post Office *(642-3213)*, Beauty Salon *(Azure Salon & Day Spa 642-4080)*, Bookstore *(Time Enough 642-7667)*, Copies Etc. *(Pacific Printing and Office Supply 642-3655)* **Under 1 mi:** Library, Laundry *(Peninsula 642-2512)*, Hardware Store *(Ilwaco Hardware 642-3104)* **1-3 mi:** Supermarket *(Sid's 642-3737)*, Liquor Store *(642-2583)*, Catholic Church *(St. Mary's 642-2002)*, Protestant Church *(St. John's Lutheran 642-4930)*, Pharmacy *(Ilwaco Pharmacy 642-3133)*

Transportation
OnSite: Local Bus *(Pacific Transit 642-9418)* **3+ mi:** Rental Car *(Enterprise 503-325-6500, 16 mi.)*, Taxi *(Royal Cab 503-325-5818, 16 mi.)* **Airport:** Astoria Regional *(18 mi.)*

Medical Services
911 Service **OnCall:** Ambulance **Near:** Chiropractor *(Ocean Beach 642-2474)*, Holistic Services *(Azure Spa 642-4080)* **1-3 mi:** Doctor *(Ocean Beach Medical Clinic 642-3747)*, Veterinarian *(Oceanside 642-2232)* **Hospital:** Ocean Beach 642-3181 *(0.5 mi.)*

Setting -- With its entrance channel at River Mile 3, Port of Ilwaco is the closest marina to the mouth of the Columbia River. Colorful fishing boats greet visitors, and a variety of businesses line the shore. A large network of docks fills the basin - hosting 800 slips. The guest docks are past the breakwater, in "East Main" on the south side of the marina. The Port office is housed in a 2-story tan building with green trim.

Marina Notes -- *Flat rate: up to 20' $12/night, 21-40' $15, over 41' $20/night. **Electricity: under 50' $2.75/day, 50-100' $4/day, over 100' $6/day. Established 1930. Home to a large number of charter, sport and commercial fishing boats. Staffed by helpful, experienced personnel. An on-site boatyard hauls out up to 50 tons, has a large area for do-it-yourself repairs, and on-the-hard storage plus a covered work area. A list of marine businesses and services is at the office. Six charter fishing boats berth here. Internet at Ole Bob's Coffee Net Café (642-2288). Well-maintained cinderblock heads with fiberglass shower stalls are located in the village. Nearby Cape Disappointment State Park (formerly Fort Canby SP) on Baker Bay has a small, 135-foot dock for visitors. Note: Use caution when entering from the ocean; over 2000 vessels have been lost trying to cross the bar at this "Graveyard of the Pacific."

Notable -- This exuberant working port is a serious draw for tourists and boaters. Explore the nearby picturesque village with shops and restaurants. Dinghy to Cape Disappointment State Park which offers fantastic views, 2 lighthouses, and the popular Lewis and Clark Interpretive Center, with exhibits and murals about the Expedition (642-3078, 10am-5pm in summer, 10am-4pm winter). Less than 4 miles north is the World Kite Museum & Hall of Fame (642-4020).

Navigational Information
Lat: 46°16.340' **Long:** 123°56.880' **Tide:** 10 ft. **Current:** 3 kt. **Chart:** 18521
Rep. Depths (*MLW*): **Entry** 11 ft. **Fuel Dock** 11 ft. **Max Slip/Moor** 11 ft./-
Access: On the Columbia River approximately 10 mi. from the mouth

Marina Facilities *(In Season/Off Season)*
Fuel: Gasoline, Diesel
Slips: 300 Total, 10 Transient **Max LOA:** 60 ft. **Max Beam:** 20 ft.
  Rate *(per ft.)*: **Day** $0.30* **Week** $75 **Month** $195
  Power: 30 amp $2, **50 amp** n/a, **100 amp** n/a, **200 amp** n/a
  Cable TV: No **Dockside Phone:** No
  Dock Type: Floating, Wood
Moorings: 0 Total, 0 Transient **Launch:** n/a
  Rate: Day n/a **Week** n/a **Month** n/a
Heads: 2 Toilet(s)
Laundry: None **Pay Phones:** No
Pump-Out: OnSite, Full Service **Fee:** n/a **Closed Heads:** Yes

Marina Operations
Owner/Manager: Daniel Todd **Dockmaster:** Same
In-Season: Year-Round, 8-9am, 4-5pm **Off-Season:** n/a
After-Hours Arrival: Call ahead
Reservations: No **Credit Cards:** No
Discounts: None
Pets: Welcome, Dog Walk Area **Handicap Access:** Yes, Heads, Docks

Port of Chinook

PO Box 185; 743 Water Street; Chinook, WA 98614

Tel: (360) 777-8797 **VHF: Monitor** Ch. 16 **Talk** Ch. 88
Fax: (360) 777-8415 **Alternate Tel:** n/a
Email: pchinook@willapabay.org **Web:** n/a
Nearest Town: Ilwaco *(7 mi.)* **Tourist Info:** (360) 642-2200

Marina Services and Boat Supplies
Supplies - Near: Ice *(Block, Cube, Shaved)*, Bait/Tackle *(Chinook Country Store 777-2248)*

Boatyard Services
OnSite: Travelift **Near:** Engine mechanic *(gas, diesel)*, Air Conditioning, Refrigeration.

Restaurants and Accommodations
Near: Restaurant *(Chinook Restaurant and Lounge 777-8233, L, D)*, *(Sanctuary Restaurant 777-8380, Scandinavian food in a remodeled church; year-round, seasonal hours)* **3+ mi:** Restaurant *(Harbor Lights 642-3196, B $6-11, L $5-11, D $12-17, 7 mi.)*, *(Pauly's Bistro 642-8447, L $7-10, D $11-16, 7 mi.)*, *(Canoe Room Café 642-4899, L $5-12, 7 mi.)*, Motel *Motel 642-8459, $30-95, 7 mi.)*, Hotel *(Eagle's Nest Resort 642-8351, (Haciendas $50-200, 7 mi.)*, Inn/B&B *(China Beach Resort 642-5660, $160-199, 7 mi.)*,*(Shelburne Inn 642-2442, $135-195, 8 mi.)*, *(Inn at Ilwaco 642-8686, $100-175, 7 mi.)*

Recreation and Entertainment
Near: Picnic Area, Grills, Playground, Galleries *(Columbia River Artists Gallery Fri-Sun Noon-6pm, Mon 11am-5pm)* **1-3 mi:** Jogging Paths, Boat Rentals, Park *(Fort Columbia State Park - hiking trails, interpretive center, observation station)* **3+ mi:** Golf Course *(Peninsula Golf Course 642-2828, 14 mi.)*, Fitness Center *(Anybody's Fitness Center 642-3649, 7 mi.)*, Bowling *(Hilltop Bowl 642-4440, 8 mi.)*, Fishing Charter *(Lucky Lady*

Charters 642-3515, 7 mi.)*, Group Fishing Boat *(Pacific Salmon Charters 642-3466, 7 mi.)*, Movie Theater *(Columbia Theatres 503-325-3516, 12 mi.)*, Museum *(Ilwaco Heritage Museum 642-3446, 7 mi.)*, Sightseeing *(Lewis and Clark Interpretive Center and 2 Lighthouses at Cape Disappointment State Park 642-3078, 8 mi.)*

Provisioning and General Services
Near: Fishmonger, Lobster Pound, Crabs/Waterman, Bank/ATM, Post Office, Protestant Church **Under 1 mi:** Market *(Chinook Country Store 777-2248)* **1-3 mi:** Copies Etc. **3+ mi:** Supermarket *(Safeway 503-325-4662, 10 mi.)*, Liquor Store *(Warrenton Liquor 503-861-1103, 10 mi.)*, Library *(Ilwaco 642-3908, 7 mi.)*, Beauty Salon *(Shear Paradise 642-3620, 6 mi.)*, Bookstore *(Time Enough Books 642-7667, 7 mi.)*, Pharmacy *(Ilwaco Pharmacy 642-3133, 7 mi.)*, Hardware Store *(Ilwaco Hardware 642-3104, 7 mi.)*, Buying Club *(Costco 503-861-1950, 12 mi.)*

Transportation
OnCall: Rental Car *(Enterprise 503-325-6500)*, Airport Limo *(Bay Shuttle 642-4196)* **Near:** Local Bus *(Pacific Transit 642-9300)* **Airport:** Astoria Regional *(13 mi.)*

Medical Services
911 Service **OnCall:** Ambulance **3+ mi:** Doctor *(Ocean Beach Medical Clinic 642-3747, 7 mi.)*, Veterinarian *(Oceanside 642-2232, 7 mi.)*, Optician *(Coastal Eye Care 642-3214, 8 mi.)* **Hospital:** Ocean Beach 642-3181 *(7 mi.)*

Setting -- Port of Chinook sits on the north shore of the Columbia River, about 10 miles from the mouth - in 15-square-mile Baker Bay. A rock pile breakwater protects the marina and the wooded hills, distant mountain vistas, and the busy commercial fishing fleet give this marina a distinctive northwest atmosphere. A small natural, vertical-sided building houses the port office.

Marina Notes -- *Flat rate: $12/night no matter the length. Liveaboards pay $100/mo. per person for any period over 5 days. Boat hoist is $12 minimum up to 19'11" then an additional $1/ft. from 20'. One-way haul out is $10. Older wooden docks are being replaced by new ones. They offer long finger slips, some with water and power on pedestals. Fuel and pump-out are on-site. A few covered picnic tables with barbeques are available. Public heads, but no showers. Note: although the controlling depth in the channel into Chinook is 9 feet, skippers will do well to mind their depth sounder when the tide is low. As little as 1 foot of water may be found in close proximity to the channel.

Notable -- Downtown is a block away, but most serious shopping is in Ilwaco or Astoria - a local bus makes both accessible. As one of three major fishing grounds on the Washington coast, seafood has always dominated Chinook's economy - in 1870 the county's first salmon cannery was built here. The Chinook Country Store carries complete salmon and sturgeon outfitting, fresh & frozen bait, licenses and groceries. Seek out Lucky Lady or Pacific Salmon Charters for some help finding that famous Chinook salmon. Or stroll the shoreline to Fort Columbia State Park (1.2 miles) and Chinook County Park.

Boaters' Notes

Add Your Ratings and Reviews at www.AtlanticCruisingClub.com

A complimentary six-month Silver Membership is included with the purchase of this *Guide*. Select "Join Now," then "Silver," then follow the instructions.

The AtlanticCruisingClub website provides updated Marina Reports, Destination and Harbor Articles and much more — including an option within each online Marina Report for boaters to add their ratings and comments regarding that facility. Please log on frequently to share your experiences — and to read other boaters' comments.

On the website, boaters may rate marinas in one or more of the following categories — on a scale of 1 (basic) to 5 (world class) — and also enter additional commentary.

▸ **Facilities & Services** (Fuel, Reservations, Concierge Services and General Helpfulness)

▸ **Amenities** (Pool, Beach, Internet Access, including Wi-Fi, Picnic/Grill Area, Boaters' Lounge)

▸ **Setting** (Views, Design, Landscaping, Maintenance, Ship-Shapeness, Overall Ambiance)

▸ **Convenience** (Access — including delivery services — to Supermarkets, other Provisioning Sources, Shops, Services, Attractions, Entertainment, Sightseeing, Recreation, including Golf and Sport Fishing & Medical Services)

▸ **Restaurants/Eateries/Lodgings** (Availability of Fine Dining, Snack Bars, Lite Fare, OnCall food service, and Lodgings ashore)

▸ **Transportation** (Courtesy Car/Vans, Buses, Water Taxis, Bikes, Taxis, Rental Cars, Airports, Amtrak, Commuter Trains)

▸ **Please Add Any Additional Comments**

19. OR – Columbia River to Yaquina Bay

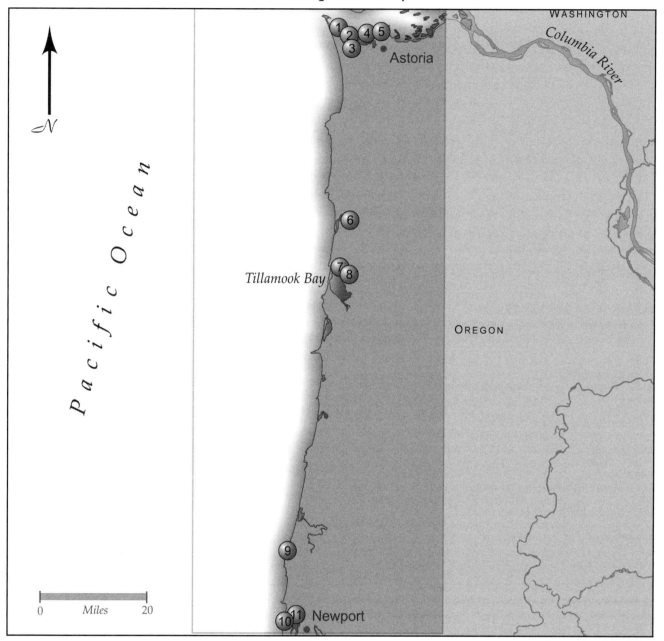

MAP	MARINA	HARBOR	PAGE	MAP	MARINA	HARBOR	PAGE
1	Hammond Marina	Columbia R./Hammond Harbor	270	7	Garibaldi Marina	Tillamook Bay	276
2	Skipanon Landing	Columbia R./Skipanon Wway.	271	8	Port of Garibaldi	Tillamook Bay	277
3	Warrenton City Mooring Basin	Columbia R./Skipanon Wway.	272	9	Depoe Bay Marina	Depoe Bay	278
4	Port of Astoria — West Basin	Columbia R./Port of Astoria	273	10	Port of Newport Marina & RV Park	Yaquina River/Yaquina Bay	279
5	Port of Astoria — East Basin	Columbia R./Port of Astoria	274	11	Embarcadero Resort Hotel & Marina	Yaquina River/Yaquina Bay	280
6	Wheeler Marina	Nehalem River	275				

▸ **Currency** — In Canadian Marina Reports, all prices are in Canadian dollars. In U.S. Marina Reports, all prices are in U.S. dollars.

▸ **"CCM"** — Denotes a Certified Clean Marina, a state/provincial award for environmental excellence. See page 298 for an explanation and page 299 for a list of Pump-Out facilities.

▸ **Ratings & Reviews** — An overview of the Atlantic Cruising Club's rating system is on page 6 and details on the content of each Marina Report are on pages 7 – 11.

▸ **Marina Report Updates** — Comments from boaters and new information from ACC reviewers and marinas are posted regularly on *www.AtlanticCruisingClub.com*.

Navigational Information
Lat: 46°12.145' **Long:** 123°57.080' **Tide:** 8 ft. **Current:** n/a **Chart:** 18521
Rep. Depths (MLW): Entry 16 ft. **Fuel Dock** 8 ft. **Max Slip/Moor** 8 ft./-
Access: 8.7 miles up the Columbia River

Marina Facilities *(In Season/Off Season)*
Fuel: Gasoline, Diesel, On Call Delivery
Slips: 180 Total, 20 Transient **Max LOA:** 90 ft. **Max Beam:** n/a
Rate *(per ft.)*: **Day** $0.38* **Week** n/a **Month** n/a
Power: 30 amp $4, **50 amp** n/a, **100 amp** n/a, **200 amp** n/a
Cable TV: No **Dockside Phone:** No
Dock Type: Floating, Wood
Moorings: 0 Total, 0 Transient **Launch:** n/a
Rate: Day n/a **Week** n/a **Month** n/a
Heads: 2 Toilet(s), 2 Shower(s)
Laundry: None **Pay Phones:** Yes, 1
Pump-Out: No **Fee:** n/a **Closed Heads:** Yes

Hammond Marina

PO Box 161; 320 Lake Drive; Hammond, OR 97121

Tel: (503) 861-3197 **VHF: Monitor** Ch. 16 **Talk** Ch. 68
Fax: (503) 861-2370 **Alternate Tel:** n/a
Email: warbasin@pacifier.com **Web:** n/a
Nearest Town: Hammond *(0.25 mi.)* **Tourist Info:** (503) 325-6311

Marina Operations
Owner/Manager: Keith Pinkstaff **Dockmaster:** Same
In-Season: Jun-Oct, 5am-5:30pm **Off-Season:** Nov-May, 7am-3:30pm
After-Hours Arrival: Use pay envelops in the mailbox
Reservations: No **Credit Cards:** No
Discounts: None
Pets: Welcome, Dog Walk Area **Handicap Access:** No

Marina Services and Boat Supplies
Communication - FedEx, UPS **Supplies - Near:** Ice *(Cube)*, Bait/Tackle *(Free Willy 861-1201)* **3+ mi:** Ships' Store *(Englund Marine Supply 325-4341, 7 mi.)*

Boatyard Services
OnSite: Launching Ramp *($5)* **Nearest Yard:** Warrenton Boat Yard (503) 861-1311

Restaurants and Accommodations
Near: Restaurant *(Coleman's Cove Restaurant 861-3547, L, D)*, *(Bouy 9 Restaurant and Lounge 861-2962, L 5-12, D 8-19)*, Motel *(South Jetty Inn 861-2500, $50-80)* **1-3 mi:** Lite Fare *(Serendipity Deli Café 861-0222)* **3+ mi:** Restaurant *(Iredale Inn 861-3574, 3 mi.)*, Fast Food *(Dairy Maid 861-2692, 3 mi.)*, *(Subway 5 mi.)*, Lite Fare *(Arnie's Café Arnie's Café 861-0520, 5 mi.)*, Pizzeria *(Fultano's Pizza 861-9367, 3 mi.)*, Inn/B&B *(Sipin'on Public House Inn Warrenton 861-2566, 3 mi.)*

Recreation and Entertainment
Near: Picnic Area, Jogging Paths, Fishing Charter *(Corkey's Charters 861-2088)*, Park *(Lighthouse Park - memorial to fishermen)*, Sightseeing *(Fort Stevens State Park 861-1671 - self-guided tours, historic sites, gun batteries, observation platform, hiking trails)* **Under 1 mi:** Group Fishing Boat *(Charlton Deep Sea Charters 338-0569)* **3+ mi:** Golf Course *(Lewis & Clark 338-3386, 8 mi.)*, Fitness Center *(Columbia Fitness 861-0688, 5 mi.)*, Movie Theater *(Columbia Theatres 325-3516, 7 mi.)*, Video Rental *(Movie Gallery 861-1229, 5 mi.)*, Museum *(Columbia River Maritime 325-2323, Children's Museum 325-8669, 7 mi.)*, Galleries *(Riversea Gallery 325-1270, 7 mi.)*

Provisioning and General Services
Near: Convenience Store *(Corkey's 4-Way Stop and Go 861-2088)*, Wine/Beer, Bank/ATM, Post Office *(861-2671)* **Under 1 mi:** Market *(Fort Stevens Shopping Center 861-1211)*, Protestant Church, Library *(Warrenton 861-3919)*, Retail Shops **3+ mi:** Supermarket *(Safeway 503-325-4662, 7 mi.)*, Delicatessen *(Serendipity 861-0222, 3 mi.)*, Liquor Store *(Warrenton Liquor 861-1103, 5 mi.)*, Farmers' Market *(Astoria Sun Market 10am-3pm May-Oct at 12th St. 325-1010, 7 mi.)*, Crabs/Waterman *(Warrenton Deep Sea Crab & Fish Market 861-3911, 3 mi.)*, Beauty Salon *(Shear Heaven 861-3866, 4 mi.)*, Barber Shop *(Clipper Station 861-2060, 5 mi.)*, Pharmacy *(Rite Aid 861-1611, 6 mi.)*, Hardware Store *(True Value Hardware 861-1161, 5 mi.)*, Buying Club *(Costco 861-1950, 5 mi.)*, Copies Etc. *(Quick Print and More 861-2650, 5 mi.)*

Transportation
OnCall: Rental Car *(Enterprise 325-6500)*, Taxi *(Old Gray Cab 338-6030)*
Near: Local Bus **Airport:** Astoria Regional *(6 mi.)*

Medical Services
911 Service **OnCall:** Ambulance **3+ mi:** Doctor *(Pacific Family Medicine 325-5300, 7 mi.)*, Dentist *(North Coast Dental 861-3707, 4 mi.)*, Chiropractor *(North Coast Chiropractic 861-1661, 4 mi.)*, Holistic Services *(Harbor Place Massage 861-0887, 4 mi.)*, Veterinarian *(Bayshore Animal Hospital 861-1621, 4 mi.)* **Hospital:** Columbia Memorial 325-4321 *(9 mi.)*

Setting -- At River Mile 8, Hammond Marina is the first Oregon facility above the mouth of the Columbia River. The main shipping channel passes close by and the Desdemona Sands, a large shoal area lies just beyond the channel. The small cove that berths the network of basic floating docks is protected by rock breakwaters. Onshore a small pale blue building hosts the heads and a nearby trailer the marina office.

Marina Notes -- *Flat rate: Mon-Thu up to 19' $12/night, 20-29' $13, 30-39' $14, 40-49' $15, 50-59' $16, 60-69' $17, 70-79' $18, 80-89' $19, 90-99' $20/night. Fri-Sun add $2/night. Pay envelopes are in the mailbox at the top of the boat launch. The marina provides wooden docks with long finger slips as well as ample alongside space, a fuel dock, and a wide launch ramp. It houses some commercial fishing vessels as well as sportfishing boats. Rogers Marine Repair (861-1527) is on-site and offers fishing charters, repair work, bait and tackle. Basic heads with Formica walls.

Notable -- The "Peacock" often docks at Hammond between deliveries of bar pilots to ships entering or leaving the river. Adjacent to the marina is a large parklike area that overlooks the river as well as the docks; a great place for picnics. Two restaurants are nearby in the village of Hammond. Take the 4.5 mi. Warrenton Waterfront Trail, which parallels the river, to Second Street Park in Warrenton. Stop at the interpretive center with its replica lighthouse in Lighthouse Park. A canoe/kayak launch is at Second Street Park. One can explore the backwaters of the Skipanon River from here. Fort Stevens State Park features lots of historical sites - the entrance is a short walk from the marina.

Navigational Information
Lat: 46°10.053' **Long:** 123°55.217' **Tide:** 10 ft. **Current:** n/a **Chart:** 18521
Rep. Depths *(MLW):* **Entry** 14 ft. **Fuel Dock** n/a **Max Slip/Moor** 14 ft./-
Access: On Skipanon Waterway off Columbia River

Marina Facilities *(In Season/Off Season)*
Fuel: No
Slips: 92 Total, 2 Transient **Max LOA:** 70 ft. **Max Beam:** 12 ft.
  Rate *(per ft.):* **Day** $0.45* **Week** n/a **Month** $10
  Power: 30 amp $0.08/kwh, **50 amp** n/a, **100 amp** n/a, **200 amp** n/a
  Cable TV: No **Dockside Phone:** No
  Dock Type: Floating, Long Fingers, Alongside, Concrete
Moorings: 0 Total, 0 Transient **Launch:** n/a
  Rate: Day n/a **Week** n/a **Month** n/a
Heads: 2 Toilet(s), 2 Shower(s)
Laundry: None, Book Exchange **Pay Phones:** No
Pump-Out: OnSite, Self Service **Fee:** n/a **Closed Heads:** Yes

Marina Operations
Owner/Manager: Bob Link **Dockmaster:** Same
In-Season: Year-Round, 8am-5pm **Off-Season:** n/a
After-Hours Arrival: Check board for available slip or call 717-3733
Reservations: Yes, Required **Credit Cards:** No; Cash only
Discounts: None
Pets: No **Handicap Access:** Yes, Docks

Skipanon Landing

PO Box 730; 200 N.E. Skipanon Drive; Warrenton, OR 97146

Tel: (503) 861-0362 **VHF: Monitor** Ch. 9 **Talk** Ch. 23
Fax: n/a **Alternate Tel:** (503) 717-3733
Email: boblink@pacifier.com **Web:** n/a
Nearest Town: Warrenton *(0.25 mi.)* **Tourist Info:** (503) 325-6311

Marina Services and Boat Supplies
Communication - Mail & Package Hold **Supplies - Near:** Ice *(Block, Cube, Shaved)*, Bait/Tackle *(Tackle Time Bait Shop 861-3693)*, Live Bait

Boatyard Services
OnCall: Compound, Wash & Wax, Interior Cleaning **Near:** Launching Ramp, Electronic Sales, Electronics Repairs, Propeller Repairs.
1-3 mi: Engine mechanic *(gas, diesel)*, Electrical Repairs, Hull Repairs, Rigger, Sail Loft, Canvas Work, Bottom Cleaning, Brightwork.
Nearest Yard: Warrenton Boat Yard (503) 861-1311

Restaurants and Accommodations
Near: Restaurant *(Kim's Kitchen 861-4314, L, D)*, *(Iredale Inn 861-3574)*, *(El Compadre 861-2906)*, Seafood Shack *(Warrenton Deep Sea 861-1233)*, Pizzeria *(Fultano's Pizza 861-9367)*, Motel *(Sipin'on Public House Inn 861-2566)* **1-3 mi:** Restaurant *(Buoy 9 Restaurant & Lounge 861-2962)*, Fast Food *(Subway)*, Lite Fare *(Arnie's Café 861-0520)*, Motel *(South Jetty Inn 861-2500)*, Hotel *(Shilo Inn 861-2181, $69-159)* **3+ mi:** Hotel *(Best Western 325-2205, $60-85, 4 mi.)*

Recreation and Entertainment
OnSite: Picnic Area, Jogging Paths **Near:** Fishing Charter *(Thunderbird Charters 861-1270)*, Group Fishing Boat *(Charlton Deep Sea Charters 961-2429)*, Video Arcade, Park **1-3 mi:** Fitness Center *(Columbia Fitness 861-0688)*, Video Rental *(Movie Gallery 861-1229)* **3+ mi:** Golf Course *(Lewis & Clark 338-3386, 6 mi.)*, Bowling *(Lower Columbia Bowl 325-3321, 6 mi.)*, Movie Theater *(Columbia Theaters 325-3516, 6 mi.)*, Museum *(Columbia River Maritime Museum 325-2323, 6 mi.)*, Sightseeing *(Fort Stevens*

State Park 861-1671 - self-guided tours available $3, 3.5 mi.)*, Galleries *(Astrovisual 325-4589, 6 mi.)*

Provisioning and General Services
Near: Convenience Store *(Mini-Mart 861-2246)*, Market *(Main St. 861-2271)*, Delicatessen *(Serendipity 861-0222)*, Wine/Beer, Fishmonger *(Oregon Coast Seafoods 861-1434)* **Under 1 mi:** Bank/ATM, Post Office *(861-7035)*, Protestant Church, Beauty Salon *(Shear Heaven 861-3866)*, Barber Shop *(Clipper Station 861-2060)*, Laundry *(Marlin Ave. 861-2182)*, Pharmacy *(Rite Aid 861-1611)* **1-3 mi:** Liquor Store *(Warrenton 861-1103)*, Synagogue, Library *(Warrenton 861-3919)*, Department Store *(Fred Meyer 861-3003)*, Buying Club *(Costco 861-1950)*, Copies Etc. *(Print Quick & More 861-2650)* **3+ mi:** Supermarket *(Safeway 503-325-4662, 6 mi.)*, Farmers' Market *(Astoria Sun Market at 12th St., May-Oct 10am-3pm 325-1010, 6 mi.)*, Catholic Church *(St. Mary 325-3671, 6 mi.)*, Bookstore *(Lucy's 325-4210, 6 mi.)*, Hardware Store *(True Value 861-1161, 4 mi.)*

Transportation
OnSite: Local Bus *(Sunset Empire Transportation 861-7433)*
OnCall: Rental Car *(Enterprise 325-6500)*, Taxi *(Old Gray Cab 861-1735)*
Airport: Astoria Regional *(3 mi.)*

Medical Services
911 Service **OnCall:** Ambulance **Under 1 mi:** Dentist *(North Coast Dental 861-3707)*, Chiropractor *(North Coast Chiropractic 861-1661)*, Veterinarian *(Bayshore 861-1621)* **1-3 mi:** Optician *(Optical Shoppe 861-9829)* **3+ mi:** Doctor *(Pacific Family Medicine 325-5300, 6 mi.)*
Hospital: Columbia Memorial 325-4321 *(6 mi.)*

Setting -- The entrance to Skipanon River is on the south shore of the Columbia - about 4 miles from Fort Stevens. The marina lies on the west shore of this quiet river, directly across from the Warrenton City Mooring Basin. Onshore, a pair of gray-sided buildings - a two-story and a single - are surrounded by lawns and trees that add a pleasant touch. The neighborhood appears largely residential, although in this region, the river is also used for moving logs.

Marina Notes -- *Flat rate: $17.50/day up to 40', $22.50/day over 40'. Formerly Skipanon Marina, new owners took the helm in Oct '05. While this facility currently welcomes overnight transients, plans are underway for the development of 53 new condominiums; the marina may close in mid '07. Fuel available at Port of Astoria West Basin and at Hammond Marina. The wide, clean docks are an usual concrete over foam. Slips can accommodate large vessels, with long, narrow fingers. There are few amenities for the boater, but the area is quiet and the fishing good. Salmon fishing begins in July.

Notable -- The marina is conveniently located about a quarter mile from "downtown" Warrenton. Several casual restaurants, motels, and small grocery stores are within walking distance. A rather easy-going bus service makes it simple to go beyond - just wave it down and name your destination. Nearby Lighthouse Park features a replica lighthouse and interpretive center. The entrance to Fort Stevens, which defended the mouth of the Columbia River from the Civil War until World War II, is about 3.5 mi. from the marina. Now a state park, the museum and remains of the fort, including well preserved gun batteries, make an interesting trip. The park boasts a whopping 579 paved campsites and 15 yurts. The wreck of the "Peter Iredale" can be seen on the beach.

Warrenton City Mooring Basin

PO Box 250; 550 NE Harbor Drive; Warrenton, OR 97146

Tel: (503) 861-3822 **VHF: Monitor** Ch. 16 **Talk** Ch. 68
Fax: (503) 861-2370 **Alternate Tel:** n/a
Email: warbasin@pacifier.com **Web:** n/a
Nearest Town: Warrenton *(0.5 mi.)* **Tourist Info:** (503) 325-6311

Navigational Information

Lat: 46°09.980' **Long:** 123°55.110' **Tide:** 8 ft. **Current:** n/a **Chart:** 18521
Rep. Depths *(MLW)*: **Entry** 17 ft. **Fuel Dock** n/a **Max Slip/Moor** 10 ft./-
Access: 8 miles up the Columbia River then 1 mile up Skipanon Waterway

Marina Facilities *(In Season/Off Season)*

Fuel: No
Slips: 370 Total, 20 Transient **Max LOA:** 90 ft. **Max Beam:** n/a
  Rate *(per ft.)*: **Day** $0.48* **Week** n/a **Month** n/a
  Power: 30 amp $4, **50 amp** n/a, **100 amp** n/a, **200 amp** n/a
  Cable TV: No **Dockside Phone:** No
  Dock Type: Floating, Wood
Moorings: 0 Total, 0 Transient **Launch:** n/a
  Rate: Day n/a **Week** n/a **Month** n/a
Heads: 2 Toilet(s), 4 Shower(s)
Laundry: None **Pay Phones:** No
Pump-Out: OnSite **Fee:** Free **Closed Heads:** Yes

Marina Operations

Owner/Manager: Keith Pinkstaff **Dockmaster:** Same
In-Season: Jun-Oct, 5am-5:30pm **Off-Season:** Nov-May, 7am-3:30pm
After-Hours Arrival: Call ahead
Reservations: Yes **Credit Cards:** No
Discounts: None
Pets: Welcome **Handicap Access:** Yes, Docks

Marina Services and Boat Supplies

Services - Docking Assistance, Trash Pick-Up, Dock Carts
Communication - FedEx, UPS **Supplies -** **Near:** Ice *(Cube)*, Bait/Tackle *(Tackle Time Bait Shop 861-3693)*, Live Bait

Boatyard Services

OnSite: Launching Ramp **Nearest Yard:** Warrenton Boat Yard (503) 861-1311

Restaurants and Accommodations

OnSite: Restaurant *(Kim's Kitchen 861-4314, L, D)* **Near:** Restaurant *(Iredale Inn 861-3574)*, Pizzeria *(Fultano's Pizza 861-9367)*, Motel *(South Jetty 861-2500)* **Under 1 mi:** Restaurant *(Ginger's Diner 861-3375)*, *(Dooger's Seafood & Grill 861-2839)*, Fast Food *(Subway, Dairy Queen L, D)*, Lite Fare *(Arnie's Café 861-0520)*, Pizzeria *(Pizza Hut 861-1616)*, Hotel *(Shilo Inn 861-2180, $69-159)* **1-3 mi:** Restaurant *(Buoy 9 861-2962)*, Motel *(Sipin'on Public House 861-2566)* **3+ mi:** Hotel *(Astoria Best Western 325-2205, $60-85, 4 mi.)*

Recreation and Entertainment

Near: Jogging Paths, Fishing Charter *(Thunderbird Charters 861-1270)*, Group Fishing Boat *(Salmon Master 861-2577, Charlton Deep Sea Charters 861-2429)*, Park *(Lighthouse Park Interpretive Center 861-7225)* **Under 1 mi:** Fitness Center *(Columbia Fitness 861-0688)* **1-3 mi:** Video Rental *(Movie Gallery 861-1229)* **3+ mi:** Golf Course *(Lewis & Clark 338-3386, 6 mi.)*, Bowling *(Lower Columbia Bowl 325-3321, 5 mi.)*, Movie Theater *(Columbia Theaters 325-3516, 5 mi.)*, Museum *(Columbia River Maritime Museum 325-2323, Flavel House 325-2563, 5 mi.)*, Sightseeing *(Fort Stevens State Park 861-1671 - self-guided tours, historic sites, gun batteries, observation platform, hiking trails/ Fort Clatsop, 4 mi.)*

Provisioning and General Services

Near: Convenience Store *(Mini-Mart 861-2246)*, Market *(Main St. 861-2271)*, Delicatessen *(Serendipity 861-0222)*, Wine/Beer, Fishmonger *(Oregon Coast Seafoods 861-1434)*, Post Office *(861-7035)*, Protestant Church *(United Methodist 861-0825)*, Barber Shop *(Clipper Station 861-2060)*, Laundry *(Marlin Ave. 861-2182)* **Under 1 mi:** Liquor Store *(Warrenton 861-1103)*, Bank/ATM *(State Farm 861-3276)*, Beauty Salon *(Shear Heaven 861-3866)*, Pharmacy *(Rite Aid 861-1611)* **1-3 mi:** Library *(Warrenton 861-3919)*, Hardware Store *(True Value 861-1161)*, Department Store *(Fred Meyer 861-3003)*, Buying Club *(Costco 861-1950)*, Copies Etc. *(Print Quick 861-2650)* **3+ mi:** Supermarket *(Safeway 503-325-4662, 5 mi.)*, Bakery *(Bakery & Deli by the Sea 861-0365, 8 mi.)*, Farmers' Market *(Astoria Sun Market at 12th St., May-Oct 10am-3pm 325-1010, 5 mi.)*, Catholic Church *(St. Francis 325-3671, 5 mi.)*, Bookstore *(Lucy's 325-4210, 5 mi.)*

Transportation

OnCall: Rental Car *(Enterprise 325-6500)*, Taxi *(Royal Cab 325-5818)* **Near:** Local Bus *(861-7433)* **Airport:** Astoria Regional *(3 mi.)*

Medical Services

911 Service **OnCall:** Ambulance **Near:** Dentist *(North Coast Dental 861-3707)*, Chiropractor *(North Coast Chiropractic 861-1661)* **1-3 mi:** Veterinarian *(Bayshore Animal Hospital 861-1621)*, Optician *(Optical Shoppe 861-9829)* **3+ mi:** Doctor *(Pacific Family Medicine 325-5300, 6 mi.)* **Hospital:** Columbia Memorial 325-4321 *(6 mi.)*

Setting -- The Warrenton Boat Basin is approximately 1 mile south of the mouth of the Skipanon, on the east shore. It is home to a large sportfishing fleet and a small number of sailboats. In contrast to the burly Columbia, the much smaller Skipanon has an intimate, less hurried feel to it. A modest network of floating docks is served by two small gray-blue buildings that house the office and bathhouse. Adjacent is a larger contemporary with a teal metal roof. This is a nononsense marina, delivering its services with a minimum of frills.

Marina Notes -- *Flat rate: varies with size and day of the week. Mon-Thu up to 19' $12/night, 20-29' $13, 30-39' $14, 40-49' $19, 50-59' $20, 60-69' $21, 70-79' $22, 80-89' $23, 90-99' $24. Fri-Sun rates are $2 higher. The wooden docks with long finger piers are showing their age, but boards are slowly being replaced. Boater amenities are limited and include pump-out, a boat ramp, a fish cleaning table, and a small book exchange in the office. The brick and sheetrock restrooms and showers are only open 7am-3:30pm.

Notable -- The Columbia River offers amazing fishing opportunities, particularly for salmon and sturgeon. Many charters and groups fishing boats are available in the area or, if you're equipped to fish off your own boat, nearby Tackle Time has supplies. Lighthouse Park, a memorial to deep sea fishermen and women, is near the marina, and the Fort Stevens State Park - the state's largest campground with miles of ocean beach - can be reached by bus or taxi. Restaurants and basic provisions are a half mile away in Warrenton. Downtown Astoria offers more urban entertainment plus additional shops and services.

Navigational Information
Lat: 46°11.430' **Long:** 123°51.300' **Tide:** 8 ft. **Current:** n/a **Chart:** 18521
Rep. Depths (*MLW*): **Entry** 16 ft. **Fuel Dock** 14 ft. **Max Slip/Moor** 14 ft./-
Access: 14 miles up the Columbia River on the south side

Marina Facilities *(In Season/Off Season)*
Fuel: Gasoline, Diesel
Slips: 335 Total, 28 Transient **Max LOA:** 109 ft. **Max Beam:** n/a
  Rate *(per ft.)*: **Day** $0.50* **Week** n/a **Month** $140-595
  Power: 30 amp $3, 50 amp n/a, 100 amp n/a, 200 amp n/a
Cable TV: No **Dockside Phone:** No
Dock Type: Floating, Long Fingers, Alongside, Wood
Moorings: 0 Total, 0 Transient **Launch:** n/a
  Rate: Day n/a **Week** n/a **Month** n/a
Heads: 4 Toilet(s), 2 Shower(s)
Laundry: 1 Washer(s), 1 Dryer(s) **Pay Phones:** Yes, 2
Pump-Out: OnSite **Fee:** Free **Closed Heads:** Yes

Marina Operations
Owner/Manager: Ken Smith **Dockmaster:** Same
In-Season: May-Oct, 8am-5pm/7 days **Off-Season:** Nov-Apr, 8am-5pm
After-Hours Arrival: Use pay envelope and display receipt where visible
Reservations: No **Credit Cards:** Visa/MC
Discounts: None
Pets: Welcome **Handicap Access:** No

Port of Astoria - West Basin

352 Industry; Astoria, OR 97103

Tel: (503) 325-8279 **VHF: Monitor** Ch. 16 **Talk** Ch. 74
Fax: (503) 325-8279 **Alternate Tel:** n/a
Email: marina@portofastoria.com **Web:** www.portofastoria.com
Nearest Town: Astoria *(1 mi.)* **Tourist Info:** (503) 325-6311

Marina Services and Boat Supplies
Services - Security, Trash Pick-Up, Dock Carts **Communication -** FedEx, UPS, Express Mail **Supplies - OnSite:** Ice *(Cube)*, Ships' Store *(Englund Marine 325-4341)*, Bait/Tackle **Near:** Propane **Under 1 mi:** Ice *(Block)*

Boatyard Services
OnSite: Travelift *(88T)*, Canvas Work *(Windward 325-5788)* **Near:** Metal Fabrication *(Columbia Pacific)*. **1-3 mi:** Launching Ramp, Electronics Repairs *(Jensen 861-2415)*, Refrigeration *(P & L Johnson 325-2180)*, Divers, Propeller Repairs *(P & L Johnson)*. **3+ mi:** Engine mechanic *(gas, diesel)*, Electrical Repairs *(Wadsworth 325-5501, 6 mi.)*.

Restaurants and Accommodations
OnSite: Coffee Shop *(Espresso Peddler 325-9012, coffee, smoothies, pastries, bagels)*, Hotel *(Red Lion 325-7373, $115-136, waterfront balconies)* **OnCall:** Pizzeria *(Fultano's 861-9367, all-you-can-eat pizza Mon nt.)* **Near:** Restaurant *(Portway Tavern 325-2651, L, D)*, *(Café Union 325-8708)*, *(Stephanie's Cabin 325-7181)*, *(Golden Luck 325-7289)*, Lite Fare *(Workers Bar & Grill 338-7291)*, Hotel *(Best Western 325-2205, $63-135)*, *(Cannery Pier 338-4702, $180-525, luxury boutique on site of Union Fish Cannery)* **1-3 mi:** Restaurant *(El Tapatio 325-1255, L $6-20, D $9-20)*

Recreation and Entertainment
OnSite: Picnic Area, Fishing Charter *(Tiki Charters 325-7818)*, Group Fishing Boat *(Tiki Charters)*, Special Events *(Astoria Regatta - early Aug 325-6311)* **Under 1 mi:** Pool *(Astoria Aquatic Center 327-7027)*, Playground, Roller Blade/Bike Paths *(Skateboard Park)*, Museum *(Columbia River Maritime 325-2323 $8/$7 seniors/$4 kids, 9:30am-5pm;*

Flavel House 325-2563 $5/$4 seniors/$2 kids, May-Sep 10am-5pm)* **1-3 mi:** Fitness Center *(Curves 338-1294)*, Movie Theater *(Columbia Theatres 325-3516)*, Sightseeing *(Astoria Column $1/car)*, Galleries *(Riversea Gallery 325-1270)* **3+ mi:** Golf Course *(Lewis & Clark 338-3386, 4 mi.)*, Bowling *(Lower Columbia Bowl 325-3321, 4 mi.)*

Provisioning and General Services
OnSite: Hardware Store **Near:** Convenience Store *(Mini-Mart 325-4162)*, Bank/ATM, Beauty Salon *(His & Her Hair Care 798-5600)*
Under 1 mi: Market *(Peter Pan 325-2143)*, Dry Cleaners *(Roy's 325-6491)*, Laundry **1-3 mi:** Supermarket *(Safeway 325-4662)*, Delicatessen *(Country Grocer 325-4031)*, Health Food *(325-6688)*, Liquor Store *(325-4784)*, Bakery *(Danish Maid 325-3657)*, Farmers' Market *(12th St. Sun 10am-3pm May-Oct)*, Fishmonger *(Fish Landing 325-1067)*, Library *(325-7323)*, Bookstore *(Lucy's 325-4210)*, Pharmacy *(Paramount 325-4541)*, Florist *(325-3571)*, Department Store *(Sears 325-3821)*

Transportation
OnSite: Local Bus *(Riverfront Trolley 325-6311-7 days, MemDay-LabDay, wknds off-season - $1/ride or $2/day / Empire Trans. 861-7433)*
OnCall: Rental Car *(Enterprise 325-6500/Hertz 325-7700, 2 mi.)*, Taxi *(Royal Cab 325-5818)* **Airport:** Astoria Regional *(5 mi.)*

Medical Services
911 Service **OnCall:** Ambulance **Near:** Holistic Services *(Columbia R. Day Spa 325.4996)* **Under 1 mi:** Dentist *(Bales 325-3230)* **1-3 mi:** Doctor *(Emergency Physicians 338-7508)*, Chiropractor *(Astoria 325-3311)*, Veterinarian *(Columbia Vet 325-2250)* **Hospital:** Columbia 325-4321 *(1 mi.)*

Setting -- The massive four-mile-long Astoria-Megler Bridge, that carries US 101 over the vast Columbia, looms directly over the Port of Astoria-West Basin - a landmark that makes finding the marina easy. Dirt-filled steel pile breakwaters marked by lights protect the basin. Five very long piers with finger slips lie perpendicular to the shore and the two-story, L-shaped Red Lion Inn hugs the whole west side of the basin. The marina office and facilities are in the tan, green-trimmed two-story building near the bridge. A picnic deck overlooks the docks and colorful restored Victorians perch on the verdant hillside above.

Marina Notes -- *Flat rates: 20-29' $12, 30-39' $15, 40-49' $20, 50-59' $25, 60-69' $30, 70-79' $35, 80-89' $40, 90-99' $55, 100-109' $60. Monthly rates vary from $140-595, with $60 additional for power; annual rates also available. This large, protected marina, managed by Port of Astoria, offers slips for boats up to 65-ft., and linear moorage for larger yachts. Guest docks are in the areas of "B" and "C" docks. On-site are a fish cleaning station, free pump-out, an 88 ton travelift, and a list of local marine services. A fish cleaning station, espresso stand, ATM, KFC, US Customs office (325-5541), and Tiki Charters (325-7818) are also in the immediate vicinity. NOTE: Seafare Restaurant at the Red Lion has closed. New bathhouse & office scheduled for early '07.

Notable -- The oldest settlement west of the Rockies, Astoria's lush vegetation benefits from an amazing 75 inches of rain a year. The revitalized 1920s-vintage downtown is less than a mile from the marina; an old, restored trolley makes the journey year-round or stroll the paved riverwalk into town and then to the East Basin. Don't miss the must-see, soaring, contemporary Columbia River Maritime Museum or Captain Flavel's 1885 elegant Victorian House.

Navigational Information

Lat: 46°11.700' **Long:** 123°48.420' **Tide:** 8 ft. **Current:** n/a **Chart:** 18521
Rep. Depths *(MLW)*: **Entry** 19 ft. **Fuel Dock** n/a **Max Slip/Moor** 17 ft./-
Access: Columbia River through Astoria Bridge on south side

Marina Facilities *(In Season/Off Season)*

Fuel: No
Slips: 82 Total, 5 Transient **Max LOA:** 109 ft. **Max Beam:** n/a
Rate *(per ft.)*: **Day** $0.50* **Week** n/a **Month** $280
Power: 30 amp $3, 50 amp $3, 100 amp n/a, 200 amp n/a
Cable TV: No **Dockside Phone:** No
Dock Type: Floating, Long Fingers, Alongside, Wood
Moorings: 0 Total, 0 Transient **Launch:** n/a
Rate: Day n/a **Week** n/a **Month** n/a
Heads: 2 Toilet(s), 1 Shower(s)
Laundry: None **Pay Phones:** Yes, 2
Pump-Out: No **Fee:** n/a **Closed Heads:** Yes

Marina Operations

Owner/Manager: Ken Smith **Dockmaster:** Same
In-Season: May-Oct, 8am-5pm/7 Days **Off-Season:** Nov-Apr, 8am-5pm
After-Hours Arrival: Use pay envelope and display receipt where visible
Reservations: No **Credit Cards:** Visa/MC
Discounts: None
Pets: Welcome **Handicap Access:** Yes, Heads

Port of Astoria - East Basin

352 Industry; Astoria, OR 97103

Tel: (503) 325-8279; (800) 860-4093 **VHF: Monitor** Ch. 16 **Talk** Ch. 74
Fax: (503) 325-4525 **Alternate Tel:** n/a
Email: marina@portofastoria.com **Web:** www.portofastoria.com
Nearest Town: Astoria **Tourist Info:** (503) 325-6311

Marina Services and Boat Supplies

Supplies - Near: Ice *(Block, Cube)*, Bait/Tackle **Under 1 mi:** Propane
1-3 mi: Ships' Store *(Englund Marine Supply 325-4341)*

Boatyard Services

OnSite: Launching Ramp **Near:** Divers. **1-3 mi:** Electrical Repairs
(Wadsworth 325-5501), Canvas Work *(Woodward 325-5788)*, Refrigeration
(P & L Johnson 325-2180), Propeller Repairs *(Northwest 325-0832)*.
Nearest Yard: Astoria Marine Construction (503) 325-4121

Restaurants and Accommodations

OnSite: Coffee Shop *(JT's Clatsop 338-6036)* **Near:** Pizzeria *(Geno's Pizza
& Burgers 325-4927)*, Motel *(Comfort Suites 325-2000, $69-109, just south
of the marina)*, Inn/B&B *(Benjamin Yung 325-6172)* **Under 1 mi:** Restaurant
(Hong Kong 325-5344, L $5-9, D $7-15), *(Annie's Uppertown Tavern 325-
1102)*, Seafood Shack *(Bowpicker Fish & Chips 458-6315)* **1-3 mi:** Restaurant *(Silver Salmon Grille 338-6640)*, *(Wet Dog Café/Astoria Brewing
Co. 325-6975)*, *(El Tapatio 325-1255, L $6-20, D $9-20)*, Fast Food
(McDonald's, Burger King), Motel *(Hotel Elliott 325-2222, $109-375,
rooms & suites)*, Hotel *(Red Lion 325-7373, $62-75)*

Recreation and Entertainment

OnSite: Picnic Area, Jogging Paths *(Riverwalk)* **Near:** Playground, Tennis
Courts **Under 1 mi:** Pool *(Astoria Aquatic Center 325-7027 - 4 pools, water
slide, lazy river, etc. $4.50/3.50 fam $10.50)*, Fitness Center *(Astoria
Aquatic)*, Museum *(Firefighters 325-2203, Flavel House 325-2563, Columbia
River Maritime 325-2323)* **1-3 mi:** Bowling *(Lower Columbia Bowl 325-
3321)*, Fishing Charter *(Tiki Charters 325-7818)*, Movie Theater *(Gateway

338-6575)*, Video Rental *(Video Horizons 325-7310)*, Cultural Attract
(Performing Arts at Historic Liberty Theater 325-2757), Sightseeing *(Astoria
Column)*, Special Events *(Astoria Regatta - 2nd wknd Aug, Astoria Music
Arts Festival, 3-day Crab & Seafood Fest - Apr)* **3+ mi:** Golf Course *(Lewis
& Clark 338-3386, 5 mi.)*

Provisioning and General Services

Near: Supermarket *(Safeway 325-4662)*, Bank/ATM, Laundry *(Uppertown
791-7765)* **Under 1 mi:** Convenience Store *(Food Mart 338-6456)*, Liquor
Store *(325-4784)*, Bakery *(Home 325-4621)*, Fishmonger *(many markets)*,
Post Office, Catholic Church, Protestant Church, Beauty Salon *(Hair Gallery
338-2988)*, Barber Shop *(Glen's 325-8099)*, Pharmacy *(Owl 325-4311)*
1-3 mi: Market *(Astoria Downtown 338-4321)*, Health Food *(325-6688)*,
Farmers' Market *(12th St. Astoria Sun Market May-Oct 10am-3pm)*, Library
(Astoria 325-7323), Dry Cleaners *(Roy's 325-6491)*, Bookstore *(Lucy's 325-
4210)*, Hardware Store *(Builders 325-3611)*, Florist *(Astoria 325-3571)*,
Department Store *(JC Penney 325-4741)*, Copies Etc. *(Lazerquick 325-1185)*

Transportation

OnSite: Local Bus *(Riverfront Trolley 325-6311 $1/trip, $2 all day/ Sunset
Empire 861-7433, near)* **OnCall:** Rental Car *(Enterprise 325-6500)*, Taxi
(Royal 325-5818) **Airport:** Astoria Regional *(7 mi.)*

Medical Services

911 Service **OnCall:** Ambulance **Near:** Dentist *(Reynolds 338-6000)*
Under 1 mi: Doctor *(Emergency Physicians 338-7508)* **1-3 mi:** Chiropractor
(Astoria Chiropractic 325-3311), Veterinarian *(Columbia Vet. 325-2250)*
Hospital: Columbia Memorial 325-4321 *(2 mi.)*

Setting -- The relatively small, quiet Port of Astoria - East Basin is located at River Mile 15.6, about 1.6 miles upstream from the impressive Astoria-Megler
Bridge. A long causeway T-bones the metal and rock breakwater that parallels the shore and protects the five main piers; tall rock jetties on both ends provide
additional protection. The basin can be entered from either side. Much of Astoria's fishing fleet, including some very large trawlers, berth here and a
considerable number of pleasure crafts moor here as well. A two-story brick and clapboard building houses the marina office and heads.

Marina Notes -- *Flat rate: 20-29' $12, 30-39' $15, 40-49' $20, 50-59' $35, 60-69' $40, 70-79' $45, 80-89' $55, 90-99' $65, 100-109' $70. Most of the docks
are concrete, with long finger piers and pedestals for water and metered electricity. Alongside wood docks provide additional space. A 1,300 ft. by 42 ft. paved
surface built on the breakwater is popular for fishing, watching the sunsets, or just observing the river scene. Basic brick heads with stainless steel fixtures.
Note: Large sea lions haul out on the docks from time to time. While they are colorful, they are both federally protected and potentially dangerous.

Notable -- Hitch a ride on the trolley that runs between the east and west basins - the picturesque downtown is about halfway between them. Or cover the
same turf on the paved 4-mile riverwalk next to the old trolley tracks. Popular tourist destinations include the Capt. Flavel House, a magnificent 1885 Victorian
built for the area's first millionaire; the spectacular Columbia River Maritime Museum for an insightful look at the river's history; the Astoria Column built on the
highest hill in town; or a stroll through the restored downtown past 75 preserved historic structures. Provision at the Sunday Market - 3 blocks along 12th Street.

Navigational Information
Lat: 45°41.425' **Long:** 123°52.983' **Tide:** 5 ft. **Current:** n/a **Chart:** 18556
Rep. Depths (MLW): Entry 25 ft. **Fuel Dock** 25 ft. **Max Slip/Moor** 25 ft./-
Access: Nehalem Bay east to Wheeler on south shore

Marina Facilities *(In Season/Off Season)*
Fuel: *Union 76* - Gasoline
Slips: 55 Total, 5 Transient **Max LOA:** 30 ft. **Max Beam:** n/a
  Rate *(per ft.):* **Day** $0.25* **Week** $50 **Month** $5/ft
  Power: 30 amp n/a, 50 amp n/a, 100 amp n/a, 200 amp n/a
  Cable TV: No **Dockside Phone:** No
  Dock Type: Floating, Wood
Moorings: 0 Total, 0 Transient **Launch:** n/a
  Rate: Day n/a **Week** n/a **Month** n/a
Heads: 1 Toilet(s)
Laundry: None **Pay Phones:** No
Pump-Out: No **Fee:** n/a **Closed Heads:** Yes

Marina Operations
Owner/Manager: Jim Neilson **Dockmaster:** Same
In-Season: May 1-Nov 1, 6:30am-Dark **Off-Season:** n/a
After-Hours Arrival: Call ahead
Reservations: Yes, Preferred **Credit Cards:** Visa/MC
Discounts: None
Pets: Welcome **Handicap Access:** No

Wheeler Marina

PO Box 72; 278 Marine Drive; Wheeler, OR 97147

Tel: (503) 368-5780 **VHF: Monitor** n/a **Talk** n/a
Fax: n/a **Alternate Tel:** n/a
Email: n/a **Web:** n/a
Nearest Town: Wheeler **Tourist Info:** (503) 368-5100

Marina Services and Boat Supplies
Supplies - OnSite: Ice *(Block, Cube)*, Bait/Tackle, Live Bait *(Shrimp)*

Boatyard Services
OnSite: Bottom Cleaning **Near:** Launching Ramp *(Free)*.
Nearest Yard: Darts Small Engine (503) 368-6519

Restaurants and Accommodations
OnSite: Seafood Shack *(Sea Shack Seafood Rest. 368-7997, L $$7-28, D $12-28, inside, deck or patio dining - cash or check)* **Near:** Restaurant *(Treasure Café 368-7740, B $3-8, L $6-10, D $13-17)*, *(Bay Ocean Café 368-6873, B, L, D)*, Hotel *(Old Wheeler Hotel 368-6000, $85-195, restored 1920s edifice across from the marina - water view rooms on the second floor)*, *(Wheeler on the Bay Lodge 368-5858, $75-135)* **1-3 mi:** Restaurant *(Currents Restaurant 368-5557, D $22-30, at Nehalem River Inn)*

Recreation and Entertainment
OnSite: Boat Rentals *(single and double kayaks: singles $22/hr., $28/4 hrs., $44/full day; doubles $28/hr., $38/4 hrs., $50/full day)* **Near:** Picnic Area, Grills, Fishing Charter, Group Fishing Boat *(Nehalem Bay Charters 368-5858)*, Video Rental *(Wheeler Grocery and Video 368-7178)*, Park *(Waterfront Park)*, Sightseeing *(Oregon Coast Explorer train ride 800-685-1719 - views of Tillamook Bay, the Pacific Ocean, Nehalem Bay & Nehalem River Valley)*, Special Events *(Crab Derby - 3rd wknd Jun; Salmon Fest - 1st wknd Aug)*, Galleries *(Ekahni Gallery 368-7959, Gypsy Fire Studio and Gallery 368-4308)* **1-3 mi:** Fitness Center *(Nehalem Bay Fitness 368-4595)* **3+ mi:** Beach *(5 mi.)*, Golf Course *(Manzanita Golf Course 368-5744, 5 mi.)*, Jogging Paths *(5 mi.)*

Provisioning and General Services
Near: Convenience Store, Market *(Wheeler Grocery and Video 368-7178)*, Delicatessen *(Bay Front Bakery and Deli 368-6599)*, Health Food *(Simples Herbal Apothecary 368-4906)*, Wine/Beer, Liquor Store *(Wheeler Liquor 368-4906)*, Bakery, Bank/ATM, Post Office, Beauty Salon *(Wheeler Beauty Shop 368-5161 Tue-Fri 9am-5pm)*, Barber Shop, Bookstore *(Ekahni Books 368-6881)*, Pharmacy *(Wheeler Pharmacy 368-5783)* **Under 1 mi:** Retail Shops **1-3 mi:** Protestant Church *(St. Catherine's Episcopal 368-7890)*, Hardware Store *(Nehalem Lumber 368-5619)* **3+ mi:** Fishmonger *(5 mi.)*, Catholic Church *(6 mi.)*, Library *(Manzanita 368-6665, 5 mi.)*, Department Store *(Fred Meyer Tillamook 815-1415, 20 mi.)*

Transportation
Near: Local Bus *(Sunset Empire 861-7433)*, InterCity Bus, Rail *(Port of Tillamook Bay RR 842-2477 "Fun Runs" $55-85)* **Airport:** Tillamook *(26 mi.)*

Medical Services
911 Service **OnCall:** Ambulance **Under 1 mi:** Doctor *(Rhinehart Clinic 368-5182)*, Dentist *(Ternus 368-7090)*, Chiropractor *(Taccogna Chiropractor 368-6050)*, Holistic Services **Hospital:** Tillamook County General 842-4444 *(23 mi.)*

Setting -- The small village of Wheeler parallels the Nehalem River and the railroad tracks - midway between Seaside and Tillamook Bay. Navigation can be difficult but the reward is a natural wonderland. A waterfront park provides a grassy picnic area and a well kept L-shaped public float and launch ramp. Just beyond are the rustic docks of Wheeler Marina. Onshore is a shingled office building, a blue residence and the weathered two-story Sea Shack.

Marina Notes -- *Flat rate: $10/day, $50/wk., regardless of LOA. Older, wooden docks are clean but somewhat uneven - slippery when wet. Finger slips and alongside space for smaller boats. No power. The small store sells fishing poles, bait, clam tubes, ice, and some clothing. A fish cleaning station and crab cooker are on-site. The marina rents single and double kayaks as well as 14-16 ft. skiffs for fishing and exploring the river. Although boater amenities are limited, the town is just across the street. Basic heads, no showers. Note: A long spit, part of Nehalem Bay State Park, extends south from the north shore, making the entrance to the river difficult. The controlling depth at the bar is only 4 ft. and can drop to 3 ft. at low tide inside the bay - the channel shifts and is only marked in the summer. An additional 120 feet of transient space is available at the Wheeler Waterfront Park.

Notable -- Wheeler has a grocery store, coffeehouses, casual restaurants, B&Bs and an assortment of antique stores, unique galleries and gift shops. It's on the edge of the Nehalem River Estuary - a great place to explore by dinghy or kayak. Bird-watching is a major pastime and, in season, the forager will find salmon, clams and crabs. Port of Tillamook Bay's Oregon Coast Explorer runs scenic trains from Garibaldi to Wheeler or Wheeler to Timber/Cochran.

Garibaldi Marina

Navigational Information
Lat: 45°33.365' **Long:** 123°54.770' **Tide:** 10 ft. **Current:** 3 kt. **Chart:** 18558
Rep. Depths (*MLW*): **Entry** 21 ft. **Fuel Dock** 17 ft. **Max Slip/Moor** 12 ft./-
Access: Northwest corner of the boat basin next to the public boat launch

Marina Facilities (*In Season/Off Season*)
Fuel: Gasoline
Slips: 75 Total, 37 Transient **Max LOA:** 30 ft. **Max Beam:** 12 ft.
 Rate (*per ft.*): **Day** $0.38 **Week** $75 **Month** $250
 Power: 30 amp n/a, 50 amp n/a, 100 amp n/a, 200 amp n/a
 Cable TV: No **Dockside Phone:** No
 Dock Type: Floating, Alongside, Wood
Moorings: 0 Total, 0 Transient **Launch:** n/a
 Rate: Day n/a **Week** n/a **Month** n/a
Heads: 2 Toilet(s), 2 Shower(s)
Laundry: None **Pay Phones:** Yes, 1
Pump-Out: No **Fee:** n/a **Closed Heads:** Yes

Marina Operations
Owner/Manager: Jeff & Val Folkema **Dockmaster:** Same
In-Season: Mar-Nov, 6am-6pm **Off-Season:** Dec-Feb, 7am-4pm
After-Hours Arrival: Call before closing for availability and slip assignment
Reservations: Yes, Preferred **Credit Cards:** Visa/MC, Amex
Discounts: Boat/US **Dockage:** 10% **Fuel:** n/a **Repair:** n/a
Pets: Welcome **Handicap Access:** Yes

Garibaldi Marina

PO Box 841; 302 Mooring Basin Road; Garibaldi, OR 97118

Tel: (503) 322-3312; (800) 383-3828 **VHF: Monitor** Ch. 19 **Talk** Ch. 19
Fax: n/a **Alternate Tel:** (503) 812-3312
Email: jeff@garibaldimarina.com **Web:** www.garibaldimarina.com
Nearest Town: Garibaldi **Tourist Info:** (503) 322-0301

Marina Services and Boat Supplies
Services - Docking Assistance, Security (*12 hrs., marina staff*), Trash Pick-Up, Dock Carts **Communication -** Phone Messages **Supplies -** **OnSite:** Ice (*Cube*), Ships' Store, Bait/Tackle **Near:** Propane

Boatyard Services
OnSite: Engine mechanic (*gas*) **Near:** Engine mechanic (*diesel*).
Nearest Yard: Darts Small Engine (503) 368-6519

Restaurants and Accommodations
OnSite: Motel (*Harbor View Inn 322-3251, $45-95*) **Near:** Restaurant (*Troller 322-3662, B $4-11, L $4-12, D $10-17*), (*Miller's Restaurant 322-0355*), (*Fisherman's Korner 322-2033, B $6-10, L $6-15, D $6-15*), (*Pirate's Cove 322-2092*), Fast Food (*Dairy Queen*), Lite Fare (*Garibaldi Pub & Eatery 322-2020*), Motel (*Inn at Garibaldi 322-3338, $55-125*), (*Comfort Inn 322-0328*) **Under 1 mi:** Motel (*Bayshore Inn 322-2552, $64-120*)

Recreation and Entertainment
OnSite: Boat Rentals **Near:** Picnic Area, Grills, Playground, Fishing Charter (*Linda Sue III Charter 355-3419, D&D Garibaldi Charters 322-0007*), Group Fishing Boat, Video Rental, Park, Special Events (*Crab Races - Mar, Garibaldi Days - last wknd Jul*), Galleries (*Classy Touch Imports 322-2030*) **Under 1 mi:** Dive Shop, Museum (*Garibaldi Museum 322-8411, Jul-Sep Thu-Sun 12-4pm; Tillamook Naval Air Station Museum 842-1130, 10 mi.*), Sightseeing (*Oregon Coast Scenic Railroad 842-7972 - departs at 12, 2, 4pm wknds in season; 1.5 hr. RT to Rockaway Beach $13/$7*) **1-3 mi:** Beach **3+ mi:** Golf Course (*Alderbrook Golf Course 842-6413, 6 mi.*)

Provisioning and General Services
Near: Convenience Store, Market (*Food Basket Market 322-3270*), Delicatessen (*Bay Front Bakery & Deli 322-3787*), Wine/Beer, Liquor Store (*Godfrey's Garibaldi Pharmacy and Liquor 322-3456*), Bakery (*Bay Front*), Fishmonger (*Bay Ocean Seafood 322-3316, 9am-5pm, to 6pm Fri-Sat*), Crabs/Waterman (*Tillamook Bay Boat House 322-3600 - live & cooked crab, fish*), Oysterman (*Bay Ocean Oyster 322-0040*), Bank/ATM, Beauty Salon (*Tina's Hair Salon 322-0373*), Barber Shop (*Tami's Barber Shop 322-2228*), Pharmacy (*Godfrey's Garibaldi Pharmacy 322-3456*) **Under 1 mi:** Post Office (*322-3675*), Protestant Church, Library (*Garibaldi 322-2100*), Florist (*Garibaldi Flowers*) **3+ mi:** Laundry (*101 One Stop 355-2923, 5 mi.*), Bookstore (*Rainy Day 842-7766, 10 mi.*), Hardware Store (*Rockaway Beach Hardware 355-2923, 5 mi.*), Department Store (*Fred Meyer 815-1415, 9 mi.*)

Transportation
Near: Local Bus (*Sunset Empire 861-7433/Tillamook Transit to Portland 815-8283, 10 mi.*) **3+ mi:** Rental Car (*E&E Auto Rentals 842-7802, 10 mi.*), Taxi (*Kenny Cab 842-5175, 10 mi.*) **Airport:** Tillamook/Portland Int'l. (*13 mi./85 mi.*)

Medical Services
911 Service **OnCall:** Ambulance (*Garibaldi Ambulance 322-3231*)
3+ mi: Doctor (*Tillamook Medical 842-5546, 10 mi.*), Dentist (*Tillamook Family Dentristry 815-1777, 10 mi.*), Chiropractor (*Tillmook Family Chiro 842-5951, 10 mi.*), Veterinarian (*Pioneer Vet. Hospital 842-8411, 10 mi.*)
Hospital: Tillamook County General 842-4444 (*10 mi.*)

Setting -- Tillamook Bay's entrance is clearly marked by jetties; about three miles in, on the north shore, is the laid-back, stopped-in-time town of Garibaldi. Unobtrusive Garibaldi Marina is tucked in just west of the Port of Garibaldi docks and the public boat ramp. The marina's set of small floating docks, on-site motel and crowded but well-supplied store cater to sport fishermen who come to Garibaldi to chase salmon, halibut, and numerous species of bottom fish.

Marina Notes -- *Flat rate for all boats $15/day, $75/week, $250/month. Family owned since the early 1950s, the current owners have been in charge since 1989. Docks are wood with short finger piers; fresh water is available slip-side, but no power. The small office building also houses a store that carries ice, fishing tackle, live bait, boat supplies, drinks and snacks. The marina rents 16 ft. aluminum fishing boats as well as clam shovels, crab rings and salmon nets; the helpful staff provides instruction on the use of the equipment they rent. A large fish cleaning station is on-site ($2). Very basic brick and cinderblock heads are located behind the marina. Note: Crossing the bar requires local knowledge, particularly in rough weather. Call the Coast Guard for conditions.

Notable -- Several rivers empty into Tillamook Bay, making this a fishermen's paradise. Just outside the breakwater you can fish for rockfish, lingcod, salmon, and halibut; crabbing is also popular throughout the year. Get updates from the marina staff. Or visit one of the fantastic fish markets located near the marina. Nearby Troller Restaurant (322-3662) serves three meals a day and provides an ATM machine and Kino-style gambling. Town is just north with a grocery store, pharmacy, and more eateries. Sunset Empire provides bus service along the coast and to Tillamook, where you can connect to an express bus to Portland.

Navigational Information
Lat: 45°33.380' **Long:** 123°54.780' **Tide:** 10 ft. **Current:** n/a **Chart:** 18558
Rep. Depths *(MLW)*: **Entry** 10 ft. **Fuel Dock** n/a **Max Slip/Moor** 8 ft./-
Access: Marina is on the left after entering Tillamook Bay

Marina Facilities *(In Season/Off Season)*
Fuel: Gasoline, Diesel
Slips: 325 Total, 15 Transient **Max LOA:** 50 ft. **Max Beam:** 10 ft.
  Rate *(per ft.)*: **Day** $0.33* **Week** n/a **Month** n/a
  Power: 30 amp n/a, **50 amp** n/a, **100 amp** n/a, **200 amp** n/a
  Cable TV: No **Dockside Phone:** No
  Dock Type: Floating, Pilings, Wood
Moorings: 0 Total, 0 Transient **Launch:** n/a
  Rate: Day n/a **Week** n/a **Month** n/a
Heads: 4 Toilet(s), 2 Shower(s)
Laundry: None **Pay Phones:** No
Pump-Out: OnSite, Self Service, 1 Central **Fee:** n/a **Closed Heads:** Yes

Marina Operations
Owner/Manager: Don Bacon **Dockmaster:** Virgil Louden
In-Season: Year-Round, 9am-5pm **Off-Season:** n/a
After-Hours Arrival: Sign in sheet at heads of gangways and office
Reservations: No **Credit Cards:** No
Discounts: None
Pets: No **Handicap Access:** Yes, Heads, Docks

Port of Garibaldi

PO Box 10; 402 South 7th; Garibaldi, OR 97118

Tel: (503) 322-3292 **VHF: Monitor** n/a **Talk** n/a
Fax: (503) 322-0029 **Alternate Tel:** n/a
Email: portofgaribaldi@oregoncoast.com **Web:** n/a
Nearest Town: Garibaldi **Tourist Info:** (503) 322-0301

Marina Services and Boat Supplies
Supplies - OnSite: Ice *(Block, Cube, Shaved)*, Bait/Tackle *(Garibaldi Bait and Tackle 322-0282)*, Live Bait, Propane

Boatyard Services
Nearest Yard: Darts Small Engine (503) 368-6519

Restaurants and Accommodations
OnSite: Restaurant *(Troller 322-3662, B $4-11, L $4-12, D $10-17)*
Near: Restaurant *(Fisherman's Korner 322-2033, B $6-10, L $6-15, D $6-15)*, *(Pirate's Cove 322-2092)*, *(Miller's Restaurant 322-0355)*, Fast Food *(Dairy Queen)*, Lite Fare *(Garibaldi Pub & Eatery 322-2020)*, Motel *(Harbor View Inn 322-3251, $45-95)*, *(Inn at Garibaldi 322-3338, $55-125)*, *(Comfort Inn 322-0328)*, Inn/B&B *(Pelican's Perch B&B 322-3633)* **Under 1 mi:** Motel *(Bayshore Inn 322-2552, $64-120)*

Recreation and Entertainment
Near: Picnic Area, Playground, Boat Rentals *(Garibaldi Marina 322-3312)*, Fishing Charter *(Garibaldi Charters 322-0007, Kerri Lin Charters 355-2439)*, Group Fishing Boat *(Troller Deep Sea Fishing 355-3419)*, Special Events *(Crab Races - 2nd wknd Mar, Garibaldi Days - last wknd Jul)*, Galleries *(Classy Touch Imports 322-2030)* **Under 1 mi:** Museum *(Garibaldi Museum 322-8411 Jul-Sep Thu-Sun 12-4pm/ Tillamook Naval Air Station Museum 842-1130, 10 mi.)*, Sightseeing *(Oregon Coast Scenic Railroad 842-7972 - departs at 12, 2, 4pm wknds in season; 1.5 hr. RT to Rockaway Beach $13/ $7 or ride in the locomotive for $30)* **3+ mi:** Golf Course *(Alderbrook Golf Course 842-6413, 6 mi.)*

Provisioning and General Services
Near: Convenience Store *(G & G Mini Market 322-2535)*, Market *(Food Basket Market 322-3270)*, Delicatessen *(Bay Front Bakery & Deli 322-3787)*, Wine/Beer, Liquor Store *(Godfrey's Garibaldi Pharmacy and Liquor 322-3456)*, Bakery *(Bay Front)*, Fishmonger *(Bay Ocean Seafood 322-3316 9am-5pm, to 6pm Fri-Sat - fish, crab, shrimp)*, Crabs/Waterman *(Tillamook Bay Boat House 322-3600 live & cooked crab, fish)*, Oysterman *(Bay Ocean Oyster 322-0040)*, Bank/ATM, Beauty Salon *(Tina's Hair Salon 322-0373)*, Barber Shop *(Tami's Barber Shop 322-2228)*, Pharmacy *(Godfrey's Garibaldi Pharmacy 322-3456)* **Under 1 mi:** Post Office *(322-3675)*, Protestant Church, Library *(Garibaldi 322-2100)*, Florist *(Garibaldi Flowers)* **3+ mi:** Catholic Church *(St. Mary's By the Sea 355-2661, 5 mi.)*, Laundry *(101 One Stop 355-2923, 5 mi.)*, Bookstore *(Main Avenue Books 842-7766, 10 mi.)*, Hardware Store *(Rockaway Beach Hardware 355-2923, 5 mi.)*, Department Store *(Fred Meyer 815-1415, 9 mi.)*

Transportation
Near: Local Bus *(Sunset Empire 861-7433/Tillamook Transit to Portland 815-8283, 10 mi.)* **3+ mi:** Rental Car *(E&E Auto Rentals 842-7802, 10 mi.)*, Taxi *(Kenny Cab 842-5175, 10 mi.)* **Airport:** Tillamook/Portland Int'l. *(13 mi./85 mi.)*

Medical Services
911 Service **OnCall:** Ambulance *(Garibaldi Ambulance 322-3231)* **3+ mi:** Doctor *(Tillamook Medical 842-5546, 10 mi.)*, Dentist *(Tillamook Family 815-1777, 10 mi.)*, Chiropractor *(Tillamook Family Chiropractic 842-5951, 10 mi.)*, Veterinarian *(Pioneer Veterinary Hospital 842-8411, 10 mi.)* **Hospital:** Tillamook County General 842-4444 *(10 mi.)*

Setting -- Once inside Tillamook Bay, visitors are greeted by a harbor surrounded on three sides by lush mountains. Commercial and sportfishing are mainstays of this coastal estuary and an extensive fleet of working boats calls Port of Garibaldi home. Casual Troller Restaurant is on-site and Harbor View Inn motel is adjacent - along with several commercial entities. Near the docks is a picnic pavilion and colorful playground.

Marina Notes -- *Flat rate: $13/day for any size boat. A wood guest dock, some distance from the main facility, offers side ties or short finger slips. The small, gray single-story harbormaster's office is located a short walk around Garibaldi Marina - past the permanent moorage. Few dockside amenities for boaters and no power - but many opportunities for outdoor types. Portable heads are on the dock. Unheated basic cement and brick restrooms are across from the marina. FYI: Garibaldi Dry Dock was purchased in 2005 and transformed into Big Tuna Marine - annual moorage and some marine services.

Notable -- Garibaldi was founded by Robert Grey in Aug 1788 when his ship, Lady Washington, entered Tillamook Bay - and was named for the Italian patriot. To learn more about the past and present of this seaside community of 1,000 people, visit the Garibaldi Museum. Tillamook Bay offers enticing opportunities for the angler: sturgeon in winter, chinook in spring, coho in summer, chinook in fall, plus crabbing and clamming. For others, there's beachcombing, birding and surfing. The seasonal Oregon Coast Scenic Railroad - a 1910 steam locomotive with open cars - makes the short trip from Garibaldi to Rockaway Beach or the longer "supper or brunch trip" to Wheeler weekends & holidays, May-September.

PHOTOS ON DVD: 10

Depoe Bay Marina

PO Box 8; Depoe Bay, OR 97341

Tel: (541) 765-2361 **VHF:** Monitor n/a **Talk** n/a
Fax: (541) 765-2129 **Alternate Tel:** n/a
Email: n/a **Web:** n/a
Nearest Town: Depoe Bay **Tourist Info:** (541) 765-2889

Navigational Information
Lat: 44°48.550' **Long:** 124°03.780' **Tide:** 8 ft. **Current:** n/a **Chart:** 18561
Rep. Depths (*MLW*): **Entry** 8 ft. **Fuel Dock** 7 ft. **Max Slip/Moor** 7 ft./-
Access: 12 miles north of Newport, right off ocean at buoy "DB"

Marina Facilities (*In Season/Off Season*)
Fuel: Gasoline, Diesel
Slips: 35 Total, 10 Transient **Max LOA:** 52 ft. **Max Beam:** n/a
　Rate (*per ft.*): **Day** $0.25* **Week** n/a **Month** n/a
　Power: 30 amp Incl., **50 amp** n/a, **100 amp** n/a, **200 amp** n/a
　Cable TV: No **Dockside Phone:** No
　Dock Type: Floating, Wood
Moorings: 0 Total, 0 Transient **Launch:** n/a
　Rate: Day n/a **Week** n/a **Month** n/a
Heads: 2 Toilet(s)
Laundry: None **Pay Phones:** No
Pump-Out: OnSite **Fee:** Free **Closed Heads:** Yes

Marina Operations
Owner/Manager: Depoe Bay **Dockmaster:** Phil Shane
In-Season: Year-Round, 8am-5pm **Off-Season:** n/a
After-Hours Arrival: Call harbormaster for moorage 765-2361
Reservations: No **Credit Cards:** Visa/MC
Discounts: None
Pets: Welcome **Handicap Access:** No

Marina Services and Boat Supplies
Communication - FedEx, DHL, UPS **Supplies - OnSite:** Ice (*Cube*)
Near: Bait/Tackle

Boatyard Services
OnSite: Launching Ramp (*$2*) **Nearest Yard:** Newport Marine (541) 867-3704

Restaurants and Accommodations
OnSite: Snack Bar (*Dockside Dogs 765-2545, in Dockside Charters*),
Inn/B&B (*Harbor Lights 765-2322, $80-135, includes B; formerly Sea Hag, new owners; harborfront balconies, some fireplaces*) **Near:** Restaurant (*Gracie Strom's Sea Hag 765-2734, B $4-11, L $5-13, D $13-24, lite fare at dinner, too; fun & famous; don't miss Gracie playing the "bottles"*), Lite Fare (*Chowder Bowl 765-2300*), (*Mick's Bay Grill 765-7658*), Inn/B&B (*Crown Pacific Inn 765-7773, $75*) **Under 1 mi:** Restaurant (*Oceanus 765-4553, reportedly the best in Depoe Bay*), (*Tidal Raves 765-2995, seafood - reasonable*), Fast Food (*Dairy Queen*), Inn/B&B (*Whale Inn 765-2789, $65-130*), (*Inn at Arch Rock 765-2560, $70-300, rooms & suites*), (*Channel House 7-65.-2140, $140-310, oceanfront rooms, balconies with hot tubs overlook the harbor entrance*)

Recreation and Entertainment
OnSite: Picnic Area, Grills, Fishing Charter (*Tradewinds Charter 765-2345, Dockside Charters 765-2545*), Group Fishing Boat (*Tradewinds; Reel Nauti 921-1628*) **Near:** Playground, Park (*Depoe Bay Sate Park*), Sightseeing (*walk the boardwalk overlooking the ocean; whale watching - Dockside*

& Tradewinds; Depoe Bay Winery 765-3311), Special Events (*Wooden Boat Show, Crab Feed & Ducky Derby - Apr, Jul 4th fireworks, Indian Salmon Bake - Sep*) **Under 1 mi:** Galleries (*Silver Heron 765-2886, Art on the Edge 765-7797*) **3+ mi:** Golf Course (*Agate Beach Golf Course 265-7331, 12 mi.*), Bowling (*Delake Bowl 994-9595, 9 mi.*)

Provisioning and General Services
Near: Convenience Store (*Whistle Stop Mini-Mart 765-2929*), Fishmonger (*Neptune's Choice 765-4000*), Bank/ATM, Retail Shops **Under 1 mi:** Crabs/Waterman (*Brite Seafood*) **1-3 mi:** Market (*Sentry Market 764-2314*), Wine/Beer, Liquor Store (*OR Liquor 765-2317*), Bakery (*Lincoln Beach Bakery 765-2629*), Post Office, Protestant Church, Beauty Salon (*Sassy Shears 765-2343*), Barber Shop (*Tom's Hair Place 765-2464*), Bookstore (*Channel Bookstore 765-2352*), Clothing Store (*T-Shirts Plus 765-3266*), Copies Etc. (*Beacon Media 764-5011*) **3+ mi:** Supermarket (*Safeway 265-2930, 13 mi.*), Library (*Lincoln City 996-2277, 12 mi.*), Laundry (*Posh Wash 994-3339, 12 mi.*), Hardware Store (*Ace Hardware 996-2131, 12 mi.*), Department Store (*Sea Towne Shopping Center 867-7710, 13 mi.*)

Transportation
OnCall: Rental Car (*Enterprise 574-1999*), Taxi (*Alaha Taxi 765-4555*)
Airport: Newport Municipal (*17 mi.*)

Medical Services
911 Service **OnCall:** Ambulance **1-3 mi:** Doctor (*Family Practice Clinics 765-3265*), Chiropractor (*Depoe Bay Chiropractic 765-3200*), Optician (*Depoe Bay Vision Clinic 765-2991*) **Hospital:** Samaritan North Lincoln 994-3661 (*12 mi.*)

Setting -- At about six square acres, Depoe Bay claims the title "world's smallest harbor." The outer channel is really a hole in the rock wall. Raging waters crash over rocks on both sides, an overhead bridge gives 42 feet of clearance in a 30-foot span and the channel turns slightly. The reward for those who risk this rite-of-passage is a calm, picturesque, diminutive bay surrounded by hills and the fishing village of Depoe Bay. Several wide wooden piers, with centered pilings, lie perpendicular to the shore opposite the bridge. At the head of the docks are Dockside Charters and the three-story Harbor Lights Inn.

Marina Notes -- *Flat rate: under 45' $10/night, over 45' $15/night. Both long finger slips and alongside moorage available. 30 amp power is included. Three fish cleaning stations are available for the lucky ones. Fishing charters and whale watching tours depart from the marina. Basic, municipal cinderblock heads with stainless fixtures are near the boat ramp; no showers. Note: Boats over 50 ft. require approval from the harbormaster before entering the Bay - the entrance channel is so narrow that turning around is nearly impossible. A Coast Guard station shares the harbor (765-2124).

Notable -- Depoe Bay is the closest point to watch the resident gray whales - usually about 50 - spend the summer right off the coast and two on-site tour companies will take you there. On either side of the harbor, stairs lead up to this tiny town of 1,000. Depoe Bay City Park is just across a footbridge at the south end of the parking lot - with a nice playground for kids and a sheltered picnic area. To the north is oceanfront Depoe Bay State Park, for panoramic views and land-based whale watch opportunities. Several upscale, oceanfront inns invite a night ashore and a variety of eateries are within walking distance.

Navigational Information
Lat: 44°37.430' **Long:** 124°03.220' **Tide:** 8 ft. **Current:** 3 kt. **Chart:** 18581
Rep. Depths *(MLW)*: **Entry** 7 ft. **Fuel Dock** 9 ft. **Max Slip/Moor** 10 ft./-
Access: Yaquina Bay south side just past bridge

Marina Facilities *(In Season/Off Season)*
Fuel: *Standard Oil* - Gasoline, Diesel
Slips: 600 Total, 200 Transient **Max LOA:** 110 ft. **Max Beam:** n/a
  Rate *(per ft.)*: **Day** $0.40* **Week** $78.50-101.93 **Month** $266-394
  Power: 30 amp Incl., 50 amp Incl., 100 amp n/a, 200 amp n/a
  Cable TV: Yes **Dockside Phone:** Yes
  Dock Type: Floating, Concrete
Moorings: 0 Total, 0 Transient **Launch:** n/a, Dinghy Dock (Free)
  Rate: Day n/a **Week** n/a **Month** n/a
Heads: 10 Toilet(s), 10 Shower(s) *(with dressing rooms)*
Laundry: 2 Washer(s), 2 Dryer(s), Book Exchange **Pay Phones:** Yes, 1
Pump-Out: OnSite, Self Service, 2 Central **Fee:** n/a **Closed Heads:** Yes

Marina Operations
Owner/Manager: Port of Newport **Dockmaster:** Chris Urbach
In-Season: Year-Round, 7am-7pm **Off-Season:** n/a
After-Hours Arrival: Check locator map at office
Reservations: Yes, Preferred **Credit Cards:** Visa/MC
Discounts: None
Pets: Welcome, Dog Walk Area **Handicap Access:** Yes, Heads, Docks

Port of Newport Marina

2301 SE OSU Drive; Newport, OR 97365

Tel: (541) 867-3321 **VHF: Monitor** Ch. 12 **Talk** Ch. 12
Fax: (541) 867-3352 **Alternate Tel:** n/a
Email: n/a **Web:** www.portofnewport.com
Nearest Town: Newport *(0.5 mi.)* **Tourist Info:** (541) 265-8801

Marina Services and Boat Supplies
Services - Security, Megayacht Facilities **Communication -** Mail & Package Hold, Phone Messages, Fax in/out *($0.50)*, Data Ports *(Wi-Fi in Laundry)*, FedEx, UPS **Supplies - OnSite:** Ice *(Block, Cube)*, Ships' Store *(Newport Marina Store or Schiewe 265-7382 2 mi.)*, Bait/Tackle, Propane

Boatyard Services
OnSite: Engine mechanic *(gas)*, Launching Ramp **OnCall:** Engine mechanic *(diesel)*, Electrical Repairs, Electronics Repairs, Air Conditioning, Refrigeration, Divers, Life Raft Service, Upholstery, Metal Fabrication
Nearest Yard: Fred Wahl Marine (541) 336-4736

Restaurants and Accommodations
OnSite: Restaurant *(Brewer's on the Bay 867-3664, L $6-14, D $8-19)*, Inn/B&B *(Newport Belle 867-6290, $100-145)* **Near:** Motel *(La Quinta 867-7727)*, Hotel *(Inn at Yaquina Bay 867-7055, $75-150)* **Under 1 mi:** Restaurant *(Newport Steak & Seafood 265-8283)*, *(Flashback's Fountain & Grill 867-6901)* **1-3 mi:** Restaurant *(Saffron Salmon 265-8921)*, *(Mo's Restaurant 265-2979, L $7-12, D $7-12)*, *(The Whale's Tale 265-8660, B $6-10, L $6-12, D $12-22)*, *(Moby Dick's Seafood 265-7847)*, *(Georgie's Beachside Grill 265-9800, ocean views, at Hallmark Resort)*

Recreation and Entertainment
OnSite: Picnic Area, Boat Rentals, Fishing Charter *(877-867-4470)*
Near: Beach, Group Fishing Boat *(on Bay front - Seagull 265-7441 crab, bottom fish, salmon $45/25 to $115; Tradewinds 265-2101 $65-165)*, Museum *(Coast Aquarium 867-3474 $12/$10 Seniors/$7 Kids, 9am-6pm summer, 10am-5pm off-season; Hatfield Marine Science Center 867-0100, donations)*,

Sightseeing *(Historic Bayfront by dinghy; Marine Tours 265-6200 - discovery $30/15; Oregon Rocket $35/20)* **Under 1 mi:** Dive Shop *(Newport Water Sports 867-3742)*, Fitness Center *(YMCA 265-3221)* **1-3 mi:** Pool, Tennis Courts, Cultural Attract *(Performing Arts Ctr 265-2787)* **3+ mi:** Golf Course *(Agate Beach 265-7331, 4 mi.)*, Movie Theater *(Newport 265-2111, 5 mi.)*

Provisioning and General Services
OnSite: Bank/ATM, Copies Etc. **Under 1 mi:** Market *(South Beach 867-7141)*, Delicatessen *(Beveins' 265-2722)*, Post Office, Protestant Church, Barber Shop *(Sportsman 265-9801)* **1-3 mi:** Health Food *(Oceana Natural Foods Co-op 265-8285)*, Wine/Beer, Liquor Store *(Newport 265-5621)*, Bakery *(Gables 574-0986)*, Farmers' Market *(Sat 9am-1pm near Chamber of Commerce)*, Catholic Church, Library *(Newport 265-2153)*, Beauty Salon *(Hair's the Thing 265-2546)*, Dry Cleaners *(Coast St. 265-5366)*, Bookstore *(Newport 265-8971)*, Pharmacy *(Newport 265-2122)*, Hardware Store *(Do It Best 265-6640)*, Florist *(The Blossom 265-8262)*, Department Store *(JC Penney 265-5369)* **3+ mi:** Supermarket *(Fred Meyer 265-8460, 4 mi.)*

Transportation
OnCall: Rental Car *(Enterprise 574-1999)*, Taxi *(Yaquina 265-5646)*, Local Bus *(Lincoln County 265-4900)* **1-3 mi:** InterCity Bus *(Greyhound 265-2253)* **Airport:** Newport Municipal *(3 mi.)*

Medical Services
911 Service **OnCall:** Ambulance, Veterinarian *(Mobile Vet 265-7448)* **Under 1 mi:** Dentist *(Newport Dental 867-3755)* **1-3 mi:** Doctor *(Family Medical 265-3772)*, Chiropractor *(Newport Chiro 265-9218)* **Hospital:** Samaritan Pacific 265-2244 *(1.5 mi.)*

Setting -- Three miles south of 93-foot Yaquina Head Light is the relatively easy entrance to the bay. Once inside, the 129-foot US 101 bridge soars above. Port of Newport lies on the south side in the small community of South Beach. Well cared for concrete piers lie perpendicular to the shore with finger piers on both sides. A wide ramp, marine store and the office are at the west end; fuel dock and pump-out are at the other end. On-site are an RV park, Rogue Brewery & its restaurant, Yaquina Bay Yacht Club, and the Newport Belle - a three-story riverboat B&B. Across the river is Historic Bayfront - the working waterfront.

Marina Notes -- *Flat rate: 24' $13.34/day, $78.50/wk.; 26' $14.56/day, $81.58/wk., 32' $14.73/day, $87.37/wk., 40' $15.77/day, $94.65/wk., 48' $16.99/day, $101.93/wk. Formerly South Beach Marina. Some piers offer cable TV & phone jacks. The store carries groceries, wine & beer, ice, fishing tackle, and clothing and can arrange fishing charters (for albacore tuna, salmon, bottom fish & crabs) and whale watching tours - or rent crab rings ($5.50/day). A large stainless fish cleaning station has running water; a crab-cooking pot is available for $6/doz. ($0.50 each). Vacuum packing is $1-1.50. A centrally-located, vertical-sided gray contemporary bathhouse hosts all-tile heads, showers and laundry. Yaquina Bay Y.C. (574-6005) offers reciprocal moorage.

Notable -- The nearby Aquarium, with over 190 species and 15,000 sea creatures, is rated among the top ten in the nation. The colorful Aquarium Village, done in a pirate motif, has shops, artist studios, and a collectibles mall. Dinghy across to the day-use landing at charming Historic Bayfront, the dungeness crab capital, which boasts the largest commercial fishing fleet on the Oregon coast plus galleries, eateries, and group fishing and whale-watch excursions.

Embarcadero Resort Hotel & Marina

Embarcadero Resort & Marina

1000 SE Bay Blvd.; Newport , OR 97365

Tel: (541) 265-8521; (800) 547-4779 **VHF: Monitor** n/a **Talk** n/a
Fax: (541) 265-7844 **Alternate Tel:** n/a
Email: See Marina Notes **Web:** www.embarcadero-resort.com
Nearest Town: Newport *(0.5 mi.)* **Tourist Info:** (541) 265-8801

Navigational Information
Lat: 47°37.770' **Long:** 124°02.468' **Tide:** 8 ft. **Current:** 3 kt. **Chart:** 18581
Rep. Depths *(MLW)***: Entry** 11 ft. **Fuel Dock** n/a **Max Slip/Moor** 10 ft./-
Access: 1 mile from Newport Bridge on north shore

Marina Facilities *(In Season/Off Season)*
Fuel: No
Slips: 232 Total, 135 Transient **Max LOA:** 40 ft. **Max Beam:** 14 ft.
  Rate *(per ft.)***: Day** $0.60 **Week** $3.15 **Month** $7.50
  Power: 30 amp Incl., **50 amp** n/a, **100 amp** n/a, **200 amp** n/a
  Cable TV: No **Dockside Phone:** No
  Dock Type: Floating, Short Fingers, Concrete
Moorings: 0 Total, 0 Transient **Launch:** n/a, Dinghy Dock
  Rate: Day n/a **Week** n/a **Month** n/a
Heads: 4 Toilet(s), 2 Shower(s) *(with dressing rooms)*
Laundry: 7 Washer(s), 7 Dryer(s) **Pay Phones:** Yes
Pump-Out: OnSite, 1 Central **Fee:** Free **Closed Heads:** Yes

Marina Operations
Owner/Manager: Shane Holland **Dockmaster:** Same
In-Season: Summer, 7am-6pm* **Off-Season:** Fall-Spr, 8am-4:30pm
After-Hours Arrival: Check in at the front desk
Reservations: Yes, Preferred **Credit Cards:** Visa/MC, Dscvr, Amex
Discounts: None
Pets: Welcome, Dog Walk Area **Handicap Access:** Yes, Heads

Marina Services and Boat Supplies
Services - Docking Assistance, Concierge, Room Service to the Boat, Boaters' Lounge, Crew Lounge, Security *(17 hrs., 7am-11pm)*, Trash Pick-Up, Dock Carts **Communication -** Mail & Package Hold, Fax in/out *($2/pg)*, FedEx, DHL, UPS, Express Mail *(Sat Del)* **Supplies - OnSite:** Ice *(Cube, Shaved)*, Bait/Tackle **Under 1 mi:** Ships' Store *(Schiewe Marine 265-7382)*, Propane **1-3 mi:** CNG

Boatyard Services
Under 1 mi: Travelift, Hydraulic Trailer, Engine mechanic *(gas, diesel)*, Launching Ramp, Electrical Repairs, Electronics Repairs, Hull Repairs, Rigger, Bottom Cleaning, Air Conditioning, Refrigeration, Divers, Life Raft Service, Painting, Awlgrip, Total Refits. **Nearest Yard:** Newport Marine and RV Service (541) 867-3704

Restaurants and Accommodations
OnSite: Restaurant *(Embarcadero 265-8521, B $6-12, L $6-14, D $14-24, Sun champagne brunch 10am-1pm, seafood buffet last Fri each mo.)*, Hotel *(Embarcadero Resort 265-8521, $77-378)* **OnCall:** Pizzeria *(Abby's 265-9336)* **Under 1 mi:** Restaurant *(Kam Meng 574-9450)*, *(Whale's Tail 265-8660, B $6-10, L $6-12, D $12-22)*, *(Sada's Sushi Bar 265-2206)*, *(Port Dock One 265-2911)*, *(Shirley's on the Bay 574-8400, B $6-11, L $7-13, D $12-25)*, *(Sharks Seafood Bar 574-0590)*, Fast Food *(Subway)* **1-3 mi:** Motel *(Days Inn 265-5767)*, Hotel *(Best Western 265-9411, $90-126)*

Recreation and Entertainment
OnSite: Heated Pool, Spa, Picnic Area, Grills, Boat Rentals *(crab boats, kayaks)*, Video Rental **Near:** Roller Blade/Bike Paths, Sightseeing

(Bayfront Walking Tours, Marine Discovery Tours 800-903-2628)
Under 1 mi: Beach, Playground, Dive Shop, Fishing Charter *(Sea Gull Charters 265-7441)*, Group Fishing Boat *(Newport Tradewinds 265-2101)*, Park *(Yaquina Bay State Park)*, Museum *(Wax Works 265-2206, Undersea Gardens 265-2206; Oregon Coast Aquarium 867-3474, 2 mi.)*, Galleries *(Forinash Gallery & Studio 867-3430)* **1-3 mi:** Golf Course *(Agate Beach 265-7331)*, Cultural Attract *(Performing Arts Center 265-2787)*

Provisioning and General Services
OnSite: Wine/Beer, Newsstand, Copies Etc. **Near:** Bank/ATM **Under 1 mi:** Market *(Bay Market 574-4364)*, Delicatessen *(Lighthouse Deli & Fish Co. 265-2400)*, Bakery *(Franz 265-9343)*, Fishmonger *(Fish Peddler's Market 265-7057)*, Beauty Salon *(Paulie's Beauty Hut 265-2046)*, Laundry, Hardware Store *(Do It Best 265-6640)* **1-3 mi:** Supermarket *(Fred Meyer 265-8460)*, Liquor Store *(Newport Liquor 265-5621)*, Farmers' Market *(Sat 9am-1pm 961-8236)*, Post Office, Library *(265-2153)*, Dry Cleaners *(Newport Dry Cleaners 265-6982)*, Bookstore *(Newport Book Center 265-8971)*, Pharmacy *(Newport 265-2122)*, Department Store *(Wal-Mart 265-6560)*

Transportation
OnSite: Bikes *($6/hr.)*, Water Taxi *($4-7)* **OnCall:** Rental Car *(Enterprise 574-1999)*, Taxi *(Yaquina Cab 265-4656)* **Near:** Local Bus *(265-4900)*, InterCity Bus *(Greyhound 265-2253)* **Airport:** Newport Municipal *(5 mi.)*

Medical Services
911 Service **OnCall:** Ambulance, Veterinarian *(Mobile Vet 265-7448)* **1-3 mi:** Doctor *(Family Medical 265-3772)*, Dentist *(Canyon Way Dental 574-0275)* **Hospital:** Samaritan Pacific 265-2244 *(1.5 mi.)*

Setting -- The Embarcadero is located upstream from Newport's historic waterfront district. Set in a residential area at the eastern end of town, the glass-fronted restaurant and gray-shingled, contemporary hotel march up the hill above the docks. A breakwater protects the upscale concrete docks that lead to a wide wooden boardwalk adorned with lampposts, barrels of flowers and benches. Private homes peek through the evergreens on the surrounding hills.

Marina Notes -- *The dockmaster is on-site Thu-Mon, 7am-6pm in summer, and 8am-4:30pm off-season (ext. no. 777). A guest in the marina is considered a guest of the beautiful hotel and its amenities, with access to a boaters' lounge, indoor swimming pool, sauna, exercise room, ping-pong table, picnic tables and grills, and a private crab dock. Room service is available to your boat and a concierge is in the hotel. Likely the most luxurious marina on the Oregon coast. The marina rents kayaks and crab boats and crab cookers are available all day. Meeting rooms can accommodate up to 150 for events. Large laundry room with ping-pong table, and a comfortable bathhouse with fiberglass shower stalls. email: information@embarcadero-resort.com

Notable -- The on-site restaurant has a wall of glass for views of the docks and the bay. Rent a bike or walk the half mile to Newport's Historic Bayfront to watch the fishing fleet return, shop, and enjoy the bustle of this quaint village. Several waterfront restaurants offer indoor or outdoor dining. Explore the Undersea Gardens and see the live diving show. Don't miss the famous Oregon Coast Aquarium and the Visitors' Center at the Marine Science Center (867-0271). State parks sit on the oceanfront on both sides of the bay. The Yaquina Head Lighthouse, built in 1872, is Oregon's tallest (574-3116).

PHOTOS ON DVD: 11

20. OR – Siuslaw River to Brookings

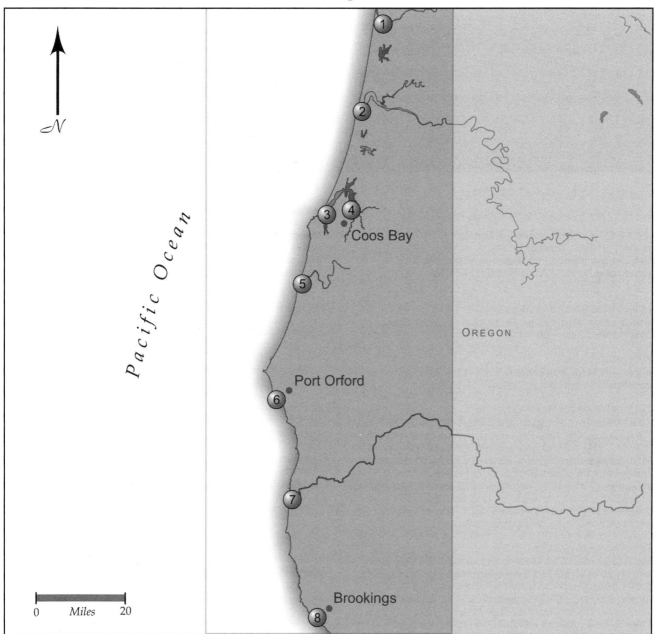

MAP	MARINA	HARBOR	PAGE	MAP	MARINA	HARBOR	PAGE
1	Port of Siuslaw RV Park & Marina	Siuslaw Bay	282	5	Port of Bandon	Coquille River/Bandon Harbor	286
2	Salmon Harbor Marina	Umpqua River/Salmon Harbor	283	6	Port of Port Orford	Port of Port Orford	287
3	Charleston Marina on Coos Bay	Coos Bay/South Slough	284	7	Port of Gold Beach	Rogue River/Gold Beach Harbor	288
4	Coos Bay City Dock	Coos Bay Harbor	285	8	Port of Brookings Harbor	Chetco River/Chetco Cove	289

▸ **Currency** — In Canadian Marina Reports, all prices are in Canadian dollars. In U.S. Marina Reports, all prices are in U.S. dollars.

▸ **"CCM"** — Denotes a Certified Clean Marina, a state/provincial award for environmental excellence. See page 298 for an explanation and page 299 for a list of Pump-Out facilities.

▸ **Ratings & Reviews** — An overview of the Atlantic Cruising Club's rating system is on page 6 and details on the content of each Marina Report are on pages 7 – 11.

▸ **Marina Report Updates** — Comments from boaters and new information from ACC reviewers and marinas are posted regularly on www.AtlanticCruisingClub.com.

Port of Siuslaw Marina

PO Box 1220; 100 Harbor Street; Florence, OR 97439

Tel: (541) 997-3040 **VHF: Monitor** n/a **Talk** n/a
Fax: (541) 997-9407 **Alternate Tel:** n/a
Email: port@portofsiuslaw.com **Web:** www.portofsiuslaw.com
Nearest Town: Florence **Tourist Info:** (541) 997-3128

Navigational Information
Lat: 43°58.060' **Long:** 124°06.010' **Tide:** 7 ft. **Current:** n/a **Chart:** 18583
Rep. Depths (*MLW*): Entry 10 ft. **Fuel Dock** 10 ft. **Max Slip/Moor** 10 ft./-
Access: Siuslaw River, through bridge on north shore

Marina Facilities *(In Season/Off Season)*
Fuel: Gasoline, Diesel
Slips: 54 Total, 5 Transient **Max LOA:** 85 ft. **Max Beam:** n/a
  Rate *(per ft.)*: **Day** $0.40* **Week** $88* **Month** $254*
  Power: 30 amp Incl., **50 amp** n/a, **100 amp** n/a, **200 amp** n/a
  Cable TV: No **Dockside Phone:** No
  Dock Type: Floating, Long Fingers, Concrete, Wood
Moorings: 0 Total, 0 Transient **Launch:** n/a
  Rate: Day n/a **Week** n/a **Month** n/a
Heads: 4 Toilet(s), 4 Shower(s)
Laundry: 1 Washer(s), 1 Dryer(s) **Pay Phones:** Yes, 2
Pump-Out: OnSite **Fee:** Free **Closed Heads:** Yes

Marina Operations
Owner/Manager: Tom Kartrude **Dockmaster:** Ken Hill
In-Season: Year-Round, 8am-5pm **Off-Season:** n/a
After-Hours Arrival: Call ahead
Reservations: Yes, Preferred **Credit Cards:** Visa/MC
Discounts: None
Pets: Welcome, Dog Walk Area **Handicap Access:** Yes

Marina Services and Boat Supplies
Communication - Data Ports *(Wi-Fi)* **Supplies - Near:** Ice *(Cube)*
1-3 mi: Propane *(Chevron 997-3351)*

Boatyard Services
OnSite: Launching Ramp, Upholstery *(Bud's 997-4856)*
Nearest Yard: Siuslaw Marina at Cushman (541) 997-3254

Restaurants and Accommodations
OnSite: Restaurant *(International C-Food Market 997-9646, L $7-13, D $9-20, & seafood market)* **Near:** Restaurant *(Traveler's Cove 997-6845, B $6-8, L $7-10, D $8-14)*, *(Fisherman's Wharf 997-2613)*, *(Florence Café 997-6851)*, *(Bridgewater Seafood 997-9405, fine dining)*, Seafood Shack *(Mo's Chowder House 997-2185, L $8-13, D $8-13)*, Fast Food *(Subway L, D)*, Lite Fare *(Lovejoy's Fish & Chips Tearoom 902-0502)*, Pizzeria *(Paisano's Pizza Deli 997-2068)*, Motel *(River House 997-3933, $69-125)*, Inn/B&B *(Lighthouse 997-3231, $75-120)*, *(Landmark 997-9030)* **Under 1 mi:** Hotel *(Best Western 997-7191, $79-129, across the bridge)*

Recreation and Entertainment
Near: Boat Rentals *(Central Coast Watersports 997-1812)*, Video Rental *(Family Video 997-8752)*, Park *(Oregon Dunes Nat'l. Rec. Area)*, Cultural Attract *(Florence Events Center 997-1994)*, Sightseeing *(Old Town)*, Galleries *(Blue Heron 997-7993, Wind Drift 997-9182)* **Under 1 mi:** Museum *(Siuslaw Pioneer Museum 997-7884, 10am-4pm Tue-Sun $3/under 16 free; Fly Fishing Museum 997-6349; Dolly Wares Doll Museum)* **1-3 mi:** Golf Course *(Sandpines Golf Links 997-1940; Ocean Dunes 800-468-4833)*, Fitness Center *(Coastal Fitness 997-8086)*, Horseback Riding *(C & M Stables 997-7540)*, Bowling *(Holiday Bowl 997-3332)*, Movie Theater *(Florence Cinemas 997-2727)*

Provisioning and General Services
OnSite: Fishmonger *(Int'l. C-Food Market)* **Near:** Convenience Store, Delicatessen *(Soup'd Up Sandwiches 997-2442)*, Bank/ATM, Beauty Salon *(Tangles 997-3160)*, Barber Shop *(Red's 997-4433)*, Dry Cleaners, Laundry *(Bob's Dry Cleaning and Laundry 997-9255)*, Bookstore *(Old Town Books 997-6205, Books N Bears 997-5979)*, Copies Etc. *(Becky's 997-5044)* **Under 1 mi:** Supermarket *(Safeway 902-1900)*, Gourmet Shop *(Grape Leaf 997-1646)*, Health Food *(Salmonberry Naturals 997-3345)*, Wine/Beer, Liquor Store, Post Office, Protestant Church, Library *(Florence 997-3132)*, Pharmacy *(Rite Aid 997-2861)*, Hardware Store *(True Value 997-8024)*, Clothing Store *(Connie's 997-8869)* **1-3 mi:** Market *(Cleawox Market 997-6435)*, Bakery *(Cheri's Bakery and Donut Shop 997-7100)*, Catholic Church, Department Store *(Bi-Mart 997-2499)*

Transportation
OnCall: Rental Car *(Enterprise 902-0808)*, Taxi *(River Cities 997-8520)* **Under 1 mi:** Local Bus *(Rhody Express 902-2067)* **1-3 mi:** InterCity Bus *(Greyhound 902-9076)* **Airport:** North Bend Municipal *(45 mi.)*

Medical Services
911 Service **OnCall:** Ambulance **Under 1 mi:** Doctor *(Family Practice 997-7134)* **1-3 mi:** Dentist *(Hayden Family Dentristry 902-8333)*, Chiropractor *(Florence Chiropractic 997-6909)*, Veterinarian *(Osburn Veterinary Clinic 902-2013)* **Hospital:** Peace Harbor 997-8412 *(1 mi.)*

Setting -- One of the most scenic ports on the Oregon coast, Florence sits at the mouth of the Siuslaw River (pronounced SIGH-oos-law) - protected by the oft-photographed 1894 Heceta Lighthouse. A wide boardwalk, topped by period lampposts, sprawls along the marina's bulkhead - stopped by a set of large pale-blue sheds. Low-slung, blue-roofed International C-Market hugs one side of the basin. The upland hosts an RV park. The Port feels a bit like a wilderness outpost - even though the docks are two blocks from charming, restored Old Town.

Marina Notes -- *Flat rate: 20' $11/day "E" dock, 24' $12 "G" dock, 28' $11 "F" dock, 40' $14 "F" dock, 40' $16 "G" dock, 48' $19; sidetie $11/ft. plus $0.25/ft. if over 20'. Power only on "G" dock. The Port Commision has authorized overflow recreational use at the West Moorage Basin, - adding another 51 concrete slips with 30 amp power. Max LOA at the Port is 48 feet, at the West Basin 85 feet. Check in at the RV office. Basic cinderblock heads with showers (and the laundry) are located in the RV park across the parking lot. Note: Crossing the entrance bar requires local knowledge.

Notable -- Unique shops, galleries, museums and restaurants populate the historic district of this pioneer fishing village. But the big attraction is the 50-mile Oregon Dunes Nat'l. Recreation Area (ODNRA) - hiking, birding, horseback or dune-buggy riding. River tours are available on the leisurely "Westward Ho!" stern wheeler (997-9691) or faster jet boats. The Siuslaw Pioneer Museum, which describes the Florence of yesteryear, is located in an old school just south of town. Ten miles north, the Sea Lion Caves (547-3111) grotto is the Stellar's only mainland rookery - it's worth renting a car.

Navigational Information
Lat: 43°40.710' **Long:** 124°10.670' **Tide:** 7 ft. **Current:** n/a **Chart:** 18584
Rep. Depths (*MLW*): **Entry** 26 ft. **Fuel Dock** 10 ft. **Max Slip/Moor** 10 ft./-
Access: Umqpua River to Salmon Harbor

Marina Facilities *(In Season/Off Season)*
Fuel: Gasoline, Diesel
Slips: 550 Total, 300 Transient **Max LOA:** 80 ft. **Max Beam:** n/a
 Rate (*per ft.*): **Day** $0.30* **Week** $72* **Month** $216*
Power: 30 amp Incl., **50 amp** Incl., **100 amp** n/a, **200 amp** n/a
Cable TV: No **Dockside Phone:** No
Dock Type: Floating, Long Fingers, Alongside, Concrete, Wood
Moorings: 0 Total, 0 Transient **Launch:** n/a
 Rate: Day n/a **Week** n/a **Month** n/a
Heads: 12 Toilet(s), 12 Shower(s) *(with dressing rooms)*
Laundry: 3 Washer(s), 3 Dryer(s) **Pay Phones:** No
Pump-Out: OnSite, Full Service **Fee:** $25 **Closed Heads:** Yes

Marina Operations
Owner/Manager: Douglas County **Dockmaster:** Jeff Vander Kley (Hbrmstr)
In-Season: Year-Round, 8am-4:30pm **Off-Season:** n/a
After-Hours Arrival: Dock "A", slips 14 & 15
Reservations: Yes **Credit Cards:** Visa/MC
Discounts: None
Pets: Welcome **Handicap Access:** Yes, Heads, Docks

Salmon Harbor Marina

PO Box 1007; 100 Ork Rock Road; Winchester Bay, OR 97467

Tel: (541) 271-3407 **VHF: Monitor** n/a **Talk** n/a
Fax: (541) 271-2060 **Alternate Tel:** n/a
Email: salmon@co.dougles.or.us **Web:** www.marinarvresort.com
Nearest Town: Winchester Bay *(0.25 mi.)* **Tourist Info:** (541) 271-4471

Marina Services and Boat Supplies
Services - Security **Communication -** Data Ports *(Wi-Fi in RV Park)*, FedEx, DHL, UPS **Supplies - OnSite:** Bait/Tackle *(or Stockade Market 271-3800, nearby)* **Near:** Ice *(Cube)*, Propane

Boatyard Services
OnSite: Travelift *(60T)*, Crane, Launching Ramp *($5 - 2)*

Restaurants and Accommodations
Near: Restaurant *(Pah Tong's Thai Restaurant 271-1750, L, D)*, Pizzeria *(Pizza Ray's and Suzy's 271-2431, B, L, D)*, Motel *(Harbor View Motel 271-3352)*, Inn/B&B *(Winchester Bay Inn 271-4871)* **Under 1 mi:** Fast Food *(King Neptune Drive-In 271-4302)* **3+ mi:** Restaurant *(Bedrock's Chowder and Grill 271-4111, B $3-9, L $4-15, D $4-15, 3 mi.)*, *(Ocean Garden 271-3590, 3 mi., Chinese)*, *(Leona's 271-5297, 4 mi.)*, *(Schooner Inn Café 271-3945, 4 mi.)*, Fast Food *(Subway 5 mi.)*, Motel *(Economy Inn 271-3671, 4 mi.)*, Inn/B&B *(BW Salbasgeon Inn 271-4831, $59-90, 4 mi.)*

Recreation and Entertainment
OnSite: Picnic Area **Near:** Playground, Fishing Charter *(Strike Zone Charters 271-9706)*, Group Fishing Boat *(Pacific Pioneer Charters 271-1967)*, Park *(Oak Rock Park)*, Special Events *(Dune Fest - early Aug, Salmon Derby - LabDay wknd, Crab Bounty Hunt Aug-Sep, Dune Musher's Mail Run - Mar)*, Galleries *(Lighthouse Gallery 271-9376)* **1-3 mi:** Cultural Attract *(Umpqua Discovery Center 271-4816)*, Sightseeing *(Umpqua Lighthouse)* **3+ mi:** Golf Course *(Forest Hills Country Club 271-2626, 10 mi.)*, Fitness Center *(Umpqua Fitness Center 271-0265, 4 mi.)*, Video Rental *(Video Wave 271-1203, 3 mi.)*

Provisioning and General Services
OnSite: Fishmonger *(Sportsmen's Cannery & Smokehouse 271-3293, Griff's on the Bay 271-2512)* **Near:** Convenience Store *(Stockade Market & Tackle 271-3800)*, Market *(Winchester Bay Market 271-2632)*, Wine/Beer, Crabs/Waterman, Bank/ATM, Post Office, Protestant Church **1-3 mi:** Dry Cleaners **3+ mi:** Supermarket *(Safeway 271-3142, 4 mi.)*, Delicatessen *(Back to the Best 271-2619, 4 mi.)*, Liquor Store *(Reedsport Liquor 271-3412, 4 mi.)*, Bakery *(Sugar Shack Bakery 271-3514, 5 mi.)*, Library *(Reedsport 271-3500, 5 mi.)*, Beauty Salon *(Better Health & Beauty Center 271-1328, 4 mi.)*, Barber Shop *(Anchor Barber Shop 271-3007, 5 mi.)*, Pharmacy *(Reedsport Pharmacy 271-3631, 4 mi.)*, Hardware Store *(Kel-Cee Ace 271-2741, 4 mi.)*, Florist *(Flower Shop 271-2270, 4 mi.)*, Retail Shops *(1 Stop Sport Shop 271-0970, Reedsport Outdoor Store 271-2311, 4 mi.)*

Transportation
OnCall: Taxi *(CB Cabs 271-3445)* **Under 1 mi:** Bikes *(Dune County ATV Rentals 271-9357)* **3+ mi:** Rental Car *(Hertz Lakeside 756-4416, 8 mi.)*, InterCity Bus *(Greyhound 464-2807, 5 mi.)* **Airport:** North Bend Municipal *(20 mi.)*

Medical Services
911 Service **OnCall:** Ambulance **1-3 mi:** Doctor *(Dunes Family Health Care 271-2163)* **3+ mi:** Dentist *(Du Val, M. DDS 271-4858, 3 mi.)*, Chiropractor *(Pacific Heights Chiropractic 271-2456, 3 mi.)*, Veterinarian *(Lower Umpqua Veterinary Clinic 271-4696, 5 mi.)* **Hospital:** Lower Umpqua 271-2171 *(3 mi.)*

Setting -- Twenty miles north of the Cape Arago Light is the entrance to the Umpqua River - the longest river between the Sacramento and the Columbia. The Umpqua Light's red and white rotating lantern pokes above the trees on a hillside just south of the entrance. A short way up river is Salmon Harbor, one of the largest recreational facilities on the Oregon coast. Two adjacent basins offer a combination of slips and alongside moorage to over 500 vessels.

Marina Notes -- *Flat rates vary by pier and boat size: 24-29' $9/night, 30-34' $10, 35-39' $11, 40-44' $12, 45-49' $13, 50-54' $14, 55-59' $15, 60-64' $16, 65-69 $17, 70-74' $18. Sample weekly & monthly rates are for a 40-ft. boat - utilities included. Vessels larger than 30-feet use "A" dock in the east basin - which is closer to town. Spacious basins - plenty of room to maneuver. After hours, deposit fee in an RV or boat launch envelope. Haulout (sling and crane) are on-site. A laundromat is in the RV park and dish-washing sinks are available on the side of the bathhouse which has tiled heads and separate heated showers ($0.50 for 3 minutes). Note: Projected depth in the entrance channel is 26 feet but shoals and breakers are present along the north jetty.

Notable -- Adjacent, facing the river entrance, is large Winchester Bay RV Resort. Several shops, cafés, galleries, and lodgings surround the harbor. Rent an ATV from nearby Dune County Rentals (271-9357) and off-road on the mountain-like Oregon Dunes. Or join a fishing boat, or crab off the docks, or dinghy to a close-by beach for clamming. The Umpqua Discovery Center (271-4816) in nearby Reedsport features exhibits about the culture and history of the area - from the early Kuuich Tribe to the present. Tours of the Umpqua Lighthouse run May-October; however, the gift shop is open year 'round (271-4631).

Charleston Marina on Coos Bay

PO Box 5409; 63534 Kingfisher Road; Charleston, OR 97420

Tel: (541) 888-2548 **VHF: Monitor** Ch. 16 **Talk** Ch. 13
Fax: (541) 888-6111 **Alternate Tel:** n/a
Email: mcallery@portofcoosbay.com **Web:** www.charlestonmarina.com
Nearest Town: Charleston *(0.25 mi.)* **Tourist Info:** (541) 269-0215

Navigational Information
Lat: 43°20.800' **Long:** 124°19.370' **Tide:** 8 ft. **Current:** 3 kt. **Chart:** 18587
Rep. Depths *(MLW)*: **Entry** 18 ft. **Fuel Dock** 12 ft. **Max Slip/Moor** 15 ft./-
Access: Once inside Coos Bay, follow South Slough to Charleston Marina

Marina Facilities *(In Season/Off Season)*
Fuel: Gasoline, Diesel
Slips: 560 Total, 15 Transient **Max LOA:** 125 ft. **Max Beam:** n/a
Rate *(per ft.)*: **Day** $0.45* **Week** n/a **Month** $5.75/ft.
Power: 30 amp Incl., **50 amp** Incl., **100 amp** Incl., **200 amp** n/a
Cable TV: No **Dockside Phone:** No
Dock Type: Floating, Long Fingers, Alongside, Wood
Moorings: 0 Total, 0 Transient **Launch:** n/a
Rate: Day n/a **Week** n/a **Month** n/a
Heads: 5 Toilet(s), 4 Shower(s)
Laundry: 5 Washer(s), 6 Dryer(s), Book Exchange **Pay Phones:** Yes, 3
Pump-Out: OnSite, Self Service, 2 Central **Fee:** Free **Closed Heads:** Yes

Marina Operations
Owner/Manager: Oregon Intl. Port of Coos Bay **Dockmaster:** Don Yost
In-Season: Year-Round, 8am-5pm **Off-Season:** n/a
After-Hours Arrival: Call ahead
Reservations: No **Credit Cards:** Visa/MC
Discounts: None
Pets: Welcome **Handicap Access:** Yes, Heads, Docks

Marina Services and Boat Supplies
Services - Security *(security patrol overnight and wknds)*, Trash Pick-Up
Communication - Mail & Package Hold, Phone Messages, Fax in/out *($1/pg)*, Data Ports *(Marina office, RV park office + Wi-Fi)*, FedEx, UPS
Supplies - OnSite: Ice *(Block, Cube, Shaved)*, Ships' Store *(Basin Tackle 888-3811)*, Bait/Tackle, Propane *(RV Park)* **Under 1 mi:** Marine Discount Store *(Englund 888-6723)*

Boatyard Services
OnSite: Travelift *(60T)*, Railway *(200T)*, Forklift *(4T)*, Crane *(12T)*, Engine mechanic *(gas, diesel)*, Launching Ramp *($5)*, Hull Repairs, Bottom Cleaning, Brightwork, Metal Fabrication, Painting, Awlgrip, Total Refits
OnCall: Electrical Repairs, Electronics Repairs, Divers, Compound, Wash & Wax, Interior Cleaning, Propeller Repairs, Woodworking, Yacht Broker
Yard Rates: $40-50/hr., Haul & Launch $125-200, Power Wash $100
Storage: In-Water $5.75/ft./mo., On-Land $1.80/ft./mo.

Restaurants and Accommodations
OnSite: Restaurant *(Basin Café 888-5227, B $3-9, L $3-7, D $8-16)*, *(Sea Basket 888-5711, L $8-12, D $8-12)* **Near:** Motel *(Captain John's 888-4041, $62-80)* **Under 1 mi:** Restaurant *(The Portside 888-5544, L $9-22, D $14-32)*, *(High Tide Café 888-3664, B $3-9, L $4-8, D $8-20)*, *(Oyster Cove 888-0703, D $12-30)* **1-3 mi:** Inn/B&B *(Talavar Inn 888-5280)*

Recreation and Entertainment
OnSite: Picnic Area, Sightseeing *(Miss Linda 888-2128 - a gleaming 76-ft. fishing boat restored for whale watching and harbor tours - 3 hrs. $48/pp)*

Near:
Near: Fishing Charter *(Betty Kay 888-9021)*, Group Fishing Boat *(Bob's Sport Fishing 888-4241)*, Special Events *(Charleston Seafood Fest - 3rd wknd Aug, Shore Bird Fest - LabDay wknd, Crab Feed - 2nd wknd Feb)*
Under 1 mi: Cultural Attract *(South Slough National Estuarine Research Reserve 888-5558; Shore Acres Boatnical Gardens 888-3732, 4 mi.)*
1-3 mi: Boat Rentals *(High Tide 888-3664)*, Park *(Bastendorff Beach; 134-acre Cape Arago State Park)* **3+ mi:** Golf Course *(Sunset Bay 888-9301, 4 mi.)*, Museum *(Coos Art Museum 267-3901, Coos Historical Museum 756-6320, 8 mi.)*

Provisioning and General Services
Near: Convenience Store *(Sunset Maket 888-3013)*, Liquor Store *(Charleston Liquor 888-4646)*, Bank/ATM, Post Office, Beauty Salon *(Beauty at the Beach 888-6421)* **Under 1 mi:** Market *(Davey Jones Locker Grocery 888-3941)*, Wine/Beer, Fishmonger *(Chuck's Seafood & Coos Bay Oyster Co. 888-5525)* **3+ mi:** Supermarket *(Safeway 267-1700, 8 mi.)*, Pharmacy *(Rite Aid 267-7240, 6 mi.)*, Hardware Store *(Farr's True Value 267-2137, 8 mi.)*, Department Store *(Bi-Mart 756-7526, 8 mi.)*

Transportation
OnCall: Rental Car *(Enterprise 266-7100)*, Taxi *(Yellow Cab 267-3111)*
3+ mi: InterCity Bus *(Greyhound Bus Lines 267-4436, 8 mi.)*
Airport: NorthBend Municipal *(9 mi.)*

Medical Services
911 Service **OnCall:** Ambulance **3+ mi:** Doctor *(North Bay 267-5151, 8 mi.)*, Dentist *(Family 269-0620, 8 mi.)*, Veterinarian *(Ocean Blvd. 888-6713, 6 mi.)* **Hospital:** Bay Area 269-8111 *(8 mi.)*

Setting -- Cape Arago Light marks the entrance to Coos Bay, the largest natural harbor between San Francisco and the Columbia River. Once inside the bar, follow South Slough two miles to Charleston. Charleston Shipyard is located on the port side past the drawbridge. At the top of the main dock is a fishing village - with eateries, marine and fishing supplies, fishing charters and an RV park. From the docks, the views are of downtown Charleston and tree-covered hills. The marina office is in a well-marked 2-story gray building with blue trim.

Marina Notes -- *To 15' $9, 16-19' $11, 20-29' $12, 30-39' $13, 40-49' $17, 50-59' $19, 60-69' $21, over 70' $25. Pay at the office. Established in the mid-1950s to serve commercial and sport fishing on Oregon's south coast. Helful, professional dockmaster. The docks are made of steel and wood; the first dock is for guests and is well marked. The 2-lane launch ramp is excellent, and the shipyard offers extensive services. The laundromat is in the RV park. Showers are in the rustic cement-floor bathhouse (3 minutes for $0.25 or 12 minutes for $1). Note: A harbor of refuge and one of the largest wood products ports in the world, The Coos Bay bar is considered the safest on the Pacific coast. Contact the Coast Guard for bar conditions (888-3102).

Notable -- This old fishing village is also a top sportfishing center as well as a busy commercial port. Rent a crab ring and try for dungeness crabs, or dig some clams - the area behind Portside is reportedly a good spot. Charter a fishing boat to search for halibut, salmon, tuna, or a wealth of bottom fish such as lingcod. Dinghy the water trails of the South Slough National Estuarine Reserve on an incoming tide; the Interpretive Center provides maps and info on tides.

Navigational Information
Lat: 43°22.115' **Long:** 124°12.625' **Tide:** 8.5 ft. **Current:** n/a **Chart:** 18587
Rep. Depths (*MLW*): **Entry** 35 ft. **Fuel Dock** n/a **Max Slip/Moor** 20 ft./-
Access: Upriver from the entrance to Coos Bay to river mile 14.2

Marina Facilities *(In Season/Off Season)*
Fuel: No
Slips: 40 Total, 40 Transient **Max LOA:** 70 ft. **Max Beam:** n/a
  Rate *(per ft.)*: **Day** $0.25* **Week** n/a **Month** n/a
  Power: 30 amp Incl., **50 amp** Incl., **100 amp** n/a, **200 amp** n/a
  Cable TV: No **Dockside Phone:** No
  Dock Type: Floating, Short Fingers, Alongside, Concrete
Moorings: 0 Total, 0 Transient **Launch:** n/a
  Rate: Day n/a **Week** n/a **Month** n/a
Heads: 1 Toilet(s)
Laundry: None **Pay Phones:** No
Pump-Out: OnSite, Self Service, 1 Central **Fee:** Free **Closed Heads:** Yes

Marina Operations
Owner/Manager: Steve Doty **Dockmaster:** Neal Lawson
In-Season: Year-Round **Off-Season:** n/a
After-Hours Arrival: See harbormaster on south dock
Reservations: No **Credit Cards:** No
Discounts: None
Pets: Welcome **Handicap Access:** Yes

Coos Bay City Dock

500 Central Avenue; Coos Bay, OR 97420

Tel: (541) 269-8918 **VHF: Monitor** n/a **Talk** n/a
Fax: (541) 267-5615 **Alternate Tel:** (541) 269-8918
Email: n/a **Web:** n/a
Nearest Town: Coos Bay **Tourist Info:** (541) 269-0215

Marina Services and Boat Supplies
Services - Trash Pick-Up **Supplies - Under 1 mi:** Bait/Tackle (*Fred Meyers*), Propane (*U-Haul 269-1333*) **3+ mi:** Ships' Store (*Englund Marine Supply 888-6723, 8 mi.*)

Boatyard Services
Nearest Yard: Charleston Shipyard (541) 888-3703

Restaurants and Accommodations
Near: Restaurant (*Gerry's 267-5600*), (*Fireside 267-7772*), (*Benetti's Italian Restaurant 267-6066, D $12-18, kids' & seniors' menu available*), (*Cedar Grill 267-7100, L, D*), (*Elizabeth's 266-7708*), (*Kosy Kitchen*), Fast Food (*City Subs 269-9000*), (*Dairy Queen, Quizon's*), Pizzeria (*Figaro's Pizza 269-7678*), (*Molly's the Pizza Parlor 267-3089*), Hotel (*Best Western 269-5111, $67-120*), Inn/B&B (*This Olde House B&B 267-5224*) **Under 1 mi:** Hotel (*Red Lion Hotel 267-4141, $68-75*) **1-3 mi:** Restaurant (*The Plankhouse Restaurant 756-8800, B, L, D in the Mill Casino; $12 early bird specials*), Hotel (*Ramada Inn 756-3191, $62-78*)

Recreation and Entertainment
OnSite: Picnic Area **Near:** Fitness Center (*Coastal Fitness Center 267-3363*), Museum (*Coos Art Museum 267-3901/ Coos Historical Museum 756-6320, 3 mi.*), Cultural Attract (*On Broadway Theater 267-2501*), Galleries (*Coos Art Museum 267-3901*) **1-3 mi:** Dive Shop (*Sunset Sports 756-3483*), Bowling (*North Bend Lanes 756-0571*), Special Events (*Oregon Coast Music Fest - Jul 897-9350*) **3+ mi:** Golf Course (*Kentuck Golf Course 756-4464, 8 mi.*), Fishing Charter (*Fishin's the Mission 297-3474, 4 mi.*), Group Fishing Boat (*Betty Kay Charters 297-3474, 6 mi.*), Movie Theater (*Pony Village Cinema 756-3447, 3 mi.*), Park (*Shore Acres State Park 888-3732, 10 mi.*)

Provisioning and General Services
Near: Convenience Store (*Shell 269-9420*), Market (*McKay's Market 267-3811*), Wine/Beer (*Oregon Wine Cellars 267-0300*), Liquor Store (*Coos Bay Liquor 267-6421*), Bakery (*Early to Rise 267-7719*), Farmers' Market (*Hwy 101 & Commercial, Wed 9am-3pm midMay-MidOct*), Bank/ATM, Catholic Church, Protestant Church, Library (*Coos Bay 269-1101*), Beauty Salon (*Salon 224 269-1414*), Barber Shop (*Driftwood Barber Shop 267-4702*), Dry Cleaners (*Wardrobe Cleaners 267-6118*), Pharmacy (*Rite Aid 267-7240*), Florist (*Checkerberry's 756-1830*), Copies Etc. (*South Coast Printing 269-5853*) **Under 1 mi:** Supermarket (*Safeway 267-1700*), Delicatessen, Post Office, Laundry (*Wash Tub Laundry 267-2814*), Hardware Store (*Farr's True Value 267-2137*) **1-3 mi:** Health Food (*Bailey's Health Food 756-3004*), Department Store (*Wal-Mart 267-4688*)

Transportation
OnCall: Rental Car (*Enterprise 266-7100*), Taxi (*Yellow Cab 267-3111*), Local Bus (*267-7111*), InterCity Bus (*Curry Public Transit 469-6822/ Greyhound 267-4436*) **Airport:** North Bend Municipal (*4 mi.*)

Medical Services
911 Service **OnCall:** Ambulance **Near:** Doctor (*Waterfall Clinic 266-0620*), Dentist (*Bay Dental 267-6673*), Chiropractor (*Bay Area Chiropractic 269-2525*), Holistic Services (*Therapeutic Touch 267-4439*), Veterinarian (*Morgan Veterinary Clinic 269-5846*) **Under 1 mi:** Optician (*Optical Shoppe 267-7877*) **Hospital:** Bay Area 269-8111 (*1.5 mi.*)

Setting -- Twelve miles upriver from the entrance to Coos Bay, this is one of the largest wood products ports in the world; commercial wood products facilities are visible on all sides. The modern Coos Bay City Dock sits right across US Highway 101 from the downtown district - in the heart of Coos Bay. A well-maintained concrete float provides side-tie moorage on the outside and long finger slips on the shoreside. Gangways lead to a beautiful, 1850-foot wooden boardwalk lined with 23 international flags and three covered pavilions - that house interpretive panels on the history of the area. The pavilion furthest from the dock contains an 80-year-old retired tugboat. A two-third mile paved walkway runs along the bay.

Marina Notes -- *Moorage free up to 12 hrs. Flat rate for overnight stays: up to 20' $5/day, 21-39' $10, over 40' $0.25/ft./day. The dock offers few amenities for boaters except its excellent location. The harbormaster lives on his boat at the dock; the port office is a block away in the gray building at the corner.

Notable -- Coos Bay has become the largest city on the Oregon Coast. Across the street at the Visitors' Center is literature on attractions and walking tours. The Coos Bay Art Museum is nearby on Anderson Ave. in a former 1930s US Post Office. July brings the Oregon Coast Music Festival, August the Blackberry Arts Festival, and September the Bay Area Fun Festival. The Coos County Historical Museum (756-6320) is a short taxi ride in North Bend - the theme of the exhibits is "Tidewater Highways." On display is an outstanding basket collection created by Native Americans, artifacts from local shipwrecks, items used in the shipbuilding trade, photographs of maritime activities, and much more.

Port of Bandon

Port of Bandon

PO Box 206; 390 1st Street SW; Bandon, OR 97411

Tel: (541) 347-3206 **VHF: Monitor** n/a **Talk** n/a
Fax: (541) 347-4645 **Alternate Tel:** n/a
Email: port@portofbandon.com **Web:** www.portofbandon.com
Nearest Town: Bandon **Tourist Info:** (541) 347-9616

Navigational Information
Lat: 43°07.250' **Long:** 124°24.750' **Tide:** 7 ft. **Current:** n/a **Chart:** 18588
Rep. Depths (MLW): Entry 12 ft. **Fuel Dock** 8 ft. **Max Slip/Moor** 10 ft./-
Access: On the right side of the Coquille River as you enter

Marina Facilities (In Season/Off Season)
Fuel: High Dock - Gasoline, Diesel
Slips: 90 Total, 5 Transient **Max LOA:** 60 ft. **Max Beam:** n/a
  Rate (per ft.): **Day** $0.30* **Week** n/a **Month** $48-143
  Power: 30 amp Incl., **50 amp** n/a, **100 amp** n/a, **200 amp** n/a
  Cable TV: No **Dockside Phone:** No
  Dock Type: Floating, Long Fingers, Alongside, Wood
Moorings: 0 Total, 0 Transient **Launch:** n/a
  Rate: Day n/a **Week** n/a **Month** n/a
Heads: 4 Toilet(s)
Laundry: None **Pay Phones:** Yes
Pump-Out: OnSite, Self Service **Fee:** Free **Closed Heads:** Yes

Marina Operations
Owner/Manager: Port Commission **Dockmaster:** Gina Dearth
In-Season: Year-Round, 7:30am-3:30pm **Off-Season:** n/a
After-Hours Arrival: Call in advance or tie up and register in the morning
Reservations: No **Credit Cards:** No, Cash Only
Discounts: None
Pets: Welcome **Handicap Access:** No

Marina Services and Boat Supplies
Communication - FedEx, UPS, Express Mail **Supplies - OnSite:** Ice (Cube), Bait/Tackle (Brandon Bait & Port of Call)

Boatyard Services
OnSite: Launching Ramp **Nearest Yard:** Charleston (541) 888-3703

Restaurants and Accommodations
OnSite: Restaurant (High Dock Bistro 347-5432), Seafood Shack (Bandon Fish Market & Chowder House 347-4282), Snack Bar (Port O' Call), (Bait Shop) **Near:** Restaurant (Wild Rose Bistro 347-4428), (Thai Talay 347-8074), Lite Fare (2 Loons Café 347-3750, L $7-9, B 7:30-11am, L 11am-3pm), Inn/B&B (Sea Star 347-9632, $55-150) **Under 1 mi:** Restaurant (The Station 347-9615, B $5-9, L $5-10, D $10-16), (Bandon Boatworks 347-2111, L, D kids' menu), Fast Food (Dairy Queen), Pizzeria (Old Town 347-3911, L, D), Motel (La Kris 347-3610, $32-72), Inn/B&B (Lighthouse 347-9316, $135-225) **1-3 mi:** Restaurant (Billy Smoothboars 347-2373), Coffee Shop (Bandon Coffee Café 347-1144), Lite Fare (Bandon Baking & Deli 347-9440), Motel (Bandon Beach 347-4430, $50-85), Hotel (Best Western 347-9441, $78-179), (Sunset 347-2453, $45-220)

Recreation and Entertainment
OnSite: Picnic Area, Boat Rentals (Port O' Call 347-2875 Kayaks, Adventure Kayak 347-3480), Fishing Charter (Port-O-Call 347-2875) **Near:** Group Fishing Boat (Prowler Charters 347-1901), Special Events (Cranberry Festival 347-9616 - 2nd wknd Sep, Celebration of Lights, Irish Festival 347-9616 - May) **Under 1 mi:** Beach (gorgeous!), Fitness Center (Bandon Fitness 347-3522), Movie Theater, Cultural Attract (Sprague Community

Theater 347-7426) **1-3 mi:** Park (West Coast Game Park 347-3106 - America's largest wild animal petting park), Museum (Bandon Historical Society 347-2164, Coquille River Museum 347-2164), Galleries (Bandon Glass Art 347-4723, Second St. 347-4133) **3+ mi:** Golf Course (Bandon Face Rock 347-3818/Bandon Dunes 347-4380, oceanfront, 3 mi./7 mi.)

Provisioning and General Services
Near: Convenience Store (Run-In Mini Mart 347-9290), Bakery (Bandon Baking & Deli 347-9440), Fishmonger (Bandon Fish Market 347-4282), Crabs/Waterman, Shrimper, Oysterman, Laundry (Samme's 347-9493), Retail Shops (Big Wheel General Store 347-3719) **Under 1 mi:** Market (Wilson's 347-3083), Supermarket (Ray's Food Place 347-2127), Delicatessen (Posh Nosh 347-1336), Bank/ATM, Post Office, Library (Bandon 347-3221), Beauty Salon (Vicki G 347-1900) **1-3 mi:** Health Food (Mother's Natural Grocery 347-4086), Protestant Church, Dry Cleaners (Bandon 347-9493), Bookstore (Winter River 47-4111), Pharmacy (Tiffany's 347-4438), Hardware Store (True Value 347-2506), Copies Etc. (Pat's 347-3553) **3+ mi:** Liquor Store (Bandon 347-2106, 4 mi.)

Transportation
OnCall: Taxi (Triple Diamond 347-1767), Airport Limo (Connoisseurs Limousine 347-5466) **3+ mi:** Rental Car (Enterprise 756-7700, 25 mi.) **Airport:** Bandon State/North Bend Municipal (3 mi./28 mi.)

Medical Services
911 Service **OnCall:** Ambulance **Near:** Dentist (Family Dental 347-4461), Veterinarian (Bandon Veterinary 347-9471) **Under 1 mi:** Doctor (North Bend 347-5191) **Hospital:** Southern Coos 347-2426 (1 mi.)

Setting -- Jetties safeguard the entrance to the Coquille River, and a seasonal light and fog signal marks the south jetty. The Port of Bandon Marina sits on the lower bend of the river, less than a mile from the entrance on the south shore. A tall rock breakwater protects the floats. A dock on the west side of the marina provides alongside moorage for transients and also houses the fuel pumps and pump-out station. Lampposts, hovering over the broad deck, light the way down the dock. A brick pathway stretches along the shore and a flower bedecked boardwalk hosts a string of eateries, picnic tables, nature and history kiosks, gardens, locally-crafted burlwood sculptures and benches and a green-roofed, glass-enclosed picnic shelter - all with great views of the harbor.

Marina Notes -- *Flat rate, cash only: 20-29' $10/day, 30-39' $12, 40-49' $14, 50-59' $16, 60' $18/day. Call during business hours for berth assignment. Fuel available at the High Dock; contact Prowler Charters (347-1901). Fish cleaning stations are within easy reach as are fresh fish markets for the unlucky angler. Bait & tackle, kayak rentals, fishing charters and casual eateries are just steps from the docks. Well-cared for heads are across the street, but no showers. Note: The channel is dredged yearly to 12 feet. A Coast Guard rescue boat is stationed here during the summer.

Notable -- Stroll the waterfront, the high-rock jetty walkway and viewing platform, the wonderful boardwalk, and charming nearby Old Town with a dozen restaurants, unique shops, galleries and a new theater. Visit the recently renovated Coquille lighthouse near the entrance of the River. Wildlife and other programs are held at the waterfront amphitheater, and Bandon hosts three major events, including the Cranberry Festival which celebrates its famous crop.

Navigational Information
Lat: 42°44.470' **Long:** 124°29.910' **Tide:** 7 ft. **Current:** n/a **Chart:** 18589
Rep. Depths (*MLW*)**: Entry** 30 ft. **Fuel Dock** 16 ft. **Max Slip/Moor** 16 ft./-
Access: Just south of The Heads, 6.5 mi. south of Cape Blanco

Marina Facilities *(In Season/Off Season)*
Fuel: Gasoline, Diesel, High-Speed Pumps
Slips: 50 Total, 20 Transient **Max LOA:** 120 ft. **Max Beam:** n/a
 Rate (*per ft.*)**: Day** $0.50* **Week** $80 **Month** $170
 Power: 30 amp Incl., **50 amp** n/a, **100 amp** n/a, **200 amp** n/a
 Cable TV: No **Dockside Phone:** No
 Dock Type: Floating, Wood
Moorings: 0 Total, 0 Transient **Launch:** n/a
 Rate: Day n/a **Week** n/a **Month** n/a
Heads: 2 Toilet(s), 2 Shower(s)
Laundry: None **Pay Phones:** No
Pump-Out: No **Fee:** n/a **Closed Heads:** Yes

Marina Operations
Owner/Manager: Port Commission **Dockmaster:** Gary Anderson
In-Season: Year-Round, 5am-10pm **Off-Season:** n/a
After-Hours Arrival: Not possible
Reservations: No **Credit Cards:** No
Discounts: None
Pets: Welcome, Dog Walk Area **Handicap Access:** Yes, Heads

Port of Port Orford

PO Box 490; 300 Dock Road; Port Orford, OR 97465

Tel: (541) 332-7121 **VHF: Monitor** Ch. 9 **Talk** Ch. 9
Fax: (541) 332-7121 **Alternate Tel:** n/a
Email: portoffice@harborside.com **Web:** portofportorford.com
Nearest Town: Port Orford *(9 mi.)* **Tourist Info:** (541) 332-3681

Marina Services and Boat Supplies
Services - Docking Assistance, Trash Pick-Up, Dock Carts
Communication - Mail & Package Hold, Phone Messages, Fax in/out
(Free) **Supplies - OnSite:** Ice *(Shaved)*, Bait/Tackle *(Dock Tackle 332-8985)* **Near:** Ice *(Cube)*, Ships' Store, Propane

Boatyard Services
OnSite: Crane *($20)*

Restaurants and Accommodations
OnSite: Lite Fare *(Dock Tackle & Gypsy Café 332-8985)* **Near:** Restaurant *(Paula's Bistro 332-9378, L, D)*, Seafood Shack *(Crazy Norwegians 332-8601, L, D)*, Motel *(Shoreline Motel 332-2903)*, Inn/B&B *(Home by the Sea 332-2855, $96-104)*, *(Holly House Inn 332-7100, $80-100)*, *(Castaway by the Sea 332-4502, $55-95)* **Under 1 mi:** Restaurant *(Pitch's Tavern 332-9313)*, Coffee Shop *(Port Orford Inn Coffee House 332-1009)*, Lite Fare *(Port Orford Breadworks 332-4022)*, Motel *(Battle Rock 332-7331, $48-75)*, Inn/B&B *(Port Orford Inn 332-0212)* **1-3 mi:** Motel *(Sea Crest 332-3040, $42-74)*, Hotel *(Tanglewood Guest House 332-6720)* **3+ mi:** Restaurant *(Wild Wind Café 332-0534, great pastries)*

Recreation and Entertainment
OnSite: Beach, Picnic Area, Dive Shop, Fishing Charter *(Pac Nor West)*, Group Fishing Boat **Near:** Park *(Buffington - with a top skateboard bowl, butterfly/hummingbird park, horseshoes, mini-golf, basketball)*, Museum *(The Black's Antique Auto Museum 9am-5pm 332-0191)* **Under 1 mi:** Fitness Center, Video Rental *(Downtown Fun Zone 332-6565)*, Special Events

(Fourth of July Celebration), Galleries *(Cook Gallery 9am-5pm daily 332-0521; Triangle Square Gallery 253-6198)* **1-3 mi:** Movie Theater *(Savoy Theater 332-3105)*, Cultural Attract *(Port Orford Lifeboat Station, Apr-Oct Thu-Mon 10am-3:30pm)*, Sightseeing *(Cape Blanco Lighthouse Apr-Oct Thu-Mon 10am-3:30pm)*

Provisioning and General Services
OnSite: Fishmonger, Copies Etc. **Near:** Convenience Store *(Circle K 332-3181)*, Market, Wine/Beer, Liquor Store *(Port Orford Liquor 332-2352)*, Newsstand, Retail Shops **Under 1 mi:** Supermarket *(Ray's Food Place 332-1185)*, Delicatessen, Health Food *(Seaweed Natural Grocery and Cafe 332-3640)*, Bakery *(Port Orford Breadworks 332-4022)*, Bank/ATM, Post Office *(332-4251)*, Catholic Church *(St. John's 347-2309)*, Protestant Church, Library *(Port Orford Library 332-5622)*, Barber Shop, Laundry *(Duds & Suds 332-6000)*, Pharmacy *(Corner Drug 247-4544)*, Hardware Store *(Coos Curry Supply 332-1818)*, Florist *(Sea Breeze Florist 332-0445)* **1-3 mi:** Beauty Salon *(New You Beauty Boutique 332-5445)*

Transportation
Near: Local Bus **3+ mi:** Taxi *(Ocean Taxi 247-8294, 17 mi.)*
Airport: Cape Blanco State Airport *(10 mi.)*

Medical Services
911 Service OnCall: Ambulance **Under 1 mi:** Doctor *(Curry Family Medical 332-3861)*, Dentist *(Port Orford Family Dentristry 332-5001)* **3+ mi:** Veterinarian *(Gold Beach Veterinary Clinic 247-2513, 28 mi.)*
Hospital: Curry General 247-6621 *(29 mi.)*

Setting -- Cape Blanco (6.5 miles north) and Humbug Mountain (4 miles south) help pinpoint Port Orford. The small town overlooks a cove that provides good protection in all but southerly weather, when it becomes exposed. Behind a rock breakwater, two hoists, painted bright yellow, sit at the water's edge and guide boaters to the marina. This is a working harbor; watch boats unloading their catch and crab and fish right off the dock.

Marina Notes -- There are no docks or slips, except for the seasonal, good-weather float. Instead, boats are lifted from the water and placed directly on trailers, which are neatly lined up on a large, concrete area with "boat pads" with water and electric hook-ups. *Flat rate: $15/day plus $5 for sling hoist or $40/day for seasonal floating dock available to pleasure boats (high fee intended to dissuade working boats from using it). Port Orford is one of a few marinas in the world that uses a "dolly dock" system. (Max LOA 44 ft., 15 ft., beam to use this system.) A few private buoys may be available to transients - inquire.

Notable -- This is the only Oregon harbor without a river bar to cross. It's known as a safe harbor, and when the seas are angry, one of the best easy-access moorages on the Oregon coast. On-site Dock Tackle and its Gypsy Café (332-8985) offers local seafood, fish and chips, clam chowder, gifts and a small nautical museum (free). Little Port Orford, the western-most city in the continental U.S., offers lots of sightseeing opportunities. Rising 200 feet above the sea, the 1870 Cape Blanco Lighthouse is the longest operating light in Oregon; climb the 64 steps for extraordinary views and visit the 105-year-old Patrick Hughes house. Shipwrecks are the theme at Battle Rock Park and the Lifeboat Station Museum (Apr-Oct Thu-Mon) just a couple of miles north.

Navigational Information
Lat: 42°25.470' **Long:** 124°25.200' **Tide:** 7 ft. **Current:** n/a **Chart:** 18601
Rep. Depths *(MLW):* **Entry** 10 ft. **Fuel Dock** 10 ft. **Max Slip/Moor** 10 ft./-
Access: On the south side of Rogue River after crossing the bar

Marina Facilities *(In Season/Off Season)*
Fuel: Gasoline, Diesel
Slips: 200 Total, 5 Transient **Max LOA:** 45 ft. **Max Beam:** n/a
 Rate *(per ft.):* **Day** $0.30* **Week** n/a **Month** n/a
 Power: 30 amp Incl., **50 amp** n/a, **100 amp** n/a, **200 amp** n/a
 Cable TV: No **Dockside Phone:** No
 Dock Type: Floating, Long Fingers, Concrete, Wood
Moorings: 0 Total, 0 Transient **Launch:** n/a
 Rate: Day n/a **Week** n/a **Month** n/a
Heads: 6 Toilet(s)
Laundry: None **Pay Phones:** No
Pump-Out: OnSite **Fee:** Free **Closed Heads:** Yes

Port of Gold Beach

PO Box 1126; 29891 Harobr Way; Gold Beach, OR 97444

Tel: (541) 247-6269 **VHF: Monitor** Ch. 16 **Talk** n/a
Fax: (541) 247-6268 **Alternate Tel:** n/a
Email: Portoffice@portofgoldbeach.com **Web:** n/a
Nearest Town: Gold Beach *(0.25 mi.)* **Tourist Info:** (541) 247-0923

Marina Operations
Owner/Manager: Pete Dale **Dockmaster:** Bruce Pinkel
In-Season: Year-Round, 8am-4:30pm **Off-Season:** n/a
After-Hours Arrival: Make prior arrangements or register in the morning
Reservations: No **Credit Cards:** Visa/MC, Dscvr, Amex, Exxon
Discounts: None
Pets: No **Handicap Access:** No

Marina Services and Boat Supplies
Supplies - OnSite: Live Bait, Propane **Near:** Ice *(Cube)*, Bait/Tackle *(Rogue Outdoors 247-7142)*

Boatyard Services
OnSite: Launching Ramp **OnCall:** Engine mechanic *(gas)*

Restaurants and Accommodations
OnSite: Restaurant *(Porthole Café 247-7411, B $4-9, L $4-7, D $12-16)*, *(Nor'Wester Seafood 247-2333, D after 5pm)*, Snack Bar *(Cone Amor 247-4270)* **Near:** Restaurant *(Wong's Café 247-7423)*, Lite Fare *(Crow's Nest Tavern 247-6837)*, Motel *(City Center Motel 247-6675, $45-65)*
Under 1 mi: Restaurant *(Spinners' Seafood & Steak 247-5160)*, *(Riverview 247-7321)*, *(Paul Bunyan Burgers 247-2785, L $7-10, D $7-10, drive-thru)*
1-3 mi: Motel *(Breakers 247-6606, $35-75)*, *(Clear Sky Lodge 247-6691)*, Inn/B&B *(Inn of the Beachcomber 247-6691, $120-150)*, Condo/Cottage *(Gold Beach Resort and Condos 247-7066, $69-119)*

Recreation and Entertainment
OnSite: Picnic Area, Sightseeing *(Jerry's Rogue Jets 800-451-3645/Rogue River Mail Boat 800-484-3511)*, Galleries *(Rogue's Gallery 247-6158 - specializing in local art)* **Under 1 mi:** Park *(Indian Creek Recreational Park 247-7704)*, Museum *(Curry County Historical Society 247-9396)*, Special Events *(Curry County Fair - Jul)* **1-3 mi:** Fitness Center *(Gold Beach Circuit Workout 247-2075)*, Group Fishing Boat *(Lower Rogue Guide Service 247-2684; Five Star Charters 247-0217)*, Cultural Attract *(Pieces of Time Gallery 247-7858)*

3+ mi: Golf Course *(Cedar Bend Golf Course 247-6911, 9 mi.)*, Fishing Charter *(Anderson's Guide Service 247-0420, 5 mi.)*

Provisioning and General Services
OnSite: Fishmonger *(Fisherman Direct Seafood 247-9494, 11am-6pm)*, Post Office, Retail Shops *(Jerry's Rogue River Museum & Gift Shop)* **Near:** Wine/Beer, Liquor Store *(Gold Beach Liquor 247-7514)*, Protestant Church, Library *(Curry Public Library 247-7246)*, Beauty Salon *(Empress Beauty Salon 247-6319)*, Florist *(Flowers By the Sea 247-7673)*, Department Store *(Rogue Outdoor Store 247-7142)* **Under 1 mi:** Market *(McKay's Market 247-7144, Ray's Food Place 247-7721)*, Bakery *(Rollin in Dough Bakery & Deli 247-4438)*, Bank/ATM *(Sterling Savings Bank 247-0478)*, Barber Shop *(Country Cuttin' 247-0760)*, Laundry *(Gold Beach Laundromat 247-4206)*, Bookstore *(Gold Beach Books 247-2495)*, Pharmacy *(Corner Drug 247-4544)*, Hardware Store *(Dan's Ace Hardware 247-6822)* **1-3 mi:** Health Food *(Savoy Natural Foods 247-0297)*, Catholic Church *(St. Charles 247-2453)*

Transportation
Near: Local Bus *(Shuttle 800-921-2871)* **Airport:** Gold Beach Municipal; Brookings State *(0.5 mi./25 mi.)*

Medical Services
911 Service **OnCall:** Ambulance **Under 1 mi:** Doctor *(North Bend Medical Center 247-7047)*, Dentist *(Gold Beach Family Dentristry 247-8000)*, Veterinarian *(Gold Beach Veteranary Clinic 247-2461)* **1-3 mi:** Chiropractor *(Gold Beach Chiropractic 247-4330)* **Hospital:** Curry General 247-6621 *(1 mi.)*

Setting -- About 30 miles north of the California border, the community of Gold Beach lies at the mouth of the Rogue River, on the south shore. Two stone jetties shelter the entrance, but breakers are often present at their outer ends. The marina is located at the north end of town, surrounded by green hills with the Route 101 bridge visible less than half a mile away. A gangway leads from an extensive network of short high-quality floating docks to the boardwalk which leads to a small number of local stores & restaurants.

Marina Notes -- *Flat Rates vary by size. Examples 20' $9/day, $45/wk., $112.50/mo., 30' $10.50/day, $54/wk., $127.50/mo., 40' $12/day, $63/wk., $142.50/mo. Register at the port office from 8am-5pm. The office is located south of the marina in a two-story blue building. Boat storage is available as well as a fish cleaning table. Basic, cement and cinderblock heads are for public use and have no showers or electricity. Note: Boaters approaching from the sea must deal with the Rogue River bar, especially since depths are often less than the 8-foot projected depth. Call for local knowledge.

Notable -- There's lots to do on-site - jet boat rides, fishing guides and charters, eateries and a free museum depicting the infamous Rogue River. Hop aboard the Coastal Treasures (Mon, Tue, Wed 8:30am) or Pacific Wonders tours (Thu-Sun 8:30am), or travel along the Rogue River (daily 2pm). Jerry's Rogue Jets offers exciting rides along the river's wild waters. The working waterfront boasts Nor' Western Seafood, Porthole Café, Fishermen Direct Seafood, Coffee Dock with fresh espresso and Rogue's Gallery.

Navigational Information
Lat: 42°02.860' **Long:** 124°16.110' **Tide:** 7 ft. **Current:** n/a **Chart:** n/a
Rep. Depths (*MLW*): **Entry** 14 ft. **Fuel Dock** 14 ft. **Max Slip/Moor** 14 ft./-
Access: The harbor lies on the south bank of the Chetco River

Marina Facilities *(In Season/Off Season)*
Fuel: Gasoline, Diesel
Slips: 657 Total, 300 Transient **Max LOA:** 105 ft. **Max Beam:** n/a
  Rate *(per ft.)*: **Day** $0.45* **Week** n/a **Month** n/a
  Power: 30 amp Incl., **50 amp** Incl., **100 amp** n/a, **200 amp** n/a
  Cable TV: No **Dockside Phone:** No
  Dock Type: Floating, Concrete, Wood
Moorings: 0 Total, 0 Transient **Launch:** n/a
  Rate: Day n/a **Week** n/a **Month** n/a
Heads: 28 Toilet(s), 6 Shower(s)
Laundry: 10 Washer(s), 10 Dryer(s), Book Exchange **Pay Phones:** No
Pump-Out: OnSite **Fee:** Free **Closed Heads:** Yes

Marina Operations
Owner/Manager: Port Commission of Brookings **Dockmaster:** Mike Blank
In-Season: Year-Round, 8am-5pm **Off-Season:** n/a
After-Hours Arrival: Pay in the morning
Reservations: No **Credit Cards:** Visa/MC, Dscvr
Discounts: None
Pets: No **Handicap Access:** No

Port of Brookings Harbor

PO Box 848; 16408 Lower Harbor Road; Brookings, OR 97415

Tel: (541) 469-2218 **VHF: Monitor** Ch. 12 **Talk** n/a
Fax: (541) 469-0672 **Alternate Tel:** n/a
Email: info@port-brookings-harbor.org **Web:** port-brookings-harbor.org
Nearest Town: Brookings *(3 mi.)* **Tourist Info:** (541) 469-3181

Marina Services and Boat Supplies
Services - Security **Communication -** Mail & Package Hold, Phone Messages, Data Ports *(Wi-Fi in RV Park)* **Supplies - OnSite:** Ice *(Cube)* **Near:** Ships' Store *(Chetco 469-6681)*, Bait/Tackle *(Four M 469-6951)*

Boatyard Services
OnSite: Travelift *(60T)*, Launching Ramp

Restaurants and Accommodations
OnSite: Restaurant *(Oceanside Diner 469-7971)* **Near:** Restaurant *(Voodoo Lounge 469-5200)*, *(Torero's Family Mexican Restaurant & Bar 469-2771, L $8-12, D $8-23, L specials $6-8)*, *(The Hungry Clam 469-CLAM, L & D $5-10)*, *(Smuggler's Cove 469-6006, L, D)*, *(Chan's Restaurant 469-7013, L $6-10, D $8-17)*, *(Sporthaven 469-5200)*, Snack Bar *(Slugs 'n Stones 'n Ice Cream Cones 469-7584)*, Lite Fare *(Feley's Café B, L - in RV Park)* **Under 1 mi:** Motel *(Ocean Suites 469-4004, $69)*, Hotel *(BW Beachfront Inn 469-7779)* **1-3 mi:** Pizzeria *(Wild River Pizza 412-7887)*, Motel *(Wild River 469-5361)*, Inn/B&B *(South Coast Inn 469-5557, $99-159)*

Recreation and Entertainment
OnSite: Picnic Area, Group Fishing Boat *(Angler's Choice Tidewind 469-0337)* **Near:** Fishing Charter **Under 1 mi:** Fitness Center *(Bayside Fitness & Aerobics 469-7118)*, Bowling *(Azalea Lanes 469-4244)*, Cultural Attract *(Performing Arts Center 469-1857)*, Special Events *(Azalea Fest - MemDay wknd, Chili Cook-off - Jul 469-0150, Kite Festival & Regatta - Jul, Arts Fest - Aug)* **1-3 mi:** Movie Theater *(Redwood 469-4632)*, Video Rental

(Blockbuster 469-8889), Park *(Azalea Park)*, Museum *(Chetco Valley Historical 469-6651)*, Galleries *(Pelican Bay 469-1807)* **3+ mi:** Golf Course *(Salmon Run 469-4888, 4 mi.)*

Provisioning and General Services
OnSite: Farmers' Market *(MidJul-MidOct Sat 9am-3pm on boardwalk)*, Fishmonger *(Dick & Casey's Seafood 469-9494;)* **Near:** Meat Market, Bank/ATM, Beauty Salon *(Tresse Fx 469-9677)*, Bookstore *(The Book Dock 469-6070)*, Retail Shops *(Beachfront Gifts 469-8025)* **Under 1 mi:** Market *(Ray's 469-3743)*, Green Market, Post Office, Barber Shop *(Gentlemen's Choice 469-4500)*, Pharmacy *(Chetco 469-2616)*, Department Store *(Sears 469-2116, Harbor Shopping Center 469-4301)* **1-3 mi:** Convenience Store *(Shop Smart 469-3191)*, Delicatessen *(Beachcombers 469-0739)*, Health Food *(Brookings Co-op 469-9551)*, Wine/Beer, Liquor Store *(Brookings 469-2502)*, Catholic Church, Protestant Church, Library *(Chetco 469-7738)*, Dry Cleaners *(Hagen's 469-3628)*, Laundry *(Old Wash House 469-3975)*, Hardware Store *(Kerr Ace 469-3139)*

Transportation
OnCall: Taxi *(412 Taxi 412-7383)* **3+ mi:** Rental Car *(Hertz Crescent City 707-464-5750, 26 mi.)* **Airport:** Del Norte County *(27 mi.)*

Medical Services
911 Service **OnCall:** Ambulance **1-3 mi:** Doctor *(Brookings Clinic 469-2330)*, Dentist *(Sims 469-9594)*, Veterinarian *(Brookings-Harbor 469-7788)* **Hospital:** Sutter Coast 464-8511 *(25 mi.)*

Setting -- Bustling Port of Brookings Harbor lies in Chetco Cove at the mouth of the Chetco River. Two adjacent basins - one used by the commercial fleet, and the other by pleasure craft - are on the river's south shore. A jetty protects the network of excellent concrete docks with long finger slips that house a mix of power and sail boats. Flags fly above a wide boardwalk that edges the waterfront; it hosts picnic tables, a seasonal farmers' market, events, and an attractive, contemporary "village" housing 34 shops and eateries. The Beachfront RV park is just south of the marina; adjacent are a chandlery, market, and restaurants.

Marina Notes -- *Flat rate: 16-21' $9.35/day, 22-26' $10.10, 27-29' $11.10, 30-39' $16.10, 40-49' $18.75, 50' $22.80/day. Pay at the port office - the gray two-story building across the parking lot from Boat Basin One. A 140-foot fishing pier with picnic tables is on-site along with a state of the art fueling facility and a six-lane boat launch. A boatyard and fishing charters round out this full service marina. The bathhouse, laundromat and Internet access are in the RV park (showers are $0.25 for 7 min.). Note: Good protection from northerly weather, but entry is not recommended during southerly blows. Bar reports are broadcast during daylight hours on KURY (910kHz). An overhead cable with a clearance of 46 feet lies 0.6 mile above the jetties and beyond the entrance to the basins.

Notable -- Surrounding the harbor are lots of unique galleries, bookstores, eateries, and other and businesses that are well worth exploring. Brookings claims to be the banana belt of Oregon; commercial flower growing, with lilies as the main harvest, gives testimony to the mild climate. Whale watching also enjoys a longer season. The Great American Smokehouse & Seafood Co. (469-6903) warrants a stop and Brandy Peak Distillery (469-0194) offers tours and tastings.

Boaters' Notes

Add Your Ratings and Reviews at www.AtlanticCruisingClub.com

A complimentary six-month Silver Membership is included with the purchase of this *Guide*. Select "Join Now," then "Silver," then follow the instructions.

The AtlanticCruisingClub website provides updated Marina Reports, Destination and Harbor Articles and much more — including an option within each online Marina Report for boaters to add their ratings and comments regarding that facility. Please log on frequently to share your experiences — and to read other boaters' comments.

On the website, boaters may rate marinas in one or more of the following categories — on a scale of 1 (basic) to 5 (world class) — and also enter additional commentary.

- **Facilities & Services** (Fuel, Reservations, Concierge Services and General Helpfulness)
- **Amenities** (Pool, Beach, Internet Access, including Wi-Fi, Picnic/Grill Area, Boaters' Lounge)
- **Setting** (Views, Design, Landscaping, Maintenance, Ship-Shapeness, Overall Ambiance)
- **Convenience** (Access — including delivery services — to Supermarkets, other Provisioning Sources, Shops, Services, Attractions, Entertainment, Sightseeing, Recreation, including Golf and Sport Fishing & Medical Services)
- **Restaurants/Eateries/Lodgings** (Availability of Fine Dining, Snack Bars, Lite Fare, OnCall food service, and Lodgings ashore)
- **Transportation** (Courtesy Car/Vans, Buses, Water Taxis, Bikes, Taxis, Rental Cars, Airports, Amtrak, Commuter Trains)
- **Please Add Any Additional Comments**

ATLANTIC CRUISING CLUB'S

GUIDE TO
PACIFIC NORTHWEST MARINAS

ADDENDA

Suggested Reading List

PACIFIC NORTHWEST REGIONAL CRUISING GUIDES

Reed's Nautical Almanac
This must-have annual covers 3,000 miles of coastline from the port of Manzanillo, Mexico north along the Pacific Coast to the Bering Strait in Alaska. It includes tide and current tables, over 2700 GPS waypoints, a coastal pilot with hundreds of harbor profiles and approach chartlets. Also extensive communications and weather resources.

The Waggoner Cruising Guide by Robert Hale
By far, the Northwest's best selling, most popular annual cruising guide, it is informative, entertaining and provides up-to-date, information on piloting, anchoring, and basic marina info for the entire region. The Guide covers all of Puget Sound, San Juan Islands, Inside Passage to Prince Rupert, B.C. the West Coast of Vancouver Island, Gulf Islands, Sunshine Coast, Princess Louisa Inlet, Desolation Sound, and the North Coast of B.C. 130 detailed maps and 250 photos.

Docks and Destinations by Peter Vassilopoulos
Clear, concise guide to cruising destinations from the San Juan Islands to Prince Rupert. Includes diagrams of some marinas and facilities, listings of services available at some locations plus local piloting information.

Anchorages and Marine Parks by Peter Vassilopoulos
A companion to Docks and Destinations, this is an easy-to-use reference to major anchorages on the B.C. coast and in the San Juan Islands. Over 200 photos, mostly aerials, and useful maps and diagrams. Includes launch ramp list and icons locating scuba and kayaking locales.

Northwest Boat Travel edited by Gwen Cole
This northwest recreational boating guide directs readers to over 2,700 places to go and things to see and do. Includes marinas, resorts, fuel docks, restaurants, secluded bays, anchorages, and marine.

Evergreen Pacific Cruising Atlas by W. Snake
Updated regional charts with 85 harbor charts and 41 photos make this popular atlas better than ever. This book covers the area from southern Puget Sound to Queen Charlotte Strait.

Evergreen Pacific San Juan Islands Cruising Atlas by Evergreen Pacific
Recent San Juan Islands chart atlas with aerial photos and a supplement for the Gulf Islands.

Evergreen Pacific River Cruising Atlas:
Columbia, Willamette & Snake Rivers by Evergreen Pacific Publishing
An atlas of charts for the rivers of the Pacific Northwest.

Gunkholing in South Puget Sound by Jo Bailey and Carl Nyberg
Detailed exploration of the cruising waters between Edmonds and Olympia including Lake Union and Lake Washington. Lots of local knowledge along with moorages, marinas, parks, facilities, restaurants, and attractions along the hundreds of miles of undiscovered shoreline in South Puget Sound. Reference charts, more than 200 photos.

Exploring the South Coast of British Columbia
by Don Douglass & Réanne Hemingway-Douglass
Routes and anchorages from Gulf Islands and Desolation Sound to Port Hardy and Blunden Harbour — with information from the authors' journeys afdsand from local skippers.

A Cruising Guide to Puget Sound by Migael M. Scherer
This is a comprehensive cruising guide to all of Puget Sound. Covers Olympia to Port Angeles including the San Juans, and Bellingham area with descriptions of more than 300 bays, harbors, anchorages, annotated charts, information on weather, history, tides and navigation. This is a first-rate book, carefully compiled by a cruiser with great depth of experience.

Exploring the San Juan and Gulf Islands
by Don Douglass & Réanne Hemingway-Douglass
Covering a thousand square miles from Deception Pass to Victoria and Nanaimo, this guidebook gives you the local knowledge you need to explore and enjoy more than 300 intimate islets and islands in this beautiful marine paradise.

Exploring the Pacific Coast — San Diego to Seattle
by Don Douglass & Réanne Hemingway-Douglass
This book covers the cruising grounds along coastal California, Oregon, and Washington, plus the Strait of Juan de Fuca and Greater Puget Sound.

Charlie's Charts: North to Alaska by Charles Wood
Sketch charts and navigation advice. Four color photographs. Victoria B.C. to Glacier Bay, Alaska, Charlie's Charts offer hand drawn sketches of approaches and harbors. Charlie is gone, now, but his wife, Margo, has carried on.

Charlie's Charts of the U.S. Pacific Coast by Charles Wood
Seattle, Washington to San Diego, California, Including the Channel Islands. This guide takes you from Seattle to San Diego offshore or harbor hopping. A GPS waypoint list is included in this updated edition.

Pacific Boating Almanac 2006: Pacific Northwest
edited by Peter L. Griffes
Covers Oregon and the Columbia River, Washington Coast, Puget Sound, San Juan Islands and British Columbia. Within the text is information about U.S. coastal piloting, tide and current tables, electronics, maps and charts, weather, navigation, and first aid.

Gulf Islands and Vancouver Island — Dreamspeaker Vol. 1
by Anne and Lauren Yeadon-Jones
In this first volume, the authors explore facilities and anchorages from Victoria to Nanaimo including the Gulf Islands and share their impressions with drawings and photos.

Desolation Sound and the Discovery Islands — Dreamspeaker Vol. 2
by Anne and Lauren Yeadon-Jones
Hand-drawn maps and photos cover approaches and way points for marinas and anchorages from the fabulous cruising grounds of Desolation Sound.

Vancouver, Howe Sound and the Sunshine Coast —
Dreamspeaker Vol. 3 by Anne and Lauren Yeadon-Jones
Featuring their unique, personal sketches that describes places to dock and anchor — including parks and some special spots — along the eastern shore of the Strait of Georgia.

The San Juan Islands — Dreamspeaker Vol. 4
by Anne and Lauren Yeadon-Jones
Destinations throughout the enchanted San Juan Islands, accompanied by the signature Dreamspeaker hand-drawn maps and striking photographs.

Northwest Marine Weather by Jeff Renner
Boater and Seattle TV meteorologist, Jeff Renner writes an educational resource book about local northwest weather basics and forecasting tools from the Columbia River to Cape Scott.

Wind Came All Ways: A Quest to Understand the Winds, Waves & Weather in Georgia Basin by Owen S. Lange
A marine weather manual for the Strait of Georgia, the Gulf Islands, the San Juan Islands, Juan de Fuca Strait, and surrounding areas — from Environment Canada.

British Columbia's Gulf Islands: Afoot & Afloat
by Marge and Ted Mueller
Information for both boaters and land-side visitors from the Afoot & Afloat series that covers more than 200 sites in the Gulf Islands.

North Puget Sound & the Strait of Juan de Fuca: Afoot and Afloat
by Marge and Ted Mueller
Coverage include Point Roberts to the south of Whidbey Island, plus the Strait of Juan de Fuca, from Neah Bay to Port Townsend. The travel guide is aimed at both boaters and land-based visitors.

The San Juan Islands: Afoot & Afloat by Marge and Ted Mueller
Whether approaching the San Juans by private vessel or ferry, this volume offers a unique twist on coverage of harbors, anchorages, and lots of quiet spots that it explores in these gorgeous islands.

Seattle's Lakes, Bays & Waterways:
Afoot & Afloat Including the Eastside by Marge and Ted Mueller
One of the only complete water guides to Seattle area, its focus is on places to explore by boat — big or small — or by foot, with maps and photos.

South Puget Sound and Hood Canal: Afoot & Afloat
by Marge Mueller and Ted Mueller
Includes the authors' favorite nooks and crannies and directs readers to parks, walks, marinas, historic sites and recreation facilities whether arriving by boat or car. Covers the South Puget Sound, from south of Seattle to Olympia, Everett to Alki Pt., west to include Hood Canal and east to Lake Union and Portage Bay.

GENERAL CRUISING GUIDES

Sensible Cruising — The Thoreau Approach
by Don Casey and Lew Hackler
A popular guide, both practical and philosophical, to cruising in a commonsense, make-do-with-what-you- have fashion. This book is full of sound, but gently humorous advice. Illustrated.

Seagoing Hitchhiker's Handbook by Greg Becker
Practical, well-researched guide to roaming the globe on other people's yachts: how to hitch a ride, where to do it, being a good crew, along with valuable tables listing most cruised routes, distances, storm and sailing seasons for those routes.

World Cruising Handbook by Jimmy Cornell
An essential reference for offshore cruisers covering all 185 maritime countries with up-to-date information on language, currency, customs and quarantine regulations, cruising permits, visas, port, medical facilities & more.

Advanced Blue Water Cruising by Hal Sutphen
What the prospective offshore sailor needs to know. Learn the pros and cons of every major equipment choice, troubleshooting systems, weather considerations, much, much more.

Advice to the Sealorn by Herb Payson
In 50 delightful, informative chapters, the author covers virtually every subject of concern to cruising sailors, from what it's REALLY like out there, to different rigs to repairs on board.

All in the Same Boat by Tom Neale
For cruisers who want to go offshore with their families for exploration and adventure but need to know how to provide for such things as home schooling, earning a living, staying in contact with home, provisioning, much more.

Bluewater Checklist by Rory Burke
A comprehensive portfolio of vital checklists covering every aspect of bluewater cruising from shopping lists and equipment inventories to safety procedures to going ashore or abandoning ship. Of great help in effectively managing any cruise.

Capable Cruiser by Lin and Larry Pardey
A classic manual for hands-on sailors. Discusses what every cruising sailor needs to know: preparation, maintenance at sea, safety aloft, seamanship, and staying healthy.

Cruising 101 by Amy Sullivan and Kevin Donnelly
Lively, real world review of cruising's pleasures and pitfalls. A good, anecdotal guide for the offshore bound.

Dragged Aboard: A Cruising Guide for the Reluctant Mate
by Don Casey
Don Casey carefully, cheerfully explains what life aboard is all about and lays to rest the questions and fears of the reluctant.

PROVISIONING GUIDES

Pacific Northwest Seafood Cookery by Stan Jones
Tested recipes for cooking the catch whether it be salmon, halibut, clams, oysters, cod, crab, crayfish and even kelp. 320 recipes include casseroles, soups, stews, salads, and spreads.

**Whale Watcher's Cookbook*: Views from the Galley* by Sharon M. Nogg
The author cooks for 20 in the tiny galley of a whale watching boat and shares the quick, simple recipes, born at sea, that she has developed.

Northwest Bounty: The Extraordinary Foods and Wonderful Cooking of the Pacific Northwest by Schuyler Ingle and Sharon Kramis
The authors offer over 275 recipes (oyster sausage, Hood River cream sauce with berries, and perfect leg of lamb among them) made with locally sourced ingredients — and add insightful commentary on local bounty.

The Northwest Best Places Cookbook: Recipes from the Outstanding Restaurants and Inns of Washington, Oregon, and British Columbia by Cynthia C. Nims and Lori McKean
A collection of the authors favorites from the region's most celebrated chefs.

Crab by Cynthia Nims
This is a tribute to the Northwest crab and describes how to prepare crab dishes of every description ranging from the old stand-bys to the daring and new. It includes over 40 recipes and 15 watercolor illustrations.

Salmon by *Cynthia Nims*
Author of <u>Crab</u>, Cynthia Nims, offers a wide range of ways to prepare another Northwest staple.

20,000 Gallons of Chowder by Ray Dunn
Inspired, time-tested, mouth-watering recipes from the legendary Colophon Cafe in Bellingham, Wash. Simple enough for any galley and loaded with fresh, sometimes surprising, ingredients, these recipes have attracted a rabid following. Author's stories are worth the price alone.

Provisioning by Dottie Haynes
Far more than a galley guide, this is a complete discussion of provisioning for every need in every situation as well as an introduction to cruising basics and living aboard for those new to sailing. Includes proven menus, advice on health, safety, seasickness, and much more.

Boat Cuisine: The All-Weather Cookbook by June Raper
This book features tested-at-sea recipes made from basic ingredients offered by an author who has cooked and cruised for many years. It is packed with information on cooking safely afloat, cooking in heavy weather, and all-important quick meals.

Guilt Free Gourmet and **Guilt Free Gourmet 2** by Sam Miles
Over 1200 low-fat recipes in each of two easily-stowable volumes — get the 70 Days of Menus, too — great for master lists and long-term cruise menu planning. The recipes are short, easily mastered and inventive.

The Care and Feeding of Sailing Crew by Lin Pardey with Larry Pardey
Practical, current information on all facets of bluewater galley provisioning, rough weather cooking, coping with differing dietary needs, seasickness, trash disposal, healthful eating, and much more.

Cooking the One Burner Way by Melissa Gray and Buck Tilton
More than 170 to simple to elaborate recipes for meals containing basic grains, rice and pasta. All recipes are tailored for the one-burner stove. Emphasis is on low-fat, high carbohydrate meals and snacks. Includes good advice on utensils, storage, and campsite (or galley) hygiene.

Cooking Under Pressure by Lorna J. Sass
Using pressure cookers in the galley saves fuel, keeps everything in the pot, and turns out delicious meals. Lorna Sass is the guru of the pressure cooker. This volume includes directions for everything from pot roast to ethnic favorites to cheesecake.

Cooks Afloat: Gourmet Cooking on the Move by David Hoar and Noreen Rudd
A well-crafted cookbook intended to teach the epicurean boater how to make gourmet meals using only a limited galley pantry and plenty of fresh-caught seafood. Includes salad alternatives, bread making, tips for canning; sprouting; harvesting shellfish, and more. Color photos throughout.

Cruising Cook by Shirley Herd Deal
The author covers how to buy, stow and prepare food for a month or more at sea. It is equally valid for an over-night cruise. Loose-leaf pages protected by an oversize plastic 3-ring binder to take abuse aboard.

From the Galleys of Women Aboard edited by Maria Russell
This is a wide-ranging collection of proven seagoing recipes from the Sea Sisters of Women Aboard. You'll find everything from salads to cookies.

Living Off the Sea by Charlie White
Charlie White offers wisdom from a master at living off the shoreline harvest. Even includes a chapter on survival living off the beach and discusses topics form red tides to exotic sea foods.

USEFUL PACIFIC NORTHWEST TRAVEL GUIDES

Exploring the BC Coast by Car by Diane Eaton & Allison Eaton
A user-friendly guide to things to see along the coast of British Columbia — with maps, directions, and insider's tips.

Pacific Northwest (Rough Guide) by Tim Jepson and Phil Lee
For the dynamic cities of Seattle and Vancouver, you'll find the stylish and critical hotel, restaurant, club and bar reviews invaluable. For those seeking adventure, the Rough Guide offers expert guidance into the great outdoors, including hiking trails and detailed maps.

The Dog Lover's Companion to the Pacific Northwest: The Inside Scoop on Where to Take Your Dog by Val Mallinson
Everything pet-friendly in the Pacific Northwest — from accommodations and off-leash parks to restaurants and pet boutiques — organized by region and paw-rated.

Beachcomber's Guide to Marine Life of the Pacific Northwest by Thomas M. Niesen, Michael K. Kunz, & David I. Wood
This comprehensive, well-illustrated guide to the marine species of the Pacific Northwest and their habitats also provides practical advice on how to spot and identify them.

Best Places Northwest by Giselle Smith
Features helpful three-day itineraries, and in-depth star-rated reviews of the best restaurants and lodgings throughout Oregon, Washington, and British Columbia. Useful for those forays from the boat.

Northwest Budget Traveler by Nancy Leson
This guide to inexpensive places to eat and stay in Washington, Oregon, and British Columbia includes deals and discounts on major attractions, shopping, concerts, theater tickets, tours, festivals, and more.

The Romantic Pacific Northwest (Romantic America) by Ken Christensen
Features romantic traveling tips, and lists unique hotels, resorts, B&Bs and restaurants in British Columbia, Washington, Oregon and northern California.

Hidden Pacific Northwest: Including Oregon, Washington, Vancouver, Victoria, and Coastal British Columbia by Eric Lucas
From the region's best-known destinations, to quirky gems, it covers Seattle, Vancouver, Victoria plus many coastal hamlets. Over 300 hikes as well.

Adventure Guide to the Pacific Northwest by Don Young and Marjorie Young
From killer whales (Orcas) in Washington's San Juan Islands to llama trekking camping trips to the high country, there are lots of ideas for excursions from or on the boat. Covers Oregon north to Victoria and Vancouver. A little old, but still useful

Kayaking the Inside Passage: A Paddling Guide from Olympia, Washington to Muir Glacier, Alaska by Robert H. Miller
The first two regions in the book mirror those areas covered in this Guide — Lower Puget Sound, Upper Puget Sound, the San Juan Islands, the Gulf Islands to Nanaimo, and then the Sunshine Coast. While most cruisers are not going to follow this itinerary by kayak, it provides many interesting excursions for local kayak rentals as well as destinations and cruises that might be equally interesting to visit in the big boat or the dinghy.

Ferry Travel Adventures in Washington, British Columbia & Alaska by Michael F. O'Malley
Short descriptions of public campgrounds; hiking and biking trails; fishing, diving, kayaking, and boating hot spots; and other locally significant attractions near ferry terminals.

British Columbia (Lonely Planet)
by Ryan Ver Berkmoes & Graham Neale
The book whole province is covered but about half the book is devoted to the same regions included in this Guide. The eastern coast of Vancouver Island, the Gulf Islands, Sunshine Coast, Vancouver and Victoria. Itineraries — some of which would also work in the boat — and maps are plentiful and very good — with lots of details.

The Seattle & Vancouver Book, A Complete Guide:
Includes the Olympic Peninsula, Victoria & More (Great Destinations)
by Ray Chatelin
A good onboard city guide evenly divided between Seattle and Vancouver with in-depth coverage of neighborhoods, hotels, eateries, shopping and attractions — with a big focus on recreation and adventures. Side trips to Tacoma and Victoria offer more options for cruisers heading to those cities.

Vancouver, Victoria & Whistler — A Ulysses Travel Guide
Most of the book is focused on thorough coverage of Vancouver with just a short — but surprisingly complete — chapter on Victoria. In addition to the usual nuts and bolts, there are ten compelling tours, with good maps, that cover every area of Vancouver.

Vancouver City Guide (Lonely Planet) by Karla Zimmerman
Detailed maps of six Walking and Cycling tours of the city's main neighborhoods are a treat. Lots of eatery reviews including "Cheap Eats," plus shopping tips, entertainment, nightlife and recreation options. Excursions to Victoria, Bowen Island and the Gulf Islands are covered, briefly, in the last chapter.

Lighthouses of the Pacific Coast: Your Guide to the Lighthouses of California, Oregon, and Washington by Randy Leffingwell
A beautifully illustrated guide that provides lots of facts and historical details about lighthouses on the Pacific coast.

50 Hikes in Washington: Walks, Hikes, and Backpacks in the Evergreen State by Kai Huschke
Some of the hikes are along the coast or within easy travel distance of it — particularly Olympic National Park, Moran State Park and Columbia River. Detailed Maps.

Northwest (Lonely Planet) by John Doerper and Greg Vaugn
Doerper knows the region well and manages to weave in apple, cheese, and oyster tours of Washington; a guide to Northwest wineries; literary and historical extracts; as well as topical essays on subjects ranging from potlatch giveaways to Columbia River explorations to the cuisine of the West.

Absolutely Every* Bed and Breakfast in Washington (*Almost)
edited by Carl Hanson
A guide to more than 500 B & B's with details of each. Includes Victorian inns, historic hotels, ranches, beach cottages, ski chalets from the mountains to the ocean beaches. Perfect for a night ashore when it's time for a shower with water pressure and a bed that doesn't rock.

Hiking the San Juan Islands by Ken Wilcox
Island hikes and walks in San Juan, Skagit and Island counties. Details of parks, viewpoints, water access and campgrounds. Photos, maps.

Essential San Juan Guide by Marge and Ted Mueller
Lively guide to accommodations, restaurants, shopping, festivals, fairs, kayak and bike rentals, and much more with descriptions, addresses, phone numbers, maps, and photos.

Seattle City Guide (Lonely Planet) by Becky Ohlsen
Compact enough to just tuck it in your pocket when you leave the boat. Overviews of each neighborhood are complemented by restaurant reviews, and one and three-day itineraries. The themed walking and biking tours include detailed maps. Excursions briefly cover the San Juan, Vashon and Bainbridge Islands.

Irreverent Guide to Seattle and Portland (Frommer's)
by James Gullo and Arthur Frommer
Insider's perspective on each destination that takes a fun, honest look at everything they review, and includes background, trivia, and local facts.

Oregon: An Explorer's Guide by Mark Highberger
About half the book is devoted to regions covered by the ACC's Guide to Pacific Northwest Marinas — North Coast, South Coast, the Lower Columbia River and the Mid-Columbia River region. The coverage has been labeled encyclopedic — and it is, which makes it a good on-board companion when cruising Oregon. Includes historic insights, including the wake of Lewis & Clark, suggested car excursions, and favorite destinations.

50 Hikes In Oregon: Walks, Hikes, and Backpacking Adventures from the Pacific to the High Desert by David L. Anderson
About the first half of the book covers hikes right along the coast and up the Columbia River but almost every one will require a car to get to the trail head from harbors or marinas.

The Photographer's Guide to the Oregon Coast: Where to Find Perfect Shots and How to Take Them by David Middleton, Rod Barbee
The same rugged beauty of the Oregon Coast that makes it harsh on us as boaters also compels us as photographers. There are specific vantage points and tips for getting the best shots at most of the landfalls covered by ACC. And most of the pointers are applicable to all coastal photography. The sample photos are impressive and inspirational.

PACIFIC NORTHWEST HISTORY and CULTURE

Native Peoples of the Northwest by Jan Halliday and Gail Chehak
More than 1,000 things to see and do with Native American people from all 54 tribes in Western Montana, Idaho, Northern California, Washington, Oregon, British Columbia and Southeast Alaska. Find out how and where to attend a "powwow", buy authentic traditional art, or raft the Trinity River with a member of the Hupa Tribe and more.

Paul Kane's Great Nor-West by Diane Eaton
Paul Kane (1810-1871) was a Toronto-born, and largely self-taught, artist who resolved early to make a lasting record of North American Indians and their traditional way of life. The central event of his career was a journey of two-and-half years through the wilderness, from the Great Lakes, through the Red River settlement (where he saw and painted one of the last great buffalo hunts), the Rockies, and into the Pacific Northwest.

Pacific Northwest — An Interpretive History by Carlos A. Schwantes
The economic and social history of the Pacific Northwest (Washington, Oregon, Idaho) from the time of the first white-Indian contact to 1987. A comprehensive and balanced history, which demonstrates that the region is indeed distinctive.

The Call of the Coast edited by Charles Lillard
A fascinating collection of essays about the waters of the Pacific Northwest, dating back to the 1850's.

A Guide to the Indian Tribes of the Pacific Northwest by Robert H. Ruby
An overview of the peoples of the Northwest and their history, from fur-trading days to modern legal battles. Includes 173 tribes from the Ahantchuyuk to the Yoncallla.

Myths and Legends of the Pacific Northwest
by Katharine B. Judson (editor)
First published in 1910 and with a new introduction by Jay Miller, this book collects the oral traditions of the Klamath, Nez Perce, Modoc, Chinook and other tribes of the Pacific Northwest. It's a wonderful introduction to the traditional mindset and importance of the land to Native Americans.

Ghost Stories from the Pacific Northwest by Margaret Reed MacDonald
Stories of friendly, funny, and mean ghosts from the Pacific Northwest, including some famous inhabitants of lighthouses and boats.

Plants Of The Pacific Northwest Coast: Washington, Oregon, British Columbia & Alaska by Jim Pojar & Andy MacKinnon
A superb, detailed, fairly comprehensive field guide, that's easy to use with clear descriptions and photos. A must for nature buffs.

The Good Rain by Timothy Egan
Seattle correspondent for The New York Times journeys through the Pacific Northwest, from manicured gardens in Vancouver, B.C., to the precipitous peaks and brooding volcanoes of the Cascade Mountains.

Phantom Waters by Jessica A. Salmonson
Water is the flowing thread in this eerie collection of Native American myths and contemporary short stories from Washington, Oregon, Idaho, Montana, and British Columbia.

Bright Seas, Pioneer Spirits by Betty C. Keller and Rosella M. Leslie
An insightful look at Sunshine Coast's transformation from the home of fishermen and loggers to the popular getaway destination of today.

The Pacific Northwest Coast:
Living With the Shores of Washington and Oregon by Paul D. Komar
This book offers a scientific exploration of natural phenomena and human activity shaping the Pacific Northwest coast.

Rains All the Time: A Connoisseur's History of Weather
in the Pacific Northwest by David Laskin
A Seattle author's historical perspective on weather in the Northwest — from early accounts to modern day radar and satellite observations.

Astoria; or, Anecdotes of an Enterprise Beyond the Rocky Mountains
by Washington Irving
A history of John Jacob Astor's Pacific Fur Company, which he started in Fort Clatsop, Oregon, and sold to British traders during the War of 1812.

A Prairie Chicken Goes to Sea by Margo Wood
This autobiography tells the story of a country girl who married Charlie Wood, the man responsible for Charlie's Charts. Her tale includes their life together and how she decided to carry on with the guides after he died.

FICTION THAT PROVIDES A SENSE OF PLACE

Onions in the Stew by Betty McDonald
The author of the Pacific Northwest classic, The Egg and I, describes how, with husband and daughters, she set to work making a life on rough-and-tumble Vashon Island in Puget Sound in the years following World War II.

Snow Falling on Cedars by David Guterson
Set in the 1950's on an island in the straits north of Puget Sound, where everyone is either a fisherman or a berry farmer, the story is nominally about a murder trial. But, lingering memories of World War II, internment camps and racism fuels suspicion of a Japanese-American fisherman, a lifelong resident of the islands.

The Pen and the Key: 50th Anniversary Anthology of Pacific Northwest Writers edited by Nigel Loring
A collection of writings from 23 celebrated Pacific Northwest authors — includes fiction, poetry, and non-fiction.

Northwest Passages: A Cascadian Anthology edited by Cris DiMarco
This compilation of science-fiction and fantasy stories was carefully selected by the editor from the work of 25 Pacific Northwest writers.

S.O.S: Chilling Tales of Adventure on the High Seas
edited by Sara Nickles
A collection of 19 stories of adventure at sea. Selected contributors: Joseph Conrad, Stephen King, Gabriel Garcia Marquez, Sebastian Junger, Paul Theroux, Ernest Hemingway.

The River Why by David James Duncan
A wonderful novel about a young man who sets out to follow his dream of fly fishing at a remote Oregon cabin, and ends up finding a new direction, romance, and ultimately his true identity.

Comfort Food: A Novel by Noah Ashenhurst
A memorable first novel that follows the adventures of six young people for more than a decade, taking the reader from Bellingham to distant places like Nepal and Budapest.

The Half Life: A Novel by Jonathan Raymond
The beauty and history of Oregon come to life in the two stories of friendship that unfold in this novel — one of two young men in the 1820's, and another of two young women in the 1980's.

Ravensong: A Novel by Lee Maracle
The challenges facing a young Native American woman in the 1950's and her struggle to find answers leads to reflection on traditional values and new customs.

To Build a Ship by Don Berry
Recently re-printed, this is the fascinating story of several Oregon pioneers who become isolated and are forced to build a schooner in the Tillamook Bay in the 1850's.

Dream Keeper: Myth and Destiny in the Pacific Northwest
by Morrie Ruvinsky
Past and present, Native myth and Western storytelling intertwine in this book about the son of a white explorer who lives for over two hundred years after the Sisters of Creation put a curse on him.

The Highest Tide by Jim Lynch
A teenager sneaks out on his kayak to explore sea creatures at night and sees a live giant squid. Instant celebrity doesn't change everyday challenges and Miles grows up as the Puget Sounds records the highest tide in fifty years.

The Sea Runners by Ivan Doig
Four Scandinavian men escape from a work camp in Alaska and travel in a canoe all the way to the Columbia River in search for a new beginning. Based on a true story from the 1850's.

Timber: A Novel of Pacific Northwest Loggers
by Roderick Langmere Haig-Brown
The story of a friendship between two men and a woman they both love, set in the Northwest woods during the heyday of steam logging.

Northern Escape by R. L. Coffield
After taking part in a robbery he didn't want to commit, Max escapes to Alaska and starts a new life in the fishing business. Eleven years later, his partner and a relentless detective both catch up with Max and everything falls apart.

Blue Poppy (Pacific Northwest Mysteries) by Skye Kathleen Moody
In this mystery novel a Fish and Wildlife agent investigates a series of murders that take place on the Olympic Peninsula, near a perfumery that creates the sought-after Blue Poppy fragrance.

Red Herring by Clyde W. Ford
After a career with the Coast Guard, Charlie Noble becomes a private investigator who navigates the familiar grounds of the San Juan Islands to solve a modern-day mystery of terrorist threats, conspiracy, and murder.

CHILDREN'S BOOKS — Pre-School

Where Do I Sleep? A Pacific Northwest Lullaby
by Jennifer Blomgren, Andrea Gabriel, Illustrator
A great bed-time book about the habitat of various animals found on the coast of the Pacific Northwest.

O Is for Orca: An Alphabet Book
by Andrea Helman, Art Wolfe Photographer
A Northwest-inspired approach to the alphabet, accompanied by Art Wolfe's beautiful nature photography.

1, 2, 3 Moose: A Pacific Northwest Counting Book
by Andrea Helman, Art Wolfe Photographer
Another book from the Wolfe-Helman team, with captivating nature photography and simple text for kids.

Northwest Animal Babies by Andrea Helman, Art Wolfe Photographer
Informative and fun, with photos and lots of interesting facts about land animals and sea creatures of the Northeast.

Ten Little Fish by Audrey Wood, Bruce Wood Illustrator
Count down from 10 with bright, boldly colored fish and easy rhymes.

The Deep Blue Sea: A Book Of Colors
by Audrey Wood, Bruce Wood Illustrator
Help little ones learn colors with memorable illustrations and short rhymes.

Can You See What I See? Seymour and the Juice Box Boat
by Walter Wick
In this whimsical search-and-find story, readers join a little toy man as he gathers together an assortment of items, builds a boat out of a juice box, and sails away on a sea of blue.

CHILDREN'S BOOKS — Ages 4 – 8

Hood River Home by Herb Marlow
Arriving in the Hood River Valley of Oregon in 1965, George Rainy, a migrant fruit worker, is hoping to find a home for his family.

Orca's Family and More Northwest Coast Stories
by Robert James Challenger
A wonderful collection of simple fables. Children learn invaluable lessons along with creatures of the West Coast.

Nature's Circle and Other Northwest Coast Children's Stories
by Robert James Challenger
Another volume of memorable lessons from writer Robert James Challenger. Illustrated by the author

Salmon Forest by *David Suzuki, Sarah Ellis, Sheena Lott* Illustrator
A biologist. takes his young daughter for a walk through a West Coast rain forest. Kate learns about the life cycle of salmon and how it relates to the life of the forest and that of the Native people in the area.

No Dear, Not Here: The Marbled Murrelets' Quest for a Nest in the Pacific Northwest by Jean Davies Okimoto, Celeste Henriquez Illustrator
Popular tourist sites of the Pacific Northwest are seen through the eyes of a couple of marbled murrelets looking for a nesting spot

Sacagawea: Northwest Explorer
by Dennis Brindell Fradin, Nora Koerber Illustrator
The story of Sacagawea, the Shoshoni girl who helped Lewis and Clark on their expedition

Northwest Coast Indians Coloring Book by David Rickman
Depicts the lifestyles and lost culture of tribal life from the late 18th to early 20th century — all 33 images meticulously researched by a museum curator.

Lewis and Clark Expedition Coloring Book by *Peter F. Copeland*
Precisely rendered, accurate depictions of the expedition in 45 illustrations.

CHILDREN'S BOOKS — Ages 9 – 12

First Salmon by Roxanne Beauclair Salonen, Jim Fowler, Illustrator
While participating in First Salmon, an annual sacred ceremony of his Native American people, Charlie realizes he misses his late uncle who taught him how to fish.

The Coast Mappers by Taylor Morrison
About mapping of the US Pacific Coast and the man who took on the assignment — George Davidson. Interwoven throughout the compelling and dramatic story cartographic methods are clearly described.

Escaping the Giant Wave by Peg Kehret
A tsunami caused by an earthquake strikes the Oregon coast where a family from Kansas is vacationing. Kyle and BeeBee, alone at the hotel, escape and even help Daren the bully make his way out.

Seaman: The Dog Who Explored the West with Lewis and Clark
by Gail Langer Karwoski, James Watlinf, Illustrator
A big, black Newfoundland dog named Seaman accompanied Merriweather Lewis — this is a fictionalized version of that true story. Some of the harsh details are softened, but Seaman, who is mentioned more than 30 times in the real diaries, provides an engaging entry into the history of this adventure.

Storm Boy by Paul Owen Lewis
In this book, inspired by a Haida Indian story, a boy is caught up in a storm on his canoe and ends up on the shore of an unknown village. He spends a year learning the traditions of a different People and teaching them about his own culture, before returning to his own town.

Living Lightly on the Water — Clean Marina & Vessel Programs

British Columbia has an estimated 400,000 pleasure craft -- most of them in the Strait of Georgia, and the Puget Sound is home to almost 200,000 power and sail vessels. The coastal waters of Washington and Oregon are also increasingly popular with recreational boaters. This creates a situation rife with obvious environmental challenges, particularly to waterfront facilities.

Keeping the waters in which we cruise clean, healthy and safe has become a major priority in North American coastal states and provinces. The maintenance, operation and storage of recreational vessels have the potential to pollute adjacent waters, impair air quality and lead to general environmental degradation. Contaminants include dust from hull maintenance operations, solvents from engine repair shops, petroleum from careless fueling practices, sewage discharges from boats, and heavy metals from antifouling paints. These pollutants may be deposited directly into waterways or they may be carried in by storm water runoff.

Certified Clean Marina Programs

The U.S. National Coastal Management Program is a federal-state partnership dedicated to comprehensive organization of the nation's coastal resources and, at the federal level, operates under the aegis of the NOAA's Coastal Programs Division. One of its major initiatives is the Certified Clean Marina program, which is interpreted and managed at the state level. The aim is to prevent pollution rather than clean it up. When a marina decides to participate, it is provided assistance in complying with environmental laws and becomes eligible for financial aid in the form of Incentive Grants Educational materials and workshops on ecologically sound management are made available both to the marina's personnel and to its clientele. After performing a self-evaluation, the facility adopts simple, innovative approaches the day-to-day operations that affect the environment, including:

- locating new or expanding marinas
- marina design, maintenance, management
- storm water management
- vessel maintenance and repair
- petroleum control
- sewage handling
- waste containment and disposal
- laws and regulations
- boater education

After a successful review by a team of the home state's Clean Marina assessors, the facility is granted a Certified Clean Marina designation by the DEP and is awarded a flag that can be hoisted at the marina, to advertise its accomplishments and status. When you see flag, you know that you're dealing with a marina which has gone the extra mile to keep fuel, sewage and hazardous chemicals out of the waters in which you cruise, fish and swim. Each U.S. marina in the *Atlantic Cruising Club's Guide to Pacific Northwest Marinas* that has met this rigid standard is recognized in Marina Notes with a **CCM** notation Canada has a growing **Pollution Prevention at Marinas** program which provides facilities with information on reducing the environmental impact of marina operations. We urge you to support the facilities that have worked so hard to ensure that they "do no harm."

No Discharge Areas, the Canada Shipping Act and the U.S. Clean Vessel Act

Both the U.S. and Canada have recognized the impact of recreational vessel overboard discharge and have taken steps to encourage boaters to behave in their own self-interest by protecting the waters they play in. While boaters contribute only a small portion of the pollution in these coastal regions, it is most often concentrated in inshore areas and small confined bays which are refreshed by larger bodies very, very slowly — if at all.

Properly "pumping out" can make a big difference in the coastal environment, because sewage microorganisms have a threefold impact: they can be visually distasteful, present a health hazard, and lower oxygen levels necessary for aquatic life in water. Although aesthetic revulsion is the most visible, it is the least important ecologically. Microorganisms contained in human sewage can cause infectious hepatitis, diarrhea, bacillary dysentery, skin rashes and even typhoid and cholera. The most common organism is coliform bacteria — found in the intestines of all warm-blooded animals. Fecal coliform, including E. coli bacteria, can increase from one bacterium to over 10 million in the 12 – 18 hour normal digestive time. Indications that a region is under stress are shellfish contamination, red tides (and other algae blooms) and declining fish populations. Biological Oxygen Demand (BOD) is the amount of oxygen that bacteria take from water to oxidize organic matter into carbon dioxide and water. The marine ecosystem goes out of balance as soon as an external source of oxygen demand is added (i.e. direct discharge of human waste).

The Canada Shipping Act's Pleasure Craft Sewage Pollution Prevention Regulations prohibits the discharge of raw sewage from pleasure boats in certain Canadian bodies of water. As of publication, eleven of British Columbia's waters have been so designated by The Ministry of Water, Land and Air (see next page for a list). Dozens of additional harbors have been nominated and are currently under review. In British Columbia, the Georgia Strait Alliance (GSA) also works to protect and restore the unique marine environment of the Strait and created a Green Boating Program.

In the US, The Clean Vessel Act was passed in 1992 and is administered by the U.S. Fish and Wildlife Service. It provides grants for the installation and operation of pump-out stations at marine facilities and to educate boaters on proper sewage disposal. Since its inception, the program has awarded states over $120 million. New grants totaling $12 million, to be awarded to 32 states, were announced in July 2006 — including $1,000,000 for the state of Washington, and over $390,000 for Oregon. Federal funds can be used for up to 75% of the cost of individual projects; State funds are often awarded to marinas as well. Since October 1994, U.S. boaters have been prohibited from discharging raw sewage into fresh water or within coastal salt-water limits (nine nautical miles for estuaries and three nautical miles on the Oceans). Currently six States have all of their waters designated as No Discharge Zones. In addition, 11 other States have segments of their waters designated as NDZs. Neither Washington nor Oregon are on either list.

Pump-Out and Clean Marina List

LOCATION	MARINAS WITH PUMP-OUT FACILITIES
BRITISH COLUMBIA	
Bowen Island	Union Steamship Company Marina
Bowser	Deep Bay Harbour Authority
Campbell River	Brown's Bay Marina
	Fisherman's Wharf
Comox	Comox Valley Harbour Authority
Cowichan Bay	Fishermen's Wharf Association
Delta	Captain's Cove Marina
Gibsons	Gibsons Landing Harbour Authority
	Gibsons Marina
Lund	Lund Harbour Authority Public Docks
Madeira Park	Madeira Park Public Wharf
Mill Bay	Mill Bay Marina
Nanaimo	Cameron Island
	Nanaimo Port Authority
Nanoose Bay	Fairwinds Schooner Cove
Parksville	French Creek Boat Harbour
Saltspring Island	Centennial Wharf
Sidney	Port Sidney Marina
	Van Isle Marina
Vancouver	Bayshore West Marina
	Coal Harbour Marina
	False Creek Yacht Club
Victoria	Coast Harbourside Hotel/Marina
	James Bay Causeway

The No Discharge Zones in British Columbia:

▸ Carrington Bay

▸ Cortes Bay

▸ Gorge Harbour & Mansons Landing

▸ Montague Harbor

▸ Pilot Bay

▸ Prideaux Bay

▸ Rosco Bay

▸ Smuggler Cove

▸ Squirrel Cove

▸ Victoria Harbour

LOCATION	MARINAS WITH PUMP-OUT FACILITIES
WASHINGTON	
Allyn	Port of Allyn North Shore
	Port of Allyn Waterfront Park
Anacortes	Cap Sante Boat Haven
	Skyline Marina
Bainbridge Island	Bainbridge Island Marina
	Eagle Harbor Marina*
	Eagle Harbor Waterfront Park
	Harbour Marina
	Winslow Wharf Marina
Bellingham	Squalicum Harbor
Blaine	Blaine Harbor
	Semiahmoo Marina
Bremerton	Port of Brownsville*
	Port Washington Marina*
Brinnon	Pleasant Harbor Marina
Chinook	Port of Chinook
Deer Harbor	Deer Harbor Marina
Des Moines	Des Moines Marina
Edmonds	Port of Edmonds
Everett	Port of Everett Marina
Friday Harbor	Port of Friday Harbor
Gig Harbor	Arabella's Landing
	Jerisich Dock
Ilwaco	Port of Ilwaco
Kingston	Port of Kingston
Kirkland	Carillon Point Marina*
La Conner	La Conner Marina
La Push	Quileute Marina
Lakebay	Penrose Point State Park
Langley	Langley Boat Harbor
Lopez Island	Islands Marine Center
Neah Bay	Makah Marina
Nordland	Mystery Bay State Park
Oak Harbor	Oak Harbor Marina
Olympia	Percival Landing Park
	Swantown Marina
	West Bay Marina
	Zittel's Marina
Orcas	West Sound Marina
Point Roberts	Point Roberts Marina Resort
Port Angeles	Port Angeles Boat Haven
Port Hadlock	Port Hadlock Marina
Port Ludlow	Port Ludlow Marina
Port Orchard	Bremerton Marina*
	Port Orchard Marina

* Certified Clean Marina ("CCM")

LOCATION	MARINAS WITH PUMP-OUT FACILITIES
WASHINGTON, con't.	
Port Townsend	Point Hudson Resort & Marina
	Port Townsend Boat Haven
Poulsbo	Port of Poulsbo
Quilcene	Quilcene Marina
Roche Harbor	Roche Harbor Resort
Seattle	Ballard Mill Marina
	Bell Harbor Marina*
	Elliott Bay Marina*
	Fairview Marinas
	Fishermen's Terminal
	Shilshole Bay Marina
Sequim	John Wayne Marina
Shelton	Jarrell Cove Marine State Park
	Jarrell's Cove Marina
	Port of Shelton
Silverdale	Port of Silverdale
Tacoma	Breakwater Marina
	Chinook Landing Marina*
	Dock Street Marina*
	Foss Landing Marina*
	Foss Waterway Marina*
	Ole and Charlie's Marina
	Point Defiance Boathouse Marina
Union	Alderbrook Resort & Spa
	Twanoh State Park
Westport	Westport Marina
OREGON	
Astoria	Port of Astoria, West Basin**
	Port of Astoria, East Basin (coming soon)
Bandon	Port of Bandon**
Brookings	Port of Brookings Harbor
Charleston	Charleston Marina**
Coos Bay	Coos Bay City Dock
Depoe Bay	Depoe Bay Marina**
Florence	Port of Siuslaw Marina
Garibaldi	Port of Garibaldi**
Gold Beach	Port of Gold Beach
Newport	Embarcadero Resort & Marina
	Port of Newport Marina
Warrenton	Skipanon Landing
	Warrenton City Mooring Basin
Winchester Bay	Salmon Harbor Marina

** Pledged to become Certified Clean Marina

Getting OnLine: Wi-Fi, Ethernet, EV-DO ... and Dial-Up

The ability to access both the web and email while cruising has become increasingly important to boaters. Marinas are providing a variety of solutions — some better than others. In each Marina Report, there is a brief description of the type and location of data ports (broadband or dial-up — either at the marina or its environs) and also notes the availability of Wi-Fi — and details the terms of use for each.

Ethernet: As in many hotels, some marinas have a stationary data center with one or more broadband Ethernet cables or plugs. This is high speed access that requires an Ethernet cable (which is a good thing to have onboard in any event); just plug in and you are on-line. Pricing for this varies from free to an hourly rate — from free to $. In the Pacific Northwest, this is usually offered in resort marinas with a major hotel component or in more remote locations that may have a satellite connection. Dial-up is offered the same way, but it is very, very slow and only good for text email downloads.

Wi-Fi or Wireless Fidelity: This is the most useful and cost-effective of the currently available remote access technologies. It allows boaters to connect via their own laptops in a line of sight coverage area extending throughout the marina and the adjacent harbor. It's wireless, provides broadband speed, doesn't require expensive equipment or a cumbersome installation, and can be used anywhere within the facility — as long as the computer (or its antenna) is within range of a base station (that area covered is called the "HotSpot" or "HotZone"). The computer must either have an internal Wi-Fi card or external USB device. And, while most laptops come with a Wi-Fi card installed, in the Northwest waters particularly, it would be prudent to upgrade to a more powerful marine-grade card and add an external antenna. Wi-Fi technology is line of sight so as long as you can see the antenna — you can usually get connected. The Wi-Fi broadcast antenna is almost always at the top of the dock perched on top of the marina office. Speed is usually 30-50 times dial-up but can run up to 100 times dial-up.

How Wi-Fi works: The marina or the community contracts with a Wi-Fi provider, which sets up HotSpots or larger HotZones that transmit and receive signals -- some of these HotSpots provide open access and are paid for by the marina or community; others are installed by a Wi-Fi provider which then sells boaters a subscription for access. (The overwhelming majority of marine hotspots in the Pacific Northwest are subscription based.) Once you notice that a hotspot has been identified by your computer, just open your Internet browser; you will either be online, compliments of the marina, or will be prompted to sign up for a subscription. At publication, subscription per diem rates were about $11 or $80 for a single month down to $25 per month based on an annual contract. The Marina Reports indicate if the Wi-Fi service is provided by a subscription service — and which one the marina has selected. That way the boater can choose a provider that services the greatest number of marinas within a given cruising area. In the Pacific Northwest, the largest provider to marinas is BroadbandXpress — with over 100 locations from Puget Sound to Alaska.

Up-Grade the Boat's Wi-Fi Device and Antenna: Built-in Wi-Fi or retail cards generally have a power of 30 Milliwatts and a range of 150-300 feet from the access tower in the marina — which works just fine if you are dock-side and close-in (ask where the marina's antenna is located). But to have a really positive experience when using Wi-Fi throughout the marina, its mooring field, and the anchorage beyond, it is advisable to trade up to a more powerful Wi-Fi device and upgrade to a separate antenna — inside the boat or permanently mounted outside. Consider upgrading the power of the Wi-Fi card or USB device to a 200 -370 Mlw card. The next step would be to upgrade the antenna to either a good quality interior antenna or a marine mounted exterior antenna. This upgrade will increase access range to 1500 feet with a boat internal antenna to 1-2 miles with an external antenna. Be sure you purchase marine grade hardware for longer durability. The cost of upgrade packages can range from $130-$400.

EV-DO — The Alternative to Wi-Fi: EV-DO is a relatively new, but pricey, technology that uses a cellular data network called C.D.M.A. 1xEV-DO (Code Division Multiple Access Evolution-Data Only). Verizon has dubbed their version the Broadband Access Plan. If you are cruising in and around cities where cell service is good, this may be a viable, albeit more expensive, option. And, you can use it "on the move" — as long as you have cell service, you have high-speed broadband of 600 kbps in major cities — akin to a cable modem. Sprint also offers this service but reports are their network and speeds are not comparable to Verizon. EV-DO requires a special PC card — Kyocera and Novatel are two manufacturers — that goes for about $50-70. The problem is the monthly fee. At publication it was $80 for 450 minutes to $170 for 4000 minutes — nights and week-ends are unlimited. If budget is not a consideration, the EV-DO should be part of your communications package. Check with the U.S.-based service providers before cruising in Canada as some plans may incur roaming charges as high as $2.00 per megabit of data transfer. In addition, there are significant cell dead zones where there is no service — phone or data.

Access Off the Boat: If the marina you've selected doesn't have a Wi-Fi system or Ethernet dataports, and you don't have EV-DO, then check out the local coffee houses, libraries and hotels. A search on www.wi-fihotspotlist.com or www.wififreespot.com will tell you where the hot-spots are and the charges. Libraries are almost always free but have a firm limit on time. T-Mobile is one of the biggest Hot-Spot providers to hotels, coffee shops (Starbucks), airports, libraries, etc. Boingo is a huge accumulator of Hot-Spots but doesn't actually install them.

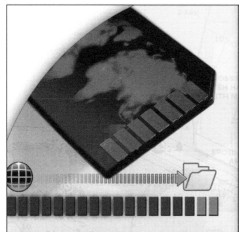

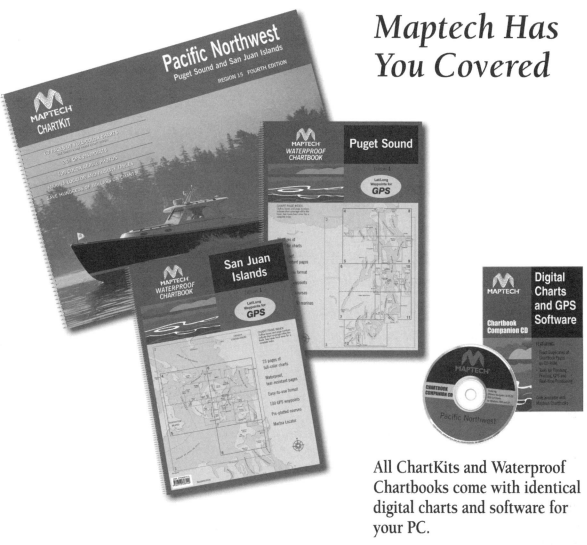

BOATERS' RESOURCES

WWW.MAPTECH.COM

Broadband Xpress

BOATERS' RESOURCES

WWW.BBXPRESS.NET

MarinaLife Concierge

BOATERS' RESOURCES

WWW.MARINALIFE.COM

Alphabetical Listing of Marinas, Harbors and Cities

About the Authors & Editors

CATHY HADEN and CHRIS HADEN

Chris and Cathy Haden have spent a lot of time "messing about in boats." They have owned a succession of boats; both power and sail, over the years. They have bare boat chartered in the San Juan Islands, gone house boating on Lake Powell and the Sacramento River Delta and cruised the California coast in their Grand Banks 48, *E Ticket*. One of their fondest memories is a weeklong cruise through southeast Alaska aboard a 65-foot Malahide trawler. Travel has always been an important part of their lives, and they have visited Germany, France, Holland, Japan, Jamaica, Thailand, Singapore, Canada and Mexico, and cruised through the Panama Canal as well. The *Atlantic Cruising Club's Guide to Pacific Northwest Marinas* is their first book together, although both have published books and/or articles individually.

Chris Haden retired from a career in the military in 2000. He saw service in Vietnam with the First Infantry Division and later joined the California Army National Guard as a full-time active duty member. His assignments led him to many far-flung destinations over the years, including Panama, Honduras, and Bangladesh. After retiring, Chris turned his hand to freelance writing and photography, with particular interest in the nature and boating fields. Catherine Haden spent her working career in the hospital patient administration field, including service on the Board of Directors of the California Association of Hospital Admitting Managers for many years. She also found time to serve as the technical editor for a medical book publisher. She followed husband Chris into retirement in 2003. Chris and Catherine have two sons, one now serving with the U.S. Army and one pursuing a career in hospital administration.

BETH ADAMS-SMITH and RICHARD Y. SMITH

Beth and Richard have also been "messing about in boats" at various times throughout their lives, and, for the last two plus decades, have been focused on "big boats." Coastal cruises have included numerous trips up and down the U.S. eastern seaboard, including the ICW, where they perfected the art of "achievable cruising" — cruise for a week or two, leave the boat, go home, come back a month or two later and continue on. Other adventures have taken them to Bermuda, the Bahamas, the Caribbean, the South Pacific, Pacific Northwest, the Baja, South America and the Mediterranean. They are firm believers in "going" even if it means taking the office along and technology has made that possible. Along the way, they've visited more than 3,000 marinas (most two, three or more times) on the east and west coasts of the U.S. as cruisers, authors and editors. They currently own and cruise a 40 foot Leopard catamaran and a 26 foot Glacier Bay powerboat.

Beth is the primary writer/photographer/editor of the *Atlantic Cruising Club's Guides to Marinas*. She is also a new media producer and has spearheaded the design of the print and electronic versions of the *Guides*. With Richard, she has co-managed the development of Jerawyn's proprietary Datastract™ publishing technology, and supervised the website programming team. In addition to her work on the *ACC Guides*, Beth is incubating the Water Lovers website (a travel resource for people with a passion for the water) and the Healthy Boat series (holistic approaches to crew lifestyle and vessel management). As the "Wandering Mariner," she writes articles on cruising, destinations and waterborne and water-oriented travel. In addition, Beth has produced numerous documentaries and multiple media projects on holistic living, ecology, peacemaking, and complementary medicine. She holds a doctorate in Technology and Media and a masters in Health Education, both from Columbia University, and a bachelors in Broadcast Journalism from Boston University.

Richard is publisher/editor of Jerawyn Publishing Inc., an umbrella company for the Atlantic Cruising Club and other publishing imprints focused on coastal lifestyles, maritime-oriented travel and holistic living. JPI's activities often involve the creation and publication of extensive databases of both factual and editorial information — including a library of over 20,000 nautical images. He is also president of Evergreen Capital Partners Inc. which undertakes merchant banking transactions both as an advisor and as a principal. Richard received a B. A. degree from Wesleyan University and an M.B.A. from Harvard Business School. When not reviewing marinas, editing books, attempting to close private equity deals or out cruising, he is playing bass in the venerable 60's rock 'n' roll band, "Gary and the Wombats."

Order Form
Please ask for the *Guides* at your local book store or chandler or order directly.

☐ Please register me as a member of the Atlantic Cruising Club so that I can benefit from even greater discounts and email specials.

☐ Please send me the following *Atlantic Cruising Club's Guides to Marinas — as soon as each is available.*

For Discounts, log onto www.AtlanticCruisingClub.com
Complimentary 6-month Silver Membership for purchasers of any ACC Guide to Marinas

_____ ***Atlantic Cruising Club's Guide to New England Marinas***; *US $24.95* (CN $29.95*)*
Bar Harbor, ME to Block Island, RI *(Including Buzzards Bay, Narragansett Bay, Martha's Vineyard and Nantucket)*

_____ ***Atlantic Cruising Club's Guide to Long Island Sound Marinas***; *US $24.95* (CN $29.95*)*
Block Island, RI to Cape May, NJ *(Including Connecticut River, New York Harbor and New Jersey Shore)*

_____ ***Atlantic Cruising Club's Guide to Florida's East Coast Marinas***; *US $29.95* (CN $35.95*)*
Fernandina, FL to Key West, FL *(Including St. John's River, St. Lucie River and the Florida Keys)*

_____ ***Atlantic Cruising Club's Guide to Chesapeake Bay Marinas***; *US $32.95* (CN $39.95*)*
C&D Canal to Hampton Roads *(Including Delmarva Atlantic Coast, Potomac, Rappahannock, Patapsco and James Rivers)*

_____ ***Atlantic Cruising Club's Guide to Pacific Northwest Marinas***; *US $29.95* (CN $35.95*)*
Campbell River, BC to Brookings, OR *(Including the San Juan Islands, Puget Sound, Sunshine Coast and Strait of Juan de Fuca)*

_____ ***Atlantic Cruising Club's Guide to Mid-Atlantic & ICW Marinas*** *(mid 2007)*; *US $29.95* (CN $35.95*)*
Norfolk, VA to St. Mary's, GA *(Including the Virginia Coast, the ICW and the North Carolina Sounds)*

_____ ***Atlantic Cruising Club's Guide to Florida's West Coast Marinas*** *(late 2007)*; *US $29.95* (CN $35.95*)*
Pensacola, FL to Key Largo, FL *(Including the Keys, Caloosahatchee River, Okeechobee Waterway and the Gulf Coast ICW)*

_____ ***Atlantic Cruising Club's Guide to California & Baja Marinas*** *(2008)*; *US $29.95* (CN $35.95*)*
Crescent City, CA to La Paz, MX

_____ ***Atlantic Cruising Club's Guide to East Coast Mega Yacht Marinas*** *(2008)*; *US $44.95* (CN $53.95*)*
Bar Harbor, ME to Key West, FL

_____ ***Atlantic Cruising Club's Guide to Northern Gulf Coast Marinas*** *(2008)*; *US $29.95* (CN $35.95*)*
Tarpon Springs, FL to Padre Island, TX *(Including Mobile Bay, the Tenn-Tom Waterway and the Gulf Coast ICW)*

_____ *Sub-Total*

_____ *Quantity Discount Percentage (see above)**

_____ *Tax (New York State Residents only add 6.75%)*

_____ *Shipping & Handling (USPS — add $4.00 for first book and $2.00 for each subsequent book)*

Final Total	Note: Credit cards will not be charged until books are shipped.

Please Charge the following Credit Card: Amex ☐   MasterCard ☐   Visa ☐   Discover ☐   Check Enclosed ☐

Number: _____   *Four-Digit Security Number:* _____

Expiration Date: _____   *Signature:* _____

Name: _____   *Email:* _____

Address: _____   *City:* _____   *State or Province:* _____   *Zip:* _____

Home Phone: _____   *Office Phone:* _____   *Fax:* _____

Boat Name: _____   *Length:* _____   *Manufacturer:* _____

Boat Type: Sail Mono-Hull ☐   Sail Multi-Hull ☐   Power ☐   Trawler ☐   Megayacht ☐

Home Port: _____   *Cruising Grounds:* _____

Please mail, fax. email, or call-in your order to: **Atlantic Cruising Club at Jerawyn Publishing, Inc.** PO Box 978; Rye, New York 10580
Tel: (914) 967-0994 or (888) 967-0994; Fax: (914) 967-5504; Email: Orders@AtlanticCruisingClub.com